REFERENCE

VOLUME 5
1800 – 1899

Science and Its Times

*Understanding the
Social Significance of
Scientific Discovery*

VOLUME 5
1800 – 1899

Science
and
Its
Times

Understanding the
Social Significance of
Scientific Discovery

Neil Schlager, Editor
Josh Lauer, Associate Editor

Produced by Schlager Information Group

Detroit
New York
San Francisco
London
Boston
Woodbridge, CT

Science and Its Times

VOLUME 5

1800-1899

NEIL SCHLAGER, *Editor*
JOSH LAUER, *Associate Editor*

GALE GROUP STAFF

Robyn V. Young, *Project Coordinator*
Christine B. Jeryan, *Contributing Editor*

Mary K. Fyke, *Editorial Technical Specialist*

Maria Franklin, *Permissions Manager*
Margaret A. Chamberlain, *Permissions Specialist*
Shalice Shah-Caldwell, *Permissions Associate*

Mary Beth Trimper, *Production Director*
Evi Seoud, *Assistant Production Manager*
Wendy Blurton, *Senior Buyer*

Cynthia D. Baldwin, *Product Design Manager*
Tracey Rowens, *Senior Art Director*
Barbara Yarrow, *Graphic Services Manager*
Randy Bassett, *Image Database Supervisor*
Mike Logusz, *Imaging Specialist*
Pamela A. Reed, *Photography Coordinator*
Leitha Etheridge-Sims *Junior Image Cataloger*

Contents

Contents

Physical Sciences

Technology and Invention

Contents

1800-1899

Preface

The interaction of science and society is increasingly a focal point of high school studies, and with good reason: by exploring the achievements of science within their historical context, students can better understand a given event, era, or culture. This cross-disciplinary approach to science is at the heart of *Science and Its Times.*

Readers of *Science and Its Times* will find a comprehensive treatment of the history of science, including specific events, issues, and trends through history as well as the scientists who set in motion—or who were influenced by—those events. From the ancient world's invention of the plowshare and development of seafaring vessels; to the Renaissance-era conflict between the Catholic Church and scientists advocating a sun-centered solar system; to the development of modern surgery in the nineteenth century; and to the mass migration of European scientists to the United States as a result of Adolf Hitler's Nazi regime in Germany during the 1930s and 1940s, science's involvement in human progress—and sometimes brutality—is indisputable.

While science has had an enormous impact on society, that impact has often worked in the opposite direction, with social norms greatly influencing the course of scientific achievement through the ages. In the same way, just as history can not be viewed as an unbroken line of ever-expanding progress, neither can science be seen as a string of ever-more amazing triumphs. *Science and Its Times* aims to present the history of science within its historical context—a context marked not only by genius and stunning invention but also by war, disease, bigotry, and persecution.

Format of the Series

Science and Its Times is divided into seven volumes, each covering a distinct time period:

Volume 1: 2000 B.C.-699 A.D.

Volume 2: 700-1449

Volume 3: 1450-1699

Volume 4: 1700-1799

Volume 5: 1800-1899

Volume 6: 1900-1949

Volume 7: 1950-present

Dividing the history of science according to such strict chronological subsets has its own drawbacks. Many scientific events—and scientists themselves—overlap two different time periods. Also, throughout history it has been common for the impact of a certain scientific advancement to fall much later than the advancement itself. Readers looking for information about a topic should begin their search by checking the index at the back of each volume. Readers perusing more than one volume may find the same scientist featured in two different volumes.

Readers should also be aware that many scientists worked in more than one discipline during their lives. In such cases, scientists may be featured in two different chapters in the same volume. To facilitate searches for a specific person or subject, main entries on a given person or subject are indicated by bold-faced page numbers in the index.

Within each volume, material is divided into chapters according to subject area. For volumes 5, 6, and 7, these areas are: Exploration and Discovery, Life Sciences, Mathematics, Medicine, Physical Sciences, and Technology and Invention. For volumes 1, 2, 3, and 4, readers will find that the Life Sciences and Medicine chapters have been combined into a single section, reflecting the historical union of these disciplines before 1800.

Arrangement of Volume 5: 1800-1899

Volume 5 begins with two notable sections in the frontmatter: a general introduction to nineteenth-century science and society, and a general chronology that presents key scientific events during the period alongside key world historical events.

The volume is then organized into six chapters, corresponding to the six subject areas listed above in "Format of the Series." Within each chapter, readers will find the following entry types:

Chronology of Key Events: Notable events in the subject area during the nineteenth century are featured in this section.

Overview: This essay provides an overview of important trends, issues, and scientists in the subject area during the nineteenth century.

Topical Essays: Ranging between 1,500 and 2,000 words, these essays discuss notable events, issues, and trends in a given subject area. Each essay includes a Further Reading section that points users to additional sources of information on the topic, including books, articles, and web sites.

Biographical Sketches: Key scientists during the era are featured in entries ranging between 500 and 1,000 words in length.

Biographical Mentions: Additional brief biographical entries on notable scientists during the era.

Bibliography of Primary Source Documents: These annotated bibliographic listings feature key books and articles pertaining to the subject area.

Following the final chapter are two additional sections: a general bibliography of sources related to nineteenth-century science, and a general subject index. Readers are urged to make heavy use of the index, because many scientists and topics are discussed in several different entries.

A note should be made about the arrangement of individual entries within each chapter: while the long and short biographical sketches are arranged alphabetically according to the scientist's surname, the topical essays lend themselves to no such easy arrangement. Again, readers looking for a specific topic should consult the index. Readers wanting to browse the list of essays in a given subject area can refer to the table of contents in the book's frontmatter.

Additional Features

Throughout each volume readers will find sidebars whose purpose is to feature interesting events or issues that otherwise might be overlooked. These sidebars add an engaging element to the more straightforward presentation of science and its times in the rest of the entries. In addition, the volume contains photographs, illustrations, and maps scattered throughout the chapters.

Comments and Suggestions

Your comments on this series and suggestions for future editions are welcome. Please write: The Editor, *Science and Its Times,* Gale Group, 27500 Drake Road, Farmington Hills, MI 48331.

Advisory Board

Amir Alexander
Research Fellow
Center for 17th and 18th Century Studies
UCLA

Amy Sue Bix
Associate Professor of History
Iowa State University

Elizabeth Fee
Chief, History of Medicine Division
National Library of Medicine

Sander Gliboff
Ph.D. Candidate
Johns Hopkins University

Lois N. Magner
Professor Emerita
Purdue University

Henry Petroski
A.S. Vesic Professor of Civil Engineering and
*   Professor of History*
Duke University

F. Jamil Ragep
Associate Professor of the History of Science
University of Oklahoma

David L. Roberts
Post-Doctoral Fellow, National Academy of
*   Education*

Morton L. Schagrin
Emeritus Professor of Philosophy and History of
*   Science*
SUNY College at Fredonia

Hilda K. Weisburg
Library Media Specialist
Morristown High School, Morristown, NJ

Contributors

Amy Ackerberg-Hastings
Iowa State University

Lloyd T. Ackert, Jr.
Graduate Student in the History of Science,
Johns Hopkins University

James A. Altena
The University of Chicago

Peter J. Andrews
Freelance Writer

Janet Bale
Freelance Writer and Editor

Kenneth E. Barber
Professor of Biology,
Western Oklahoma State College

Bob Batchelor
Writer,
Arter & Hadden LLP

Charles Boewe
Freelance Biographer

Scott Bohanon
Freelance Writer and Historian

Kristy Wilson Bowers
Lecturer in History,
Kapiolani Community College, University of Hawaii

Sherri Chasin Calvo
Freelance Writer

Geri Clark
Science Writer

Catherine M. Crisera
Freelance Writer

Guillaume de Syon
Assistant Professor of History,
Albright College

Thomas Drucker
Graduate Student, Department of Philosophy,
University of Wisconsin

H. J. Eisenman
Professor of History,
University of Missouri–Rolla

Lindsay Evans
Freelance Writer

Loren Butler Feffer
Independent Scholar

Keith Ferrell
Freelance Writer

Randolph Fillmore
Freelance Science Writer

Mark R. Finlay
Associate Professor of History,
Armstrong Atlantic State University

Richard Fitzgerald
Freelance Writer

Maura C. Flannery
Professor of Biology,
St. John's University, New York

Donald R. Franceschetti
Distinguished Service Professor of Physics and
*   Chemistry,*
The University of Memphis

Jean-François Gauvin
Historian of Science,
Musée Stewart au Fort de l'Ile Sainte-Hélène,
Montréal

Jim Giles
Freelance Writer

Sander Gliboff
Ph.D. Candidate,
Johns Hopkins University

Phillip H. Gochenour
Freelance Editor and Writer

Brook Ellen Hall
Professor of Biology,
California State University at Sacramento

Gerald F. Hall
Writer and Editor

Robert Hendrick
Professor of History,
St. John's University, New York

Jessica Bryn Henig
History of Science Student,
Smith College

Mary Hrovat
Freelance Writer

Philip Johansson
Senior Editor,
Earthwatch Institute

Matt Kadane
Ph.D. Candidate,
Brown University

P. Andrew Karam
Environmental Medicine Department,
University of Rochester

Evelyn B. Kelly
Professor of Education,
Saint Leo University, Florida

Rebecca Brookfield Kinraide
Freelance Writer

Israel Kleiner
Professor of Mathematics,
York University

Judson Knight
Freelance Writer

Lyndall Landauer
Professor of History,
Lake Tahoe Community College

Mark Largent
University of Minnesota

Josh Lauer
Freelance Editor,
Lauer InfoText Inc.

Lynn M. L. Lauerman
Freelance Writer

Garret Lemoi
Freelance Writer

Adrienne Wilmoth Lerner
Division of History, Politics, and International
*   Studies,*
Oglethorpe University

Brenda Wilmoth Lerner
Science Correspondent

K. Lee Lerner
Prof. Fellow (r), Science Research & Policy Institute,
Advanced Physics, Chemistry and Mathematics,
Shaw School

Stephen A. Leslie
Assistant Professor of Earth Sciences,
University of Arkansas at Little Rock

Carolyn Crane Love
Freelance Writer

Eric v. d. Luft
Curator of Historical Collections,
SUNY Upstate Medical University

Elaine McClarnand MacKinnon
Assistant Professor of History,
State University of West Georgia

Lois N. Magner
Professor Emerita,
Purdue University

Marjorie C. Malley
Historian of Science

Jim Marion
Freelance Writer

Ann T. Marsden
Writer

Megan McDaniel

William McPeak
Independent Scholar,
Institute for Historical Study (San Francisco)

Lolly Merrell
Freelance Writer

Leslie Mertz
Biologist and Freelance Science Writer

Kelli Miller
Freelance Writer

J. William Moncrief
Professor of Chemistry,
Lyon College

Heather Moncrief-Mullane
Masters of Education,
Wake Forest University

Stacey R. Murray
Freelance Writer

Ashok Muthukrishnan
Freelance Writer

Brid C. Nicholson
Drew University

Lisa Nocks
Historian of Technology and Culture

Stephen D. Norton
Committee on the History & Philosophy of Science,
University of Maryland, College Park

Shawn M. Phillips
Burial Sites Archaeologist,
State Historical Society of Wisconsin

Brian Regal
Historian
Mary Baker Eddy Library

Sue Rabbitt Roff
Cookson Senior Research Fellow,
Centre for Medical Education,
Dundee University Medical School

Michelle Rose
Freelance Science Writer

Steve Ruskin
Freelance Writer

Martin Saltzman
Professor of Natural Science,
Providence College

Elizabeth D. Schafer
Independent Scholar

Morton L. Schagrin
Emeritus Professor of Philosophy and History of
Science,
SUNY College at Fredonia

Neil Schlager
Freelance Editor,
Schlager Information Group

John B. Seals
Freelance Writer

Brian C. Shipley
Department of History,
Dalhousie University

Tabitha Sparks
Graduate Student, English,
University of Washington

Keir B. Sterling
Historian, U.S. Army Combined Arms Support
Command,
Fort Lee, Virginia

Gary S. Stoudt
Professor of Mathematics,
Indiana University of Pennsylvania

Zeno G. Swijtink
Professor of Philosophy,
Sonoma State University

G. Ann Tarleton

Todd Timmons
Mathematics Department,
Westark College

David Tulloch
Graduate Student,
Victoria University of Wellington, New Zealand

Contributors

1800-1899

Julianne Tuttle
Indiana University

Stephanie Watson
Freelance Writer

Karol Kovalovich Weaver
Instructor, Department of History,
Bloomsburg University

Richard Weikart
Associate Professor of History,
California State University, Stanislaus

Giselle Weiss
Freelance Writer

A.J. Wright
Librarian,
Department of Anesthesiology,
School of Medicine,
University of Alabama at Birmingham

Michael T. Yancey
Freelance Writer

Introduction: 1800–1899

Overview

The nineteenth century brought the world telephones, telegraphs, steamboats, electric lights, movies, sewing machines, cars, electric motors, the railroad, Ferris wheels, and aspirin. It was the age of invention, ending with the famous pronouncement in 1899 that "Everything that can be invented has been invented" (Charles H. Duell, Commissioner, U.S. Office of Patents). There are many candidates for the century's greatest invention, but the winner may be the future itself. While history has seen individuals, such as Francis Bacon, who imagined a world different from that of their parents, most people throughout history did not. They have expected their professions, tools, and entertainments to be essentially the same as those of their parents and grandparents. In the nineteenth century this changed, as inventors and their inventions captured the public imagination.

It is no coincidence that two important literary genres were born in the 1800s: the mystery story and science fiction. Edgar Allan Poe's Auguste Dupin was arguably the first detective in fiction, the precursor of Sherlock Holmes. Both characters used reason and deduction to understand the world. The popular audiences for their stories accepted this; they were confident that a deliberate and systematic approach would reveal the truth. Meanwhile, the heroes of Jules Verne and H. G. Wells used inventions to fly to the moon, explore the depths of the ocean, and travel through time. The public welcomed these stories, and many saw them as more than diversions. They experienced so many changes in their lives that, often, these fictions looked like predictions.

Looking Back to the Eighteenth Century

Of course, many changes came in the eighteenth century, but these were chiefly political. When Americans rebelled and created a new political philosophy, Thomas Jefferson could imagine freedom and equality. But even though he was an inventor, he believed America would remain a simple agrarian society. The French Revolution executed a king and founded a republic, but it also beheaded Antoine Lavoisier (1743-1794), known as "the Newton of chemistry." Isaac Newton (1642-1727) may have stood on the shoulders of giants to achieve a revolution in physics, but in 1800 most of his successors were still standing in his shadow. In fact, physics and mathematics stagnated, particularly in England, as Newton's accomplishments came to be seen as the final word.

In the new century, however, perhaps because revolutions had loosened conventions and shaken the social order, it became acceptable to challenge established dogmas. The emerging sciences of biology, chemistry, and archeology extended Newton's methods into new realms. Engineers and physicians carried the resulting technologies into everyday life. And, in Newton's own disciplines—physics and mathematics—people of courage broke free of his mechanical, clockwork universe to discover radiation, probability, imaginary numbers, and other original concepts that would shape the next century. The eighteenth century transformed our view of humans. It put the power of change into our hands, then built, and eventually shattered, a confidence in certainty and truth.

The Nineteenth Century: Building Blocks

Nineteenth-century scientists strove to rationalize the universe. Physics and astronomy led the way, but much of chemistry was still inured in alchemy. "Vitalism" and other mystical points of view dominated biology, and archeology had little standing in the Western world, where most educated people believed that the world was only 6,000 years old and humans were a separate creation from animals.

Chemistry provided many of the early triumphs of the rationalization process. In 1803 John Dalton (1766-1844) postulated the existence of atoms and began working on the proof. This added force to discovery of the elements, and by 1869 63 elements were known. Later in the century Dmitri Mendeleyev (1834-1907) saw a pattern to the elements when he looked at their masses and chemical characteristics. He organized them in an original way that made sense of chemicals and their reactions. This organization, the periodic table of the elements, allowed Mendeleyev and later scientists to predict the existence of such elements as gallium, neon, krypton, and radon—all of which were discovered later. By the end of the century chemistry, particularly synthetic chemistry, had become an essential and profitable part of society. Dyes made the world more colorful; patent medicines and synthetic fertilizers provided for human health; explosives moved mountains, made great engineering projects possible, and caused mass annihilation in war. Chemistry had created a vital role and a new identity for itself, with the periodic table as its icon.

Biology took a different path, perhaps because it touched more directly on humanity's view of itself. Classification and cell descriptions were at the leading edge of activity at the beginning of the century. These helped to provide a sense of order without making a strong challenge to accepted beliefs that viewed the world in a static way. Since the core of biology is process—e.g., growth, differentiation, competition, synergy, reproduction—its progress had to await a new insight.

In 1831 22-year-old Charles Darwin (1809-1882) undertook a voyage as a naturalist on the HMS *Beagle*. His findings shattered ideas about the age of the universe, the origin of humans, and the nature of biology. The heart of his thesis, evolution, was so disturbing that he did not publish his findings for 27 years. Variation and natural selection, or "survival of the fittest," were explained in Darwin's landmark 1859 work *On the Origin of Species*. Evolution required a much older world. Species were no longer fixed, in fact they were related. Darwin's next book went further. *The Descent of Man* (1871) joined humans to the rest of the biological world and challenged their special place. This upset many deeply held religious beliefs and demystified all of nature.

While it liberated science, it also spawned social darwinism, which was used to justify colonialism, racism, and the abuse of workers.

Archeology and paleontology took advantage of the doors opened by Darwin. Dinosaurs captured the fancy of the public, and digging fossils became a popular endeavor. Pierre Broca (1824-1880) determined that Neanderthal man was part of a prehuman species, setting off the search for the "missing link" connecting human and apes. What might have been a basis for understanding the common nature of humans and their shared relationship with animals was sometimes turned to demonstrate "scientifically" the inferiority of certain races. Phrenology and other pseudosciences made claims about white superiority, and the idea of eugenics was popularized.

A deeper understanding of genetics, the work of a humble Austrian monk, was unrecognized in its own time. Gregor Mendel (1822-1884) methodically investigated genetic inheritance by growing peas. His work provided a foundation for the twentieth century's icon for biology, DNA.

Inventing the Future

At the same time that people were coming to appreciate change in the natural world, they found themselves with unprecedented power to create change. The railroad may have been the first popular example. For the first time in history, people could travel faster than a galloping horse could carry them. The railroad extended cities, connected communities, fueled the Industrial Revolution, and changed concepts of time and space.

The sewing machine brought another kind of change. It freed time, since prior to its invention people spent fully one-third of their working hours creating and mending clothes (not to mention sails, curtains, and shoes). Sewing machines also increased productivity. Since tailors, who were generally men, resisted their introduction, manufacturers marketed them to women, allowing them to participate in the economy, and giving them independence that helped them secure their political and legal rights.

The most famous inventions of the nineteenth century are associated with equally famous inventors. Alexander Graham Bell (1847-1922) invented the telephone. Robert Fulton (1765-1815) invented the steamship. Thomas Edison (1847-1931) invented the electric light, the phonograph, and the motion picture. All of these inventions, thanks to the emerging methods of mass production and distribution, had a profound effect upon the daily lives of ordinary people. But this only partially explains their inventors' fame. When beset by patent battles and competing technologies, inventors found that they could brand their inventions, secure their wealth, and become celebrities with self-promotion. There was an economic value to Edison providing quotable quotes like "Genius is 1 percent inspiration and 99 percent perspiration." Edison used public demonstrations of technology to his advantage, and even ran a negative campaign against Nikola Tesla's (1856-1943) alternating current (AC) that included the electrocution of animals. (The advantages of AC for transmitting electricity over long distances were significant enough, however, that Edison's direct current technology lost out.) Thus, myth and reality were interwoven to create an age of invention.

Unexpected Truths and Consequences

Electricity was the darling of nineteenth-century physicists. It made them close collaborators with the inventors of the era and pushed the bounds of experimental science. Understanding electrical theory was essential to James Maxwell's (1831-1879) work, which helped unify concepts of electricity and magnetism. Such syntheses were aimed not just at explaining and taming nature, but at revealing its absolute truth. Mathematicians were engaged in the same pursuit, developing new tools and methods, and finding underlying consistencies that made their discipline more rigorous. From the early days of the century, however, there were indications that the precise truth they sought was unattainable. Even as public confidence in science reached its height, its limits were becoming apparent. People used scientific discourse to deceive themselves and each other and to confirm prejudices. One such

"proof," for example, showed that education was unhealthy for women. Just as importantly, probability emerged as a discipline in the 1800s. First used for error checking, it developed later into an expression of the statistical, intrinsically uncertain nature of the universe.

The Legacy of Nineteenth-Century Science

Society has come to rely on chemistry for plastics, fuels, fertilizers, and medicines. The houses we live in, the clothes we wear, and the food we eat are often, if not usually, the product of a deep understanding of chemistry that began with the periodic table. By the beginning of the twentieth century the understanding of bacteria and, by extension, sanitation, that came from nineteenth-century advances in biology, helped fuel tremendous population growth. Biology also increased understanding of fertility and led to artificial means of birth control. This essentially stopped the rise in population for developed countries by the end of the twentieth century.

In the twentieth century the flow of new inventions continued, reinforcing popular expectations of change. Many nineteenth-century inventions evolved into improved, but still-recognizable, forms. The car and the electric light, two of the most notable nineteenth-century inventions, created essential change in human cultures.

Pathological use of science, both to facilitate and to excuse brutality, left an indelible mark on the twentieth century and reduced confidence in science as a source of truth and progress. Kurt Gödel (1906-1987) and Werner Heisenberg (1901-1976) demonstrated how incomplete and uncertain scientific truths are. Even so, science remains the touchstone for rational discussion.

PETER J. ANDREWS

Chronology: 1800–1899

1804 Napoleon Bonaparte crowned emperor of France; launches a decade of conquests in which he subdues virtually all of Europe except Britain and Russia.

1807 French mathematician Jean-Baptiste-Joseph Fourier announces his famous theorem concerning periodic oscillation, which will prove invaluable to the study of wave phenomena.

1814-1815 Congress of Vienna sets the boundaries of European states, boundaries that will remain virtually intact for 99 years following Napoleon's defeat at Waterloo in 1815.

1822 Jean François Champollion deciphers the Rosetta Stone, thus making possible the first translations of ancient Egyptian hieroglyphics.

1823-1824 United States declares the Monroe Doctrine, ordering an end to European colonization of the Western Hemisphere; a year later Spain vacates the New World after its defeat by forces under Simon de Bolívar and others at the Battle of Ayacucho.

1829 Russian mathematician Nicolai Ivanovich Lobachevski discovers non-Euclidean geometry, paving the way for the mathematics of curved surfaces.

1837 French artist Louis Jacques Mandé Daguerre makes the first photograph, or daguerreotype, a still life taken in his studio.

1844 Having earlier patented his telegraph machine, Samuel Morse successfully transmits the first Morse code message over a telegraph circuit between Baltimore and Washington: "What hath God wrought?"

1848 Revolution breaks out in numerous European cities; Karl Marx and Friedrich Engels publish the *Communist Manifesto.*

1854-1856 Britain, France, Turkey, and Sardinia fight Russia in the Crimean War, a conflict noted for the nursing reforms of Florence Nightingale and for the fact that it was the first war covered by photojournalists.

1859 English naturalist Charles Darwin publishes *On the Origin of Species,* setting forth natural selection as the mechanism governing evolution.

1861-1865 Civil War and emancipation of slaves in the United States.

1864 French chemist Louis Pasteur invents pasteurization, a process of slow heating to kill bacteria and other microorganisms.

1865 Laying the groundwork for antiseptic surgery, English surgeon Joseph Lister uses phenol to prevent infection during an operation on a compound fracture.

1865-1876 Nain Singh, an Indian "pundit" employed by the British, leads several expeditions into the Himalayas and Tibet, including Lhasa, the capital city of Tibet, forbidden to Westerners.

1866 Austrian botanist Gregor Johann Mendel discovers the laws of heredity, presenting data that would not gain wide recognition until 1900.

1869 First periodic table, which arranges the elements in order of atomic weight and predicts the existence of undiscovered elements, created by Russian chemist Dmitri Ivanovich Mendeleyev.

1870-1871 Franco-Prussian War results in defeat of France; establishment of world's first communist state, the short-lived Paris Commune; and unification of Germany.

1873 James Clerk Maxwell publishes *Treatise on Electricity and Magnetism*, a landmark work that brings together the three principal fields of physics: electricity, magnetism, and light.

1875 Alexander Graham Bell first transmits sound over electric cable; in the following year he demonstrates his new telephone.

1879 Thomas Edison produces the first practical incandescent lightbulb.

1884-1885 Conference of Berlin effectively divides Africa into various European colonial spheres of influence.

1894-1895 Defeat of China in Sino-Japanese War marks rise of Japan as a world power.

1894-1906 Dreyfus Affair in France, involving false charges against Jewish army officer Alfred Dreyfus, exposes undercurrents of European anti-Semitism, creates sharp and lasting divisions between political left and right.

1898 Victory in Spanish-American War establishes United States as a colonial power, with possessions including Cuba, the Philippines, and Guam.

1899-1902 Second Anglo-Boer War; first systematic use of concentration camps.

Exploration and Discovery

Chronology

1804-1806 Meriwether Lewis and William Clark explore the American West on their way to the Pacific Ocean.

1822 Jean François Champollion deciphers the Rosetta Stone, thus making possible the first translations of ancient Egyptian hieroglyphics.

1831-1836 The HMS *Beagle*, a British vessel, explores both coasts of South America; on board is Charles Darwin, who begins forming his theory of evolution while in the Galapagos Islands.

1840 American explorer Charles Wilkes and French explorer Jules-Sébastien-César Dumont d'Urville simultaneously discover the continent of Antarctica.

1845-1851 British archaeologist Austen Henry Layard excavates the ruins of ancient Assyrian cities Calah and Nineveh.

1848 Gold is discovered at Sutter's Mill in California, beginning the California Gold Rush.

1850 British naval officer Robert McClure, on board the HMS *Investigator*, discovers the Northwest Passage between the Atlantic and Pacific Oceans.

1853-1856 British missionary David Livingstone becomes the first European to cross the entire African continent, from south to north; along the way, he discovers Victoria Falls (1855).

1862 British explorer John Hanning Speke discovers the source of the Nile River at Lake Victoria.

1873 German amateur archaeologist Heinrich Schliemann discovers the ruins of Troy, long thought to be a purely legendary city.

1872-1876 An expedition aboard the HMS *Challenger* systematically explores the ocean depths, temperature, and underwater life of the Atlantic and Pacific Oceans.

1874-1877 British explorer Henry Morton Stanley conducts extensive exploration of the African continent.

Overview:
Exploration and Discovery
1800-1899

Overview

One of the greatest thrills man can experience is the discovery of something that no one has ever seen. The exhilaration of traveling a wild unexplored locale, facing hazards natural and native while discovering the hitherto unknown, has attracted explorers of the world for thousands of years. By the eighteenth century man's quest for the unknown led explorers such as Captain James Cook (1728-1779) on scientific voyages around the globe. Some attempted, unsuccessfully, to reach the farthest corners of the globe—such as a 1773 British Admiralty expedition to the North Pole. Others, such as surveyor and explorer Alexander Mackenzie (1763-1820), who traveled across land to the Pacific Ocean, explored a single continent—North America. Organizations developed such as the African Association founded in June 1788, whose main objective was the exploration of Africa. By the end of the eighteenth century, man's hunger for knowledge of the world had become insatiable, leading to the most active period of Earth exploration: the 1800s.

The expeditions of the 1700s were limited in scope and significance when compared to the amazing accomplishments of explorers in the 1800s. Never before or since has so much of Earth been discovered in such a brief period of its history. In all, man's compulsion to discover, describe, and catalog his world—as well as conquer it—resulted in a flood of exploration in the 1800s. There were expeditions to solve unanswered geographical questions, such as the existence of a Northwest Passage and the source of the Nile. There were expeditions to expand scientific knowledge, such as the first deep-sea exploration of the HMS *Challenger* (1872-6) and voyages to South America that led to new discoveries in the fields of zoology, botany, and geology. Meanwhile, other explorations, especially those sponsored for political purposes, were expanding national boundaries—in America and Australia, for example—as well as imperial domains, as was the case in Africa. Adventure in the nineteenth century was not only for explorers, however, as archaeological discoveries in the Middle East and Mediterranean were also significant.

Exploration for Scientific Purposes

The first class of nineteenth-century exploration, for scientific purposes, could accurately describe nearly every expedition undertaken in the period. The information brought back by explorers stimulated a new perspective on man and his environment. New, more accurate maps and geographical reports resulted from the journeys and voyages of topographical engineers and surveyors. New discoveries were made in the fields of botany, zoology, ornithology, marine biology, geology, and cultural anthropology. Especially significant were expeditions to South America. From 1799-1802 Alexander von Humboldt (1769-1859) and Aimé Bonpland (1773-1858) explored the Orinoco River and most of the Amazon River system in northwest South America, identifying plant and animal life and studying climatology, meteorology, and volcanoes. Humboldt used his discoveries to create an encyclopedic work entitled *Kosmos*, which cataloged his own extensive scientific knowledge and much of the accumulated knowledge of geography and geology of his time. In northeast South America Robert Schomburgk (1804-1865) explored the interior of Guyana from 1835 to 1839 as one of the first funded expeditions of Britain's Royal Geographical Society, which was founded in 1830. In addition to extensive mapping of rivers and geographical features, Schomburgk collected hundreds of botanical, zoological, and geological specimens for study. Along the coast of South America, the voyage of the British ship HMS *Beagle* (1831-6), with Charles Darwin (1809-1882) aboard, made scientific discoveries that inspired Darwin's theory of evolution, one of the titanic achievements in modern science.

While explorations were penetrating the hot jungles and rivers of South America, other scientific expeditions were braving the frosty regions of the North Pole, Antarctica, and Tibet and discovering, at last, both the Northwest and Northeast Passages. In 1831 James Clark Ross (1800-1862) was the first to discover the Magnetic North Pole. The first major voyage of exploration undertaken by the young United States was the U.S. Exploring Expedition led by Charles Wilkes (1798-1877), which sighted the Antarctic mainland early in 1840. Several American scientists accompanied Wilkes on the voyage and returned with thousands of scientific specimens from the lands visited, as well as important information on weather, sea conditions, and safe sea passages, bringing distinction to the expedition. Two more firsts were accomplished by the discoverers of the Northwest and Northeast Passages, sought by 300 years of explorers. In 1854 Irishman Robert McClure (1807-1873) completed a four-year journey of the Northwest Passage to Asia—by ship, by foot, then by ship again. Likewise, in 1879 Nils Nordenskiöld (1832-1901), a Finnish scientist, completed the first transit of the Northeast Passage, a sea route from Europe across the northern coast of Asia to the Pacific.

Exploration to Expand National Boundaries and Imperial Terrain

The second class of nineteenth century exploration, for political purposes, includes expeditions sent out for the express political goal of expanding national boundaries as well as those intended to expand imperial terrain. Continental/national boundaries were addressed by expeditions in Australia, Siberia, and North America. In 1802 Matthew Flinders (1774-1814) was the first to circumnavigate Australia and to chart its southern coast. The Australian interior was explored by numerous teams of scientists, surveyors, and discoverers. These included Edward Eyre (1815-1901), the first to explore central Australia and the first to traverse the continent, and the ill-fated transcontinental explorers Robert O'Hara Burke (1820-1861) and surveyor William John Wills (1834-1861), who, after traversing the continent from Melbourne to present-day Normanton near the Gulf of Carpenteria, both died of starvation on their return journey.

While Australia was eagerly exploring its continental boundaries, Russia was rapidly expanding its borders, annexing Siberia and other central Asian provinces. Thanks to the extensive explorations of men such as Nikolay Przhevalsky (1839-1888), who traveled throughout central and eastern Asia, mapping, collecting biological specimens, and surveying future travel routes, Russia was able to lay claim to considerable natural resources and valuable winter ports and to consolidate its territories in the Far East.

Like their counterparts in Australia and Russia, nineteenth-century American explorers played no small part in the rise of its Manifest Destiny—the expansion of its boundaries to the

Pacific Ocean. One of the most significant feats of American exploration was that of Lewis and Clark's Corps of Discovery. From 1804-6 Meriwether Lewis (1774-1809) and William Clark (1770-1838) explored the uncharted American Far West on their way to the Pacific Ocean, helping cement the United States' claim to parts of the Pacific Northwest. Another American expedition that spurred interest in western expansion was that of Zebulon Pike (1779-1813), whose discoveries led to the conquest and settlement of lands in the Southwest. American expansion was further aided by the expeditions of John Frémont (1813-1890), whose dramatic account of western adventures excited the American public to a greater level of enthusiasm for the West.

National boundaries weren't the only lines expanding due to nineteenth-century exploration. Explorers were both the forerunners and forefathers of European imperialism, especially on the African continent. The "Dark Continent" was traversed in 1855-6 by David Livingstone (1813-1873), the first known European to do so, covering much uncharted African territory. Another important African discovery, made in 1858, was the source of the Nile found at Lake Victoria by John Speke (1827-1864). From 1874-7 Henry Stanley (1841-1904) explored the entire length of the Congo. The southern and central African expeditions of Livingstone, Speke, and Stanley resulted in a frenetic race between European nations to colonize Africa and introduce so-called "civilized" European ways into the continent's peoples. This included an infusion of Christian missionaries and enterprise-oriented merchants and traders, many of whom exploited the African natives.

Archaeological Exploration

The final class of nineteenth-century exploration, while not technically of that classification, hinges closely on the spirit of romanticism tied to the exploration of the time. Nineteenth-century romanticism stressed not only an interest in the remote and an appreciation of external nature; it also emphasized an exhaltation of the primitive and an idealization of the past. The subsequent rising interest in antiquities produced several significant archaeological discoveries such as the uncovering of the Egyptian temple of King Ramses II in 1813 by Jean-Louis Burckhardt (1784-1817), the deciphering of the Rosetta Stone in 1822 by Jean-François Champollion (1790-1832), and the locating of the ancient Greek city of Troy in 1873 by Heinrich Schliemann (1822-1900).

Conclusion

Fundamental developments in technology changed the character of exploration after the 1800s. Most significant were the evolution of the aviation and aeronautics industries and the revolution of photography and film. Computers, telephones, and global positioning satellites have also "technified" the business of exploring. With the assistance of such technology, twentieth-century explorers have been able to make more detailed surveys of Earth's surface, explore the depths of the ocean and Earth's interior, and voyage to the moon and stars, as the quest for the unknown has extended beyond Earth.

ANN T. MARSDEN

Humboldt and Bonpland's Landmark Expedition to the Spanish Colonies of South America (1799-1804)

Overview

Alexander von Humboldt (1769-1859), a German geologist and naturalist, and Aimé Bonpland (1773-1858), a French botanist, engaged in a new sort of scientific travel involving systematic measurement and observation of a remarkable range of organic and physical phenomena with dozens of sophisticated scientific instruments. Humboldt's ultimate goal for these researches was to understand nature as an interconnected whole. Humboldt and Bonpland inspired a generation of scientific explorers and established new methodologies and new instrumentation standards.

Background

The eighteenth-century expeditions of Charles Marie de La Condamine (1701-1774), Louis Antoine de Bougainville (1729-1811), and Captain James Cook (1728-1779) provided the model of scientific exploration followed by Humboldt and Bonpland. In all of these earlier instances scientific travelers bravely explored mysterious lands and oceans while continuously collecting specimens and measuring astronomical and geological phenomena. Upon returning home these explorers published popular and scientific accounts describing heroic adventures and exotic sights and, especially in the case of Cook, presenting a wide range of botanical, geological, oceanographical, and anthropological findings.

While mostly adhering to this model, Humboldt's efforts in particular were inspired by a range of scientific interests and a commitment to comprehensive empirical observation surpassing those of any scientific explorer before or after. Natural objects, Humboldt insisted, can be understood only within the full range of their environment: rainfall, humidity, temperature, barometric pressure, electrical charge of the air, chemical composition of the atmosphere and soil, geomagnetism, longitude, latitude, elevation, surrounding geological formations, surrounding plants and animals, and nearby human activity and culture must all be measured or observed. Humboldt called his scientific enterprise a *physique du monde,* or terrestrial physics. Inspired by the philosophy of Immanuel Kant (1724-1804), Humboldt was seeking to discover amid the geographical distribution and variation of phenomena nature's constant and most simple laws and forces.

Towards this end Humboldt and Bonpland carried with them an unprecedented array of instruments, all financed by Humboldt himself. Telescopes, sextants, theodolites, compasses, a magnetometer, chronometers, a pendulum, barometers, thermometers, hygrometers, a cyanometer, eudometers, a rain gauge, leyden jars, galvanic batteries, and chemical reagents were carried and used across the continent. Scientific instruments had been greatly improved in recent years both in accuracy and in portability. Humboldt had gained expertise in using these instruments through years of scientific study and travel in Europe. The expedition of Humboldt and Bonpland to the Spanish colonies, then, was truly at the frontiers of science.

For almost five years, from July 1799 to April 1804, as the Napoleonic Wars raged in Europe, Humboldt and Bonpland traveled throughout what is now Venezuela, Cuba, Colombia, Peru, Ecuador, and Mexico mapping, collecting, measuring, sketching, describing, and observing all the way. It was a tremendously arduous journey accomplished on foot, canoe, and horse with equipment carried by a caravan of as many as 20 mules or by numerous canoes assisted by Indian guides. Not surprisingly, glass jars and instruments broke. Despite the hardships Humboldt, in particular, thrived in the tropical climate, displaying tremendous energy and strength and, unlike Bonpland, rarely falling ill.

In Venezuela their primary goal was to explore the Orinoco River and discover its connection to the Amazon watershed. After trekking through Venezuelan mountains and plains they canoed the Orinico's vast system for 75 days. Humboldt performed calculations upon observations of Jupiter's moons and other celestial objects in order to map the Orinico's course. Humboldt and Bonpland also systematically collected plants while carefully measuring every possible environmental factor. Through global studies in "plant geography" Humboldt hoped to eventually be able to infer the diversity and density of vegetation at any point on Earth. Vegetation for Humboldt represented an organic force as measurable as heat or magnetism.

Upon reaching the southern border of the Spanish colonies, the explorers traveled back through Venezuela. After visiting Cuba they explored Colombia, Ecuador, and Peru for 21 months. Humboldt, an expert in geology and minerals, was particularly interested in studying volcanoes of the Andes and sites of major seismic activity for clues as to Earth's formation. Crossing the Andes four times (and setting a mountaineering record of 19,289 feet) Humboldt and Bonpland carefully measured the magnetic axes of mountains and the inclination of strata in order to understand the forces that had generated the volcanic range. By carefully attending to all the data, especially data deviating from the general north-south orientation, Humboldt hoped to develop a comprehensive dynamical theory of mountain ranges to replace what he considered simplistic explanations of his predecessors.

In January 1803 the explorers sailed to Mexico. During the voyage Humboldt charted the course of the cold coastal current that now bears his name. Humboldt and Bonpland paid special attention to Mexico's mining districts in relationship to the geology, economy, and anthropology of the country. After a year in Mexico Humboldt and Bonpland sailed to Cuba and then to Philadelphia. They met with President Thomas Jefferson, an ardent scientist himself, in Washington and Monticello. In June 1804 Humboldt and Bonpland departed Philadelphia for home carrying 30 large crates of collected materials. For all their tremendous successes they were disappointed in one thing. Originally they had planned to travel to the Philippines and other Spanish possessions throughout the globe. War and bad luck had frustrated those plans.

Impact

Upon return to Europe Humboldt and Bonpland were celebrated as heroes. Humboldt went on to write numerous books recounting the rigors of the trip and the beauty and strangeness of the mysterious continent. These books, which were widely translated and widely read, portrayed the scientist as a fearless, virile adventurer who was willing to endure any hardship for the pursuit of knowledge.

Through his voluminous popular and scientific writings on the South American expedition, Humboldt became the most famous naturalist of his day and inspired a generation of scientific explorers. He and Bonpland had proven the possibility of a sophisticated inland scientific expedition employing a vast range of the best instruments. Humboldt's quantitative, technical methodology was quickly taken up by many American explorers of the western United States and by British, German, French, and Scandinavian explorers. His model of plant geography greatly inspired, for one, Charles Darwin (1809-1882) in his studies of the geographical distribution of species. Humboldt's style of scientific travelogue, in which he vividly recounted sights, sensations, and scientific observations from a personal viewpoint, was adopted by Darwin, Alfred Russel Wallace (1823-1913), Louis Agassiz (1807-1873), and other scientific explorers. The extent of Humboldt's influence on later explorers is indicated by the number of towns, counties, rivers, and mountains bearing his name in the western United States.

Humboldt's influence extended well beyond scientific exploration. His work on plant geography became a basis of the field of plant ecology at the end of the century. His "political geography" of Mexico, which incorporated social, economic, and manifold environmental factors, was quickly emulated by other geographers. His technique of "iso-maps," which connected with lines geographical points of equal mean temperature, magnetic intensity, rainfall, and so on, was adopted by researchers in many sciences and, notably, is retained in the isobars and isotherms of our weather maps. Even painters such as the American F. E. Church responded to Humboldt's writings by journeying to the tropics to faithfully portray exotic plants amid their sublime, tangled environment.

More generally, Humboldt's scientific writings contributed to a new vision of science and nature. Under Humboldt's influence, any science centered around the isolated specimen in the laboratory had come to be branded as out-moded or even false. Nature was complex and science must attend to the myriad of interconnected factors contributing to this complexity. Humboldt was not the first to conceive of nature or science this way. Nor did all natural scientists embrace Humboldt's goal of discovering nature's unity through measurement. But through his and Bonpland's exploits in a difficult terrain with dozens of sophisticated instruments and through his extensive writings presenting data and explaining their significance, Humboldt demonstrated how such a science could be pursued.

Inspired by Humboldt's vision, many scientists turned their attention to complex phenomena such as the tides, the weather, and geomagnetism, which required heterogeneous empirical investigations across the globe. In order to study these phenomena researchers invented better, more accurate instruments and carried their instruments to diverse locations. They also adopted Humboldtian tables, graphs, and isomaps as tools for organizing and understanding data. Perhaps the most successful developments in Humboldtian science came in the field of geomagnetism. Humboldt himself had urged governments to establish global stations for observing magnetic and other phenomena. This idea gained impetus after famed German mathematician Carl Friedrich Gauss (1777-1855) successfully analyzed Humboldt's measurements in terms of spherical harmonics in 1833. Observational stations were established around the globe in the 1830s and 40s by several European nations. Especially in Britain this enterprise was motivated as much by colonial and navigational concerns as by a commitment to knowledge or international cooperation in science. In 1852 British astronomer Edward Sabine (1788-1883), comparing data tabulated at the stations in

Toronto and Tasmania, determined that statistical variations in geomagnetic disturbances corresponded to the recently discovered sunspot cycle. With this discovery the science of solar-terrestrial physics was born. The success of the magnetic stations encouraged the establishment of similar observational networks, most notably in meteorology. That network is, of course, still with us today on a much larger scale.

By the middle of the nineteenth century Humboldt's mode of universal science, in which an individual single-handedly seeks to integrate understanding of a vast range of organic and physical phenomena, had become untenable. In an era of scientific specialization Humboldt was indeed the last scientific polymath. In the meantime Humboldt and Bonpland's scientific accomplishments in South America, which had so astonished their contemporaries, had been overshadowed by the work of new generations of investigators using better instruments and pursuing geographically broader investigations.

JULIANNE TUTTLE

Further Reading

Books

Botting, Douglas. *Humboldt and the Cosmos.* London: Joseph, 1973.

Cannon, Susan Faye. "Humboldtian Science." In *Science and Culture: The Early Victorian Period.* New York: Dawson, 1978.

Dettelbach, Michael. "Global Physics and Aesthetic Empire: Humboldt's Physical Portrait of the Tropics." In *Visions of Empire: Voyages, Botany, and Representations of Nature,* edited by David Philip Miller and Peter Hanns Reill. Cambridge: Cambridge University Press, 1996.

Humboldt, Alexander von. *Personal Narrative of a Journey to the Equinoctial Regions of the New Continent.* Abridged and translated by Jason Wilson. London: Penguin Books, 1995.

Nicolson, Malcolm. "Alexander von Humboldt and the Geography of Vegetation." In *Romanticism and the Sciences,* edited by A. Cunningham and N. Jardine. Cambridge: Cambridge University Press, 1990.

Periodicals

Cawood, John. "Terrestrial Magnetism and the Development of International Scientific Cooperation in the Early 19th Century." *Annals of Science* 34 (1977): 551-87.

Nicolson, Malcolm. "Humboldtian Plant Geography after Humboldt: The Link to Ecology." *British Journal for the History of Science* 29 (1996): 289-310.

The Discovery of Australia and Tasmania Greatly Expands the British Empire

Overview

Long after the northern regions of the world were known and occupied, the Southern Hemisphere was still unexplored and obscure. When the Europeans finally were able to build ships that could safely make long voyages, men were sent on arduous and difficult expeditions to explore the area and gain a foothold there. In the East Indies and Southeast Asia, riches, resources, and raw materials abounded that European nations not only needed but wanted. As nations expanded their power and extent, they found a new continent, new islands, and opened new trade routes. The new continent, settled by the British and named for an ancient, non-existent land called *Terra australis*, became a far flung extension of the British Empire called Australia.

Background

When Europeans began to realize the extent of their own continent, they presumed that a land mass of similar size and weight must lie in the Southern Hemisphere to balance the globe. It was called *Terra australis incognita,* or the unknown southern land. The idea lay dormant for centuries, for Europeans had no means to reach it. When ship building improved, they ventured out into the seas.

. The Chinese, Arabs, or other Asians may have seen Australia, but references are unclear. The Portuguese may have discovered the West Coast in the sixteenth century. The Spanish found land in the same area but did not follow up on it. In 1615 a Dutch explorer reached Cape York, the northernmost point on the Australian continent, but didn't connect it with Spanish or Portuguese discoveries. Another Dutch explorer landed on a barren coast in the West and called it New Holland, but he found no gold, rich civilizations, spices, or other resources. Another Dutch captain discovered another wild and barren place he called Van Diemen's Land; today it is Tasmania.

Explorers were sent to find the elusive southern continent with the hope of riches and civilizations worth the time and effort. Hopes were fading when nothing concrete was found. In the 1770s Captain James Cook (1728-1779) sailed past 40 degrees south latitude and found only ocean and the tip of a frozen land called Antarctica. There was no *Terra australis*. This achievement answered a centuries-old question and put the idea of a large continent to rest. Ships and sailors could now travel these latitudes without the fear of encountering a large land mass. By the end of the eighteenth century, New Holland was still unsettled, uninviting, and unexplored. The maps of some explorers showed New Zealand, Tasmania, and New Guinea as part of this unknown land. The Europeans knew something was there but were not exactly sure what or where it was.

Several motives led European governments to underwrite the expense of these expeditions. They needed new lands and resources to keep their governments strong. Overseas colonies brought prestige and power, as well as resources. More and more raw materials like cotton, wheat, wool, gold, spices, and new foods were needed to satisfy the growing population in Europe.

By 1800 Europeans, especially the British, were at war with French dictator Napoleon Bonaparte and were alarmed at renewed French interest in the southern continent. Hoping for another land of infinite resources like North America, the British mounted an expedition to claim the whole southern land, however large it was. Matthew Flinders (1774-1814), in the ship *Investigator,* was chosen to survey the coast. He began in the Southwest, sailed eastward past the Great Australian Bight and Spencer Gulf to Port Philip (Melbourne). Every few miles, he landed, noted the people and animals, recorded the topography of the land, made maps, and charted the bays, rivers, and reefs. After resting in Sydney, he resumed his exploration sailing along the coast of Queensland. He noted the presence of the Great Barrier Reef, the Coral Sea, and various features of the tropical peninsula of Cape York. He sailed west to Arnhem Land until his ship was in such bad shape that he could not go on.

In his papers Flinders championed the name Australia for this new continent soon to be claimed and occupied by the British. The British Empire had not reached its full extent, though it controlled Canada and was making inroads in South Africa and India. Australia would be the first British colony in the Pacific Ocean.

Impact

In the early 1800s, European nations were competing all over the world for trade, markets, and resources. The European population was growing, people were living longer and better lives, and they were expanding their industrial development and beginning to need more space and resources. Europeans had used up their raw materials and had to find new ones in the far corners of the world. They roamed the seas searching for resources and new markets for their products. A strong sailing fleet was the most vital part of this business. A country had to build seaworthy ships and have the skill to sail them anywhere in the world. The British were masters of these activities.

The basis of this enterprise was called "mercantilism," a philosophy born in France in the seventeenth century. The national government controlled all economic activity in its own nation. It made sure more goods were sold than bought to keep a favorable balance of trade—that is, more money should come in than go out. It owned colonies in far corners of the world, each of which existed to produce goods for the mother country. The system ensured the nation power, security, and self-sufficiency. Most European nations followed this philosophy in one form or another. Spain and Portugal had colonies in South America and Asia, the British were in Canada, and the Dutch were in Southeast Asia. The Dutch and French had landed in eastern Australia, but neither had settled there. It was so barren they doubted crops would grow, and the natives did not seem willing to work. The British and Dutch engaged in several armed conflicts over trade as the Dutch had a monopoly on the commodities in the East Indies or spice islands (Indonesia). A shooting war erupted between Britain and the Dutch in the Indies in 1780. The Dutch were strong in trade but militarily weak, and the British had little difficulty subduing them.

William Pitt the younger was Prime Minister of England from 1783-1800. With the French Revolution and the subsequent war against Napoleon, Pitt had to make sure France would not gain access to Britain's eastern trade routes. Because the U.S. was no longer available as a place to send undesirable people from England, he championed the idea of using Australia as a penal colony. He had considered Africa and discarded it. Joseph Banks had reported the area around Sydney to have rich soil and lots of vegetation. Native inhabitants posed no problem, as they did not challenge the newcomers but hid in the vast deserts of the interior. So in 1788 the English government sent 1,000 convicts to Botany Bay, south of Sydney. These men and women were convicted criminals, many transported for minor crimes like stealing a loaf of bread, plus some Irish political prisoners. They ranged in age from children of twelve or thirteen to men and women of seventy years or more. In Australia they served a seven-year sentence. After that they could become free settlers, and many did. This export of prisoners lasted from 1788 to 1840 in New South Wales and continued elsewhere until 1868. Free immigration, passage, and settlement was encouraged after the Napoleonic Wars were over in 1815.

Australia was a strategic outpost. Having a base here helped the British keep the seas open around India, New Guinea, the East Indies, and the Pacific Ocean for their ships and commerce. A governor was appointed to administer each area in Australia, and he was the employer of the convicts. Many governors were autocratic and harsh, but they generally maintained strict British codes of ethics and law. Later, civil liberties were gradually introduced to Australian settlers. By 1800 the population of New South Wales and Norfolk Island numbered 5,000 people, 3,000 sheep, and 500 cattle and other animals.

The settlement of Australia effectively expanded the national boundaries of England and set it on a path to the creation of a huge empire. Great Britain took over India beginning in 1757, Australia in 1788, South Africa in 1814, and New Zealand in 1840. Australia was a colony with no pretensions to independence, peopled by citizens whose origins were in the lower classes of England. Many early settlers had been convicted of minor crimes and taken to the continent without their consent. They were ruled by the aristocratic elite and the British government in the first years. In the twentieth century Australia became an independent nation and part of the British Commonwealth. The culture that developed in this remote frontier was as far from European tradition or class distinctions as the United States had been. The circumstances of its beginning have colored its national character ever since.

The settlement of Australia gave the British a base in the Pacific Ocean, relief from overcrowding at home, and a place from which to gather resources like gold, wool, and food staples and to sell manufactured goods. While the continent did not contain the resources that North America had, Australia was strategically more important, and it became a part of the largest empire in the world. In the nineteenth and early twentieth centuries, the English people and the crown boasted that the sun never set on the British Empire. Australia was one of the reasons that boast could be made.

LYNDALL B. LANDAUER

Further Reading

Flinders, Matthew. *A Voyage to Terra Australis.* Adelaide: Libraries Board of South Australia, 1966.

Hughes. Robert. *The Fatal Shore.* New York: Vintage Books, 1988.

Ingleton, Geoffrey C. *Matthew Flinders, Navigator and Chartmaker.* Victoria, Australia: Genesis Publications, 1986.

Martin, Ged., ed. *The Founding of Australia.* Sydney: Hale and Iremonger, 1978.

Terrill, Ross. *The Australians.* New York: Simon and Schuster, 1987.

American Far West:
The Lewis and Clark Expedition

Overview

One of the greatest feats of exploration in North America was that undertaken by Lewis and Clark's Corps of Discovery from 1804-1806. During their travels, the Corps of Discovery explored the Mississippi and Missouri river basins, made scientific discoveries about many plant and animal species new to science, contacted Native American tribes, and helped cement the United States' claim to parts of the Pacific Northwest, formerly claimed by Great Britain and Russia.

Background

In a move of questionable legality and constitutionality, President Thomas Jefferson (1743-1826) purchased the Louisiana Territory from the French in 1803. The territory, stretching from New Orleans to Canada and encompassing the majority of the drainage basins of the Missouri River and west of the Mississippi River, increased the size of the United States dramatically. Although the Louisiana Territory proved a boon economically and provided a windfall of scientific knowledge, the primary reasons for the purchase were political and commercial. Jefferson found it intolerable that a foreign power (first Spain, then France) should control New Orleans, through which the commerce of the western boundary states passed. Jefferson was also interested in learning more about the Native American tribes that he planned to add to the United States, hoped to find an easy, mostly water route to the Pacific Ocean, and wanted to take much of the lucrative fur trade from the British of Canada and the Pacific Northwest.

In 1802, shortly after the Spanish transferred the Louisiana Territory to the French, Jefferson sent an ambassador to Paris to meet with Napoleon. Conveying the message that the United States was interested in purchasing Louisiana from the French and would take it by force otherwise, Napoleon agreed to sell the territory for nearly $10 million, earning much needed funds for his government. The deal was approved by Congress amid a great deal of controversy, closely followed by approval of $2500 to fund a Corps of Exploration. Jefferson had already been in discussion with his friend and personal secretary Meriwether Lewis (1774-1809) about leading an expedition to explore the Louisiana Territory; with the approval of Congress, Jefferson gave Lewis permission to make preparations for a journey of exploration, mapping, and diplomacy with Native American tribes. Lewis recruited former military officer and friend William Clark (1770-1838) to serve as the expedition's co-leader.

At the time the Corps of Discovery left there was a tremendous amount of erroneous information about the lands west of the Mississippi. About all that was known with any degree of certainty was the latitude and longitude of the mouth of the Columbia River and other landmarks on the west coast, based on measurements taken by Captain James Cook (1728-1779) and other oceanic explorers. Jefferson, one of the leading intellectuals of the day, firmly believed that in the American interior would be found wooly mammoths, giant ground sloths, active volcanoes in the Badlands of the upper Missouri, and other misconceptions. Most

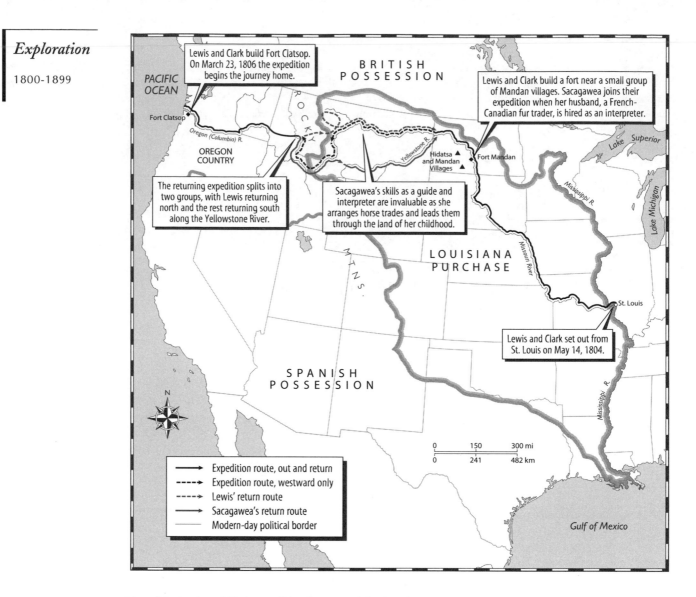

Map of the Lewis and Clark expedition that crossed the American West.

importantly, Jefferson was convinced that the highest mountains in North America were the Blue Ridge Mountains and that an easy route to the Pacific would be found with, at most, a low and short portage. This last was among the most important of Jefferson's mistaken ideas; easy access to the Pacific figured importantly into Jefferson's commercial plans for the United States.

Finally, Jefferson was concerned about the possibility of an imperialist, expansionist France with territories in North America. If France sought to settle the Louisiana Purchase, Jefferson anticipated the need to seek an alliance with the British, a politically unpalatable prospect given the recently ended Revolutionary War. Jefferson wanted North America for the United States, not for European powers.

The typical American at this time had little interest in Louisiana, except for those few who trapped for a living. The existing United States was sparsely settled at that time, so there was little population pressure to move westward, and the economy was largely agrarian, so the need for raw materials was similarly low. At that time, too, the typical American was concerned about survival; farming, avoiding attacks by Native Americans, and staying healthy. They had little time to ponder the political implications or the scientific curiosities of Louisiana.

Impact

The Lewis and Clark expedition had both immediate and long-term impacts on most Americans. These can be summarized as follows:

1. This expedition was the first major organized survey of the interior of a major continent. The Spanish had similarly explored much of South America, but with an eye towards exploitation of resources and little regard for scientific or geographic knowledge. The interiors of Africa, Asia, and Australia were still largely unknown to Western civilization.

2. The findings of this expedition encouraged the rapid settlement of the Louisiana Territory by farmers and trappers. This, in turn, was a step on the path towards the American concept of a "Manifest Destiny" to fill and rule most of the North American continent.

3. Lewis and Clark's positive and negative contacts with Native American tribes helped set the stage for conflicts to come. They alienated some powerful tribes, befriended others, made arbitrary decisions regarding official dealings with others, and encouraged settlers to move into tribal lands.

4. They ruled out the possibility of rapid and easy travel to the Pacific, confirming that cross-continental travel would be, for some time, long and risky. This, in turn, meant that communication and trade across the expanding United States would become increasingly cumbersome until improved travel and communications (unforeseen in Lewis and Clark's day) were invented.

The Lewis and Clark expedition was launched for political, strategic, scientific, and commercial aims. This made it unlike most other major exploration efforts. The Spanish in South America sought riches and converts to Christianity. The British around the world sought commerce, raw materials, and strategic advantage, as did the Dutch. Some voyages had been previously launched for scientific gain, but these tended to ignore nonscientific aims. The Corps of Discovery was virtually unique in attempting so much and succeeding so well in virtually all areas. This success also helped to vindicate Jefferson's purchase of the Louisiana Territory as well as his insistence on launching the expedition. In addition, the knowledge returned by the expedition helped to bring the continental interior into better focus, replacing many myths with hard-earned fact.

Upon their return, the members of the Corps of Discovery lost little time in publishing memoirs, giving public lectures, and talking about the rich lands and plentiful herds they had seen. Lewis presented his specimens, jour-nals, and scientific discoveries to the government and to the leading intellectuals of the day, winning great acclaim. All of this encouraged settlers to continue pushing westward, even though lands in the existing states could support far greater populations than they then had. Less than 20 years after their return, sailing ships were making regular voyages around South America to trade with the West Coast. Forty years after their return, gold was discovered in California, launching the California Gold Rush. These events would have occurred with or without Lewis and Clark, but their reports likely accelerated the settling of the American West with all of the good and bad that accompanied the process.

Another long lasting impact made by the Corps of Discovery was in the area of relations with the Native American tribes west of the Mississippi. Jefferson was deeply interested in establishing political and trade relations with these tribes for strategic advantage over the French and British, as well as for economic gain for the United States. Unfortunately, Lewis (who took the lead in most of the interactions with Native Americans) was condescending, treating many of the people with whom he dealt as children. This engendered resentment and animosity among some tribes, a few of which attacked the expedition at various times. At other times, the behavior of the men towards the natives they encountered was less than exemplary, causing further problems. Finally, Lewis was instructed to encourage tribal leaders to visit Jefferson in Washington and succeeded in persuading several to do so. Unfortunately, some of these men died during their travels and others were treated poorly when they arrived. These negative experiences, along with the American government's tendency to make and break treaties, caused many problems over the next century.

The final major impact made by the Corps of Discovery was to lay to rest the hope of an easy passage between oceans. Jefferson was certain that an easy path existed for travel across the North American continent. He had no idea that the Rocky Mountains were as high or as rugged as they turned out to be, just as he was sure that a short and easy canoe portage would suffice to take one from the headwaters of the Missouri River across the Continental Divide to the headwaters of the Columbia River. In this, he was mistaken, as were many of the day's top thinkers. Lewis and Clark showed that any travel across North America was going to be long, difficult, and risky for many years to come. In effect, their explorations helped to make North America a larger place.

Although the expedition was ostensibly sent to explore the Louisiana Purchase and to try to find an easy route to the Pacific, Lewis and Clark were also instructed to explore as far north as possible while remaining within the Missouri River drainage basin. The Louisiana Territory extended throughout this drainage basin and Jefferson, as well as many in Congress, hoped that a major tributary would be found that ran primarily to the north, giving the United States a valid claim to much of Canada. Needless to say, such a river was not found and the national boundary was eventually fixed at its current location. However, the Corps' explorations beyond the Missouri River basin and into the Columbia River basin were clearly beyond the boundaries of the Louisiana Purchase and, therefore, outside the borders of the United States. Although they could not justify a northward extension of U.S. territory, they did help to extend the country's borders west to the Pacific.

P. ANDREW KARAM

Further Reading

Ambrose, Stephen E. *Lewis & Clark: Voyage of Discovery.* Washington, DC: National Geographic Society, 1998.

Ambrose, Stephen E. *Undaunted Courage: Meriwether Lewis, Thomas Jefferson, and the Opening of the American West.* New York: Simon & Schuster, 1996.

Moulton, Gary, ed. *The Journals of the Lewis and Clark Expedition.* Lincoln, NE: University of Nebraska Press, 1988.

Zebulon Pike and the Conquest of the Southwestern United States

Overview

In late October 1806, Zebulon Montgomery Pike (1779-1813) led an expedition that professed its main goal as mapping the Arkansas and Red Rivers. In reality Pike's explorations may have been designed to gauge the military strength of a potential enemy, Spain, and possibly even provoke an international incident which would lead to war. Nonetheless his journey, although fraught with error and controversy, proved to be influential on the development and conquest of the region and had an impact on settlement patterns throughout the western United States in the eighteenth century.

Background

The United States in 1806 was a growing country. Just three years previous in 1803 the country had secured the Louisiana Purchase, one of the largest land deals in western history, from France. That same year President Thomas Jefferson sent Meriwether Lewis (1774-1809) and William Clark (1770-1838) to make a survey of the newly acquired land. Before this time, most of what is currently the western United States belonged to Spain. After Spain ceded large parts of the territory to Napoleon, the French leader wasted little time in selling the land to the United States to raise money to finance his campaigns in Europe.

The rest of what is now the southwestern United States remained in Spanish hands. This includes present-day Texas, New Mexico, Arizona, Utah, Nevada, Colorado, and California. The Spanish prohibited American traders from operating in the areas under their control, as they were extremely wary of the United States Government's designs on the region. Even after the large acquisition of the Louisiana Purchase territories, many in the United States government coveted the rest of the Spanish lands. Not the least of these was the United States Army's ranking officer at the time, and Governor of Upper Louisiana, General James Wilkinson.

Pike may have been the commander of the 1806-1807 expedition, but General Wilkinson was the mastermind. Wilkinson was a complex character at best and a traitor at worst. Wilkinson had been in the pay of the Spanish government for years, referred to as Number 13 in Spanish diplomatic correspondence, not for political or ideological reasons, but simply as a means to supplement his lifestyle. At one time he received $12,000 by supplying fake invasion plans of the southwest to the Spanish. Pike's expedition probably was a key element in one of Wilkinson's biggest plots.

One of Wilkinson's partners in this scheme was then Vice President of the United States Aaron Burr. Burr and Wilkinson hoped, using Pike as a willing or unwilling dupe, to instigate a war with Spain. Then he and Burr would lead an army of their own against the Spanish with the goal of securing a piece of the area for a private empire.

It is not conclusively known if Pike was aware of Wilkinson and Burr's scheme before he began his expedition. Pike was an ambitious man and he felt that a peacetime army left him little room for advancement and fame. Upon seeing the accolades given to Lewis and Clark upon their return from their great explorations, Pike was able to secure a position as the leader of an expedition to locate the source of the Mississippi River. He failed, in that he missed the true headwaters by 25 miles (40 km), but returned to St. Louis in the spring of 1806 and by the fall had left on his trip to map the Arkansas and Red Rivers.

Yet Pike admitted to Wilkinson that he had a plan for reconnoitering Spanish territory and reaching Sante Fe. He would move into Spanish-controlled lands and when confronted by Spanish authorities claim to be lost and then offer to visit the government in Sante Fe to offer explanations and apologize.

Pike could not have been considered the most experienced or the best man for the job at hand. The journal of his expedition was confiscated by the Spanish, and the notes he was able to hide from them are described as "patchy." Many of the facts he reported were reviewed as "for the greater part very inaccurate," and he is reputed to have stolen and copied the map he made of the area from a German mapmaker named Alexander von Humboldt (1769-1859).

In addition, Pike and his men suffered through an ill-advised and poorly planned winter crossing of the Rocky Mountains. Besides this they really did get lost several times. At one point Pike realized that they had been traveling in a circle for over a month. After venturing into Spanish territory, he and his party were arrested just as he had "planned" at the trip's outset. The Spanish confiscated nearly all of Pike's journals and notes written until that time, and what reports Pike was able to bring out after that had to be smuggled. Taken to see Spanish officials in Sante Fe, and eventually deep into present-day Mexico, Pike returned nine months after he left, on what he had foreseen as a four-month's travel, to a nationwide scandal.

Burr and Wilkinson's scandal broke and the results of this severely tainted Pike's expedition. It turned out that Wilkinson had no authority to even order such a mission and Burr was brought to trial over the plot. Burr was eventually acquitted. Pike, for his part, refused to say anything bad about Wilkinson and even wrote passionate pleas in the General's defense. Wilkinson was able to save his own career by trying to pin everything on Burr and in the end it is Pike, and what positive work his expedition accomplished, that suffered the most as the result of Wilkinson's machinations.

Pike died, in the Battle of York, (now Toronto) Canada, during the war of 1812.

Impact

Pike's expedition directly led to the conquest and settlement of the Spanish, and later Mexican, lands of the Southwest by the United States. By 1846 war did finally come to the area but it was not between the United States and Spain, as Mexico had won independence from the Spanish in 1821. This was the war Wilkinson, Burr, and possibly Pike had wished to start in 1806. The end result was a treaty in which the United States was "sold" New Mexico, Arizona, Utah, Nevada, Colorado, and California. In addition Texas seceded from Mexico and joined the United States.

This annexation of the lands, of which Pike was the first American to explore, served to ultimately drive any European influence from what is now the continental United States. If Spain and/or Mexico had kept possession of this region, the balance of power in the western hemisphere would have evolved down a much different path.

While it is highly likely that this "purchase" of the Southwest would have occurred even if Pike had never set foot on his travels through the area, his journey also served to publicize this part of the country and brought it to the forefront of the national scene. Later expeditions, notably those of Stephen Harriman Long in 1820, built on the groundwork, however shaky, laid down by Pike and brought a more accurate and scientific study of the area.

Pike's journals and descriptions of the area made note of the abundant wildlife. As soon as it was realized that fur and valuable minerals existed in the area in great abundance, the fur trade thrived, and trappers and fortune seekers moved into the region, hastening the time when more and more permanent settlements would rise up. This was just one of the many reasons that lead to the large-scale displacement, exploitation, and oppression of the Native American population in the area.

Pike's expedition also had great influence on settlement patterns in the United States throughout the 1800s. The idea of the American Great Plains as the "Great American Desert" may seem like a fallacy today but in Pike's time, a place so empty of timber must have seemed practically inhospitable to someone raised in the then heavily forested eastern United States. This image of the Great Plains as desert persisted throughout nearly all of the 1800s. As a result initial western settlers concentrated on finding a route through the Rocky Mountains to California and the Northwest, leaving most of this area uninhabited until late in the homestead movement.

Pike described the city of Sante Fe as vitally important as a trade center in the area and that it had tremendous potential. He noted that it was lightly defended, and his descriptions of the city were to lead others in the future to seek their fortune by trading in the region. William Becknell (1796?-1865) pioneered a route to Sante Fe from Pike's starting point in Saint Louis, which later became the famous Sante Fe Trail.

Later, Missouri Senator Thomas Hart Benton proposed that a federal road be constructed along this route. The Sante Fe road was finished in 1827. By 1831 traffic on the road was heavy, and there are estimates that over 130 wagon trains a year traveled the road from St. Louis to Sante Fe. Benton went on to become one of the main proponents of the Manifest Destiny movement by the 1840s. This movement culminated with the passing of the Homestead Act, which became law in 1863. With free land in the West available for anyone with the will to take it, it wasn't long before the West was teeming with settlements.

While not all of the results of the expedition of Zebulon Pike were positive, and the motives for his journey controversial, the impact of this venture in the history and development of the southwestern and western United States cannot be ignored.

JOHN B. SEALS

Further Reading

Abernathy, Thomas. *The Burr Conspiracy.* New York: Oxford University Press, 1954.

Frazier, Ian. *Great Plains.* New York: Penguin, 1989.

Hollon, W. Eugene. *The Lost Pathfinder: Zebulon Montgomery Pike.* Norman: University of Oklahoma Press, 1969.

Jackson, Donald. *The Journals of Zebulon Montgomery Pike.* Norman: University of Oklahoma Press, 1966.

Stallones, Jared. *Zebulon Pike.* New York: Chelsea House, 1992.

The Temples at Abu Simbel

Overview

The great temples of the pharaoh Ramses II at Abu Simbel had been unknown to the West until 1813, when they were visited by the Swiss explorer Johann Ludwig Burckhardt (1784-1817). In 1817 Giovanni Belzoni (1778-1823) began the process of digging away the sand that had hidden most of the site. Copies of its hieroglyphic inscriptions were used by Jean-Francois Champollion (1790-1832) as he completed deciphering the Egyptian script a few years later. In the 1960s the temples at Abu Simbel were moved to prevent their submersion as a result of the construction of the Aswan High Dam.

Background

The famous temples at Abu Simbel, in the southern Egyptian region of Nubia, were built by the pharaoh Ramses II in about 1260 B.C. to celebrate the thirtieth year of his reign. Building ornate tombs and temples to their own memory was always a priority for pharaohs, who were regarded as "god-kings." To the Egyptians, these statues represented extensions of their being. But Ramses II may have built more monuments to himself during his 67-year reign than any other ruler in antiquity. Abu Simbel was among his most ambitious undertakings. "His Majesty commanded the making of a mansion in Nubia by cutting in the moun-

The colossal Egyptian temple of Abu Simbel, built by King Ramses II during the thirteenth century B.C. *(AP/Wide World Photos. Reproduced by permission.)*

tain," reads an inscription. "Never was the like done before..."

A horde of artisans was required to carve these massive monuments in stone into the rose-colored sandstone cliffs beside the Nile. Its presence would help to remind the pharaoh's distant Nubian subjects of his might. The remoteness of the site also kept it out of the way of the priestly hierarchy, which might have taken issue with the degree of self-aggrandizement Ramses intended.

Abu Simbel was already holy ground, with shrines dedicated to the local gods Horus of Meha, and Hathor of Ibshek. When Ramses appropriated it, he took precautions against incurring the wrath of these deities by including images of them in his newer, larger monuments. While other Egyptian gods were honored in the

temples as well, none loomed larger than Ramses himself.

The complex at Abu Simbel consisted of two temples. Their basic designs were similar to those built in the open. At the entrance to the Great Temple were four huge seated statues of Ramses. Each was 67 feet (20 m) high, and weighed 1,200 tons. The adjacent Small Temple was dedicated to his favorite consort, Nefertari, "for whose sake the very sun does shine." On its façade were six giant statues, four of the pharaoh and two of his queen, and smaller images of their children. Like most ancient Egyptian monuments, the statues were painted with red ochre and other pigments that wore off over the long centuries. Just inside the entrance to the larger temple, in the Great Hall, were eight 30-foot-high standing statues of Ramses, four on each side. The halls and chambers of the temples continued 160 feet (49 m) into the cliff.

The ancient Egyptians' knowledge of astronomy was evident in the construction of the Great Temple. It was oriented so that twice each year, in February and October, the rising sun would stream all the way into the innermost sanctum and wash over two of the statues seated there, one of Ramses and one of Amun, the god of the southern Egyptian capital of Thebes. The first October date may have been chosen 3,200 years ago to correspond with the temple's opening ceremonies. The other seated statues in the inner sanctum were those of the gods Ptah and Re-Harakhti. The walls were covered with inscriptions and bas-reliefs.

The priests probably continued to maintain the temples at Abu Simbel for a few hundred years after Ramses's death. Eventually, though, Egypt's hold on Nubia began to loosen. In the sixth century B.C., Greek mercenary soldiers scratched a paragraph of graffiti into Ramses's shin, among the oldest Greek inscriptions known. After this, Abu Simbel appeared to be forgotten. It was not mentioned along with the pyramids in the Seven Wonders of the World known to the ancient Greeks and Romans. For centuries the temples sat unvisited, slowly being covered by drifting sands.

Impact

In 1813, the Swiss explorer Johann Ludwig Burckhardt was traveling up the Nile in an attempt to reach the interior of Africa. He turned back just before the Third Cataract, in what is now Sudan. On his return journey he decided to stop and look for the "temple of Ebsambal," of which he had heard rumors. These stories referred to the Small Temple, most of which was still exposed. Inhabitants of nearby villages sometimes hid there when nomadic Bedouin raided their homes.

As Burckhardt was leaving the Small Temple, he stumbled upon a line of colossal buried statues. Only one was exposed to the extent that he could see its face. "A most expressive youthful countenance," he wrote in *Travels in Nubia*, "approaching nearer to the Grecian model of beauty than of any ancient Egyptian figure I have seen." He guessed that the statues guarded the entrance to another temple, cut into the rock cliff and hidden under the sand.

Four years later, Giovanni Battista Belzoni, an Italian adventurer sent to collect antiquities for the British Museum, spent three weeks digging away enough sand to proceed past the entrance of the Great Temple. "Our astonishment broke all bounds," he wrote in *Voyages in Egypt and Nubia*, "when we saw the magnificent works of art of all kinds, paintings, sculptures, colossal figures, etc., which surrounded us."

Belzoni could not read the hieroglyphic inscriptions on the walls, because the ancient Egyptian writing would not be deciphered for another few years. In fact, copies of inscriptions from Abu Simbel, sent to Jean-Francois Champollion in 1822, provided some of the clues that, along with the famous Rosetta Stone, helped him decode the script.

Inside the Great Temple, Belzoni made sketches of what he saw, but it was so hot that perspiration made the paper wet and drawing difficult. Perhaps as a result, a number of errors were made in his finished artwork. For example, he depicted all the Great Hall statues of Ramses wearing the double crown of Upper and Lower Egypt, when in fact the row of statues on the south side wore only the single crown of Upper Egypt.

The English were in competition with the French in searching for great finds in Egypt, and Belzoni's triumph scored one for England. This rankled the French, and Belzoni received poison pen mail and death threats. A rival claimant even shot at him, but fortunately missed.

In 1819, another expedition cleared away enough sand to reveal that the statues on the façade were seated. The exposure of the four huge statues coincided with an increase in Western interest in Egypt, and tourists began coming to Abu Simbel.

The sands continually threatened to encroach and blanket the temples all over again. In 1892 diversion walls were built to hold them at bay. These walls were reinforced in 1910, and the temples were no longer in danger of being swallowed up by the desert. Ironically, it would be water that would next endanger them, more than half a century later.

In the 1960s, the Aswan High Dam was built to control the floodwaters of the Nile. It was 180 miles (290 km) downstream from Abu Simbel, and what was to become Lake Nasser began to accumulate behind it. In a major international project initiated by the United Nations Educational Scientific and Cultural Organization (UNESCO), the threatened temples were cut from the cliffs and moved in 950 huge blocks to a safer spot 212 feet (65 m) up and 690 feet (210 m) back from the shore. Today, their original site, along with much of the ancient land of Nubia on the banks of the Nile, is underwater.

SHERRI CHASIN CALVO

Further Reading

Beaucour, Fernand, Yves Laissus, and Chantal Orgogozo. *The Discovery of Egypt: Artists, Travellers and Scientists.* Paris: Flammarion, 1990.

Brown, Dale M., et al. *Ramses II: Magnificence on the Nile.* Alexandria, VA: Time-Life Books, 1993.

MacQuitty, William. *Abu Simbel.* New York: Putnam, 1965.

The Rosetta Stone: The Key to Ancient Egypt

Overview

The Rosetta Stone was found in Egypt in 1799. Inscribed upon it was the same text in hieroglyphs, another Egyptian form of writing called demotic, and Greek. Since Greek was well understood, the stone provided a key to deciphering the others. With all the inscriptions that still existed in stone, the ability to understand hieroglyphs vastly increased our knowledge of the civilization of ancient Egypt.

Background

Hieroglyphs, the stylized pictures and other symbols used in ancient Egyptian texts, were used for almost 3,500 years. They have been beautifully preserved over the millennia, etched into many stone walls and tablets. Yet the knowledge of their meaning was lost after the fourth century, when Egypt came under Byzantine rule. The latest known hieroglyphic inscription was carved at the Philae Temple in 394 A.D.. This temple dedicated to the goddess Isis was one of the last surviving centers of Egyptian religious ritual, owing its longevity to the popularity of the Isis cult among Greeks and Romans.

The word hieroglyph comes from the Greek for "priestly carving." For many hundreds of years, people didn't realize that hieroglyphs were simply the way the ancient Egyptians wrote. They thought the symbols were a type of magical code, and that one had to be an initiate of the esoteric arts in order to understand them. A few scholars did make attempts to decipher them, but with little success. What passed for standard reference works, such as the fourth- or fifth-century *Hieroglyphika* of Hor-Apollo, an Egyptian who wrote in Greek, and the translations of the seventeenth-century German linguist Athanasius Kircher, were little more than fantasies.

In July 1799, a year after Napoleon's armies had captured Egypt, a group of French soldiers would come upon the key to understanding the ancient civilization of Egypt. Working on a fort near Rosetta in the Nile Delta, they found a black basalt slab about the size of a desktop covered with inscriptions, which would come to be called the Rosetta Stone. The inscriptions were unusual in that they were in three different scripts. The French officers realized the stone might be important, and sent it on to Alexandria.

The inscription at the top of the stone was in Egyptian hieroglyphs. At the bottom there was an inscription in Greek. Between them was *demotic* script, a later cursive form of hieroglyphs that came into use around 600 B.C., derived from an earlier cursive script called *hieratic*. The word "demotic" comes from the Greek *demos*, meaning "of the people," indicating that this was the script used for everyday purposes by that minority of the population that was literate.

Looking at the stone, the French began to realize that the three inscriptions might say the same thing. This would make the Rosetta Stone an incredible gift to scholars, because the Greek version of the inscription could be readily understood, providing a key to the other two.

The text of the stone was a royal edict dating from 196 B.C., during the time of the

The Rosetta Stone. *(Corbis Corporation. Reproduced by permission.)*

Ptolemies. The Ptolemies were a Macedonian family who ruled as kings of Egypt from 323 through 30 B.C., after the death of Alexander the Great. During this time, the official language of the court was Greek, but demotic script was used to communicate with the people, and hieroglyphs were used to impress them and appease their gods. The Rosetta Stone commemorated the reign of Ptolemy V Epiphanes on the occasion of the anniversary of his coronation. It proclaimed the devotion of the king to the gods and to the Egyptian people.

Impact

The Rosetta Stone was shown to Napoleon, who was very impressed. He arranged for printers to come from Paris and make copies of the inscriptions by inking the stone and laying paper upon it. The copies were sent to the best linguists in Europe. The Greek text was translated in 1802 by the Rev. Stephen Weston. Next, work began on the demotic text. In 1803, a Swedish diplomat named Johan Akerblad published his initial results, identifying the proper names in the text and a few other words.

The linguist whose work was most instrumental in understanding the hieroglyphic text was Jean-Francois Champollion (1790-1832). Champollion had been fascinated by ancient languages since childhood. He began working on the Rosetta Stone inscriptions in 1808 when he was 18 years old.

Champollion made three basic assumptions in his effort to decipher the hieroglyphs. He looked for hints in the script used by the early Egyptian Christians, or Copts, assuming that this represented the last remnants of the language of the pharaohs. Soon after he began working on the Rosetta Stone inscriptions, he identified correspondences between the Coptic alphabet and 15 signs of the demotic script.

Then, Champollion realized that although the hieroglyphs obviously included *ideograms*, symbols intended to represent objects or ideas, there were also *phonograms*, symbols representing sounds. In most written languages, ideograms were gradually discarded as phonetic symbols took hold, but the ancient Egyptians retained them both.

Finally, he recognized that the groups of hieroglyphs encircled by an oval loop, or *cartouche*, were phonetic symbols for the pharaohs' names. Champollion found the name Ptolmys in Greek and demotic, and so was able to decipher the cartouched hieroglyphic characters for the name as well. An obelisk found by Giovanni Belzoni and sent to England also bore both Greek and hieroglyphic texts. From this inscription Champollion was able to pick out the name Kliopadra, defining the sounds of a few more hieroglyphic signs. Champollion had realized that since the names of these Ptolemy rulers were Greek in origin, they would have no meaning in the Egyptian language. Therefore they would be represented only with phonetic symbols. A copy of an inscription from the temple of Ramses II at Abu Simbel afforded additional clues.

During the Late Period (712-332 B.C.) there were as many as 6,000 different hieroglyphs in existence, although no more than about 1,000 were in general use at any one time throughout most of ancient Egyptian history. As in Hebrew and Arabic, the phonograms corresponded only to consonants. Vowels were simply omitted. In English this would correspond to writing "brk" for "brook," "break," and "brick." In hieroglyphic text, a special ideogram called a *determinative* would be added to remove the ambiguity. In our example, a determinative for water would be added to "brk" to convey the meaning "brook."

Hieroglyphic text is also like other Middle Eastern languages in that it was generally written from right to left. Unlike them, there was also the alternative of going from left to right. The pictures of people and animals in the text always face toward the beginning of the line. If the inscription is written from top to bottom, as is also common, the signs face toward the beginning of the series of columns.

Champollion took 14 years to solve the puzzle of the hieroglyphs. In 1822, he wrote a letter to the French Royal Academy of Inscriptions, explaining his results. He defined an alphabet of 26 letters and syllabic signs, of which about half turned out to be correct. He also included an explanation of determinatives. In 1824 Champollion published his book *Precis du Systeme Hieroglyphique*, in which he expanded upon the information in the letter, as well as correcting some of his own mistakes and a few of the English physicist Thomas Young. Young had been working on the Rosetta Stone inscriptions and made substantial progress, independently coming to some of the same conclusions as Champollion. His work had been published in 1819, in a supplement to the fourth edition of the *Encyclopedia Britannica*.

Champollion died of a stroke in 1832 when he was only 41; his *Egyptian Grammar* and *Egyptian Dictionary* were published posthumously. In 1897, an exhaustive reference called the *Berlin Woerterbuch* was begun, including all the words in all the known Egyptian manuscripts and inscriptions.

Additional copies of the Ptolemy V text were later found in other locations, allowing Egyptologists to fill in sections of hieroglyphs that had been missing where the top of the Rosetta Stone was broken off. The stone itself had changed hands soon after its discovery, when Napoleon's forces were routed by the English. Today it is displayed in the British Museum.

SHERRI CHASIN CALVO

Further Reading

Andrews, Carol. *The British Museum Book of the Rosetta Stone.* New York: Peter Bedrick Books, 1981.

Davies, W. V. *Reading the Past: Egyptian Hieroglyphics.* Berkeley: University of California Press, 1987.

Scott, Joseph and Lenore Scott. *Egyptian Hieroglyphics for Everyone.* New York: Funk and Wagnalls, 1968.

Wilson, Hilary. *Understanding Hieroglyphics: A Complete Introductory Guide.* Passport Books, 1995.

James Clark Ross and the Discovery of the Magnetic North Pole

Overview

James Clark Ross (1800-1862), commander in the British Navy and England's most experienced and successful Arctic explorer, discovered the Magnetic North Pole in June 1831. During the eighteenth century, explorers wanted to find a Northwest Passage connecting the Atlantic and Pacific Oceans. It was one of the most important exploration goals of the time. While Ross did not discover the Northwest Passage, his discovery was judged a significant achievement both in science and in Arctic exploration. Finding the Magnetic North Pole advanced knowledge of the Earth's magnetic field. Knowing its location allowed mariners, sailing in any part of the world, to better fix their position.

While the scientific discovery of the Magnetic North Pole had little social or political impact, Ross raised the obsession with Arctic exploration to a fevered pitch. For the rest of the century, explorers continued to seek, but with greater passion, the Northwest Passage and the geographic North Pole. Their quest also became a test and affirmation of human ingenuity, stamina, and desire.

Background

When nineteenth-century explorers sought a Northwest Passage that would connect the Atlantic and Pacific Oceans, they expected to find it by sailing through the icy waters west of Greenland and tacking around frozen landmasses in the far north of Canada. Finding the passage became an obsession, similar to the quest for the Holy Grail (the chalice Christ was thought to have used during the Last Supper) in the Middle Ages, or the race for the moon during the twentieth century.

Some Arctic explorers were driven by their blind expectations and desires for fame rather than by scientific proof that such a passage might exist. As a result, many expeditions proceeded unwisely and were trapped when their ships become lodged in ice as winter closed in. Many Arctic explorers died.

Arctic explorers also sought the Magnetic North Pole and the geographic North Pole. The geographic North Pole is at the top of the world, but the Magnetic North Pole, some distance from the geographical North Pole, is the polar part of the northern hemisphere's magnetic field. Earth's magnetic field can be measured by several means: a compass that points horizontally north or a wire instrument, called a dip circle, that dips due south at the Magnetic North Pole (the same instrument at the magnetic equator would register horizontally).

By 1830 scientists knew that Earth's magnetic field varied in intensity from location to location and that Earth's magnetic field could be measured by three elements: horizontally (by a compass that points to the north magnetic pole), vertically (by the dip circle that dips either at an angle or, on the north magnetic pole, dips straight down), and what is called the "total magnetic field." All three measurements intersect in a certain way at the Magnetic North Pole.

The fact that the Magnetic North Pole and the geographical North Pole do not coincide is important to mariners charting their position who must know not only the horizontal direction of north via the compass, but the angle between the direction north, as indicated by the compass, and the angle of the magnetic dip, called the "declination" of Earth's magnetic field. Declination varies around the world and accurate measurements of declination in various places in the world are maintained and charts drawn from measurements. From these charts and a compass, a mariner can know his or her position.

The precise location of the Magnetic North Pole, however, has not always been known.

Impact

In 1831 British Navy commander, naturalist, and explorer James Clark Ross searched for the Magnetic North Pole on Boothia Peninsula, north of Canada and west of Greenland. At that time, Ross was Britain's most experienced Arctic explorer. He joined the British Navy at the age of 12, served on his first Arctic voyage at age 18 under his uncle John Ross (1777-1856), and was promoted to commander in 1827. Ross had earlier searched for the Magnetic North Pole during his 1827-31 Arctic voyage with John Ross.

Interest in Arctic expeditions reached its height in Great Britain in the 1830s and 1840s. Part of the excitement was tied to the notion of a British sense of superiority and the necessity to prove it. Success in the harsh Arctic climate was regarded as a way to demonstrate the worth and power of the British people, whose colonial empire was still extensive, but also beginning to decline.

During the winter of 1831, when James Ross's ships were trapped in the Arctic ice, he led a series of overland expeditions aimed at mapping Arctic features and, if it could be located, finding the Magnetic North Pole. Expeditions could not start until the long Arctic winter of darkness was over. By March 1831 Ross and his party, accompanied by Inuit (Eskimos) with sleds and dogs, were able to set out to chart land features and explore.

Ross, unlike many Arctic explorers before him, had become friendly with the Inuit. He learned their language, learned about their culture, and learned how to cope with the environment as they did, using clothing made of furs and seal skins. He also learned to travel by dog sled and to live in native igloos. These adaptations not only allowed him to explore and hunt for food more efficiently than his predecessors, but also prepared him for his ultimate successes in discovery.

After the expeditions, each lasting weeks, the Ross party returned to the iced-in ship. On his third foray away from the ship, in June 1831, Ross took the proper instruments that would tell him if he had found the Magnetic North Pole, which he thought was nearby. At the time, he was traveling in a large area of land British explorers had named "Boothia," after the British gin manufacturer Felix Booth, who had financed exploration.

At 8 o'clock in the morning on June 1, 1831, Ross reckoned that he had reached the Magnetic North Pole. To make sure, he used a "dip circle," an instrument from which he suspended horizontal delicate needles. When his dip circle registered a dip in the needle of 89 degrees and 59 minutes, due south, close to 90 degrees, Ross knew he was at the Magnetic North Pole. Their compass and chart readings also told him that he was at latitude 70 degrees, 5 minutes, and 17 seconds north, and longitude 96 degrees, 46 minutes, and 45 seconds west. This was then the first time the Magnetic North Pole had been located and stood upon. But there was nothing there to be seen.

Ross wrote in his log that "even I could have pardoned anyone among us who had been so romantic or absurd to expect that the magnetic pole . . . was a mountain of iron, or magnet . . . but nature had here erected no monument to denote the spot which she had chosen as the centre of one of her great and dark powers."

Ross and his party claimed the barren but important spot for England and King William. They erected a pile of stones, called a "cairn," on the spot after placing a canister bearing record of the discovery at the bottom.

Ross was surprised to find that the Magnetic North Pole moved even as he measured its location. A century and a half later, scientists said that Ross was fortunate to be able to measure the Magnetic North Pole because it was later found that the location could change as much as 30 minutes in an hour; Ross took an hour to measure his location. Since Ross first found it, the Magnetic North pole has moved to latitude 75 degrees, 5 minutes north, and longitude 100 degrees, and 5 minutes west.

Earth is not uniformly magnetized over its entire surface. Scientists also now know that not only does the Magnetic North Pole move, but also the intensity of Earth's magnetic field varies from place to place and, over time, diminishes.

When they returned to England after four winters in the Arctic and the achievement of finding the Magnetic North Pole, James Clark Ross and his men were cheered as heroes. Ross enjoyed fame and was even the subject of a poem recited by school children:

Sir James Clark Ross, the first whose sole / Stood on the North Magnetic Pole!

Ross was offered knighthood with the title "Sir," but he turned it down. He was later knighted after his expeditions to the Antarctic in an unsuccessful bid to locate the Magnetic South Pole.

Excitement over Arctic exploration and the search for the Northwest Passage became less important in the 1850s after Ross's fellow explorer and noted British Navy commander John Franklin (1786-1847) and his crew disappeared in the Arctic. Not even Sir James Clark Ross could find them. Ross, who sailed in 1848 to search for them, eventually returned tired and sick from the ill-fated and unsuccessful rescue voyage.

Using many of Ross's charts and techniques, including his practice of relying upon the technology and assistance of Arctic native peoples, the Northwest Passage between the Atlantic and Pacific Oceans was successfully navigated by Norwegian explorer Roald Amundsen (1872-1928) in 1903-5. Amundson used the Magnetic North Pole, which Ross charted, to help locate his positions and find his way. American Robert Peary (1856-1920) was the first to set foot upon the geographical North Pole in 1909. The Arctic "grail" had finally been retrieved.

After the Arctic was successfully explored and the magnetic and geographic North Poles located and stood upon, interest in Arctic exploration waned. However, the successes of nineteenth-century Arctic exploration, both scientific and heroic, inspired subsequent adventurers to seek out new places to explore. The same spirit of adventure that motivated Ross and other Arctic explorers would later drive twentieth-century efforts to travel into space and to visit the moon.

RANDOLPH W. FILLMORE

Further Reading

Berton, Pierre. *The Arctic Grail.* New York: Viking, 1988.

Dodge, Ernest. *The Polar Rosses.* New York: Harper and Row, 1973.

Lee, E. W. *Magnetism, An Introductory Survey.* New York: Dover Publications, 1970.

Nanton, Paul. *Arctic Breakthrough.* Toronto and Vancouver: Clarke Irwin, 1970.

Ross, M. J. *Polar Pioneers.* Montreal: McGill-Queens University Press, 1994.

The Voyage of the HMS *Beagle*

Overview

Charles Darwin (1809-1882) was among the most influential scientists who ever lived. He began his career as a naturalist aboard the HMS *Beagle*, on its five-year surveying mission around South America and across the Pacific. Darwin's work was to make the *Beagle's* journey one of the best documented surveys of its time. His observations would eventually result in his theory of evolution by natural selection. This theory holds that species change gradually because individuals best-suited to their environments are more likely to survive, reproduce, and pass their desirable traits to their offspring.

Background

Charles Darwin launched the greatest revolution in the history of biology with his theory of evolution by natural selection. However, like many original thinkers, he did not show extraordinary promise in school. Darwin came from a prominent and wealthy family. His grandfathers were the famous physician Erasmus Darwin (1731-1802) and Josiah Wedgwood (1730-1795), the manufacturer of the fine pottery that bears his name. Erasmus Darwin was a radical freethinker. Wedgwood, a Unitarian, was only slightly more orthodox. Yet their children Robert Darwin and Susannah Wedgwood, who met through their fathers' friendship, led a conventional, upper-class English life, and raised their own family in the established Anglican Church.

Charles Darwin was a born naturalist, and enjoyed tramping around the woods and collecting rocks and insects. Being an outdoorsman was acceptable for a nineteenth-century English gentleman; in particular, hunting was a popular pastime. But science was considered a hobby, not a suitable occupation. At school, the emphasis was on Greek and Latin, and young Darwin was scolded for wasting his time on chemistry experiments.

Given his scientific bent, a medical career seemed like a possibility, and so Darwin was sent to Edinburgh University in 1825. It was several decades before the advent of anesthesia, and watching an operation being performed on a screaming child put an end to his interest in medicine. At that point, Darwin's father Robert decided that the best place for his tender-hearted and somewhat eccentric son was in the Anglican clergy, and sent him to Christ's College at Cambridge.

Darwin was not particularly interested in becoming a clergyman, but neither was he alarmed by the prospect. At the time, many English scientists were clergymen. Gentlemen of independent means were able to arrange for a small country parish and a quiet life that gave them both a respectable occupation and time to indulge their own interests. In any case, Darwin was happy to go to Cambridge. He enjoyed university life. Most of his time was spent socializing, or in the outdoor pursuits he loved: hunting, fishing, and riding horseback. He became fascinated by beetles, and went to great lengths to collect them. He undertook many hiking expeditions with botany professor John Henslow (1796-1861) and geology professor Adam Sedgwick (1854-1913).

At about this time, Darwin had two experiences that greatly influenced the future direction of his thought. The first was a re-reading of a scientific text, *Zoonomia*, written by his grandfather Erasmus. Erasmus Darwin had noted that animals such as dogs and horses were bred for desired traits, changing the characteristics of their breeds over time. If that were possible, then perhaps many characteristics of species had changed during the course of their history, and possibly a number of species had come from a single root, or "living filament." Charles Darwin was very impressed with his grandfather's ideas, but realized he had been theorizing ahead of the evidence. No facts were presented that directly supported the theories.

While hiking with Sedgwick in Wales, Darwin observed what he considered to be the opposite error. A local workman had discovered the fossilized shell of a tropical mollusk in a nearby gravel pit. Fossils had been known since the seventeenth century; they were thought to be remnants of a time before the biblical flood. The tropical shell fascinated Darwin, but it gave Sedgwick pause. It was impossible, he declared, because Wales was not a tropical place. He seemed annoyed to have encountered something that did not fit into his established theories.

In combination, these experiences molded the way Darwin was to do science. He was wary about theorizing without sufficient evidence, and he recognized the importance of looking carefully at evidence, even (or perhaps, especially) that which seemed anomalous. As a result, he was inclined to collect as much data as possible before beginning to draw conclusions.

The opportunity to do so would arrive unexpectedly. In 1831, Captain Robert FitzRoy of the Royal Navy was about to undertake a surveying trip to South America with his ship the HMS *Beagle*. He already had hired one naturalist to study the plants, animals, and minerals of the South American coast and the islands nearby. He wanted a second naturalist, one who could afford to pay the expenses of gathering, storing, and shipping his specimens, and who was of the appropriate aristocratic social class to be good company for him. Professor Henslow recommended Darwin, who fit the bill in all respects. A man of orthodox religious beliefs, FitzRoy was especially pleased that Darwin had studied for the clergy. He hoped that he could find evidence that supported the literal truth of the Bible's account of Creation.

Impact

In preparing for his trip, Darwin read the latest works on natural history and geology. One of these was *Principles of Geology*, by Charles Lyell (1797-1875). Lyell proposed a new explanation for the extinct creatures that appeared in the fossil record. Rather than being evidence of a biblical catastrophe, he wrote, gradual changes in the Earth modified conditions in such a way that some unprepared creatures slowly died out. What most alarmed many of his readers was that this process would take many thousands or even millions of years. This was contrary to religious teachings that the Earth was about 6,000 years old.

Darwin was not by nature a religious iconoclast. He had carefully reviewed the Anglican theology before agreeing to study for the clergy, and found nothing there to object to. Yet Lyell's ideas were interesting to him. And, with his grandfather's book in mind, he took the reasoning one step further. If species were constantly dying out, yet there was still a huge variety of different plants and animals on Earth, new types must have arisen, and be arising still, to replace those that are lost. Was God continuing to intervene with new, separate acts of creation? With these questions in mind, Darwin set off on his journey.

The HMS *Beagle* was a relatively small wooden sailing vessel, only 90 feet (27 m) long, with 74 people on board. It sailed from Plymouth on December 27, 1831. Darwin was seasick throughout the trip, particularly when the weather was bad. The sociable young man was generally able to get along with the opinionated and rather bad-tempered Captain FitzRoy, but relations were occasionally tense. At one point, an argument over slavery, which Darwin opposed, led to his temporary banishment from the captain's table.

When they arrived at the coast of South America, near Salvador, Brazil, most of the crew stayed with the ship gathering survey data. Darwin was free to explore as he chose. The lush forests, full of plants and animals he had never seen before, were like a paradise to him. In general, he worked alone. As was the custom of the time, he studied birds by shooting and collecting them, as many as 80 different species on a single morning. He could accumulate dozens of insect specimens in a day. He found many fossils in Brazil and Argentina. Soon the *Beagle* began to fill up with Darwin's plants, animals, and rocks, hauled on board by the sackful. His shipmates nicknamed him the "Flycatcher."

Darwin studied, preserved, and labeled his finds, packed them up, and shipped them back to Henslow in England. He took such detailed notes that the voyage of the *Beagle* was one of the best-documented scientific expeditions of its time. His careful work made his reputation back home. On the ship, his courage and vigor spared him from the scorn sometimes leveled at learned passengers. He took many side trips, such as a journey on horseback across the pampas with South American cowboys, or *gauchos*. At the southern tip of South America, Tierra del Fuego, Darwin went mountain-climbing. During the course of the voyage, however, his health deteriorated, especially after the bite of a disease-carrying insect left him bedridden for seven weeks.

During the *Beagle's* years of travels in South America, Darwin gave a great deal of thought to Lyell's ideas about geology and the history of the Earth. Near the Straits of Magellan, he experienced an earthquake, and then noticed that the shape of the land around him had changed. Mussels that were on rocks underwater were now exposed, and left to die. He found fossilized seashells up in the mountains, and discovered a petrified forest at 7,000 feet (2,134 m) above sea level. He uncovered the fossils of prehistoric animals and, near Buenos Aires, the fossilized tooth of a horse. Yet he knew that there had been no horses in South America when the Spanish had arrived and re-introduced them in the sixteenth century. All these discoveries influenced Dar-

win's ideas about how changes had occurred on the Earth and in animal species.

In 1835, the *Beagle* sailed from Lima, Peru, on the western coast of South America, to a group of Pacific islands called the Galapagos. The name came from the Spanish word for giant tortoises, which were among the unusual animals Darwin observed there. He noticed that many of the animals and plants were unique to the islands. In fact, there were even many differences between the species on each individual island. The small size of the living community coupled with its geographic isolation made it an ideal laboratory for studying changes in species.

For example, each island had species of small birds called finches. Although these birds all looked similar, their beaks were of different shapes, depending on what type of food was available on the island. The differences were not confined to birds. Some types of lizards could swim well and ate seaweed. Others lived on the land and ate cactus. It seemed clear that, in some way, plants and animal species adapted to their environments.

After the Galapagos, the *Beagle* continued into the Pacific, visiting Tahiti, New Zealand, and Australia. It then skirted the southern tip of Africa. Off Africa's western coast, on the Cape Verde Islands, Darwin noticed that the animals, though distinct, were similar to those of Africa. He remembered that the distinct species of the Galapagos were more similar to those of South America. Why should these isolated species be the same in general form, but differ in details, from those of a nearby continent?

The *Beagle* sailed into the port of Falmouth on October 2, 1836. Darwin never left England again. Nor was he ever again in completely good health, although he married and had 10 children, and lived to the age of 73. Within a few years he had published an account of his travels, and then continued to study his data. Darwin, the careful observer, was loath to jump to conclusions. He would sift through his evidence for 20 years before publishing *On the Origin of Species*, the book that was to change the face of science.

SHERRI CHASIN CALVO

Further Reading

Darwin, Charles Robert. *Journal of Researches into the Geology and Natural History of the Various Countries Visited by HMS Beagle, Under the Command of Captain FitzRoy, R. N., from 1832 to 1836.* London: H. Colburn, 1839.

Darwin, Charles Robert. *On the Origin of Species by Means of Natural Selection.* London: J. Murray, 1859.

Dibner, Bern. *Darwin of the Beagle.* New York: Blaisdell Pub. Co., 1964.

Marks, Richard Lee. *Three Men of the Beagle.* New York: Alfred A. Knopf, 1991.

Moorehead, Alan. *Darwin and the Beagle.* New York: Harper and Row, 1969.

Robert H. Schomburgk Explores the Interior of British Guyana, Brazil, and Venezuela and Is the First European to Visit Mount Roraima

Overview

Because dense jungles and dangerous rivers made exploration difficult, little was known about the interior of South America in the early and mid nineteenth century. Despite the many dangers and difficulties in this region, Robert Hermann Schomburgk (1804-1865), a German-born explorer and naturalist, hired by the British government, traveled the rivers of present-day Guyana, Brazil, and Venezuela. He mapped geographical features and collected geological and botanical specimens. After discovering how major rivers connected, Schomburgk marked the boundaries of what became the modern nations. His exploration and mapping also opened the rivers for transport and commerce. Schomburgk's efforts also had the unintentional impact of paving the way for the twentieth-century development of the tropical rainforests and the continued extermination of South American native peoples in Venezuela and Brazil.

Background

When the European powers—Great Britain, France, Spain, and Holland—began to colonize northeastern South America in the sixteenth century, little was known about the interiors of what became the modern nations of Brazil, Venezuela, and Guyana. Because rivers, mountains, and other geographic features needed to be explored and charted before economic development could begin, explorers were often hired by European governments to investigate unknown lands.

The era between 1830 and 1860 was a time during which scientists, especially naturalists, were trying to understand the world of nature as well as understand the tribal peoples they encountered, whom they considered "primitive." It was during this time that naturalist Charles Darwin (1809-1882), who later postulated the

theory of evolution through the process of natural selection, sailed on the HMS *Beagle* off the west coast of South America. In the Arctic, explorers searched for a Northwest Passage between the Atlantic and Pacific Oceans. European explorers pushed into the African interior, into what was called the "dark heart" of Africa.

What is today the modern South American nation of Guyana was colonized by the French, British, and the Dutch in 1815. In 1831, the British gained control of three coastal settlements that became British Guyana. In 1835, and again in 1841, the British government sent expeditions to explore the interior Guyana. Each expedition was led by Robert Hermann Schomburgk, a German-born explorer and naturalist.

When Schomburgk explored the tropical rainforest wilderness of Guyana, Venezuela, and Brazil, he fought against great natural obstacles, yet traveled extensively and mapped never-before-seen land features and charted many rivers. In his travels, he also made first-time contact with many isolated tribal groups of South American native peoples, whom he attempted to count.

Impact

In October 1835, on his first expedition, Schomburgk led 22 men in canoes up what he later determined was British Guyana's longest river, the Essequibo. Not only was he interested in charting the rivers, he also wanted to retrieve animal and plant specimens for the British government and for his own scientific interests as well.

Dangerous rapids and frequent waterfalls slowed his travel. Not far into the journey, Schomburgk's party began suffering from dysentery and had to rest and regain their health during the month of November. The rainy season started in December, making further progress difficult when they finally continued pushing on up the Essequibo.

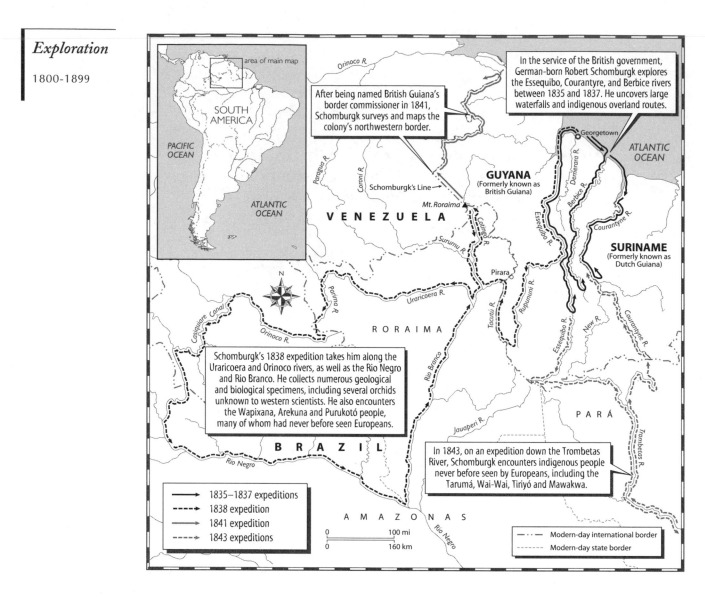

Map of Robert H. Schomburgk's journey.

By February 1836, Schomburgk's party reached the point where the Essequibo met the Rupunumi River, but could go no farther. Here they encountered an impassable waterfall. Since it had no native name, he named the 24-foot (7 m) falls "King William's Cataract," in honor of Great Britain's King and his employer. Unable to go down the falls and unwilling to spend months cutting a land path around it, he had to turn back. As they headed back to Georgetown, biological and geological specimens were lost when a canoe overturned in the rain-swollen, white water.

In September 1836, Schomburgk made a second river trip, this time traveling up the Courantyne River, the easternmost river in Guyana which now marks a boundary between Guyana and neighboring Surinam. Again he ran into an impassable waterfall. Schomburgk had heard from natives that there was a footpath in the wilderness around the falls, but was unable to locate it. Disappointed, Schomburgk turned back, but with the knowledge that this was a major river.

Not discouraged, Schomburgk tried to navigate another river in Guyana—the Berbice. Unlucky again, Schomburgk's foray up the Berbice was a near disaster. First, native guides deserted them. Then the expedition was set upon by thousands of ants, followed by packs of dangerous wild hogs, which sent his party scurrying up trees to escape. Dense jungle made the progress slow, and they nearly ran out of food before completing the trip. Despite his difficulties, over five years Schomburgk mapped the three great rivers of Guyana.

By the early 1840s, coastal South America was already developed agriculturally, with sugar cane the main crop. Sugar plantations flourished by using slave labor. When Schomburgk traveled into what is now Brazil along the Guyana-Brazil border, he happened upon Brazilian slave traders who had kidnapped native children for slave labor. In his account, Schomburgk reported that children as young as five or six years old were taken prisoner and sold into slavery. During this same time, as native peoples were being wiped out in coastal areas, Schomburgk observed that most native Guyana Indians had been exterminated. As the native source of slaves disappeared, plantation owners began importing African slaves, natives from the Caribbean islands, and indentured servants from Asia to work in the sugar cane fields.

Continuing his expeditions, Schomburgk traveled the Serum River, stopping at a mountain called Mount Roraima. He was the first Westerner to visit the mountain, which now divides Brazil, Guyana, and Venezuela. Schomburgk's brother, Richard, also active in South American exploration, tried to climb the mountain soon after Robert first arrived at Mount Roraima. Richard, Robert reported, was unable to scale the sheer walls that rise from half-way up to the flat top.

During this trip, as during all of his expeditions, Schomburgk met native peoples who had never before been seen by Europeans. He recorded their locations and, when he could, their numbers. Eventually, he reached the great Orincoco River that runs through now what is Venezuela. This river took him back to the Essequibo and to Georgetown, the main British settlement in Guyana.

Schomburgk's extensive travels served to establish a boundary between Guyana and Venezuela. He mapped what became known as the "Schomburgk Line," a mark between the two countries. In 1840, when he received an award from the Royal Geographic Society, Schomburgk recommended that a settlement be started on the Guyana boundary. The British government accepted the recommendation and awarded him the post of Guyana's "border commissioner."

Not content with an administrator's duties, in 1841 Schomburgk once again set out on a river quest, this time following the Essequibo to its source on the Brazilian border. All along the way he encountered tribal peoples who had never been seen by Europeans. As Schomburgk identified and counted many groups of native South Americans in Brazil and Venezuela, he

also reported that most of the Guyana natives had been severely reduced by disease and warfare. His was one of the earliest voices to draw attention to the human tragedy. "Their present history is the finale of a tragical drama; a whole race of men is wasting away," he wrote of the disappearing Indians. "The system of Brazilians of hunting Indians for slaves exists to this day in all its atrocities." Schomburgk concluded that in 1838 "the aggregate number of Indians in British Guyana" was only 7,000.

When he returned to England in 1844, Schomburgk was knighted and given the title "Sir." He became a naturalized British citizen.

Schomburgk's work at charting unexplored rivers and collecting botanical and geological specimens clearly aided in the economic development of the region, and the impact of his expeditions helped to increase trade and the wealth of exploitive colonial governments that extracted minerals, sugar, and other commodities called "cash crops." Much of this development was, however, at the further expense of remaining native peoples when, for their own gain, colonial governments and growing international corporations removed, annihilated, or enslaved native populations for the rest of the nineteenth century and on into the twentieth century.

Schomburgk's efforts were hailed as a great triumph for exploration and trade. However, since his exploration, extensive twentieth-century plantation farming, mining operations, and other development in parts of these areas have threatened to continue to destroy the region's tropical rainforests and the existence of the area's remaining native peoples. Developers seeking oil are the most recent group threatening dwindling numbers of South American native peoples. Schomburgk's "firsts" in meeting with South American native peoples have had an unintentional but lasting impact of making native peoples more vulnerable to more than a century of development.

Because South American rivers flowed through dense, almost impassable, jungle, many native groups were not contacted until a century later following Schomburgk's explorations. Unlike the development process in North America that both assimilated and exterminated Native American groups in the northeastern parts of North America, and then slowly pushed large groups of native peoples westward where they faced further extermination and reservation life, European developers in South America often destroyed isolated, relatively small groups of native populations as they found them.

Today, remaining native peoples in this part of South America continue to fight developers for their lives and try to slow the development in the tropical rainforests, in order to save the resources on which they depend.

RANDOLPH FILLMORE

Further Reading

Bodley, John. *Victims of Progress*. Palo Alto: Mayfield Publishing, 1982.

Bohlander, Richard E. *World Explorers and Discoverers*. New York: Da Capo Press 1992.

Goodman, Edward J. *The Explorers of South America*. Norman: University of Oklahoma Press, 1992.

Pear, Nancy and Daniel B. Baker. *Explorers and Discoverers, From Alexander to Sally Ride*. Detroit: UXL, 1998.

Edward Eyre Explores the South and Western Territories of the Australian Interior and Helps Open the Territories to the Transport of Goods and Animals

Overview

Australia during the nineteenth century was the site of extensive settlement and political and cultural expansion by the British colonial governments in the continent. In particular, the period between 1830 and 1860 saw a tremendous surge in the continent's population. One aspect of this expansion was the effort to explore the vast, sparsely populated land. Numerous explorers made arduous, brutal, and sometimes fatal efforts to explore Australia. Edward Eyre's 1840-41 journey through the South and Western territories of the Australian interior—a journey marked by thirst, starvation, freezing cold, and murder— epitomizes these expeditions. The journey assisted in opening up the South and Western territories to the transport of goods and animals.

Background

Edward John Eyre (1815-1901) arrived in Australia when he was 17 years old. He was born in England, where his father was a minister and had the benefits of classic schooling. Early on, he showed an interest in both government affairs and the businesses that were beginning to flourish in the newly colonized continent. When he was only 31 he served as lieutenant governor of New Zealand and held this post for seven years. (Later, he would serve as governor of Jamaica for one year in 1864-65.)

Eyre's main interests were sheep and cattle, and he spent much of his life not only building his flocks and herds of livestock but trying to find new ways of transporting them to the markets

Edward John Eyre. *(The Granger Collection LTD. Reproduced by permission.)*

where his wool and surplus stocks could be sold. His primary interest was sheep, since they could survive more easily in sparse vegetation zones.

Eyre pioneered the government's efforts to open up South Australia for both settlement and agrarian development. Not content with limiting his business interests to the coastline, Eyre was determined to explore the possibilities of routes to the central part of the continent, where sheep stations would encourage depots or townships where settlers could migrate.

Accordingly, in 1839 he organized a group on his own to reach the center of the continent. When he encountered what he thought would be a likely refuge—Lake Torrens—he discovered that it was entirely covered with salty mud. Finding saline swamps on one side and insurmountable sandhills on the other, he took the only other route open to him: the Flinders Ranges, which lay approximately 250 mi. (400 km) north of Adelaide and covered about the same distance due north. He got as far as the aptly named "Mount Hopelee" before turning back.

While Eyre was eyeing central Australia, the government in Adelaide was planning an entirely different direction for new growth. It believed that staying fairly close to the coast, where it hoped to establish a cattle trail, would be the most practical move as well as the safest. There were other considerations as well. While exploring the terrain for movement of cattle and sheep, the explorers were to keep their eyes open for good land that could be profitably farmed. (Many still had visions of the lush English countrysides, where lots of water made for bountiful crops.)

Because of his personal skills in working livestock as well as surviving in the bush, the government selected Edward Eyre to lead the expedition. With an eye to the future, Eyre not only accepted the responsibility, he volunteered to pay half the cost of the venture.

The Eyre party, which left in June 1840, consisted of six men, including Baxter, his station manager, and an aboriginal friend named Wylie who recruited two other aborigines. They anticipated it would take three months to reach Spencer Gulf and loaded their supplies with that time frame in mind. In addition, they took 13 horses and 40 sheep. All this would be augmented by the government, which was to send a ship with the remaining supplies and would be found at anchor in Spencer Gulf.

As many intrepid adventurers have found, exploration on paper seldom translates into actual field reality. When Eyre traveled across what is now called the Eyre Peninsula, he stayed on the coastal edges as planned. The unbelievably harsh conditions were compounded (and probably caused) by a total lack of drinkable water in any direction. Eyre was a practical man. He sent most of the members of the party back to Adelaide and kept only Baxter, Wylie, and the other two aborigines. His thought was that a smaller group would have a better chance of moving more rapidly, needing fewer supplies and livestock. When the four remaining men left Fowler's Bay, they took only 11 pack horses and 6 sheep. They estimated a trek of approximately 800 mi. (1,300 km) through some extremely rough country. Crossing the Nullarbor Plain (another aptly named area where no trees could be found) was especially difficult, with absolutely no shade from the fierce rays of the sun. There was little water and, because of the huge cliffs, the trekkers couldn't cool themselves in the waters of the bay.

They were almost certainly doomed by the time they reached the top of the Great Australian Bight and would likely have perished but for some friendly aborigines who taught them how to dig behind the sand dunes and occasionally find water.

In spite of this advice and help, they went without water for five days. Finally, they were saved by chancing upon some wells that had been found and perpetuated by some aboriginal tribe at the present site of Eucla, which is on the border between South Australia and Western Australia. In an effort to restore themselves physically and mentally, they remained at the well for six days before resuming their arduous journey.

They continued to move westward, keeping as close to the beach as possible. Once again, water was next to impossible to find and they followed the example set by the aborigines in the party of breaking off the roots of the gum trees and sucking on them for drops of moisture.

Walking on the sand was difficult for the pack horses and, to lighten their burdens, Eyre abandoned firearms, horseshoes, spare water bags, and even clothing. In spite of these efforts, the horses had to be left behind, one by one. By this time the party was using sponges to collect early morning dew from leaves as their only source of water. Food was getting very low and when one of the horses was obviously sick, the party killed the animal for food. Both Eyre and Baxter became very ill from this desperate move and could not go any further. The aborigines went ahead but came back in a few days, almost starving.

At this point Eyre and his men were about halfway to the West Australian coast. The killing heat of the summer sun was replaced by the cold of winter—especially noticeable at night. Having left much of their clothing behind with the other items abandoned earlier, the travelers were unprepared for the winter conditions. One night while Eyre was on watch, he was surprised to hear a gunshot. Wylie came running back with the news that two of the disgruntled aborigines had shot and killed Baxter, stolen most of the supplies and firearms, and fled. Wylie refused to go with them and remained with his longtime friend, Eyre. The ground was solid rock and they could not even provide a decent burial for Baxter, only covering him with a blanket and leaving him on the surface.

Wylie and Eyre plodded on, going without water for three days, with very few supplies left and almost 600 mi. (1,000 km) in front of them. After seven grueling days they found a waterhole that had been discovered earlier by the natives. The only food they had was kangaroos, which they killed and ate—with absolutely no enthusiasm. Wylie even consumed a dead penguin he found on the shore.

The indomitable pair continued walking for over a month in the direction of Western Australia. Finally, in June 1841 they chanced upon a French whaling ship that lay at anchor in the bay. Thanks to its Captain Rossitor (an Englishman), Eyre and Baxter were given food, water, and even some wine and brandy.

Eyre and Wylie recovered their strength and fitness in about two weeks. Since they now had good clothing, plenty of food and water, and renewed enthusiasm, they pushed on to their goal. It was, by far, the easiest part of the whole expedition. Even though they still had to contend with heavy rains and cold weather, they reached Albany in July. The entire trip had taken more than a year.

Impact

Eyre's success was hailed by the government, and the expedition's impact was substantial. It furthered the goal of opening South and Western Australia to the transport of animals and goods and assisted with the expansion of colonial settlement into all parts of Australia. Although the various territories were not yet politically unified, the enormous expansion of settlements throughout the remainder of the nineteenth century would eventually lead to the country's political unification in 1901. Nonetheless, the difficulty of Eyre's journey is reflected in the fact that even today vast areas of arid, rocky plains in South and Western Australia remain virtually uninhabited.

Eyre received many honors during his long life, including a gold medal from the Royal Geographic Society for his incredible journey. The Australian government named a large lake above the Flinders Ranges and just southeast of Oodnadotta for him. Further, the peninsula that bears his name is strategically located west of Adelaide and, since it is triangular in shape, covers many square miles. Unlike many other fearless explorers, Eyre lived a long, prosperous life, eventually returning to England, where he remained until his death in 1901.

As for Wylie, he was awarded a government pension for life. He returned to Albany, where he lived among his own people until the end of his life. His contribution to the opening of the Western Australia Territory is highly regarded and valuable.

In retrospect Eyre, along with Robert O'Hara Burke (1820?-1861), Charles Sturt (1795-1869), and at least a dozen other Australian explorers all came to the same final conclusion: lots of land with very little water! Parts of the central lowlands find their way into the ocean—mostly through the Murray River and its tributaries. A large percentage of the pitifully small rainfall in much of the 65% desert or semi-desert areas evaporates or dissipates into such salty land depressions as Lakes Eyre, Torrens, and Gairdner. Fortunately for the Australians, they eventually found the largest artesian basins in the world beneath the arid lowlands. To this day, a limited amount of bore water is pumped to sustain grazing animals in the arid zone.

Thanks to Edward John Eyre and his stalwart companion Wylie, South and Western Australia were opened to the eventual successful transport of livestock and other goods overland. The horrors they endured (and survived) paved the way for the booming economic future of "the land down under."

GERALD F. HALL

Further Reading

Estensen, Miriam. *Discovery: The Quest for the Great South Land.* New York: St. Martin's Press, 1999.

Eyre, Edward John. *Autobiographical Narrative of Residence and Exploration in Australia 1832-1839.* Caliban Books, 1984.

Hughes, Robert. *The Fatal Shore: The Epic of Australia's Founding.* New York: Vintage Books, 1988.

Ryan, Simon. *The Cartographic Eye: How Explorers Saw Australia.* New York: Cambridge University Press, 1996.

The Wilkes Expedition and the Discovery of Antarctica

Overview

In 1838 Charles Wilkes (1798-1877), a United States Naval officer, set sail on an exploratory mission to the far reaches of the southern seas with six small and barely adequate ships, 82 officers, 342 sailors, and nine scientists and artists. This expedition, the United States Exploring Expedition, was charged with exploring waters in the extreme south to learn more about weather, sea conditions, uncharted lands, and other information of scientific and economic interest. The Wilkes expedition succeeded in discovering the continent of Antarctica, mapping large sections of Australia, and gathering a wealth of scientific and commercial information.

Background

Although the existence of a southern continent had been proposed by the ancient Greeks, who felt a large southern landmass must exist to balance the land north of the equator, no confirmed sightings of the purported continent had been made. Captain James Cook (1728-1779) and the crew of the *Endeavour* had approached the continent in 1773, but were turned back by heavy ice and never sighted the continent. Later, in 1821, American sealer Nathaniel Palmer discovered the Antarctic Peninsula (also called the Palmer Peninsula). Russian explorer Fabian Gottlieb von Bellinghausen (1778-1852) discovered some small islands near Antarctica and an American named Davis actually landed on the continent. None of these men, however, realized they had discovered the southern continent and, because of this, none are credited with the discovery. In fact, by 1840 many were beginning to suspect that there was no southern continent at all.

During the first part of the nineteenth century, American whaling and sealing ships were actively sailing every ocean, gradually depleting stocks of these animals near the United States and, eventually, around South America, too. As the ships pushed further south they began to encounter strong weather, uncharted regions, and had trouble locating new hunting grounds. Commercial sailors were loath to share information that might take away their competitive advantage, so a comprehensive understanding of the far southern oceans did not exist.

In 1836 Congress passed a bill authorizing the president to "send out a surveying and exploring expedition to the Pacific Ocean and the South Seas." The expedition's goals would be to fill in the gaps in knowledge about the seas, weather, and lands in the extreme south to give American sailors an advantage over those of other nations when sailing those seas. The goals of the expedition were stated thusly:

"Although the primary object of the expedition is the promotion of the great interests of commerce and navigation, yet you will take all occasions not incompatible with the great purposes of your undertaking to extend the bounds of science and promote the acquisition of knowledge."

Unbeknownst to Congress, Great Britain and France had reached similar decisions and were sending squadrons of ships south at the same time.

After offering the leadership position to a number of well-qualified candidates (all of whom turned the opportunity down for one reason or another), Wilkes was asked to lead the Expedition of Exploration, in spite of his relative lack of sea experience. Wilkes accepted and, on August 18, 1838, the expedition set sail. They were to return four years later, having accomplished their goals in the Southern Ocean as well as mapping parts of the Australian coast, a number of South Pacific islands, and large parts of the West Coast of North America.

Impact

The Wilkes expedition was somewhat of a nautical coup for the United States, as it was the first major voyage of exploration undertaken by the young nation. (Though the 1804-06 expedition of Lewis and Clark was, granted, a major expedition, it was entirely land-based and mostly within the confines of the United States.) There were several ways in which Wilkes's expedition had an impact on society:

1. Wilkes returned with a great deal of scientific knowledge, much of which could be used to enhance the position of American sailing ships (especially whalers and sealers) when operating in extreme southern waters. Several American scientists—among them, James Dwight Dana

(1813-1895), a geologist and biologist who rose to prominence because of his work on this expedition—got their start with Wilkes.

2. Wilkes, in a race with the French and British to confirm or deny the existence of a southern continent, was able to claim for the United States the distinction of being the first person to sight the last continent discovered on Earth. Other discoveries in geography included mapping large sections of Australia for the first time and helping to chart many of the islands of the South Pacific.

3. The information returned by Wilkes on weather, sea conditions, and commercial animals helped American sailors gain a foothold in the lucrative whaling and sealing grounds of the Southern Ocean.

4. As the last continent discovered, this expedition can be said to have completed the initial phases of mankind's exploration of the Earth. With all of the major land masses now discovered, attention shifted towards exploring the interiors of the lesser known continents as well as well-publicized races to the north and south poles of later years.

Although Wilkes's primary charge was to return with information that could give the United States an advantage over other nations in hunting whales and seals, the longest lasting benefit of this expedition was the scientific information brought back to the United States. At least three scientists accompanying Wilkes achieved international renown for their work and the specimens returned to the United States provided many years of fruitful work for many more researchers. In fact, at one point Charles Darwin (1809-1882), when writing a monograph on barnacles, requested the loan of "some of the species (of barnacles) collected during your great expedition." These specimens were later donated to become the foundation of the National Museum of Natural History, a part of the Smithsonian Institution complex.

During his expedition, Wilkes became aware that the Frenchman Jules-Sébastien-César Dumont d'Urville (1790-1842) and Englishman James Ross (1800-1862) were both seeking the southern continent at the same time. All were seemingly aware of Palmer's and Bellinghausen's sightings and all were attempting to claim for their nations the honor of discovering the last continent on Earth. In one of the incidents that led to scandal, Wilkes initially logged that he first sighted the Antarctic continent on January

19, 1840. Later, realizing that Dumont d'Urville had logged the same date, Wilkes changed his logs to indicate the sighting took place on January 16th. In fact, Wilkes ended up charting more of the Antarctic continent than either other captain and was the first to be able to prove he had sighted a continent rather than a long archipelago encased in ice. This incident generated such controversy, though, that as late as 1910 there was considerable acrimony between the English, French, and Americans regarding who should be properly credited with this discovery and who it was appropriate to name various features for. Although generally a debate among the upper classes and intelligentsia, the debate was followed sporadically by larger segments of the population when various arguments were reported in the popular media.

While most of the plaudits received by this expedition are for its scientific discoveries, Wilkes's work did benefit whalers, fishermen, and sealers greatly. Where, previously, many ships were lost or damaged by storms, uncharted islands, or uncharted reefs, several safe transit lanes were identified. However, to some extent, the commercial impact of Wilkes's work was shorter-lived than the scientific impact because, once discovered, the stocks of whales and seals were quickly depleted. This made the long trip south less profitable, leading fewer and fewer whalers in that direction. In addition, many individual captains had fairly detailed knowledge about parts of the Southern Ocean, but were unwilling to share it for fear of losing their competitive advantage. So, in a sense, Wilkes managed to recreate, consolidate, and make public knowledge that already existed in small, scattered bits and pieces.

The final major impact of this expedition was that it signaled an end to the age of discovery, if that age is defined as mankind's learning of the large details of our world. As the last continent to be discovered, attention turned increasingly towards exploring continental interiors and exploiting the wealth found there. To be sure, the mapping of the Earth was far from complete, but its broad outlines were now known.

Wilkes returned to the United States in 1842, having spent four years on his expedition. The expedition covered a total of 87,000 miles (139,000 kilometers) through some of the most dangerous waters on Earth, charting over 1,000 miles (1,600 kilometers) of Australian coast, 1,500 (2,400 kilometers) miles of the Antarctic coast, and several hundred islands and reefs. He also charted and explored large sections of the North American Pacific coast, the Philippine Islands, Hawaii (then called the Sandwich Islands), and Fiji. Expedition scientists returned thousands of specimens of insects, plants, fossils, minerals, corals, seashells, and artifacts from the native peoples of the lands they visited. During this time, he lost only one ship and 15 men through disease, drowning, or injury.

P. ANDREW KARAM

Further Reading

Gurney, Alan. *Below the Convergence: Voyages Towards Antarctica, 1699-1839.* New York: Penguin Books, 1998.

Headland, Robert. *Chronological List of Antarctic Expeditions and Related Historical Events.* Cambridge: Cambridge University Press, 1989.

Stanton, William. *The Great United States Exploring Expedition of 1838-1842.* Berkeley and Los Angeles: University of California Press, 1975.

The Buried Cities of Assyria

Overview

During the nineteenth century, archaeological discoveries in the Middle East changed the way scholars thought about the history of Western civilization. The translation of the ancient languages of Mesopotamia coincided with the spectacular excavations of the palaces of Assyrian kings and the biblical city of Nineveh. Eventually Mesopotamia was to be recognized as the location of the world's first urban civilization, even more ancient than Egypt. Some of the writings uncovered there parallel accounts found in the earliest sections of the Hebrew Scriptures.

Background

Mesopotamia is the name of an ancient region including parts of what are now Syria, Turkey, Iraq, and Iran. Its heart was the rich "Fertile Crescent" between the Tigris and Euphrates Rivers, long known as a cradle of civilization. A Mesopotamian city called Ur of the Chaldees is said to be the birthplace of the biblical patriarch Abraham, whom Jews and Arabs regard as their ancestor.

Abraham's people were Semites, nomadic tribes whose remote origins were probably in Arabia. The Sumerians, among whom they lived in Mesopotamia, stood poised at the dawn of history. They built the world's first cities and invented writing, using wedge-shaped symbols called *cuneiform*, about 5,000 years ago.

By about 2300 B.C., the Sumerian civilization was weakening. The population of their cities had increased faster than their ability to provide housing and sanitation. The rich soil of their farmlands had become depleted as irrigation leached nutrients from the over-cultivated

fields. Finally, they were overrun by another Semitic people known as the Akkadians. The descendants of the Akkadians and other invaders formed kingdoms and empires, and ruled parts of Mesopotamia for the next 2,000 years. They included the Babylonians, the Assyrians, and the Amorites.

The influence of the Sumerians lived on in Mesopotamia, as their successors adopted much of the Sumerian culture as the basis of their own. It was also a major contributor to the civilization of the Hebrews, who took it with them to the eastern Mediterranean almost 4,000 years ago. Through the Bible, it helped to form the foundation of Western culture and thought.

Some of the great Mesopotamian cities such as Nineveh were mentioned in the Bible, and also in the writings of ancient Greek historians including Herodotus and Xenophon. In the twelfth century, the rabbi Benjamin of Tudela (b. 12th century), traveling from Spain to visit Jewish communities in the Middle East, correctly identified the ruins of Nineveh. But in medieval Christian Europe, the lost civilizations of Mesopotamia were all but forgotten.

The Renaissance revived interest in learning about the ancient world. After 1530, Mesopotamia was controlled by the Ottoman Turks. Over time, trade and diplomatic relations increased, and more Europeans started traveling to what was now a remote and neglected provincial backwater. With renewed scholarship, understanding of the historical importance of the region grew. In an era when every educated European knew the Bible well, the places mentioned in the Scriptures held special significance.

Impact

The first archaeological excavations in Mesopotamia were made in 1780 by Abbe Beauchamp, an emissary of the Pope. His finds included a number of inscribed bricks, but at the time no one could read them. Still, his memoirs became quite popular, and increased interest in the area.

In 1756, the king of Denmark sent a scientific expedition to the Middle East. Tragically, five of its six members were felled by disease. However the sixth, the German Carsten Niebuhr (1733-1815), explored the ruins of the 2,000-year-old palace of the Persian kings of Persepolis. There he found and copied inscriptions in three different languages, all written in the same cuneiform script. He published an account of his discoveries in 1772.

The first of the three languages to be deciphered, in 1802, was Old Persian. This was accomplished by a German schoolteacher named Georg Friedrich Grotefend. The key was his being able to pick out repetitive phrases used to honor Persian kings, as well as the kings' names, from his knowledge of Greek historical texts.

Henry Creswicke Rawlinson (1810-1895), a British diplomat and scholar first posted to Persia on military duty in 1835, climbed sheer cliffs at Behistun to copy another multi-lingual set of 1,200 cuneiform lines. The most inaccessible inscriptions were reached with the help of a "wild Kurdish boy" slung from a rope taking papier-mache casts.

This larger body of text helped Rawlinson decipher the second language, Babylonian. The Babylonian language was a later dialect of Akkadian, a Semitic language related to Hebrew and Arabic. Rawlinson published a preliminary translation of the Babylonian text from the Behistun cliffs in 1851. The third language, Elamite, belonged to a people centered at Susa in southwestern Iran. They preceded the Persians and seemed to be native to that area. Elamite is not similar to any other known language. It was not deciphered until much later, and is still incompletely understood.

Rawlinson's linguistic breakthrough fortuitously occurred just as the great age of Mesopotamian archaeology was beginning. In 1843, Paul Emile Botta (1802-1870), while serving as French consular agent in the Iraqi city of Mosul, discovered the city of the Assyrian king Sargon II, dating from the eight century B.C.

Sargon, which means "declared king," was the name of three rulers of ancient Assyria. The first, Sargon of Akkad, was the first great emperor in history, and ruled for 61 years. It was he who conquered the Sumerians and began the period of Akkadian rule in Mesopotamia 23 centuries ago. He extended his empire as far as Persia to the east, and the Mediterranean Sea and Asia Minor to the west.

Sargon I reigned about 500 years later. Little is known about him, although his kingdom was active in forming trade colonies. Sargon II was one of Assyria's last and most powerful kings, although his relatively short reign lasted only from 722 to 705 B.C. He conquered Babylonia, Armenia, and Philistia, and destroyed one of the two adjacent Jewish kingdoms on the shores of the Mediterranean, scattering the famous 10 lost tribes.

Sargon II built a new capital to celebrate his victories, but soon died in a battle in Persia. Eventually his city crumbled into a raised mound of earth at the Iraqi town of Khorsabad. Botta had been having little success digging for a year at a site in nearby Kuyunjik that was later to yield spectacular discoveries. However, acting on a tip from a villager, who said that antiquities were underfoot in Khorsabad, he moved there and was almost immediately successful. Botta's finds included huge statues of winged bulls and lions with human heads, and many exquisite bas-reliefs. Most were shipped to Paris and exhibited at the Louvre.

The English archaeologist Austen Henry Layard (1817-1894) found eight more Assyrian palaces a few years later. In 1849, he excavated a mound at Kuyunjik and uncovered the famous city of Nineveh. These discoveries earned him the title "Father of Assyriology." Like Botta, he shipped most of the artifacts he found back to his own country, where they were displayed at the British Museum. There were similar statues and about two miles (3.2 km) of bas-reliefs. Most significant of all, there was a library of about 24,000 cuneiform tablets assembled by the learned king Ashurbanipal, who reigned in the seventh century B.C.

The Crimean War of the 1850s put a damper on archaeological expeditions. Attention turned to studying the cuneiform languages, with Ashurbanipal's library now providing a large amount of material to work with. The script was primarily syllabic; for example, there were seven different "r" symbols, corresponding to ar, ir, er, ur, ra, ri, and ru. However, some symbols were shorthand for complete words, similar to the way we sometimes use the symbol "&" for "and."

Linguists quickly noted the similarity of Akkadian to other Semitic languages such as Hebrew and Arabic. But the cuneiform script didn't seem adapted particularly well to the Semitic language family. For example, there were vowel patterns in Semitic tongues that had no corresponding expression in the script. At the same time, Akkadian did not have a symbol for "lion," although lions were common in ancient Mesopotamia and had even been known to eat the guard dogs of nineteenth-century archaeologists. This suggested that Akkadian had originated elsewhere, presumably at a time before the appearance of the Semitic Akkadians in Mesopotamia. The uncomfortable fit of the written symbols to the language prompted scholars to surmise that the Akkadians had, after their arrival in Mesopotamia, borrowed the script of the earlier, non-Semitic inhabitants to express their language.

Layard had found tablets in Ashurbanipal's library upon which Akkadian words were listed next to unknown words in another language, written in the same script but unfamiliar. Rawlinson proposed in 1855 that this unknown language was that of the originators of the cuneiform writing. The mysterious people were identified as Sumerians in 1869 by Jules Oppert, after he studied Akkadian inscriptions referring to "king of Sumer and Akkad."

The most spectacular discovery to emerge from Ashurbanipal's library was found by George Smith (1840-1876), a young scholar reading through the tablets at the British Museum. Suddenly he came upon an Akkadian version of the biblical story of the great flood. He read how a flood had come to punish mankind. Like Noah, a king named Utnapishtim, along with his family and animals, survived the deluge in a ship of his own making. The rain stopped after six days, and the ark came to rest on a mountaintop. Just as in the Genesis account, a bird was sent out, and its failure to return indicated the re-emergence of land. Another fragment of text contained a story resembling that of Adam and Eve.

The fascination with the ancient land of Mesopotamia and the light it sheds on the earliest days of Western civilization was sparked by the many discoveries of the nineteenth century, and continues to this day.

SHERRI CHASIN CALVO

Further Reading

Baumann, Hans. *In the Land of Ur: The Discovery of Ancient Mesopotamia.* New York: Random House, 1969.

Brown, Dale, et al. *Sumer: Cities of Eden.* Alexandria, VA: Time-Life Books, 1993.

Cottrell, Leonard. *The Quest for Sumer.* New York: G.P. Putnam's Sons, 1965.

John C. Fremont and Exploration of the American West

Overview

John C. Fremont's explorations of the West in the 1840s were undertaken with the sponsorship of the United States government to expand the boundaries of the country, to make maps for Americans who wanted to settle in the area, and to notify Great Britain and Mexico that the U.S intended to expand its borders all the way to the Pacific Ocean.

Background

Fifty years after the United States was created, its citizens and leaders began to look westward in earnest. Twenty-six states made up the United States in 1840, but only three were west of the Mississippi River. The land from the river to the Pacific Ocean covered more square miles than the existing United States, but few people knew anything about it and the only confirmed knowledge on the area had been brought to Washington, D.C., by famed explores Meriwether Lewis (1774-1809) and William Clark (1770-1838) 40 years before. The government knew that England could take over the Oregon territory, which included all of the present day states of Oregon, Washington, and some of Idaho, whenever it wished. In addition, Mexico had long claimed the territory from Oregon to the Mexican border and from the Pacific Coast to Colorado and New Mexico. If Mexico gained enough strength, it might gain control of the region.

The United States long considered that all lands to the west belonged to individual states that wished to claim them. Few paid any attention to the rights of indigenous native tribes who lived there. England was the most powerful nation in the world and owned all of Canada. If England took Oregon, too, it would be represent a further threat to the sovereignty of the United States. It would also make England even more powerful and expand Canada to an uncomfortable size. Mexico, traditional owner of the land south of Oregon, was weak but threatened armed conflict if the United States tried to move in.

Land was power in European society, and early American settlers brought that idea with them to North America. For 20 years mountain men, adventurers, and fur traders had traversed the land beyond the Mississippi River, bringing to the East tales of towering mountains, fierce native peoples, raging rivers, and endless plains. But they had no proof of any of it as they made neither maps nor scientific observations. Americans were eager to move into this unknown land, and the government wanted to encourage them. But doing so without knowing what lay beyond the frontier was not wise.

Thus, there were three motives for westward expansion that all grew critical at once; 1) the need for the United States to gain control of Oregon and remove the threat of British power on its doorstep; 2) the desire to obtain the land south of Oregon from a very weak Mexico, by peaceful means, if possible (there were, however, already signs and rumors that a war would be necessary); and 3) a desire to facilitate American settlers who wished to move west by providing them with maps. Their presence in many areas of the West would help the government secure the land by justifying its presence and dominance.

Impact

With his training, talent, and interests, John C. Fremont (1813-1890) was the perfect man to help fulfill the ambitions of the government in the West. In 1837, when he was 25 years old, he applied for a commission in the United States Army Corps of Topographical Engineers. He was approved as a Second Lieutenant because he had good training and a strong background in mathematics, exploration, surveying, and geographical field work. He also had a reputation as a careful observer and mapmaker.

His first task with the army engineers was to travel west with a respected scientist and surveyor named Joseph Nicollet (1786-1843) to survey the territory between the Mississippi and Missouri rivers. From Nicollet, Fremont learned to make astronomical observations, record the lay of the land accurately, and sketch accurate maps. He also learned to observe and identify plants and animals, soil, geographical formations, and minerals. Along the way, he became proficient in techniques of surviving in the wilderness. This training, and his later actions, would lead to a great deal of notoriety and make him stand out in a field in which amateurs outnumbered professionals.

Back in Washington after his first expedition, he set about writing up his travels and making accurate maps. At this time he met a United States Senator from Missouri, Thomas Hart Benton. He was the man who chose Fremont to lead several important expeditions and was the father of Jessie, who would soon became Fremont's wife. Benton was powerful in the government. He was a strong supporter of exploration of the West and an advocate of westward expansion. Fremont spent a great deal of time at the Benton house and became friends not only with Benton but also with the rest of the family—especially Jessie's brother, whom he took on his next expedition west.

The trans-Mississippi West had been explored by Zebulon Pike (1779-1813), Stephen Harriman Long (1784-1864), and Benjamin-Louis Bonneville (1796-1878), but they had not produced usable maps. There was a great clamor for information, and the expansionists in Washington were pushing for a two-ocean nation. Some were opposed to it because they did not want to antagonize either Britain or Mexico. One of the latter was President John Tyler. So in the 1840s Fremont led three major expeditions to the West. They were officially not for the purpose of securing the land but were to gather scientific information only. The underlying purpose to expand west, however, was understood by Fremont. He was to map mountains, rivers, and trails and observe conditions, wild life, and natives as far west as the Missouri River, but he pushed all the way to South Pass—the only viable way to cross the Rocky Mountains.

By the time Fremont left on another expedition in May of 1842, he had met Charles Preuss, a German cartographer who would become Fremont's mapmaker, and Kit Carson (1809-1868), who guided all of his journeys. Fremont wrote the first of many reports on this expedition. It was well received and had a good circulation when it was published. Soon, a more ambitious expedition was proposed by Benton and other expansionists. This time Fremont was to map the Oregon Trail all the way to the Pacific Ocean and coordinate his findings with Navy Captain Charles Wilkes (1798-1877), who had already explored and mapped the western coast.

The expedition arrived at the Dalles in Oregon in October of 1843, completed its ordered tasks, and was supposed to start back eastward. Fremont decided that instead of going over land already surveyed, he would investigate the Great Basin. This was a huge depression that covers the modern states of Nevada and Utah and lies between the Rocky Mountains and the Sierra Nevada. His stated object was to map and gather data on plants, trees, and natives and to look for legendary rivers and lakes in the area. In reality, he wanted to go to California, because things were happening there and he wanted to be at the center of the action. He was too far from the seat of power to ask permission or to be deterred, so he took it upon himself to take his expedition west.

Fremont mapped and took observations all over the Great Basin heading west toward California. He discovered Pyramid Lake in Nevada, the American River in California, which he named the Salmon Trout, and Lake Tahoe, which he called Lake Bonpland, a name no one ever used. He and his hardy band of explorers crossed the Sierra Nevada in the dead of winter without losing a man. In California he subsequently took part in the Bear Flag Revolt to free California from Mexico, and he inspired Americans in the area to work for California's independence. When war broke out between the United States and Mexico in 1846, he served in the army and later as military governor of California. At the end of the war in 1848, when California was transferred to the United States in the Treaty of Guadalupe Hidalgo, Fremont took part in the convention that wrote the first state constitution. When California became a state in 1850, he was elected one of its first senators. In 1856 he became the first candidate for President of the United States selected by

the brand new Republican Party. He lost to James Buchanan.

Fremont was an explorer, a scientist, a politician, an army man, and sometimes a controversial hothead. He was also a catalyst. Always in the right place at the right time, his exploration of the West was the first to examine, observe, and record the vast reaches of the West with scientific accuracy. The United States eventually claimed all of the land he explored. He gained great notoriety for his actions and accomplishments and was called the "pathfinder." His controversial behavior and subsequent court martial for refusing to follow orders given by General Stephen Watts Kearney following the war with Mexico did nothing to dim his reputation. His support of California independence from Mexico and then statehood was in line with expansionist motives. Fremont had fulfilled the expansionists' goals. After the 1842-1843 expedition and later the Mexican War, there was never any doubt that the United States would conquer and claim all land from the Mississippi River to the Pacific Ocean.

LYNDALL B. LANDAUER

Further Reading

Benton, Thomas Hart. *A History of the American Government for Thirty Years from 1820 to 1850.* New York: D. Appleton, 1854-56.

Billington, Ray Allen. *Westward Expansion.* New York: MacMillan, 1949.

Egan, Ferol. *Fremont: Explorer for a Restless Nation.* Reno: University of Nevada Press, 1977.

Fremont, John Charles. *The Expeditions of John Charles Fremont.* 2 vols. Edited by Donald Jackson and Mary Lee Spence. Urbana: University of Illinois Press, 1970.

Limerick, Patricia Nelson. *The Legacy of Conquest.* New York: W.W. Norton, 1987.

Nevins, Allan. *Fremont, Pathmarker of the West.* Lincoln: University of Nebraska Press, 1939. Reprinted by Bison Books, 1992.

Unruh, John D. *The Plains Across.* Urbana: University of Illinois Press, 1982.

Robert McClure Discovers the Elusive Northwest Passage

Overview

The Northwest Passage was the persistent vision of many early explorers. They traveled across the ocean and up rivers and lakes searching for a waterway to the Orient and into the New World. The explorers were lured westward with hopes of finding immediate tangible wealth. The fastest mode of travel at that time was by waterways, so they followed all they could find. The search for the Northwest Passage, a route extending from the Atlantic Ocean to the Pacific Ocean by way of the Arctic Archipelago (now Canada), required centuries of effort and inspired one of the world's most competitive maritime challenges. Sir Robert John Le Mesurier McClure (1807-1873), an Irish naval officer, was the first explorer to complete the crossing of the Northwest Passage, doing so in 1850. For 450 years a history described by self-sacrifice, tireless curiosity, the wealth of the fur trade, and personal tragedy lured men to find what some had called the mythical waterway. This is the history of the Northwest Passage.

Background

The actual route of the Northwest Passage is located fewer than 1,200 miles (1,900 km) from the North Pole and 500 miles (800 km) north of the Arctic Circle. The route extends approximately 900 miles (1,400 km) east to west, from north of Baffin Island to the Beaufort Sea (above Alaska) and consists of a series of deep channels that cut through Canada's Arctic Islands. (It is amazing to think how early exploration crews must have felt as they sailed through the hazards of giant icebergs. Imagine a stream of approximately 50,000 icebergs up to 300 feet (90 m)in height constantly drifting by the sailing vessels that searched for something that might not be there.) The Northwest Passage is equally difficult to cross when returning. Much of year, as the polar ice cap presses down on the shallow northern coast of Alaska, it funnels masses of ice between Alaska and Siberia and into the Bering Strait. Most of the attempts to find the passage followed one of two hypotheses: first, a waterway around North America or, second, a waterway through the continent.

The personal and political desires to find the elusive Northwest Passage were numerous.

The British Navy was relatively unemployed at the end of the Napoleonic Wars. The British government became excited by the enthusiasm of Sir John Barrow to the potential of discovering the passage. As second secretary to the admiralty, he put together a large naval expedition for the project.

Supported by Spain, Christopher Columbus's (1451-1506) voyages had provided maps of the coasts of Central and South America. Even with this information, however, people still had little knowledge of the unvisited North. Some thought that the continents of North America and Asia leaned together. Theoretical geographers reasoned, then, that the coast of Newfoundland was simply the edges of a peninsula that extended eastward from the Orient. With the lucrative trade in furs and other natural resources in mind, a faster passage to the Orient would give English merchants the opportunity to locate trade routes and markets not yet dominated by Spain or Portugal.

The impact of opening the Northwest Passage would be enormous. Not only would it allow regular commercial ocean traffic, proving to be of worldwide economic significance as it pertained to trade relations among nations, modes of transportation would also dramatically change. It was thought that the passage would permit the use of much larger vessels and shorten travel time. If the passage existed, the distance between Tokyo and London would be shortened to under 8,000 miles (12,800 km). At the time, the primary route was around the tip of Africa. This route was 14,670 (23,470 km) miles long.

Unfortunately, the many different attempts to discover the Northwest Passage lured opportunists of questionable character. It seems that in the excitement of discovery, some people always found "evidence" for believing what they wished to believe. The intensity of the search was so great that some travelers claimed to have discovered the strait when in fact they had not. Juan de Fuca was the most notable of these opportunists. Fuca was from Greece, adopting a Spanish pseudonym to please the government of Mexico. In reality, he only sailed the northern waterways through the passage for 20 days, giving standard reports that described very fruitful land rich in gold, silver, pearl, and many other natural resources. History has given him much credit, however, naming a famous strait of water after him.

Tragedy and failure are among the realities of exploration. Sometimes tragedy, however, provides the opportunity to learn. Of all the recorded histories of the Northwest Passage, the worst tragedy came in 1845. Sir John Franklin (1786-1847) and his crew of 128 men vanished. The result of this loss was a 12-year search that, in turn, contributed greatly to geographic knowledge. At the search's peak in 1850, as many as 14 ships were looking for this unlucky crew at the same time and a further expedition was at work from the mainland. The mystery of Franklin and his party was finally pieced together. Franklin and his crew had wintered at the western end of Lancaster Sound. He turned south in the spring of 1846 to Peel Sound, a stretch of water that had not been navigated, and then to Victoria Strait, off the northern tip of King William Island, in 1859. It is here that his beleaguered ships eventually were abandoned. There were no survivors.

Sir Robert John Le Mesurier McClure was born in Ireland and became an explorer for the British. In 1848 McClure entered the navy and accompanied Sir James Clark Ross (1800-1862) to the Arctic. In 1850 McClure took command of the ship, the *Investigator*. This was one of two ships sent to search for Sir John Franklin in the western part of the Arctic Archipelago. Captain Richard Collinson was the commander of the other ship, the *Enterprise*. They became separated in the Pacific and later discovered that Collinson had spent three years in Victoria Island and reached the Victoria Strait. This was a short distance from where Franklin's ships had been abandoned. Collinson found no additional clues with which to continue his search so he eventually returned home.

McClure's search for the passage was from the West. He discovered two entrances to the Northwest Passage. From the Pacific he entered the Bering Strait. (The U.S.-Russian boundary extends through the strait). McClure sailed along the Alaskan and Canadian coasts and discovered the Prince of Wales Strait. The strait is an arm of the Arctic Ocean and extends 170 miles northeastward from Amundsen Gulf to Viscount Melville Sound. It separates Banks and Victoria islands and now is part of the Northwest Territories of Canada. The Strait was named after Albert Edward, then the Prince of Wales. While wintering here, McClure continued his exploration by

sledge (a strong heavy sled) along the shore, reaching Barrow Strait. Here he discovered McClure Strait, also known as M'Clure Strait. Approximately 60 miles (96 km) wide and 170 miles (270 km) long, this strait is an eastern arm of the Arctic Ocean and the Beaufort Sea. West of Viscount Melville Sound, it lies between Melville Sound Banks Island (to the south) and Eglinton Islands (to the north). The sounds and straits of the Northwest Passage are navigable only under favorable weather conditions.

The *Investigator* became icebound for nearly two years near Banks Island in Mercy Bay. McClure finally abandoned his ship to travel over land by sledge. Captain Henry Kellett rescued McClure and his crew at Melville Island. Kellett was also making the expedition from the West. Soon Kellett's ship also became icebound, in this case for a year or so, and both crews abandoned their ships, moving on to a third ship. McClure and Kellett together managed to proceed on foot to Beechey Island to become the first men to cross the Northwest Passage. McClure is credited with completing the entire journey and proving the existence of the Passage. Eventually both parties were rescued by Sir Edward Belcher's (1799-1877) expedition, which brought the men home. McClure was censured for returning without the *Investigator*. Nevertheless, he was highly commended for his discoveries and knighted in 1854.

The story continues as it was not until 1906 that the Northwest Passage was conquered completely by sea. Roald Amundsen (1872-1928?), a Norwegian explorer, completed the three-year voyage on a boat named *Gjoa*, which had been converted from a 47-ton herring boat. Interestingly, Amundsen's excursion began secretly as he sailed to escape creditors who wanted the expedition stopped.

Impact

The Northwest Passage captured the imagination of many famous explorers. The prestigious list includes Captain James Cook (1728-1779), Sir Francis Drake (1540?-1596), Henry Hudson (1565?-1611), Sir Martin Frobisher (1535?-1594), Sir John Ross, and Jacques Cartier (1491-1557). Many met with disaster and all but one met with failure. Today, the Northwest Passage serves not only as a highway for commerce but also functions as a means to protect the North American continent.

One of the great long-term results of the discovery of the Northwest Passage was the Distant Early Warning (DEW) line. The DEW began in 1954 as a primary line of air defense, its role being to warn of an "over the Pole" invasion of the North American continent. (It is remarkable to think that the Northwest Passage made possible the gigantic task of bringing the weighty and vast supplies for the DEW Line and its later equipment.) The early explorers had established that the Northwest Passage existed. It took, however, an air reconnaissance mission headed by Vice-Admiral Will, U.S.N, to establish that the Bellot Strait was a possible passage for cargo ships and tankers. The DEW Line is, notably, near the spot where Sir John Franklin and his men perished in the icy—and still unconquered—Arctic.

SCOTT BOHANON

Further Reading

Delgado, James P. *Across the Top of the World: The Quest for the Northwest Passage.* New York: Checkmark Books, 1999.

Graf, Miller. *Arctic Journeys: A History of Exploration for the Northwest Passage.* New York: Peter Lang, 1992.

Savours, Ann. *The Search for the North West Passage.* New York: St. Martin's Press, 1999.

David Livingstone Traverses the African Continent

Overview

David Livingstone (1813-1873) began exploring Africa in 1841 and spent most of the next 32 years there, until his death in 1873. In his travels he discovered or traced some of Africa's major rivers and lakes, elucidating much of the drainage system of the central and southern continent. As a missionary, he fought against the African slave trade and the exploitation of African natives, especially by the Portuguese and the Boers (Dutch settlers in South Africa). He also wrote extensively of his travels and discoveries, winning much acclaim and helping to influence Western attitudes toward Africa and its inhabitants.

Background

Africa had been known to Europeans since the time of the Greeks, who had built cities along the African coast. However, with the exception of a small distance along the Nile River, Europeans had little conception of the size of the continent until Bartholemeu Dias's voyage in 1487 and 1488 in which Africa's southern-most point was rounded for the first time. Over the next three centuries, although the coastline and some of the major African rivers were mapped, virtually nothing of the African interior was known.

Crucial to understanding the central and southern African interior was the pattern of lakes and rivers. These were the key to travel and trade as well as providing food, water, and transportation for the native populations. The Greeks knew of the existence of major rivers, but never learned their locations or even in which direction they flowed. The source of the Nile was unknown, and the presence of the Great Lakes of the African Rift Valley was just a topic of speculation.

The African slave trade was in full swing in the early part of the nineteenth century. The chance to combat the slave trade, as well as the opportunities to do missionary work and to explore Africa for potential commercial activity led governments and private organizations to sponsor trips of exploration into the African interior. David Livingstone made several such journeys during the three decades he spent exploring Africa. During these journeys, Livingstone explored, mapped, conducted some scientific investigations, preached, and more. And, during his infrequent returns to Britain, he wrote and spoke of his work in Africa, helping to spark public interest and appreciation of the continent and its inhabitants.

Impact

Livingstone's work in Africa can be roughly broken into three major areas: 1. Scientific and geographic studies of the African interior, including teasing out the details of its waterways. 2. Work with the African people, including fighting the slave trade, his missionary work, and his studies of the Africans. 3. Teaching the public about

Africa and Africans through books, lectures, newspaper accounts, and commercial possibilities.

These will each be discussed in further detail in the rest of this essay.

During most of his travels, Livingstone spent a great deal of time exploring the great rivers and lakes of Africa. In addition to his attempts to find the source of the Nile River, Livingstone discovered or explored the Zambezi, Congo, Shire, Lualaba, Rovuma, and Zouga rivers as well as Lakes Tanganyika, Nyasa, Ngami, Mweru, Bangweulu, and others. In fact, he spent most of his time traveling these waterways as he pursued his other aims, including those of trying to drive out the slave trade with a combination of religion and commerce.

One of Livingstone's most spectacular discoveries was Victoria Falls, one of the world's greatest waterfalls. Ignoring the native name, which translated as "smoke does sound," Livingstone renamed the waterfall after his queen, the name it retains to this day. Livingstone took a small steamship to Africa on one of his journeys, although it turned out to be as much a hindrance as an asset. However, as one of the first steamships on the great African lakes, it did help him to accurately map the shores, visit natives, and travel some of the rivers. It is possible that the understanding of African hydrology fostered by these travels is the greatest geographic legacy Livingstone left because of the complexity and importance of these waterways and lakes.

An unintended consequence of his understanding of the hydrology of central and eastern Africa turned out to be the pattern made by the great African rivers and lakes, a pattern that eventually turned out to owe much to plate tectonics. It turns out that the African Rift Valley is where the continent of Africa began to rip apart from tectonic stress, creating great faults along which lakes and rivers formed; the great rivers and lakes Livingstone spent three decades exploring.

In addition to his journeys, Livingstone spent several years living in Africa, during which he studied local geology, botany, and zoology. Although his scientific discoveries pale in comparison with his other accomplishments, Livingstone did send back information that was assessed to help determine the economic potential of those parts of Africa.

In spite of Livingstone's explorations, his primary reason for traveling to Africa in the first place was to help fight the slave trade, to help convert the natives to Christianity, and to better understand the African natives. He did all of these in abundance.

Livingstone's primary aim was to introduce and push Christianity, commerce, and civilization into the heart of Africa. As a missionary, he tried to find the most populous parts of the interior, the better to convert natives who could then help him to convert others. In addition, through his explorations of Africa's rivers, he was helping to open a path for others to follow him, spreading religion, commerce, and civilization behind them. He also hoped that he could encourage the economic development of central and southern Africa, creating sufficient economic incentive to end slave trade. While the slave trade did eventually slow down greatly, this may have owed as much to changing laws in the Americas and elsewhere as it did to Livingstone's efforts.

Livingstone did make great progress in learning the African languages, many of which are related to one another. This helped greatly because several of his expeditions were solitary and required the help of Africans to be successfully completed. In fact, Livingstone traveled a great deal with Africans, many of whom became close friends. Establishing this level of rapport and understanding not helped Livingstone survive some of the travails of his journeys (including having his left arm and shoulder mauled by a lion), but gave many of the native inhabitants an initially favorable impression of Europeans in general and the English in specific. And, by transmitting information about the natives to England, Livingstone helped future travelers to better understand the land and the people they were about to encounter.

This, of course, leads to what may be Livingstone's longest-lasting legacy; his journals, books, and lectures and the impression they left on the public. Livingstone wrote and lectured extensively about his experiences in Africa. These books sold well and were widely read and quoted. Newspaper accounts of his travels were frequent and popular, including those by Henry Stanley (famous for his line "Dr. Livingstone, I presume?") who helped save Livingstone's life after a series of mishaps left him ill and without medicine or supplies. Livingstone was primarily a kind man with a deep compassion for those about whom he wrote. His books and lectures inspired others to try to understand, to help, and to follow in his footsteps. As public interest in Africa grew, so did public involvement. Livingstone himself made enough money through his books and lectures that he was able to conduct many of his expeditions without external funding. In addition, several private and public organizations were established with the goal of improving knowledge and understanding of both Africa and its inhabitants.

At this time, too, the British Empire was still in its ascendancy. One of Livingstone's stated goals was to introduce commerce to the African interior so that the slave trade would wither and die. To do this, he felt it important to arouse public and governmental interest in Africa's commercial possibilities and, to do that, he explored and wrote. Although no British colony was established in areas he explored and the few missions he founded either failed or were destroyed, the British did turn parts of central and southern Africa into colonies, although without the same success they enjoyed in other colonial ventures.

Unlike many explorer-authors, Livingstone's accounts were neither self-aggrandizing nor needlessly sensationalistic. They nonetheless sold well, leading readers to a better understanding of the continent, its geography, its natives, and the evils of the slave trade. By showing in his writing that the African interior was accessible, fertile, and populated, readers could see the potential for development and the possible rewards. And, by writing of the excitement of discovery, the wonders to be seen, and the plight of some of the natives, Livingstone's books

helped convince many to try to help. All in all, his writings may have had a greater influence on European understanding of and sentiments towards Africa than those of any other man. From that standpoint, it was through Livingstone's eyes that the world saw Africa and, to some extent, we still do.

David Livingstone died in Africa at the age of 60. He was found dead, kneeling as though in prayer at the side of his bed. Following local custom, his heart was removed and buried beneath a tree and his body was packed with salt and carried to the coast to be returned to England for burial. His legacy includes not only his maps and books, but also his belief that Africa could advance into the modern world. While his explorations may have helped to advance European colonialism in Africa, his belief in Africans helped to advance feelings that later grew into African nationalism and led to the independence of many African republics in the second half of the twentieth century.

P. ANDREW KARAM

Further Reading

Martelli, George. *Livingstone's River: A History of the Zambezi Expedition, 1858-1864*. New York: Simon & Schuster, 1969.

Seaver, George. *David Livingstone: His Life and Letters*. New York: Harper, 1957.

Robert O'Hara Burke Traverses the Australian Continent from North to South

Overview

The period between 1830-60 in Australia was marked by a tremendous expansion of British colonial interests, from people to economic interests. It is also notable for a series of expeditions that helped the British discover what the continent's physical terrain was like. Closing this period was a tragic but ultimately influential expedition by Robert O'Hara Burke (1820?-1861).

Background

Robert Burke's commission by the Royal Society of Victoria to traverse the Australian continent was an odd one: Burke had no experience in explo-

ration, knew nothing about the native inhabitants he would most likely encounter, and knew close to nothing about survival methods in unknown territories. Fortunately for him, Burke teamed up with William John Wills (1834-1861), a knowledgeable bushman and more experienced traveler.

The government plan was to cross the continent from south to north, with the final destination the Gulf of Carpentaria. The straight line distance (we now know) is approximately 1,480 mi., (2,370 km) a trek that would be turn out to be incredibly arduous due to difficult terrain.

Since the party anticipated long stretches of desert, 24 camels were included in the 28 horse-

and-wagon train. Thanks to an obscure government accounting clerk, we still have a partial list of the supplies that were in the wagons when they left Melbourne: 20 camp beds, 57 buckets, 80 pairs of shoes, 30 cabbage tree hats, preserved fruits, brandy, local vegetables, firearms, and 6 tons of firewood. They believed their provisions would last at least two years and would be sufficient for the 18 members of the trek. Perhaps they would have—but only in the hands of a competent leader. The expedition left Melbourne in August 1860.

Robert Burke was not the man for the job. He was impatient, unwilling to accept advice from those who were (as time would tell) far more experienced and careful than he, and a headstrong decision-maker. Early on in the journey, Burke became frustrated by the slow pace of travel, a pace in part due to the size of the party. His first mistake was to speed up travel by abandoning supplies and good equipment. He had been warned by an experienced explorer named Augustus Charles Gregory not to travel until the middle of the summer (December through February in Australia) so as to have cooler weather in the middle of the continent. However, Burke had knowledge of another party headed by Robert Stuart (1785-1848), who represented the South Australian government (the various British colonies in Australia had autonomous governments at this time) and was also headed for the Gulf of Carpentaria. When the Burke party reached the Darling River in New South Wales at a settlement named Menindie, he made another one of his errors in judgment. He chose eight men to accompany him in setting up a depot at Cooper's Creek, which was thought to be about halfway between Melbourne and the Gulf. The rest of the men were to follow in a few days.

Burke and his eight followers reached Cooper's Creek early in November, and they set up a major camp and waited for the rest of their companions to arrive. When six weeks had passed with no sign of the travelers, Burke made another disastrous decision. He elected to take three of the men (Wills, Charles Gray, and John King) and make a mad dash to the Gulf. When he left a man named Brahe and the other men behind (along with some of the animals), he instructed them to "stay put" for as long as possible.

The next four months were hot and hellish. Burke and his small group actually came very close to reaching the Gulf of Carpentaria in early February but were unable to traverse the swamps and brush lying next to the coast. In any case the men had to start back immediately because their supplies were running dangerously low. There was little comfort in knowing that they had been the first expedition to cross Australia from south to north when they might not return alive to tell anyone about their feat.

As with the trip to the Gulf, the return trip was extremely difficult, what with the incredible heat, thunderstorms, marshy terrain, very slim rations, and exhausted animals. When Gray's hunger got the best of him and he was caught stealing a bit of food, Burke gave him a beating. Eventually, Gray could no longer walk and died in his tracks. The energy it took to bury him cost the others a whole day's walk. When Burke, Wills, and King made it back to Cooper's Creek on April 21, 1861, they found it deserted with the only sign of former occupants in a message carved into a tree: DIG—3 FEET N.W. Burke and company had missed their colleagues by a matter of hours. The message carved into the tree directed the men to food that was buried nearby, along with directions on how to get to the nearest town.

Due to Burke's imprudence and some fateful circumstances, the group did not get to the aforementioned town. Rather, the party ate some of the food left for them, and then Burke and King set out for Adelaide on the southern coast. Wills stayed behind. Burke and King continued on, but Burke died of exhaustion shortly thereafter. King then returned to camp to find Wills dead. Luckily, King received help from local aborigines—native inhabitants of Australia—and managed to stay alive. When he was later found by a search party, he described his adventures living with a friendly group of aborigines who taught him how to survive in what we know today as "the outback."

Although the expedition ended so tragically, it had all the appearances of a comedy of errors. If any of the men had had the good sense to make friends with the aborigines (who obviously knew how to survive under the worst possible circumstances), they could have "lived off the land" and taken adequate breaks to recover their strength. Also, if Burke had been even mildly experienced in survival techniques, they could have survived on the banks of creeks that held fish for the taking.

Impact

The main impact of the expedition was the added knowledge it gave to colonial British authorities about the route traveled by Burke and his men. Although Australia is a small continent, two-thirds of it is desert or semi-desert. In the 1860s, when the colonial authorities wanted to know more about the continent's economic possibilities, it was essential to learn firsthand whether the interior of the continent would support the kind of life they had lived in the British Isles. They surmised that the entire future of the continent hinged on water—still the most precious of Australia's natural resources.

Although memorial statues were erected in Melbourne in honor of Burke and Wills, it seems little enough for the magnitude of their accomplishment. In light of what we know today, it was monumental. We are grateful to Wills for his diary and notes, which detailed parts of the ill-fated but world-famous expedition. He looked to a future well beyond his own years and, in the tradition of the genuine scientist, passed along the information that helped others forge new trails in the interior of the Australian continent.

GERALD F. HALL

Further Reading

Colwell, Max. *The Journey of Burke and Wills.* Sydney, Australia: P. Hamlyn, 1971.

Estensen, Miriam. *Discovery: The Quest for the Great South Land.* New York: St. Martin's Press, 1999.

Hughes, Robert. *The Fatal Shore: The Epic of Australia's Founding.* New York: Vintage Books, 1988.

Ryan, Simon. *The Cartographic Eye: How Explorers Saw Australia.* New York: Cambridge University Press, 1996.

Exploration of the Nile River:
A Journey of Discovery and Imperialism

Overview

The nineteenth-century efforts to find the source of the Nile River, one of the great rivers in the world, can be seen both in the light of genuine scientific exploration and discovery and in the light of naked imperial expansionism. On the one hand, John Hanning Speke's discovery that Lake Victoria is the source of the Nile was the culmination of centuries of curiosity. On the other hand, the discovery reflected the efforts of European colonial powers to explore, conquer, and control the African continent and its vast natural resources.

Background

Our understanding of rivers and river systems has increased greatly in the last few decades. Satellite images have shown us more about the geography of the world's major rivers than we ever knew from land exploration. What we now understand is that the Nile River is not a single flowing body of water that snakes through hundreds of miles of terrain. In fact, what we call the Nile River of today is actually a major drainage source of regions deep into the African continent in an area called the Great Rift Zone. The many beautiful tributaries and lakes create a chain of waterways that empty into a final passageway to the sea at the Great Nile Delta.

The Nile River began to carve its passage through the continent about 30 million years ago. Some believe the main headwaters were the section of the present-day system called the Atbara River. The current understanding of the Nile includes two rivers as the actual source. They are the White Nile and the Blue Nile. The length of the Nile from the major tributary, the White Nile at Lake Victoria, Uganda, is approximately 4,100 mi (6,695 km). The Nile River system flows through nine countries: Uganda, Sudan, Egypt, Ethiopia, Zaire, Kenya, Rwanda, and Burundi.

Travel along the Nile River has a history that starts long before Europeans ever witnessed its shores. The early Egyptians, who lived along its banks, used its waters to carry a wide variety of goods, including the limestone blocks used to build the pyramids. The records of their commerce are the first pictures we have into the character and size of this mighty river. The Egyptians limited their use of the river, however, and never ventured farther than the known empires that thrived along its coast. As far as we know, there was never any attempt made by the Egyptians to discover the actual source of the river.

Greek history describes several explorations of varied length and success. Many travelers, like Herodotus, found the waterfalls near present-day Aswan an impenetrable barrier for boat voyage and turned back. Later, Grecian investigators described a variety of lakes and side rivers as possible sources, but they did not actually explore these territories. The suggested sources some of these writers proposed were extraordinarily precise considering they did not actually follow the routes to their supposed destination.

The first recorded exploration for the source of the Nile occurred with the Romans in 66 A.D. Emperor Nero sought to expand the reach and wealth of his empire and launched a failed expedition up the White Nile. The unlucky adventurers met additional imposing waterfalls along the route that made the heartiest of them weary with the constant trek through the surrounding wilderness. The Roman expedition was finally stopped by a tough and resistant people of the As-Sudd.

It was not until the seventeenth century that exploration of the Nile began in earnest. By the late 1600s both the Spanish and British empires were supporting expeditions around the world in search of new territory. A little-mentioned Spanish expedition, led by a Jesuit priest, Pedro Pàez (1564-1622), was successful in locating the source for the Blue Nile in 1618. The Blue Nile is the source of almost 70 percent of the floodwater measured at Khartoum. Pàez found a small spring above Lake Tanganyika in northwestern Ethiopia. It empties into the lake at about 6,000 ft (1,800 m) above sea level. The path following the outflow of the lake to its meeting with the White Nile follows a tortuous route down deep gorges, over high waterfalls, and through dense rain forest.

The quest for the source of the Nile during the nineteenth century was driven by the desire to find waterways across the region. The control of water has always been considered to be of economic importance, since the country that controls the water controls the commerce along its waterways. Motivated by potential wealth and fame, expeditions began across the region.

In northern and central Sudan, Muhammad 'Ali (1769-1849), viceroy of Ottoman Egypt in 1821, sent his sons in search of the source of the White Nile. Turkish expeditions vied with the Egyptians for geographical exploration. Missionaries from Austria and other European countries began to traverse the region as well, bringing Christianity to tribal people. Their journeys were recorded and catalogued in great detail. From these journals the world began to accumulate much of its knowledge about the physiography of the Nile region. Records of undiscovered mountains and lakes sparked a flurry of expedition that

was to become one of the most famous quests in history: the search for the source of the Nile.

One of the most prestigious and influential societies of the nineteenth century was the Royal Geographical Society of London. It was established in 1830 in Westminster near the Royal Albert Hall. The society, composed of wealthy gentlemen, sponsored expeditions around the world, including to the Arctic and Antarctica. In particular, great financial support was lent to the exploration of Africa. Famous expeditions led by members such as David Livingstone (1813-1873) and Sir Richard Francis Burton (1821-1890) were financed by the society. It was the expedition led by John Hanning Speke (1827-1864) that is credited with the actual discovery of the source of the Nile. It was the continued pursuit of this exploration by both Burton and Speke that has become the stuff of legend.

Burton was raised by his father in England, France, and Italy, where he showed an amazing talent for language and dialect. After expulsion from Oxford in 1842, he joined the army in British India, where he learned to speak Arabic and Hindi. In his travels around the world he eventually learned to speak 25 languages, a feat that made him an invaluable member of any expedition. In 1855 his fascination with the White Nile and its source compelled him to form an exploration party with three additional officers of the British East India Company. One of these officers was John Hanning Speke, a friend with whom he would have a lifelong relationship.

Speke is credited as the first European explorer to reach Lake Victoria, the identified source of the Nile. He was a British officer commissioned in the British Indian Army in 1844. He served in Punjab and traveled extensively throughout the region. He joined with Burton in the 1855 attempt to locate the source of the Nile. Together they planned a route that would take them along a route used by Arabs for trading, one that would take them along the shores of Lake Tanganyika. On this expedition Speke was critically wounded by the Somalis at Berbera. Another officer of the party died in the attack, and Burton was pierced by a spear through the jaw. The expedition was dismantled and Burton returned to England.

Undaunted, Speke and Burton continued to pursue their plans for discovering the Nile source. Speke and Burton teamed up in Zanzibar for another attempt at the elusive discovery during 1857-58. This attempt proved to be as devastating as the first. Their goal was to find a lake described in many journals and texts, from the ancient

Greeks to the seventeenth-century missionaries. They believed it to be in the heart of Africa and were confident that it would be the elusive source of the mighty river. Their journey, however, floundered for six months along the shores of the African coast, where they had hoped to find a route inland towards the lake. The physical strain of the expedition took its toll on both men. Burton contracted malaria and Speke became nearly blind. They came, exhausted, upon the shores of Lake Tanganyika, which they discovered was not the source of the mighty river after all. Dismayed, they discovered that an important tributary, the Rusizi River, flowed into and not out of the lake.

Burton was physically unable to continue the journey at this point. As he made plans to return to England, Speke regrouped. Burton wished Speke to return with him to prepare another expedition, but Speke's refusal and organization of another trip created a rift between the two men that would never heal.

In 1860 Speke asked James Augustus Grant (1827-1892), a loyal lieutenant in the British Indian Army, to accompany him on another expedition towards a great lake that Speke believed to be the Nile's source. Speke read many descriptions of this unrecorded lake and believed it to be the true source of the Nile. Again, the men endured the terrible hardships of illness, conflict with native peoples, and fatigue. At times Speke was so ill that command of the expedition was handed to Grant, who stalwartly pushed the weary band forward.

On July 28, 1862, the two men found themselves at the outlet of a tremendous lake that they undoubtedly believed to be the true source of the Nile River. They named the Lake after the reigning Queen of England: Victoria. Lake Victoria still stands as one of the most majestic lakes in the world and is considered by all to be the source of the Nile River. While exploring the extent of the lake, Speke and Grant separated and conducted independent surveys around the vast body of water. On one of these forays Speke discovered the famous and beautiful waterfalls we now know as Ripon Falls.

Impact

Although the goals of the expedition were in part scientific, mention must be made of the expansionist agenda of the British colonial empire that financed and carried out the trip. The southern and central African expeditions of Livingstone, Speke, Burton, and others resulted in a frenetic race between European nations to colonize Africa and introduce so-called "civilized" European ways into the continent's peoples. This included an infusion of Christian missionaries and enterprise-oriented merchants and traders, many of whom exploited the African natives. The subsequent carving up of Africa by the European colonial powers—with artificial "nations" created where none had existed before—left a legacy of violence and conflict that continues to the present day.

Speke and Grant returned to England with news of their discovery only to find controversy. Burton refused to believed they had found the source. He discredited their achievements and tried to gain support for his own travels back into the wilderness. The members of the Royal Geographic Society were divided in their opinions of Speke and Grant's discovery. However, in a hotly debated controversy, it was decided that since the society had financially supported Speke's expedition it would support his claim. Burton was devastated. He could find no financial support for his own mission and created a public feud over the incident.

Speke returned with Grant to Lake Victoria and began a mapping project that was to continue for some time. When he finally returned to England he challenged Burton's criticism of his achievements and agreed to a public debate over his findings. Unfortunately, a few days before the scheduled debate, Speke accidentally shot himself in a hunting accident and the two men, Burton and Speke, were never reconciled.

Today, Speke is credited with the discovery of the Nile source. The history of the many hearty adventurers who attempted to discover the wonders of this famous river and the land that contains it is long and colorful. The Nile deserves its place as one of the natural wonders of the world, and it will continue to inspire the imaginations of generations to come. The discovery of its source will also continue to inspire historians to understand the legacy of European colonialism in Africa.

BROOK ELLEN HALL

Further Reading

Erlich, Haggai and Israel Gershoni, eds. *The Nile: Histories, Cultures, Myths.* Lynne Reinner Publications, 1999.

Harrison, William. *Burton and Speke.* New York: St. Martin's Press, 1982.

Nomachi, Kazuyoshi. *The Nile.* (Photographic book.) Odyssey Publications, 1998.

Speke, John Hanning. *Journey of the Discovery of the Source of the Nile River.* New York: Dover, 1996.

The Nain Singh Expeditions Describe Tibet

Overview

Any attempt to establish a chronological opening and closing of Tibet's borders during the eighteenth century will produce conflicting reports—depending on whether they originate in England, India, Nepal, China, or Tibet itself. One fact remains constant, however. No one disputes the enormous contributions of Nain Singh, an Indian Pundit who explored the secret world of Tibet and gave the world valuable geographical and cultural information about one of the most tightly guarded and protected countries in the world.

Background

Early in the seventeenth century, both the Jesuits and Capuchins were thrown out of Tibet. However, in 1846, two French Lazarist priests, Evarist Huc and Joseph Gabet, entered the gates to Lhasa and became the first missionaries to see the Tibetan capital for almost a full century. Lhasa had long been the official seat of government for Tibet but, because of strong religious ties to central China, they were not entirely autonomous.

From the initial subjugation of India by the British in the seventeenth century, the new rulers had been aware of the commercial possibilities that awaited them and their goods in the neighboring Himalayan states. They remained passive until they were finally in control of territory that encompassed all the land from the Bay of Bengal to the foothills of the mighty mountain barriers to the north. Coincidentally, the Ghurkas were engaged (at the same time) in taking over the smaller Newar states which now form Nepal: Katmandu, Bhatgaon, and Patan. These were particularly desirable to the Ghurkas because they controlled the all-important trade routes between the Gangetic plain and Tibet. This control caused a marked decline of the early trading that the English had established through the East India Company and forced them to consider alternate routes to Lhasa and the rest of Tibet.

The British were not alone in seeking this valuable exchange of goods. Russia was showing a marked interest in central China and had already made several exploratory forays into the interior. Records from the late eighteenth century show lists of goods which the British believed would attract the interest of the Tibetans: sec-

ond-quality cloth, cheap watches, clocks, trinkets, pocket-knives, scissors, large (but imperfect) pearls, shells, smelling-bottles, and similar barter items. In exchange for this questionable quality merchandise, the British expected to receive gold dust, silver, musk from the musk deer, yak tails (which they used as fly whisks), and precious wool for their mills. Later, it was discovered that there were gold-producing mines in Tibet, along with other primitive and precious metals.

In the early 1800s, there were several explorers and dilettantes who ventured into Tibet, Nepal, and other exotic territories, but the notes they brought home were—in the words of John MacGregor, author of *Tibet: A Chronicle of Exploration* (1970)—short on fact, long on trivia.

Under the rule of Queen Victoria, the British Empire had prospered and grown. It stretched so far around the world that it gave rise to the motto: "The sun never sets upon the British Empire." With all its acquisitions, India remained their most important territory. It truly was the "Jewel in the Crown" and the British had every intention of exploiting both the land and its people to their greatest advantage.

Impact

Accordingly, in 1852, the government commissioned Thomas George Montgomerie, a British army engineer, to prepare maps of the entire country—a daunting task for any team. (A look at a world atlas will show the enormous amount of territory occupied by the Indian states.) Twelve years later, Montgomerie had mapped the domains of the maharajah of Jammu and Kashmir, located in northwestern India. It was a monumental accomplishment, but Montgomerie then encountered a problem. He found that many parts of India were forbidden to outsiders and opted not to risk his men or equipment in defying the border guards. Further, the land itself was a formidable barrier in that India is bordered on the north by the snow-capped Himalayas, which were difficult in good weather and impassable when snows were present. This natural barrier provided good protection for the small states of Nepal, Sikkim, and Bhutan and kept the foreigners out. In addition, these states, along with Tibet, were highly suspicious of British power in any form.

Montgomerie was a curious man. He wanted to know what was on the other side of the mountain. Sending an Englishman was not an option so he sent a Muslim clerk in his office (Abdul Hamid) to Yarkand, a small village in what was then called Chinese Turkistan. After six months, Hamid had accurately mapped the location of Yarkand and, as an added bonus, had used some of his time to spy on Russian activities in the surrounding areas. Unfortunately, Hamid died while returning to India but when his notes were delivered to Montgomerie, he was so pleased that he set up a training center for Indians who would penetrate Tibet and report on any activities they observed. The recruits were called "pundits," a word meaning teacher or educated person.

Searching for reliable Indians led Montgomerie to a family known to have worked previously for both English and German explorers. His first pundits were the headmaster of a school in the village—Nain Singh—and his cousin, Mani Singh. Both proved adept and showed considerable skill in using surveying instruments, navigational astronomy, and altitude calculation methods. But these men were not just surveyors or geographers, they were trained as spies and traveled with secretly compartmented luggage and hidden pockets in their clothing.

One of their most interesting achievements was a particular method of walking. They were trained to perfect their stride so that each pace was the same size as the previous one. With this system, they were able to measure distances by keeping count of the number of paces in any given period. Also, they shortened their Buddhist prayer beads from 108 to 100 to aid them in their counting. Both were given code names but Nain Singh was always the "chief pundit."

The two men began their first venture into Tibet in January of 1865 and arrived in Katmandu, Nepal, two months later. A suspicious governor at the Tibetan border refused to let them enter and the cousins separated to devise other ways of getting through the gates. Nain joined a caravan and posed as a merchant and was able to get through to Jih-k'a-tse, the second largest city in Tibet.

In January of 1866, Singh found himself in Lhasa where his real work was to begin. Records show that he completed 20 solar and stellar observations that came amazingly close to the exact latitude of Lhasa. Also, by using the boiling point of water, he was able to calculate Lhasa's altitude at 11,700 feet (3,566 m) above sea level. This figure is also very close to what we know presently.

Although Singh was invited to a group audience with the twelfth Dalai Lama, he decided to forgo the opportunity since he believed some of the Tibetans were becoming suspicious of his activities. Since capture would mean execution, he went back to his caravan and resumed their course along the main east-west trade route of Tibet. He used this portion of the trip to chart the Tsangpo River from its source to its confluence with the Kyi-Chu River. The caravan was attacked and Singh was held captive but finally found his way back to India by July of 1867. His cousin, Mani, also returned about the same time. They had been away for more than two years and had walked 1,200 miles (1,920 km). They deduced this by converting the number of steps they had taken: 2,500,000.

After a short rest, Singh began his second expedition into Tibet. He had reported seeing large golden Buddhas, confirming what the British had already suspected: Tibet held large quantities of gold. This time he was accompanied by his brother, Kalian, who had also been trained as a pundit. After numerous difficulties, both from foul weather and suspicious Tibetans, the two men split up as they had planned. Nain found the goldfields which were in a hostile, bleak plain at 16,000 feet (4,877 m), possibly the highest inhabited spot in the world. Although the man in charge of the field was suspicious, Singh was able to gain his confidence by giving him some coral jewelry for his wife. He reported seeing one gold nugget that weighed two pounds (.91 kg).

While on his way home, he managed to free his cousin, Mani, who had been held prisoner by some nomads and also located his brother, Kalian. All three returned safely to India.

In 1874, the British government asked Singh to make one more expedition to Lhasa. His ultimate goal was to reach Peking (now known as Beijing), China. Singh and four companions disguised themselves as Buddhist monks, rounded up a flock of sheep as additional cover and set out on their last adventure. They never reached Peking but were able to gather much valuable information by mapping mountain lakes and finding some that had never been mentioned previously.

The group reached Lhasa in 1875 but stayed only two days. Rumors about British spies were circulating and fearing apprehension,

Singh sent the others to Kashmir. He then headed for the fastest route to India but was captured and held in prison until February 1876. He was able to escape and returned home via Calcutta.

Nain Singh was an impressive example of loyalty and resourcefulness. His feats were reported in the *Geographical Magazine* of the Royal Geographical Society, who awarded him its coveted gold medal for his accomplishments. He also received a gold watch from the Paris Geographical Society. The government granted him a parcel of land and a life-time pension for his contributions to the crown.

After his historical discoveries, Singh was active in training other pundits to assist the government in its continuing efforts to expand trade routes. He ultimately retired to his land where he lived a quiet life. Conflicting reports about his demise state that he died in 1882 from cholera, but another report indicates that he died in 1895 from a heart attack. His contributions to his country and the rest of the world regarding the Asian continent remain a benchmark for all who followed.

GERALD F. HALL

Further Reading

Lamb, Alastair. *Britain and Chinese Central Asia; The Road to Lhasa, 1767 to 1905.* London: Routledge and Kegan Paul, 1960.

Saari, Peggy and Daniel Baker. *Explorers and Discoverers; From Alexander the Great to Sally Ride.* Detroit: Gale Research, 1995.

The Discovery of Troy

Overview

The ancient Greek poet Homer wrote of the city of Troy, but in medieval times its location was forgotten, and many doubted that it existed at all. An enthusiastic amateur, Heinrich Schliemann (1822-1890), was determined to find the fabled city. He actually did find the site, and a great treasure trove besides. However, his rather unscientific approach to archaeology led to mistakes and misinterpretations that continue to provoke controversy today.

Background

The poet Homer lived almost 3,000 years ago, and his *Iliad* is considered to mark the dawn of Greek literature. Several of the plays of Aeschylus, Sophocles, and Euripides were based upon it. The *Iliad* is an epic woven together from shorter oral poems, and tells the story of a time when Greece was at war with a city on the coast of Turkey, across the Aegean Sea. The name of the city was Troy.

The civilizations of the ancient Mediterranean often fought over trading alliances, wealth, and territories. According to legend, however, the Trojan War, which took place a few hundred years before Homer's time, was a 10-year conflict over a beautiful woman named Helen. Helen was the wife of Menelaus, the king of the Greek city of Sparta. Paris, the son of the Trojan King Priam, carried Helen off to Troy with the assistance of the goddess Aphrodite. Then Agamemnon, the brother of Menelaus and ruler of another Greek kingdom called Mycenae, led an army to Troy to bring Helen back to Sparta.

The *Iliad* covers a period of 54 days during which a conflict develops between Agamemnon and the Greek hero Achilles, who is under his command. Achilles withdraws from the battle, until his best friend Patroclus is killed by Hector, another of Priam's sons. Determined to have revenge, Achilles kills Hector outside Troy, and his funeral is the last scene in the epic. The fall of the city of Troy is described in Homer's later work, the *Odyssey*.

Homer himself is shrouded in mystery; as with Shakespeare, scholars debate whether a man by that name actually wrote any or all of the works attributed to him. The location of Troy was unknown after about 400 A.D., and many doubted that it had existed at all outside the realms of fable. But for centuries, upper-class schoolchildren were taught a curriculum that stressed classical literature, including Homer.

Heinrich Schliemann was born in 1822, the son of an impoverished German clergyman. According to his autobiography, he was introduced to stories of Troy by his father, and was captivated by the romance of the world Homer portrayed. Schliemann's formal schooling was cut short in early adolescence by the need to support himself. But he continued to study, and was especially proficient in languages, eventually teaching himself 18 of them. Schliemann was also a very successful businessman, and at the age of 41 was able to retire to pursue his interest in archaeology, and especially his dream of finding the city of Troy.

Impact

In the 1860s, Schliemann turned his life completely around. Not only did he quit his business, he divorced his first wife. In a letter to a Greek friend, he put in his order for a new one. She must be poor, but well educated, someone who shared his passion for Homer. He wanted someone beautiful, who had black hair and looked Greek. The friend managed to find a woman who met these specifications, and later that year Schliemann and Sophia Engastromenos were married. She was to be his partner in archaeology as well as in life.

Schliemann toured the Aegean with Homer as his guide. From the descriptions in the *Iliad*, he became convinced that Hissarlik, Turkey, was the site of ancient Troy. It took him two years to get a government permit to excavate there, and he had to finance the dig himself. He also had to promise to turn half his finds over to a Turkish museum, and leave any ruins in the same condition in which he found them. When the digging finally began in 1871, these promises were completely ignored.

Schliemann worked in an era when archaeology was mostly treasure-hunting. Some of the most advanced archaeologists were beginning to understand that excavation is a destructive process. It must be done slowly and carefully, while recording detailed information, to learn as much as possible. But Schliemann was not among these pioneers of scientific archaeology. He attacked the Hissarlik site as enthusiastically as he did everything else.

Ruins were uncovered soon after the excavation began. But Schliemann reasoned that the ancient city he was looking for would be deep under the layers of eons. He had his workers dig down to the lowest level and push aside the intervening rubble. It wasn't until a year later that he realized he had gone too far. Settlement at the Hissarlik site predated the Troy he sought by as much as 1,700 years. And his excavation had damaged or destroyed everything in the path of his deep trenches.

In 1873, digging in a newer layer, he found a palace he was sure was of the right period. He also found a paved road and a sacrificial altar. This, he was convinced, was the real Troy of the *Iliad*. But the archaeological community, not convinced of Schliemann's credibility, paid little attention.

Then, one day Schliemann's pick struck a shiny object. Immediately he dismissed his workers, and continued digging. He unearthed jewelry of gold and silver, goblets, plates and vessels of gold and copper, and a shield. This hoard, which Schliemann called "Priam's treasure," attracted a great deal of publicity. Schliemann became a famous archaeologist almost overnight. He smuggled the treasure out of Turkey and displayed it in his house in Athens. A later photograph of Sophia Schliemann draped in the heavy jewelry became one of the most famous images of its time.

Schliemann eventually gave the treasure to the Berlin Museum. More than 50 years after his death, during World War II, the entire hoard disappeared. A long-standing rumor that it was hidden away deep in Russian vaults was proven true when these began to open up at the end of the Cold War. In April 1996, the treasure went on display in Moscow.

Yet scholars agree that "Priam's treasure" never belonged to the king for which it was named. It was a remnant of a much earlier culture. The bedrock layer in which Schliemann dug first, called Troy I, dates from about 3,000 B.C., the Early Bronze Age. Its ruins include brick walls and crude pottery. After Schliemann realized he had excavated too deeply, he found the treasure in the next layer up, now called Troy II, a city of stone walls with artifacts of finely worked metal. But this was still about 1,000 years before the events of the *Iliad*.

After Schliemann's death in 1890, his widow vowed that his work would continue. She funded further excavations by Wilhelm Dörpfeld (1853-1940), who was more scientific in his orientation. He found nine separate cities, one atop the other, at the Hissarlik site. He believed that the sixth of these was the Troy of the *Iliad*. It was larger than its predecessors, with high limestone walls protecting its perimeter.

In 1932, a University of Cincinnati expedition led by Carl Blegen (1887-1971) studied the site. Like Dörpfeld, Blegen found nine layers, but recognized that Troy VI had been destroyed by an earthquake. This meant it wasn't Priam's city, fallen in a war or raid. The city now believed to be the Troy of legend is the next layer, Troy VIIa. It is built of similar materials, as if rebuilt after the earthquake. But it lasted only about 100 years before being destroyed by fire and looting.

Troy VIII, which stood while Homer actually lived, was a small Greek village. Troy IX was the city of Ilium, ruled by the Greeks and later by the Romans. Alexander the Great held athletic games there in the 300s B.C. to honor Achilles, from whom he believed himself to be descended. The city lasted until the reign of the Roman Emperor Constantine the Great in the 300s A.D.

SHERRI CHASIN CALVO

Further Reading

Braymer, Marjorie. *The Walls of Windy Troy: A Biography of Heinrich Schliemann.* New York: Harcourt, Brace, 1960.

Caselli, Giovanni. *In Search of Troy: One Man's Quest for Homer's Fabled City.* New York: P. Bedrick Books, 1999.

Duchêne, Hervé. *Golden Treasures of Troy: The Dream of Heinrich Schliemann.* New York: Abrams, 1996.

Edmonds, I. G. *The Mysteries of Troy.* Nashville: Thomas Nelson, 1977.

Schliemann, Heinrich. *Ilios: The City and Country of the Trojans: The Results of Researches and Discoveries on the Site of Troy and Through the Troad in the Years 1871-72-73-78-79.* Including an autobiography of the author. With maps, plans, and about 1,800 illustrations. London: J. Murray, 1880.

Traill, David A. *Schliemann of Troy: Treasure and Deceit.* New York: St. Martin's Press, 1995.

Wood, Michael. *In Search of the Trojan War.* New York: Facts on File Publications, 1985.

Deep-Sea Exploration:
The HMS *Challenger* Expedition

Overview

The HMS *Challenger*, a 200-foot warship converted for scientific use, left the English port of Sheerness on December 7, 1872, for a four-year voyage of exploration. Unlike previous expeditions, *Challenger* left to explore the sea itself, the first scientific expedition of oceanographic exploration. During *Challenger's* four years away from home, her crew and scientists founded the era of modern scientific study of the ocean and its creatures. *Challenger* returned to England on May 24, 1876, having traveled 68,890 nautical miles (79,277 miles or 127,584 km) with scientific specimens from over 350 sampling locations. These specimens included samples of bottom materials, samples of bottom animal life, water samples at a variety of depths, and samples of aquatic life from various depths. In addition, seawater temperature at different depths was measured, as were surface weather conditions, measurements of ocean currents, and other information. Following her return, a total of 50 volumes of scientific reports were published by a team of over 100 scientists that described and explained the voyage's findings. The voyage was said to have been "the greatest advance in the knowledge of our planet since the celebrated discoveries of the 15th and 16th centuries."

Background

The great age of exploration and discovery that began with the Portuguese and Spanish in the fifteenth century had largely run its course by the latter part of the nineteenth century. All the major landmasses of the world had been discovered by that time, the coastlines charted, and explorers were making major inroads towards exploring the continental interiors for geographic, scientific, and commercial purposes. Throughout this time the oceans served as highways between nations and continents, highways upon which ships sailed carrying raw materials, finished goods, people, and money from port to port. Fishermen, whalers, sealers, and others depended on the sea for their livelihood, and the products of their work fed and employed countless people in nearly every country. Yet, in spite of this dependence on the oceans, no systematic scientific study had been made of them until the *Challenger* set sail in 1872.

Although explorations on land were progressing well, land encompasses only about 30% of the Earth's surface. No matter how thoroughly the continents were mapped and explored, there was a limit to our understanding of the Earth unless serious, scientific exploration of the oceans took place. This is the role that *Chal-*

lenger, under the scientific direction of Charles Wyville Thomson (1830-1832) and the military leadership of Captain George Nares (1831-1915) was to fill.

The *Challenger* voyage was a logical progression from other scientific voyages sponsored by Britain. Captain James Cook (1728-1779) made three voyages of discovery with the *Endeavour* between 1768 and his death in 1779, Charles Darwin (1809-1882) accompanied the *Beagle* in 1831 on a voyage of nearly five years, and other lesser known scientists and explorers made similar voyages. But England's empire on land was held together by her dependence on the sea. So, in 1870, Thomson, a professor of natural history at Edinburgh University, persuaded the Royal Society to recommend a voyage of oceanographic exploration and study, a recommendation that was granted.

At the time *Challenger* set sail, geologists were relatively certain that there were vast expanses of featureless plains at the bottom of the oceans, probably covered with all the sediments washed from the continents. They were also sure that life could not exist at the ocean bottoms and that any sediments found would likely be unremarkable, fine-grained sediments that varied little from place to place. All of these suppositions were shown to be wrong. Finally, the land-based scientists had some facts with which to work.

Impact

The major impact of the *Challenger* expedition was scientific in nature, but these discoveries either were staggering in and of themselves or later led to other discoveries that fundamentally changed the way in which we view our world. Some of these were:

1. Discovery of undersea mountains, particularly along the Mid-Atlantic Ridge.

2. Discovery of life at the bottom of the oceans.

3. Charting of surface and subsurface ocean currents, salinity differences, and temperatures.

4. Returning samples of sea-floor sediments, including manganese nodules, from a variety of locations.

From the viewpoint of long-term scientific impact, the discovery of the Mid-Atlantic Ridge is perhaps the most significant discovery made by the *Challenger,* though this was not appreciated until more than 80 years after the voyage.

Not only did this dispel the conception of the abyssal plains as uniformly flat and featureless, but the Mid-Atlantic Ridge turned out to be crucial to understanding the concept of plate tectonics. It is there that geologists first understood how the tectonic plates upon which continents ride are formed and travel around the globe. The theory of plate tectonics, in turn, has proved to be a unifying force in the fields of geology, evolutionary biology, and climatology, among others. The presence of undersea mountains helped to validate the concept of the earth as a dynamic geologic environment, a living planet in the geological sense that mountains and islands are born, age, and are eroded away. Another discovery, a deep-sea trench 8,515 meters (nearly 27,000 feet) deep, turned out to be due to plate tectonics, too. This trench is actually a subduction zone where one plate is diving beneath another, completing the journey begun at a mid-ocean ridge such as the Mid-Atlantic Ridge.

At each observation station, the *Challenger* attempted to dredge the ocean floor for any animal life they could find. At first it was widely thought that no life could survive the high pressures and low temperatures at the sea floor. This notion was proved wrong when *Challenger's* nets brought up living creatures from the greatest depths. Finding life under such conditions led, in turn, to a reevaluation of the conditions under which life can exist, suggesting that living beings are not as fragile as previously thought. This concept has lately been taken to even greater extremes with the discovery of "extremophiles"—organisms that survive temperatures above boiling, extreme pressures, highly acidic solutions, vacuums, and extreme radiation levels. All told, *Challenger* scientists discovered over 4,700 species that were new to science.

Adding to this already impressive legacy, *Challenger* also obtained a suite of samples at each observation station that gave information about water temperature, water chemistry, and current direction at a number of depths down to the ocean floor. At the same time, surface conditions were noted, including weather, the direction of surface currents, and any other information that could be gathered. The immediate impact was the start of global maps of surface currents and weather conditions. The measurements obtained at various depths set in motion studies that later led to a better understanding of the manner in which ocean and atmosphere interact to cause weather, the interdependence of ocean and climate, and the impact that changes in ocean conditions can have on terrestrial climate over longer periods of time. For example, scientists have recently discovered the intimate link between wind patterns over the Pacific Ocean and the onset of El Niño weather conditions that affect parts of Asia, the Americas, Australia, and Africa. For the first time, scientists began to understand that conditions in the oceans were not uniform, that the oceans could affect the weather (and vice versa), and that these facts could be significant scientifically, economically, and militarily.

The discovery that sea floor sediments were varied also came as a surprise to many. In fact, *Challenger* brought back many surprises. Parts of the sea floor were covered with large nodules that turned out to be primarily manganese and other metals. Recently, technology has been developed that will allow mining of these areas for the manganese, a process that, once started, could have significant economic impacts as well as potentially serious environmental effects on the areas of ocean that are mined. In addition, it was found that large parts of the ocean floor were covered with biogenic ooze; sediments derived mainly from microscopic shells composed of calcium carbonate and silica. In fact, in many places siliceous shells predominate because calcium carbonate dissolves. These shells helped demonstrate the major role played by microscopic organisms in the ocean ecosystem and assisted with proper identification of rocks composed of similar sediments found on land.

Like any good research, *Challenger's* discoveries led to as many questions as were answered. These, in turn, led to more questions as knowledge of the oceans increased, and this process continues today. As important as *Challenger's* discoveries were, equally important is the branch of scientific study that was thus established. Today, oceanographic exploration is a legitimate form of scientific inquiry practiced by thousands of scientists on over 100 oceanographic research vessels in all major bodies of water on Earth. All of this was started by *Challenger,* the first ship to purposely set out to explore the 70% of our planet covered by water.

P. ANDREW KARAM

Further Reading

Saari, Peggy, and Daniel B. Baker. "HMS *Challenger.*" In *Explorers & Discoverers: From Alexander the Great to Sally Ride.* Detroit: UXL, 1995.

Thurman, Harold, and Alan Trujillo. *Essentials of Oceanography.* Englewood Cliffs, NJ: Prentice Hall, 1999.

Henry Morton Stanley Circumnavigates Africa's Lake Victoria and Explores the Entire Length of the Congo River

Overview

The nineteenth century introduced a period of European-led African exploration unlike any other in history. Most significant were the travels of Henry Morton Stanley (1841-1904), a British-born American journalist who was the first person to travel and record the entire length of the Congo River. Stanley was also the first European to circumnavigate Lake Victoria and the man responsible for opening parts of central Africa to transportation. Stanley's discoveries answered some of the main questions about the geography of Africa's interior waterways. His observations became the foundation for Belgian King Leopold's violent Congo Free State and inspired a period of imperialism whose effects continue today.

Background

One of the first European sightings of the Congo River occurred in the late fifteenth century, when Portuguese slave traders glimpsed the river's outlet at the Atlantic Ocean. By the late eighteenth century interest in Africa and its navigable waterways intensified as slave trading reached a terrible high, and a transportation route through the continent was much sought after.

In 1795 Scottish physician Mungo Park (1771-1806) explored the Niger River and first spoke of the immensity of the Congo, which he assumed originated from a large lake in the center of Africa. Other geographers theorized that the Congo was actually a smaller waterway feeding a larger, as yet unnamed, river. In 1816

James Kingston Tuckey (1776-1816) attempted to navigate the Congo from the Atlantic Ocean but moved no farther than Cooloo, 200 miles (320 km) upriver, because of disease, attacks from tribes, and enormous waterfalls.

By 1836, when more than 10 million Africans had already been shipped out of their homeland as slaves, the major European powers declared slave trading illegal and thus removed a large commercial interest in African exploration. This shifted the focus of exploration to geographical science and Christian missionary work. Several European explorers were already underway by the early-to-mid-nineteenth century, led mainly by the desire to find the source of the Nile River, which was the most used river in Africa at the time. Richard Francis Burton (1821-1890) and John Hanning Speke (1827-1864) explored part of Lake Victoria and a section of the Nile, and theorized that either Victoria or Lake Tanganyika, southwest of Victoria, was the river's source. Sam and Florence Baker (1821-1893 and 1841-1916, respectively), who discovered Lake Albert, guessed that it was the source of the Nile.

By 1872 there was still no factual evidence supporting which body of water was the true source of the Nile River. British missionary David Livingstone (1813-1873), while partly on a quest to seek the elusive source of the Nile, discovered the Zambezi River and Victoria Falls. Livingstone's expedition went on to discover parts of the main network of Africa's largest rivers, including the Congo, but his work remained unfinished, leaving many questions that Stanley would soon answer.

Henry Morton Stanley's first African expedition was in 1871, on assignment for *The New York Herald* to find Livingstone, who was assumed dead. Stanley's famous question upon finding him, "Dr. Livingstone, I presume?" made Stanley a household name in the explorer frenzy that followed Livingstone's journeys. Although not a scientist, Stanley was sent back out to answer the geographic questions left following Livingstone's death in 1873. Among these, Stanley set out in 1874 to circumnavigate the enormous Lake Victoria to see if it was a single body of water, and—more importantly—to see if it was the much-sought-after source of the Nile River. Stanley also planned to circumnavigate Lake Tanganyika, to see if it was the source of the Nile, as Burton had suggested. Finally, Stanley planned to finish Livingstone's work of mapping the Lualaba River. Livingstone had theorized that the Lualaba, which flowed from Lake Bangweolo, was quite possibly the Nile itself. (Others thought that the Lualaba was the same as the Congo River, not the Nile.)

Considering the obstacles faced by explorers of the age, which included deadly disease, bloody tribal attacks, and starvation, Stanley's expedition around Lake Victoria was relatively uneventful. He was well funded, had designed his own boat that could be disassembled into four sections, had a crew nearly 200-men strong, and encountered few hostile tribes. He finished the circumnavigation in under two months and continued on to Lake Tanganyika.

It took four months for Stanley to meet the banks of Tanganyika, but he circumnavigated it successfully in 51 days. His trip concluded that Lake Tanganyika had no outlet river, so it could not be the source of the Nile, disproving Burton's theory. On his way to look for the Lualaba, loosely following Livingstone's last route, he spotted the confluence of the Lualaba and Luama Rivers, adding to the theory that if the Lualaba were actually the Congo, it was fed periodically by smaller streams. On October 17, 1876, he departed on the Lualaba toward Nyangwe.

While Stanley was traveling toward Nyangwe, British explorer Verney Lovett Cameron (1844-1894) had already arrived. He, too, had planned to uncover the Lualaba/Congo mystery; he suspected that the Lualaba was a river that fed the Congo. Cameron wrote his theory in his journal: "...for where else could that giant among rivers, second only to the Amazon in its volume, obtain the two million cubic feet of water which it unceasingly pours into the Atlantic?" Stanley would soon discover that Cameron's theory was right. Unfortunately for Cameron, he never left Nyangwe to see for himself. He tried for several weeks to secure men and canoes for the journey, but stories from Arab traders of the dreadful cannibals who lined the thick jungle, and the dangerous rapids along the route, prevented his departure.

Stanley, however would not be dissuaded. Like Cameron, he negotiated with Tippoo Tib, a wealthy, ruthless Arab slave trader who had a strong and relentless troop of men who could provide a safe escort for Stanley through the Congo's wilds. While Tippoo Tib never agreed to Cameron's negotiations, he made a deal with Stanley to accompany him through the first 90 days of the expedition. Tib's escort provided another 400 men, of which many died from smallpox, pleurisy, malaria, and typhoid. In the first 50 days the expedition encountered several suspected cannibal tribes that flocked toward the slow-moving Stanley caravan in fast canoes, shooting poisoned arrows and sounding war cries. After more than 15 of these encounters, Tib retreated, leaving Stanley to fend for himself.

The tribal attacks grew more ferocious and frequent. At each attack, however, Stanley acquired more canoes, until he had a flotilla large enough to keep his entire party on the river. Soon they came to a series of dangerous falls, unsure of how long or treacherous each rapid was. They portaged the boats, carrying them through dark rainforests filled with well-camouflaged attackers who frequently pounced on the anxious expedition. The Stanley Falls, as it was named, took three weeks to descend, contained seven cataracts, and proved the team's first geographic obstacle. At the sixth cataract, Stanley determined that they had crossed the equator. He also found that their altitude had dropped considerably below that of the Nile at its lowest measured point, so in order for the Lualaba to be the Nile, as Livingstone thought, it would have to flow uphill from the Stanley Falls. This flow was an improbable feat, making it impossible for the Lualaba to be the Nile River.

Soon after the Falls, the river swelled to immense widths, sometimes nearly nine miles across, and provided the longest stretch of navigable waters on the river. The expedition reached a pool, which Stanley named the Stanley Pool, during this period of continued tumultuous travel. At this point the team counted a total of 32 battles with hostile, allegedly cannibalistic, tribes. The remaining tribes that the expedition encountered from the Stanley Pool until the end of the journey were peaceful, but the river was not. The Congo, as Stanley had now surmised that the Lualaba and the Congo were the same river, would have nearly 200 miles (320 km) of the most severe rapids he would encounter. These were the same cataracts that had defeated Tuckey, only Stanley faced them at their onset, not their end. He lost several men to the falls, moving as little as 500 yards a day while carrying the boats and confiscated canoes along a steep river bank. He finally stopped at Embomma, only a day's journey from the end of the river, satisfied that he had completed his goal to explore the unknown sections of the Congo.

Impact

Stanley's success was received with wild admiration. He had proven that the Lualaba was indeed the Congo River, that the Nile did not originate from Lake Victoria or Lake Tanganyika as previously thought, and that a European could survive the wilds of the Congo—albeit with some luck—by navigating the enormous Congo River from the heart of the continent. Stanley's journey also concluded what we know about the character of the Congo River: from its source, just south of Lake Tanganyika, the river begins as the Lualaba, heads southwestward to Lake Bangweolo, then turns north to the Zambia/Zaire border to Lake Mweru, where it becomes the Congo. The mighty river crosses the equator twice, placing it in both the Northern and Southern hemispheres. After 3,000 miles (4,800 km) of a wild path through extreme landscapes, it reaches the Atlantic Ocean.

Beyond solving one of the nineteenth century's geographical unknowns, Stanley's expedition created—and widely publicized—the sentiment that the interior of Africa was open and accessible to all of Europe. For almost ten years following his famous journey, a period referred to as "the Scramble" took effect. In short, it was a mad race between European nations to colonize Africa and introduce so-called "civilized" European ways into the continent's peoples. This infiltration included money-seeking traders and missionaries, who convinced hundred of tribes to convert to Christianity and adopt its social code.

Stanley was soon recruited by King Leopold of Belgium to build stations along the Congo and bypass the unnavigable rapids with roadways and railroads. With these structures in place, Leopold soon turned the Congo into the Congo Free State. This era under Leopold's covert control proved to be a period of the worst abuse and exploitation of Africans in history. His reign of terror, which included routine executions and punishment by amputation, was finally discovered by the rest of Europe by 1907. Leopold was removed from power, and the Congo Free State became the Belgian Congo in 1908. The introduction of European influence, however, changed Africa forever. Commerce thrived, but tensions grew and oppression of native Africans was commonplace. Gold and diamond mining changed the landscape, big game hunting became a common vacation pursuit, and the railroads and road systems removed the geographic boundaries that once defined the various tribes throughout the continent. In 1960 the Belgian Congo was given its independence and became the country known as Zaire. Today, renamed the Congo, it continues to have periodic bouts of politically motivated violence.

LOLLY MERRELL

Further Reading

Books

Anstruther, Ian. *Dr. Livingstone, I Presume?* New York: Dutton, 1957.

Bennett, Norman R. *Stanley's Dispatches to "The New York Herald."* Boston: Boston University Press, 1970.

Bierman, John. *Dark Safari: The Life Behind the Legend of Henry Morton Stanley.* New York: Knopf, 1990.

Forbath, Peter. *The River Congo.* New York: Harper and Row, 1977.

Stanley, Henry M. *Through the Dark Continent.* New York: Harper and Brothers, 1878.

Other

The Journal of African Travel Writing. http://www.unc.edu/~ottotwo

Stanley and Livingstone (film). Directed by Henry King, starring Spencer Tracy, Cedric Hardwicke and Richard Greene, 1939.

Nikolay Przhevalsky and Russian Expansion: The Exploration of Central and East Asia

Overview

Between 1869 and 1884, Russian explorer Nikolay Przhevalsky (1839-1888) conducted a number of expeditions of exploration into central Asia. During these expeditions he helped to chart the interior of Siberia, Mongolia, and China. He made a number of important natural history discoveries, including the last discovered species of wild horse and wild camel. He became the first European since Marco Polo (1254-1324) to visit Lop Nor in China and he was the first European to explore the upper portions of the Huang Ho River, a river important in Chinese history and folklore for over three thousand years.

Background

During the first part of the nineteenth century Russia was not a large nation. However, like the United States, it had a large expanse of land, stretching to the Pacific Ocean, that was occupied by nontechnological people. In the middle of the century, following the sale of Alaska to the United States, defeat in the Crimean War, and the sale of some of the Kuril Islands to Japan, Russia began to flex its muscles in central Asia, in part to help compensate for these other losses.

At this time, too, the Chinese were distracted by foreign incursions and internal rebellion (the Taiping rebellion), drawing their attention away from the northern border. This combination of circumstances opened the door for Russian annexation of Siberia and other central Asian provinces, as well as Russian seizure of Chinese territories that were later formalized in treaties signed in the 1860s. Some of the apparent Russ-

Nikolay Przhevalsky. *(The Library of Congress. Reproduced by permission.)*

ian imperialism occurred because military officers greatly overstepped their authority, while other territorial annexations followed governmental policy, albeit at a more rapid rate than intended. In addition, the Russians pointed out that they were simply following the lead set by the Americans in their West, by the British in India, the French in Africa, and the Dutch in their colonies. In 1861, partly to consolidate their hold over the Russian Far East, the city of Vladivostok was founded on the coast of the Sea of Japan.

By 1879 Russia had absorbed eastern provinces to the Pacific Ocean and the Sea of Japan. However, a great deal of this land was largely or completely unexplored, except as was necessary to annex it or to wage military campaigns. Among those who set out to alleviate this lack of knowledge was Przhevalsky, who traveled throughout central and eastern Asia, mapping, collecting biological specimens, and surveying future travel routes. By the time of his death in 1888, Przhevalsky had visited and written about many of the recently acquired territories, as well as some that Russia hoped to acquire.

Przhevalsky made a number of discoveries during his travels. In addition to describing Przhevalsky's horse and the wild bactrian camel, he collected widely from both plant and animal kingdoms, preserving his specimens for later scientific study. As the first European in centuries—and, in some cases, even longer—to enter many parts of central Asia, he either discovered or rediscovered important features such as the salt lake at Lop Nor, the Tien Shan mountains, the Takla Makan desert, and more. He also crossed the Gobi Desert, was turned away from Lhasa, Tibet, and explored parts of what were to become Turkistan, Kazakhstan, Kyrgyzstan, and Uzbekistan, to name a few. His books relating his travels were published in both Russian and English, becoming popular for a time.

Impact

Przhevalsky's explorations and discoveries must be set in the larger context of Russian and Chinese history to ascertain their full impact at that time and later. The Russian conquest of Siberia and central Asia had largely ended by 1700, though some areas remained free of annexation until the mid-nineteenth century. These annexations took place, for the most part, rapidly and without terrible loss to Russia. What makes this unusual is that the Mongol hordes that repeatedly overran European and Chinese empires for centuries came from precisely this part of the world. By the time Przhevalsky set out on his journeys, not only had central Asia been largely annexed to Russia, but its inhabitants had fallen under the control of either Russia or China, apparently never again to wield military power. The mere fact that he could explore this area for several years was evidence that centralized governments now controlled central Asia.

In addition to signaling the end of nomadic might, Przhevalsky's explorations gave indication of Russia's continuing interest in consolidating its territories and seeking access to warm water ports for commercial and military purposes. Russia was concerned about British incursions to the north from their colony of India, as well as possible Chinese claims to these lands. By mapping these regions first, Russia hoped to not only lay claim to them, but possibly to extend its borders closer to ports that would remain open even during the cold Russian winters. The first of these goals was not unlike America's purchase of the Louisiana Territory and expansion to the West; by so doing, it was thought, the possibility that European powers might compete with the United States in the New World was diminished. So, too did Russia's claims in the Far East help reduce the chance of competition in its own backyard. Along those same lines, Przhevalsky's explorations of Chinese territories helped to bolster Russian claims to them and, by opening routes from northern China into, for example, Turkistan, Russia helped to encourage migration and trade between these areas.

These expeditions also helped to compensate for Russia's humiliating loss to the British-French-Turkish alliance in the Crimean War. This war was started by Russia, ostensibly to protect southern Slavs and the Orthodox Church from the Ottoman Empire. Although Russia initially seized parts of what is now Romania and destroyed many Turkish ships in the Black Sea, most of the war was fought on the Crimean Peninsula, on Russian territory. It ended in October 1856, after two and a half years, with Russia suing for peace. The embarrassment of this loss was partially offset by subsequent territorial gains and military excursions in the East, including the annexation and exploration of these territories.

It must also be noted that, although already annexed, some of the Eastern territories remained not only unexplored, but largely untouched by their annexation. Although the annexed territories paid Russian taxes, their situation was not very different from previous political arrangements under which they paid tribute to their nomadic conquerors. By exploring these territories and writing of his travels, Przhevalsky was able to excite some popular interest in the Russian people. This, in turn, helped to encourage some to move east, bringing with them the Russian language and some degree of education and culture. From that perspective, the "Russification" of the Russian Far East was assisted by Przhevalsky's travels and writings.

One of the unintended consequences of Russia's exploration of and claims to portions of central Asia may be seen in lingering border disputes between Russia and China during most of the twentieth century. As noted above, Russia occupied and annexed parts of Chinese territory while China was engaged in rebellions and other military distractions. This allowed Russia to more or less force China to relinquish some territories in the Treaties of Aigun (1858) and Peking (1860). Some of these areas and others along the Russian-Chinese border became the subject of disputes that periodically escalated into armed conflict (always stopping short of actual war) between Russia and China on a number of occasions. Such tensions persisted well into the 1990s, though easing somewhat towards the end of that decade.

Not least among the impacts of annexing, exploring, and opening Siberia and the Russian East was that of natural resources. This part of the world has become increasingly important in the production of petroleum, diamond, gold, timber, and more. The mineral resources of Siberia and the Central Asian republics are still not fully realized, and seem likely to increase in importance to the several nations that now comprise this area. By chance, because of Russian claims and exploration, these resources are Russia's (and the central Asian republics') to exploit and not China's. This, in turn, could have a significant impact on the future economic development of China, Russia, and the republics in which these resources now lie. Resources from this region helped to finance and build the Soviet Union, including (ironically) its competition with China, and they may now help the new republics of the former Soviet Union if political stability permits their development and exploitation.

Although hardly a household name, even in Russia, Przhevalsky contributed greatly to the exploration and opening of the Russian Far East. It is tempting to compare him to the American explorers Meriwether Lewis (1774-1809) and William Clark (1770-1838), but this comparison would be inaccurate. Lewis and Clark were sent to explore, map, and discover central and western North America. In addition, a very important part of their mission was to make contacts with the natives of these areas to try to build commerce for the fledgling United States. By comparison, Przhevalsky's trips were primarily for the purpose of exploration and consolidation. Commerce was not stressed, the rough boundaries of the areas to be explored were known, and the inhabitants were already technically subjects of an established and ancient Russian nation. In addition, while Lewis and Clark's expedition lasted for a much longer time than any of Przhevalsky's, they made only a single voyage, while Przhevalsky returned for a total of four trips covering a much greater geographical extent. Though similarities exist, the differences of these journeys of exploration are more apparent.

Przhevalsky died in 1889 on the shores of Ysyk-Köl in Turkistan, one of the largest mountain lakes in the world, first discovered by him in 1883 (and briefly named after him). Although his work was instrumental in the exploration and exploitation of the Russian Far East, he remains a relatively obscure figure who is chiefly known by Przhevalsky's horse.

P. ANDREW KARAM

Further Reading

Dukes, Paul. *A History of Russia: Medieval, Modern, Contemporary, c. 882-1996*. Durham, NC: Duke University Press, 1998.

Przhevalsky, Nikolay. *From Kulja, Across the Tian Shan to Lop-Nor*. Translated by E. Delmar Morgan. New York: Greenwood Press, 1969.

Przhevalsky, Nikolay. *Mongolia, the Tangut Country, and the Solitudes of Northern Tibet*. London: S. Low, Marston, Searle, & Rivington, 1876.

Rayfield, Donald. *The Dream of Lhasa: The Life of Nikolay Przhevalsky*. London: P. Elek, 1976.

Luigi Maria D'Albertis Explores Unknown Interior Regions of New Guinea

Overview

In his two voyages up New Guinea's Fly River, Luigi Maria D'Albertis (1841-1901) became one of the first Europeans to explore the interior of the world's second-largest island. D'Albertis returned with a great number of scientific specimens, relatively accurate maps of the areas he visited, and scarcely credible stories of adventure. He also returned to accusations of disreputable collecting practices (particularly with respect to ethnographic specimens), poor leadership qualities, and needless conflict with the island's natives.

Background

Now that modern scientists have mapped virtually all of the planets and major moons in our solar system, it is difficult to imagine a time in which large portions of our planet remained unmapped, unexplored, and unknown except to the local inhabitants. This was, however, the case through virtually all of human history. By the middle of the nineteenth century European explorers had visited the interiors of Africa, Australia, South America, and much of Asia. There were, of course, many areas left unmapped, but the single largest area not yet visited by Europeans was the island of New Guinea, the second largest island (after Greenland) on Earth.

New Guinea was first visited by Europeans in 1545, and its coastline was mapped in a haphazard fashion over the next 300 years. Early expeditions that attempted to land for provisioning were usually met with fierce attacks by the natives, and stories of cannibalism were quickly spread by survivors of these expeditions. It was not until the latter part of the nineteenth century that an expedition undertook to explore the New Guinea interior in a scientific manner.

The first scientific expeditions concentrated on the coast and near-coast parts of New Guinea. Russian, Dutch, English, and others spent from a few days to several months in several locations. During this time, D'Albertis made a relatively short trip on foot across a narrow isthmus connecting the northernmost part of New Guinea (the "bird's head") to the main body of the island. One expedition in 1874, led by an explorer named McFarlane, had traveled 150 miles by boat up a river, but no other journeys into New Guinea's interior had yet occurred. D'Albertis was intent on changing this.

D'Albertis was a larger-than-life Italian explorer who had become interested in New Guinea after hearing stories about it from a visiting French naturalist. In his early adulthood, he had earned a good scientific reputation, which he used to convince the government of New South Wales to give him a small steamship with which to ascend the Fly River. On his first voyage in 1875, amid many confrontations with natives, he and his crew penetrated 580 miles

(930 km) along the Fly, returning with a great many specimens and no small number of adventures. His second voyage took place a few years later, but was much less successful because of more constant trouble with natives (who probably remembered confrontations during the 1875 voyage) and significant conflicts with his crew. Amid such hindrances, D'Albertis traveled more slowly on his second voyage and only made it as far as 530 miles (850 km) along the Fly. Following his return from New Guinea, D'Albertis retired to Italy to write of his adventures in New Guinea, published as *What I Did and What I Saw.*

Impact

The impact of D'Albertis's explorations into New Guinea can be roughly divided into two aspects—gains in scientific knowledge and the excitement of the popular imagination.

There is no disputing the large amount of scientific and ethnographic information returned by D'Albertis. He brought back hundreds of specimens of plants and animals new to science, as well as geographic information from New Guinea's interior and detailed information about the island's natives. All of this added immeasurably to the scientific knowledge of the day. This scientific haul was met with acclaim upon his return. Unfortunately, D'Albertis may have erred in choosing to return to the Fly River for his second voyage. During his 1875 voyage, he traveled about as far as was possible before the river became too shallow to proceed further. By choosing to retrace his previous path, D'Albertis was unlikely to penetrate further into New Guinea and, in fact, added little new knowledge because he broke no new ground. Had he instead chosen to ascend another major river, he might have accomplished more.

D'Albertis also returned to a fair degree of criticism about his methods of dealing with New Guinea's natives and the manner in which he collected some of his specimens. A volatile man, he was frequently questioned about his tendency to assume he was under attack any time he saw natives approaching. He would frequently set off rockets to frighten or impress the natives and, on at least a few occasions, launched rockets at villages for no apparent reason. On other occasions, he took bodies from funeral sites, took ethnographic artifacts from huts, and even bought heads from natives for his collections. All of these practices earned him controversy and no small degree of criticism.

Because of his apparent misdeeds, D'Albertis's scientific legacy must be viewed as a mixed success. On the one hand, as the first European to explore any significant portion of the New Guinea interior, every new fact and specimen he returned was a gain for science. On the other hand, his dubious collecting practices served to taint many of his ethnographic discoveries. Finally, by choosing to ascend the Fly a second time, he not only revisited areas previously described, but made things more difficult for himself by having to fight off natives who remembered his previous voyage. This, in addition to ethical questions, no doubt limited his ability to make scientific observations during this voyage.

Scientific and ethical issues notwithstanding, D'Albertis's expeditions were of great interest to the public. New Guinea was among the last places on Earth where exciting new discoveries could be made. Though the polar regions still awaited intensive scientific scrutiny, there was convincing evidence that there was little to be discovered except ice, rock, and cold. No new animals or exotic tribes awaited explorers. New Guinea, however, was subject to literature that was closer to fantasy than to reality. Writers spoke of cannibals, tall mountains, volcanoes, and strange animals, all of which was true. They also wrote of cities of gold, striped ponies, birds with 18-foot wingspans, and marsupial pachyderms, none of which was true. However, in the absence of any actual explorers, there was no way to reliably separate fact and fantasy. D'Albertis's trips began to change that and, in the process, managed to rekindle public excitement for exploration at a time when it was beginning to wane.

Public interest in exploration had begun to subside during this time for reasons similar to those behind flagging support for American lunar exploration after the first few Apollo landings. In July 1969 landing on the moon was the most exciting thing ever done by mankind, and the moon landing commanded a greater audience than any other event in human history. By 1972 public interest was so low that Apollo rockets already built were turned into museum pieces because nobody wanted to fund another trip to the moon. In the case of terrestrial exploration in the late nineteenth century, there was the tendency to feel that, once a few initial forays into the unknown had returned with a great deal of information and little incident, there was little reason to return. The excitement of breaking new ground by visiting a place for the first time vanished and the public had much less interest in the longer and less exciting process of consolidating those first tentative journeys. Or, in other words, the first few visits were a lot of fun, but after that, exploration became more like work, which was neither fun nor interesting to the public. As the last major landmass to be explored beyond the coastline, only New Guinea still excited the public imagination as Africa, Asia, and South America once had, and D'Albertis had a major role in that process.

D'Albertis's journeys had residual effects on subsequent exploration into parts of New Guinea and the discovery of New Guinea's economic potential. His stories of conflict with the natives tended to reinforce previous stories of danger, encouraging future expeditions to approach New Guinea with loaded weapons and the willingness to use them. In the Fly River area, the natives certainly remembered his visits and made life difficult for future expeditions along the Fly. However, this information and resulting animosity was not easily transmitted to other tribes because of the thick jungles, high mountains, and generally impassable terrain around New Guinea.

It was eventually discovered that New Guinea has significant economic resources, including gold and petroleum. While neither of these resources are found in the areas visited by D'Albertis, there are now mining camps in the upper reaches of the Fly River basin, as there are in many other places in New Guinea. Though D'Albertis discovered neither gold nor petroleum during his travels, his trips did help begin the process of mapping the entire New Guinea interior, setting the stage for the initial exploration upon which recovery of mineral resources depends. In this sense, D'Albertis's explorations in New Guinea can be considered the first of many that had the end result of opening the New Guinea interior for commercial development.

P. ANDREW KARAM

Further Reading

Souter, Gavin. *New Guinea: The Last Unknown.* New York: Taplinger Publishing Company, 1963.

Nils A. E. Nordenskiöld Discovers the Northeast Passage

Overview

In 1879 Nils Adolf Erik Nordenskiöld, a Finnish scientist and explorer, completed the first transit of the Northeast Passage, a sea route from Europe across the northern coast of Asia to the Pacific. This passage, like the Northwest Passage to Asia, had been long sought and speculated about. At the time of Nordenskiöld's voyage aboard the *Vega,* it was considered certain that an all-water route existed, but most felt it to be impassable due to heavy ice. By making the passage safely, Nordenskiöld proved such speculation to be groundless. Following his return to Europe, Nordenskiöld wrote and published a popular book, *The Voyage of the Vega,* in which he described his journeys and discoveries.

Background

European cartography in the late nineteenth century was an interesting mix of the known, the speculated, and the unknown. While mapmakers were moving away from the fantastic speculation of previous centuries, there were still large parts of the world that were simply unknown and about which there was no verified information. By the 1870s, most of the world's coastlines had been well mapped and European explorers were staging what may be described as assaults on the continental interiors of Africa, Australia, Asia, and South America, while American explorers were busily investigating the American West. One of the few coastlines left to be explored was that of northern Asia, along the Arctic Ocean.

For several centuries, a number of explorers had sailed in search of a Northwest Passage across the top of the Americas to the Orient. They hoped to find a relatively short and easy route from Europe because the existing routes around the Cape of Good Hope or Cape Horn were both long and dangerous. In addition, there was the hope that economic benefit might be found in the Arctic Ocean itself, in the form of fishing, resources in yet-to-be-discovered lands, or in some other area. And, to some degree, national pride was at stake. One of the great peacetime adventure-seeking activities of the eighteenth century was exploring the world. Nations that added significantly to world geographic knowledge could claim territories, trade routes, and recognition.

At the same time the search for the Northwest Passage was underway, a smaller group was searching for the Northeast Passage across the top of Asia. These explorers felt that it should be possible to sail a ship north, around Scandinavia, past the Russian Arctic, through the Bering Strait, and down to Japan, China, and the rest of Asia. Confirmation of such speculation was hindered by ever-present ice, unknown sea conditions, and the possibility that Asia and North America might actually be joined, preventing passage by sea.

By the 1870s, much of the Russian Arctic coast had been mapped, but this had been done in bits and pieces. It was by then known that Siberia and North America did not touch across the Bering Strait, but it was not definitively known that they did not meet further to the north. For many years, dating back to the time of the Vikings, the far northeastern part of the world had been explored by Scandinavians. It was not a surprise that the final assault on the Northeast Passage would be made by a Finn living in Sweden.

By the time Nordenskiöld set out from Tromsö, Sweden, in 1878, he had already established a reputation for Arctic research and exploration, earned through his many expeditions to Greenland, Spitzbergen, and the European Arctic. Before embarking on this trip, Nordenskiöld thoroughly researched existing maps in an effort to find evidence for a Northeast Passage. Concluding that it existed, he first had to sail around the Scandinavian Peninsula—no small feat in itself—and head to the east. He made it to within about 120 (190 km) miles of the Bering Strait before ice forced him to winter over. He completed the journey the following spring. In so doing, he had proved it possible to reach the Pacific by sailing through the Arctic Ocean, but this route proved economically impractical because it was only open for a short time each year.

Impact

Nordenskiöld's voyage had both direct and indirect effects on late nineteenth-century society. The mapping of uncharted coastlines was wind-

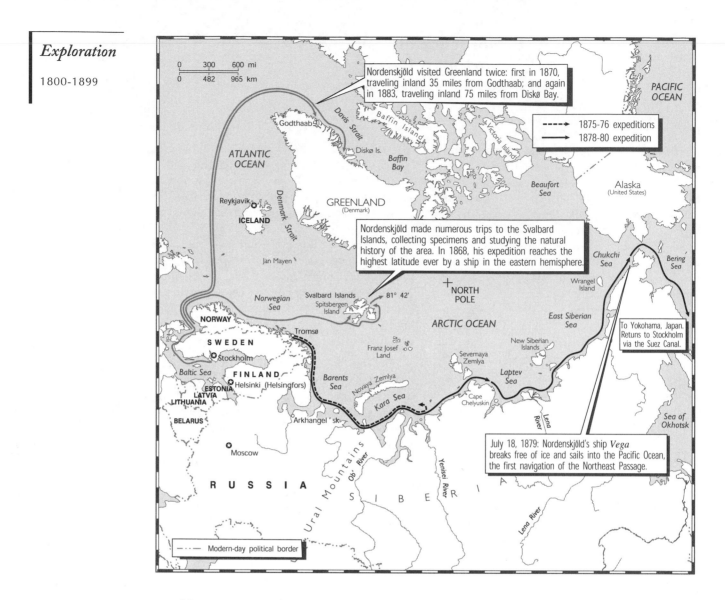

Map of the Arctic region and Greenland showing the explorations of Nils Nordenskiöld.

ing down at this time and, from that perspective, Nordenskiöld's voyage was, in a sense, symbolic of the end of this sort of exploration.

The direct impact of the *Vega's* voyage primarily involved the scientific and geographic knowledge returned by the expedition. First and foremost was the knowledge, as opposed to previous speculation, that the Northeast Passage actually did exist, though it provided no economic boon to anyone. Nordenskiöld showed that the Americas were really completely separated from Asia, something about which there was some debate in spite of the voyages of James Cook (1728-1779) and Vitus Bering (1681-1741) during the previous century. This, in turn, proved definitively that Russia had no geographic claim to the portions of Russian America they had explored and, in some cases, settled. Nor-

denskiöld also returned a large number of scientific samples for study upon his return. These samples, which included geologic, oceanographic, and biological materials, proved to be of great interest to scientists. Finally, there was the added proof that there was no large northern island or continent yet to be discovered, claimed, and explored. In one sense, the *Vega* expedition helped to place a northern limit to colonial ambitions. The indirect impacts of this expedition are somewhat more difficult to describe.

From an economic standpoint, while the search for an easier route to Asia was important, doing so via either a Northwest or Northeast Passage was rapidly losing its importance. The advent of the clipper ships in mid-century had cut the passage from China down from six months to less than three, making it easier to transport perishables (such as tea) from the Orient to Europe and America. In 1870 the Suez Canal opened, shortening the voyage by nearly 6,000 miles (9,600 km) and several weeks. And, as steamship technology became better known, there was the promise of fast, efficient transit to any part of the globe in vessels not dependent on wind or weather. So, by the time Nordenskiöld sailed, what he hoped to find was hardly as important to the world as it had once been. His expedition, and virtually all subsequent expeditions, were driven less by hopes for commercial gain and more by hopes for scientific gain and national pride.

Another impact of the *Vega* expedition involved the historical information discovered—or rediscovered—by Nordenskiöld during his preparations for the voyage. Seeking to learn as much as possible before going into the unknown, he was tireless in his quest for old maps and travelers' accounts of Arctic sea travel. He managed to turn up maps thought lost for decades or centuries that provided interesting information about the knowledge, theories, and cultures of those times. Surprisingly, he found one map dating to 1507 that was surprisingly accurate in its portrayal of North America and other continents. Although Nordenskiöld wanted to believe this showed some unexpected knowledge on the part of the mapmaker, it turned out to be pure speculation and, at one point, the mapmaker had referred to this map as being a mistake. However, through his researches, Nordenskiöld assembled an impressive collection of ancient maps that were donated for academic study upon his death.

As noted above, Nordenskiöld's voyage was one of the last to map a previously unexplored coast in its entirety. True, most of the coast had been mapped earlier, but these maps were not assembled into a coherent whole; many were incomplete or inaccurate, and they had not been verified. Because of this, the maps that did exist could not be fully trusted. In mapping the northern coast of Asia, Nordenskiöld was helping to usher in an era in which very few coastlines were still unknown. The emphasis in exploration was thus turning increasingly to exploration of continental interiors rather than coastlines, evidenced by new, emerging exploration styles.

In mapping a coast and discovering islands, the explorer, often as not, will land briefly to collect samples, meet local natives, and, if possible, reprovision the ship. The emphasis was as much on the time spent at sea as on the time spent ashore, and if trouble with the native populations arose, it was usually possible to return to the ship and sail on, leaving the problem behind. This was obviously not always the case as the deaths of Ferdinand Magellan (1480?-1521), Cook, and others attests, but these were in many respects exceptions to the general rule.

In overland exploration, by comparison, explorers were not as able to leave their troubles behind because they could generally travel no more quickly (and usually less quickly) than the native populations. In addition, overland journeys were often smaller and less self-sufficient. Charles Wilkes (1798-1877) sailed for three years to the Antarctic and South Pacific with six ships and 433 officers, men, and scientists. By comparison, Heinrich Barth (1821-1865) went with two others and a handful of servants on a five-year journey through northern Africa. Wilkes spent no more than a few months at a time in any single port, while Barth spent his entire journey surrounded by African tribes. In sea voyages, the emphasis was on mapping coasts and making a quick landing to collect a handful of scientific specimens, then moving on to the next island, while in land-based exploration a more detailed and in-depth knowledge of local geology, ecology, and peoples could be gained.

As the style of exploration changed, so, too, did society. The emphasis began to shift from claiming new lands for empire building to exploiting the lands already added. While there were still unknown places to explore on the globe, most of future exploration would use ships for transportation rather than as a primary base of operations.

P. ANDREW KARAM

Further Reading

Mansir, A. Richard. *Quest for the Northeast Passage.* Montrose, CA, and Los Angeles: Kittiwake Publications, 1989.

Nordenskiöld, Nils Adolf Erik. *Facsimile-Atlas to the Early History of Cartography With Reproductions of the Most Important Maps Printed in the XV and XVI Centuries.* Translated by Johan Adolf Ekelöf and Clements R. Markham. 1889. Reprint, New York: Kraus Reprint Corp., 1961.

Raurala, Nils-Erik, ed. *The Northeast Passage: From the Vikings to Nordenskiöld.* Helsinki: Helsinki University Library, 1992.

A Race Around the World

Overview

In 1890, inspired by Jules Verne's novel *Around the World in Eighty Days*, one of the first woman reporters set out to make the trip even faster. Elizabeth Cochrane, who went by the pen name Nellie Bly (1867-1922), was traveling for over a month before she found out that another reporter, Elizabeth Bisland, had been dispatched by a rival publication in the hope of getting around the world first. Beating out both the fictional Phileas Fogg and her real-life competitor, Nellie Bly completed the journey in 72 days, a record that would not be bested until the days of air travel.

Background

In the 1880s, women reporters were all but unknown. There was one female writer in a New York City newspaper office, J.C. Crody, who used the name Jennie June. A young woman named Sally Joy worked for the *Boston Post*. Elizabeth Cochrane, a Pennsylvania teenager, got her first newspaper job at the *Pittsburgh Dispatch* because she didn't like an essay the paper ran.

The opinion piece in question, "What Girls Are Good For," railed against the employment of women, who were beginning to get jobs in business. Cochrane wrote a letter to the editor opposing the essay, but didn't sign her own name. In the Victorian era, young ladies weren't expected to write letters to newspapers. So she signed it "A Lonely Orphan Girl."

The editor, George Madden, was impressed by the letter. He assumed it was written by an enterprising young man who wanted a job at the newspaper. Taking the "wrong side" of the issue, he reasoned, was simply a way to stand out from the crowd. He advertised for the writer to contact him, and was startled when Elizabeth appeared. Still, after he bought a few of her articles, including a controversial piece on unfair divorce laws, she persuaded him to give her a regular job on the staff. He liked her work, and besides, controversy sold newspapers. The mores of the time argued against an 18-year-old girl writing under her own name, so Elizabeth Cochrane became Nellie Bly, after a girl in a popular song by Stephen Foster.

Bly often used her writing to expose social injustice. She did stories on deplorable working

Nellie Bly. *(Corbis-Bettmann. Reproduced by permission.)*

conditions in Pittsburgh factories, and children living in slums. She was not afraid to name names of factory owners and landlords. Before long, businessmen threatened to pull their advertising from the *Dispatch*. Madden was on the horns of a dilemma, which he attempted to resolve by giving Bly a raise and hiding her away for a year in the society section. There she was to write about parties held by rich matrons. Bored and restless, she eventually persuaded him to let her go back to writing serious articles, including a series on conditions in the Pennsylvania jails. Soon she went undercover to write about women working on assembly lines, and the factory owners' protests against Bly recommended. Back to the society page she went, with another raise.

After seeing pictures of Aztec ruins in an art gallery, Bly came up with the idea of traveling to Mexico and writing about the sights to be seen there, as well as the Mexican people and the way they lived. Surely the factory owners could not object to that. Her stories ran in the *Dispatch*, and were picked up by other newspapers all over the country. Bly herself was picked up by the newspaper she most wanted to write for, the prestigious New York *World*, published by Joseph Pulitzer.

Impact

Like Bly, Pulitzer was reform-minded. Unlike Madden, he had the resources to stand up to business interests. Bly was able to go back to writing undercover exposes. In one famous case, she arranged to be committed to the notorious Blackwell's Island insane asylum in order to write about the abuses there. Disguised as the wife of a businessman, she laid a trap for a lobbyist who was bribing politicians.

Meanwhile, the success of the Mexican trip was still in the back of her mind. According to family legend, her mother's uncle had taken a three-year trip around the world. In 1872, the French novelist Jules Verne had published a story about a traveler named Phileas Fogg, who raced around the globe in 80 days. If she made the trip even faster, Bly realized, not only would she be able to write fascinating stories, but it would be great publicity for the *World*. On this basis, she pitched her idea to Pulitzer.

Pulitzer, concerned about the size of the undertaking and the danger to a woman traveling alone, considered it for almost a year. Then he decided to go ahead with it, but to send a man instead. Bly threatened to quit, finance her own trip, and beat out the *World* reporter. Finally Pulitzer agreed to let her go, but insisted that she leave within a few days. If she took any longer, word might get out and someone else could try to make the trip first.

In fact, that is exactly what happened. On November 14, 1889, the same day that Nellie Bly sailed east from Jersey City, Elizabeth Bisland of *Cosmopolitan* magazine had taken a train west from New York. Traveling in the opposite direction, her aim was to beat Bly around the world and make the first triumphant return to New York. But Bly was already a celebrity, and railroad companies and steamship crews made special efforts on her behalf. She was also extremely resourceful and persistent in her own right, valuable qualities for anyone, particularly a woman, circling the globe alone in her day.

When Bly made her trip, not only were there no automobiles or airplanes, but telecommunications was in its infancy. Telephones had been recently invented, but the first inter-city lines, between Boston, Providence Rhode Island, and New York, had opened only a few years before. Bly was able to send telegraph messages back to the *World*, via coast-to-coast and trans-Atlantic cables laid in the 1860s. But this required being within range of a telegraph office.

Marconi's invention of wireless communication would come five years after Nellie Bly's journey. Frequently during her trip her newspaper office had no idea where she was.

Bly set off with two small bags, wearing a fashionable plaid coat that would become her trademark. Her first stories did not arrive back at the *World* until 17 days later, when she had crossed the ocean and arrived in London. After crossing the English Channel, she risked a detour for what was to be one of the high points of her trip: a visit to the aged Jules Verne at his home in Amiens, France.

Next Bly caught a mail train across Europe to Brindisi, Italy. There she picked up a ship that steamed through the Suez Canal and into the Indian Ocean to Ceylon (now Sri Lanka). Other ships took her to Singapore, Hong Kong, Japan, and across the Pacific to San Francisco. While in Asia, she tried other methods of travel, including mule cart and sampan, and acquired a pet monkey. A large crowd awaited her in San Francisco, along with a special train, re-routed to avoid a blizzard that might have delayed her homecoming. There was a brief layover in Chicago, where Nellie Bly became the first woman reporter ever admitted into the Chicago Press Club building.

Bly's train pulled into Jersey City on January 25, 1890, marking the end of a round-the-world journey that had taken 72 days, six hours, and 11 minutes. Not only had Nellie Bly beaten the fictional milestone of 80 days, her real-life competitor, Elizabeth Bisland, was still on a ship in the middle of the Atlantic, having missed connections and experienced bad weather. Four days later, Bisland's ship steamed into the New York harbor. She too had beaten the 80-day mark. But the crowds had gone home, and the dignitaries were done with their speeches.

Nellie Bly became a national hero. Her book about the trip was a bestseller. Fast trains and racehorses were named after her, and songs were written about her. Internationally famous, she commanded a salary of $25,000 a year, an enormous amount for a reporter at the time. She could cover any story she wanted, and continued her crusades for social reform. As she had in Pittsburgh, she exposed slum conditions in New York, and she described her horror after witnessing an execution. She also debunked the claims of a popular "mind-reader," went up in a balloon, and down in a diving bell. But her influence extended beyond her writing, because of the example she set. By the turn of the century, woman reporters were working at newspapers around the country.

SHERRI CHASIN CALVO

Further Reading

Bisland, Elizabeth. *In Seven Stages: A Flying Trip Around the World.* New York: Harper & Brothers, 1891.

Bly, Nellie. *Nellie Bly's Book: Around the World in 72 Days.* New York: Pictorial Weeklies Company, 1890.

Kroeger, Brooke. *Nellie Bly: Daredevil, Reporter, Feminist.* New York: Random House, 1994.

Marks, Jason. *Around the World in 72 Days: The Race Between Pulitzer's Nellie Bly and Cosmopolitan's Elizabeth Bisland.* New York: Gemittarius Press, 1993.

Verne, Jules. *Around the World in 80 Days.* New York: Bantam, 1984.

Biographical Sketches

Heinrich Barth
1821-1865
German Explorer

Heinrich Barth spent most of his adult life exploring Saharan and sub-Saharan Africa in the company of James Richardson (1806-1851) and on his own. Originally sent to help gather information on the slave trade, Barth made some of the most reliable maps of North Africa and devoted a great deal of attention to studying the customs of the tribes he encountered in his travels.

Barth was initially assigned to assist James Richardson, an English explorer sent by a British religious society to learn about the North African slave trade. Wanting to make the expedition an international one, Richardson accepted the recommendation of Prussia's ambassador and asked Barth to join him. The expedition, which left in 1850, included Richardson, Barth, and the German geologist Adolf Overweg (1822-1852). Records suggest it may well have been the best-equipped and best-organized expedition to the Sahara launched up to that time.

Unfortunately, Barth and Richardson did not get along well and there was a considerable amount of tension between the two of them. The tension escalated to the point that they set up separate camps for each man and his servants at the end of each day. Nevertheless, they continued exploring together for several months before deciding to split up in late 1850.

After going their separate ways, Richardson decided to head directly for Lake Chad while Barth and Overweg took a longer, more westerly route. Barth and Overweg, in turn, split up with Barth exploring the regions south and east of Lake Chad, agreeing to meet Overweg and Richardson both at the lake in April 1851. Richardson, however, never made the rendezvous, dying of a tropical fever just three weeks before Barth's arrival. Overweg did reach the lake, only to die of malaria 15 months later at the age of 29. Before his death, he and Barth conducted extensive studies of the Lake Chad area, using a boat they had brought with them across the desert.

Like the others, Barth fell ill, but he recovered and pressed on with his explorations. By this time he had been in Africa for nearly three years and realized that reaching Timbuktu, his next destination, would likely take another two years. He decided to continue anyway and reached the city in September 1853.

In early 1854 Barth, hearing that a party of Europeans was at Lake Chad searching for him, left Timbuktu to return to the lake. Meeting up with them, he discovered that he had been presumed dead. Barth eventually made his way back across the Sahara to Tripoli, traveling during the hottest part of the year. From there, he returned to London, arriving in 1856 after an absence of over five years. In a report totaling five volumes, Barth contributed the first reliable maps of North Africa, detailed scientific observations, and extensive notes on the customs of the tribes he had observed during his travels. According to an account published in the *National Geographic* in 1907, Barth "...gave the first definitive account of the Saharan region after a journey of great extent and importance...Barth's journeys were of great value, for he not only made known to the world the existence and accessibility of hundreds of thousands of square miles of fertile territory, but he also gave in five volumes an enormous amount of geographical information, in which he treated quite thoroughly the ethnology of the various tribes of the Central Sudan."

After returning from Africa, Barth was appointed a professor of geography at Berlin University, where he remained until his premature death at the age of 44.

P. ANDREW KARAM

Paul Emile Botta
1802-1870
French Archaeologist and Diplomat

Paul Emile Botta was among the first archaeologists to study the ruins of the ancient civilizations of Mesopotamia. In 1843, he discovered the palace of the Assyrian king Sargon II near what is now Khorsabad, Iraq.

Botta was born in Turin, Italy. His father, the Italian historian and physician Carl Botta, became a French citizen in 1814. Paul Botta also studied medicine, but then entered the French diplomatic corps.

Initially assigned to Alexandria, Egypt, Botta secured an appointment to the city of Mosul in Mesopotamia, then controlled by the Ottoman Empire. Julius Mohl, a well-known scholar of Middle Eastern civilization, had convinced the French government that someone interested in archaeology would be right for the consulate post, and Botta fit the bill. He was the son of a historian, he spoke Arabic, and he was fascinated by the prospect of discovering the lost cities of Assyria. Their locations had been forgotten; the cities were known only from biblical references and other ancient documents, some of which contradicted each other.

For the first year, Botta dug at Kuyunjik, close to the Tigris River. This was a site that Austen Henry Layard (1817-1894) would later realize was the biblical city of Nineveh. But Botta wasn't digging deep enough, and his results there seemed unpromising. Meanwhile, he had heard that the nearby village of Khorsabad sat on a mound of inscribed bricks and sculpted stone, a rumor that he had at first refused to credit.

When an assistant returned from a scouting mission talking of carvings lying on the ground, Botta immediately moved his entire entourage from Kuyunjik. After only a week of digging at Khorsabad, the team began to unearth the palace of Sargon II, who reigned from 722 to 705 B.C. It was part of Sargon's ancient capital of Dur Sharrukin, and was the first Assyrian site ever uncovered. Its riches included fabulous statues of winged animals with human heads, relief sculptures, and a number of inscriptions in the characteristic cuneiform, or wedge-shaped, writing of ancient Mesopotamia.

Botta mistakenly believed that he had found Nineveh, and reported as much back to Paris. The French government immediately declared itself the leader in the study of antiquities and further financial support was quickly forthcoming, allowing Botta to continue excavating.

An artist, Eugene Flandin, was sent over to make on-site drawings. This was particularly important because under the dry earth of the Middle East, the ruins had been protected from air as well as moisture. Some relief carvings and other artifacts began to crumble as soon as they were exposed, particularly those made brittle from ancient fire damage.

Botta shipped hundreds of statues and other antiquities back to Paris. Unfortunately, one shipment sunk to the bottom of the swiftly flowing Tigris, but statues of Sargon and the winged bulls from his palace did make it safely to the Louvre. They went on display in 1847, in a newly established Assyrian Museum there.

Later, Botta served diplomatic posts in Jerusalem and Tripoli. He also wrote accounts of his discoveries and studied cuneiform writings. He died in 1870 in Acheres, France.

SHERRI CHASIN CALVO

Johann Ludwig Burckhardt
1784-1817
Swiss Explorer

Johann Ludwig Burckhardt was the first European in modern times to visit the ancient city of Petra in what is now Jordan, and the great temples of the ancient Egyptian pharaoh Ramses II at Abu Simbel.

Burckhardt was born in 1784 in Lausanne, Switzerland. In 1806 he went to England, where he studied at Cambridge University. He traveled to the Middle East in 1809 under the auspices of an English organization called the Association for Promoting the Discovery of the Interior Parts of Africa.

As did other great nineteenth-century explorers of the Middle East, Burckhardt adopted many local ways. He became fluent in Arabic, and learned in Islamic doctrine. He often wore

Muslim garb, and even took an Arabic name, Ibrahim ibn Abdullah.

After he had spent almost three years in Aleppo, Syria, Burckhardt set off for Cairo with the goal of joining a caravan across the Sahara to Timbuktu. His route took him through southern Jordan, where he rediscovered the ruins of Petra.

Petra was an ancient trading city of reddish stone, with houses and temples cut into the surrounding cliffs. It was settled by an Arabian people called the Nabataeans by 500 B.C., and conquered by the Romans in 106 A.D. Eventually other nearby cities took over Petra's economic role, and it became primarily a religious center. A Christian city around the year 300, it was controlled by the Muslims in the 600s, and taken by the Franks during the Crusades. It was abandoned by the thirteenth century, and fell into ruins. While biblical and Roman accounts of Petra kept its memory alive in the West, its location was forgotten until Burckhardt's 1812 visit.

When he arrived in Cairo, Burckhardt found no suitable caravan forming, so he decided to approach the interior of Africa by traveling up the Nile. South of Aswan, in the ancient Nubian region of southern Egypt, he became the first European to see the 3,000-year-old temples of Ramses II at Abu Simbel.

The two temples of Abu Simbel were carved into the side of a sandstone cliff. The larger, dedicated to the chief gods of the cities Heliopolis, Memphis, and Thebes, has become famous for the four huge stone statues of a seated Ramses II on its facade. Each is over 65 feet (20 m) tall. The smaller temple is dedicated to Ramses's queen, Nefertari.

Next, Burckhardt planned a pilgrimage to the Arabian city of Mecca, the holiest site in Islam. Non-Muslims are not allowed into Mecca. However, he was assisted by the viceroy of Egypt, who knew of his reputation and was able to arrange for him to be admitted by having him declared a Muslim. As Burckhardt was probably the only non-Muslim ever to travel openly to Mecca, his published account is still notable for its unique point of view.

After his journey through Arabia, Burckhardt returned to Cairo. He died of dysentery before his 33rd birthday, still waiting to join a caravan to Timbuktu. His five travel journals were published posthumously: *Travels in Nubia* (1819), *Travels in Syria and the Holy Land* (1822), *Travels in Arabia* (1829), *Notes on the Bedouin and Wahabys* (1830), and *Arabic Proverbs* (1830).

SHERRI CHASIN CALVO

Sir Richard Francis Burton
1821-1890
British Explorer and Scholar

An adventurer with a gift for languages, Sir Richard Francis Burton was the first European to reach many of the once forbidden areas of the world, including many ancient Muslim cities, as well as Lake Tanganyika during his quest for the source of the Nile River. He wrote extensively of his travels, publishing 43 volumes and providing invaluable insight into the religion, lives, and customs of the people with whom he came in contact. In addition, he translated many texts of the East for Westerners, including the *Kama Sutra* and the definitive translation of *Arabian Nights*.

Born to a retired officer of the British army and his wealthy British wife, Burton was the oldest of three children. He and his brother and sister grew up in England, France, and Italy. By all accounts an unruly child, Burton nevertheless showed promise in his studies, including a great gift for languages. When he arrived at Trinity College, Oxford, in 1840, he was already fluent in Greek, Latin, Italian, and French. (He would eventually master 40 languages and dialects.)

He was expelled only two years later, however, due to disciplinary problems. He decided to join the English army, and Burton became an officer during Britain's campaigns in Afghanistan and what is now Pakistan. Having now mastered the Arabic and Hindi languages, he became an accomplished and valuable spy.

After his army service, Burton began his exploration of ancient Muslim cities. In 1853, disguised as an Afghanistani Muslim, he visited Cairo, Egypt, and the holy cities of Islam, Mecca, and Medina in Saudi Arabia. Though not the first non-Muslim to visit Mecca, the birthplace of the prophet Mohammed, Burton provided the most insightful and comprehensive information to date about the place and its people in his book, *Pilgrimage to El-Medinah and Mecca*. In 1854, he led an expedition to Harar in what is now Ethiopia, and became the first European to enter that city. He wrote of this adventure in *First Footsteps in East Africa*.

Harar was Burton's first step toward his journey to find the source of the Nile, specifically the White Nile, which joins the Blue Nile to form the Nile River. British army officers, including John Hanning Speke (1827-1864), accompanied Burton in 1855, but their

Sir Richard Francis Burton. *(The Granger Collection NY. Reproduced by permission.)*

tated with essays and footnotes regarding customs and sexual practices.

Due to society's view of her husband's scandalous work, Burton's wife burned most of his diaries and journals after he died, attempting to protect her husband's reputation and place in history.

JIM MARION

Jean-François Champollion
1790-1832
French Egyptologist, Linguist and Historian

Jean-Francois Champollion was one of the founders of Egyptology as it became a scientific discipline in the nineteenth century. His most important contribution was the deciphering of hieroglyphic writing after the discovery of the Rosetta Stone.

Champollion was born in 1790 in Figeac, France. He is sometimes called "Champollion le jeune," the younger, because his early education proceeded under the supervision of his older brother, the archaeologist Jacques Joseph Champollion. At 10 he was sent to the Lyceum in Grenoble. He was a precocious student of languages from an early age, teaching himself Hebrew, Arabic, Syriac, Chaldean, and Chinese as well as the Latin and Greek that were part of the standard curriculum of the day.

As a 16-year-old, Champollion presented a paper to the Grenoble Academy asserting that the language spoken by the contemporary Egyptian Copts was the same language spoken in the days of the pharaohs. This turned out to be incorrect, but the realization that certain signs in Late Egyptian script corresponded to letters of the Coptic alphabet was instrumental in his later work.

After furthering his education in what were then called Oriental languages at the College de France, Champollion began teaching history and politics at the Grenoble Lyceum when he was only 18. There, he married Rosine Blanc and had a daughter, Zoraide. He also began writing books about ancient Egypt: *Introduction to Egypt Under the Pharaohs* (1811) and *Egypt of the Pharaohs, or Researches in the Geography, Religion, Language and History of the Egyptians Before the Invasion of Cambyses* (1814).

An 1818 appointment to a chair in history and geography at the Royal College in Grenoble gave Champollion more freedom to pursue his

expedition was cut short by an attack by Somali natives, who seriously injured Burton and Speke. After returning to England to recover, Burton and Speke began a new expedition in East Africa in 1857. Both near death, they saw Lake Tanganyika on the border of what is now Zaire and Tanzania, the first Europeans ever to do so. Burton believed he had found the White Nile's source, but Speke pushed on, discovering Lake Victoria and claiming it as the source. Burton and Speke remained at odds over their discoveries, though Lake Victoria was eventually proven to be the true source of the White Nile.

In 1861, Burton married Isabel Arundell and began his career with the British Foreign Office, which included assignments in Brazil, Syria, and Italy. These posts allowed him time to continue his writing and his translations, including the erotic literature of the East, with which he became increasingly fascinated. He was determined to expose his repressed and conservative Victorian society to the views of these works, and to advocate the sexual liberation of women, as well as men. His controversial work continued with his translation of the Eastern sexual manual, the *Kama Sutra*, and culminated in the 16-volume translation of *Arabian Nights*, a collection of Persian-Indian-Arabian stories (including "Sinbad the Sailor"), which he anno-

studies of the ancient Egyptian language. In 1799, after Napoleon's armies took the Nile Delta, a French soldier had found a black basalt slab with an inscription carved in hieroglyphs, a later Egyptian script called demotic, and Greek. The Greek translation provided an overall meaning for the Egyptian inscriptions, and the most distinguished European linguists competed to decipher them.

Champollion was the first to do so. His earlier studies of the Coptic alphabet and its relationship to the demotic script enabled him to see that some of the Egyptian hieroglyphs were strictly ideograms; that is, picture signs, while others also represented sounds. He published his work in 1822.

In 1826, Champollion became curator of the Egyptian department at the Louvre. It was during this period that he made his only trip to Egypt. The 1828-1829 expedition undertook a systematic survey of Egyptian monuments and their inscriptions. Champollion took voluminous notes and made sketches, from which his protégé Ippolito Rosellini produced finished engravings. In 1832, while back in Paris and still organizing the material, Champollion suffered a fatal stroke at the age of 41. His work was published posthumously and formed the basis for field studies by Karl Richard Lepsius (1810-1884) and John Gardner Wilkinson (1797-1875).

SHERRI CHASIN CALVO

William Clark
1770-1838
American Explorer

William Clark is best known as co-leader of the Corps of Discovery, leadership he shared with Meriwether Lewis (1774-1809). In this role Clark helped explore the Louisiana Purchase and western territories stretching to the Pacific Ocean, becoming in the process one of the greatest American explorers.

Born in 1770 in the state of Virginia, Clark's family soon moved to Kentucky. Clark was the younger brother of General George Rogers Clark, a Revolutionary War hero and a friend of Thomas Jefferson (1743-1826). Clark joined the United States Army in 1790 and first met Lewis in 1795 when both were assigned to the same rifle company for six months. Prior to their meeting, Clark fought in the Battle of Fallen Timbers.

William Clark. *(The Library of Congress. Reproduced by permission.)*

Though they spent only a short time together, Lewis and Clark developed a great respect for one another that lasted despite infrequent contact over the next several years. Resigning his commission because of ill health and responsibility for his family's business, Clark moved to Indiana, where he lived until contacted by Lewis in 1803 and asked to help lead the expedition.

In many ways, Clark complemented Lewis's abilities. While both were skilled woodsmen, Clark was an accomplished surveyor, an excellent mapmaker, and a talented waterman—all skills that would be needed for their expedition. Like Lewis, Clark was a born leader. Unlike Lewis, however, Clark kept a regular journal, though without the eye for detail or the literary flair that Lewis exhibited in his entries.

All evidence indicates that Clark shared leadership of the expedition equally with Lewis, though documents show that Lewis had the higher military rank throughout. At times, especially on the return journey, the men split up, each taking part of the expedition in separate directions in order to explore as much as possible. It appears as though Clark worked with Lewis's full confidence and did a superb job of leading his part of the expedition.

After their return from their journey, Clark was recommended for promotion to Lieutenant

Colonel, a recommendation rejected by Congress because of Clark's lack of seniority. However, Congress did approve naming him Superintendent of Indian Affairs in the Louisiana Territory, a post to which he was assigned in 1807. A few years later, Clark courted and married Julia Hancock, and in 1809 fathered a son whom he named after his friend Meriwether Lewis.

In spite of his accomplishments with the Corps of Discovery, Clark's leadership and political skills were so impressive that, for nearly a century, they overshadowed his role as part of the Lewis and Clark expedition. In fact, a massive history of the Jefferson administration written towards the end of the nineteenth century made scant mention of the Corps of Discovery at all. Part of the reason for this relative lack of attention may have been the stigma associated with Lewis's suicide in 1809. The full reason, however, is not fully known. In any event, the early part of the twentieth century saw a fuller realization of the role played by both Lewis and Clark in exploring and opening the American West, led in part by Theodore Roosevelt's enthusiasm for their journals and journey.

Clark died in 1838 at age 68. He left a legacy of accomplishments as an explorer, leader, and political appointee. His undisputed skills as a woodsman, waterman, and soldier further round out his reputation as a remarkable man who made a deep and lasting impression on his nation.

P. ANDREW KARAM

Luigi Maria D'Albertis
1841-1901
Italian Explorer

Luigi D'Albertis led two unprecedented trips of exploration up the Fly River into central New Guinea. During these expeditions, he made a number of important scientific and ethnographic discoveries, but also fought many battles against the inhabitants, alienating virtually all of them. He was severely criticized for the manner in which he dealt with the local natives during these trips, as well as for his methods in collecting ethnographic specimens.

D'Albertis was born in preunification Italy. In his youth, he fought at the side of Giuseppe Garibaldi in the struggle for Italian independence and unification. D'Albertis was an energetic, larger-than-life person—lusty, fatalistic, excitable, and flamboyant. He also seemed to have remarkably few scruples when it came to

protecting his expedition or gathering samples. Unfortunately, his lack of tact in dealing with New Guinea natives and his questionable collection methods turned the natives against him, making the return from his first trip difficult and his entire second voyage dangerous.

From his youth, D'Albertis was convinced that his life would reach a pinnacle during his thirty-fifth year, a conviction that proved true. This premonition—or self-fulfilling prophecy—came to fruition after D'Albertis fell under the spell of New Guinea. He decided that his first journey of exploration would be bold. He would follow the Fly River (named for the ship that discovered it) as far inland as possible, with the goal of being the first to penetrate more than a few tens of miles into the island. To do this, he bought a small steamship, the *Neva* and, in 1875, set forth on his first voyage to New Guinea. With its shallow draft and small size, the *Neva* was not a comfortable vessel, but it proved able to steam 580 miles (930 km) up the Fly, allowing D'Albertis to collect many specimens new to science. On this trip, D'Albertis took a crew of nine, a sheep, a python, and a pet dog, as well as a large number of rockets, guns, and ammunition. This voyage, as it turned out, was the peak experience of D'Albertis's life, as he would never again accomplish so much or penetrate so far into New Guinea.

Upon his return, D'Albertis came under increasing criticism for his treatment of his crew, his willingness to open fire at the slightest provocation, and his tendency to pilfer villages, graves, and other sites for specimens. This criticism did not, however, stop him from launching a second expedition in 1877. One incident on this voyage, which occurred after an attack on the *Neva* that ended in the fatal shooting of a New Guinea native, sheds some light on the origin of the criticism leveled against him. In D'Albertis's words:

"Should they return [referring to the dead man's companions], they will perceive our guns are weapons not to be despised, and they will learn to moderate their desires for the heads of strangers. But, meanwhile, I shall preserve this man's head in spirits, for Bob [a native guide], in remembrance of his younger days, did not hesitate to cut it off the poor savage's body."

With that, he pickled the head in alcohol. His second voyage did not accomplish nearly as much as his first, and was marred by severe problems with his new crew as well as by nearly constant skirmishes with the natives. Returning to Italy, D'Albertis wrote a two-volume account of his voyages and never again left his native land. He spent most of the rest of his life at his home in Rome and died 24 years later of mouth cancer.

P. ANDREW KARAM

Jules-Sébastien-César Dumont d'Urville
1790-1842
French Naval officer and Explorer

Jules-Sébastien-César Dumont d'Urville spent a great deal of his life on voyages of discovery for France. He is perhaps best known for his explorations near Antarctica, some performed in competition with the American vessel command-ed by Charles Wilkes (1798-1877). He was also responsible for the French purchase of the statue known as the Venus de Milo and made several voyages of discovery to the southern Pacific.

Dumont d'Urville was born in rural France and by the age of 17 had entered the French Navy. He showed early talent, graduating at the top of his class and learning several languages as well as showing a talent in some of the sciences.

After several years of serving as an officer with French expeditions to the Black Sea and the South Pacific, Dumont d'Urville was given com-mand of a voyage to the southern ocean. During this three-year voyage he charted many new islands, surveyed parts of the New Zealand coast, and circumnavigated the globe. However, on his return he was accused of undue harsh treatment of his crew and other faults. This led to his removal from command for seven years, assigned to shore duty until 1837. During this time, too, he returned with charts, maps, and large num-bers of scientific specimens and sketches.

Dumont d'Urville was given command of a French expedition to the southern ocean by King Louis-Philippe, who was interested in countering English and American interests and voyages in that area. He charged Dumont d'Urville with voyaging south "as far as the ice permits" and then turning westward to continue exploring New Zealand, New Guinea, Western Australia, and the South Shetland Islands. Hop-ing to reach further south than any previous expedition, Dumont d'Urville promised his crew 100 francs each for reaching 70° south latitude and an additional 20 francs for each additional degree they penetrated southward.

Jules-Sébastien-César Dumont d'Urville. *(Corbis Corporation. Reproduced by permission.)*

After reaching pack ice, Dumont d'Urville and his ships followed it, attempting to pene-trate within sight of the continent they felt lay to the south. They failed and, in fact, did not even reach as far south as had Weddell several years earlier. On a subsequent attempt, they did man-age to penetrate some distance into the ice pack, only to have the ice close behind them. After five days of intense effort the crew succeeded in free-ing the ships and their voyage continued. Final-ly, on January 19, 1839, he logged sighting land.

The rest of the voyage was a trial for the crew, who were beginning to develop scurvy, intense seasickness, fever, and dysentery. Several crew members died of these diseases before they could return to port and more died in Hobart, Tasma-nia, as Dumont d'Urville prepared for yet another attempt to reach the mainland in late 1839.

Dumont d'Urville finally succeeded in setting foot on land south of the Antarctic Circle on Janu-ary 20, 1840, an island he claimed for France and named for his wife, Adélie. After this, he and his crew spent another eight months exploring the southern ocean before returning to Hobart, New Zealand, New Guinea, Timor, and, finally, France. When they pulled into the Toulon harbor, the expedition had been gone for over three years, during which they had accomplished an enor-mous amount of exploration and scientific obser-vation for France. In recognition of this, Louis-

Philippe promoted Dumont d'Urville to the rank of Rear Admiral in the French Navy, awarded him the highest honor of the French Geographical Society, and gave 15,000 francs to be distributed among the 130 surviving crew members.

Dumont d'Urville continued serving the French navy for another two years. In 1842, while on vacation with his wife and son, Dumont d'Urville was killed in a train accident near Versailles.

P. ANDREW KARAM

Edward John Eyre
1815-1901
English Explorer and Government Official

Usually, explorers are known for their relatively short life spans. Not so with Edward Eyre. He enjoyed an adventurous, somewhat illustrious career that lasted until he was 86 years of age.

Born in England as the son of a clergyman, Eyre and his family migrated to Australia when he was just 17. Young but ambitious, he showed strong interests in both governmental affairs and the budding world of commercial enterprise. As a result, he was appointed lieutenant governor of New Zealand at age 31 and remained in this post for seven years. (Later in life, he served as governor of Jamaica for one year in 1864-65.)

It was apparent to the young Eyre that the lush farm crops of his native Britain were never going to be possible in the deserts of Australia. Instead, he turned his efforts toward sheep and cattle, which were the principal commercial activities of the time. Although cattle were somewhat profitable as food sources, he looked more favorably on sheep since they yielded good wool, which sold well in European markets.

Initially, Eyre's properties were limited to South Australia, where there was sufficient water for his flocks and for the vegetation they needed to survive. It was soon obvious, however, that new ranges had to be found both for animal forage and for transporting them to wider markets, and thus Eyre set out on expeditions to discover the new ranges. His first effort was directed toward the central part of the continent and the Lake Torrens area in the Flinders Ranges. This expedition was a fiasco in that the so-called "lake" proved to be an impassable, salty marsh, and the sandhills around it were

equally impossible to cross. After reaching the aptly named "Mount Hopeless," Eyre returned to Adelaide, where he discovered that the government was planning an entirely different approach to transporting livestock and goods: a route along the coast of Fowler's Bay all the way to Albany at the southwestern portion of the continent.

This seemed like a good plan and—because of his experience and leadership abilities—Eyre was selected to guide the party on the proposed expedition. He was so optimistic about the venture that he volunteered to pay half the costs involved.

The party included six men, including his sheep station manager, Baxter, an aboriginal friend named Wylie, and two other aborigines selected by Wylie because of their experience traveling in the Australian bush.

When they started out in June 1840, they anticipated a three-month leg that would take them to Spencer Gulf, where a government ship would await them with additional supplies for the remainder of the trek. Their initial outlay included 12 horses and 40 sheep as well as food, firearms, and clothing for the hot and cold seasons they would be on the trail. In a very short time, however, Eyre realized that they were in serious trouble. The horrendous conditions they encountered were compounded by a total lack of drinkable water in any direction. Always practical, Eyre sent most of the party back to Adelaide and pushed on with fewer supplies and livestock. During the next few months, the group suffered hardships that brought them close to death on several occasions. When two of the aborigines murdered Baxter and fled with over half of the pitifully small supplies and firearms, Eyre and Wylie escaped certain death when they finally sighted a ship in the bay with a kindly English captain who restored their health and fitness and gave them supplies to continue. Eyre and Wylie eventually reached their destination in July 1841.

Eyre received many honors in his adventurous life, including a gold medal from the Royal Geographic Society for his incredible journey. The Australian government also named a large lake after him as well as a large peninsula at the eastern end of Fowler's Bay in South Australia. He returned to England where he lived until his death in 1901.

GERALD F. HALL

Matthew Flinders
1774-1814
English Sea Captain, Explorer and Cartographer

Matthew Flinders explored the coast of Australia, charted it more carefully than any explorer had before, and gave it the name Australia. He sailed around both Australia and Tasmania, proving they were islands, and accurately located many coastal features as well as nearby islands, reefs, bays, and rocks on British maps.

Born in England in 1774, Matthew Flinders studied navigation and cartography so he could go to sea. In 1789 his uncle got him on a ship as a servant. He was fifteen, five feet six inches tall, wiry with black hair and dark eyes. A year later he was a midshipman with a berth on *Bellerophon*, a British, 74-gun man of war, where he learned important sailing skills and navy operations.

In 1791 Flinders was on the ship *Providence* with Captain William Bligh (1754-1817). *Providence* reached Tahiti in April of 1792 and sailed west with 600 bread fruit trees for the West Indies. Landing for water and food on New Guinea and Timor, Flinders saw new lands, people, and animals. The breadfruit was planted in the West Indies to augment food for British colonial workers.

Returning to Australia, Flinders and a friend named George Bass, in a small ship called *Norfolk,* charted the coast south of Sydney for the governor. Then they explored the Furneaux Islands and proved Tasmania was not attached to Australia. Needing a new ship by then, Flinders went home. In April of 1801 he and Ann Chappell, daughter of a sea captain, were married. Flinders by then was a lieutenant and a respected explorer.

A small English penal colony had been established in Australia a few years before, but the British had done nothing to secure the land. Now the government authorized an expedition to Australia and chose Flinders as Commander. The 334-ton sloop *Investigator* was old and not very sound, but it was all the navy could spare for exploration, since it was embroiled in a war with French dictator Napoleon Bonaparte.

Flinders reached the south coast of Australia in August of 1801 and began at once to make a complete and accurate survey of the south shore. He saw wombats, kangaroos, and aborigines—native inhabitants of Australia. He and his crew carefully described the birds, animals, plants, and insects they saw. Flinders observed

Matthew Flinders. *(The Library of Congress. Reproduced by permission.)*

tides, made soundings of bays and capes, and noted topographic features. After a rest in Sydney the expedition left again to chart the eastern and then the northern coasts of the island. By December *Investigator* was in such poor shape that Flinders left her and took a smaller ship home. When he stopped for supplies at Mauritius, a French controlled island east of Africa, Matthew was imprisoned and his papers, charts, journals, and letters were confiscated by the governor. After six years of captivity, an order from Napoleon set Flinders free. His maps and other papers were returned.

Back home, he was honored in scientific and naval circles, but his health had been damaged by the years in a tropical prison, and he never went to sea again. He spent years readying charts and notes for publication in his book, *A Voyage to Terra Australis.* It was a semi-official publication and he was not paid. He did not finish the personal account of his voyages. He was promoted to post-captain in 1814, but was not given full pay. When he died that same year, there was no pension for his wife. New South Wales and Victoria in Australia, however, both gave her a pension in honor of her husband's work. Flinders had shown that Australia was a continent, and by the time of his death its possibilities were drawing the British to its shores.

LYNDALL B. LANDAUER

Sir John Franklin
1786-1847
English Explorer and Naval Officer

Famed arctic explorer John Franklin was born in Spilsby, Lincolnshire. Following elementary and grammar school education, he went to sea over his father's objections. When he was 14, Franklin's father, a cloth merchant, was persuaded to help his son secure a place in the Royal Navy as a volunteer on HMS *Polyphemus*, and the lad saw action during the Battle of Copenhagen (1801). This was the first of a series of naval assignments that took him to Australia, China, Europe, South America, and the Caribbean during the next 15 years. At the end of the Napoleonic wars, during which he had been wounded and mentioned in dispatches, he was discharged in the standard manner as a lieutenant on half pay. He was not yet thirty.

What saved his career was the decision by the Royal Navy in 1818 to resume Arctic exploration, specifically the search for a northwest passage from the Atlantic Ocean to the Pacific Ocean through the islands lying to the north of the Canadian mainland. Franklin was selected for this duty in part because of his excellent war record and also because early in his career he had accompanied his uncle, Captain Michael Flinders (1774-1814), in an exploration of Australia.

Franklin commanded several expeditions, some by land, to the Canadian Arctic in 1818, 1819-1822, and 1825-1827. Franklin's weaknesses as a leader and explorer, notably his insistence on adhering to the admiralty's instructions when they did not apply to existing conditions, and his inability to adapt his objectives to changing circumstances in the field, were for the most part overlooked. The new geographic information he brought back did not, in the view of some critics, justify the personnel he'd lost owing to starvation or exposure and the cost in terms of effort, ships, and equipment. On the other hand, he won great respect for his efforts to be self sufficient while on the ice and for his courage in the face of daunting adversities. Promoted to commander early in 1821 and post captain in November 1822, he was made a fellow of the Royal Society in the latter year. He published narratives of two of his expeditions in 1823 and 1828. He was awarded a gold medal by the Geographic Society of Paris and an honorary degree from Oxford University. He was knighted in 1829.

Sir John Franklin. *(The Library of Congress. Reproduced by permission.)*

During the 1830s, the British Admiralty once again discontinued Arctic exploration, and Franklin served as a peacekeeper during Greece's struggle for independence. He was later (in 1837) appointed lieutenant governor of Van Diemen's Land (present-day Tasmania). His efforts at reform in that penal colony, in the face of political opposition and his administrative inexperience, resulted in his dismissal in 1843. In 1842 he published an account of some of his experiences.

Renewed interest on the part of the Royal Navy to find the northwest passage led to Franklin's appointment as commander of his final expedition, and he left England in May 1845 with two well-equipped Royal Navy vessels, *Erebus* and *Terror*, 129 men, and sufficient supplies for three years. Last observed by some whalers in Baffin Bay several months later, Franklin and his crew were never seen alive again. In the 14 years after he failed to return, more than 30 search expeditions were sent out by the admiralty, the Hudson's Bay Company, and by Lady Jane Franklin, his second wife, most between 1847 and 1859. In the meantime, Franklin was promoted to rear admiral in 1852. Lady Franklin attracted much sympathy, particularly in Britain and the United States, where several search expeditions were mounted. While looking for Franklin, would-be rescuers themselves mapped much new territory. Confirmation of his death came only in 1854, when Inuit residents in the region of King William Island provided information concerning the final days of Franklin and his men. Not until 1859, however, were the human remains of Franklin's party and their belongings finally located there. Franklin's vessels had evidently been caught in the ice near that point, and he died there at age 61 in June 1847. Several parties of survivors themselves subsequently died while trying to reach the mainland of Canada. In reconstructing the routes they had taken, searchers confirmed the Northwest Passage route. It was not until the Norwegian explorer Roald Amundsen's (1872-1928?) voyage of 1903-06, however, that any vessel successfully negotiated the passage.

Through his considerable efforts, Franklin, a figure of great persistence and winning personality, contributed much new cartographic information to what had been known about Canada and Alaska's Arctic coastlines.

KEIR B. STERLING

John Fremont. *(The Library of Congress. Reproduced by permission.)*

John Charles Fremont
1813-1890
American Explorer and Mapmaker

John Charles Fremont was an American explorer of the eighteenth century who parlayed skills as a mapmaker and explorer to earn a huge fortune and to achieve both high military rank and political office.

Born in Savannah, Georgia, in 1813, Fremont's family moved to Charleston, South Carolina, after the death of his father. He enrolled at Charleston College in 1829 but was expelled in 1831 for poor attendance. Despite this Fremont was still recognized for his skill in mathematics. By 1833 he was a mathematics instructor on the United States Navy war sloop *Natchez*. He prospered in this position and was named a professor of mathematics by the Navy in 1835.

At this time his interests turned to civil engineering. In 1836 he surveyed a potential railroad line in the western Carolinas and eastern Tennessee. This seemed to whet his appetite for further exploration, and by 1838 he assisted the noted French scientist Joseph Nicollet (1786-1843) in surveying the upper Mississippi and Missouri rivers.

In 1841 Fremont headed his own expedition to survey the Des Moines River for Nicollet, and later in that year married 17-year-old Jessie Benton, the daughter of influential Missouri Senator Thomas Hart Benton. Senator Benton, a prime proponent of the "Manifest Destiny" movement, at first refused to sanction his daughter's marriage to Fremont, so the young couple eloped. Later on the Senator would change his opinion about Fremont and become one of his most ardent backers and supporters. Using the backing provided by his father-in-law, Fremont was to lead three major expeditions in the western United States during the 1840s, eventually mapping much of the territory between the Mississippi valley and the Pacific Ocean.

An 1845 trip to California, whether purposely or accidentally, led Fremont into the military and political machinations taking place before and during the United States's war with Mexico. He threw his support behind the dissident founders of the "Bear Flag Republic" who were revolting against Mexico. After being named a Major in the United States Army by General Robert F. Stockton (1795-1866), Fremont went on to assist in the conquest of California, actually receiving Mexico's formal capitulation of the territory in 1846.

At this time United States Army General Stephen Watts Kearney entered California with orders to set up a new government. Instead, professing his loyalty to General Stockton, Fremont refused to obey direct orders from Kearney. This resulted in his court martial in 1847-1848. President James K. Polk set his penalty aside but Fremont resigned his commission nonetheless.

By the end of 1848 Fremont had amassed a huge fortune in the California gold rush and had remained influential in California politics. In 1850 he was elected one of the state's first two senators.

Fremont was a fierce abolitionist and his strong anti-slavery stance led to his being the fledgling Republican party's first presidential candidate in 1856. He lost a narrow decision to Democrat James Buchanan but did much to solidify the northern and western states against the south, paving the way for the election of Abraham Lincoln four years later.

When the Civil War broke out in 1861 Lincoln appointed Fremont "Commander of the West." His brief tenure in this post was extremely controversial. He enacted an unpopular martial law in the city of St. Louis, Missouri, where he was headquartered and then arbitrarily issued an order freeing all slaves in the state. This proved to be a political embarrassment for Lincoln at the time, and he removed Fremont from his post in 1862.

Fremont later ended up losing most of his previous fortune, and the early 1870s was a difficult time for him and his family. At this time his wife's writings and sketches, many about her husband's previous adventures, supported the family. Arizona Territory named him their Governor in 1878, a position he held until 1883.

In 1890 he was named an Honorary Major General in the United States Army by a vote of Congress. He died later that same year in New York City.

JOHN B. SEALS

Alexander von Humboldt
1769-1859

German Naturalist, Traveler and Statesman

Alexander von Humboldt gained fame for his adventurous scientific travels and highly accurate empirical researches throughout the Spanish colonies of South America. He wrote extensively on his findings and sought to establish a geographical understanding of nature.

Born in Berlin, Humboldt received an excellent education from private tutors as a boy and later at the great universities and academies of

Alexander von Humboldt. *(Archive Photos, Inc. Reproduced by permission.)*

Germany. He mastered several languages including French, Spanish, Italian, and English and studied literature, history, philosophy, mathematics, and the sciences. He excelled in everything but was particularly fascinated with the life and earth sciences. He studied botany, geology, mineralogy, chemistry, electricity, and plant physiology, always mastering each subject he took up.

Tales of the eighteenth-century expeditions Louis Antoine de Bougainville (1729-1811), Captain James Cook (1728-1779), and others enthralled the young Humboldt. After Humboldt befriended Georg Forster, who had sailed with Captain Cook, the two new friends took a botanizing tour through parts of Europe in 1790. Back in Germany Humboldt continued his studies and in 1792 entered the Prussian mining service. All the while he continued his scientific travels whenever possible. In keeping with the German Romantic science of this era, Humboldt held that nature must be understood as an interconnected whole. In particular, he believed that plants could be understood only within their environment. As he traveled the Alps in 1795, studying the effects of climate and altitude on plants, he became convinced of the necessity of more comprehensive researches involving astronomical and geomagnetic observations, researches that could indicate the broad range of factors affecting plant life.

In 1796 Humboldt received a large inheritance, which allowed him to leave the mining service and pursue his dreams of scientific travel. In the next few years he learned astronomical mapping techniques and other scientific skills needed for an extended expedition. The burgeoning Napoleonic wars thwarted his plans time and again for expeditions to Egypt, the South Pole, and the volcanoes near Naples, Italy. In 1798, however, Humboldt traveled with Aimé Bonpland (1773-1858), a French botanist, to Marseilles, France, and then Spain on a mapping and botanizing expedition.

In Spain he and Bonpland were granted permission to explore the Spanish colonies. Financed by Humboldt's inheritance, the two explorers tirelessly traversed the mountains, rivers, and deserts of South America for almost five years while painstakingly collecting, measuring, mapping, and observing a remarkable range of phenomena.

Humboldt spent most of the next 22 years in Paris writing scientific and popular publications, including a 33-volume series, based upon the vast collections and observations from the South American expedition. His vivid and detailed accounts of strange specimens, dramatic terrain, and exotic Indians, as well as the political and economic circumstances of an almost unknown territory, won him a vast readership in many languages. Humboldt frequently presented his data in tables, graphs, or maps in order to reveal patterns or correlations that, he expected, would point to nature's most simple laws and forces. Humboldt's engraving of the Andes mountain range in South America offers one example of his approach. In it he synthesized for the reader "the general results of five years in the tropics" by mapping and tabulating "all the phenomena that the surface of our planet and the surrounding atmosphere present to the observer." All conceivable factors that varied according to geographical position were discussed and charted by Humboldt.

Humboldt's plans for other major expeditions were foiled by war or other circumstances. In 1827, having spent his inheritance on the publication of his findings, Humboldt moved to Berlin to serve the Prussian court. In 1829, at age 60, he traveled through Russia, including Siberia, at the Czar's request. As in South America he collected biological and mineralogical specimens and measured geological, meteorological, and magnetic phenomena. Based upon his earlier researches he was able to predict the existence of diamonds in the Ural Mountains.

Humboldt's last years were devoted to writing *Kosmos,* a popular scientific work that beautifully expressed his far-ranging vision of the interconnectedness of the universe and of the human place within that universe. Humboldt died while writing volume five of *Kosmos.*

JULIANNE TUTTLE

Meriwether Lewis
1774-1809
American Explorer

One of the greatest explorers in United States history, Meriwether Lewis shared command of the United States Corps of Discovery that explored the newly acquired Louisiana Purchase for President Thomas Jefferson (1743-1826). Sharing command with William Clark (1770-1838), Lewis helped lead his men on a journey of exploration, scientific discovery, and diplomacy that met and exceeded all expectations.

Born into one of Virginia's leading families in 1774, Lewis led a childhood that mingled privilege with frontier hardships. His father, a plantation owner, served in the colonial army until his death from pneumonia when Lewis was five years old. Lewis's mother, a strong-willed woman, ran the plantation, sending Lewis away for schooling. Following his formal education, Lewis joined the army and was eventually assigned as secretary to President Jefferson, a family friend.

While serving Jefferson, Lewis was personally tutored by the president and stayed as a guest in Jefferson's home quite frequently. After consummating the Louisiana Purchase, Jefferson assigned Lewis as commander of a Corps of Discovery whose mission it would be to explore the newly acquired lands, collect scientific specimens, make initial diplomatic contact with the Native American nations along their path, and try to determine the economic potential of the lands they were to traverse.

Although unusual, Lewis chose to share command with his friend, William Clark. Lewis and Clark selected a group of frontiersmen and soldiers to join their expedition and set out from St. Louis in May 1804. In general, Lewis proved himself a capable commander, an excellent explorer, and a fine observer of nature. He gathered thousands of scientific specimens and

Meriwether Lewis. *(The Library of Congress. Reproduced by permission.)*

frontier. He repeatedly put off transcribing his journals for the scientific community, concentrating on publishing his memoirs instead. He was assigned several posts during this time, filling them adequately, but not with the same talent he had used to lead the Corps of Discovery. Under circumstance never fully understood, Lewis committed suicide in 1809, following a series of arguments and squabbles with several government officials. His later problems notwithstanding, he was remembered by Jefferson in 1813:

"Of courage undaunted, possessing a firmness & perseverance of purpose which nothing but impossibilities could divert from it's direction, careful as a father of those committed to his charge, yet steady in the maintenance of order & discipline, intimate with the Indian character, customs & principles, habituated to the hunting life, guarded by exact observation of the vegetables & animals of his own country, against losing time in the description of objects already possessed, honest, disinterested, liberal, of sound understanding and fidelity to truth so scrupulous that whatever he should report would be as certain as if seen by ourselves, with all these qualifications as if selected and implanted by nature in one body, for this express purpose, I could have no hesitation in confiding the enterprize to him."

P. ANDREW KARAM

wrote up detailed descriptions equal to what any trained specialist would have written. However, Lewis was a mediocre diplomat who was often condescending and patronizing to both his Native American advisors and the tribal chiefs he met during his expedition.

It is worth noting that very little along the route matched the pre-expedition expectations. The Rocky Mountains were thought to resemble the Appalachians and a relatively easy crossing was anticipated. It was fully expected that an all-water route across the continent would be found or, at least, one that had only a short and easy portage. Jefferson warned Lewis to expect to find wooly mammoths, giant ground sloths, and other such animals along the way, as he believed them to exist in the continental interior. In fact, most of what was "known" about the continent turned out to be wrong. That Lewis and Clarke succeeded despite this lack of knowledge is even more impressive.

Lewis and Clark reached the Pacific in the autumn of 1805, wintered over, and returned to St. Louis in late September 1806 after an absence of nearly three years. From there Lewis returned to Washington where he was received with acclaim by both Jefferson and the academic world.

In the following few years, Lewis seemed to have difficulties adjusting to life away from the

David Livingstone
1813-1873
Scottish Missionary and Explorer

Born into a devoutly religious and hard-working family in Blantyre, Scotland, David Livingstone's faith and work habits would become the map and vessel for a lifetime of exploration. His parents impressed upon him the importance of spreading the Christian gospel. At age 10, he was put to work at a cotton mill, where he labored from six in the morning until eight at night. Afterwards, he went on to study, sometimes late into the night.

In 1836 he began medical school at Anderson's College in Glasgow, intending to become a medical missionary. It was there that he heard fellow Scotsman, Dr. Robert Moffat (1795-1883), speak of having seen, "The smoke of a thousand villages, where no missionary has ever been." In 1840 Livingstone set sail for Africa and arrived at Capetown on March 14, 1841.

David Livingstone. *(The Library of Congress. Reproduced by permission.)*

He spent the next 15 years moving in and about the interior of this remarkable continent. It was a geographical delight to him and he learned all he could about its native population and their cultures. In time, he had occasion to meet the Boers and the Portuguese and developed an intense dislike for their inhumane treatment of the native Africans.

Eventually, he began working with Moffat as a medical missionary in the port town of Kuruman. His work took him into more remote parts of the country and, by the summer of 1842, he had already ventured farther north into Kalahari country than any other white man in history. He decided to leave Kuruman in hopes of spreading the gospel in new, untried areas. Apparently unhindered by typical needs for the companionship of countrymen, he moved to an inland village to learn African languages and customs. During a trek to Mabotsa where he hoped to establish another Christian mission, he was attacked and badly mauled by a lion. Although he recovered generally, he was never again able to fire a gun with any accuracy and was often at the mercy of jungle animals.

In 1845 Livingstone married Moffat's daughter, Mary. A combination of droughts and Livingstone's desire to spread Christianity caused them to relocate three times over the next five years. Eventually, Livingstone sent her

and their four children back to Britain out of concern for their safety and their need for security and education.

Between 1853 and 1856, Livingstone traversed the entire continent, traveling an incredible 4,000 miles (6,400 km) of land along the Zambezi River, none of which had been touched by Europeans. On November 17, 1855, Livingstone laid eyes upon the awesome Victoria Falls. He had discovered most, but not all, of his "highway" when he returned to Britain and recorded his journeys in his first book, *Missionary Travels*.

Tragedy filled the next decade. In 1862 he embarked on what was later called the Zambezi Expedition, on behalf of the British government. Plans to navigate the Zambezi proved to be impractical and the expedition was finally recalled. Not in time, however, to avoid the death of Mary Livingstone who died during the journey. Two years later, the discovery of the impassable Murchison Cataracts shattered his hope of finding a waterway to central Africa. To Livingstone, this did not mean the end of his dream or his travels and he continued to explore Africa on his own terms.

After disappearing from the public eye for several years, a challenge to find Dr. Livingstone was issued. *New York Herald* staff reporter Henry Morton Stanley (1841-1904) took the challenge. He spent a year traveling through Africa before he had the opportunity to pose the famed understatement—"Dr. Livingstone, I presume?"—upon encountering Livingstone.

Stanley offered to help Livingstone back to the coast, but he refused the offer and continued his travels. Livingstone died in 1871 at the age of 60.

BROOK HALL

Sir Robert John Le Mesurier McClure
1807-1873
Irish Explorer

The name of Sir Robert McClure is inextricably linked with the Northwest Passage because of his determination to find the elusive waterway linking the Atlantic and Pacific Oceans.

Born in Wexford, County Wexford, Ireland, on January 28, 1807, Robert McClure completed his apprenticeships and training to become an Irish naval officer. He later settled in England and was given command of the *Investigator*.

For many centuries, seafaring nations have tried to overcome what is probably the world's severest seagoing challenge: the navigable crossing of a series of deep, treacherous channels that pass through Canada's Arctic Islands. Reaching the passage from the Atlantic side meant threading a watery way between, around, and past about 50,000 giant icebergs (many up to 300 ft [91 m] high), with a constant drift toward the south between Greenland and Baffin Island. The alternate approach from the Pacific coast is no less daunting because of the masses of ice that are funneled into the Bering Strait. Despite the dangers, many explorers have tried to find the entrance to the Northwest Passage.

Beginning in the late 1400s, there are records of explorers attempting to find this route. Christopher Columbus (1451-1506), Vasco da Gama (1460?-1524), and Ferdinand Magellan (1480?-1521) laid the groundwork for some of the Dutch captains who later attempted to reach the passage around Russia. However, the discovery that intrigued many prominent explorers during the 1500-1600s was the possibility of a northwest route through the icy waters of northern Canada. This intrepid group included Jacques Cartier (1491-1557), Sir Francis Drake (1540?-1596), Sir Martin Frobisher (1535?-1594), and Captain James Cook (1728-1779). None of them succeeded and several lost their lives in this discouraging pursuit. Even Sir Humphrey Gilbert (1539?-1583), who wrote a rosy treatise of this endeavor, which inspired others to set sail, was lost at sea when he attempted the voyage in 1583. Later, in 1611, Gilbert's young son, Henry Hudson (1565?-1611; for whom the famous Bay is named), took his own shot at "the prize" but when the waters of the bay turned into an icy trap, his crew became mutinous and set Hudson and seven crew members adrift to their eventual demise.

In 1845, two sailing vessels named the HMS *Erebus* and HMS *Terror* disappeared in the North American Arctic. Captain John Franklin (1786-1847) and 128 crew members were lost at sea. McClure was sent to find Franklin. Starting from the Pacific entryway, McClure found his way into the Bering Strait and eventually found two entrances to the Northwest Passage. They were both near Banks Island, which became part of the Northwest Territories of Canada. His ship fell victim to the ice floes in Mercy Bay, and, after nearly two years in the ice, he was forced to abandon the vessel. Fortunately, he and his men were rescued by Captain Henry Kellett, but the Kellett ship was also icebound for an additional year before being rescued by a third expedition. McClure died in England on October 17, 1873, after a long, illustrious career in the British Navy.

GERALD F. HALL

Baron Nils Adolf Erik Nordenskiöld
1832-1901
Finnish Explorer and Scientist

Nils Adolf Nordenskiöld made a total of 10 voyages of exploration to the polar regions, exploring the areas around Greenland, Spitzbergen, and discovering the Northeast Passage from Europe to Asia. He later became a professor of mineralogy at the University of Stockholm and director of the Mineralogy Department at the Swedish Museum of Natural History. In his later years, Nordenskiöld studied the history of cartography, assembling a fine collection of maps of historical significance.

Nordenskiöld was born in Finland to a wealthy family. He attended the Royal Alexander University of Finland (now the University of Helsinki), graduating in geology and mineralogy. In 1858, he set out on his first voyage of exploration, the first of 10 that he would complete between 1858 and 1883.

At this time, Finland was still under Russian domination as the Archduchy of Finland. As such, and unlike most other European voyages of exploration, Nordenskiöld did not set out to expand an empire. Nordenskiöld also did not set out to "fill in the white spaces on the maps"; rather, he tried to add to the body of scientific knowledge about the regions he visited and, on later expeditions, to prove the existence of a water route from Europe to Asia. This was also a time of increasing pride on the part of Nordic people, spurred on by the (then) recent discoveries suggesting that Vikings may have visited North America before Columbus, a fact not proved conclusively until the twentieth century.

During Nordenskiöld's explorations, he conducted detailed mapping of the eastern coast of Greenland, the southern coast of Spitzbergen, and the northern coast of Russian Asia. In addition to writing about his expedition's scientific, navigational, and geographic achievements, Nordenskiöld also wrote a book based on his journey

Nils Adolf Nordenskiöld. *(The Library of Congress. Reproduced by permission.)*

this was in his viewing of a sixteenth-century map of the world in which America is shown separated from Asia, in spite of the widespread belief of that time that the continents were joined. Nordenskiöld attributed this to superior knowledge of the mapmaker when, in fact, the mapmaker had stated in separate documents that he had erred and the two continents should have been joined. All in all, however, Nordenskiöld contributed tremendously to science, geography, and the history of cartography, and even his errors cannot detract from these accomplishments. Nordenskiöld died in 1901 at the age of 69.

P. ANDREW KARAM

Zebulon Montgomery Pike
1779-1813
American Explorer and Army Officer

Zebulon Pike was born in New Jersey, the son of a professional army officer. Beginning in 1794, when only 15, Pike served in his father's infantry unit as a cadet and entered the army as a second lieutenant of infantry in 1799. Aware of his very limited formal education (Pike had attended country schools in his native New Jersey and in Pennsylvania), and impatient with the slow pace of peacetime promotions, he taught himself mathematics, the rudiments of science, French, and Latin. He also studied the European army tactics of his day. In 1801 he married the daughter of an army general. Rising steadily through the ranks, he achieved the grade of brigadier general at the comparatively early age of 34, shortly before his death. After spending the years 1799-1805 in a variety of routine frontier assignments, he was selected, while still a first lieutenant, to take a party of men and find the source of the Mississippi River.

On reaching Leech Lake in what is now northern Minnesota early in February 1806, Pike believed that he had accomplished his mission. He returned to St. Louis after a challenging 9-month, 5,000-mile (8,000 km) trip with his 20-man command. During this journey, which was intended to supplement the western expedition being led by captains Meriwether Lewis (1774-1809) and William Clark (1770-1838), Pike encountered several Indian tribes and secured a large tract of land near Leech Lake from the Sioux for a future army post. His efforts were not altogether successful, however. The U.S. Senate did not approve Pike's treaty with the Sioux.

through the Northeast Passage called *The Voyage of the Vega* (the name of his ship), in which he showed himself to be an excellent writer and popularizer. This book combining a travelogue with scientific observations, ethnographic studies, and some history was the first of its kind to be published in Scandinavia and was later translated into a number of foreign languages.

In preparation for his voyages with the *Vega* in 1878 and 1879, Nordenskiöld had spent much time studying maps of the far north. This led him, after his return, to the study of the history of maps and the influence of maps on the European view of the world. Virtually abandoning his studies in geology and mineralogy, he spent most of the rest of his life in this pursuit, making some significant discoveries, but also making some serious blunders.

Nordenskiöld's most serious mistake was accepting too readily any map he found as authentic. By so doing, he ended up purchasing several forged maps, basing some statements on documents later found to be fakes. This detracted from his credibility and caused others to take many of his writings less seriously than would otherwise have been the case. In addition, Nordenskiöld on several occasions erred in his interpretation of old maps, usually attributing knowledge of intent to ancient mapmakers that they could not have had. One particular instance of

Zebulon Pike. *(The Library of Congress. Reproduced by permission.)*

There were many mistakes in Pike's maps, which were not well drawn, and later explorers determined that Lake Itasca, rather than Leech Lake, was the true source of the Mississippi. Nonetheless, Pike's efforts would later buttress American boundary claims following the War of 1812.

Despite Pike's deficiencies as a mapmaker, General James Wilkinson, the Army's controversial commanding general, next directed him to explore and make maps of the Red and Arkansas Rivers to the west. Pike and his two dozen-man party left St. Louis in July 1806. He and his men got lost owing to the poor maps with which they had been supplied. He discovered but did not climb the mountain that today bears his name. He and his men located the source of the Arkansas River and continued their search for the Red. During the winter of 1806-1807 Pike built a fortified camp for some of his sick men near the present-day site of Canyon City, Colorado. With his other men, he continued southward, reaching the Conejos River in what is now southern Colorado. Wrongly believing this to be his objective, he erected another stockade and sent word for the other men to join him there. Realizing that the Spanish colonial authorities were looking for him, Pike agreed to go with a larger Spanish force that was sent to apprehend the members of his command. Having examined documents Pike had in his possession, the Spanish governor at Santa Fe concluded that he was a spy and insisted that Pike be sent to Chihuahua, Mexico, for further interrogation.

After some months in Spanish custody, and in response to demands from President Thomas Jefferson, Pike and his men were released by the Spanish authorities, and he finally reached Louisiana at the end of June 1807. There, instead of receiving praise for his exploits, Pike found himself accused of being party to a plot by former U.S. Vice President Aaron Burr and General Wilkinson to create an empire in the Southwest by combining some American territory with parts of Spain's colonial possessions in northern Mexico. With difficulty, Pike cleared his name, and resumed his regular military duties.

Pike's papers from his 1806-1807 expedition were not returned to American authorities by the Mexican government for nearly a century. Despite their seizure, Pike published *An Account of Expeditions to the Sources of the Mississippi and through the Western Parts of Louisiana* (1810), which was later republished in Europe. In the work he commented that the Great Plains might in time become as famous as the deserts of North Africa. Pike also described some of the mammals and birds he and his men had encountered, including the grizzly bear. He sent several cubs east, which were displayed at Charles Willson Peale's (1741-1827) museum in Philadelphia, first as part of a menagerie, later as mounted specimens. An engraving of the bears appeared in John Godman's *American Natural History* (1826-1828), the first original treatise on American mammals by an American.

Pike was an active military commander during the War of 1812. He was killed in an explosion following a successful attack on the Canadian city of York (now Toronto), six weeks after his promotion to brigadier general.

KEIR B. STERLING

Nikolay Mikhaylovich Przhevalsky
1839-1888
Russian Explorer, Geographer and Zoologist

In the late nineteenth century central and eastern Asia remained a mystery to most of the world beyond the region's inhabitants. Russia, partly because of it's geographic proximity but largely because of it's desire to expand its rule, dispatched Nikolay Przhevalsky on four separate expeditions through east-central Asia. During these trips, Przhevalsky collected thousands of plant and animal specimens, mapped unrecorded stretches of land, rivers, and lakes, and recorded and named previously unseen plants and animals. Of these rare finds, Przhevalsky is best known for discovering the wild camel and the *Equus przhevalskii,* a stocky, resilient horse he found in Mongolia.

Born in a well-to-do family of Russian and Polish descent, Nikolay Mikhaylovich Przhevalsky grew up in the Siberian wilderness, where he learned to hunt in thick forests filled with wild boar and black bear. Even though he was said to have a photographic memory, Przhevalsky's passion for outdoor activity made him a restless student. He eventually joined the Russian military in Moscow. In order to secure an assignment back in the wilds of Siberia, he studied diligently to gain entry into the Academy of the General Staff in St. Petersburg. He was accepted, and soon thereafter he published *Memoirs of a Sportsman,* the first of many books on his outdoor experience.

Inspired by the African travels of British explorer David Livingstone (1813-1873), Przhevalsky devoured books about botany and zoology while studying navigation at the Academy. He became a history and geography teacher and wrote a geography textbook in 1865. After years of proposing a geographic expedition through central Asia, and funding his own expedition to the Manchurian and Korean borders, Przhevalsky received funds and support from the Imperial Geographic Society to make a longer journey through central Asia.

In 1870 he left Siberia and traveled south through Peking, across the A-la-shan Desert, around Lake Baikal, and crossed the formidable Gobi Desert to reach Kalgan, China. During the trip, Przhevalsky collected sub-alpine and subtropical botany samples; recorded hundreds of species of birds; and, since his barometer broke early in the expedition, measured and mapped altitude by heating water and recording the temperature at which it boiled. As he approached the Tibetan plateau, a 14,000-foot expanse of frozen tundra, he spotted wild camels, while his own camels succumbed to the lack of oxygen.

Przhevalsky returned to Russia a celebrity, and money and prestigious titles soon followed this sudden fame. He published a book on his travels, *Mongolia and the Country of the Tanguts,* lectured widely on his plant and animal discoveries and prepared for his next trip. He set off again in 1876, this time through the peaks of

Tien Shan, across the ancient city of Lop Nor, and through the Taklan Desert. He spotted wild camels again in the Altyn Tag, and he mapped out the "missing link" of the mountains between Pamir and the Nan Shan. He discovered the wild Mongolian horse in the Dzungarian Desert, known to Mongolians as the "Kurtag," but which he named *Equus przhevalskii.* He traversed passes higher than 16,000 feet, and he came within 160 miles of Lhasa, the Tibetan spiritual capital and home of the Dalai Lama, before officials turned him back. He returned to Russia through the Jahar Mountains and arrived with carts of rare specimens and news of spotting *Equus przhevalskii.*

For Przhevalsky's third journey, the Geographic Society asked him to conduct more scientific research, so he searched for—and discovered—the source of the Huan Ho River, while hunting snow leopards and black bear. He found a new, safer trade route from Turkestan to Tsaidan, traveled through the 60-ft-high sand dunes of the Takla Makan, and mapped Khotan. Upon his return he was promoted to Major General, wrote more volumes of his expedition, and prepared for his final journey.

In 1888 Przhevalsky headed off again for Lhasa. He camped and hunted along the River Cher, where a typhoid epidemic had wiped out the local population. He rode on to Lake Karakol, where he quickly became sick and eventually died. His staff buried him on the shores of Lake Karakol, which has been periodically renamed Lake Przhevalsky and where a monument now stands in memory of his accomplishments.

LOLLY MERRELL

Baron Ferdinand Paul Wilhelm von Richthofen
1833-1905
German Explorer and Geographer

Baron Ferdinand von Richthofen conducted important explorations of China and eastern Asia. In addition, as a professor of geography, he launched the study of geomorphology (the study of the Earth's surface features) and he had a significant influence on the field of physical geography.

Richthofen was born into an aristocratic family in Karlsruhe, Germany, in 1833. He attended universities in Wroclaw, Poland (at that time, Breslau, Germany), and in Berlin. He left on his first journey of exploration in 1860, joining a Prussian expedition to East Asia and on to Java, Siam, Burma, and finally California in 1863. In 1868 he left for Asia again, exploring parts of China and Japan, and returning to Germany in 1872.

Upon his return, Richthofen became a professor of geology at Bonn in 1875. He later became a professor of geology at Leipzig in 1883, and Berlin from 1886 on. In these posts, he helped to found the study of landforms, called geomorphology, and helped to make physical geography into a scientific study by investigating the effects of the atmosphere, underlying geology, hydrology, and other factors at work upon the earth's surface.

Richthofen's first major contribution to the science of geography was his 1883 lecture, "On the Problems and Methods of Modern Geography," given at his inauguration as Chair of Geography at the University of Leipzig. In this address he set forth the limits of geography—that it is the study of the surface of the Earth. From there, he pointed out that one could look at the Earth's surface in one of two ways—mathematically or physically. Viewed as a mathematical surface, geographers are interested in measuring distances, drawing borders, determining elevations, and so forth. When viewed as a material surface, on the other hand, geographers must also consider the forces acting on that surface that can cause it to alter its shape and properties. In this view, the surface interacts with the atmosphere (through weathering and weather-related phenomena), the hydrosphere (erosion by rivers, formation of lakes, and so forth), living organisms (what we now call the biosphere, which can act to shape the surface), and with the underlying geology, or lithosphere (earthquakes, landslides, volcanic eruptions, etc.).

Finally, Richthofen also pointed out that humans were beginning to shape the face of the Earth through their activities. One example of such human impact may be seen in the mountains of the Philippines, where the land is so heavily terraced for rice cultivation that the mountain's original outlines have been altered for at least a thousand years.

In this address and his subsequent writings and teaching, Richthofen helped to open what was described as a new epoch in geographical studies. He provided the first systematic survey of the entire field of geography and the first critical assessment of what it could—and should—accomplish. In a sense, Richthofen helped to turn geography into a true science, rather than simply a set of maps and descriptions of observations.

Although the idea of "multidisciplinary" research is largely considered a product of the late twentieth century, it is obviously something that Richthofen realized the importance of a century earlier. By recognizing and drawing attention to the need to integrate diverse studies of geological, hydrological, meteorological, biological, and human factors at work on the Earth's surface, he was ahead of his time. Richthofen died in 1905 in Berlin at the age of 72.

P. ANDREW KARAM

James Clark Ross
1800-1862
British Naval Commander

James Clark Ross is remembered for his extensive experience and successes exploring both the Arctic and Antarctic. His most notable achievement was the discovery of the Magnetic North Pole in June 1831. In his time, Ross was recognized as the world's most successful and accomplished Arctic explorer.

Born in London, England, on April 15, 1800, Ross entered the British Navy at the age of 12. He participated in his first Arctic voyage in 1818, serving under his uncle John Ross (1777-1856), who was searching west of Greenland and north of Canada for an Arctic northwest passage between the Atlantic and Pacific Oceans. Later, between 1819 and 1827, James Ross sailed under British explorer Sir William Edward Parry (1790-1855). In 1823 Ross was promoted to lieutenant and, because of his skills as a naturalist, was elected to the Linnaean Society. In 1827 he was promoted to the rank of commander.

From 1829-33 James Ross sailed on John Ross's second Arctic voyage. James Ross was more popular with the crew than his uncle, who it is said often tried to claim credit for the discoveries and work accomplished by his nephew. Uncle and nephew are believed to have been often at odds.

During this voyage, when the ship became ice-bound and stranded in the winter of 1831, James Ross led a series of overland expeditions. He had become friendly with native Arctic peoples, the Inuit, and used their assistance and their knowledge of the Arctic to survive his overland expeditions.

On one expedition, during June 1831, he discovered the Magnetic North Pole, then located on the Boothia Peninsula, west of Greenland. The Magnetic North Pole is different from, and distant from, the geographic North Pole. Earth has a magnetic field that shifts and exists in varying intensity in different locations. Ross discovered that the Magnetic North Pole shifted constantly, even as he measured its location. Used by mariners to help plot their location, the Magnetic North Pole has since undergone a major shift in location since Ross discovered it.

In 1839 the British government placed James Ross in command of an expedition to the Antarctic to try to find the Magnetic South Pole. Although he failed to find it, he did get farther south than any other British navigator at that time and surveyed and named major Antarctic landmasses. For these accomplishments, and his past successes, he was knighted and given the title "Sir" when he returned to England in 1843.

After Ross married to Anne Coulman in 1847, he vowed never again to go to sea. However, when the famous British Arctic explorer Sir John Franklin (1786-1847) and his ships and crew disappeared on an Arctic expedition seeking the Northwest Passage, Ross broke his promise. He set out in May 1848 to find them, though failed to locate the missing men, who were three years overdue. Ross followed Franklin's route through Barrow Strait west of Baffin Island, but soon found himself blocked by ice. Ross and his crew, several of whom died and many more became seriously ill, endured a harsh ordeal until they escaped the ice. Ross himself suffered illness through the ordeal and returned exhausted and defeated.

During his explorations, Ross discovered and named many Arctic and Antarctic locations. On his voyage to find the Magnetic South Pole he discovered and named Ross Island in 1841, and sighted and named Victoria Land. He also discovered and named Mount Erebus, an active Antarctic volcano, after one of his ships. The Antarctic Ross Sea was subsequently named for him.

Ross's book, *A Voyage of Discovery and Research to Southern and Antarctic Regions,* was published in 1847, at the height of his fame. During his lifetime, he participated in six expeditions to the Arctic and Antarctic, where he spent nine winters and 16 summers.

His wife died in 1857 at the age of 40, and Ross himself died five years later in 1862.

RANDOLPH W. FILLMORE

Heinrich Schliemann
1822-1890
German Archaeologist and Businessman

Heinrich Schliemann, a self-educated German businessman turned archaeologist, unearthed the ruins of ancient Troy and other lost cities mentioned in the *Iliad* of Homer. This accomplishment stunned many who had not believed the cities existed at all, let alone that Schliemann would find them.

Schliemann was born in the Mecklenburg region of northern Germany in 1822. The son of an impoverished minister, he had little formal education before being compelled to leave school and earn a living. Yet from his father he inherited an interest in ancient history. He studied the *Iliad* and other classical literature, and eventually taught himself 18 languages.

Working his way up through the ranks of a trading company in Amsterdam, Schliemann became its representative in St. Petersburg, Russia, in 1846. He imported sugar, coffee, and indigo for dye, and became a wealthy man before he was 30. He spent a few years in California, establishing a successful bank in Sacramento. Schliemann appreciated the frontier spirit of the American West, and would later become a United States citizen.

In the 1860s Schliemann decided he had made enough money. He would be well off for the rest of his life, and was free to devote his time to the study of the ancient world. He visited the Middle East, North Africa, and Asia, and enrolled at the university of the Sorbonne in Paris at the age of 44.

When he visited Greece, Schliemann was captivated. He was steeped in the great works of Homer, the *Iliad* and the *Odyssey*. He explored the Greek island of Ithaca using Homer as a guide. Yet just as with Shakespeare, the figure of Homer is riddled with mystery. There is no historical record of the poet, and although it now seems likely there was indeed a man named Homer, he may not have written everything ascribed to him. In any case, the *Iliad*, which describes the Trojan War, could hardly be taken as an accurate historical record. The war took place in the eleventh or twelfth century B.C. between Greece and Troy, a rival trading power on the Turkish coast. The *Iliad* was written in about 850 B.C., from stories passed along over three or four centuries. Schliemann believed in the existence of Troy and wanted to find it. But he incorrectly identified many of the artifacts he found, and actually destroyed much of the layer corresponding to the Trojan War era as he continued to dig.

After his excavations at Hissarlik, Schliemann returned to Greece, and uncovered more treasures at the site of Mycenae, another city mentioned in the *Iliad*. He died in 1890, on a visit to Naples to see the excavations at Pompeii.

SHERRI CHASIN CALVO

Heinrich Schliemann. *(Archive Photos. Reproduced by permission.)*

his reliance on Homer, coupled with his lack of formal credentials, discouraged the archaeologists of the day from taking him seriously.

In 1869, Schliemann married a young Greek woman named Sophie, who was to be his companion in his archaeological endeavors. The next year, he began excavating at Hissarlik, a rocky Turkish plain that famous ancient soldiers like Julius Caesar and Alexander the Great had believed to be the site of Troy. The surroundings fit all the descriptions given in the *Iliad*. Soon, he had uncovered a palace and a temple. While this caused a sensation in newspapers around the world, Schliemann's work continued to be ignored by scientists and museums. Finally, the discovery of a vast hoard of gold and jewels made them take notice.

Schliemann did indeed discover Troy. However, he made a number of mistakes as well. In his time, the most scientifically advanced archaeologists were just beginning to realize the necessity of careful excavation in order to extract the most knowledge from a site. And Schliemann, who was certainly among the most enthusiastic of archaeologists, was not among the most scientifically advanced.

In particular, Schliemann did not fully understand how the layers of ruins at a site correspond to the time in which they were built, with the oldest layers farthest down. As a result,

Sir Robert Hermann Schomburgk
1804-1865
German-British Naturalist and Explorer

German-born naturalist Robert Hermann Schomburgk explored the interior of Guyana for Britain's Royal Geographical Society between 1835 and 1939, mapping rivers and other geographical features and collecting hundreds of botanical, zoological, and geological specimens for study. In 1841, Schomburgk was commissioned by the British government to return to South America to explore, survey, and establish boundaries along the Guyana-Venezuela frontier. The resulting "Schomburgk Line" was significant in the final boundary settlements of the 1890s.

On June 5, 1804, Schomburgk was born the son of a minister in the Prussian Saxony town of Freiburg. In the late 1820s, after receiving a Prussian education which included lessons in geology and the natural sciences, Schomburgk moved to the United States, where he settle in Richmond, Virginia, as a tobacco merchant. When a fire destroyed his tobacco business in 1830, his early love for botany and natural history and a desire to travel led Schomburgk to the West Indies, where he surveyed the coasts of Anegada in the British Virgin Islands. His surveys were published in the journal of Britain's Royal Geographical Society, which had been founded in 1830.

In 1834, Schomburgk was commissioned by the Royal Geographical Society to lead one of its first funded explorations—into the interior of British Guyana in northeastern South America. The expedition set out in October 1835, and traveled to the upper Essequibo River, the longest river in Guyana. The month of November saw the expedition delayed when all members fell ill with dysentery, and they were forced to turn back in December when the onset of the rainy season made river travel too hazardous. In February 1836, the expedition resumed their charting of the Essequibo River, and Schomburgk became the first European to visit many of its waterfalls and steep rapids, the largest of which he named King William's Cataract for Britain's king. Along the journey, Schomburgk and his team collected numerous geological, zoological, and botanical specimens, including a giant water lily he named the *Victoria regia* in honor of Britain's monarch.

In September 1836, Schomburgk's team ventured up the Courantyne River in eastern Guyana (now the boundary with Suriname) but was again impeded by waterfalls too difficult to bypass. In November, he began exploring the Berbice River, where he was deserted by his native guides, ran out of food, and was attacked by swarms of ants as well as a herd of stampeding wild hogs. In January 1837, Schomburgk forged an overland path back to the Courantyne to finish his commission to explore and survey all of Guyana's great rivers. Wanderlust struck again and, before his return to Europe, he traveled into the Brazilian territory of Roraima, becoming the first European to see Mount Roraima, and meeting many native tribes who had never encountered a European.

Schomburgk returned to England, where he published his minor classic of nineteenth-century exploration, *Description of British Guiana* [sic] (1840). Also that year, he was awarded a gold medal by Britain's Royal Geographical Society for his work in Guyana. While in England, he met with government representatives to recommend further development of Guyana, emphasizing the need to map its boundaries. As a result, he was soon named Guyana's boundary commissioner, returning in 1841 to explore, survey, and establish boundaries between Guyana and Venezuela.

From 1841 to 1843, Schomburgk marked the frontier region and established the boundary that was eventually named the "Schomburgk Line." He continued his travels in Guyana until his return to England in 1844, where he became a naturalized citizen and was knighted by Queen Victoria for his contributions in opening up British Guyana and providing vital statistical surveys and data for Britain's boundary conflicts with Venezuela as well as Brazil.

In 1848, Schomburgk joined the diplomatic service as British consul to Santo Domingo. In 1857, he was appointed consul in Bangkok, Siam (now Thailand), where he supplemented his diplomatic duties with a survey of the Isthmus of Kra during explorations into Southeast Asia. Schomburgk finally retired in 1864 due to declining health and returned to his native Germany, where he died the following year in Berlin.

ANN T. MARSDEN

Charles Sturt
1795-1869
Australian Explorer

Charles Sturt was an Australian explorer best known for his expeditions down the Murrumbidgee and Murray Rivers in Australia. Born in India on April 28, 1795, and educated in England, Sturt entered the British Army at the age of 18. His military career took him to such places as Spain, Canada, France, and Ireland. In 1827, he traveled to Australia to become the military secretary to the governor of New South Wales, Ralph Darling. The following year, Darling commissioned Sturt to explore this new land.

Calling on his military field experience, Sturt organized and outfitted his group for anticipated hardships and difficult terrain. In December of 1828, he departed Wellington and soon came upon Macquarie, Bogan, and Castlereagh Rivers. He also discovered a new river, which he named Darling River in honor of his governor. On his second major trek, Sturt ventured along the mighty Murrumbidgee River, where the riparian growth was abundant and productive in many areas. He observed the dominant tree species along the banks, which were the native river red gums (a type of Eucalyptus). The bountiful trees and shrubs supported numerous species of birds, mammals, and creatures which had never before been seen or recorded. The fish in the river provided sustenance for the group and were a welcome change from the "salted meat" diet.

Following the Murrumbidgee River, Sturt discovered another huge river, which he named the Murray River to honor the Colonial Secretary Sir George Murray. This river is the principal waterway of Australia, flowing 1,609 miles (2,589 km) across southeastern Australia from the Snowy Mountains all the way to the Great Australian Bight of the Indian Ocean. Its size can be gauged by the number of rivers which flow into it: the Darling, Murrumbidgee, Mittta Mitta, Ovens, Goulburn, Campaspe, and Loddon.

Charles Wilkes
1798-1877
American Naval Officer

Charles Sturt. *(The Library of Congress. Reproduced by permission.)*

Sturt followed the Murray all the way to its source near Adelaide. Along the way he met and dealt peaceably with many aborigines along the banks. However, poor diet and physical hardships took their toll, and nearly blind and totally spent, Sturt returned to England, where he wrote *Two Expeditions into the Interior of Southern Australia, 1828-31* (1833). This publication inspired the choice of South Australia for a proposed new British settlement.

Sturt returned to Australia in 1834, where a grateful British government rewarded him with a 5,000-acre land grant. He could have easily lived out his life on a profitable sheep station or other agrarian pursuits, but Sturt was still an intrepid explorer. Once again, in 1844-46, he led an expedition into northern Australia, departing from Adelaide and arriving at the edge of the Simpson Desert. This time there were no new discoveries to report, although his group was the first party to penetrate the center of the continent. They were finally driven back by the paralyzing heat and an outbreak of scurvy.

Once again, he served his country as registrar general and colonial treasurer before he left Australia to settle permanently in England (1847). It was there that he wrote *Narrative of an Expedition into Central Australia* (1849). He lived in England on an annual pension of 600 pounds ($3,500). Sturt died in 1869.

GERALD F. HALL

Charles Wilkes spent his entire working life in the United States Navy. He is best known for leading a four-year voyage of exploration that circled the globe, mapped large parts of the Pacific and Australia, and charted over 1,500 miles (2,400 kilometers) of the Antarctic coast. He also constructed and opened the forerunner of the U.S. Naval Observatory near Washington, D.C.

Wilkes was born in New York City in 1798 to John and Mary Wilkes. He joined the Navy in 1818, specializing in oceanography. One of his first assignments was to take charge of the recently established Depot of Charts and Instruments, upon which he began construction of a simple astronomical observatory. This grew to become the U.S. Naval Observatory, an important center for astronomical research for many years.

In 1838 Wilkes was given command of a six-ship expedition of discovery, the U.S. Surveying and Exploration Expedition. This expedition, which was to last four years and covered 87,000 miles (139,000 km), mapped large tracts of the Pacific, Australia, and Antarctica, endured severe weather, and returned thousands of scientific and anthropological specimens for further study.

Wilkes was actually the fourth or fifth person asked to lead this expedition, but those asked before him either refused or left. An officer with very little time at sea, Wilkes proved himself to be a strict disciplinarian, driving both himself and his crews rigorously throughout the expedition. Leaving the United States with six ships, Wilkes returned from the voyage having lost only one ship and 15 men.

Following his return, Wilkes found himself court-martialed for inaccurate records (one British ship was logged as sailing across a stretch of what Wilkes recorded as solid land), excessive discipline, and possible falsification of records. All of the charges brought against him were eventually dropped, with the exception of one—he ordered more than the allowed 12 lashes for six crewmen found guilty of theft. Angry and disappointed, Wilkes spent the next three years writing a five-volume narrative of his voyage, of which 100 printed copies were distributed.

At the start of the Civil War, Wilkes returned to active duty and was given command at sea. In 1861, in command of the *San Jacinto,* Wilkes intercepted the British steamer *Trent* in

Charles Wilkes. *(The Library of Congress. Reproduced by permission.)*

National Cemetery, his tombstone commemorating his discovery of the Antarctic continent.

P. ANDREW KARAM

the Caribbean and apprehended two Confederate agents, James Mason and John Slidell. The "Trent Affair" very nearly brought Great Britain into the Civil War on the side of the Confederacy and gained Wilkes more notoriety. Further commands followed as did several ill-advised comments against Gideon Welles, the Secretary of the Navy. These comments led to Wilkes's court-martial for disobedience, disrespect, insubordination, and conduct unbecoming an officer. Found guilty on these counts, Wilkes was subjected to a public reprimand and was suspended from the Navy for a year.

Wilkes died in 1877 leaving a mixed legacy. His naval career was checkered, to say the least. Court-martialed, forced out of the Navy, and disliked by his men, Wilkes was hardly a model officer. Besides founding the Naval Observatory, Wilkes's only major accomplishment was leading the U.S. Surveying and Exploration Expedition. However, this was such a resounding success that it more than compensated for the rest of his career. Several American scientists earned international recognition because of their work during those four years and the collections Wilkes returned to the United States became a major part of the original holdings of the Smithsonian Museum when it opened. Based in large part on this expedition, Wilkes was promoted to Rear Admiral in 1866 and is buried in Arlington

Biographical Mentions

Carl John Andersson
1827-1867

Swedish hunter who explored portions of central and southern Africa both in the company of Francis Galton and alone. Andersson was primarily interested in traveling to South Africa to hunt big game and, hearing of Galton's upcoming journey, asked to join. When a local chief refused to let the party pass through his territory, Andersson continued on alone, eventually reaching his goal. He later died of dysentery and is buried in South Africa.

William Henry Ashley
1778-1838

United States Congressman and fur trader whose "rendezvous" system of trading is credited with revolutionizing the fur trade and encouraging explorations of the American West. In 1820 Ashley became the first lieutenant governor in the state of Missouri. Ashley cofounded the Rocky Mountain Fur Company in 1822, and established an annual rendezvous—an open-air wilderness market where trappers could bring their furs to the company at the end of the season and purchase from the company the supplies needed for the coming year. Ashley served as a congressman from 1831 until his death.

William Balfour Baikie
1825-1864

Scottish explorer, naturalist, and linguist who explored West Africa, studying its botany, anthropology, and languages. Baikie became a naval surgeon in 1848 and, between 1854 and 1864, served the British government on the Niger, opening navigation of the river, constructing roads, and eventually becoming Consul of British Nigeria. Before his death Baikie collected native vocabularies of 50 dialects and translated portions of the Bible and Book of Prayer into Hausa.

Florence Baker
1841-1916

Transylvania-born explorer who was the first woman to explore the Nile River. Born Florence von Sass, she had fled the Ottoman Empire in the Balkans to arrive in Turkey in 1860, where Sam Baker, a British explorer of Africa, rescued her from a Turkish slave auction. She accompanied him on his journey to look for the source of the Nile River in Africa. She earned a reputation as an excellent shot, a keen negotiator, and critical member of all of Baker's expeditions throughout Africa in the late nineteenth century. She and Baker eventually married.

Samuel White Baker
1821-1893

British explorer who tried to discover the source of the Nile River. Baker left in 1860 on an expedition of exploration. He met up with John Speke and James Augustus Grant at Gondokoro in 1863 and, in 1864, reached Lake Albert (now Lake Mobutu Sese Seko) on the Uganda border. Although Baker never succeeded in reaching the source of the Nile, he was knighted in 1866 upon his return.

Pedro João Baptista
1800?-1900?

Angolan slave and explorer who, in the company of Amaro José, made the first recorded transcontinental transit of Africa. Returning from their journey in 1815, Baptista was appointed captain of the Portuguese "Company of Pedestrians," with the charge of "keeping open the communication that had been discovered between the two coasts of Western and Eastern Africa." However, it seems likely that this company did much because very little Portuguese exploration of Africa occurred subsequently.

James Pierson Beckwourth
1798-1866?

African-American fur trader and mountain man who discovered a popular trail for gold miners through the Sierra Nevada mountains in California. Today, that trail is known as the Beckwourth Pass. Born a slave in Virginia, Beckwourth was freed at the age of 26. He headed west and lived with Crow Indians for about six years before setting off for California during the gold rush. Accounts of Beckwourth's death, which occurred during a visit to the Crow, vary. It is generally believed he was poisoned—either by an ex-wife or by Crow tribesmen—when Beckwourth refused their request to make him their chief.

Fabian Gottlieb Benjamin von Bellinghausen
1778-1852

Estonian explorer who made a number of discoveries in the Pacific Ocean and may have discovered Antarctica. Commanding an around-the-world voyage of exploration that lasted from 1819-1821, Bellinghausen reached as far south as 70° latitude and was probably the first person to actually sight the Antarctic continent. However, not realizing the significance of this, he is not generally credited with the discovery. The Bellinghausen Sea was named in his honor.

Isabella Bird
1831-1904

English explorer who traveled widely and wrote numerous books. Bird was born at a time when women's activities were severely restricted. At the age of 40, after studying science and literature, she began to travel alone. Following trips to Australia, Hawaii, the United States, the Orient, Tibet, and Morocco, she went home, gave lectures, and wrote of her adventures. Her books showed that a woman could live an unorthodox life and still contribute to society. Many of her books are still in print.

Gregory Blaxland
1778-1853

British explorer who emigrated to Sydney, Australia, in 1806, becoming the first in a long line of adventurers in Australia. In 1813 Blaxland explored the Blue Mountains in southeastern Australia. In 1819 he settled on a farm in Parramatta, near the Blue Mountains, where he experimented with various forms of agriculture, investigated several fodder-type grasses, and tried out vine growing and winemaking. Blaxland's wine won silver (1823) and gold (1828) medals from London's Royal Society of Arts.

Anne Blunt
1831-1917

English explorer and Arabian horse breeder. Blunt was the granddaughter of the poet Lord Byron. Although she was an accomplished artist and chess player, Blunt's real love was the outdoors. In 1869 she married Wilfrid Blunt, a diplomat, and the couple made several trips to the Middle East. At the end of one expedition, they brought back Arabian horses and began to breed them at her home, Crabbet Park. Blunt also bred salukis, native Arabian dogs. She spent her final years in Egypt, and was buried there soon after her eightieth birthday.

Wilfrid Scawen Blunt
1840-1922

British poet and traveler who traveled extensively in the Middle East and India, and championed anti-imperialist measures on behalf of Egyptian, Irish, and Indian nationalists. From 1859-70 Blunt served in seven countries for Britain's Diplomatic Service. Between 1875 and 1904 he made many visits to the Middle East. His adventures included traveling dressed as a Bedouin, visiting desert tribes, following ancient caravan routes, and journeying through hundreds of miles of uninhabited desert.

Nellie Bly
1865?-1922

American newspaper reporter who set a speed record for round-the-world travel in 1889-90. Born Elizabeth Jane Cochrane, she wrote under the name Nellie Bly, taken from popular song. Considered a pioneer in journalism and investigative reporting, Bly is heralded as the first reporter to "go behind the scenes" for a firsthand look at a wide range of controversial subjects, notably the mistreatment of patients in mental institutions. Bly made headlines in 1889 when she sailed from New York to beat the record of Phileas Fogg, hero of Jules Verne's *Around the World in Eighty Days*. She completed her trip in 72 days, six hours, 11 minutes, and 14 seconds.

Sidi Mubarak Bombay
1820?-1885?

African guide who accompanied John Hanning Speke, Richard Francis Burton, and Henry Morton Stanley on several expeditions throughout Africa. Bombay, a resilient man who survived disease and constant battle during his journeys through Africa, traveled with Speke and Burton to discover the source of the Nile River, and he assisted Stanley on his journey to find David Livingstone. Bombay was most faithful to Speke, who mentions him frequently as a sometimes temperamental yet charismatic companion and servant through his two great expeditions to find the source of the Nile.

Aimé Jacques Alexandre Bonpland
1773-1858

French botanist who traveled on a scientific exploration to South America, Cuba, and Mexico (1799-1804) with Alexander von Humboldt.

During the journey, Bonpland collected and described over 6,000 new species of plants. In 1846 he was named professor of natural history at Buenos Aires. While a professor, he undertook a journey up the Paraná River, but was arrested by José Francia, the dictator of Paraguay, who imprisoned Bonpland for nine years.

Nathaniel Bowditch
1773-1838

American mathematician and astronomer best known for his 1802 work, *The New American Practical Navigator,* a theoretical and practical guide to navigation at sea that is still in use today. Bowditch is also remembered for his translation of Pierre Laplace's *Celestial Mechanics.* The four volumes of this translation, appearing between 1829 and 1839, included extensive commentary making the very difficult text more accessible to English-speaking scientists.

Pierre Paul François Camille Savorgnan de Brazza
1852-1905

Italian-born French explorer who undertook three great journeys of exploration in the Congo beginning in 1875. His explorations of the Ogowé River, from the coast of Gabon to the interior and to the north of the Lower Congo, laid the foundations of the future colony of French Equatorial Africa, which extended over 1.25 million square miles. From 1886-87 de Brazza served as commissioner-general of French Equatorial Africa.

Robert O'Hara Burke
1820?-1861

Irish explorer best known as the leader of an ill-fated expedition to cross Australia from north to south in 1860-61. The Royal Society of Victoria sponsored the trip, which included 18 men besides Burke. Midway through the trek, Burke became impatient when supplies did not arrive on schedule and decided to go ahead accompanied by William John Wills, Charles Gray and John King. Unfortunately, his rash actions resulted not only in his own death, but also that of Wills and Gray. Valuable information of the expedition was found in a journal that had been kept by Wills and discovered with his body. It indicated that Burke had died of exhaustion around June 28 in 1861. Later, a statue honoring Burke and King was erected in Melbourne.

René Auguste Caillié
1799-1838

In 1827 the French Geographic Society offered a reward for the first European to reach Timbuktu and return. Already accustomed to the hardship of transatlantic voyages, Caillié—a French explorer known in Senegal as "Abd Allah"—embarked on a long desert crossing, after which he finally reached Timbuktu on April 20, 1828. The recollections of his journey were published in 1830. A stamp was released in 1999 by the French government to commemorate the two hundredth anniversary of his birth.

Verney Lovett Cameron
1844-1894

British explorer who was the first European to cross the African continent from east to west. From 1857-83 Cameron served in the Royal Navy. In 1873 he was sent to Africa to relieve David Livingstone. After learning of Livingstone's death, Cameron continued to explore the region, documenting valuable geographical, political, and anthropological information about central Africa and, by 1875, crossing equatorial Africa. In 1882 Cameron returned to Africa with Sir Richard Burton's Gold Coast mining expedition.

Menigildo de Brito Capello
1841-1917

Portuguese soldier and explorer who, in the company of Roberto Ivens, made one of the first trips across Africa from west to east. Although much of the route they followed was already known to Europeans, they made new explorations between the Zambesi and Luapula Rivers. In the process, they corrected geographic errors reported by previous explorers and investigated the basins of these rivers.

Charles Robert Darwin
1809-1882

English naturalist whose theory that species evolve through natural selection launched a revolution in the biological sciences as well as a storm of scientific and religious controversy. His 1859 book *On the Origin of Species* outlined his theory, and in 1871 *The Descent of Man* applied it to humanity, describing human evolution from ape-like ancestors. Darwin began his scientific career studying plants, animals, fossils, and geology on the famous five-year voyage of the British vessel HMS *Beagle*, which explored the coast of South America and the Galapagos Islands. See long biography on p. 159.

Paul Belloni Du Chaillu
1835-1903

French-American traveller and anthropologist who explored West Africa extensively between 1855 and 1859. He was probably the first white man to see gorillas, which were only known to scientists of the time through a few skeletons. A second journey to that region (1863 to 1865) enabled him to confirm the presence of a long forgotten pygmy tribe in the African forests. Although his first published narrative (1861) was bitterly attacked, the validity of Du Chaillu's statements was subsequently verified by numerous independent investigations.

John Forrest
1847-1918

Australian politician and explorer who became his country's first premier. In his various official capacities, he spearheaded both harbor and railroad projects along with a plan for supplying water to the valuable gold fields. In addition to promoting substantial land settlements, he worked for women's suffrage and protected the interests of smaller states within Australia. He was the first native Australian to be honored by a title in the British peerage—Baron Forrest of Bunbury—and he was also knighted. His writings reflect his lifelong interests and love of country: *Explorations in Australia* (1875) and *Notes on Western Australia* (1884).

Simon Fraser
1776-1862

American-born Canadian fur trader and explorer best known for his harrowing exploration of a trade route from Canada's far west to the Pacific Ocean in 1808. His path west followed along an unknown river, now known as the Fraser River in his honor. As a partner in the North West Company for much of his career, he was charged with establishing trading relations with the Indians in the undeveloped interior of New Caledonia (now British Columbia).

Louis Claude de Saulses de Freycinet
1779-1842

French naval officer and explorer who led a round-the-world scientific voyage from 1817-20. Freycinet entered the French navy in 1794. In 1800 he was chosen by Napoleon for an expedition to complete surveys and natural science observations along the coast of Australia. In 1817 Freycinet embarked on an expedition around the world. His extensive account of the voyage comprised 13 volumes and was fully published in 1844, two years after his death.

Francis Garnier
1835-1873

French naval officer and explorer who graduated from the Ecole Navale and then went to China, where he took part in the 1860 Anglo-French Pa-li-chau victory. Transferred to Cochin China, Garnier was to become the first officer and later the commander of the Mekong River exploration (1866 to 1868). His life ended while defending Hanoi's citadel against an attack by the Pavillons Noirs rebels. Garnier's military feats and scientific achievements served as a catalyst for the French penetration into Indochina.

Ernest Giles
1835-1897

Australian explorer who forged a route from central Australia near Ayres Rock to the west coast. Although a highly skilled adventurer who knew how to survive in the daunting Australian bush land, Giles's first expedition (1872-1874) ended in failure, as he ran out of water and supplies. A second attempt, sponsored by the Victorians and the government of South Australia, in 1876 was successful. Unfortunately, his companion, Gibson, succumbed to the hardship of the trip and died before reaching the west coast.

James Augustus Grant
1827-1892

British soldier and explorer who, with John Speke, set out to discover the source of the Nile River. As a soldier, Grant saw action during the Indian Mutiny and the Abyssinian campaign. As an explorer, Grant is best known for his book *A Walk Across Africa*, in which he published his observations about the native customs and geography of those parts of Africa he explored. Grant retired from the British Army as a colonel.

Charles Francis Hall
1821-1871

Canadian explorer who conducted three research expeditions to the Arctic. His first two expeditions were in search of survivors and clues from a failed 1845 voyage led by Sir John Franklin. The map of the west coast of Canada's Melville Peninsula, which Hall prepared on his first voyage, was so exact that it was not improved upon until the application of aerial photography. His final adventure, begun in 1871, was to lead a government sponsored United States expedition to the North Pole. He made it to 83°N but died suddenly on the return trip.

Matthew Alexander Henson
1866-1955

American explorer who, with Matthew Peary, was the first American to reach the North Pole. Henson ran away from home at age 12, lying about his age to go to sea as a crew member. He received an informal education and, later, became a vital member of several expeditions of exploration to Greenland and the Arctic. In 1909, with Matthew Peary, he reached the North Pole. Unfortunately, his accomplishment was not recognized for several decades, in part because Henson was an African American.

Wilson Price Hunt
1782-1842

American explorer of the trans-Mississippi West. In 1810-12 Hunt lead a party called the Astorians—named after their sponsor, John Jacob Astor—up the Missouri River and across the continent in an effort to establish a fur trading station at the mouth of the Columbia River. This was the second group, after Lewis and Clark, to attempt such a trek. The Astorians suffered incredible hardships while pioneering what would become the Oregon Trail, proving that there was no easy way to reach the Pacific coast via the river systems.

Roberto Ivens
1850-1898

Portuguese explorer who, with Menigildo Capello, explored portions of southern Africa and traversed the continent from west to east. Their expedition, whose purpose was to survey the hydrographic basins of the Congo and Zambesi Rivers, ran into difficulties when some local tribal chiefs refused to provide assistance, fearing their men would not return. However, Ivens and Capello pressed onward, making many new discoveries and correcting many erroneous reports of previous explorers.

Amaro José
1800?-1900?

Angolan slave and explorer who, with fellow slave Pedro João Baptista, was sent by Portuguese army officers on the first recorded transit of Africa from west to east. This journey (1802-1811) followed many unsuccessful efforts and was hampered by inadequate supplies and other difficulties. Part of their journey was spent in the company of Arab traders who acknowledged making the journey previously. However, these earlier travels were never recorded in a reliable manner and could not be verified.

Wilhelm Junker
1840-1892

Russian explorer whose journeys of exploration included travels to Iceland, Tunis, Egypt, and the Sudan. Junker conducted most of his explorations in north and central Africa, where he explored the lower Sobat River and portions of the White Nile and its tributaries. Junker also made numerous important ethnographic observations in central Africa. At one point, prevented by an uprising from crossing Sudan, he returned to Europe by traversing Africa to the east through what is now Tanzania.

Elisha Kent Kane
1820-1857

American arctic explorer who in 1850 led an unsuccessful mission to northwest Greenland to search for British explorer Sir John Franklin, missing since 1845. A trained physician, Kane became a naval surgeon in 1843. He led a second, very difficult arctic expedition in 1853, during which he and his team accomplished much geographic, meteorological, geologic, and other scientific research. Icebound for a year and a half, Kane and his crew abandoned ship, were rescued by a relief expedition, and returned to New York City. Kane died the following year.

Mary Henrietta Kingsley
1862-1900

English traveler who ventured to remote areas of Africa and nearby islands, collecting information to complete an unfinished book by her deceased father, George Henry Kingsley. While visiting the French Congo, she was brave enough to pass through a tribal zone inhabited by the Fang, a tribe well known for its cannibalism. In spite of many narrow escapes, she managed to collect an assortment of beetles and freshwater fishes, which she donated to the British Museum. Her writings were very sympathetic toward the problems and lack of medical care for the black Africans she had encountered. Kingsley died in Simonstown (now in South Africa) at the time of the Boer War, during which she was nursing sick prisoners of the conflict.

Kintup
1849?-?

One of the better known and dedicated Indian "Pundit" explorers of the nineteenth century.

When European travelers were banned from Tibet, Kintup trained as a geographer and slipped into the region in 1880 under a false identity. His mission was to explore the Tsangpo River, the possible source of the famed Brahmaputra River. He survived four years of enslavement and imprisonment while continuing to seek new geographical information. When he was eventually freed he returned to India, only to have his findings discounted. His disillusionment caused him to leave his country and no further record of his life is known.

Otto Astavich von Kotzebue
1787-1846

Estonian naval officer and explorer who completed three round-the-world voyages, including the first circumnavigation of the world under the Russian flag (1803-06) with Baron Adam Johann von Krusenstern. From 1815-18, Kotzebue headed a successful circumnavigation of the globe, discovering several uncharted islands in the Marshalls, visiting California and Hawaii, and exploring Russia beyond the Bering Strait. From 1823-26 he headed another globe-circling expedition, continuing his explorations.

Johann Ludwig Krapf
1810-1881

German explorer who was the first European to discover Mount Kenya. His explorations, along with those of Johann Rebmann, proved important in opening East Africa to exploration, especially from the Indian Ocean. Krapf worked with Rebmann in several of his more productive expeditions, including those to the volcanoes of eastern Africa.

Adam Johann von Krusenstern
1770-1846

Russian admiral and explorer who led the first Russian expedition to circumnavigate the globe. Krusenstern was accompanied by the German/Estonian naval officer Otto von Kotzebue during this journey, which took place from 1803-1806.

Austen Henry Layard
1817-1894

English archaeologist who unearthed the ruins of ancient Assyrian cities from beneath ancient mounds in what is now Iraq. There he discovered alabaster palaces, gigantic statues of winged animals with human heads, and many tablets of wedge-shaped cuneiform writing. Among the cities was Nineveh, a city mentioned in the Bible but which had been regarded by many as legend. Layard became known as the "Father of Assyriology." Later in life he served as a British diplomat.

Friedrich Wilhelm Ludwig Leichhardt
1813-1848?

German naturalist and explorer who made several expeditions into the Australian interior. Setting forth from Brisbane in 1844, Leichhardt was presumed lost until he appeared in Sydney two years later. In 1848 he set forth for his third journey across the continent but disappeared with his entire party without a trace. Although searches were made for over a century afterwards, they were never found. Leichhardt was considered a national hero at the time of his disappearance.

Karl Richard Lepsius
1810-1884

German Egyptologist who made important studies of hieroglyphic writing and the Egyptian language. On his two expeditions, he excavated the Hawarah labyrinth in the Faiyum, discovered the Canopus Decree at Tanis, and collected antiquities for the Berlin Museum, where he headed the department of Egyptology. He also brought back a large collection of detailed drawings of temples, monuments, and inscriptions. His extensive writings, along with almost 900 illustrations, are preserved in his 12-volume work, *Denkmaler aus Aegypten und Aethiopien* (1849-1859).

Thomas Manning
1772-1840

British traveler who was the first Englishman to visit Tibet (1811-12). During the nineteenth century, Europeans were strictly prohibited from entering Tibet, and speedily expelled if found there. In order to bypass this policy, Manning lived in Canton, China, where he worked as a doctor from 1806 until his journey to Tibet in 1811. From Canton, he visited Lhasa in Tibet, staying there for several months and enjoying several interviews with the Dalai Lama.

Jean Baptiste Marchand
1863-1934

French soldier and explorer who investigated the African interior. Marchand was best known for his explorations into Niger, West Sudan, and the Ivory Coast—all in West Africa. During a journey into Central Africa in 1898, he precipitated a French-British crisis by hoisting the French tricolor flag at Fashoda, on the White Nile. Marchand reached the rank of general in the French army, serving with distinction in World War I.

Carl Friedrich Phillip von Martius
1794-1868

German explorer and botanist whose 1817-20 journey to South America yielded over 6,500 plant species, which he presented to the Munich herbarium. Upon returning from his journey, Martius, who had studied at Erlangen University and the Royal Bavarian Academy, published a joint account of the Brazilian expedition with Dr. J. B. von Spix. Martius was appointed a member of the Royal Bavarian Academy, became a professor in Munich (1826), and was made conservator of the Royal Bavarian Academy's botanical gardens (1832).

Thomas Livingstone Mitchell
1792-1855

Scottish explorer and surveyor known for his extensive surveying journeys into the interior of Australia as Surveyor-General of New South Wales. From 1831-46 Mitchell conducted four expeditions in eastern Australia, exploring and surveying areas such as the Darling River (1835), the Murray River (1836), Discovery Bay (1836), and the Barcoo River (1846). Mitchell kept detailed journals of his explorations and published several books, including *The Australian Geography* (1850).

Gustav Nachtigal
1834-1885

German physician and explorer who served as personal physician to the ruler of Tunisia and helped Germany establish protectorates in western equatorial Africa. Nachtigal was sent by Kaiser William I to the Kingdom of Bornu, in what is now northern Nigeria. He reached his goal by traveling a route never before taken by a European, crossing through the central Sahara. He later served as German Consul in Tunis.

Fridtjof Nansen
1861-1930

Norwegian explorer best known for deliberately freezing his ship into Arctic pack ice in an attempt to reach the North Pole. Nansen also journeyed across Greenland and led an arctic voyage of discovery in the sealing vessel *Viking*. Nansen was appointed keeper of the Oslo Natural History Museum after this voyage and was later the first Norwegian ambassador to Britain. In

1922 he was awarded the Nobel Peace Prize for his relief work in Russia following the revolution.

Peter Skene Ogden
1794-1854

Canadian fur trader and explorer for the Hudson's Bay Company who led five trapping expeditions between 1824 and 1829 to the "Snake River Country" (the upper reaches of the Columbia River) with the aim of discouraging American trappers from coming into the area. The northwestern United States and Canada were Ogden's only home and "playing field." In fact, his impact upon the fur trade and explorations of the region were such that schools, parks, and even the third largest city in Utah bear Ogden's name today.

William Edward Parry
1790-1855

British explorer who made several unsuccessful attempts to reach the North Pole and to find the Northwest Passage. During his explorations across the North American continent he was able to map a considerable amount of new territory. Some of the many new areas he described were Prince Regent Inlet, Wellington Channel, Barrow Strait, and Melville Sound. In 1819-20 Parry was the first person to sail beyond 11° west longitude mark in the Arctic. Although his primary purpose was land discovery, he developed an interest in native people and their culture. He studied the Inuit Eskimo for two winters in 1821-23. His last effort to find the Northwest Passage was in 1824-25, also unsuccessful. Parry was later knighted.

Julius von Payer
1842-1915

Austrian explorer who made discoveries in the Russian Arctic and Greenland. Payer graduated with honors from the military academy, becoming an officer in the Austrian army. At the age of 27 he joined an unsuccessful German attempt to reach the North Pole, trying again the following year with a similar lack of success. He subsequently joined or led several successful expeditions to the Russian Arctic and Greenland, culminating in 1912 with a submarine expedition to the Arctic.

Robert Edwin Peary
1856-1920

American explorer of the Arctic credited with leading the first successful expedition to the North Pole. Peary, along with assistant Matthew Henson, explored much of the Arctic and mapped routes to the North Pole from Greenland and the Northwest Territories of Canada. It was believed that in 1909 he reached the pole with Henson and four Eskimo assistants. In the 1980s, however, Peary's route was examined, and, due to possible navigational errors, it is now uncertain if his 1909 trek actually attained the North Pole.

Flinders Petrie
1853-1942

English archaeologist who was among the first to emphasize preserving scientific knowledge by the careful excavation of an entire site, rather than simply hunting for spectacular finds. Early in his career, Petrie investigated the ancient ruins of Stonehenge in England. In the 1880s, he began an important series of excavations in Egypt, including the predynastic cemetery at Naqada. He established the British School of Archaeology in Egypt and the *Journal of Egyptian Archaeology*.

Ida Pfeiffer
1797-1858

German explorer and writer who made a noted trip to present-day Indonesia. Pfeiffer was born in Austria as Ida Reyer. After a childhood in which her father encouraged her to learn and become physically strong and self-sufficient, she married in 1820 and raised two sons. In 1835, when her sons established their own homes and she was free of family obligations, she separated from her husband and began traveling. Pfeiffer published many books; her most famous was *A Lady's Voyage 'Round the World*. She traveled alone and on a tight budget. Pfeiffer shocked the established world with her visit to the East Indies (now Indonesia). While there, she met with the Batak tribe, renowned for cannibalism. Perhaps no previous European had ever been allowed into their territory, and she was the first person to report on the Batak way of life. The Bataks treated her as a curiosity and passed her from tribe to tribe. She became uneasy after the Batak gestured that they wanted to kill and eat her. Pfeiffer replied in broken Batak that she was too old and tough to make good eating. They were amused and eventually let her escape unharmed.

Henry Creswicke Rawlinson
1810-1895

English archaeologist who deciphered the cuneiform, or wedge-shaped writing used in ancient Mesopotamia. Rawlinson's work made it possible to identify ruins uncovered in Iraq as Assyrian capitals, including the famed Nineveh of the Bible. A huge library of clay tablets found in Nineveh yielded evidence that the Mesopotamian civilization was even older than that of Egypt. Among the most important works to emerge from the Nineveh library was the *Epic of Gilgamesh,* an account of the creation of the world and the great flood similar to those found in the biblical Book of Genesis.

Johann Rebmann
1820-1876

German missionary who explored the interior of East Africa and, in 1848, was the first European to see Mount Kilimanjaro. Rebmann's explorations provided geographic information that proved important to future English expeditions led by Sir Richard Burton, David Livingstone, and others. Some of his expeditions in conjunction with fellow German, Johann Krapf, were especially valuable to subsequent explorers.

Susie Carson Rijnhart
1868-1908

Canadian-born physician whose ardent Christian faith led her to exploration and missionary work. In 1898 during a missionary expedition deep into the land of Tibet, she lost her child and husband. In danger herself, she embarked on a two-month hazardous journey over mountain passes and across river torrents to reach safety in China. Returning to America in 1899, Rijnhart vividly related her Asian experience in an acclaimed book entitled *With the Tibetans in Tent and Temple* (1901).

Friedrich Gerhard Rohlfs
1831-1896

German explorer who became the first European to make a complete Trans-Sahara journey from the Mediterranean to the Gulf of Guinea (1865-1866). In the 1860s, Rohlfs used his extensive knowledge of Arabic to explore North Africa disguised as an Arab, compiling geographical data and acquiring valuable information about Trans-Sahara trade routes. In 1876, he was selected to represent Germany at a conference in Brussels of African geographers and explorers. In 1885, Rohlfs was appointed German Consul-General in Zanzibar.

May French Sheldon
1847-1936

American author and explorer who was one of the first white women to visit Central and East Africa. Born in Pennsylvania, Sheldon worked in England in the publishing business. She married an American, Eli Lemon Sheldon, in 1876. She later became a best-selling author and renowned traveler. Her husband financed her journeys and she was typically accompanied by more than 100 porters, guides, and servants. Her most popular book was *Sultan to Sultan,* (1892), inspired by her trip to Central and East Africa, where she described her experience with more than 35 different tribes. Sheldon gave lectures on her travels in Africa to raise money for the Belgian Red Cross during World War I.

Philipp Franz von Siebold
1796-1866

German physician, botanist, and explorer who introduced Western medicine to Japan and brought Japanese plants to European gardens. Siebold worked for the Dutch for many years, first as a physician in Batavia (now Jakarta) and later in Japan. He was expelled from Japan for learning too much about the country, although not before gathering enough scientific information to write important papers about Japanese plants and animals with Dutch and German colleagues.

Nain Singh
1832?-1882?

Indian "Pundit" explorer who began his geographical training with the British when Tibet was closed to foreign travel. He is credited with being the first geographer to discover the famous city of Lhasa, the home of the Dali Lama. Nain Singh began his training in 1863 and, with his brother, spent many years recording important geographical information about Tibet. Some of his discoveries included the semi-prosperous gold fields in the northern region of the Himalayas.

Jebediah Strong Smith
1799-1831

American fur trader and explorer best known for his explorations into the central Rocky Mountains and the Columbia River basin. Smith was the first white man to travel overland to California through the Sierra Nevada Mountains and the Great Basin. He conducted two missions of exploration into the American Southwest before being killed by Comanche Indians while leading a wagon train to Santa Fe at the age of 32.

John Hanning Speke
1827-1864

English explorer who discovered the source of the Nile, the longest river in Africa. Speke served in the army in India before setting out with Richard Francis Burton (1821-1890) to explore the African interior. Together they discovered Lake Tanganyika and then, traveling alone, Speke discovered Lake Victoria, which he recognized as the source of the Nile. When, upon his return to England, his claim about the Nile's source was doubted, Speke made another trip to confirm his discovery, only to have his discovery questioned again. He was killed in a shooting accident before he could set forth a third time.

Johann Baptist von Spix
1781-1826

German explorer who explored large areas of the Amazon River basin. Spix was among the first northern Europeans to explore South America. With fellow Germans von Eschwege, von Wied, Saint-Hillaire, and Carl Martius, Spix used the Amazon as a gateway into the heart of South America, exploring large areas along the river's path and those of some tributaries. In doing this, he helped pave the way for more extensive future explorations of central South America.

Ephraim George Squier
1821-1888

American archaeologist and journalist who explored and excavated prehistoric earthworks throughout the United States and Central and South America. Squier held diplomatic posts in Central America and Peru, but is best known for his *Ancient Monuments of the Mississippi Valley* (1847). This profusely illustrated book, co-authored with archaeologist Edwin Hamilton Davis, provided accurate maps and drawings of prehistoric mounds built by Native Americans. However, it also popularized the myth that these sites were the work of a vanished race of Mound Builders.

Henry Morton Stanley
1841-1904

Welsh-born American considered among the greatest explorers of Africa. Having served on both sides in the American Civil War, Stanley got his start in journalism as a war correspondent. He lead several major expeditions into Central Africa, during which he explored the entire length of the Congo River, circumnavigated Lake Victoria, and collected significant topographic information while searching for the source of the Nile River. His encounter with Scottish explorer David Livingstone, whom Stanley had been dispatched to locate in Tanzania, is immortalized in the famous greeting—

"Dr. Livingstone, I presume?" Stanley also coined the phrase "Darkest Africa."

John Lloyd Stephens
1805-1852

American traveler and author of popular archaeological accounts. Educated as a lawyer, Stephens was a successful executive in transportation companies. He is remembered for his travel narratives about Central America, the Near East, Middle East, and Central Europe. Though not an archaeologist himself, he brought the Mayan monuments to the world's attention through his illustrated *Incidents of Travel in Central America* (1841) and *Incidents of Travel in Yucatan* (1843).

Robert Stuart
1785-1843

Scottish trader and explorer who made way for permanent settlement by opening commercial fishing in the Great Lakes. Born in Scotland, Stuart immigrated to Canada in 1807 and became a fur trader. He eventually became a partner with John Jacob Astor in the American Fur Company. This was important because the American Fur Company became one of the first great American trust companies. The history of the frontier was strongly influenced by Stuart's economic development of Michigan. He was the director of trade for the area called Mackinac, and he was responsible for fostering trade in lead. Steamboat transportation in the region was also under his influence.

Annie R. Taylor
1855?-1920

English missionary and traveler who explored China and Tibet. At 13 she decided to dedicate her life to missionary work. At this tender age she began to visit the slums in London and Brighton. When she was 28, against her family's wishes she sailed to Shanghai, China. For three years she traveled up and down the Yangtze River and also ventured near the forbidden Tibetan border. Taylor was accepted by and moved freely among the Chinese women. She was one of the first foreigners who urged them to study the Bible. One of her most passionate causes was ending the ancient tradition of foot binding, in which a young girl's feet were tied in tight cloth strips and never permitted to grow naturally. After her return to London she formed a "Tibetan Pioneer Band" consisting of nine men. They traveled to Darjeeling but the effort failed. She then opened a shop in Yatung. Taylor's travels aided the British government in trade relations with Tibet.

Joseph Thomson
1858-1895

Scottish explorer who conducted several expeditions into Africa, adding greatly to the general knowledge of African geography. A geologist by training, Thomson joined the Royal Geographical Society's African expedition, which left England in 1878. He took command of the expedition when the leader died. During this and a subsequent journey, he became the first European to see Lake Baringo and Mount Elgon. He also explored along the Upper Congo River and the Sokoto region in Nigeria.

Johann Jakob von Tschudi
1818-1889

Swiss doctor and explorer who made numerous expeditions to South America. Tschudi compiled valuable information on the natural history, geography, agriculture, antiquities, and commerce of South America, and wrote extensively about his travels. His published works include five volumes of *Travels to South America* and his *Travels in Peru* (1847), which was praised for its comprehensive portrait of Peru. From 1866 to 1883, Tschudi was the Swiss Ambassador to Austria.

Edward Whymper
1840-1911

British mountaineer, writer, and artist who is the best known of the nineteenth-century mountaineers because of his many "firsts" in Europe and South America. From 1860 to 1869, Whymper conquered several unscaled peaks of the Alps, including being the first to find a viable route up the Matterhorn (1865). He described these adventures in *Scrambles Amongst the Alps* (1871). In 1880, Whymper was the first climber to reach the summit of Chimborazo in the Andes.

William John Wills
1834-1861

British-born Australian explorer who made the first overland crossing of the Australian continent in 1860 as second-in-command and surveyor on Robert O'Hara Burke's (1821-1861) ill-fated Great Northern Expedition, on which he and Burke died of starvation. After beginning medical studies, Wills migrated to Victoria, Australia, where he accepted work as a Crown Lands surveyor (1852). In 1858, he became a staff member

of a meteorological observatory on Flagstaff Hill in Melbourne. In 1863, Wills and Burke's remains were reburied in Melbourne with a memorial to their transcontinental adventure.

Bibliography of Primary Sources

~

Books

Belzoni, Giovanni Battista. *Voyages in Egypt and Nubia.* 1819. This book summarized Belzoni's many expeditions in Egypt and described the numerous treasures he found—and took back to Europe with him.

Bly, Nellie. *Nellie Bly's Book: Around the World in 72 Days.* 1890. Bly summarized her famous trip in this 1890 work.

Burckhardt, Johann Ludwig. *Travels in Nubia.* 1819. This book included Burckhardt's account of his discovery of the famous temple at Abu Simbel.

Burton, Richard Francis. *First Footsteps in East Africa.* 1856. This book is Burton's account of an 1854 expedition that he led to Harar in what is now Ethiopia; he became the first European to enter that city.

Burton, Richard Francis. *Pilgrimage to El-Medinah and Mecca.* 1855-56. In 1853, disguised as an Afghanistani Muslim, Burton visited Cairo, Egypt, and the holy cities of Islam, Mecca, and Medina in Saudi Arabia. Though not the first non-Muslim to visit Mecca, the birthplace of the prophet Mohammed, Burton provided the most insightful and comprehensive information to date about the place and its people in this book about the trip.

Champollion, Jean-François. *Précis du système hiéroglyphique.* 1824. Champollion, who also wrote important works about Egyptian history, here offered a comprehensive study of Egyptian hieroglyphics.

D'Albertis, Luigi Maria. *New Guinea: What I Did and What I Saw.* 2 vols. 1880. A memoir about the author's noteworthy expedition to New Guinea.

Darwin, Charles Robert. *Journal of Researches into the Geology and Natural History of the Various Countries Visited by HMS Beagle, under the Command of Captain FitzRoy, R. N., from 1832 to 1836.* 1839. In this book Darwin offered an early account of his trip on the *Beagle.* Darwin's experiences and observations from the trip would later form the basis of his theory of evolution.

Darwin, Charles Robert. *On the Origin of Species by Means of Natural Selection.* 1859. The landmark work in which Darwin offered his theory of evolution through natural selection.

Fourier, Jean-Baptiste Joseph. *Description d'Egypte.* 1809-28. Although best known as a mathematician, Fourier was also a noted Egyptologist who spent several years in Egypt with Napoleon. After his return to France, he was responsible for publishing the massive body of scientific and literary discoveries from Egypt, a project that became the *Description d'Egypte.* This work included Fourier's significant historical preface on Egypt's ancient civilization.

Livingstone, David. *Missionary Travels.* 1857. Livingstone's account of his first expedition to Africa was an enormous commercial success.

Przhevalsky, Nikolay. *From Kulja, Across the Tian Shan to Lop-Nor.* 1879. An account of the author's pioneering second expedition through central Asia, during which he spotted the rare wild horse *Equus przhevalskii,* commonly known as Przhevalsky's horse.

Przhevalsky, Nikolay. *Mongolia, the Tangut Country, and the Solitudes of Northern Tibet.* 1876. The Russian explorer's account of his first expedition through central Asia.

Schliemann, Heinrich. *Ilios: The City and Country of the Trojans; the Results of Researches and Discoveries on the Site of Troy and through the Troad in the Years 1871-72-73-78-79.* This work comprised Schliemann's accounts of his numerous discoveries in Asia Minor.

Stanley, Henry Morton. *In Darkest Africa.* 1890. A memoir of Stanley's final African expedition (1889).

Stanley, Henry Morton *Through the Dark Continent.* 1878. An account of Stanley's famous 1874-77 journey through central Africa in search of the source of the Nile and the key to the central African lakes.

Sturt, Charles. *Narrative of an Expedition into Central Australia.* 1849. This is Sturt's account of his 1844-46 expedition into the heart of Australia, an unsuccessful attempt to discover fertile land in the center of the continent.

Sturt, Charles. *Two Expeditions into the Interior of Southern Australia, 1828-31.* 1833. Sturt's memoir of his travels in South Australia inspired the choice of South Australia for a proposed new British settlement.

Life Sciences

Chronology

1817 French naturalist Georges Cuvier publishes a foundational work in paleontology, *The Animal Kingdom,* in which he compares fossil animals with living species.

1827-1838 American ornithologist John James Audubon publishes *Birds of America.*

1828 German chemist Friedrich Wöhler first synthesizes urea and, with others, lays the foundations for organic chemistry and later biochemistry.

1830 English optician Joseph Jackson Lister devises a method of eliminating chromatic aberration from microscope lenses, thus ushering in modern microscopy.

1838-1839 German botanist Matthias Jakob Schleiden and German physiologist Theodore Schwann apply cell theory to the plant and animal kingdoms, respectively.

1859 Discovery of living concretions on a telegraph cable laid in very deep waters explodes the widely held belief that life cannot exist below a depth of 600 meters.

1859 English naturalist Charles Darwin publishes *On the Origin of Species,* setting forth natural selection as the mechanism governing evolution.

1864 French chemist Louis Pasteur invents pasteurization, a process of slow heating to kill bacteria and other microorganisms.

1866 Austrian botanist Gregor Johann Mendel discovers the laws of heredity, presenting data that would not gain wide recognition until 1900.

1872 German botanist Ferdinand Julius Cohn presents the first of four papers that mark the beginnings of bacteriology as a distinct field.

1883 German biologist August F. Weismann begins work that will culminate six years later in his germ-plasm theory, a precursor of modern ideas concerning DNA.

1898 Dutch botanist Willem Beijerinck, describing the causative agent of the tobacco mosaic disease, first identifies what comes to be known as a virus.

Overview:
Life Sciences 1800-1899

Natural history, the description and classification of natural forms, had been the main occupation of life scientists in the eighteenth century, and it expanded dramatically in the nineteenth. Voyages of exploration brought thousands of new plants and animals back to Europe from Africa, the Americas, Asia, and Australia. Improved microscopes revealed hitherto unseen microorganisms and details of anatomy. In addition, geologists began studying fossils systematically, bringing many strange and new organisms to light and giving natural history a truly historical dimension. The organic world proved to be more extensive, old, and diverse than previously imagined.

Despite the exciting findings to come, many naturalists at the turn of the nineteenth century were dissatisfied with their field. A true science, they felt, should not just describe and classify, it should explain things as well. Form, or morphology, needed explaining: Why were there so many plant and animal forms? Why did certain forms exist and not others? Why did some species have internal structures in common? Also, how did species develop and maintain their forms? Function, or physiology, needed explaining, too: How did living things work? How did function relate to form? Further important questions included why species were found in some places and not others (biogeography), how to interpret the fossil record (paleontology), and whether biological theories applied to humans.

This movement to go beyond descriptive natural history gave rise to modern biology. Indeed, the word "biology" was invented in 1800 for the new approach and the problems it addressed. Here is a sampling of the answers and explanations upon which biologists built their science.

Biologists first took Newtonian physics as a model for what scientific explanations should be like. Around 1800 men such as Johann F. Blumenbach (1752-1840) and Carl F. Kielmeyer (1765-1844) tried to explain life processes by means of special forces and laws, the way Isaac Newton had explained the movements of falling objects with the gravitational force and the laws of motion. To explain function, for example, Kielmeyer used such forces as "irritability," which made muscles contract in response to

stimuli, or "sensitivity," which acted on the nerves and conveyed sense perception.

In the study of form, the old theory that embryos were preformed and only had to expand and unfold gave way to the idea of a formative force, advanced by Blumenbach, that made embryos take shape in stages. The force obeyed laws dictating the changes the embryos had to go through. Similar reasoning inspired early theories of evolution, which assumed that similar laws and formative forces produced the series of forms in the fossil record.

Evolutionist Jean Baptiste de Lamarck (1744-1829) borrowed his concept of force from the theory that heat and electricity were carried by invisible fluids. (Even today we speak of heat "flow" and electrical "juice.") Some force-bearing fluid, he suggested in 1809, made animals active and changed their forms in response either to the environment or to internal causes.

An influential alternative to the rule of laws and forces was the "idealism" of Karl E. von Baer (1792-1876) and other early morphologists. Idealists assumed that every species or natural group corresponded to an unchanging "idea" that was the model for its form. In his landmark work on embryology (1828), von Baer declared that the idea of the adult animal was present in the embryo and guided its progress. Somehow the end product caused its own development—a doctrine known as "teleology."

Biologists did not remain satisfied with special laws, forces, ideas, and teleology as explanatory devices. There were too many exceptions to the laws and questions about how the forces acted or how the idea caused its own development. They began looking inside the organism for structures and mechanisms that might cause biological processes more directly.

The cell theory of Matthias Schleiden (1804-1881) in 1838 and Theodor Schwann (1810-1882) in 1839 taught that cells were the building blocks of life and that biological explanations should involve observable cellular activity. The movement, growth, and division of cells were to explain the development, growth, and maintenance of plants and animals forms. Gregor Mendel (1822-1884), the founder of genetics, was part of this trend. He did not merely propose laws that predicted what forms would result from cross-breeding and in what ratios. He also showed how the behavior of reproductive cells could explain why the laws worked, if they contained different hereditary factors.

Physiologists studied physical and chemical processes going on in organs, tissues, and cells, and by mid-century, leading physiologists, including Hermann Helmholtz (1821-1894) and Emil DuBois-Reymond (1818-1896), called special, biological laws and "vital forces" unnecessary and unscientific. Studies of respiration and energy metabolism, begun by Helmholtz and others in the 1840s, eventually showed that processes in cells were no different, in principle, from what was observed in test tubes. Research on the electrical nature of nerve function, on fermentation, photosynthesis, and other biochemical processes helped made it clear, by the 1890s, that the laws of physics and chemistry were the only laws biology needed.

A most decisive blow against laws, forces, and teleology was dealt by Charles Darwin (1809-1882). His 1859 theory of natural selection provided a concrete biological mechanism of evolution. The tendency of living things to vary in form, the inevitability of a struggle for life among the variants, and their unequal success in producing offspring, were all observable processes that combined to cause species to change and become better adapted to their environments. The results of variation and selection were unpredictable and therefore could not be explained by laws or teleology. They had to be studied and explained like historical events.

History was the missing piece in many biological puzzles. In biogeography, it explained why similar environments often were populated by different species. Environment alone did not determine which species would live where; the

movements and evolutionary changes of species also played a role. In morphology, Darwinism explained similarities among living species and between living and fossil forms by their common family histories, and a generation of morphologists, led by Ernst Haeckel (1834-1919), worked out the basic family trees.

Haeckel and Thomas Huxley (1825-1895) were among the promoters of Darwinian evolution who brought its social and theological consequences before the public. Were the human mind and moral sense products of natural selection like the body? If so, could there be objective moral standards, a soul, or any basis for religion? Could, or should, humans shape their own evolutionary future? Would it improve mankind if society and international relations were as ruthlessly competitive as Nature? Answers varied. Some "social Darwinists" argued that if struggle and selection could improve animals, then it was good for humans, too, and justified economic competition, colonialism, racism, and war. Some engaged in polemics against religion.

Others thought that the ability to cooperate would free us from natural selection and make Darwinism irrelevant in human affairs.

At the end of the century, new research directions re-emphasized the experimental methods of the physiological laboratory and used them to explore the biological questions that Darwinism had made urgent. Chief among them were the physical causes of variation, heredity, and development, as well as the interactions between organisms and their physical environment. Research in these areas laid the groundwork for the twentieth-century disciplines of developmental biology, genetics, cytology, and ecology.

SANDER GLIBOFF

Johann Blumenbach and the Classification of Human Races

Overview

Johann Friedrich Blumenbach (1752-1840) was a prominent German anatomist and early anthropologist who played a major role in elevating science above racial prejudice and toward scientific objectivity. His dissertation *On the*

Unity of Mankind (1795), still recognized for its quality and sound scientific approach to the study of human variation, is considered the starting point of anthropology.

As a professor of medicine at the University of Göttingen, Blumenbach was an influential teacher and respected researcher. His publication *Textbook of Comparative Anatomy* (1805) was an unprecedented and superior comparative anatomy and physiology text. Through his natural history studies and his research with fossils, Blumenbach developed revolutionary ideas about life on earth, including the recognition that a number of fossil species had become extinct and other species had emerged more recently.

Blumenbach believed the earth and all its plants and animals had an ancient history, and this idea led to the geological-paleontological time line he devised based on these new ideas. Blumenbach presented both unprecedented concepts about humanity and nature, and the protocols and techniques needed for objective scientific research to study these phenomena.

Background

One of the first classifications of mankind was made by Carl Linnaeus (1707-1778), the father of taxonomy, in *Systema Natura* (1758). In this work Linnaeus followed both continental geography and a color scheme that divided man into white European, dark Asiatic, red American, and black Negro. In the style of the times, many of the "characters" used by Linnaeus to classify his races were quite subjective and unscientific, such as "hopeful" Europeans, "sad and rigid" Asiatics, "irascible" American natives, and "calm and lazy" Africans.

In 1749 French naturalist Georges Buffon (1707-1788) also formulated a classification scheme to distinguish between types of humans. He envisioned six varieties, adding the Lapps, or polar group, and the Tartars, or Mongolians, to the European, American, south Asian, and Ethiopian divisions, but similarly included derogatory adjectives as descriptors of these races.

Other attempts at classification were also simplistic and subjective, such as Meiner's (1793) reference to all nations being derived from two stocks: the handsome, white peoples, including the Celts, Sarmatians, and oriental nations; and the ugly, dark peoples, which consisted of all the rest. These early publications, though earnest approaches to scientific investigation, were flawed by their inclusion of cultural bias and subjectivity, and ignorance of genetics and psychology. During this time scientists often had to rely on the observations and descriptions of people, animals, and phenomena made by adventurous travelers, and had only meager collections of actual organisms to study, including humans.

By the nineteenth century scientists were becoming more aware of the need for objectivity and the importance of using actual physical characteristics and measurements to study and classify animals and humans. However, cultural comparisons and prejudice were still part of some classification schemes, typically elevating the status of Europeans and lowering other groups to the position of primitive peoples.

When Blumenbach began his university studies, he was greatly influenced by Christian Buttner's natural history lectures and discussions of exotic places and people. Blumenbach developed a passion for the natural sciences, including anatomy and the variations of the human race. He began his renowned private collection of biological and ethnographic objects and articles, which included the skulls of peoples from around the world and their art works and literature. His *On the Unity of Mankind* is still seen as a very objective and scientifically sound approach to studying the human animal, while avoiding the use of subjective behavioral characteristics and cultural bias. While Blumenbach incorporated basic differences in skin pigmentation and hair color in his study, he also relied heavily on facial features, shape of teeth, and skull morphology to identify five human races consisting of Caucasian, Malaysian, Ethiopian, American, and Mongolian.

Though similar in some aspects to a few previous publications, Blumenbach's work was more scientifically objective and based on physical data. He also introduced important ideas and terminology such as "Caucasian" to refer to the average appearance of the white, or European, stock. Blumenbach considered the Caucasian race as a middle ground of humanity, with two divergent lines in either direction: the Malaysian people situated intermediate to the Ethiopian lineage, and the American native as intermediate to the Mongolian people.

His detailed study of skull morphology and his considerable cultural awareness lead Blumenbach to categorically support a single species status of all humans, as well as the basic equality of all races and peoples. Blumenbach strongly opposed the subjectivity and perceived cultural superiority held by Europeans, and fought against social and political abuses of anthropological ideas. His adamant support and defense of the "Negro" race as equal in capability and intelligence is pointed and well advanced for his time. Regarding the unity of mankind, Blumenbach concluded that the "many varieties of man as are at present known to [be] one and the same species."

As a preeminent research anatomist, Blumenbach confirmed the anatomical similarities of vertebrates, and detailed the differences among the various animal groups. His *Textbook of Comparative Anatomy* (1805) is considered to be an unprecedented text of comparative anatomy and physiology, as well as a thorough investigation of the anatomical relationships between the major animal groups.

Blumenbach also promoted some revolutionary concepts and ideas in the area of natural history in his essays *Beytrage zur Naturgeschichte* (1811) and *Handbuch der Natureschichte* (1830), both of which exerted a significant influence on scientific study and the advancement of natural history. The advances made by Blumenbach during the early nineteenth century preceded the other great naturalists of that time, Charles Darwin (1809-1882) and Alfred Russel Wallace (1823-1913).

Impact

On the Unity of Mankind was a ground-breaking publication because it forcefully stated that mankind was but a single species and reinforced the need for objective research in science. Blumenbach recognized that all such research should be based only on relevant anatomical characters, such as the skull morphology he used in his study. Blumenbach illustrated clearly that the various races of people did not significantly differ from each other in the measured values of these important characters. He saw humanity as a flow of variation that did not exhibit any marked differences and did not merit any separation at the species level.

Blumenbach believed that this species unity was demonstrated by the fact that all races have accomplished equal cultural development, including writing, art, and scientific investigation. Blumenbach also considered it an important fact that all races are capable of reaching the human perfectibility of form, and these races can and do produce individuals who would be seen by any other human as beautiful.

Blumenbach's dissertation presented ideas that the scientific world recognized as valid and clearly reflective of the true nature of mankind. His work was hailed for its objectivity and scientific merit, and accepted as part of the new science that would study mankind as a historical member of the natural world. Both the scientific and cultural basis of anthropology can be traced back through Johann Blumenbach and his impressive publication *On the Unity of Mankind*. Blumenbach also created a vast private ethnography collection, which included the skulls and other items used in his study, and a range of cultural materials and artifacts from distant places and peoples, including a collection of African literature.

After Blumenbach, there were numerous attempts to classify the races and variations of humans, and race classification became part of a serious pursuit toward a systematic body of knowledge about mankind. Based on the careful measurements of various physical characteristics, three main groups or races were eventually recognized—Negroid, Mongoloid, and Caucasoid. These terms were typically used to represent: the Negroid people of Africa, Melanesia, and New Guinea; the Mongoloid people of Asia, native America, and the Eskimos of the Arctic; and the Caucasoid people of the Indo-European areas. Some researchers elevated these three main groups to the level of a variety, and then divided them into races, including Joseph Deniker (1852-1918), a racial anthropologist of this time, who categorized 10 different races within Europe alone.

Rearranging and scrambling these various classifications occurred repeatedly through the nineteenth century. During this time, the works of Darwin, Wallace, and Gregor Mendel (1822-1884) began to influence anthropology significantly, as the sciences of evolution and genetics gained attention and acceptance. Gradually it became obvious that human variation was based not on different genes, but rather on varying frequencies of the same genes shared by all populations of humans. No classification system that separated humans into distinct species, or even sub-species, could hold up under the light of these sciences.

Still, by the end of the nineteenth century, the idea of a hierarchy of races that elevated some nations and people above others, was widely accepted by many in the upper classes of Europe, Great Britain, and the United States. The wealthy and powerful were often smug in the belief that their superior station and position was justified and secured by nature. Theological dogma was interwoven into this debate and many attempts were made to sort the various races into a pattern purportedly designed by the Creator.

In this way, races were incorrectly seen as preordained, pure, and rigidly fixed in their current form. This erroneous belief has survived into the present, despite the advances of evolution, genetics, and psychology. Blumenbach and the anthropologists he influenced helped to clear away this ignorance and prejudice, and gradually the recognition of racial equality and human rights grew into a force that helped end the practice of human slavery and other such injustices.

Genetic studies of the twentieth century have confirmed the single species status of mankind, and indicate that human races are genetic pools always in flux, forming and blending under the effects of isolation, migration, and genetic exchange. Modern anthropology and evolutionary science continue to study mankind, eager to learn the origins of our species, *Homo sapiens*. Important new discoveries are made almost daily, as new fossils of human ancestors continue to provide clues about where and when humans developed.

Recent genetic research comparing the DNA contained in human mitochondria indicate that all living people may be derived from a very small ancestral population that lived in east Africa as recently as 200,000 years ago. This small group of modern humans, called a "mitochondria Eve" by some scientists, grew into a larger population that spread throughout Africa and, about 100,000 years ago, groups of these humans began to migrate out of Africa to the Middle East, Europe, Asia, and the Americas.

There is still a great deal to be learned about the humans that migrated out of Africa and when these migrations occurred. This area of scientific research remains as charged and exciting as ever, and many critical discoveries about mankind have yet to be made.

KENNETH E. BARBER

Further Reading

Bendyshe, Thomas. *Life and Works of Blumenbach*. London: Longman, Green, Longman, Roberts, and Green, 1865.

Boyd, W. C. *Genetics and Races of Man*. Boston: Little, Brown, 1950.

Dobzhansky, T. *Mankind Evolving*. New Haven: Yale University Press, 1962.

Garn, S. M. *Human Races*. 3rd ed. Springfield, IL: Charles C. Thomas, 1971.

Mayer, Ernst. *Animal Species and Evolution*. Cambridge: Harvard University Press, 1963.

Osborne, Richard. *The Biological and Social Meaning of Race*. San Francisco: W. H. Freeman and Co., 1971.

Population Theory: Malthus's Influence on the Scope of Evolution

Overview

Approximately 60 years before the now historic publication of Charles Darwin's *On the Origin of Species* in 1859, Reverend Thomas Robert Malthus (1766-1834) penned a commentary on what he perceived to be the destiny of the human population in eighteenth-century Britain. Malthus's *Essay on the Principles of Population* profoundly impacted the evolutionary theories of Charles Darwin (1809-1882) and Alfred Russel Wallace (1823-1913), and continues to resonate through social, political, and environmental issues that affect the lives of people today.

Background

Prior to 1750, agricultural practices were as they had been since the Middle Ages. Great Britain was an agricultural society in which farmers worked long hours using simple tools—wooden plows, hoes, and scythes—to produce scanty crops. Using these tools and techniques, farmers were scarcely able to eke out a living. The mid-1700s saw a shift from an agricultural society to an industrial society as new farming practices and mechanization, the growing use of machines, reduced the number of workers needed to produce food for Europe's growing population. As a result, unemployed farmers migrated to the new industrial towns to seek employment in factories.

The Industrial Revolution radically transformed the economic structure of British society from a system of feudalism—a hierarchical system of lords and serfs that concentrated wealth at the top—to one of capitalism. Free enterprise and cost efficient machines caused factory owners, bankers, and entrepreneurs to gain significant wealth and power. The middle and upper classes prospered from the labors of the poor who filled the factories and toiled long hours for little pay.

Malthus's *Essay on the Principles of Population* was written in response to William Godwin's *The Enquirer.* In *The Enquirer* Godwin (1756-1836) promoted population growth as the stimulus for attaining equality among men. Godwin described population growth as a positive force that paves the way to greater wealth and improvement for all. Malthus attempted to point out weaknesses in Godwin's philosophy by way of simple mathematics. Rather than seeing increasing population size as a way of improving the standard of living for all Britons, Malthus viewed it as a limiting factor that reduced the opportunity for the poor to escape from the miseries and hardships of their daily lives.

Malthusian principle states that populations grow geometrically while resources grow only arithmetically. Simply stated: Population grows much more quickly than the food supply. Malthus believed that the inability of available resources to keep pace with ever increasing population size ultimately results in a continuing struggle for survival by the lower economic classes. His *Essay on the Principles of Population* describes the outcome of mankind's struggle to obtain increasingly limited resources as a life filled with misery and vice. It was this concept of a struggle for survival within large populations that was adopted by evolutionary biologists, forever changing evolutionary thought.

Impact

The impact of Malthus's *Essay on the Principles of Population* on Charles Darwin as he sought the mechanism for evolution has never been understated. Darwin himself recorded in his 1876 autobiography the following:

"In October 1838, that is, fifteen months after I had begun my systematic inquiry, I happened to read for amusement Malthus on *Population,* and being well prepared to appreciate the struggle for existence which everywhere goes on from long-continued observation of the habits of animals and plants, it at once struck me that under these circumstances favourable variations would tend to be preserved, and unfavourable ones to be destroyed. The results of this would

be the formation of a new species. Here, then I had at last got a theory by which to work."

Malthus's work caused Darwin to refocus on a bigger picture. While Darwin's predecessor, Jean Baptiste de Lamarck (1744-1829), and contemporaries such as Jean Agassiz (1807-1873) and Richard Owen (1804-1892) focused on individual organisms and the belief that something was driving the evolution of organisms in a direction toward perfection independent of environmental influence, Darwin dismissed this popular theory and continued his quest for another explanation.

Upon reading *Essay on the Principles of Population,* both Darwin and Alfred Russel Wallace independently adapted Malthus's "struggle for existence" principle and applied it to plant and animal species, thereby arriving at the theory of natural selection. Darwin's theory of natural selection stretches Malthus's principle beyond the boundaries of the human population and political economy. Darwinian theory of natural selection made the connection between organisms and their environments stronger than they had ever been.

Prior to Darwinian theory, many believed that individual plants and animals changed in response to their environments. According to Lamarkian theory, then the dominant school of thought with regard to evolution, a giraffe's neck grows longer in its lifetime to reach leaves on taller trees. This individual, in turn, passes this acquired trait on to its offspring. Thus, it was believed, organisms change in response to their environment. However, under this model it would seem that the environment exerts little direct pressure on a species.

Darwin's model challenged the passive role of the environment. He proposed that the environment set the stage for a competition for survival. In economics, as in biology, those that are best able to compete are best able to survive. In a capitalist society, those that are most successful, as measured by accrued wealth, are those who are able to best utilize their resources. Similarly, environment or factors within the environment select for those traits that enable plants and animals to survive long enough to reproduce.

Darwin recognized that many plant and animal species produced far more offspring than could survive. Upon reading Malthus, he surmised that the large numbers of offspring are produced in a biological gamble. According to the selection theory, within the large numbers of offspring some will have the traits that will aid in their survival and will pass desirable traits on to their offspring. Those lacking the traits needed to successfully utilize resources and reproduce do not survive.

The idea that organisms and species survive as a result of characteristics selected for by the environment did not sit well with orthodox religion. The automated nature of the selection process precludes the involvement of a divine figure. Additionally, Christian philosophy holds all life as a gift. In light of Darwinian theory, the essential randomness and apparent wastefulness of the selection process argued against any form of divine intervention. This significant difference in philosophy resulted in battle lines being drawn between Creationists—those believing that the hand of God guides changes in species—and Evolutionists—those contending that species change as a function of natural selection.

Malthus's influence is still felt today as the battle rages on between Creationists and Evolutionists, affecting the way Evolution is taught in schools across America. Similarly, as global population approaches six billion, concerns continue to mount over the Earth's ability to support the growing human population, while at the same time preserving essential ecosystems and maintaining biological diversity.

MICHELLE ROSE

Further Reading

Fedoroff, Nina V., and Joel E. Cohen. "Plants and Populations: Is there time?" *Proceedings of the National Academy of Science* 96 (25 May 1999): 5903-5907.

Ghiselin, Michael T. "Perspective: Darwin, Progress, and Economic Principles." *Evolution* 49 (December 1995): 1029-1037.

Thomson, Keith Stuart. "1798: Darwin and Malthus." *American Scientist* 86 (May-June 1998): 226-229.

Invertebrate Zoology, Lamarckism, and Their Influences on the Sciences and on Society

Overview

The ideas of naturalist and systematist Jean Baptiste de Lamarck (1744-1829) influenced the notions surrounding evolution and also sparked social Lamarckism, which developed years after his death. He was also responsible for making a respectable field out of the study of invertebrates. On the evolutionary front, Lamarck propounded evolution and the mutability of species, but is best known for his "use and disuse" hypothesis. This hypothesis states that traits acquired during an individual's life span can be passed from generation to generation. After his death, many writers and philosophers rallied behind and expanded upon Lamarck's equivocal belief that animals have control over their evolutionary course. The product was social Lamarckism, an unorganized but influential movement.

Background

Lamarck died some thirty years before the publication of Charles Darwin's theory of evolution by natural selection in 1859. While many of Lamarck's ideas were later found to be misguided, his belief in evolution and the mutability of species helped set the stage for their future scientific consideration.

His scientific career took a turn from botany, in which he had gained a solid reputation, when his employer, France's Jardin des Plantes, reorganized into the National Museum of Natural History (Muséum National d'Histoire Naturelle). Lamarck became the museum's professor of "insects and worms," a new field for this man who was nearing 50 years of age. In the years to come, he would coin the term invertebrates for this group of organisms and make great contributions in their classification.

While Lamarck was developing his considerable skills as an invertebrate systematist, he spent many hours considering the method of evolution and devised a "use and disuse" hypothesis. He suggested that organisms over their lifetime could incur alterations to their organs and structures simply by using them more or using them less. For example, a bird might develop stronger lungs to account for the heightened energy demands of flight, whereas a burrowing mole's eyes might degenerate and become smaller, because the animal had no need for them. Any changes acquired during the animal's lifetime would pass down to its descendants. Over time, the animal would evolve to enhance these altered traits.

Lamarck also considered whether animals could "will" these changes, occasionally propounding that they did indeed have control over their evolutionary paths. His thoughts on this idea shifted, but they nonetheless spurred considerable debate, particularly in the century following his death.

Impact

Lamarck's work in systematics became a foundation for future invertebrate classification. Before he took on this formidable task, the invertebrates were considered lower life forms that required and deserved little attention, compared to other animals and to plants. When he accepted his position at the National Museum of Natural History, he quickly became intrigued by these neglected animals and recognized that the group contained an enormous diversity too large to be lumped into one classification. He divided the invertebrates into such groups as the arachnids (spiders and their allies) and annelids (earthworms and their allies)—classifications that persevere today. In many scientists' eyes, he is the father of invertebrate zoology.

Lamarck also received acclaim as a naturalist for his earlier botanical work, particularly by way of his well-received 1778 publication *Flore Française,* which described the plants of France.

It was Lamarck's ideas on evolution, however, that would bring the most notice to the French naturalist. At the turn of the century, Lamarck began to speculate about the method of evolution, and in 1809 outlined them in *Philosophie Zoologique.* While other evolutionists were suggesting that the environment directed the varied alterations in an organism, Lamarck felt that environmental modification triggered an organism's "need" or "urge" to change. The organism responded to that need by developing behaviors that led to increased or decreased use of various organs or structures. If no corresponding organs existed, the organism would generate one. This "First Law," as described by Lamarck, allowed that organs and structures could enlarge or shrink based on their use. His "Second Law" stated that these acquired traits could be passed on to descendants. In other words, organisms changed gradually over time

because their needs ultimately drove their responses to environmental changes and those acquired responses were heritable. Over the years, Lamarck also espoused that organisms have control over the "needs" and "urges," and could thus direct their evolution.

Lamarck's ideas remained largely unknown during his life, because general opinion held that evolution was simply incorrect, and species instead existed unchanged over time. In the mid- to late-1800s, however, the scientific side of Lamarck's "use and disuse" hypothesis began to generate considerable discussion and argument. That changed when the study of genetics began in earnest in the early 1900s as Gregor Mendel's experiments on heritability were rediscovered. Eventually scientists confirmed that genes transfer traits from generation to generation, and that acquired traits were not heritable. The "use and disuse" hypothesis was torn asunder. Additional blows to Lamarck's ideas accumulated as scientists began to understand that evolution was driven by random, nondirected mutations. Evolution had no defined, progressive path as Lamarck suggested.

Lamarck's ideas gained a brief resurgence around the time of World War I when Austrian biologist Paul Kammerer (1880-1926) reported experimental results that seemed to support the idea of heritable acquired traits. His experiments centered around a species of toad that lived on dry land. While the males of most toad species living in moist habitats have a pad on their forelimbs to assist in gripping the female during mating, this dry-land species no longer carried the trait. Kammerer reared the toad in an artificially moist habitat, and reported that the toad developed the formerly absent pads. He also reported that the pads were heritable.

After worldwide publicity about his findings, including media speculation that the experiments opened the door to the creation of modified and enhanced human beings, skeptical scientists examined a preserved specimen of the padded males to find that the dark pads were a result of an application of India ink. Kammerer claimed that the ink was merely an attempt by a student to retain the pad's coloration, which had faded during preservation. With his credibility in ruins, Kammerer killed himself.

Despite the decline in the scientific value of Lamarck's ideas, his hypotheses generated a significant impact in the social arena. In the late nineteenth century and early twentieth century in particular, many philosophers began to select portions of Lamarck's ideas, augment them, and create what came to be called Lamarckism. In particular, they singled out his beliefs that a species may have some control over its future state and that evolution encompasses a trend toward perfection.

In *Philosophie Zoologique,* Lamarck submitted that evolution is a progressive path with lower, imperfect forms at the bottom of the evolutionary tree and higher, more perfect forms at the top. Man was the most complex and perfect. Evolution, then, was a means toward perfection. Because all organisms would over time progress to the top of the tree, he reported that nature must continuously add new organisms to the bottom of the tree via spontaneous generation.

As with a number of his other ideas, Lamarck remained flexible in his opinions. He alternated between convictions that organisms could or could not control their evolutionary destiny, and that evolution was progressive versus solely a response to local environmental conditions. His lack of scientific evidence to back up or refute any of his claims gave him considerable reign to explore philosophical questions and put forth various hypotheses. Contemporary and later critics of Lamarck noted this lack of scientific credibility as a major fault. Darwin was one of Lamarck's greatest critics, and often made a point of distancing his ideas from those of Lamarck. Other scientists, however, looked back on Lamarck as an important thinker of his time.

Supporters of Lamarckism ignored the naturalist's misgivings about some of his own ideas, along with the absence of a scientific basis for them, and developed a rather unorganized social movement that embraced the idea that organisms, including humans as the highest form, could continue to direct the betterment of themselves in succeeding generations. Lamarckians did not accept that evolution was driven by random, nondirected mutations. Lamarckism was a philosophy filled with optimism, but not for all

human races. Lamarckians believed that among humans, white races ranked at the top of the evolutionary tree, and non-white races fell somewhere below. Lamarckians felt that humans could progress most effectively through cooperative social reform and grand educational programs. They believed that learned behavior would be passed along to subsequent generations, a notion which had been refuted by biologists and scientists.

British social philosopher Herbert Spencer (1820-1903), who is often mentioned in connection with Lamarckism, actually limited his connection to this movement by taking up its use-inheritance pretense as the foundation of evolution, but maintaining that nature instead of nurture guided evolution. Spencer is more aptly described as a social Darwinist, who helped promote the "survival of the fittest," a phrase he coined. He held that the individuals who were most well adapted were also most likely to survive and reproduce. Like Lamarckism, social Darwinism fell from popularity in the early 1900s.

LESLIE A. MERTZ

Further Reading

Bowler, Peter J. *Evolution: The History of an Idea.* Rev. ed. Berkeley: University of California Press, 1989.

Bruno, Leonard C. *The Tradition of Science: Landmarks of Western Science in the Collections of the Library of Congress.* Washington, DC: Library of Congress, 1987.

Byers, Paula K. *Encyclopedia of World Biography.* 2nd ed. Detroit: Gale Research, 1998.

Greene, John C. *The Death of Adam: Evolution and Its Impact on Western Thought* Ames, IA: Iowa State University Press, 1959.

Lamarck, J. *Philosophie Zoologique.* New York: Lubrecht and Cramer, 1960.

Lamarck, J. *Zoological Philosophy: An Exposition with Regard to the Natural History of Animals.* New York: AMS Press, 1963.

Singer, Charles. *A Short History of Scientific Ideas to 1900.* London: Oxford Press, 1959.

Advances in Plant Classification and Morphology

Overview

Beginning in the sixteenth century, as botanists explored more areas of the world and identified more and more species, the problem of how to

best classify all these species became critical. By the mid-nineteenth century a number of botanists had devised classification schemes that were based on using a large number of traits. As the century progressed there was also greater interest not only in plants' external form or morphology, but also in internal structures and in microscopic examination of plant tissue. With this trend came more work on plant development, on how plant structures arose, enlarged, and changed over time. This increased morphological knowledge improved classification of plants by providing more information on which to base categorizing decisions.

Background

The basic classification system used by nineteenth-century botanists grew out of the work of botanist Carl Linnaeus (1707-1778) in the mid-eighteenth century. He developed a system for giving each species a two-part name, the first part being for the genus or group to which the species belonged and the second for the species itself. Linnaeus's classification system was called artificial in that he chose a particular characteristic as the basis of classification; in the case of many plants the flower was the structure used. All plants having similar flowers would be put into the same group even though this group would include plants that were otherwise very different from each other.

This system worked well in allowing botanists to identify plants, but to many botanists it seemed inadequate. A number of botanists, most notably Antoine-Laurent de Jussieu (1748-1836) in France, set out to create a more effective system at the end of the eighteenth century. These systems were called "natural" because they were not as arbitrary as that of Linnaeus; they were based on the examination of a large number of plant characteristics, not just those of the flower.

In the first half of the nineteenth century the greatest contributions to the study of plant classification were made by Swiss botanist Augustin de Candolle (1778-1841). He coined the term taxonomy to describe the theoretical study of classification—the investigation of the different ways to describe and categorize a species. Candolle's goal was to classify and organize information on all plant species. To achieve this goal he began work on what would become the 17-volume *Prodromus*, which provided classifications and descriptions of all known plants. Augustin de Candolle did not live to see this work completed. The last of the volumes were written by his son, Alphonse de Candolle (1806-1893), also a noted botanist.

In his studies on plant morphology Augustin de Candolle was influenced by the botanical work of the great German poet, Goethe, who was very interested in botany. Goethe (1749-1832) argued that there was a general plan or form that underlay all plant form, with particular species being variations on this general theme. Candolle, too, saw unity underlying the diversity of plant form. He also agreed with Goethe that the parts of the flower were all related to the leaf form—that, for example, petals and other flower parts could be seen as modified leaves. This idea was very influential in the nineteenth century and reflected an interest in finding a way to simplify or unify the study of plant form.

The diversity of plant species and even the great number of forms within a single plant—stems, leaves, and intricate flowers with many different parts—made the study of plants a complex business; any underlying simplicity would have been welcome. By the end of century, however, interest in this approach to morphology had waned because a unifying theme had come from another direction, not from the study of external morphology but from investigation of internal anatomy and development.

Impact

After the publication of Charles Darwin's *Origin of Species* in 1859, with its presentation of the theory of evolution or change in species over time, natural classification came to mean attempting to organize species in terms of their evolutionary relationships. Species that had evolved from a common ancestor, for example, would be grouped together. The theory of evolution thus provided a new concept on which to base classification, though it frequently proved difficult to figure out evolutionary relationships, and the idea of an entirely natural classification system remains an unfulfilled goal in biology.

One aspect of botany that contributed a great deal to evolutionary theory was the study of biogeography, that is, finding relationships between plant form and geographical characteristics such as climate and terrain. This field was first investigated at the beginning of the century by Alexander von Humboldt (1769-1859), a German geologist who spent several years exploring South America with botanist Aimé Bonpland (1773-1858). Humboldt was impressed with the way vegetation varied with altitude and with the amount of rainfall in a region.

Later, a number of botanists including Asa Gray (1810-1888) in the United States and Augustin de Candolle and his son Alphonse in Switzerland deepened this investigation. Alphonse de Candolle in particular studied the geographical distribution of a large number of plant species and came to the conclusion that most plants seem to have originated in one location. This provided support for the theory of evolution, since the theory includes the idea that the organisms most likely to survive are those that are best adapted to a particular environment. So it makes sense that plants in different kinds of climates would have different traits. Desert plants, for example, would have adaptations to prevent loss of moisture, while those growing on the shaded rain forest floor would be adapted to surviving in low-light conditions.

One criticism of many plant morphologists such as the Candolles was that they focused almost exclusively on what could be seen with the naked eye or with a simple magnifying glass. The microscope figured little in their studies, and, in fact, microscopic work was considered suspect by many botanists because they were uncomfortable with information obtained in this way; they were worried about what they were really seeing through the lenses of the microscope and how it was related to directly visible form. But as the century progressed, botanists became more comfortable with the use of this instrument, particularly as it became apparent that there was a complex world of structure at the microscopic level and that different botanists using different microscopes were seeing similar forms—in other words, that one botanist's results were reproducible by another, an important hallmark of scientific inquiry in general.

Use of the microscope led to a greater interest in plant development—in how the embryonic plant arose, grew larger, and became more complex. Much of the focus of morphological research had been on angiosperms, flowering plants, which are considered the most advanced plants and those that evolved most recently. But concern with development brought a broadening of botanists' investigations, with more work being done on non-flowering plants, (mosses, ferns, and gymnosperms, e.g. pines and fir trees).

British botanist Robert Brown (1773-1858) discovered that development in flowering plants began with the fertilization of the female cell or ovum by the male cell from a pollen grain; this process took place within a structure called the ovule at the base of the flower. Later, German botanist Wilhelm Hofmeister (1824-1877) found that the embryo then develops from the ovum and eventually becomes encased in protective layers to form the seed. Hofmeister was one of the leading plant morphologists of the nineteenth century. He studied development not only in flowering plants but in a variety of other plants including ferns, and this is how he came to discover the great unifying principle underlying plant form.

Ferns had been studied by a number of noted botanists before Hofmeister; they were a fascinating group of plants because they were quite different from flowering plants, producing no flowers or seeds but rather spores that lacked the protective coatings of seeds. The spore develops not into a mature plant, but into a tiny, leaf-like structure called a prothallus or gametophyte. German botanist Carl von Nägeli discovered a set of cells within the gametophyte called the antheridium that produced sperm and was thus comparable to the pollen-producing anther of flowering plants. It was Polish Count J. Leszczyc-Suminski who later found the archegonium, the structure in the gametophyte that produces eggs. Fertilization takes place on the gametophyte and a new structure, the sporophyte, develops on it. It is the sporophyte that eventually grows into the mature fern plant on whose leaves or fronds arise the sporangium, the spore-producing structures.

Hofmeister saw the fern as having two parts to its life-cycle, with one part, the gametophyte, producing sex cells, eggs and sperm, and the other producing the sporophyte or mature plant. We now know that the gametophyte is haploid, that is, it has half the number of chromosomes as the diploid sporophyte, but Hofmeister made his observations before the role of the chromosomes as carriers of genetic material was understood. He did see, however, that this alternation of generations, between gametophyte and sporophyte, between haploid and diploid structures, was found in all plants, not just in ferns, though the size and structure of the two forms varied widely.

In flowering plants the sporophyte structure is clearly dominant, with the pollen being the male gametophyte and the embryo sac found within the ovule at the base of the flower being the female gametophyte. The sporophyte is also dominant in gymnosperms such as pines, but in seaweed or green algae the gametophytes and sporophytes are often so similar in size and structure as to be indistinguishable from each other when examined with the naked eye. In moss, the visible form of the plant is the gametophyte, with the sporophyte usually forming small structures arising from the gametophytes. Such wide variation in structures makes Hofmeister's discovery that much more impressive, and it provided the first really significant unifying principle in botany, an idea that related the structure and development of all plants.

MAURA C. FLANNERY

Further Reading

Isely, Duane. *One Hundred and One Botanists.* Ames, IA: Iowa State University Press, 1994.

Morton, A. G. *History of Botanical Science.* New York: Academic Press, 1981.

Pilet, P. E. "Augustin de Candolle." In *Dictionary of Scientific Biography,* edited by C. C. Gillispie. New York: Scribner's, 1974.

Proskauer, J. "Wilhelm Hofmeister." In *Dictionary of Scientific Biography,* edited by C. C. Gillispie. New York: Scribner's, 1974.

Sivarajan, V. V. *Introduction to the Principles of Plant Taxonomy.* 2nd ed. Cambridge: Cambridge University Press, 1991.

Stevens, Peter. *The Development of Biological Systematics.* New York: Columbia University Press, 1994.

Georges Cuvier Revolutionizes Paleontology

Overview

In the wake of the French Revolution (1789-99), young naturalist Georges Cuvier took up arms against conventional beliefs calling into question the accepted view of the history of Earth as well as the relatedness of fossil organisms to living species. Through extensive field collection and meticulous study of fossil specimens, Cuvier began amassing the evidence that would establish extinction as a biological reality. His methodologies, field practices and procedures, and groundbreaking work paved the way for future paleontologists.

Background

Fossils have fascinated and mystified humans for thousands of years. Collected, discussed, and described by such historical figures as Aristotle (384-322 B.C.), Leonardo da Vinci (1452-1519), and Martin Luther (1483-1546), fossil origins served as a source of speculation and debate. Once believed to form in Earth's crust by processes similar to the crystallization of minerals, by the late 1600s fossils were generally recognized as the remains of plants and animals. The realization that fossils were neither randomly distributed throughout the crust of Earth nor found in all rock types led naturalists to examine fossils more closely as they sought answers to age-old questions.

While many of his colleagues offered explanations for the occurrence of fossils and other natural phenomena, providing cursory observational evidence to support these explanations, Cuvier (1769-1832) carried out exhaustive field studies collecting evidence to support his position that fossils were the remains of species that no longer existed. He became a bold innovator offering what few before him had: physical evidence. For him, fossils became more than natural oddities but rather served as the crux for a different view of the history of Earth.

Cuvier entered the infant field of geology with questions and concerns. The predominant philosophy at the time held that fossil remains were the remnants of animals that still existed elsewhere on Earth. To suggest that they were the remains of animals that no longer existed was counter to religious creationist doctrine. Cuvier, a devout creationist, suspected that Earth was much older than suggested by the biblical interpretation. He challenged current Earth history view on the basis of the physical evidence provided by the fossils themselves.

A gifted comparative anatomist, Cuvier carefully reconstructed ancient animals, thereby documenting the past existence of large mammals that resembled no living species. Through detailed studies of elephant anatomy, he demonstrated beyond a shadow of a doubt that African and Indian elephant species were two distinct species. Additionally, he demonstrated that the fossil remains of elephant-like organisms recovered from Europe and Siberia belonged to neither species but rather to a species that no longer existed—woolly mammoths. In addition to mammoths, Cuvier published articles describing giant ground sloths, Irish elk, and American mastodons in addition to a number of invertebrate species.

Cuvier's work extended beyond the proof and naming of extinct species, as he proposed the mechanism for the demise for these species—most of which existed before man, a notion hotly debated in and of itself. His proposition that extinction resulted from geological "catastrophes" or sudden changes in Earth's surface provided a causal agent for mass extinction and a foundation for species transformation. Cuvier was a steadfast opponent of organic evolution, and his work significantly influenced future scientists.

Impact

With the advent of Cuvier's work and publication of his catastrophe theory, fossils were viewed with heightened interest. Students and disciples of Cuvier continued to seek fossil evidence and scientific explanations that supported Cuvier's theories of extinction and Earth history.

A renowned student of Cuvier and glaciologist, Louis Agassiz (1807-1873) employed both Cuvier's methodologies and his philosophies as he explored living and fossil fishes in addition to the potential effects that glacial episodes must have had on Earth's ancient populations and topography (land surface features). Agassiz made lasting contributions to evolutionary biology and systematics through his construction of classification based on morphological characteristics. Similar to Cuvier, Agassiz opposed evolutionary thought even as his work was used by biologists to support it.

Cuvier's influence extended well beyond his own students, impacting those who would throw the gates to the field of vertebrate paleontology wide open. While not the first to discover a reptilian fossil—that distinction belonged to William Buckland (1784-1856)—Gideon Mantell's discovery of fossil iguana-like teeth in Sussex, England, in 1822 and his subsequent discovery of Hylaeosaurus in 1833 promoted continued interest in fossils. Close examination of his finds led Mantell (1790-1852) to suggest that based on physiologic differences Iguanodon and Hylaeosaurus belonged to different genera.

Sir Richard Owen (1804-1892), a comparative anatomist in the style of Georges Cuvier and a staunch anti-evolutionist, studied the relationships between groups of organisms. Upon examination of the known reptile fossils, Megalosaurus, Iguanodon, and Hylaeosaurus, Owen concluded that these organisms were members of a group distinctly different from living lizard species. In 1842 Owen published the results of his study, naming this new grouping of large, extinct reptilian animals "Dinosauria"—which meant "terrible lizard" or "fearfully great reptile."

In 1863, as curator of the British Museum of Natural History, Owen acquired, named, and described Archaeopteryx lithographica, a crow-sized animal possessing morphologic features characteristic of both lizards and birds. Regardless of how much Archaeopteryx looked like the missing link between reptiles and birds, Owen rejected evolution as the causal agent of the bizarre appearance of this fossil specimen. In fact, in response to evolutionists claiming Archaeopteryx as a missing link between animal phyla, Owen demanded evolutionists produce more "missing links."

In 1868 the nearly complete skeleton Hadrosaurus foulkii, discovered 10 years earlier in Haddonfield, New Jersey, went on public display at the Academy of Natural Sciences of Philadelphia. Through comparative anatomy techniques, Joseph Leidy concluded that Hadrosaurus was closely related to the Iguanodon discovered 50 years earlier on the other side of the Atlantic Ocean.

Completion of the Transcontinental Railroad in 1869 geared the world of paleontology up for the first of several eras of dinosaur discoveries. Discovery of dinosaur fossils in Colorado in the late 1870s initiated the "First Great Dinosaur Rush" as paleontologists scrambled to identify previously unknown species of the extinct beasts. The public's fascination with dinosaurs grew with each new discovery and public display. Two paleontologists, Edward Drinker Cope (1840-1897) and Othniel Charles Marsh (1831-1899), became bitter rivals as each sought to be the first to unearth the most spectacular dinosaur fossils. Thus began one of the most well-known rivalries in paleontological history. The fierce competition between Cope and Marsh resulted in the discovery and identification of no less than 28 new genera of dinosaurs.

Paleontology today still bears some resemblance to the science founded on Cuvier's principle of hard evidence. Fossil organisms are still identified using morphological characteristics. However, modern paleontologists strive to do more than identify the organisms to which fossils belong. Modern paleontology often involves the fields of paleobotany—the study of fossil plants—and paleoecology—the study of ancient habitats. Anatomical studies and reconstruction of extinct species excavated from the Rancho La Brea tar pits, for instance, provides extensive insight into the population structure of animals that lived 10,000 to about 400,000 years ago, a period termed the "Rancholabrean Land Mammal Age."

As to the role fossils play in settling the age-old debate between creationists and evolutionists, fossils are proving to be invaluable resources for determining the relatedness between extinct and modern species. Multiple copies of ancient DNA obtained from fossils have been made through the duplicating process of polymerase chain reaction, or PCR. A determination and comparison of the nucleotide base sequences between extinct and modern species often shows that which Cuvier refused to consider—a relationship between extinct and modern species of animals.

MICHELLE ROSE

Further Reading

Books

Hellman, Hal. *Great Feuds in Science: Ten of the Liveliest Disputes Ever.* New York: Wiley, 1998.

Rudwick, Martin J. S. *Georges Cuvier, Fossil Bones, and Geological Catastrophes: New Translations and Interpretations of the Primary Texts.* Chicago: University of Chicago Press, 1997.

Periodicals

Janczewski, D.N., et al. "Molecular Phylogenetic Inference from Saber-Toothed Cat Fossils of Rancho La Brea." *Proc. Natl. Acad. Sci.* 89 (October 1992): 9769-9773.

Watching as Life Begins:
The Discovery of the Mammalian Ovum and the Process of Fertilization

Overview

The discovery of the mammalian ovum brought the realization that human reproduction occurred in the same way as did that of other animals. In the second half of the nineteenth century, microscopic techniques improved enough to allow scientists to observe the nuclei of cells. Turning this new ability to the study of sexual reproduction, they were able to see that fertilization involved the merging of cell nuclei from the male and female parent. This superseded the previous view of the sperm stimulating the ovum to develop by physical or chemical means, and led the way to the modern understanding of reproduction and genetics.

Background

From ancient times, people have understood that traits may be inherited, and bred animals in order to reproduce desirable attributes. But they had no idea of how heredity actually worked. Aristotle (384-322 B.C.) and other Greek philosophers believed that hereditary traits were passed in the blood, and we still use terms like "blue-blood" and "blood relative" today. While it was clear to the ancients that sexual intercourse resulted in pregnancy, the mechanism by which a new being was conceived and developed was also a mystery. The general view was that the male contributed the "seed" of a new organism, while the female's role was to nourish it as it grew.

In the seventeenth and eighteenth centuries, another school of thought arose, which held that the essence of the new being was concentrated in the egg. These were understood as containing perfect miniature organisms, which needed only the stimulation of the male in order to trigger their growth. Each tiny being carried its own even tinier offspring in miniature, and so on, like a set of nested dolls going all the way back to Creation and forward for all time. Some scientists argued against this theory, called preformation, because it did not explain the inheritance of traits from both parents. Opponents of preformation got a boost in 1651, when the English physician William Harvey (1578-1657) published his studies of deer embryos. In their early stages, the embryos looked nothing at all like fully formed deer.

The invention of the microscope in the mid-1600s by Anton van Leeuwenhoek (1632-1723) allowed scientists to see individual cells for the first time. The term "cell" was coined in 1665 by Robert Hooke (1635-1703), upon observing holes encased by walls in a sample of cork. Leeuwenhoek first described his observations of sperm cells in 1677. This renewed the debate as to whether these tiny "animalcules" contained preformed organisms, or served to awaken those in the egg.

In 1759 Kaspar Friedrich Wolff (1733-1794), a German physiologist, dealt preformation another blow. His experiments with chick embryos showed that specialized organs develop from undifferentiated tissue. This is the same process we can readily observe in the growth of a plant, as a new shoot differentiates into stem, flowers, and leaves. Better microscopic and experimental techniques of the nineteenth century would shed light on the actual mechanisms of reproduction and development.

Impact

It was by no means obvious to early natural historians that eggs had anything to do with the reproduction of humans and other mammals. Many types of animals lay their eggs, but in mammals, fertilization and embryonic development take place internally. The first known claim to have observed the human ovum was made by the Dutch physician Reinier de Graaf (1641-1673) in 1672, but there is some dispute as to what he actually saw. For the next 150 years attention turned to studying the more easily accessed eggs of birds and frogs. The first undisputed observation of the human ovum was made by Karl Ernst von Baer (1792-1876), and reported in a letter in 1827.

Early nineteenth-century scientists watching the first stages of the development of frogs' eggs saw them divide repeatedly, but they did not at first connect this with the concept of cells. The idea that cells are the building blocks of all life was advanced in the 1830s by the German scientists Matthias Schleiden (1804-1881) and Theodor Schwann (1810-1882). It was not until 1858 that Rudolf Virchow (1821-1902) proposed that cells could arise only from other cells. Meanwhile, one of the most important developments in the history of biology was quietly taking place in the garden of an Austrian monk.

Gregor Mendel (1822-1884) discovered the laws of heredity while growing peas, observing red and white flowered plants and how the colors appeared in hybrids. In an 1866 paper he explained that heredity is caused by informa-

tion-carrying entities we now call genes. The genes exist in pairs, one from each parent. Mendel's ideas appeared before their time, and no one knew what to do with his work. Developments over the next few decades resulted in its rediscovery at the turn of the century. Similarly, DNA, the molecule that codes the genetic information, was discovered in 1869 by Swiss biochemist Friedrich Miescher (1844-1895), but dismissed at the time as unimportant.

In the 1850s the German botanist Nathanael Pringsheim (1823-1894) was among the first to observe the union of egg and sperm, in his experiments with freshwater algae. He watched the sperm force its way through the outer layer of the egg and seem to dissolve within its protoplasm. His studies demonstrated that the sperm cell actually combined with the egg, rather than simply acting upon it.

Microscopy improved in the 1870s and 1880s with the development of immersion lenses, optics corrected for spherical and chromatic aberration, and new specimen fixing and staining techniques. Suddenly the details of cells were revealed. The study of reproduction was revolutionized when scientists could begin to observe the structure of the cell nucleus.

Oskar Hertwig (1849-1922) and Hermann Fol (1845-1892) independently observed the penetration of a sea urchin eggs by sperm cells in the late 1870s. Hertwig was able to discern that fertilization involved the merging of two cell nuclei, one from the egg and one from the sperm cell. This was a major advance in thinking about reproduction. Even after Pringsheim and others had observed the sperm entering the egg, the assumption had been that the "dissolution" of the sperm resulted in some type of chemical stimulation. Hertwig's work showed that the embryo developed from a union of the two cells. He also described how once fertilization has occurred, additional sperm are prevented from entering the egg.

The nucleus was observed to contain a substance called chromatin. During ordinary cell division, or mitosis, the chromatin could be seen to condense into discrete chromosomes. Each cell had a fixed number of chromosomes, depending on the type of organism. Cells divided to form new cells with the same number of chromosomes as those that preceded them. Human cells have 46 chromosomes.

The German biologist August Weismann (1834-1914) realized that if egg and sperm cells were produced by the same mechanism, their merging would result in a doubling of the hereditary information in each generation. The solution to this problem is key to understanding the assortment of hereditary traits we see in sexual reproduction. The gametes, or sex cells, are generated in a process called meiosis, in which the parental gene pairs are split. The gametes are haploid cells, with half the normal complement of chromosomes. Two gametes merge in sexual reproduction, resulting in the embryo's having diploid cells with the normal amount of genetic material, half of which derives from each parent.

In 1892 Weismann wrote that the determinants of hereditary traits were located in the chromosomes, an assertion that was regarded as speculative at the time, and which was not proven until some years later. He called this hereditary information the "germ plasm," a com-

ponent of the cell nucleus passed along in such a way that it remained isolated from anything that happened to the adult organism. This contradicted the Lamarckian view that traits acquired during an organism's lifetime could be inherited, and provided support to the theory of evolution by natural selection.

SHERRI CHASIN CALVO

Further Reading

Farley, John. *Gametes and Spores: Ideas About Sexual Reproduction, 1750-1914*. Baltimore, MD: Johns Hopkins University Press, 1982.

Harris, Henry. *The Birth of the Cell*. New Haven, CT: Yale University Press, 1999.

Pinto-Correja, Clara. *The Ovary of Eve: Egg and Sperm and Preformation*. Chicago: University of Chicago Press, 1997.

John James Audubon Publishes His Illustrated *Birds of America* (1827-1838)

Overview

Prior to 1827 the few published descriptions and illustrations of North American birds had been done in workmanlike but not definitive fashion. French-American artist-naturalist John James Audubon (1785-1851) changed this with the publication of his *Birds of America,* first in a limited elephant folio edition for wealthy subscribers and institutions and later (1840-44) in a smaller popular edition. This work forever changed the way in which Americans perceived their avian fauna.

Background

From the mid-eighteenth century on, visitors to the British colonies in North America, and later the United States, wrote books and articles describing the birds and other fauna and flora they found there. English naturalist Mark Catesby (c. 1679-1749), in his *Natural History of Carolina, Florida, and the Bahama Island* (1731-1743), described and illustrated the animals and plants he encountered. He is considered a pioneering ornithologist, having depicted 109 birds. Catesby also enjoys a reputation as America's first ecologist. He was, however, primarily a botanist. Nevertheless, more than 75 modern

A watercolor of a golden eagle by John James Audubon. *(The Library of Congress. Reproduced by permission.)*

bird species are based in whole or in part on his descriptions. His illustrations, however, while reasonably accurate, lacked inspiration.

English naturalist Thomas Pennant, who never visited North America, described, in his two-volume *Arctic Zoology* (1785), 17 birds collected by Captain James Cook (1728-1779) in the Bering Strait and western Alaska region and another 83 found by various collectors in and around Hudson Bay. Pennant's illustrations were often romanticized and not always accurate. He also included many species from northern Europe and some previously described from the United States, while giving coverage to other vertebrates. Additional comments were contained in a 1787 *Supplement* volume. In subsequent years, naturalists and explorers from several European nations added considerably to these lists.

Scottish-born Alexander Wilson (1766-1813) published seven volumes of his *American Ornithology* between 1808 and his death in 1813. Two final volumes were brought out by his friend, Philadelphia businessman-naturalist George Ord. Wilson carefully described 278 birds, 48 of them new to science, but his travels were for the most part limited to the eastern seaboard of the United States, and his illustrations, though workman-like and generally accurate, lacked distinction.

The English explorer-naturalist Sir John Richardson and his colleague William Swainson, in their *Fauna Boreali-Americana,* part second, *The Birds* (1831), admirably described and illustrated 240 Canadian species, many from the western provinces, setting a high standard not equaled until later in the nineteenth century.

Several factors combined to make Audubon's project a defining work for its time. The Elephant Folio edition of his *Birds of America* depicted 489 birds (later increased to 500 in the popular edition), nearly twice as many as had appeared in Wilson's books. He covered more ground than any of his predecessors in the United States, visiting the south central and midwestern states and territories, as well as the eastern seaboard. He also traveled to Quebec and Labrador and went up the Missouri River. His artistic talents were far superior to the skills of those who had come before him. Also, the geographical distribution of many of the birds he described ranged into the Middle West and Great Plains areas, then little known to the general public. Finally, he later made a point of bringing his work to the attention of interested Americans and Europeans by selling subscriptions to his book.

Growing numbers of explorers, travelers, military men, and settlers were moving into the central and western territories during the 1820s and 1830s. Birds and other wildlife were the subject of curiosity to these and other Americans. American publishers were becoming more numerous and some of their efforts more ambitious. From this time on, prompted in part by Audubon's work, the great variety and range of American animal and plant life began to draw more than casual attention from American and European naturalists. Increasing numbers of these scientists were doing their own research in the field.

Audubon seems not to have had much formal art training, though he possessed a natural genius. For many years, it was thought that he studied with French portrait painter Jacques Louis David in Paris, but Audubon's modern biographers do not believe this is true. For nearly 20 years, from his first arrival in Pennsylvania in 1803, Audubon gradually perfected his technique, shooting birds and animals, arranging his specimens with the use of wires and pins, drawing them skillfully, and sometimes practicing his taxidermic skills. His first trip to the central plains took place in 1807. Unfortunately, limitations of time and money prevented him from ever seeing and depicting the birds of the Far West. Audubon's efforts to support himself, his wife, and their two sons (two daughters died in childhood) by means of various business ventures ultimately resulted in bankruptcy. Lucy Audubon became a teacher to help support the family, and Audubon himself made and sold portraits of various persons to bring in needed income. But not all blame for his business reverses can be attributed to Audubon. He and his partners also suffered from commercial interruptions brought about by President Thomas Jefferson's embargo on American exports between 1807-09, the War of 1812, and from periodic worldwide and domestic economic adversities. Brief employment as a taxidermist at a museum in Cincinnati ended when the director admitted he could not pay Audubon's salary, though he praised Audubon's abilities as an artist and taxidermist.

Audubon's chance meeting with several contemporaries may have influenced his decision to become a full-time bird artist. Alexander Wilson and Constantine S. Rafinesque, two active naturalists, had at different times visited Audubon's store in Kentucky. Wilson was peddling his bird books, which Audubon did not buy, while Rafinesque sought information concerning local wildlife. Audubon thought his artistic abilities superior to Wilson's, although he was intrigued by the amount of information his visitor had collected and the literary manner in which he wrote about birds. Rafinesque was a scientist, not an artist, who spread himself too thinly across many disciplines. He frequently published descriptions of new species of plants and animals he had identified, some on tenuous grounds. Many scientists of his time thought this activity somewhat irrational and did not take Rafinesque seriously.

Audubon, who had gradually been accumulating several hundred of his drawings and paintings, began to see that his fascination with birds might be translated into a project that would interest a wider audience, while also bringing in needed income. In 1820 Audubon and his wife decided that he should take the plunge and compile a book about American birds. Several years followed, during which Audubon created many new paintings, and in 1824 he took a portfolio of them to Philadelphia, the nation's center of scientific research. There and in New York, a number of prominent naturalists praised his work and encouraged him to get it published. But Audubon also needlessly antagonized several key individuals, among them friends and supporters of Wilson's who in turn influenced Philadelphia engravers and printers not to help Audubon with his book. Audubon concluded that he had no alternative but to get the project published in England.

With high hopes Audubon scraped funds together for a trip to England in 1826. He soon secured the backing of scientists and socially prominent persons there, many of whom subscribed to his Birds of America. Their advance payments made publication possible, though some subscriptions ultimately lapsed or were not paid for. Audubon had to support himself by painting, often of extraneous subjects. After several false starts, Robert Havell Jr. of London was engaged to engrave the plates, and he worked rapidly and well to complete the project.

While he was in Europe, Audubon received much praise for his work, and he was elected to a number of scientific societies. Several Atlantic crossings followed over thirteen years, as Audubon went back to America to find and paint more of the birds he needed for his project. Some were derived from study skins procured by others; ultimately, 489 birds were depicted in 435 plates.

In 1830 William MacGillivray, a young Scots naturalist, was hired to polish Audubon's

English and help the older man prepare copy in the proper scientific manner for his *Ornithological Biography,* the five-volume text that was to accompany Audubon's plates. Audubon also had to be reminded of the necessity of preparing museum study specimens of his birds as a basis for his bird descriptions. The elephant folio edition of the *Birds of America* was finally completed in 1838. A *Synopsis,* with corrections and additions, appeared in 1839. Audubon and his sons, both competent artists in their own right, then turned their attention to the publication of a smaller, "popular" American edition (7 volumes, 1840-1844), in which Audubon's text was finally combined with his plates. Ultimately, the 435 plates in the elephant folio edition rose to 500, matching the number of birds described. Background details were very much simplified. Though still expensive, nine editions of this smaller edition were published, the last in 1871.

Impact

Baron Georges Cuvier (1769-1832), the eminent French anatomist and paleontologist, termed Audubon's *Birds of America* "the greatest monument ever erected by Man to Nature." Audubon's pictures and text take the reader back to a time when American birdlife was in many ways different. There were many more birds a century and a half ago; some species then numerous have since become extinct.

Though much of the scientific information accompanying Audubon's *Birds* has been revised over the past 150 years, his bird drawings, increasingly in demand, have often been reprinted. Sets of the original engravings have been broken up and the individual prints sold for very large sums. Audubon's reputation as a distinguished early American artist is secure. Many persons enjoy the dramatic spirit and color of his birds and mammals. For others, his work exemplifies their nostalgic feelings about American wildlife in the early years of the Republic. Audubon also enjoys some standing as a pioneering conservationist, though his concern with preserving American wildlife developed relatively late in life.

KEIR B. STERLING

Further Reading

Books
Audubon, John James. *Birds of America.* 7 vols. New York: Dover Publications, 1967.

Audubon, John James, and John Bachman. *The Quadrupeds of North America.* 3 vols. New York: Arno Press, 1974.

De Latte, Carolyn. *Lucy Audubon.* Baton Rouge: Louisiana State University Press, 1982.

Ford, Alice. *John James Audubon.* Norman: University of Oklahoma Press, 1964.

Fries, Waldemar. *The Double Elephant Folio: The Story of Audubon's Birds of America.* Chicago: The American Library Association, 1973.

Herrick, Francis H. *Audubon the Naturalist.* Rev. ed. New York: D. Appleton Century, 1938. Reprint in two volumes, New York: Dover Publications, 1968.

Low, Suzanne. *An Index and Guide to Audubon's "Birds of America."* New York: American Museum of Natural History and Abbeville Press, 1988.

Shuler, Jay. *Had I The Wings: The Friendship of Bachman and Audubon.* Athens: The University of Georgia Press, 1995.

Tyler, Ron. *Audubon's Great National Work: The Royal Octavo Edition of the Birds of America.* Austin: University of Texas Press, 1993.

Periodicals

Allen, Elsa Guerdrum. "History of Ornithology before Audubon." *Transactions* 41 (1951): 387-591.

Energy Metabolism in Animals and Plants

Overview

Metabolism focuses on the sum total of all physical and chemical changes that take place within an organism. Derived from the Greek word *metabole,* meaning "change," it includes all energy and material transformations that occur within a living cell.

Energy is a fundamental feature of life. Processes at the cellular level are the same whether they are in animals, plants, fungi, or bacteria. Living matter is made up of large molecules called proteins, which are assembled from some 20 amino acids.

At the beginning of the nineteenth century, however, people held to traditional beliefs that special biological laws or forces controlled life. This belief, called vitalism, was well entrenched in the romantic thinking prevalent in that period. It is understandable that people who had no knowledge of atoms and molecules would explain life by adopting such ideas to account for the apparent miracle of life.

Scientists began to apply the principles of chemistry and physics to biological functions during the nineteenth century. Friedrich Wöhler (1800-1882) shocked the academic world by making an organic compound out of inorganic materials. This direct challenge attacked the precepts that non-living things are made only from non-living things and living things come from only living things. Henri Dutrochet (1776-1847) studied a variety of living things and argued that processes are similar in all living things. He believed life can be explained in terms of physi-cal and chemical forces. German physiologist Max Rubner (1884-1932) determined that certain foods have energy value. He opened the door for the study of comparative nutrition. Max von Pettenkofer (1818-1907) and Carl von Voit (1831-1908) created measuring devices to study the physiology of metabolism.

As the century progressed, these investigators, along with many others, helped render vitalism obsolete and established the foundations of modern biology.

Background

It is hard for people today to put themselves in the state of mind of scientists at the beginning of the nineteenth century. Modern scientists agree that both living and non-living materials are made of molecules. But in the 1800s, many prominent scientists did not accept molecules as real. Likewise, they believed living and non-living things were made differently. Living things were endowed with a special, or "vital," life force that was completely separate and different from inorganic things. The legacy of alchemists—pseudoscientists who sought to turn lead into gold—also left a magical or mystical aura around chemistry.

John Dalton (1726-1824), Amadeo Avogadro (1776-1856), and others changed how chemicals were made in the laboratory. Dalton had proposed that every element is made of tiny particles called atoms, which can neither be divided nor destroyed. Every atom of each element is identical. Scientists began to argue that living things also were made of the same substances as non-living things. A great debate of "mechanism versus vitalism" ensued. During the nineteenth century the debate played out in experiments that determined the fundamental chemistry and physics of life. It seemed amazing that the real vital force could be found in chemistry, determined by the various ways in which certain atoms are arranged. In fact, it was shown that these arrangements of atoms determine organic molecules such as proteins, amino acids, and nucleic acids.

Metabolism involves the chemical activities essential to life and is a process of building up and breaking down complex substances. These substances are needed by the living organism to grow, repair or replace cells, and give energy. Today, the term "organic" is used to refer to chemistry of carbon compounds, but in the early nineteenth century it referred to all things made of living matter.

Impact

Fredrich Wöhler, a German physician, lived in Heidelburg when he came under the influence of Leopold Gmelin, one of Germany's most prominent physicians. Gmelin recognized Wöhler's talent and encouraged him to go into chemistry. Wöhler went to Stockholm to study with the leading chemist of Europe, Jöns Berzelius (1779-1848). There he absorbed his professor's new techniques and enthusiasm for finding new elements. While working in the municipal technical school in Berlin, Wöhler made two of his major discoveries. In 1828 he accidentally synthesized urea—the chief compound used by the body to excrete nitrogenous wastes. Fascinated by the synthesis, he pursued and found that urea had the same chemicals as ammonium cyanate. The formula for ammonium cyanate is NH_4CNO, while the formula for urea is $CO(NH_2)_2$. The two compounds contain exactly the same number of atoms of the same elements, though their properties are very different. This was the first discovery of an isomer. The significance of the discovery was that he was able to synthesize urea, the product of a living process, from the inorganic compound. This showed inorganic materials could be made into organic substances. This realization stunned the scientific world.

After the death of his wife, Wöhler befriended German chemist Justus von Liebig (1803-1873). The two investigated the theory of radicals—groups of atoms that work together as one atom—and struggled to comprehend the nature of organic compounds. Wöhler has been credited as the father of organic chemistry. With the synthesis of urea, he discredited the notion of the vital force once and for all. Older historians hailed this as a deliberate attempt by Wöhler to smash vitalism; however, most recent historians contend that he was much more interested in urea and its compounds than in philosophical statements.

French physiologist Henri Dutrochet noticed that the chemical and physical processes in plants and animals were very similar. While investigating cellular processes, he discovered and named the process of osmosis. Osmosis is the passage of substances from a place of high concentration to low concentration across a barrier or membrane. He was the first to investigate respiration in plants and found that green pigment in plants used carbon dioxide. He studied light sensitivity and geotropism, the response to gravity. Constructing an osmometer, a device that measures osmotic pressure, he recognized the role of internal plant transport by diffusion across these semipermeable membranes. He was one of the first to reorganize that individual cells are very important to the functioning of the organism.

Dutrochet insisted that physical and chemical forces explain the processes of living things and that all living things are similar. In 1824 he advanced the understanding of the cell principle when he declared that all organic tissues are actually globular cells of exceeding smallness. These precepts are his most important contribution to science.

Justus von Liebig made several exciting discoveries relating to organic chemistry. He organized the compounds into the various groups. He was one of the first to apply chemistry to biology in the field of biochemistry. He became a university professor in 1824 and was very dedicated to chemistry education, setting up a design for laboratory instruction that made possible the great advances in chemistry in Germany in the nineteenth century. He had many outstanding students, such as Max Rubner and Max von Pettenkofer. Liebig's work with isomers of cyanic and fulmeric acids influenced Wöhler's work. In 1838 Liebig shifted his attention to the chemistry of animals and plants, studying their metabolism. He rejected the idea that plants get their food from humus and made great contributions to practical agriculture.

Rubner added to the knowledge of metabolism by discovering that the rate is proportional to the surface of the body. He made a respiratory apparatus into a calorimeter. Placing a dog inside, he measured the heat production of the dog, relating size to diet. The experiment showed that heat in warm-blooded animals is related to the energy supply of nutritive materials. In 1884 he extended the work of Liebig by making quantitative determinations of the energy values of certain foods. We now know these values as calories. His work made possible a scientific explanation for metabolism and a basis for the study of comparative nutrition.

Max von Pettenkofer, a German hygienist and chemist, is recognized as the father of experimental hygiene. He used specially designed chambers for analyzing food and air consumed by an animal, and also the products exhaled and excreted. He added much information to the process of metabolism.

Carl von Voit, a German physiologist, measured metabolism in humans and other mammals and founded modern nutritional science. Voit was a student of both Liebig and Wöhler at the University of Munich. He later served as a professor there and conducted experiments to determine how animals use proteins, fats, and carbohydrates. In 1882 he collaborated with Pettenkofer to build a respiration chamber that could hold human subjects. They tested subjects during various states of activity, rest, and fasting by measuring how much food was taken in and how much waste was excreted. They also measured consumption of oxygen and how much carbon dioxide and heat were given off. For 11 years they measured calorie intake in relation to energy requirements. Their experiments illustrated the laws of conservation of energy in living animals and showed that metabolism is actually carried out in the cells rather than in the blood, as some had proposed.

Together, the work of these scientists made important strides toward revealing the complex processes of molecular activity and metabolism. They laid the foundation for the amazing breakthroughs in cell respiration and molecular biology in the twentieth century.

EVELYN B. KELLY

Further Reading

Brock, William H. *The Norton History of Chemistry.* New York: W. W. Norton, 1993.

Morange, Michael. *A History of Molecular Biology.* Cambridge: Harvard University Press, 1990.

Partington, James R. *A Short History of Chemistry.* Mineola, NY: Dover Publications, 1990.

Advances in Cell Theory

Overview

In the 1830s Matthais Jakob Schleiden (1804-1881) and Theodor Schwann (1810-1882) established the basic principles of modern cell theory. Cell theory allowed scientists to see the cell as the fundamental unit of life. Schleiden and Schwann realized that the entity Robert Brown (1773-1858) had named the "nucleus" was an essential feature of cells in animals as well as plants. As cell theory developed it provided a powerful new framework for understanding the structure of the body, the mechanism of inheritance, the development and differentiation of the embryo, the unity of life from simple to complex organisms, and evolutionary theory. According to cell theory, the body is composed of cells and cell products.

Background

Tissue doctrine, elaborated by Marie François Xavier Bichat (1771-1802), represents an ambitious attempt to analyze the fundamental structural and vital elements of the body. According to Bichat, the human body could be classified into 21 different kinds of tissue. Obviously, Bichat's simple tissues were themselves complex combinations of vessels and fibers rather than the most basic unit of structure and function. Many of Bichat's followers, however, thought of the tissue as the body's ultimate level of resolution even after the microscope revealed a world of previously invisible entities. When microscopic investigations began in the seventeenth century, naturalists reported seeing various corpuscles, vesicles, sacs, and globules. In 1665 the word "cell" appeared in Robert Hooke's book *Micrographia,* along with pictures of little boxes seen under the microscope in freshly cut sections of cork. While the cells in cork appeared to be empty, Hooke (1635-1703) noted that living plants contained cells full of streaming green juices.

By the beginning of the nineteenth century, various theories had been advanced about the "globules" found in biological materials. Some of these globules may have been cells, but others were probably optical illusions and artifacts. Plants clearly contained cell walls enclosing some fluid, but it was difficult to see a basic relationship between the microscopic structure of plants and that of animals. One of the keys to understanding the fundamental similarity of plant and animals cells was the discovery of the nucleus in 1831 by Scottish botanist Robert Brown. Brown is also famous for his studies of the phenomenon now called Brownian motion. Brown described the nucleus of plant cells as a dark circular area that appeared to be more opaque than the cell membrane. The discovery of the nucleus provided a focus for studies of plant cells, but it was more difficult to understand the basic structure of animals. By the end of the 1830s, however, botanists and zoologists had developed the concept of animal and plant life known as the cell theory.

Impact

Credit for the formal statement of cell theory is generally attributed to Schleiden and Schwann. Schleiden, however, noted his interest in the work of the French botanist Charles Brisseau-Mirbel (1776-1854), who believed that cells were found in all parts of the plant. In 1838 Schleiden published "Contributions to Phytogenesis" in Müller's *Archives for Anatomy and Physiology*. Taking Brown's work on the nucleus as a starting point, Schleiden focused his attention on the distribution of this structure. Soon he came to regard the nucleus, which he renamed the cytoblast, as a universal characteristic of all plants. Schleiden believed that all plants were aggregates of cells. Although the cells were actually independent entities, each cell also acted as an integral part of the plant. Thus, all aspects of plant physiology were fundamentally manifestations of the vital activity of cells.

Schleiden described several possible methods of cell formation in *Principles of Botany*. He thought that the hypothesis known as "free-cell formation" was most likely. According to this hypothesis, cell growth was rather like the process by which crystals grew. A fluid called the cytoblastema, rich in sugars and mucus, supported the formation of a nucleolus. As mucus particles aggregated around the nucleus, the cytoblast formed and grew into the young cell. Eventually, a complete cell developed within a rigid cell wall.

Schleiden thought that plants could also grow by the formation of cells within cells so that the contents of a cell were divided into two or more parts. Although the process of cell formation was unclear, Schleiden rejected the possibility that life arose through spontaneous generation.

By clearly expressing the idea that a plant was a community of cells, Schleiden provided the key principle of the cell theory. His description of this hypothesis led Schwann to consider the possibility that animals might also be composed of cells. Animals appeared to be more complicated than plants, even at the microscopic level. Also, animal cells were difficult to analyze under the microscope because they are usually transparent and do not have the cell walls found in plants. Schwann had noticed cell nuclei in preparations of notochord, but until he talked with Schleiden he did not grasp the importance of these observations. After Schleiden described the importance of the nucleus in plant cells, Schwann realized that the cell nucleus was the unifying factor in the most diverse forms animal tissues. In his book *Microscopical Researches into the Accordance in the Structure and Growth of Animals and Plants* (1839) Schwann attempted to prove the basic unity of the plant and animal worlds. Schwann provided evidence that cells are the basis of all animal tissues, and that animal tissues originated from cells which were analogous to the cells of plants. The generation of animal cells seemed to be similar to the processes described by Schleiden in plant cells. Building on the work of Brown and Schleiden, Schwann argued that true cells could be defined in terms of the presence of the nucleus. The whole animal was composed of cells and cell products. In *Microscopical Researches* Schwann summarized his research and explicitly stated the powerful generalization known as cell theory.

In a section called "Theory of the Cells" Schwann attempted to deal with the question of whether cellular phenomena should be considered mechanical or vital. Schwann discussed the problem of chemical changes that took place within the cell and in the surrounding cytoblastema. He called these cellular activities "metabolic phenomena." He coined the word "metabolic" from the Greek to describe "that which is liable to occasion or to suffer change." According to Schwann, metabolism was a universal property of cells and, therefore, of life. He elaborated his theory by analyzing fermentation in yeast as an example of the metabolic activities of cells. Schwann's metabolic theory led to a battle between the German chemists Justus von Liebig (1803-1873) and Friedrich Wöhler (1800-1882) and the great French chemist and microbiologist Louis Pasteur (1822-1895).

While Schwann and Schleiden had provided a powerful new framework for understanding the structure of plants and animals, their theory was different from modern cell theory in several important respects. The major problem was their concept of free-cell formation and the cytoblastema. In subsequent years various botanists, zoologists, and microscopists attacked this concept, but it was Rudolf Virchow (1821-1902) who established the principle that every cell is the product of a pre-existing cell. Virchow was, therefore, the author of cell theory in its modern form. His ideas were summarized in a series of lectures later published under the title *Cellular Pathology* (1858).

LOIS N. MAGNER

Further Reading

De Duve, C. *A Guided Tour of the Living Cell*. 2 vols. New York: Scientific American Library, 1984.

Hughes, A. *A History of Cytology*. New York: Abelard-Schuman, 1959.

Rudnick, D., ed. *Cell, Organism and Milieu. Society for Study of Development and Growth*. New York: Ronald, 1958.

Schleiden, M. J. *Principles of Scientific Botany; or, Botany as an Inductive Science*. Translated by E. Lankester. 1849. Reprint, The Sources of Science, no. 40. New York: Johnson Reprint, 1968.

Schwann, T. *Microscopical Researches into the Accordance in the Structure and Growth of Animals and Plants*. Translated by Henry Smith. 1847. Reprint, New York: Kraus Reprint, 1968.

Virchow, R. *Cellular Pathology*. 1860. Reprint, New York: Dover, 1971.

Wilson, E. B. *The Cell in Development and Heredity*. New York: Macmillan, 1925.

The Agricultural Sciences Flourish and Contribute to the Growing Size, Health, and Wealth of Western Nations

Overview

During the nineteenth century the agricultural sciences flourished in the Western world. When the century began there was not one person who devoted a career to the scientific investigation of agriculture; by the century's end, however, at least two thousand scientists worked in one of the agricultural colleges, experiment stations, and laboratories devoted to the agricultural sciences found around the globe. In the course of one hundred years, scientific discoveries fundamentally reshaped the way farmers selected seed, fertilized fields, fed livestock, controlled pests, and processed foods. In an era of industrialization and urbanization, the application of science to agriculture contributed to the growing size, health, and wealth of many Western nations.

Background

For much of the nineteenth century agricultural chemistry was virtually synonymous with agricultural science. Beginning with the French chemist Antoine Lavoisier (1743-1794), who operated a model farm of his own, scientists began to recognize that chemical compounds and reactions explained many issues related to plant and animal growth. Britain's Board of Agriculture commissioned Sir Humphry Davy (1778-1829) to give lectures on connections between chemistry and agriculture, leading to his influential 1813 text, *Elements of Agricultural Chemistry*. Meanwhile, demographic pressures in the German states, overused soils in the United States, and depressed farm prices in Britain caused agricultural productivity and profitability to stagnate and provided further incentive for agricultural scientific developments.

Rapid improvements in chemical education contributed to the greater reliability of laboratory analyses and increasing popular interest in chemical discoveries. In this context, in 1840 German scientist Justus von Liebig (1803-1873) published the book that is generally considered the central work in nineteenth-century agricultural chemistry. In *Organic Chemistry and Applications to Agriculture and Physiology*, Liebig explained that the exact requirements for plant growth could be measured through laboratory combustion analyses, so that scientists could tell farmers precisely what they needed to maintain a balance of chemical inputs and outputs. More than a theorist, Liebig also vocally promoted and popularized his teachings, insisting that governments recognize chemistry's importance in agricultural and economic development.

The sudden interest in agricultural chemistry induced landowners, farmers' organizations, and various governments to establish educational and research facilities devoted to the agricultural sciences. Jean-Baptiste Boussingault (1802-1877), for example, operated an experimental estate in France after 1834. In 1843 English landowner and fertilizer manufacturer John Bennett Lawes (1814-1900) established a larger, more influential facility at Rothamsted that is still in operation. Meanwhile, German governments devoted public monies to similar institutions, subsidizing nearly 80 state-supported agricultural experiment stations in Germany by the turn of the century. In addition, several German universities established well-funded institutes devoted to agricultural science teaching and research. Similar institutions came more slowly to the United States, but a series of insect plagues, animal disease outbreaks, and chronic problems with fertilizer fraud and adulteration, combined with glowing reports of the German stations, led to the passage of the Hatch Agricultural Experiment Station Act of 1887. Under this bill the federal government provided funds that permitted each state and territory to establish facilities devoted to basic and fundamental agricultural research.

By the end of the nineteenth century the agricultural sciences were unusually well-respected and well-funded and were a high priority in the budgets of most of the industrialized nations of the world. In addition, growing urban populations were able to pressure agricultural scientists to devote attention to issues such as nutrition, public health, and food adulteration that carried this work beyond the farm. Yet the agricultural sciences were also entering a period of transition at the end of the century, as many questioned the centrality of chemistry in the study of agriculture. With the emergence of sub-disciplines such as physiology, nutrition, bacteriology, biochemistry, genetics, mycology, meteorology, mineralogy, applied entomology, and other sciences, the agricultural sciences eventually lost some of their significance and distinctiveness.

Impact

Perhaps the most significant aspect of the agricultural sciences was its status as a symbol of the usefulness of science in general. Early in the nineteenth century agricultural writers trusted the experienced farmer to make sound judgements about applying science to the agricultural enterprise. By the middle of the century, however, advances in the agricultural sciences shifted the notion of expertise to the trained scientist who had access to the specialized knowledge of the disciplines. To demonstrate their value to the rural economy, agricultural chemists conducted soil analyses and investigations of fertilizer fraud that also revealed the legitimacy of the laboratory as the locus for scientific work. Through organizations such as the Association of Official Agricultural Chemists in the United States and similar groups overseas, chemists developed coordinated and standardized methods that increased their usefulness. Soon, farmers' groups pressured scientists to expand their role in regulating the agricultural marketplace through analyses of seed quality, the values of animal feeds, and butterfat content of dairy products.

The emergence of the agricultural sciences also represented a democratization of the scientific enterprise. Publicly funded research and inexpensive publications displaced the elitist agricultural groups that controlled knowledge at the beginning of the century. Most scientists who entered the field gained a reputation as objective and knowledgeable public servants, and they thereby contributed to the rising status of professional scientists overall. Ample job opportunities also meant that the field was more open to the sons of farm and middle-class families than other branches of science and academia.

Though scientists debated, often heatedly, which chemical elements and compounds were most vital to plant nutrition, their enthusiasm for chemical solutions to agricultural problems helped to create the artificial fertilizer industry. Scientists and bureaucrats did not simply investigate fertilizers, they promoted them, linking them with both national prosperity and personal entrepreneurial opportunities. The ensuing search for bones, rock phosphates, guano, and potash salts, all of which had chemical constituents useful as fertilizers, became central issues in international political and economic history. By 1860 Britain consumed 50% of the world's production of fertilizers, reflecting the empire's international economic clout as much as any other survey of imperialism. Other nations also embraced fertilizer chemistry; in Japan, for example, the Meiji government recognized that expansion of fertilizer chemistry and food production was essential before further industrialization could occur. In general, the fertilizer industry predated the growth of other branches of the chemical industry.

Agricultural researchers were also on the forefront of developments in fields that had impact far beyond the agricultural sector. German botanist Julius von Sachs (1832-1897), for example, developed research methodologies while at the isolated agricultural experiment station in Tharandt, Saxony, that shaped the discipline of plant physiology. Wilbur O. Atwater (1844-1907), Oskar Kellner (1851-1911), and other founders of nutrition science began their work with agricultural subjects and produced detailed analyses of the metabolism of fats, proteins, and carbohydrates. Serious efforts to control food adulteration originated at Wilhelm König's agricultural experiment station in Mönster. Research on the role of soil bacteria in the nitrogen cycle sparked further research in symbiosis, ecology, and nitrification, and the subsequent discovery of a number of useful antibiotics. American plant breeder Luther Burbank (1849-1926), working without the aid of the genetics theory worked out by Gregor Mendel (1822-1884), created scores of useful plant varieties and hybrids that reshaped commercial fruit and vegetable industries. Developments in dairy science were also notable, as control of milk-borne diseases permitted the commercialization of dairy farming and brought important changes to Western foodways.

The agricultural sciences contributed to significant boosts in farm productivity, but not everyone benefited from these developments. Virtually everything that scientists recommended added to the costs of farming. Although relatively few farmers could afford fertilizers, pesticides, commercial seeds and feeds, and other capital expenditures, those who did faced fewer

risks of crop failure. Declining risks permitted greater specialization in fewer and fewer profitable farm commodities, thus reducing the value of tending a variety of crops and animals. A decline in genetic diversity among the remaining varieties of commercial plants and animals has been another consequence. In brief, the advancements in the agricultural sciences helped quicken the pace of farm consolidation and rural depopulation in the nineteenth century. These developments also dramatically affected the colonial nations of the non-western world. European imperialists invested heavily in agricultural research in their colonies, where the commercial development of cash crops came at the expense of the food crops that were the basis of traditional agricultural practices.

MARK R. FINLAY

Further Reading

Browne, Charles A. *A Source Book of Agricultural Chemistry.* Waltham, MA: Chronica Botanica, 1944.

Knoblauch, Frieda. *The Culture of Wilderness: Agriculture as Colonization of the American West.* Chapel Hill: University of North Carolina Press, 1996.

Marcus, Alan I. *Agricultural Science and the Quest for Legitimacy: Farmers, Agricultural Colleges, and Experiment Stations, 1870-1890.* Ames: Iowa State University Press, 1985.

Rosenberg, Charles E. *No Other Gods: On Science and American Social Thought.* Baltimore, MD: Johns Hopkins University Press, 1976.

Rossiter, Margaret W. *The Emergence of Agricultural Science: Justus von Liebig and the Americans, 1840-1880.* New Haven, CT: Yale University Press, 1975.

True, Alfred C. *A History of Agricultural Experimentation and Research in the United States, 1607-1925.* USDA Miscellaneous Publication 251. Washington: GPO, 1937.

Wines, Richard A. *Fertilizer in America: From Waste Recycling to Resource Exploitation.* Philadelphia: Temple University Press, 1985.

Cell Division and Mitosis

Overview

Cell theory, the cornerstone of biology, states in modern form that (1) living matter is composed of cells; (2) chemical reaction takes place within cells; (3) cells arise from other cells; and (4) cells have information that is passed from parent to daughter cells. In the nineteenth century three

scientists advanced the knowledge of cell division and mitosis. German botanist Hugo von Mohl (1805-1872), using improved microscopic techniques, studied cell division in algae and determined that cells divide. German biologist Walther Flemming (1843-1905) applied aniline dyes to observe the processes of cell division in

An early illustration of cell division by Hugo von Mohl (1839).

the nucleus and how chromosomes behave. Belgian zoologist Edouard van Beneden (1846-1910) first described the process of fertilization and stages of embryonic division.

Background

At the beginning of the nineteenth century, scientists knew nothing about the existence of cells. When they looked at a whole cat or bird or human being, they thought they were looking at the basic unit. Some proposed that body parts were made of strings and tubes that grew like crystals and enlarged until the person attained adult proportions. The internal organs were pots or pipes of different shapes.

The general orderliness of living things had led to a system of thought called "vitalism."

Because organs were basically structured in the same way, vitalists contended that the breath of life that God gave Adam was the "vital force of life." Absence of the vital force was death—when the force or spirit or ghost is given up. It was an ancient system of thinking, but eighteenth-century German philosopher George Stahl (1660-1734) put a stranglehold on biology for the next two centuries by applying vitalism to life sciences. His related idea, called "animism," dominated early cell biology by insisting that all body parts came from the soul. Before there was knowledge about cells, vitalism was mainstream science. Even in the early days of scientific method, it was not unusual to mix natural investigation with supernatural explanation.

Another popular precept was the idea of spontaneous generation—that living things could develop from nonliving things. Even though Francesco Redi (1626-1697) had performed his famous experiment disproving spontaneous generation in the seventeenth century, the controversy still raged in 1860. In fact, the conflict was so spirited that the Paris Academy of Science conducted a contest inviting members to shed new light on this subject. It was Louis Pasteur (1822-1895) who, in 1864, devised an experiment to show that microorganisms did not appear spontaneously. However, many stubbornly clung to the old ideas until the last few decades of the nineteenth century.

Other events had taken place to set the stage for the discovery of cell division. During the seventeenth century, Robert Hooke (1635-1703) observed cork and called the blocks that he saw cells because they reminded him of jail cells. In 1673 Anton van Leeuwenhoek (1632-1723) ground lenses to look at a small world of "cavorting beasties," but cell studies were hampered by the poor quality of magnifiers and by beliefs in spontaneous generation and other ideas of biological structure.

Another philosophical turn was developing at the end of the eighteenth and early nineteenth centuries. The rise of natural philosophy led to the proposal that a "chain of being" connects all living organisms as well as inanimate objects. This chain, it was thought, started with minerals and progressed from sponges to plants, to birds, to mammals, and, finally, to man. For these philosophers, all species were interrelated and the ideal was to return to nature and the natural man. The German poet Johann Wolfgang von Goethe (1749-1832) and philosopher Friedrich Schelling (1775-1854) were leading proponents of this movement, which has been dubbed Romantic philosophy. However, these ideas of interrelations and connectedness with nature spawned interest in zoology and botany as well as comparative development, including embryology.

Impact

During the nineteenth century, the development of the microscope became critical in determining mitosis and cell division. Lenses were improved so that higher powers could be seen with accuracy. Illumination made it possible to see intricate details.

In 1839 two Germans—Matthias Schleiden (1804-1881) and Theodor Schwann (1810-1882)—advanced a radical theory that proposed that plants and animals were made of cells. They showed all cells to have a nucleus, cell body, and cell membrane.

Mohl made his most important contributions in the field of microscopic botany. Using his knowledge of optics, he wrote a manual on the use of the microscope in 1846. His great claim to fame was the development of microscopic study of plant cells. In the 1850s German pathologist Rudolph Virchow (1821-1902) suggested that all cells are formed from existing cells by cell division. Drawing upon this insight, in 1853 Mohl published in a classical reference of the time a description of the cell as the elementary organ. Looking closely at the cell under his developed microscope, he was able to show how aligned cells fused and how cells moved within. He saw how the cell generated by dividing and understood how the cell obtained nourishment, moved, and reproduced. His meticulous studies led to a description of cell structures of the membrane, protoplasm, nucleus, and cellular structure. He was the first to delve into cytochemistry and named the intracellular contents "protoplasm."

Mohl was the first to determine that new plant cells are formed by cell division. Working with the alga Conferva glomerata, he noted how cells around the edges of the colonies began to divide into two. Although it was unique for the day, his view of division and the fibrous walls of plants cells is well confirmed today.

Flemming became a professor at the University of Kiel, Germany, where he carried out his great work on cell division. While the knowledge that cells divide into two daughter cells was becoming accepted, exactly how the division occurred was still a mystery. Several years after Mohl proposed the idea of algal cell division, Flemming used the oil immersion lens to study cells and stains to reveal their structure. He had heard of the aniline dyes developed by Paul Ehrlich (1854-1915) and used by Robert Koch (1843-1910), and he used a special class of these dyes to reveal small, thread-like structures in the nucleus. The thread-like structures would later be dubbed "chromosomes," meaning "colored bodies."

Flemming found he could apply these aniline stains to cells killed at different stages of dividing. He was able to make slides that revealed a clear and established sequence of what occurs during cell division. According to his observation, the threads shortened, then split lengthwise in two. Each of the halves went to a separate end of the cell. The cell then divided into two daughter cells. Flemming found nine stages in the processes of cell division and gave names to each stage. He also coined the term "mitosis" for the process of cell division.

Flemming insisted correctly that animal and plant cells have the same process of division. Even with the high-power electron microscopes of modern science, Flemming's descriptions of the phases are very current today, with only a few exceptions. The nine phases, which correspond to the drawings and plates produced by Flemming, are now known as interphase (when the cell is not dividing), early prophase, mid-prophase, early metaphase, metaphase, early anaphase, late anaphase, early telophase, and late telophase.

Flemming then focused on nuclear division in the testes and concluded that sperm are formed from cells that have divided twice. In 1887 he saw the sperm precursor cells—spermatocytes—divide and then divide again. These are now known as meiotic divisions. He also traveled to Naples to study cell division in the egg cells of echinoderms (starfish). He concluded that the egg also divided twice. At the end of the process of meiosis, the egg and sperm have only half of the chromosome information of the original cell.

The idea of sperm and egg as the mechanism of reproduction was not well developed at the beginning of the nineteenth century. Flemming had spent some time studying the formation of egg and sperm. It was Beneden, though, who in 1883 published a work on fertilization. Studying the egg of *Ascaris megalocephala,* a nematode or roundworm, he described the essential process of fertilization. One-half nuclei from the female joins with one-half from the male and forms pronuclei—which are now called chromosomes. In drawings and plates he accurately represented the chromosomes. He described how the nuclear membrane of the egg and sperm break down. The sets of chromosomes then line up at the equator. The variety of *Ascaris* Beneden studied had only two chromosomes, with each parent contributing one to the zygote, or fertilized egg.

In 1887 Beneden published a paper describing the centrosome as a permanent cell organ, which was in the cell during resting period and divided into two parts before the beginning of the next mitosis.

The work of these three scientists—Mohl, Flemming, and Beneden—had advanced the concepts of mitosis, meiosis, and fertilization. They moved cellular investigation away from spontaneous generation, romantic natural philosophy, and vitalism to an era of scientific method. By the end of the nineteenth century it was generally accepted that cells divide in phases, sperm and egg are formed in a special division process, and the union of sperm and egg make a new cell. By 1900 cell division or mitosis had become an integral part of cell theory, presenting a coherent view of how an organism—both simple and complex—arises and reproduces.

EVELYN B. KELLY

Further Reading

Harris, Henry. *The Birth of the Cell.* New Haven: Yale University Press, 1999.

Hughes, A. *A History of Cytology.* London: Abelard Schuman, 1959.

Rensberger, Boyce. *Life Itself: Exploring the Realm of the Living Cell.* Oxford: Oxford University Press, 1999.

Evolution, Natural and Sexual Selection, and Their Influences on the Sciences

Overview

Although evolution was not an altogether new topic, Charles Robert Darwin (1809-1882) brought it into the limelight in 1859 with the publication of his *On the Origin of Species by Means of Natural Selection.* His hypothesis that current species evolved from past species by the flexible and non-rigid pathway of natural selection brought him enormous praise, but also much criticism. Since then, scientists have learned about genetics and the role that genes and mutations play in natural selection, and have continued to provide confirmation of Darwin's ideas. Despite the considerable scientific evidence, evolution is still under fire from some present-day critics, who believe the theory flies in the face of religious beliefs.

Background

Many scientists had a hand in developing what is now termed the theory of evolution. Evolution is essentially the transformation of organisms over time, where new species evolve from previously existing species. The scientifically accepted method of evolution, which was proposed simultaneously and independently by Darwin and fellow scientist Alfred Russel Wallace (1823-1913), is natural selection. Natural selection is the process that occurs when a heritable trait within a species has a favorable enough effect that those individuals with the trait have a better chance of surviving over time. In other words, those individuals with the trait are more likely to reproduce successfully, and they therefore pass the trait down through the generations. Eventually the species carries the trait as part of its general makeup.

Unlike many previous hypotheses, including the notion proposed by Jean Baptiste de Lamarck (1744-1929) that animals could control the course of their evolution, Darwin believed an animal's "will" had nothing to do with natural selection.

According to the theory of evolution by natural selection, new species can arise under various conditions. For example, a geographical barrier, such as the ocean separating two islands, may split a bird species into two separate populations. On separate islands, the birds may find different foods. Perhaps one island has an abundant population of insects that live in the bark of trees. The birds with the strongest, most chisel-like beaks would have the greatest opportunity to find food by breaking through the bark and reaching the insects. They would also be more likely to survive to reproductive age. If that trait is heritable, descendants would become more likely to have that bill. If a strong bill continues to be beneficial, the birds generation-by-generation would continue to develop a more and more chisel-like bill. Likewise, birds on the second island might benefit from a bill more conducive to sipping nectar, and the population might eventually carry that trait. Eventually the two populations may become so different that they don't recognize individuals from the other population as potential mating partners. At that point, the two become different species by definition.

Sexual selection can also give rise to new species by reason of aesthetic preferences: individuals prefer mates with one trait over another. Perhaps females of a bird species prefer males with longer feathers. Those with the longer feathers mate more often, even to the exclusion of males with shorter feathers. Eventually the species will contain males with longer and longer feathers. Aesthetics, then, can also drive evolution.

Many other conditions and a wide variety of traits can occur to split one species into two.

Darwin developed his detailed hypothesis of evolution and natural selection over many years, beginning with his voyage on the HMS *Beagle*. The five-year trip (1831-36) was a surveying mission, and Darwin served as its naturalist. On the voyage, Darwin spent day after day observing, collecting specimens, and writing elaborately detailed notes on the flora and fauna. Drawing on his interest in geology, he also spent considerable time describing fossils and geological formations.

Once he returned, Darwin spent the next several decades poring over his notes, studying the sciences, particularly biology, and developing his ideas on "descent with modification." In 1859, when his manuscript for *Origin of Species* was well under way, Wallace sent Darwin an essay describing ideas very similar to those Darwin was also writing about. The two presented a joint paper, which received little notice, and the following year Darwin published his best-known work. The full title was *On the Origin of Species by Means of Natural Selection, or the Preservation of Favoured Races in the Struggle for Life.*

Impact

The publication of *Origin of Species* caused an uproar. Religious leaders, newspaper editorial writers, and many members of the general public exclaimed that the hypothesis suggesting that man is but an evolutionary end-product of previous organisms was preposterous and even blasphemous. In fact, Darwin had sidestepped the issue of the evolution of man in *Origin of Species,* but more than a decade later acknowledged man's place in evolution with *The Descent of Man, and Selection in Relation to Sex.*

The religious battle against evolution continued to roil in the United States and came to a head with the 1925 Scopes trial, when a Tennessee high school teacher was convicted for teaching evolution in his class. The U.S. Supreme Court overturned the decision in 1968, finally removing any legal barriers to teaching what had become a fundamental scientific view and the basis for biological sciences. About a decade later, other states tried again to temper the theory of evolution by requiring educators to teach creationism along with evolution. The Supreme Court stepped in again in 1987 to deny those stipulations.

As late as 1999, creationists in the United States had again succeeded in turning back the clock by taking another tack. This time, the Kansas Board of Education forbade teachers from testing students on the subject of evolution. Scientists immediately decried the action, with some even calling for colleges and universities to deny admission to students from Kansas.

In response to the board's decision, acclaimed biologist and author Stephen Jay Gould wrote in *Time:* "No scientific theory, including evolution, can pose any threat to religion—for these two great tools of human understanding operate in complementary (not contrary) fashion in their totally separate realms: science as an inquiry about the factual state of the natural world, religion as a search for spiritual meaning and ethical values." He added, "The major argument advanced by the school board—that large-scale evolution must be dubious because the process has not been directly observed—smacks of absurdity and only reveals ignorance about the nature of science."

On the scientific front, Darwinian evolution took on a revolutionary air in the late nineteenth century. Collections of fossils were now views into the family history of life on Earth. The relationship of animals and plants gained an added depth with the understanding that one may have given rise to another, or that two separate species had a common ancestor. Scientists had yet to understand the role of genes in heredity, and a new scientific field began to form as questions arose about the mechanisms behind natural selection.

When Gregor Mendel's experiments on heredity in pea plants was rediscovered at the turn of the twentieth century, and the field of genetics slowly began to develop, evolution had already become a mainstay in the biological sciences. Genes became the previously unknown gears that turned the wheels of natural selection and that, in turn, drove evolution. Late in the twentieth century, the field of developmental biology became entwined with genetics and evolution when scientists found that essentially the same genes control specific stages of development. These control genes not only occur in different animal species but even cross the barrier between the plant and animal kingdoms.

Studies also began on the impact of evolution on animal behavior. Scientists used the ideas put forth by the theory of evolution to begin to understand kin selection, in which an individual in an animal population looks out for a relative, even at that individual's own expense. For example, some animals take on the role of look-out for a group and will issue a predator-warning call even though that call will likely draw the attention of the predator to the calling individual. The idea that kin selection would ultimately favor the related individuals' ability to reproduce and thus the continuation of their genetic composition, had its essence in evolution and natural selection.

Sexual selection provided explanations particularly about aesthetic changes that evolved in animals. For example, the presence of large, cumbersome antlers on male white-tailed deer or the persistence of colorful plumage on some male birds, even when such plumage makes them noticeable to predators. Sexual selection explains that if a female preferentially mates with males carrying the heritable trait, that trait will continue and possibly even become enhanced in future generations. Sexual selection can work both ways. Certain male frogs, for example, appear to prefer larger females.

Scientists have also raised questions regarding the potential battles between sexual and natural selection. For example, when does the benefit of larger and larger antlers become a big enough detriment to the health and survival of the male deer that the latter outweighs the former? In other words, under what conditions does sexual selection bow to natural selection?

This research has spurred studies in other areas as well. Work on the evolution of animal behavior became a paradigm for understanding human behavior. During the twentieth century, scientists in many fields began to consider the influence of evolution on many aspects of human behavior. Evolution has become ingrained in psychology, the social sciences, and even political science. Studies are not only considering how human behavior has evolved to its current state, but how humans are continuing to adapt to changing social and political conditions.

In summary, the theory of evolution by natural selection has not only changed the way biological scientists view the world, but has become a pillar of scientific research in general.

LESLIE A. MERTZ

Further Reading

Books

Bowler, Peter J. *Evolution: The History of an Idea.* Rev. ed. Berkeley: University of California Press, 1989.

Bruno, Leonard C. *The Tradition of Science: Landmarks of Western Science in the Collections of the Library of Congress.* Washington DC: Library of Congress, 1987.

Byers, Paula K. *Encyclopedia of World Biography.* 2nd ed. Detroit: Gale Research, 1998.

Darwin, Charles R. *On the Origin of Species by Means of Natural Selection, or the Preservation of Favoured Races in the Struggle for Life.* London: John Murray, 1859.

Darwin, Charles R. *The Descent of Man and Selection in Relation to Sex.* London: John Murray, 1871.

Gould, Stephen J. *Wonderful Life: The Burgess Shale and the Nature of History.* New York: W. W. Norton and Company, 1989.

Greene, John C. *The Death of Adam: Evolution and Its Impact on Western Thought.* Ames, Iowa: Iowa State University Press, 1959.

Periodicals

Gould, Stephen J. "Dorothy, It's Really Oz: A Pro-creationist Decision in Kansas is More Than a Blow Against Darwin," *Time,* 154 (23 August 1999): 59.

Social Darwinism Emerges and Is Used to Justify Imperialism, Racism, and Conservative Economic and Social Policies

Overview

Social Darwinism was a sociological theory popular in late nineteenth-century Europe and the United States. It merged Charles Darwin's theory of natural selection and Herbert Spencer's sociological theories to justify imperialism, racism, and *laissez-faire* (i.e. conservative) social and economic policies. Social Darwinists argued that individuals and groups, just like plants and animals, competed with one another for success in life. They used this assertion to justify the status quo by claiming that the individuals or groups of individuals at the top of social, economic, or political hierarchies belonged there, as they had competed against others and had proven themselves best adapted. Any social or political intervention that weakened the existing hierarchy, they argued, would undermine the natural order.

Background

Darwin's theory of natural selection and the subsequent arguments by social Darwinists were based heavily on the work of Thomas Malthus (1766-1834), an early nineteenth-century British clergyman who wrote *Principles of Population.* Malthus predicted that food resources increased arithmetically while human populations, unchecked by war, disease, or famine, increased geometrically. The disparity between resources and population meant a constant struggle among members of a given population for scarce resources. Darwin (1809-1882) applied the Malthusian principle to the natural world and posited his theory of natural selection. In *Origin of Species* (1859) he argued that the scarcity of natural resources led to competition among individuals, which he called "the struggle for survival." Through this competition, the best-adapted members of a given population were most likely to be successful, reproduce, and pass their beneficial adaptations on to their offspring. Poorly adapted members, he asserted, probably would not survive and therefore would not pass their lower quality traits to the next generation.

Social Darwinists argued on the basis of Darwin's theory of natural selection that the best adapted humans naturally rose to the top of social, political, and economic strata. Therefore, they argued, those members at the top of society, either by virtue of hard work or birth, were the best-adapted citizens. They used this rationale to argue against welfare policies that would help the poor by redistributing resources from the most fit members to the least fit, which they claimed would violate the natural order and allow the perpetuation of less fit members. Darwin himself did not promote social Darwinism and probably would have opposed many of the claims of social Darwinists.

Social Darwinism was the product of late nineteenth-century economic and political expansion. As the European and American upper class sought to extend its economic and political power, it employed scientific explanations to justify the increasingly obvious gap between rich and poor. The social Darwinists' reliance on natural laws allowed social, political, and scientific leaders to dismiss those who sought to redistribute wealth and power by claiming that reformers were violating the natural hierarchy. By extending their arguments to address entire nations, some social Darwinists justified imperialism on the basis that the imperial powers were naturally superior and their control over other nations was in the best interest of human evolution. The increasing public interest and respect for the sciences also contributed to the success of social Darwinism, as policies that had the stamp of scientific legitimacy were accepted as above political interest or influence.

Impact

While Darwin coined the term "struggle for survival," it was Herbert Spencer (1820-1903) who invented and popularized the concept of "survival of the fittest," and Spencer is widely considered the chief proponent of social Darwinism. Spencer's synthesis of evolutionary thought with sociology, psychology, and philosophy provided the stamp of scientific justification to the social and political leaders who sought to preserve the status quo and promote unrestrained competition. Originally trained as an engineer, Spencer developed an increasing interest in the natural and social sciences and proposed theories that linked them under the umbrella of evolution. He believed that biological evolution had brought about human intellect, which in turn produced society. Therefore, he argued, human intellect and social activities were products of biological evolution, and all three operated on natural laws. His work was a clear reflection of English industrialism, which was dedicated to promoting competition, exploitation, and struggle in the human social realm. He asserted that all aspects of life, be it human, plant, or animal life, were guided by the constant struggle in which the weak were subjugated by the strong. However undesirable humanitarians might find this process, he argued, it was the natural order of things and could not be altered by charity, welfare policies, or legislative actions.

Spencer was well known in Europe, but he was especially popular in the United States because his work provided Americans with a scientific justification for free competition, which was widely recognized as the most effec-

tive path to economic progress. Between the 1860s and 1900, Americans purchased more than 350,000 copies of Spencer's books, and his influence on late-nineteenth century figures such as Henry James, John Dewey, and Josiah Royce was significant. Andrew Carnegie (1835-1919) was one of America's most prominent followers of Spencer and popularized social Darwinism in America. He called Spencer his "teacher" and visited him in England. In 1882 Spencer arrived in the United States for a widely publicized tour that brought together American writers, scientists, politicians, theologians, and businessmen around the doctrines of social Darwinism.

William Graham Sumner (1840-1910), a Yale sociologist, was another prominent American social Darwinist. In *What the Social Classes Owe to Each Other,* Sumner argued against governmental and private charity attempts to improve the conditions of the lower classes. Like Spencer, Sumner believed that society evolved and operated in a deterministic fashion and any attempt to alter social hierarchies was doomed to failure. Using his authority as a scientist, Sumner legitimated aggressive competitive practices of American businessmen by declaring their activities to be the source of human evolutionary progress.

The best known American opponent of social Darwinism was Lester Ward (1841-1913), a paleontologist and one of the founders of sociology in America. Ward argued against the social Darwinists' natural justifications for the status quo and posited the theory of telesis, or planned social evolution. While social Darwinists focused on the role of competition in the natural and social worlds, Ward highlighted the importance of cooperation and marshaled historical evidence against Sumner to argue that human progress was the product of cooperative activities and intelligence, not merciless competition. He used this analysis to urge social and political leaders to adopt measures deliberately aimed at social improvement.

By the turn of the century social Darwinists were attacked and their credibility undermined by reform Darwinists, who used the same scientific theories about the natural world to uphold opposite conclusions about society. Reform Darwinists asserted that the scientific knowledge of evolution allowed social and political leaders to intervene in the natural order to better the human condition. Using Darwin's theory of natural selection and Gregor Mendel's recently rediscovered theories of inheritance, reform Darwinists argued that humans could control their own evolutionary destiny by adopting interventionist policies such as public sanitation and eugenics.

Individuals who have been labeled "social Darwinists" did not use the term to describe themselves. Reform Darwinists and other critics of *laissez-faire* economic policies invented the label in the early twentieth century as a derogatory term to describe their opponents' position. In doing so, they highlighted the influence of Darwin's theory of natural selection on social and political activities by emphasizing the social Darwinists' use of Darwin's work. More recent historians have emphasized the social influences that went into Darwin's theories, such as the nineteenth-century British tendency to emphasize competition and overlook cooperation and altruism in the natural world. Taken together, the work of early and late twentieth-century scholars illustrates the reciprocal influence between science and society, as social concerns affected the development of evolutionary theory and then that evolutionary theory influenced later social developments.

MARK A. LARGENT

Further Reading

Books

Bannister, Robert. *Social Darwinism: Science and Myth in Anglo-American Social Thought.* Philadelphia: Temple University Press, 1979.

Degler, Carl. *In Search of Human Nature: The Decline and Revival of Darwinism in American Social Thought.* New York: Oxford University Press, 1991.

Hofstadter, Richard. *Social Darwinism in American Thought, 1860-1915.* Philadelphia: University of Pennsylvania Press, 1945.

Spencer, Herbert. *An Autobiography.* New York: D. Appleton and Company, 1904.

Spencer, Herbert. *Principles of Sociology.* New York: D. Appleton and Company, 1880.

Sumner, William Graham. *What Social Classes Owe Each Other.* New York: Harper & Brothers, 1883.

Ward, Lester Frank. *Dynamic Sociology; or, Applied Social Science as Based upon Statistical Sociology and the Less Complex Sciences.* New York: D. Appleton and Company, 1883.

Young, Robert. "Herbert Spencer: and 'Inevitable' Progress." In *Victorian Values: Personalities and Perspectives in Nineteenth-Century Society,* edited by Gordon Marsden. London: Longman, 1990.

Periodicals

Young, Robert. "Charles Darwin: Man and Metaphor."*Science as Culture* 5 (1989): 71-86.

Louis Pasteur's Battle with Microbes and the Founding of Microbiology

Overview

In 1800 the origins of infectious diseases were unknown and as mysterious as they had been in the Middle Ages. By 1900 the causes of many of these diseases that ravaged humans throughout history had been discovered, and methods such as vaccination and improved sanitation were being developed to deal with them. The main figure in this achievement was Louis Pasteur (1822-1895), a French scientist who first demonstrated the crucial role microbes (microscopic organisms) play in the life process. He established the germ theory of disease and was the first to show that vaccines against infectious diseases can be manufactured. Vaccinations against viral diseases, antibiotics, infection-free surgery, safe milk and food, effective sanitation systems: all these developments owe much to Pasteur, the founder of modern microbiology (the study of unicellular microorganisms and their activities).

Background

In the early nineteenth century science and medicine seemed no farther advanced than they had been half a millennium earlier. Aristotle's concept that many plants and small animals were spontaneously generated from soil or decaying animal matter still influenced scientists. In medicine many still believed that human health was determined by internal "humors" that could be treated by drawing off the patient's blood. Blood-letting was so widespread that in 1833, 41 million leeches for this purpose were imported into France alone. Most importantly, although the existence of microbes had been known for two hundred years, very few scientists suspected, and no one had proven, their role in causing diseases.

As a result, infectious diseases such as tuberculosis, cholera, post-operative infection, typhoid, diphtheria, syphilis, puerperal (childbirth) fever, and malaria killed tens of thousands of people in Europe and America annually; millions more died worldwide from these and other infectious diseases such as yellow fever and plague. Cholera, for example, killed 145,000 people in France in 1854. Microbes also had a devastating economic impact by causing diseases that killed animals. Infectious diseases periodically struck the poultry, sheep, silkworm, and swine industries. In addition, problems caused by microbes in the vinegar, wine, and beer industries often spoiled production.

Although many nineteenth-century scientists struggled with these diseases, Pasteur is most often considered the founder of microbiology, in part for his remarkable success in dealing with many of these problems. Also, in science such credit often goes to the individual who convinces the world. Pasteur was very ambitious and vigorously sought recognition for his work, missing few opportunities to make it known to as wide an audience as possible. One of his assistants said Pasteur "neglected nothing . ... [in trying] to attract attention." It was these public relations activities that convinced the world of the importance of microbes.

Pasteur also benefited from the support of the vast hygiene movement that developed in Europe to combat diseases such as cholera and typhoid. Not knowing the causes of these diseases, the hygienists used scientifically worthless terminology such as "contagion," "miasma," and "morbid spontaneity." Since Pasteur's germ theory convincingly explained everything they could not, the hygienists threw the support of their powerful movement behind him, a major reason for Pasteur's deification in French and European public opinion.

Impact

In 1855 Pasteur was teaching chemistry in Lille, an industrial city in northern France. Microbiology got its start when he became interested in fermentation problems at a local distillery where sugar beets were fermented to alcohol through the addition of yeast. This led to his first encounter with microbes. Anton van Leeuwenhoek (1632-1723) had discovered microorganisms when, using microscopes, he found that

Louis Pasteur at work in his Paris Laboratory. *(The Library of Congress. Reproduced by permission.)*

uncountable billions of minute "animalcules" existed everywhere. But until Pasteur microorganisms had been studied only to establish their morphology (form and structure) and taxonomy (relationship to other microbes). Few nineteenth-century scientists suspected that microbes could cause disease; the idea that invisible organisms could kill humans and cattle seemed preposterous. Improved microscopes, Pasteur's skill in using them, and his intuitive genius would change that view. He would need these tools; microorganisms are extraordinarily diverse (some need air to live, others die in it; some thrive in cold environments, others succumb when deprived of heat). The fact that many microbes, particularly viruses, were too small to be seen with existing microscopes created further difficulties.

Pasteur's work on sugar beet ferment led to his announcement in 1857 that, contrary to the opinion of leading chemists, the yeast used in the process was a living microbe and necessary for successful fermentation. He found that the vats where the process failed contained no yeast, but instead had smaller microbes that killed the yeast. He concluded that microbes were living, and sometimes harmful, organisms. He also discovered the microbes that caused milk to turn sour and expanded his research into fermentation problems in the wine industry (and later into the brewing industry). He again discovered that microbes existed that could spoil the fermentation process. He found that if wine was heated for a brief period, the harmful microbes would be destroyed but the wine remained sound. Because of this, the heating of milk, wine, juices, and other liquids and food products to prevent the growth of harmful microorganisms is called "pasteurization."

Pasteur's work on fermentation brought him into a debate over spontaneous generation. While no scientist still believed that animals such as flies and mice spontaneously developed out of decaying matter, many assumed that microbes did spontaneously generate; in 1859 Félix Pouchet published a paper "proving" it. Pasteur disagreed. One of his talents was his ability to design clever experiments validating his contentions. In the controversy that followed, Pasteur's experimental genius led him to devise several methods to show that the microbes were carried in the air and did not arise spontaneously on nutrient-rich surfaces. In the most famous, he made flasks with long, narrow, horizontally curving "gooseneck" openings. Nutrient solutions were heated in the flasks to purify them: then Pasteur let unfiltered air enter through the gooseneck openings. Microbes settled in the curve of the goosenecks, but none appeared in the solutions, proving that air carried microbes but did not cause them to develop spontaneously.

Pasteur, now living permanently in Paris, announced in 1862 that he was going to apply his germ theory of fermentation to the study of disease. The few other scientists who also believed infectious diseases were caused by microorganisms could not prove it with conclusive experimental evidence. Pasteur's first chance to do so came in 1865, when the government asked him to investigate why large numbers of silkworms were dying. After several years of experiments, he concluded that the silkworms were suffering from diseases caused by two different microbes. By 1866 he had developed a method allowing growers to isolate the healthy from the diseased worms, thus saving the industry.

Evidence was accumulating that Pasteur's germ theory was correct. A cholera epidemic (1854) in London let John Snow (1813-1858) demonstrate that proper sanitation was essential in fighting contagious diseases. In the 1860s Joseph Lister (1827-1912) developed antiseptic methods that killed bacteria during and after operations, greatly reducing deaths in surgery. In 1874 Lister credited Pasteur's germ theory with furnishing him with the "principle upon which alone the antiseptic system can be carried out." In Hungary Ignaz Semmelweis (1818-1865) announced in 1861 that he had greatly reduced deaths of women in labor through the use of antiseptics. Pasteur infuriated French physicians by constantly lecturing them that microbes on their unclean clothing, hands, and unsterile instruments were killing women in childbirth. By the 1880s the medical profession was won over to the importance of inhibiting the growth of microorganisms with antiseptics.

Meanwhile, a German doctor, Robert Koch (1843-1910), was also using the microscope as a weapon against disease. Staining his samples for better visibility and identification, Koch isolated the microbes that cause anthrax, cholera, and tuberculosis. While Koch was isolating and identifying various microbes of disease, Pasteur was heading in a more pragmatic direction. Edward Jenner (1749-1823) had used the fact that milkmaids who got cowpox (a mild disease) never contracted smallpox (a deadly disease) to develop the practice of vaccination. He injected small amounts of cowpox fluid into people, who were then protected against smallpox. No one knew what caused smallpox, but Jenner had shown that vaccination could prevent the disease. Pasteur wondered if he could develop vaccines for animal diseases in his laboratory. This had never been done before; Jenner's smallpox vaccine occurred naturally as cowpox.

The breakthrough came in 1880 when Pasteur was working on the microbe that causes fowl (chicken) cholera. He discovered that if he kept cultures of the microbe for long periods of time and then inoculated chickens with them, the fowls recovered from the disease. When he inoculated the same chickens with the most virulent pure culture, the animals were not affected—they were immune. He had invented a vaccine for fowl cholera. This discovery opened a whole new branch of microbiology, the study of immunity. Pasteur realized that weak microbes had stimulated the host (the chickens) to produce substances (antibodies) that protected them in the future. He now sought to apply this process of attenuation (weakening) to other microbe-caused diseases.

Fowl cholera was a rare disease. Anthrax, on the other hand, was a common and costly disease that killed thousands of animals annually. So in 1881, when Pasteur announced he had developed an anthrax vaccine on the same principle as his fowl cholera vaccine, he attracted international attention to a public test of it at Pouilly-le-Fort. Twenty-four sheep and a few cows were injected with Pasteur's vaccine; a similar number of sheep and cows were not. Two weeks later both groups were given lethal doses of anthrax culture. All the vaccinated animals lived, all the unvaccinated animals died, creating a milestone in the history of microbiology. Pasteur marketed the vaccine, and deaths caused by anthrax became rare. He also developed a vaccine against swine fever that was sold throughout Europe, further increasing his fame and spreading the concept of immunization.

Pasteur now decided to try to develop a vaccine for hydrophobia (rabies in humans). He could not identify the microbe causing rabies because it was a virus, too small to be seen by the microscopes of the period. Nor could he cultivate it in his laboratory, because viruses can only multiply in living cells. He injected the invisible microbe, which was known to be in the saliva of rabid animals, into rabbits. Since rabies attacks the nervous system, after the rabbits died he dried their spinal cords for several weeks. From these spinal cords he prepared a new solution that contained the microbe and injected it into a second series of rabbits. He repeated the procedure a dozen times, then vaccinated some dogs with the attenuated microbes. When they were bitten by rabid dogs, they were immune to the disease. But since all the dogs in France could not be vaccinated, Pasteur decided to try the vaccine on unvaccinated dogs after they had been bitten by rabid animals, but before the symptoms appeared. The procedure worked: the bitten dogs did not develop rabies.

The question was whether the same procedure would work on humans bitten by rabid animals. In 1885 Pasteur treated a boy badly bitten by a rabid dog with his vaccine; the boy survived, showing no symptoms of rabies. This was the first successful human vaccination with a manufactured vaccine. After successfully treating a second bite victim later that year, Pasteur became internationally famous. Within a few months victims of rabid animal attacks from as far away as Newark, New Jersey, and Smolensk, Russia, came to Paris for his vaccine. Contributions poured in, allowing him to build the Pasteur Institute; a building in which, because of failing health, he never worked. But he had made his contributions to science. He had convinced the world that microbes were alive, were connected to disease, and could be successfully battled by vaccination.

ROBERT HENDRICK

Further Reading

Books

Debré, Patrice. *Louis Pasteur,* trans. by Elborg Forster. Baltimore: Johns Hopkins University Press, 1998.

De Kruif, Paul. *Microbe Hunters.* New York: Blue Ribbon Books, 1926.

Dubos, René. *Louis Pasteur: Free Lance of Science.* New York: Da Capo, 1960.

Duclaux, Emile. *Pasteur: The History of a Mind,* trans. by Erwin Smith and Florence Hedges. Metuchen, NJ: Scarecrow, 1973.

Geison, Gerald L. "Louis Pasteur." In *Dictionary of Scientific Biography,* edited by C.C. Gillispie. New York: Scribner's, 1974.

Geison, Gerald L. *The Private Science of Louis Pasteur.* Princeton: Princeton University Press, 1995.

Grant, Madeleine P. *Louis Pasteur: Fighting Hero of Science.* London: Ernest Benn, 1960.

Latour, Bruno. *The Pasteurization of France,* trans. by Alan Sheridan and John Law. Cambridge: Harvard University Press, 1988.

Vallery-Radot, René. *The Life of Pasteur,* trans. by R.L. Devonshire. New York: Doubleday-Page, 1923.

Periodicals

Hendrick, Robert. "Biology, History, and Louis Pasteur." *American Biology Teacher* 53 (November/December 1991): 467-478.

Gregor Mendel Discovers the Basic Laws of Heredity while Breeding Pea Plants (1866)

Overview

In a monastery garden rather far removed from the rest of the scientific community, Gregor Mendel studied the transmission of physical characteristics from one generation of pea plants to the next, thereby deciphering the basic principles governing heredity. Mendel was not the first person to study heredity, but he was the first to carefully study the inheritance of traits with planned experiments, carefully recorded data, and statistical analysis of results. His quantitative approach allowed him to translate his findings into a coherent and reproducible theory of how traits are passed from one generation to the next. Mendel's contribution was not appreciated during his lifetime but became the foundation for our understanding of genetics in the twentieth century.

Background

Gregor Mendel (1822-1884) was not the first scientist to question how physical characteristics are transmitted from one generation to the next. Centuries before Mendel began breeding pea plants, humans grasped the idea of inheritance despite having no idea how it worked. Throughout history, inheritance of "familial" traits in humans has been important in social organization. That children often resemble parents or grandparents was noted far back in history. The protruding bottom lip of the Hapsburg family (ca. thirteenth century), for example, was a distinct physical characteristic that helped define members of this royal clan. Historically, inheritance played a large role in agriculture as well. Farmers learned that by crossbreeding and inbreeding animals (and plants) with different traits, they could improve on nature and create hybrids with desirable characteristics.

Recorded theories about inheritance date back to the days of Aristotle (384-322 B.C.). For example, Aristotle suggested that a mixture of semen from a male and menstrual fluid from a female combine to generate offspring. When Anton van Leeuwenhoek (1632-1723) developed the microscope and discovered sperm in 1677, people believed that they saw in the sperm cell a miniature person (a homunculus) that was ready to be incubated in the female womb. Others, called ovists, believed it was the egg that harbored the next generation. The concept of spontaneous generation, in which simple life forms could take shape from substances such as ooze and mud, was another very popular theory until Louis Pasteur (1822-1895) disproved it in 1864. The theory of inheritance of acquired characteristics, which states that changes (through use and disuse) in an organism's body that occur during its lifetime are passed to offspring, was incorporated into Jean Baptiste Lamarck's theory of evolution in 1809 and was widely accepted through most of the 1800s.

In 1856, when Mendel started breeding pea plants to try to understand heredity, he knew of some of these theories and was interested in them. He knew of gametes and fertilization, but

mitosis, meiosis, chromosomes, genes, and DNA had yet to be discovered. Mendel also was well aware of the history of breeding plants and animals to create hybrids. He knew that hybrids from the same kinds of plants looked similar but that when those hybrids mated they could produce offspring with traits different from those of their parents.

Mendel had good scientific instinct. His approach to the study of heredity was new and different (and unappreciated during his lifetime). He was successful because he framed his question in a scientific manner, he chose a good model (the pea plant), and he analyzed his data quantitatively. The pea plant was easy to grow and it self-pollinated, which meant that he could control reproduction between individuals. He began by spending two years breeding more than thirty varieties of pea plants to make sure that they bred true (meaning that offspring had the same physical characteristics as the parents) and to define distinct physical characteristics. Mendel then chose seven traits for further study and carefully designed experiments in which he crossed parents with different traits. He began by crossing parents with different variations of a particular trait (e.g. height) and counting the number of offspring bearing each form (e.g. short and tall). He later crossed those offspring to get a second generation, again counting individuals that were tall or short. Later still, Mendel crossed parents with two and then three distinct traits.

A mathematical approach to biology was unprecedented in the 1800s, but Mendel took this approach and analyzed his data statistically. His analysis of the numbers of offspring with particular traits that came from crosses of parents with particular traits illustrated that the results of breeding could be predicted based on mathematical probabilities. Based on his data, Mendel deciphered what came to be called Mendel's law of segregation: hereditary units that determine a particular trait occur in pairs that separate during gamete formation so that a gamete receives half of the pair. This means that two units (now known to be genes), one from each parent, combine to determine which form of a trait an organism will have, and the factors can be distributed in different ways in each generation. These factors are discrete (meaning that they do not blend), and one factor is dominant over (or masks) the other. Mendel's examination of multiple traits led to the law of independent assortment, which says that members of each pair of genes are distributed independently when gametes are formed.

In the twentieth century scientists have questioned Mendel's methods because his data seemed to match statistical probabilities too closely. Historians of science wonder if, perhaps, Mendel already knew what to expect prior to actually analyzing the data or if he already had his theories in mind when he did his experiments. Regardless of his methods, Mendel's pea plants and the write-up of his results form the foundation for modern genetics.

Impact

Mendel's discoveries had virtually no impact on the scientific community or on life in general in the nineteenth century. He reported his results at a meeting of the Brünn Natural History Society in 1865, but no one was interested in his work. He published the results in a manuscript, titled *Experiments with Plant Hybrids,* in the Proceedings of the Brünn Natural History Society in 1866. The journal was distributed to 120 societies and libraries throughout Europe and America, but it was cited in only four papers before it was found and recognized as an explanation for inheritance in 1900.

During the years between Mendel's publication and its rediscovery in 1900, several others proposed theories to explain inheritance (although none have stood the test of time). For example, Charles Darwin (1809-1882) developed the theory of pangenesis to explain how variation arises in a population. He proposed that each tissue in the adult buds off particles (gemmules or pangenes) that concentrate in the reproductive organs; thus, an acquired trait (such as enlarged muscle resulting from weightlifting) could be passed to offspring because it left a lasting imprint on the cells of the body. Like Lamarck, Darwin thought that use and disuse would affect heredity and provide the source of variation on which natural selection would act. Likewise, Ernst Haeckel (1834-1919) proposed that the basic units of living matter have a memory. Blending inheritance, in which the information for a trait donated by each parent would blend together to create an intermediate trait, was another popular theory. August Weismann (1834-1914), on the other hand, rejected pangenesis and instead proposed "continuity of the germplasm" (gamete-producing cells) in which he recognized the importance of gametes and the cell nucleus in heredity. He believed that the body was simply a host for the germ cells; this theory explained how cells could be reproduced generation after generation and remain unchanged, thus supporting the popular notion of the fixity of species.

While these theories were being disputed in scientific circles, Mendel gave up trying to push his theory of heredity. However, it was his work (if not all of his laws) that stood the test of time. The first golden age of genetics was kicked off in 1900 when botanists Hugo De Vries (1848-1935) in Holland, Carl Correns in Germany, and Erich von Tschermak in Austria independently found Mendel's paper and realized that it explained much of their own research and described the mechanism for heredity. At this point in time, the old notions about heredity were being disproved and the world was ready to gradually accept Mendel's work.

William Bateson (1861-1926), who is considered one of the creators of modern genetics, championed Mendel's work and coined the terms "Mendelism" and "genetics." A host of other scientists took an interest in heredity, and advances in understanding came quickly. Cytology, or the study of cells, also had burgeoned in the late 1800s and provided an understanding of how cells divide. Together, Mendel's theories and research at the cellular level could explain inheritance. For example, in the early 1900s Mendel's two laws were explained by the finding that chromosomes carry genes, that chromosomes segregate at meiosis, and that there are genes on different chromosomes.

Mendel's work also impacted Darwin's theory of evolution. One of Darwin's problems in explaining inheritance was that he thought inheritance and variation were opposite processes that could not coexist. In the early 1900s, however, De Vries fixed this problem when he proposed that mutations (changes) in genes produce variation. Scientists in the 1930s then generated the synthetic theory of evolution that incorporated natural selection, the principles of genetics, the concept of mutation, and the idea that populations are the units on which natural selection acts.

By 1910 Mendel's work had been amplified, and it was clear that inheritance often was not as straightforward as predicted by Mendelian genetics. However, Mendel's basic laws provided the framework that allowed researchers to determine how exceptions were produced. For example, while the law of segregation always held true, scientists often did not observe dominance and found that blending (codominance) seemed to occur. Independent assortment was found to be limited to genes on different chromosomes. In addition, researchers found that multiple genes influence many traits, genes can be linked, traits can be linked to sex chromosomes, lethal genes exist, and many times a trait has multiple states rather than just two (e.g. tall and short).

From 1910 to the turn of the twenty-first century, our understanding of genetics has advanced quickly and tremendously. Classical genetics gave rise to molecular biology and biotechnology. We are in the midst of a second golden age of genetics, in which biotechnology offers the promise of better crops, fixing damaged genes during gestation, and curing disease with "gene therapy." Mendel's basic findings discovered in a monastery garden, however, resonate throughout modern genetics. His work has positively affected the agriculture and livestock industries. His laws are used daily in genetic counseling to help parents determine the risk of having babies with various diseases and birth defects. And Mendel's work led to genetic terms sprinkled throughout every basic biology and genetics textbook currently published.

LYNN M.L. LAUERMAN

Further Reading

Books
Corcos, Alain, and Floyd V. Monaghan. *Gregor Mendel's Experiments on Plant Hybrids: A Guided Study.* New Brunswick: Rutgers University Press, 1992.

Darwin, Charles R. *The Variation of Animals and Plants under Domestication.* 2 vols. London: John Murray, 1868.

Mayr, Ernst. *The Growth of Biological Thought: Diversity, Evolution, and Inheritance.* Cambridge: Harvard University Press, 1982.

Mendel, Johann Gregor. "Experiments in Plant Hybrids," 1866. Translation in Stern, Curt and Eva R. Sherwood, eds., *The Origins of Genetics: A Mendel Source Book.* San Francisco: Freeman & Co., 1966.

Periodicals
Bateson, William. "The Facts of Heredity in the Light of Mendel's Discovery."*Report of the Evolution Committee of the Royal Society* 1 (1902): 125-160.

Other
http://www.netspace.org/MendelWeb/

Ferdinand Cohn and the Development of Modern Bacteriology

Overview

Ferdinand Julius Cohn (1828-1898) is recognized as one of the founders of modern bacteriology. He contributed to the creation of this discipline in two important ways. First, he invented a new system for classifying bacteria, which provided microbiologists with a more standardized nomenclature with which to work. Secondly, his drive to understand the life cycles of microorganisms, combined with his research into the effect of heat on bacteria, contributed to the downfall of spontaneous generation arguments. These achievements took place in the context of advances in experimental techniques and laboratory equipment.

In the nineteenth century the terms bacteriology and microbiology were used in often confusing and contradictory ways. For the purpose of this essay, the term bacteriology is limited to mean simply the study of bacteria, although it often refers to their relationship to disease or other medical questions. Microbiology is the general study of bacteria and other microbes with a focus on their morphology and physiology.

Background

The discipline of bacteriology originated with the recognition that bacteria are organisms in their own right—that they are different from algae, fungi, and other single-celled microorganisms. This idea is central to Cohn's belief in the constancy of bacterial species and his creation of an extensive classification system for microorganisms, in which bacteria are given their own place.

He was not the first to develop a classification system of microorganisms that included bacteria. Otto Friedrich Müller (1730-1784) did so in the late eighteenth century and Christian Ehrenberg (1795-1876), later Cohn's teacher, built upon Müller's system in the early nineteenth century. Ehrenberg described several species of what we would recognize today as bacteria. Advances in microscope technology and staining methods improved the ability to observe these very small and quickly moving organisms. As a result, better descriptions, in the form of words and illustrations, appeared in the zoological and botanical literature of the nineteenth century. Cohn drew upon these in making his own system. His most important contribution to the founding of bacteriology, was in recognizing that bacteria, because of their peculiar shapes, motions, and methods of reproduction, represented separate genera and species.

He first addressed these issues in his *On the Developmental History of Microscopic Algae and Fungi* (1853), in which, opposing previous investigators, he placed bacteria in the plant rather than animal kingdom. Drawing upon his botanical education, he made the novel suggestion that, although they differed in color, bacteria were very similar in shape and method of reproduction to microscopic algae. His ability to draw analogies between larger (and thus easier to see) organisms, and the nearly invisible bacteria, allowed him to interpret his observations in a new way. Over the next 20 years, culminating in the 1870s with three articles entitled *Untersuchungen der Bakterien* (*Researches on Bacteria*), his influence on the scientific community grew steadily.

During this period, a debate raged over the nature of microbe species. Some contended that the many different types of organisms they observed were actually only a few species that could adopt many shapes. On the opposing side of the argument, others contended that each species was constant and thus defined by one form. Cohn supported the second group, and argued during the 1860s and 1870s for clarity in the usage of nomenclature. He, for example, criticized the French chemist Louis Pasteur (1822-1896) for using several names, such as cryptogames, animalcules, infusorien, monads, and more, for the same microorganisms.

Cohn made his classification system public in 1872. For him, bacteria were "chlorophyllless cells of spherical, oblong, or cylindrical shapes, which are now and then rotated or crooked, which multiply exclusively through transverse division and are either isolated or vegetate into colonies." He arranged the bacteria into four "tribes" based on their external characters, and each tribe contained one or more genera. The first tribe, *Sphaerobacteria*, were round and contained the genus *Micrococcus*; the sec-

ond, *Microbacteria*, were rod-shaped and contained the genus *Bacterium*; *Desmobacteria*, the third tribe, were thread-like and contained the genera *Bacillus* (staff) and *Vibrio* (vibrating); the final tribe, *Spirobacteria* were coil-shaped and contained the two genera *Spirillium* and *Spirochaete*. Cohn continued to make order of the different organisms, describing many species for each genus. For example, in *Micrococcus* alone he placed 10 different types.

During the 1870s, microscope technology and staining methods were still quite undeveloped. This forced Cohn to build his system around morphological traits. He did, however, emphasize the necessity of using physiological characters in order to fully understand the relationships between the different species of bacteria. He challenged microscope builders to design instruments that would clear the fog away from the microscopic world he and his fellow microscopists were exploring.

Impact

The notion, promoted by Cohn and others, that bacterial species were constant, led to methods of growing pure cultures. Pasteur was using pure cultures to support his claims that different types of fermentations were caused by specific microorganisms. German physician Robert Koch (1843-1910) would later apply similar reasoning in developing the germ theory of disease, which suggests that each disease is caused by a specific bacterium. The pure culture methods coming to the fore in the 1870s required the use of sterile organic solutions (infusions), which were made from a wide variety of plants and animals. Sterilizing these infusions usually required only boiling them for five minutes; however, at times this did not prove sufficient. In an effort to discover the reasons for this, researchers such as William Roberts, John Tyndall (1820-1893), and Cohn conducted elaborate investigations on how heat effected the growth of bacteria.

Cohn, drawing on Robert's work, was the first to discover that some bacteria possess a life cycle that includes a spore stage. This life cycle included an incubation period, a stage of progressive growth, a fastigium (the period of maximum development of a disease), and a period of remission. During this last stage, the bacteria are resistant to very high and low temperatures. On a visit to England, Cohn visited Tyndall and presented him with a copy of his report on this matter. Tyndall incorporated Cohn's discovery to invent a process of discontinuous heating, which successfully killed bacteria and their spores. This process came to be known as "Tyndallization" and is still used in microbiology laboratories.

By contributing to Tyndall's work, Cohn contributed to ending any serious scientific debate concerning the origin of life by spontaneous generation. In France, by 1870, spontaneous generation had become a dead issue for most serious scientists. In England, however, the influential Henry Charlton Bastion (1837-1915) kept the issue active. Using what were considered at the time to be sound experimental procedures, he observed that bacteria would appear in turnip-cheese infusions that had been sterilized by boiling and then sealed. Since boiling should have killed everything in the infusion, the bacteria must have, in Bastion's view, been produced by spontaneous generation. Through the 1870s, a series of extensive investigations on turnip-cheese were conducted by supporters of the germ theory. They strove to prove that Bastion's bacteria had somehow found its way into his infusions after they were boiled, or that his investigative procedures were faulty and the microbes had not been completely killed off in the first place.

Among those challenging Bastion were Pasteur, Tyndall, and Cohn. Throughout the 1860s Pasteur had challenged many supporters of spontaneous generation on the basis of their faulty procedures. In the next decade, joined by Roberts and Tyndall, the criticisms came from a new direction. Tyndall had shown in 1877 that five minutes of boiling was insufficient to sterilize some types of organic infusions. It was Cohn's paper on *Bacillus subtilis* (1876) that inspired this line of investigation for Tyndall. Although Bastion continued to argue for spontaneous generation into the first decade of the twentieth century, his claims were mostly ignored by the scientific community.

In 1872 Cohn published "Bacteria: The Smallest of Living Things," an article written for a popular audience. Here he discussed the possibility that the presence of bacteria in blood and other bodily tissues during epidemics and in cases of disease could mean that bacteria are the conveyors and originators of infection and contagion. He drew upon the work of several earlier researchers, including that of Casimar Davaine (1812-1882), who had shown that anthrax was always accompanied by the presence of the same microorganism, to build a case for the importance of bacteria. Cohn's article, with its poetic language and its exploring questions, introduced the bacteria and their still unknown, yet crucial role in all life, to a wide public audience.

Cohn contributed to bacteriology, not only through his popular literature and scientific investigations, but also by using the influence his institutional powers gave him. As director of the plant physiology institute at the University of Breslau and editor of its journal, he promoted the people and ideas that he respected. One such person was Robert Koch. Koch visited Cohn for several days in 1876, in order to demonstrate his claim that the bacteria *Bacillus anthracis* was the cause of the disease anthrax. Cohn, convinced of the correctness of Koch's ideas and procedures, published his work in the 1876 volume of the institute's journal. Koch's work and the students he later trained ushered a "Golden Age of Bacteriology" during the decade of the 1880s. This era of great productivity was built, in part, upon the foundations of Cohn's numerous discoveries in bacterial life cycles and the classification system he constructed for them.

LLOYD ACKERT

Further Reading

Bulloch, William. *The History of Bacteriology.* London: Oxford University Press, 1960.

Cohn, Ferdinand J. *Bacteria: The Smallest of Living Organisms.* Baltimore, MD: Johns Hopkins University Press, 1939.

Vandervliet, Glenn. *Microbiology and the Spontaneous Generation Debate During the 1870s.* Kansas: Coronado Press, 1971.

The Discovery of Viruses

Overview

By the late nineteenth century, the work of Louis Pasteur (1822-1895) and other scientists had established the germ theory of disease and identified the bacteria that caused many ailments. But they found that some diseases were caused by invisible agents that could not be filtered out, agents that came to be called viruses. The experiments by Martinus Beijerinck (1851-1931) and Dmitri Ivanovsky on the tobacco mosaic virus in the 1890s are generally thought of as the beginning of the science of virology, but it was not until 40 years later that viruses could be isolated with extra-fine filters and imaged using electron microscopes.

Background

Infectious diseases have always been humanity's lot. Throughout history, plagues were not uncommon, and many people died of diseases that would be preventable or easily curable today. Early explanations of disease included possession by evil spirits and imbalance of "humours," or bodily fluids. The name "influenza," coined in the fifteenth century, is Italian for "influence"; the influence, that is, of the heavenly bodies.

During the Renaissance, scholars began to suggest that illnesses could be spread by invisible particles. But until the microscope was invented by Anton van Leeuwenhoek (1632-1723) in the 1600s, scientists had no idea that microorganisms existed. Even then, it was another 200 years before they were proven to be associated with disease.

Meanwhile, contagious illnesses continued to run rampant. In eighteenth-century Europe, about 95 percent of the population contracted smallpox at some point in their lives, and as many as one person in 10 died of it. Carried to the New World by Europeans, it wiped out millions of Native Americans. By 1796 Edward Jenner (1749-1823), an English country doctor, had seen many people die of smallpox. Yet the milkmaids he knew rarely got sick with this dread disease. They did often contract cowpox, an illness with much milder symptoms. Could exposure to cowpox somehow protect against smallpox as well?

It was well known that people who had survived smallpox were thereafter immune. The 1700s had seen widespread adoption of exposure to what was hoped would be a mild dose of smallpox in order to achieve this immunity. But the strategy often backfired, as there was no real way to control the severity of the resulting disease.

Jenner hoped that inoculation with cowpox might be a way to achieve immunity without being exposed to smallpox at all. He tested his theory on a healthy eight-year-old child named James Phipps. He took pus from a milkmaid's cowpox sore and scratched it into the boy's arm, where it caused a small infected spot that soon subsided. Six weeks later, Jenner took the riskier

step of inoculating the boy with smallpox in the same way. The child did not get smallpox then, or ever. Jenner had developed the first vaccine for a viral disease, long before anyone knew what a virus actually was.

Impact

The great nineteenth-century French scientist Louis Pasteur was a proponent of the germ theory of disease, the idea that illnesses were caused by living microorganisms, or microbes, that could spread between people and multiply in the body. Pasteur was able to identify and study a number of the microbes called bacteria that were involved in diseases, and also in other processes such as fermentation and the souring of milk. He disproved the then-popular idea that microbes could arise out of nothing by "spontaneous generation," and began to investigate how a human or animal "carrier" could spread a disease without contracting the illness itself.

But there were some diseases, such as rabies, for which Pasteur could not find the germ, although he insisted it must be there. We now know that these diseases are caused by even tinier agents called viruses. Other examples of viral diseases that would have been familiar to Pasteur include the common cold, mumps, measles and polio.

Rabies was of particular concern to Pasteur because it was almost always fatal. In his experiments, he found that rabies-infected tissue transferred from animal to animal gradually became weaker and less likely to cause the dreaded disease. This caused him to hypothesize that a weakened extract of tissue infected with rabies might be protective against the disease, and usable as a vaccine even after a person had been bitten. By 1885 he had prepared such an extract, although he was by no means ready to try it on humans. But when nine-year-old Joseph Meister was brought to Pasteur after having been bitten by a rabid dog, there was no alternative for the boy but a painful death, and Pasteur was persuaded. With great trepidation, he administered the vaccine, and it worked. Joseph Meister did not develop rabies, and lived another 55 years.

Although Pasteur had found a way to defeat rabies, he never did find the germ he was looking for. He assumed that it was simply another type of bacteria, a variety too small to be seen with the microscope. Pasteur was half right; most viruses were much too small to see with nineteenth-century microscopic techniques. The

Scottish surgeon John Buist did manage to see the smallpox virus in 1887, but didn't recognize the tiny dots he observed. Even the largest viruses are about one-tenth the size of typical bacteria, and most are much smaller than that. But in fact they are not simply miniature bacteria, but something else altogether.

In 1892 a Russian scientist named Dmitri Ivanovsky was studying tobacco mosaic disease, which destroys the leaves of tobacco plants. The disease was clearly infectious; plants that came into contact with the sap from diseased plants were damaged as well. This ability to reproduce itself eliminated the possibility that the damaging agent might be a simple toxin. Hoping to find the bacteria responsible for the infection, Ivanovsky ran an extract of diseased leaves through a very fine filter, with pores small enough to trap any known type of bacteria. But he found that whatever caused the disease went right through his filter. No matter how many times he tried to strain out the microbe, the liquid still retained the power to infect other plants.

Ivanovsky published his findings, but little attention was paid. He himself thought that there might have simply been a problem with his filters. Six years later, however, not knowing about Ivanovsky's work, the Dutch botanist Martinus Beijerinck performed the same experiments and got the same results. While the infectious agent could not be filtered out, it seemed to be destroyed when the liquid was heated.

Beijerinck concluded that the infectious agent was not a microbe at all, but a "contagious living fluid." Just as Jenner had a century before when writing about his smallpox vaccine, Beijerinck used the term "virus" from the Latin word for poison or pestilence. Other scientists soon showed that hoof-and-mouth disease, yellow fever, and other infectious ailments were also caused by these "filterable viruses."

Late nineteenth-century scientists attempting to study viral diseases were in a sense groping around in the dark. In the 1930s filters could finally be manufactured with pores tiny enough to prove that viruses are particulate after all, rather than being fluid in nature. The earliest electron microscopes also appeared in the 1930s, and viruses could at last be seen. Today we know that viruses are not living cells like bacteria, but rather tiny packets of genetic material that must infect the cells of their unwilling host in order to reproduce. They mutate quickly, and our ever-increasing knowledge of them is always under challenge by the appearance of new viral diseases such as AIDS and the Ebola virus.

SHERRI CHASIN CALVO

Further Reading

Nourse, Alan E. *The Virus Invaders.* New York: Franklin Watts, 1992.

Oldstone, Michael B. A. *Viruses, Plagues and History.* Oxford: Oxford University Press, 1998.

Radetsky, Peter. *The Invisible Invaders.* Boston: Little, Brown, 1991.

Middle-Class Victorian Men and Women Collect, Identify, and Preserve Plant and Animal Species, Broadening Human Knowledge of the Natural World and Transforming Biology into a Mature Science

Overview

In the nineteenth century the science of biology, including botany and zoology, was just beginning to mature while astronomy, physics, and math were already established sciences. Victorian members of the middle class were affluent and interested in learning about the natural world. They had leisure time to attend lectures, visit museums, and collect specimens of plants, insects, animals, and fossils. Many of these enthusiasts provided specimens for naturalists and added to scientific knowledge and understanding of the natural world in a material way. By the end of the century, enthusiasm for amateur scientific enterprise waned as the increasing complexity of science and the need for professional training and expertise to keep up with advances in these disciplines overwhelmed the formally inexperienced.

Background

The natural world has been observed and commented on since Aristotle. Species and types of plants and animals were recognized, but classification went no farther than obvious divisions. Each natural historian drew and arranged specimens his or her own way. Understanding of the enormous varieties of organisms in the natural world could not become a science until an orderly system of classification and identification was devised. Thousands of species were collected and described, but all was confusion.

Swedish naturalist Carolus Linnaeus (1707-1778) recognized the need for order and created a system based on groups of species and subspecies. The tenth version of his *Systema Naturae,* published in 1758, classified mammals, birds, reptiles, and fish in two groups—vertebrates (with backbones) and invertebrates (without backbones). Plants were divided into parts and arranged according to sexual characteristics, identified by a binomial formula of two Latin names designating genus and species. Linnaeus also devised the Latin name *Homo sapiens* (wise man) for humans. Although this system has been revised, refined, changed, and augmented since its original appearance, it is still essentially the one in use today.

England and other European nations had been exploring the rest of the world for 200 years with an eye to exploiting its commerce. On board each ship was a naturalist whose task it was to collect examples of plants and animals wherever the ship went and an artist to draw accurate images of the curious and fantastic life in the four corners of the earth. When the ships returned home in the 1800s, these images were published and became popular with a new audience.

This audience was made up of a new middle class who gained importance in early nineteenth-century Europe. Victorian society is defined as middle-class Europeans and Americans who lived when Queen Victoria (1819-1901) ruled England from 1837 to 1901. It was made up of men of wealth and their families. These men created, exploited, and were expanding a new industrial world. They included shipbuilders, factory owners, architects, builders, insurance company operators, and bankers who owned the means of industrial production and supported businesses employing thousands of people. They also profited handsomely from their investments.

This was a time of rapid change in Western society. New inventions made life easier and required industrialization. Victory over disease brought a population explosion, which led to urbanization. Workers were needed in factories, steel mills, mines, and to build bridges and railroads. They moved into cities, which became crowded slums. However, the standard of living of the middle class rose as did their education, and they had ample leisure time to pursue their interests.

Impact

Collecting, arranging, theorizing, and discussing the natural world became a popular activity engaged in by men and women with the leisure time to pursue intellectual activities. Many had no goal other than to satisfy their personal interest and desire for self-improvement. Attending public lectures and visiting museums augmented this interest. Museums, an old idea for preserving scientific and antiquarian accumulations, usually located in private homes, were opened to the public. The first public museum was the Ashmoleon in Oxford, England, begun in 1683. The British Museum opened in 1759 and the American Museum of Natural History in 1869. Museums became popular places to visit and learn. Hundreds of thousands of specimens of plants, flowers, animals, insects, and fossils were gathered, prepared for display, and studied by amateur naturalists. Printed explanatory cards enabled visitors to understand what they were seeing and derive pleasure from the visit.

The importance of the activities of these middle-class seekers after knowledge and diversion was threefold. First and foremost, their collections and discoveries added to the body of knowledge of the natural world, though this was not necessarily the goal of the enthusiast. Current knowledge of plants, animals, and insects was augmented by new specimens and insights.

Second, to collect specimens with discrimination, a natural historian, professional or amateur, had to know what to look for. This helped popularize the importance of and need for education, recognized as fundamental for self-fulfillment and for individual contributions to society.

Thirdly, the popularity of natural history helped create professional biologists by extending current knowledge. Many professional biologists began as amateur naturalists. Because of the enormous varieties of plants, animals, insects, and fossils, naturalists began to focus on a single species or variety of organism.

Those who collected, especially botanical species, did so for their own edification and not for the advancement of a particular science. They often belonged to networks of collectors who contributed to magazines and journals in which they wrote up their discoveries. Naturalists like Asa Gray (1810-1888), American professor of natural history at Harvard, encouraged amateurs to collect in their own geographical area and to send specimens to him. He was also generous in helping to identify species for them and occasionally paid them to collect species he wanted.

Natural historians—American, English, and European—appreciated the work of prolific, accurate, amateur collectors from various countries. Such amateurs often worked on an equal footing with the professionals and some became illustrious scientists, others knowledgeable enough to write texts on botany or zoology for general consumption.

The case of Captain Charles Scammon, a whaling captain, serves as an example. Zoologists studied whales for centuries, but were limited to specimens of bones only. Whalers saw cetaceans alive and breathing. Captain Scammon, who whaled off the west coast of the United States, was untrained in science but recognized the California Gray whale (*Eschrichtius robustus*) as a new species unknown to the biological world. In 1876 his *Marine Mammals of the West Coast of North America* contained the first published drawing and description of the physical properties of the species. He was also one of the first naturalists to note behavior, routes of migration, and food sources of the subject of his study.

The discovery and collection of fossils was another area where amateurs contributed to knowledge of the natural world. In 1822 the first recorded dinosaur was discovered in a backyard in Sussex, England, by Gideon Mantell (1790-1852). Later more were found in Oxford and soon specimens were being found all over the world. Richard Owen (1804-1892) wrote a basic treatise on the new fossils in 1841 and coined the name dinosaur. Plant fossils had been known since the Greeks, but in the nineteenth century they were shown to be prints of living plants. In 1856 the first early human, Neanderthal Man, was discovered in a cave by a German quarry man. It was reported in print in 1863 just as controversy over evolution arose.

In 1859 one of the most famous of amateur naturalists, Charles Darwin (1809-1882), turned biological knowledge on its head with his book *On the Origin of Species*, which contained his famous ideas about evolution. Based on observations during a voyage around the world in 1830-35, ideas from his grandfather Erasmus Darwin (1731-1802), and philosophical and geological works by Thomas Malthus (1766-1834), Charles Lyell (1797-1875), and others, Darwin's treatise stunned the world. His work was of enormous importance to biology and to man's understanding of himself and his world, and its impact on society and educated men and women cannot be understated.

Some people applauded, some objected in utter indignation, but people noticed Darwin's work. The educated Victorian who had been collecting and studying the natural world for years was able to understand this new idea. Many protested, many vilified the writer, but they could comprehend it. Some began to study particular plants or animals in order to refute or corroborate the Darwin's ideas. The age of the earth was also pushed back because of the startling discovery of Neanderthal Man, as well as older fossils. Evolution required ages to work and that was now seen as possible. Religious objections were vocal and vitriolic, but did not stop advances in biology.

At the beginning of the nineteenth century emphasis was on observation, collecting, and classifying, and the word "biology" was just beginning to be used. By the end of the century the focus was on creating experiments, theoretical deductions, and was turning for answers to microscopic structures of organisms. Science had also begun to unseat orthodox theology as the principal source of explanation of the natural world and its workings. Biology had acquired a huge body of knowledge, a system of classification, and a theoretical model that allowed practitioners to interpret and manipulate knowledge. The Victorians had played a large part in propelling biology into a modern science.

LYNDALL LANDAUER

Further Reading

Darwin, Charles. *Descent of Man.* New York: American Home Library, 1902.

Darwin, Charles. *On the Origin of the Species.* London: Ward, Lock, 1910.

Ewan, Joseph. *A Short History of Botany in the United States.* New York: Hafner, 1969.

Heyck, T. W. *The Transformation of Intellectual Life in Victorian England.* London: Croom Helm, 1982.

Himmelfarb, Gertrude. *Darwin and the Darwinian Revolution.* Garden City, NY: Doubleday, 1962.

Keeney, Elizabeth. *The Botanizers*. Chapel Hill: University of North Carolina Press, 1992.

Lyell, Charles. *Principles of Geology*. Chicago: University of Chicago Press, 1990-91.

Singer, Charles. *A History of Biology*. 1931. Reprint, London and New York: Abelard-Schuman, 1959.

Scientists in Europe and the United States Lay the Foundation for the Modern Science of Ecology

Overview

Ecology is the branch of science that deals with the interrelationships of plants, animals, and the environment. The world has a great variety of living things ranging from simple one-celled organisms, such as bacteria and fungi, to man. No organism lives by itself. In some way, each depends on the living and non-living things in the environment. This perception of interdependence did not evolve until late in the twentieth century. Nineteenth-century researchers on expeditions, plant and animal scientists in Europe, and American naturalists all contributed bits and pieces. Their efforts would continue into the early part of the twentieth century. Work on various elements of ecology was fragmented and not connected, and because of communication problems, research efforts were not well known or publicized. Several scientists studied different populations, communities, and parts of an ecosystem. Both plant and animal ecology developed separately until American biologists and environmentalists in the twentieth century began to emphasize the biotic whole.

Background

No one can point to the exact place where ecology began. There is no founder or father. The Greek Theophrastus (c. 370-285 B.C.), one of Aristotle's students, described the morphology or make up, natural history, and therapeutic use of plants. He even made the effort to develop a scientific naming system or nomenclature. More than 500 plants were divided into what is known as genera. He also described seed germination and sexual reproduction of flowering plants. The first use of the term "ecology" was by a German zoologist and evolutionist, Ernst Haeckel (1834-1919). He declared that all nature is explained in a unified theory of evolution and used the term "oekologie" in his book *Generalle Morphologie der Organismen* ("General Structure of Organisms") in

1866. He coined the term ecology for interrelatedness relating to evolution. Scientists today have specific meanings for the term ecology. Major precepts of ecology at the end of the twentieth century follow: 1) All living and non-living things are interconnected. 2) The world is divided into separate units known as ecosystems, for example a forest; populations are factors in an ecosystem. 3) An area where all organisms live together is called a community. 4) Organisms have energy roles as producers, consumers, and decomposers. These links are represented by food chains or food webs. To arrive at these precepts, animal and plant physiologists of the nineteenth century laid the foundations through many separate contributions. However, it was not until late in the twentieth century that the strong conceptual bases for ecology were focused.

Impact

A force that began in the eighteenth century and continued in the nineteenth was the biological expedition sponsored by the government of Great Britain. Such endeavors had interesting names and went to exotic places to study plants and animals. Early travelers were no more than curious adventurers, but as the expeditions continued, they became more scientific. The biologists became trained observers. From an expedition called the "Investigator" in 1801, botanist Robert Brown (1773-1858) wrote a classic work on the plants of Australia and New Zealand, describing how certain plants adapt to differing environmental conditions. In 1831 the *H.M.S. Beagle* landed on the Galapagos Islands off the coast of Ecuador. On board was geologist Charles Darwin (1809-1882), who began to ponder the significance of the strange animals that had developed on the isolated islands. He also collected specimens and took many notes on the living things in South America and Australia. He noted the variations of animals that had adapted to different environments and presented his ideas of how organisms evolved at a scientific meeting in 1858. His theory of evolution became an exploding bomb throughout the world. These expeditions piqued the development of zoogeography, or the study of animals, and phytogeography, the study of plant distribution.

In 1858 English ornithologist (one who studies birds) Philip L. Sclater based Earth divisions on the kinds of bird inhabitants. Not long after, biologist Adolf Engler (1844-1930) did the same type of world classification for plants. Many individual adventurers made a contribution. From 1854 to 1862 English naturalist Alfred Russel Wallace (1823-1913), along with British naturalist Henry Walter Bates (1825-1892), traveled to the Amazon and later to the Malay Archipelago to collect specimens. Here, he grouped the animals according to an imaginary line that became known as Wallace's line. The animals west of the line were like those in Asia; those east of the line were like the Australian animals. After reading Thomas R. Malthus's (1766-1834) "Essay on Population," he reported that he was struck with the idea that natural selection is a process by which historical changes occur in a plant or animal species. The story is told how he arrived at his ideas, independent of Charles Darwin. Wallace was in the Molluccas or Spice Islands and became ill with malaria. While he was struggling with the terrible fevers and chills of the infection, he perfected his ideas of the survival of the fittest and evolution.

Both the work of Wallace in zoogeography and Sir Joseph Dalton Hooker (1817-1911), a plant collector and systematist, laid the foundation for the emerging biogeography of the nineteenth century. They had an influence on the work of Charles Darwin, who mentioned their works in his seminal book, *The Origin of Species.*

In the early and mid-1900s biologists in Europe and America approached the study of living things from two different points of view. Europeans botanists concentrated on structure, composition, and distribution of plant communities. America botanists were fascinated with the development or changes in plant communities, called succession. Study of zoology likewise developed as comparative zoology, and it was completely dominated as the century progressed by evolution and population dynamics.

The early botanists of Europe began with studies of the geography and structure of plants. However, transition from phytogeography to plant ecology progressed throughout the century.

Austrian botanist Gottlieb Haberlandt (1854-1945) was the first person to study plant tissue culture. At the University of Tübingen Haberlandt studied under Simon Schwendener, who emphasized that structure and function must be studied together. Haberlandt later succeeded Schwendener as chair of plant physiology at the University of Berlin. Haberlandt wrote a book in 1884 in which he classified plants based on the function of twelve tissue systems, such as mechanical, absorptive, or photosynthetic. His system was not accepted by other botanists, but his studies of the relationships of structure and environment led to later developments in ecology.

Andrea Franz Wilhelm Schimper (1856-1901) was one of the first to divide the continents into floral regions. He traveled extensively in Brazil, Java, and Africa and wrote about the climatology and physiology of the world's vegetation. He described how plants spread to new areas and how they adapt to these new territories. Another European, Danish botanist Johannes Eugenius Warming (1841-1924), related living plants to their surroundings. Beginning his studies in Greenland, he described how the structures of plants adapted to their surroundings. In his 1895 study called *Oecology of Plants,* he grouped and characterized plant communities as a group of species growing in the same locality and subject to the same external conditions.

U.S. botanists took a more pragmatic approach. Charles Edwin Bessey (1845-1915) introduced the systematic study of plant morphology and experimental laboratory at the college levels at the Iowa State University Agricultural College and the University of Nebraska. George Perkins Marsh (1801-1882) was a U.S. diplomat and congressman who had the foresight of government's role in natural resource management. His book *Man and Nature* (1864) was one of the most significant of the nineteenth century in ecology and resource management.

The science of oceanography also developed in the last half of the nineteenth century. However, several scientists laid the foundation. From 1872 to 1876 under the leadership of Charles Wyville Thomson, scientists aboard the British ship HMS *Challenger* studied oceanography, meteorology, and natural history. The teams made vast collections of sea life and were the first to determine the importance of plankton, the free-floating plants and animals that are the beginning of the food chain for other animals. The science of marine biology was born with this trip. Karl August Möbius (1825-1908) was a

German zoologist who greatly contributed to marine biology. He conducted research on corals and foraminifera, (tiny one-celled animals with calcium shells), leading to the discovery of symbiotic relationships in marine invertebrates. He was also interested in fishery biology, mussels, oyster breeding, and possibly pearl cultivation.

Louis Agassiz (1807-1873) was a Swiss-born U.S. naturalist who developed an interest in ichthyology when he inherited a collection of Brazilian fishes from the Amazon. He traveled throughout Europe, studying both live species and fossils of extinct species of fish and the animals that were found with them, such as echinoderms and mollusks. In 1836 he began a study of the glaciers in Switzerland and concluded these were great sheets of ice that covered large areas. In 1846 he moved to the U.S., settling as a professor of zoology at Harvard. He published many papers and worked to assist a complete revolution of the study of natural history in the U.S. Almost every notable U.S. naturalist in the nineteenth century was a pupil of Agassiz.

American naturalists began the rudiments of the environmental movement by popularizing nature. John James Audubon (1785-1851) became known for his drawings of North American birds. In the summer of 1869 John Muir (1838-1914) made his first long trip to Yosemite and became known as the most romantic and sublime of the naturalists. Taking the American experience an additional step was Clinton Hart Merriam (1855-1942), a biologist and ethnologist who helped found the National Geographic Society (1888) and the U.S. Biological Survey (1896), which is now the Fish and Wildlife Service.

Scientists of the nineteenth century made unique contributions to the science of ecology. From the adventurous expeditions to the pure botanists and zoologists of Europe to the American practical and romantic naturalists, who encouraged government resources management, the whole science emerged. Today's ecological studies and the environmental movement are powerful forces that will continue to be refined in the twenty-first century.

EVELYN B. KELLY

Further Reading

Darwin, Charles and Greg Suriano. *The Origin of Species.* Modern printing. New York: Random House, 1998.

Daws, Gavin and Mary Fujita. *Archipelago: The Islands of Indonesia for the Nineteenth Century Discoveries of Alfred R. Wallace.* Berkeley: University of California Press, 1999.

Muir, John. *My First Summer in the Sierra.* Modern printing. New York: Viking Penguin. 1996.

Severin, Tim. *The Spice Islands Voyage: The Quest for Alfred Wallace, the Man Who Shared Darwin's Discovery of Evolution.* New York: Carroll and Graf, 1998.

Neanderthals and the Search for Human Ancestors

Overview

During the nineteenth and early twentieth centuries, the search for man's evolutionary past lead to dramatic new conclusions about life on Earth and the biological history of mankind. The search for our fossil ancestors, once initiated, began to answer questions about where humans first appeared, what they looked like, and most importantly, how old they are. The first discovery of fossil humans threw surprising new light on our past, though challenged established views about society, race, and Christian religious beliefs concerning man's special place in the universe.

Background

During the Enlightenment, Western thinkers disillusioned with the Genesis story of creation began to speculate about how human beings had originated. The problem that plagued these researchers was that there was no physical evidence to support their theories. Some fossils had been discovered that looked human, but these were passed over as the remains of *Homo Diluvinii*—giants killed off during Noah's flood. These early discoveries not withstanding, the science of paleoanthropology (the study of early people) began in earnest in the late eighteenth and early nineteenth centuries when naturalists began finding curious artifacts and bones. Frenchmen Isaac de la Peyrere (1596-1676) and Jacques Boucher de Crevecoeur de Perthes (1788-1868), and Englishman John Frere (1740-1807) claimed that stones they found in caves in France and England were actually stone tools created by people in the distant past.

Though they looked man-made, it could not be proven whether these tools had been fashioned by the hand of man or formed naturally. As such, the work of these men was largely ignored.

The first significant fossil evidence for the antiquity of humans was found in the Neander valley of Germany in 1856. The finds—a skullcap and a few limb bones—were curious because while they had obvious human characteristics, they were also very primitive. Neanderthal Man, as the fossils were called, quickly became a subject of controversy. Some said they were the remains of an ancient man, while others said it was only the skeleton of a deformed modern human who died of bone disease. A few years later when Charles Darwin's *Origin of Species* was published, supporters held up Neanderthal man as proof that humans had evolved from more primitive, primate forms. Early critics of Neanderthal man were mostly anti-evolutionists who disagreed with this idea.

The work that really brought the man/ape link to public attention was published by Darwin's protege, Thomas Henry Huxley (1825-1895). Inspired by Darwin, Huxley's *Man's Place in Nature* (1863) argued the close affinity between humans and primates. In the first such systematic study, Huxley compared human and ape anatomy and found them startlingly similar, and therefore closely related. This closeness, he argued, was due to evolution. This finding lead to the concept of the "missing link." If man and apes were related by evolution, the idea went, then there should be a creature which was a transitional form, or link, between the two. Inspired by Huxley, Darwin then published *The Descent of Man* in 1873. Darwin's book also argued for the close relationship of humans and primates, but was more concerned with how various human characteristics were the result of evolution and sexual selection. Both Huxley's and Darwin's work was misinterpreted as saying humans evolved out of monkeys, when they actually said that humans and primates share a common ancestor.

In 1868 road workers at Les Eyzies, France, found skeletal remains and tools which were ancient yet very modern looking. Dubbed Cro-Magnon man (found in a "large shelter"), these fossils were more advanced than the Neanderthals and thus fell somewhere between them and modern humans. Near Spy, Belgium, more Neanderthal fossils were discovered in 1886. This proved that the first Neanderthal fossil was not a freak of nature, but part of a large population of human-like beings who lived in the very distant past.

German anthropologist Ernst Haeckel (1834-1919) lectured widely in the 1870s that man had descended from an ape-like ancestor. Haeckel was the leading supporter of Darwin and evolution in Germany. Since there were few Neanderthal or Cro-Magnon fossils known at the time, Haeckel postulated a hypothetical human ancestor he called *Pithecanthropus* which was more primitive than the Neanderthals and closer to primates on the evolutionary line. His work inspired Dutchman Eugene Dubois (1858-1940) to go searching for the remains of this creature. Haeckel said that since humans were related to the apes and apes live in the tropics, that would be the place to look (Darwin suggested that Africa was the cradle of humankind). Dubois, a professor of anatomy in Amsterdam, set off to Sumatra to find the missing link in 1887. After several fruitless years working a site along the Solo river near the village of Trinil, Dubois found what he was looking for in 1891. Consisting of a skullcap, femur, and a tooth, the creature seemed to be a man-like ape much more primitive than the Neanderthals. Using Haeckel's term, the find was labeled *Pithecanthropus,* but was commonly called Java Man (the modern designation is *Homo Erectus,* or the man who stands up straight). Dubois insisted that Java Man was the missing link between humans and apes. Haeckel claimed that Dubois's find proved that humans had descended from primates. Not everyone agreed, however.

A debate began over how to interpret what Dubois had found. Problems arose with the fossils themselves. Though he tried to work systematically, Dubois did not keep careful records of the rock layers, or strata, in which the fossils were found. Not knowing the strata makes it difficult to accurately date fossils. Also, the fossils had not all been found in the same spot. If they were not all found together, one could not be sure they belonged to the same skeleton. Dubois frantically showed his fossils to everyone he could, trying to convince other scientists that his discovery was genuine. The pressure began to show in Dubois. He took any criticism personally and eventually became so frustrated that he went into seclusion, refusing to show Java Man to anyone. For decades after, no one was allowed to view the fossils and it was rumored that Dubois hid them under the floorboards of his home.

In the continuing controversy, the Neanderthals were seen either as the ancestor of modern humans and a separate species, as an evolutionary dead-end not related to modern humans at all, or as a variation on the human line. The most consequential opinion on the status of the Neanderthals was that of French anatomist Marcellin Boule (1861-1942). In 1908 an almost complete Neanderthal skeleton was found near a site in France called La Chapelle-aux-Saints. The fossils were sent to Boule, an evolutionist who did not favor the straight-line, hierarchical evolutionary ladder popular at the time, but instead preferred to see evolution as a branching bush with many shoots leading off the main trunk. His study of the La Chapelle fossils, as well as other Neanderthal remains, was published as *Fossil Men* (1921). Though Neanderthal had been characterized as a dim-witted brute, Boule took the model to the extreme. He fashioned an image of Neanderthals as beast-like creatures, powerful, stoop-shouldered, animalistic in their behavior, communicating by grunts, and violent. For Boule the Neanderthals were more ape than man. It is from Boule that we get the classic image of the "cave man." Boule, as well as many others, were uncomfortable with the idea of the apparently degenerate Neanderthals as the ancestors of modern humans, especially modern Europeans. Using the branching approach to creating a human family tree, the Neanderthals could be safely pushed off to the side and out of the direct line of descent from modern humans. However, American paleoanthropologist Ales Hrdlicka (1869-1943) argued that Neanderthals were human, that they were *Homo Sapiens* (the smart man), and that they were variants caused by local environmental and dietary conditions. Hrdlicka and like-minded anthropologists saw human evolution as progressing slowly and smoothly. They argued that all the fossil humans from Java Man to modern people were one species and thus closely related.

Impact

The Neanderthals and other human-like fossils created problems for many in the Victorian world of the nineteenth century. Christian tradition held that God created Adam and Eve in a state of perfection six thousand years ago, but the Neanderthals were obviously far more ancient and primitive. The discovery of their fossils coincided with a wave of intellectual turmoil. Darwin's *Origin of Species,* the growing criticism of the literalness of the Bible, unrest brought on by the Industrial Revolution, and advances in physics rocked Victorian society. These new developments implied to many people that life was out of their control and spinning randomly through time and space without the guidance of an all-powerful creator. Man was not the epitome of all life or the product of divine design as once thought, but merely the result of the unpredictable movements of the cosmos. This convinced many Victorians that life was descending into chaos, society was in peril, and moral certainties were in jeopardy. If the Neanderthals were our ancestors, what were we, they wondered. The sure, unmoveable bedrock of Western society no longer seemed so solid.

The existence of fossil humans supported not only the general concept of evolution, but that people themselves had evolved. Even those who accepted evolution sometimes balked at accepting the brutish, ape-like Neanderthals as their ancestors. The cave men were used by various people to further their political and social agendas. The powerful head of the French scientific establishment, Georges Cuvier (1769-1832), an ardent anti-evolutionist, stated flatly that there could not be any human fossils because man had been recently created by God. If the Neanderthals and Java Man were our ancestors, it suggested modern humans were not the creation of God and were therefore closer to apes than angels. Western society's economic, corporate, and political claim of cultural superiority rang less true if the blood of the Neanderthals ran through its veins. Men were not monkeys, critics claimed, because that idea undermined social order. In a famous 1861 debate over evolution held at Oxford University, Bishop Samuel Wilburforce (1805-1873) chided T. H. Huxley by asking if he was related to a monkey on his mother's or father's side. Huxley, who had agendas of his own, fired back by saying he would rather be related to a monkey than a man like Wilberforce who introduced ridicule into such an important discussion.

Others saw the hulking Neanderthals not as ancestors of Europeans, but of Africans and other people of color. In the early twentieth century the American Henry Fairfield Osborn (1857-1935) commissioned Charles R. Knight (1874-1953) to do a series of illustrations for his book *Men of the Old Stone Age* (1916). Under Osborn's direction, Knight depicted Neanderthals as dark-skinned, slow-witted cave men (in the mold of Boule), while the more modern Cro-Magnons were shown as light skinned, intelligent, and creative. The pictures were reprinted over and over and helped establish the view of the "cave man" in the popular imagination that still holds. The illustrations were a subtle attempt to extend racial stereotypes to the past. If, as the evolutionists claimed, all humankind descended from a single origin, then all people were related and part of the same family. As such, oppression of one group of people was not the oppression of some "other," but of themselves. The Victorian world was unwilling to accept such an idea.

The problem of the Neanderthals and other fossil human ancestors for nineteenth- and early twentieth-century society was that they called into question beliefs about divine creation and our relationship with God, as well as our relationship to each other. If man was just an intelligent ape who evolved over tens of thousands of years or more, then he might not be quite as special as he thought himself to be. To the Victorians (and many still today) human descent from

animals denied man's central place in the universe as God's special creation and questioned his place in nature.

BRIAN REGAL

Further Reading

Boule, Marcellin. *Fossil Men.* London: Gurney and Jackson, 1923.

Huxley, Thomas Henry. *Man's Place in Nature.* New York: D. Appleton and Co., 1896.

Lewin, Roger. *Bones of Contention.* New York: Simon & Schuster, 1988.

Osborn, Henry Fairfield. *Men of the Old Stone Age.* New York: Charles Scribner's Sons, 1915.

Shreeve, James. *The Neandertal Enigma.* New York: William Morrow and Co., 1995.

Van Riper, A. Bowdoin. *Men Among the Mammoths: Victorian Science and the Discovery of Human Prehistory.* Chicago: University of Chicago Press, 1993.

Biographical Sketches

John James Audubon
1785-1851
French-American Naturalist and Artist

John James Audubon is famous for the brilliant artistry of his *Birds of America* and his *Viviparous Quadrupeds of America,* completed between 1827 and 1854. Audubon was the son of his father's Creole mistress, who died when he was seven months old. He was formally adopted by his father, a French naval officer, and his father's French wife, who educated him at home. He briefly attended Rochefort-Sur-Mer Naval Academy in 1796.

Audubon's earliest drawings date from the year 1805. His claims that he had studied with the French painter Jacques Louis David in 1802-03 are now regarded as invalid by modern biographers. He briefly visited the United States in 1803, and in 1806 began a business career in New York, Kentucky, Missouri, and Louisiana that ended in bankruptcy in 1819. He married Lucy Bakewell in 1808. They had two daughters (both of whom died in infancy) and two sons, Victor Gifford (b. 1809) and John Woodhouse (b. 1812), who later helped their father in his artistic work. Lucy Audubon was a teacher

whose earnings provided much of the family's income for years.

Beginning in 1819-20, with a short-lived position as a taxidermist in Cincinnati, Audubon was an itinerant artist, clerk, and tutor. By 1820 he decided to devote full time to science and art. He planned a large scale set of paintings depicting all the known American birds in elephant folio size and began painting them. Between 1826 and 1839, he visited England four times, supervising engravings of his paintings while soliciting subscriptions from well-to-do individuals and institutions in Europe and the United States. The plates for his *Birds of America* were published in four volumes (1826-1838), followed by a summary and index volume, *A Synopsis of the Birds of America* (1839). The *Ornithological Biography* (published in five volumes between 1831 and 1839), designed as the accompanying text to his paintings, was combined with the pictures in a smaller octavo edition between 1840 and 1844. It was the most commercially successful of any natural history book published anywhere in the world to that time, and it was frequently reprinted. The income from this series provided Audubon's family with their first permanent home, in New York City, in 1842.

Audubon's last project was a study of American land mammals, the *Viviparous Quadrupeds of North America,* with the paintings published in two volumes (1845-46). The three-volume text, by North Carolina minister and mammalogist John Bachman (1790-1874), appeared between 1846 and 1854. Text and pictures were combined in later editions. Audubon's eyesight began to fail in 1846, and by the next year he had become senile. His sons completed the paintings, and with Bachman's help brought the project to a conclusion. John Woodhouse Audubon was the more gifted animal artist, while Victor painted backgrounds and served as business manager for the family publishing activity. They were assisted by the Reverend Bachman's second wife, Maria, who completed many of the insect and flower backgrounds for both the *Birds* and the *Quadrupeds.* Today, it is difficult for experts to tell which mammals were painted by John Woodhouse and which were completed by his father. Both sons married daughters of the Reverend Bachman, but tragically, both brides died young. Victor and John both remarried but died not too long after their father, in 1860 and 1862, respectively.

After her sons died, Lucy Audubon resumed teaching to help support herself and her fifteen grandchildren. In 1863, short of income, she sold all but three of the original bird paintings done by her husband to the New York Historical Society for $4,000. Most of the copper plates used in printing the Elephant Folio engravings were melted down for the value of the metal; several dozen survive today at the American Museum of Natural History in New York City and other institutions. Audubon's fame rests on the lifelike appearance and the graceful artistry evident in the *Birds of America* and the *Quadrupeds,* which have been many times reprinted. Audubon, however, sometimes had to work rapidly, before his specimens decayed, and in some cases the birds and mammals were painted in somewhat stiff and unnatural poses. His work, however, has been very influential in familiarizing Americans with the fascinating variety of North American birds and mammals. Toward the end of his life, he expressed concern about the declining numbers of certain species, and he has become identified with the beginnings of America's conservation movement.

KEIR B. STERLING

Karl Ernst von Baer
1792-1876
Estonian Biologist

Karl Ernst von Baer was an Estonian biologist who discovered the mammalian ovum—the reproductive egg in female mammals. He made significant contributions to the study of the embryonic development of animals.

Born in Piep, Estonia, to parents descended from Prussian nobility, Baer studied medicine at the University of Dorpat in Estonia. He continued his studies in Vienna, Austria, and later in Würzburg, Germany. He then accepted a position to teach anatomy, anthropology, and zoology at the University of Königsberg.

From his investigations at Königsberg arose one of Baer's most important discoveries. Scientists had long been trying to determine the nature and location of the mammalian egg. In 1672 Regnier de Graaf (1641-1673) discovered follicles in the ovaries and believed that the follicles themselves might be eggs. When he found structures in the uterus that were even smaller than the follicles, the role of follicles was thrown into doubt. During his research, Baer, using a microscope, found a structure within the follicle. He concluded that this structure was the mammalian egg. Baer's discovery, which he expanded on in his 1827 treatise, called in English *On the Origin of the Mammalian and Human Ovum,* confirmed that mammals develop in a manner essentially like that of other animals.

Baer's other great achievement was his theory on embryonic development. In *On the Development of Animals* (1828-1837), he argued that the embryos of all animals develop from a simple and homogeneous stage to a complex and heterogeneous one. Suggesting that the younger the embryos of different species are the closer the resemblance between them, he showed that during embryonic development general characteristics of animals appear before traits specific to a species. This idea came to be known as the "biogenetic law." In addition, Baer argued that vertebrate embryos form distinct layers that later differentiate into all the organs or structures of the animal. These layers are the ectoderm, mesoderm, and endoderm. This theory is called the germ layer theory of development.

In his studies of the embryo, Baer was the first to show that mammalian reproduction involves the fusion of an ovum with sperm, not the mingling of female and male seminal fluids as had been previously thought. He discovered and described the functions of the extraembryonic membranes—the chorion, amnion, and allantois. Baer also discovered the notochord in early vertebrate embryos as well as the neural folds. (The notochord in vertebrates develops into the vertebrae, or backbone, while the neural folds are concerned with formation of the nervous system.)

Following his tenure at Königsberg, Baer accepted a position as librarian at the St. Petersburg Academy of Sciences. He was also appointed Professor of Comparative Anatomy and Physiology at the Medico-Chirugical Academy in St. Petersburg. While in St. Petersburg he took part in various scientific expeditions, including many to Novaya Zemlya, in the Russian Arctic, where he was the first naturalist to collect plant and animal specimens. Baer's greatest accomplishments, however, would remain the discovery of the mammalian egg and the publication of his theory on embryonic development.

GARRET LEMOI

Spencer Fullerton Baird
1823-1887
American Zoologist and Government Official

Spencer Baird is best known as a skilled biological investigator who shaped government agencies to popularize and advance professional zoological science and who applied much of that knowledge for practical ends.

Born in Reading, Pennsylvania, the son of an attorney, Baird was an 1840 graduate of Dickinson College. After briefly studying medicine in New York, he returned to Dickinson, secured his M.A. degree, and taught natural history and chemistry there for four years.

In 1850 Joseph Henry, the first secretary of the recently formed Smithsonian Institution in Washington, invited Baird, then age 27, to serve as assistant secretary. Baird spent the rest of his life at the Smithsonian. Early in his tenure he created and organized what would become the United States National Museum (now the National Museum of Natural History), the first federal government agency concerned with the study of America's animal resources.

Unable for reasons of health and work responsibilities to undertake much active field research on his own, Baird identified a number of junior military officers and young civilians with an affinity for natural history. These workers went out into various parts of the country and brought animal specimens and information back to Washington. Baird arranged to have many of the workers assigned or attached to the Pacific Railway Survey expeditions of the 1850s, which investigated various routes for an intercontinental railroad to the West Coast. Baird supplied his young associates with the necessary instructions, relevant Smithsonian publications, and collecting supplies, but little money, as he was chary about spending the limited funds at his disposal.

By the late 1850s Baird had sufficient specimens and other data on hand to compile volumes on American mammals (volume 8 of the *Reports of Explorations and Surveys to Ascertain the Most Practicable and Economical Route for a Railroad from the Mississippi River to the Pacific Ocean,* 1857) and birds (volume 9 in the same series, 1858). Both were later published in popular editions (*Mammals of North America,* in 1859 and *The Birds of North America* in 1860), which summarized what was then known about their subjects. Baird also co-authored *A History of North American Birds* (3 volumes, 1874) and *The Water Birds of North America* (2 volumes, 1884). In addition, he contributed to several other volumes concerning the reptile collections brought back by the Railroad Surveys. He continued to find places for younger zoologists on the various geographical and geological surveys of the western territories in the late 1860s, 1870s, and 1880s. Most of these men contributed additional specimens to the National Museum's collections, which, with many subsequent additions, are today among the world's most comprehensive.

In 1871, following some years of increasing interest in marine biology, Baird organized—and, to the end of his life, financed out of pocket—the United States Fish Commission. This agency encouraged the conservation and propagation of food fish and research in marine biology while providing support to the commercial fishing industry. It served as the foundation on which most present-day federal fishery efforts are based.

Baird was a pioneer among American biogeographers, who were concerned with the biological and geographic factors that influence the distribution of life on Earth. He was an early exponent of conservation and also worked to make scientific information available to the general public. For example, he edited eight volumes in a series, the *Annual Record of Science and Industry,* which were commercially published between 1872 and 1879.

When Joseph Henry died in 1878, Baird became the second Secretary of the Smithsonian, serving until his death nine years later. Baird's accomplishments as a builder of several federal scientific agencies, as founder of the National Museum collections, and as mentor of many younger American naturalists were all of considerable importance. In these roles, and as a synthesizer of information, his work was crucial, particularly to the later development of American vertebrate zoology and museum administration.

KEIR B. STERLING

Augustin Pyramus de Candolle
1778-1841
Swiss Botanist

Augustin de Candolle is considered one of the most important botanists of the nineteenth century. His major contributions were in the fields of plant classification and morphology, the study of form, and in the geographical distribution of plants.

Candolle was born in Geneva and his early studies were in Switzerland. In 1796 he moved to Paris, where he studied medicine and the natural sciences and where he often visited the natural history museum, which was a center for research. He came to know leading scientists of the day, including zoologist Georges Cuvier (1769-1832) and the early advocate of evolution, Jean-Baptiste de Lamarck (1744-1829). These connections increased Candolle's interest in biology to the point where he abandoned medicine and devoted himself to botany, focusing on the investigation of plant structure.

Candolle remained in Paris until 1808, when he was made chair of botany at the College of Medicine in Montpellier in southern France. From there, he moved back to Switzerland in 1816, when he became chair of natural history and director of the botanical garden at the Academy of Geneva. He retained these positions until 1835, when he retired and his son, Alphonse, also a noted botanist, took over his work. While at the Academy, Augustin Candolle created an impressive herbarium, a collection of dried plant specimens, that continues to be studied by botanists to this day.

The primary focus of Candolle's research was in taxonomy, a word that Candolle himself coined. It refers to the theoretical study of classification, of how organisms are sorted into categories. Candolle's work was influenced by that of the great French botanist, Antoine-Laurent de Jussieu (1748-1836). Both favored a natural system of classification as opposed to the artificial system of Carl Linnaeus (1707-1778), who had created the first widely accepted classification system. In this system a single feature, the structure of the flower, was used as the basis of classification. Plants with similar flowers would be placed in the same category even though they might differ greatly in other characteristics. A natural system, on the other hand, is based on decisions about overall similarity, with many characteristics being examined and considered.

Candolle was also influenced by the botanical work of the great German poet, Goethe (1749-1832), who was an enthusiastic student of botany. Goethe argued that there was a general plan or form that underlay all plant structure, with particular species being variations on this general theme. Candolle, too, saw unity underlying the diversity of plant form. He also agreed with Goethe that the parts of the flower were all related to the leaf form—that petals, for example, could be seen as modified leaves. This idea was very influential in the nineteenth century and reflected an interest in finding a way to simplify or unify the study of plants.

Candolle studied many plant families and wrote books on such plants as lilies and cacti. He also published several impressively illustrated works, including one with paintings by perhaps the most famous botanical illustrator of all time, Pierre Redouté (1759-1840). In terms of contributions to botany, Candolle's most significant work was called the *Prodromus*, which ultimately ran to 17 volumes, several of which were written and published after his death by his son Alphonse. This work was a description of the whole range of plants and attempted to include all known species. It presented information not only on classification but on ecology, agriculture, and phytogeography, which is the study of how climate and terrain influence the distribution of plant species. This last area is where Candolle made his most original contributions, and he is considered one of the founders of this field.

MAURA C. FLANNERY

Ferdinand Julius Cohn
1828-1898
German Botanist and Bacteriologist

Ferdinand Julius Cohn, a German botanist, is recognized today as a founder of bacteriology. He was adept at observing and describing the life cycles of microorganisms. This talent led him, in the 1870s, to construct the first classification system for bacteria.

Cohn was the first of four sons born to Isaac and Amalie (Nissen) Cohn in Breslau, Lower Silesia (now Wroclaw, Poland). His parents lived in Breslau's Jewish ghetto with few financial means until they set up a business selling rapeseed oil for lamps.

Cohn was a precocious child who learned to read by the age of two and began to study natural history at the age of four. He studied at the School of Master Weber from the age of four until six, then entered the Gymnasium of St. Maria Magdelena in 1835. He advanced steadily until the age of 10, when a hearing deficiency arose. Due to this ailment, he suffered from emotional retardation that he did not begin to overcome until 1840. His partial deafness accompanied him throughout his life in spite of the advice and "miraculous remedies" of many doctors. A six-month trip to Berlin with a friend bolstered his ego and upon graduation in 1844, he felt prepared for the freer and more scientific life of the university.

Cohn first attended the University of Breslau as a student in the philosophical faculty, where he studied a wide variety of subjects including astronomy, philosophy, and chemistry. Influenced in part by his professors Heinrich Göppert and Christian Nees von Esenbeck, Cohn chose botany to be his main course of study.

Although Cohn grew up in a period of partial liberation of earlier restrictions on Jews, he was nevertheless barred from the degree examination at Breslau. After several of Cohn's teachers, and his family's friends, failed to obtain a governmental waiver, he transferred to the University of Berlin in October 1846. He eventually regarded this as a fortunate turn of events, for it was in Berlin that he was first introduced, by Johannes Müller (1801-1858), to a new style of natural sciences. During this period, Cohn found a patron in Eilhard Mitscherlich (1794-1863), a chemist who also investigated botanical questions, and met Christian Ehrenberg (1795-1876), who introduced him to the study of microbes. On November 13, 1847, at age 19, Cohn graduated from the University of Berlin with a doctorate in botany.

Cohn remained in Berlin during the turmoil of the 1848 revolution, after which he made his permanent home in Breslau. There he joined the Silesian Society for the Natural Sciences and later directed its botanical section. After 1850, Cohn researched a wide variety of topics including the sexuality of algae and fungi, plant tissues and organs, and the effect of light on microscopic plants. His most lasting influence, however, was his research on bacteria.

In 1850 he was promoted to *Privatdocent* at the University of Berlin and gained recognition for his work on the microorganism *Protococcus pluvialis*. This work represents a shift away from the cell theory of Matthais Schleiden (1804-1881) and its focus on organelles located in the protoplasm. In 1854 Cohn published "On the Development of Microscopic Algae and Fungi," in which he reorganized the classification systems of Otto Friedrich Müller (1730-1784), Ehrenberg, and Felix Dujardin (1801-1860). In 1859 Cohn was promoted to extraordinary professor of botany. He married Pauline Reichenbach in 1867.

In 1866 Cohn convinced his university to organize an institute of plant physiology, which he directed. He edited the institute's journal *Studies on the Biology of Plants,* in which he published his "Investigations on Bacteria." Cohn's research and publications, particularly his proposal that bacteria can be arranged in genera and species, helped to found the study of bacteriology.

LLOYD ACKERT

Georges Cuvier
1769-1832
French Anatomist and Paleontologist

Georges Cuvier was France's leading naturalist and the father of paleontology and comparative anatomy. Born in the Jura mountain region of France on August 23, 1769, Cuvier attended the Carolinian Academy in Stuttgart, Germany. Cuvier had always collected natural objects and was fascinated by the study of plants and bugs. A job as a personal tutor to a noble family on the Northern coast of the country kept him clear of the turmoil of the French revolution. It also sparked his interest in marine life. He then received a minor government job where he began his career as a naturalist. In 1795 he traveled to Paris and was appointed professor of animal anatomy at the new National Museum of Natural History. He was later made Inspector-General of Public Education by Napoleon. He eventually held positions under several different and often opposing governments.

Georges Leopold Cuvier. *(The Library of Congress. Reproduced by permission.)*

As a naturalist Cuvier developed several important scientific ideas. He argued that the individual parts of an organism operated as a single unit, with each unit playing an integral part in the whole. No part could alter its function without adversely affecting the entire organism. This idea of biological integration convinced Cuvier that evolution—or transmutation as it was then known—was a fallacy because if one part of the organism changed through evolution, it would throw the entire organism out of alignment. Mummified cats and tomb artwork brought out of Egypt by the French army convinced Cuvier that since the ancient Egyptian cats were no different from modern ones, no evolution had taken place. He also studied the bones of elephants discovered in and around Paris. He saw that, though they were elephants, they were not exactly like any known species. In 1796 he began to suggest his theory of extinction to account for this discrepancy. A devout Christian, Cuvier thought living things came into being through the special creation of God. Cuvier had become a powerful political figure in French science and used his power to discredit those who supported the idea of evolution there. Several French naturalists—Georges Louis Leclerc Buffon (1707-1788), Geoffroy Saint-Hilaire (1772-1844), and Jean Baptiste de Lamarck (1744-1829)—had put forward rudimentary theories of evolution. Due to Cuvier's intransigence and active opposition, their work, especially that of Lamarck, was kept from wide attention and fell into obscurity. Cuvier also opposed the antiquity and evolution of humankind.

While Cuvier staunchly opposed evolution, he did establish the concept of extinction. Fossils had long been recognized as the remains of ancient life forms, but few had any idea of what had happened to these animals. Some suggested they still existed in a remote part of the world where they waited to be discovered. In his 1812 book *Researches on the Fossil Bones of Quadrupeds,* Cuvier argued that the earth had gone through a series of titanic geologic upheavals he called revolutions. These catastrophes, as others called them, occurred on a worldwide scale and rearranged the face of the earth, obliterating many species of plants and animals. These upheavals, Cuvier claimed, caused the extinction—or complete eradication—of species. The planet was then repopulated by special creation, not evolution. By doing studies of these extinct fossil forms, Cuvier created the science of paleontology.

Cuvier's other contribution to biology was the technique of comparative anatomy. He separated living things into various "branches," like the vertebrata (back boned animals) and the mollusca (symmetrical invertebrates), which he said were unrelated. If they looked alike, he claimed, it was due to having similar functions, not because they evolved from one another or a common ancestor. By comparing the parts of different organisms he could learn about their function. While it was not his intention for it to do so, Cuvier's discipline of comparative anatomy was used by others for the study of how organisms are related and how they evolved. This approach also allowed for the reconstruction of what an animal looked like from fragmentary fossil evidence. If, for example, an animal's tooth structure is known, a good deal can be extrapolated from the teeth to give a rough idea of what the entire animal looked like. These are basic techniques still in use by biologists and paleontologists today.

BRIAN REGAL

Charles Robert Darwin
1809-1882
English Naturalist

Charles Robert Darwin is best known for his hypothesis that natural selection is the driving force behind evolution. Although the idea

Charles Darwin. *(The Library of Congress. Reproduced by permission.)*

that species evolve over time was not a new one, Darwin's famous work, *On the Origin of Species by Means of Natural Selection* (1859), brought public attention to the hypothesis and presented a field of evidence for natural selection. Simultaneously and independently, fellow scientist Alfred Russel Wallace (1823-1913) presented similar findings.

Darwin grew up in Shrewsbury, England, the fifth of eight children to Robert and Susannah Darwin. His grandfathers were quite famous—Erasmus Darwin (1731-1802) was a noted physician and poet, and Josiah Wedgewood was known for his porcelain—although both died before his birth. Although he initially enrolled for his higher education at Edinburgh University in medicine, his interests as a naturalist won out when he joined Cambridge and became involved in entomology, geology, and botany.

With a bachelor's degree from Cambridge newly in hand, Darwin accepted the position of unpaid naturalist on the HMS *Beagle*, a small cruiser that embarked on a five-year voyage to South America at the end of December 1831. During the trip, Darwin took meticulous notes on the geology, flora, and fauna of the coastal locations they visited. He collected many animal specimens, and supplemented them with descriptions of their location, range, and habits.

During the long voyage, he also began to notice slight differences between animals living on various islands, including the Galapagos, and also between living animals and fossils, raising initial questions that would later become the subject of his landmark research concerning evolution.

After Darwin returned to England, he published *Journal of Researches* (1939) and *The Structure and Distribution of Coral Reefs* (1842). He also edited the four-volume *Zoology of the Voyage of the Beagle* (1839-1843), which included material from systematists who studied the collections made during the voyage. The well-received books gave Darwin status as a top scientist. He married first cousin Emma Wedgewood in 1839, and four of their children became distinguished scientists in their own right.

Over the next two decades, Darwin began to formulate ideas about evolution and natural selection by combining the data he gathered on the *Beagle* voyage, his work with specimens he harvested on the trip, studies of selective breeding in domesticated animals, and previously proposed ideas from scientists like Thomas Malthus (1766-1834) and Charles Lyell (1797-1875). Most of this work Darwin did at his home, called Down House, in the village of Downe. Chronic fatigue and intestinal ailments forced him to work in seclusion. Although it was undiagnosed at the time, Darwin was likely suffering from the insect-borne Chagas' disease, which he had probably contracted during the *Beagle* voyage.

Darwin began to compose his thoughts about a method for evolution in the early 1840s, but withheld his notes after somewhat similar views by Robert Chambers (1802-1871) in 1844 drew fierce criticism. In the 1850s he worked out additional details, and in 1858 learned that Wallace was developing basically the same hypothesis of evolution by natural selection. Wallace and Darwin wrote a joint paper and presented it to the Linnaean Society in 1858.

It wasn't until 1859, however, and Darwin's publication of *On the Origin of Species by Means of Natural Selection, or the Preservation of Favoured Races in the Struggle for Life* that the ideas drew widespread interest. With that interest came a good deal of misinterpretation and even condemnation. Darwin remained secluded at Down House, fielding criticisms and releasing additional editions of *Origin of Species*.

For the rest of his life, Darwin continued to develop his hypothesis, and in 1871 published

The Descent of Man and Selection in Relation to Sex, which presents the topic of sexual selection and the place of humans along the evolutionary path. In addition, Darwin began to place an increasing focus on domesticated plants and animals in his later years. From 1860-80 he published works on orchids, climbing plants, and insectivorous plants.

Charles Darwin died on April 19, 1882 at the age of 73, and was buried at Westminster Abbey.

LESLIE A. MERTZ

Ernst Haeckel
1834-1919
German Embryologist

Ernst Haeckel was among the first German biologists to adopt Charles Darwin's ideas about evolution in the latter half of the nineteenth century. He popularized evolutionary thought on the European continent and added many novel concepts to biology and the social sciences. Haeckel is most often remembered for his "law of recapitulation," the belief that the evolutionary history of any given animal is repeated during its embryological development.

Haeckel was born in Potsdam, Germany, to middle-class parents, Carl and Charlotte. He was a curious naturalist from the start, studying botany in grammar school and cultivating his artistic skills to produce sketches and watercolors of living things. He began medical studies at the age of 18 and was licensed as a general practitioner, surgeon, and obstetrician at the age of 25. But soon after Charles Darwin's *Origin of Species* was published in 1859, Haeckel read a German translation of it and abandoned his medical practice. After three years of intense study he became a professor of comparative anatomy at the University of Jena.

The chief objects of study for Haeckel were invertebrates—small animals without backbones, including sponges and segmented worms. He traveled widely in search of unusual forms, exploring throughout Europe, North Africa, and Asia. In one trip to the Mediterranean he named almost 150 new species of radiolarian, marine protozoa with spiny shells. Haeckel's careful embryological studies of invertebrates formed the core of his evolutionary theories.

Haeckel met Charles Darwin (1809-1882) in 1866, and they corresponded for many years. The German embryologist was one of the most

Ernst Haeckel. *(The Library of Congress. Reproduced by permission.)*

prolific supporters of evolution in continental Europe, but he diverged from Darwin's thoughts in many important respects. While Darwin proposed natural selection as the mechanism of evolution, each species changing gradually through the selective survival of advantageous traits, Haeckel's concept of evolution was steeped in his embryological discoveries. He asserted that new species evolved through novel steps at the end of their embryos' development. This integral role of embryology in the evolution of animal life was at the base of Haeckel's law of recapitulation.

As Haeckel put it, and has been quoted ever since, "ontogeny recapitulates phylogeny." Ontogeny, being the development of an individual animal, is an instant replay in fast-motion of phylogeny, the evolution of that animal's lineage through time. This was not a completely new idea; it was common among embryologists who had long observed the early worm-like or fish-like embryos of higher animals. But Haeckel did more than anyone to popularize and refine the concept. He even proposed an ancestral organism, a "gastraea," similar to the first hollow ball of cells that make up an embryo, as the hypothetical ancestor of all multi-celled animals, stimulating other hypotheses in this direction.

Haeckel's vision of evolution also diverged from Darwin's in its overall shape. While Darwin imagined a tree or bush, with all species more or less connected by spreading branches, Haeckel proposed an unbranching line. Each animal was related to the next up the chain, he reasoned, by added changes in embryonic development. Haeckel's concept was a scientific rendering of the Great Chain of Being, a ranking of living things that had occupied much of Western thought for centuries and inevitably put human beings literally at the top.

The law of recapitulation gained some popularity in Europe during Haeckel's generation but was largely discredited by the time of his death. Haeckel's thoughts have had a more lasting effect on the public perception of evolution, and on the social sciences. His linear vision of evolution not only supported a ranking of species, with humans at the top, but a ranking of humans from the more "primitive" (typically non-Caucasian) to the more "evolved" European races. Haeckel and his followers gave scientific validity to racial prejudices, ultimately leading to the tragic genocidal policy of Adolf Hitler and the Third Reich during the 1930s and 1940s. Despite its flaws, obvious to us now, recapitulation remained a central paradigm of biology into modern times.

PHILIP JOHANSSON

Thomas Henry Huxley
1825-1895

British Anatomist, Paleontologist and Zoologist

T. H. Huxley was a major figure behind the propagation of Darwin's theory of evolution and a noted advocate of science education. Huxley contributed to the growing study of the classification of organisms by studying fossils. He was instrumental in shifting the emphasis in paleontology away from geology (the study of the rocks in which fossils were found) to biology and the study of the structure of the fossil itself.

Huxley had little formal education as a youth. Instead, he taught himself by reading scientific and religious works, studying nature, and doing experiments on his own. In 1841 he became an apprentice to his physician brother-in-law, and the next year was awarded a scholarship to London's Charing Cross Hospital, allowing him to gain intensive study of anatomy and physiology as well as the scientific method. After graduating in 1845, Huxley enlisted in the Royal Navy and was eventually assigned to the HMS *Rattlesnake*, as ship's surgeon, for a survey trip to

Australia. Though it was not his job to do so, Huxley spent much of his time on the journey studying the zoology of the ocean by collecting specimens and dissecting them. He became fascinated by the tiny plankton that swarmed the sea and was able to clear up the somewhat muddled classification system other scientists used on these animals. Some of Huxley's findings were published back in London and he was elected to the prestigious Royal Society. Eventually, he left the navy and entered the British scientific world. In 1854 he was appointed professor of natural history at the School of Mines and started his study of paleontology and geology. From a working class family himself, Huxley wanted to extend the opportunity of an education to others who would not normally have access to one. In 1885 he began his so-called workingman's lectures in science for the general public. He gave popular talks on biology and other aspects of science to enthusiastic crowds.

Huxley was involved in several important scientific feuds in his career. The first was with Richard Owen (1804-1892), the established head of the British Natural Science community. The argument was over whether the newly discovered dinosaurs were closer to mammals, as Owen argued, or reptiles as Huxley did (we now know that Huxley was right). Owen and Huxley also fought over the relationship between man and the primates. Owen said that the similarities between humans and apes were only superficial, while Huxley argued that the two were closely related due to evolution. Huxley defended Charles Darwin (1809-1882) with such zealousness that he was dubbed "Darwin's Bulldog." By the time Darwin's *Origin of Species* was published in 1859, he and Huxley were already friends. Darwin sent him an advance copy and Huxley was immediately taken with the elegance of the work and became convinced that organic evolution was a reality. As a major early defender of Darwin, Huxley gained wide renown for his famous (1805-1873) encounter with Oxford bishop Samuel Wilberforce. In a public debate in 1860 Wilberforce asked Huxley through which side of his family he was related to the apes. Huxley replied that he would rather be related to an ape than a man who introduced ridicule into a scientific discussion.

Though a supporter of Darwin and evolution in general, Huxley was not initially convinced that Darwin's idea of natural selection as the agent of evolution was supportable. As an empiricist (someone who requires verifiable facts before accepting a theory) Huxley was troubled by the lack of fossil evidence to prove Darwin's idea. He wanted to be able to do experiments or otherwise check to see if natural selection worked. Traveling to America in the 1870s, he met with paleontologist O. C. Marsh (1831-1899), who had put together an impressive fossil collection showing the progression of extinct to modern horses. This was the proof Huxley needed. He proclaimed Darwin's theory to be vindicated by physical evidence and accepted it completely.

Huxley added his voice to the debate about the social aspect of evolutionary theory. Many were troubled by Darwinian evolution because it suggested that life was without direction and thus meaningless. Huxley countered these evolutionary pessimists by saying that if there was no direction or purpose given to life by God, then man must take the initiative and give himself purpose and a reason to live. His optimistic view—and distrust of religious dogmatism—led him to coin the word "Agnostic" in 1870 to describe someone who might believe in God, but who saw no proof for His existence.

BRIAN REGAL

Ida Henrietta Hyde
1857-1945
American Physiologist

Ida Hyde was a pioneering woman scientist whose academic and professional successes established a foundation for future female researchers. Earning an esteemed foreign doctorate, Hyde set precedents for other women who were interested in studying with international experts. She also enabled women scientists to have access to more laboratories with superb equipment and facilities. By securing university recognition for her expertise, funding scholarships, evaluating scholarly candidates, and establishing groups supportive of female researchers, Hyde assured opportunities for talented, dedicated women scientists to achieve desired educations and careers.

Born in Davenport, Iowa, Hyde was the daughter of Meyer H. and Babette (Loewenthal) Heidenheimer. Her parents changed their surname to Hyde when they emigrated from Wurttemberg, Germany, to Chicago. Hyde worked in a millinery shop and attended night classes at the Chicago Athenaeum before enrolling at the University of Chicago in 1881. After one year,

she taught Chicago elementary students, initiating science courses in public schools.

By 1888 Hyde had enrolled at Cornell University, completing her bachelor's degree in biology within three years. She earned a scholarship to study biology at Bryn Mawr College, where she studied jellyfish embryos. A professor at the University of Strassburg, Alexander Wilhelm Goette, asked Hyde to collaborate with him. Although women were unwelcome as students at the university, Goette insisted Hyde have laboratory privileges. When the faculty refused to grant her a Ph.D., Hyde contacted the University of Heidelberg, asking if she could receive a doctorate from that institution. A group of administrators said that school's rules did not prohibit female doctoral students. Hyde participated in zoology and chemistry classes but was denied access to the physiology courses of Wilhelm Kuhne (1837-1900) at the medical college. Despite this obstacle, Hyde was the first female to earn a Ph.D. at Heidelberg, finishing with honors in 1896.

The university paid for Hyde's research trip to the Naples Zoological Station, where she studied octopus salivary glands. She also conducted investigations at the University of Bern, where she was introduced to Henry P. Bowditch (1840-1911) of the Harvard Medical School. Bowditch arranged for Hyde to become the first woman employed to research at Harvard's medical laboratories. From 1896 to 1898 she analyzed heart blood flow. In 1898 Hyde accepted the position of assistant professor of zoology at the University of Kansas. She was selected as the first female member of the American Physiological Society in 1902. Three years later, Hyde was named head professor of the new Department of Physiology and also received a university medical school position. She initiated a program to detect schoolchildren with contagious diseases and presented public health lectures. She studied several summers at Chicago's Rush Medical School, the University of Liverpool, and the Marine Biological Laboratory in Woods Hole, Massachusetts. Hyde wrote two textbooks—*Outlines of Experimental Physiology* (1905) and *Laboratory Outlines of Physiology* (1910).

Angered by conflicts with students concerning academic performance, Hyde resigned in 1921. She returned to Heidelberg to study radium's biological effects. She also conducted the first scientific studies of how music influences physiology. Hyde invented the microelectrode, a device for stimulating a single cell which enabled researchers to further understand cellular properties and behavior. She published an article in the 1921 *Biological Bulletin* about her device. Concerned about assisting female scientists, Hyde established a group that accrued funding for American women to research at the Naples Zoological Station. She also secured laboratory space for women at Woods Hole and served on an accreditation board to insure women applicants were qualified to pursue graduate science work at foreign universities. Hyde financed scholarships for women scientists. She wrote about her professional frustrations in an article, "Before Women Were Human Beings." Hyde moved first to San Diego then Berkeley, California, where she died.

ELIZABETH D. SCHAFER

Jean-Baptiste Pierre Antoine de Monet, Chevalier de Lamarck
1744-1829

French Naturalist and Systematist

Jean-Baptiste de Lamarck. *(The Library of Congress. Reproduced by permission.)*

Jean-Baptiste de Lamarck did much to clear the path toward the current theory of evolution and natural selection. While some of his ideas about evolution were later proven false, others led later scientists, including Charles Darwin (1809-1882), to develop the concepts that now form the basis of the biological sciences. Another of Lamarck's great contributions was the classification of invertebrates, a term he coined for animals lacking a backbone.

Lamarck was born in Bazentin-le-Petit in Picardy, north of France, the youngest of 11 children. His family fell into the upper class socially but not financially, and he would struggle with money matters throughout his life.

After a few years in a Jesuit seminary, Lamarck decided to continue his family's history of military service and joined the French army in 1761. When, seven years later, an accidental injury ended his military career, he began to pursue scientific studies with an emphasis on botany.

His botanical work led to the 1778 publication of the greatly successful *Flore Française,* which described the plants of France. The book helped Lamarck win admission to the French Academy of Sciences in 1779 and to land a position as assistant botanist at the Jardin des Plantes, which specialized in its botanical gardens, along with studies of medicine, chemistry, and biological sciences.

A reorganization of the Jardin des Plantes into the Musée National d'Histoire Naturelle (National Museum of Natural History) in 1793 eliminated the need for Lamarck as a botanist and instead catapulted him into the position of professor of the natural history of "insects and worms." Although he was nearly 50 years old and lacked a background in the field, Lamarck quickly switched gears and began a study of these animals, for which he introduced the term invertebrates.

At the museum, Lamarck became a renowned systematist. He recognized the differences between insects, spiders and other arachnids, and worms and other annelids, and he classified them into separate groups. He explained his classification system in *Systéme des animaux sans vertébres* (*System of Invertebrate Animals*) published in 1801, and the seven-volume *Histoire naturelle des animaux sans vertébres* (*Natural History of Invertebrate Animals*), published from 1815-1822.

Shortly after the turn of the century, Lamarck began to voice his views on evolution. He believed that species changed over time, and that current species were the descendants of earlier forms. He also helped shift the widely accepted evolutionary depiction of a "chain of being," which was a straight line, into a branching tree.

He is best known, however, for what has come to be known as the "use and disuse" hypothesis. He felt that species mutated over time to adapt to changes in the environment. If an animal found itself in a drier habitat, for example, it might over the generations develop thicker skin to avoid desiccation. Other organs followed this pattern. Lamarck also often suggested that organisms could "will" the change in themselves. For example, a deer-like animal that wanted to eat the upper-level leaves on tall trees could stretch its neck, which eventually would lengthen. Over the generations, this continued activity would give rise to giraffes. He said that if humans followed this path, their desire to become the dominant species might result in the development of bipedal locomotion and larger brains.

Lamarck's "use and disuse" hypothesis drew increasing fire, and the emerging field of genetics in the 1900s sealed its fate as a false notion.

Despite his fine reputation as a systematist, Lamarck remained a pauper most of his life. Nonetheless, he was married four times and had several children, two of whom cared for him when he lost his sight late in life. He died in poverty in 1829. The lease on his rented gravesite ran out a few years after his death, and the location of his remains is unknown.

LESLIE A. MERTZ

Justus von Liebig
1803-1873
German Chemist

Justus von Liebig was one of the most influential chemists of the nineteenth century. As a scientist, Liebig offered bold theoretical frameworks and trained a generation of students and colleagues who led developments in organic chemistry, pharmaceutical chemistry, physiological chemistry, agricultural chemistry, and industrial chemistry for decades. As a popularizer of science, Liebig explained to academics, bureaucrats, farmers, and monarchs chemistry's utility as the basis for modernization, improvements in public health and nutrition, and greater international cooperation. As an entrepreneur, Liebig demonstrated the commercial potential of applied chemistry through enterprises that manufactured fertilizers, mirrors, baking powders, infant formulas, and meat extract.

Liebig was born in Darmstadt in 1803, where his parents operated a small shop that sold hardware and useful chemicals like paints

and varnishes. Unable to complete a formal secondary education for financial reasons, Liebig moved to nearby Heppenheim to apprentice as a pharmacist, then the most important step for someone with a career interest in chemistry. Liebig went on to study chemistry at the University of Bonn and the University of Erlangen, where he received a doctorate with little more than the promise to do research on the topic of plant chemistry. He then earned a grant that allowed him to study in Paris under some of the leading chemists of the day, especially Joseph-Louis Gay-Lussac (1778-1850).

In 1824, as part of the grand duchy's broader initiative to modernize its infrastructure, Liebig received an appointment at the small University of Giessen in his home state of Hessen-Darmstadt. Liebig soon transformed Giessen into a center for chemical education that earned an international reputation for its innovative teaching methods and successful graduates. Liebig's approach stressed the development of laboratory skills. Through use of standardized techniques and equipment like the "potash bulp apparatus" and the "Liebig condensor," students learned how to efficiently perform routine analyses. As their skills developed, students tackled independent research projects that led to the synthesis and identification of countless new compounds and reactions. Liebig's own research focused on organic chemistry theory, during a crucial period in its development. In particular, he refined the theory that certain groups of organic compounds, called radicals, remain unchanged through a series of reactions that produce related compounds. Liebig also edited *Annalen der Pharmacie und Chemie,* which became the most influential journal in the field.

In 1840 Liebig published two works that established his reputation as an influential commentator on the major scientific issues of his day. The first, an article sharply critical of the quality of chemical instruction offered in Prussia, prompted many European governments to revamp their approach to scientific education. The second, *Organic Chemistry and Its Applications to Agriculture and Physiology,* laid the foundation for a generation of research in the agricultural sciences. Liebig argued that farmers should be cognizant of the role that chemical compounds play in every aspect of farm operations. Illustrating that mineral nutrients from the soil leave the farm with every harvest and livestock sale, Liebig clearly outlined a concept of chemical cycles that stressed a balance of chemical inputs and outputs. He also endorsed artificial fertilizers as an appropriate means to maintain soil fertility. In a related book on animal chemistry, published in 1842, Liebig announced his theories on the protein radical, the metabolism of fats, and the relationship between digestion and respiration, laying the groundwork for future research on nutrition and biochemistry.

In the latter half of his career, Liebig distanced himself from laboratory teaching and his own research projects, concentrating instead on efforts to promote and popularize chemistry in the German area and beyond. His *Chemical Letters,* eventually published in 11 languages, taught the significance of chemistry in a popular language and format. In 1852 he took a post at the University of Munich that required little in the way of teaching or research, allowing him to concentrate on his roles as popular lecturer and public savant. His later career also included efforts to recycle London's sewage for agricultural purposes, to market the byproducts of South American beef as Liebig's Extract of Meat, and to address international issues of science policy and philosophy of science.

Though many of the specific theories that Liebig promoted have since been proven incorrect, his career furthered the emergence of chemistry as a central branch of scientific inquiry.

MARK R. FINLAY

Thomas Robert Malthus
1766-1834
English Economist

Thomas Malthus was one of the most important English economic theorists of the early nineteenth century. He is best known for his influential work, *An Essay on the Principle of Population,* first published anonymously in 1798, but reissued in six editions by 1826. Malthus further developed the *laissez-faire* (French for "let things alone") views of economist Adam Smith (1723-1790), who promoted the capitalist principles of self-interest and free competition in the marketplace. Malthus's book proved valuable for Charles Darwin (1809-1882) as he began to formulate his concept of natural selection through the struggle for existence. Though Malthus's population principle has been largely discredited, fears of overpopulation periodically stir new interest in his ideas.

Malthus studied at Cambridge University, graduating with honors in math. Thereafter

Thomas Malthus. *(The Library of Congress. Reproduced by permission.)*

Malthus joined the Anglican clergy and became priest over a parish in Surrey in 1796. His book on population brought him sufficient renown to be named the first professor of political economy in Britain (and probably in the entire world) at the East India College in 1805. Malthus later published other economic works, including *Principles of Political Economy* (1820).

Malthus's *Essay on the Principle of Population* was a polemical work directed against the overweening optimism of two Enlightenment thinkers: William Godwin (1756-1836) and Marquis de Condorcet (1743-1794). Condorcet argued in his work *Progress of the Human Mind* (1793) that human progress would ultimately result in universal happiness, virtue, and economic equality. Godwin's *Enquiry Concerning Political Justice* (1793) was similarly utopian. Godwin was an anarchist who believed that humans would progress in their struggle against vice and sickness to such an extent that they would ultimately become immortals who ceased reproducing.

These views were, according to Malthus, completely unrealistic. He set out to refute them by showing that poverty, misery, and many evils in society are not the result of irrational institutions, as Condorcet and Godwin believed, but are rooted in nature. Malthus's population principle rested upon two statements: 1) human population (as with other organisms) tends to increase geometrically, that is, in a series 1, 2, 4, 8, 16, 32, etc.; 2) the food supply can at best only increase arithmetically, that is, in a series of 1, 2, 3, 4, 5, 6, etc. Malthus provided statistics to try to establish these statements. The deduction is then obvious: The increase in food supply can never keep pace with the increase in population.

Overpopulation is, according to this view, an ever-present specter in human society that can never be overcome. If there is not some other check on population growth, such as war or disease, poverty and starvation are inevitable. Faced with the stark horror of this picture of society, Malthus suggested only late marriages of the prudent as means to overcome poverty caused by overpopulation. He remained pessimistic about ultimate solutions to the problem.

Malthus's essay was instrumental in leading both Darwin and Alfred Russel Wallace (1823-1913) to their idea of natural selection as the driving force behind the evolution of biological organisms. Darwin recognized that if organisms reproduced far beyond their abilities to survive, this would cause intense competition, especially between members of the same species. Since the organisms better adapted to their environments would survive at a greater rate than those less well adapted, this would allow better variants to propagate their characteristics.

Darwin acknowledged Malthus's influence on his theory in *Origin of Species* (1859); Darwin had read Malthus's *Essay on the Principle of Population* some twenty years earlier as he was developing his initial ideas about natural selection. Later, in *Descent of Man* (1871), Darwin explicitly espoused the validity of the Malthusian population principle for human populations. Malthus's ideas spread widely because of the spread of Darwinism, and many social theorists in the late nineteenth century, often referred to as Social Darwinists, built their view of human society on the Malthusian population principle embedded in the Darwinian theory of the struggle for existence.

Two factors have discredited Malthus's population principle. First, population growth has been checked in most industrialized societies through the use of birth control. Secondly, agricultural productivity has increased much more rapidly than Malthus thought possible, sustaining a population today that Malthus would have considered impossible.

RICHARD WEIKART

Matthew Fontaine Maury
1806-1873
American Oceanographer

The first real scientist of the sea and the first oceanographer, Matthew Maury was called "Pathfinder of the Oceans." Scientific navigation did not exist before Maury, and his work improved the safety of every ship at sea. He wrote the first serious text on oceanography, *The Physical Geography of the Sea* (1855), and inspired many students to study oceanography. He compiled charts of wind and wave patterns, currents, hazards of all oceans and harbors, and was the first to show that meteorology could become a science.

Matthew Maury was born in Virginia. When he was five years old, his father moved the entire family, five boys and four girls, to a cotton farm in Tennessee. The children worked on the farm and were educated in a small country school. Injured when he was 12 years old, Matthew was sent to the Harpeth Academy for further schooling. Though his father wanted him to study medicine, Matthew was especially drawn to science. His oldest brother was a naval midshipman whose adventures in the South Pacific may have enticed Matthew to a military career. It was a fateful day when, without his father's knowledge, Matthew secured appointment as a midshipman in the navy with the help of Sam Houston, then congressman from Tennessee. Matthew was 19 years old when he reported for active duty in August 1825.

The next four years shaped his education and his life's work. Sailing as a midshipman on three ships, he went all the way around the world, learned to sail and navigate, met all kinds of people, studied mathematics, geology, and astronomy and learned many languages. While visiting various countries, he compiled a lunar table to be used by navigators. When he returned home in 1829, the table was published, the first of many publications. He wrote more than 200 articles on navigation, oceanography, meteorology, astronomy, naval reform, and the need for a naval academic institution; the latter article spurred the construction of the U.S. Naval Academy in 1845. Maury also compiled a number of nautical charts. He collected information for his charts by communicating with all naval and merchant ship captains. He asked them to send him any information they had on currents, winds, and weather wherever they were in the world. The resulting charts made navigation less hazardous in all oceans and harbors of the world for anyone involved in merchant shipping or naval operations.

He settled in Virginia, married in 1834, had two daughters, and worked for extra money as superintendent of a gold mine. In the 1850s he charted the floor of the Atlantic Ocean between Europe and the United States, demonstrating that a telegraph cable could be laid to connect the two continents. He was soon promoted to the rank of Commander in the navy. His book *The Physical Geography of the Sea* was published in 1855.

Maury supported the Confederate States of America during the Civil War and was sent to England on a mission for his government. There he remained, receiving accolades from many nations on his work. He wrote a number of articles on his favorite subjects not only in English but also in French, Spanish, and German.

When the Civil War ended, Maury returned to Virginia and became superintendent of the Department of Charts and Naval Oceanography, which later became the U.S. Naval Observatory. He was a lecturer on meteorology at Virginia Military Institute when he died in Virginia in 1873.

LYNDALL BAKER LANDAUER

Gregor Johann Mendel
1822-1884
Austrian Monk, Biologist and Botanist

In the quiet setting of a monastery garden, Gregor Mendel bred pea plants in an attempt to understand heredity, or how characteristics are passed from parent to offspring. After more than seven years of research, Mendel deciphered the basic principles governing heredity (now called Mendelian laws). Although Mendel's work was not appreciated until 16 years after his death, the results of his experiments and his interpretation of those results ultimately provided the foundation for the field of genetics, and he is considered the father of the discipline.

Gregor Mendel was born Johann Mendel in Heinzendorf, Moravia, in 1822. He was the second child of Anton and Rosine Mendel, who were peasant farmers. In order to be educated, Mendel joined an Augustinian monastery in Brünn, Austria (now Brno, Czech Republic) in 1843, and there he took the name of Gregor and became an ordained priest in 1848. Mendel left the monastery briefly to study mathematics,

Gregor Johann Mendel. *(The Library of Congress. Reproduced by permission.)*

physics, and science at the University of Vienna from 1851 to 1853.

Despite Mendel's intellect and the lasting significance of his future findings, he repeatedly failed the tests required to obtain a teaching certificate (although he did serve as a substitute teacher at local schools and taught experimental physics and natural history at the Brünn Modern School for 14 years). This failing, however, did not discourage his love of the natural world around him. He was an amateur scientist with interests as diverse as meteorology, botany, and theories of evolution, and he read the scientific literature of the day. He was an astute observer and grew curious about how plants in the monastery garden obtained unusual characteristics. With questions about inheritance in mind, Mendel began in 1856 the experiments for which he is now famous.

That plants and animals could be bred to produce hybrids with desirable characteristics was known long before Mendel came along. However, Mendel's quantitative approach to studying heredity, in which he carefully designed experiments and mathematically analyzed the data, was new; it was this approach that allowed him to recognize the patterns underlying heredity and to develop a theory to explain them. His decision to study pea plants also was crucial to the success of his work. The pea plant normally self-pollinates, meaning that male gametes fertilize female gametes from the same plant; this feature meant that Mendel could carefully control reproduction and artificially cross plants with different characteristics. Mendel's pea plants also had a number of clearly defined and measurable traits (e.g. seed color, stem length), seven of which he studied in a logical progression of experiments in which he crossed plants with different sets of traits.

After spending years studying more than 21,000 pea plants, Mendel recognized that he could predict the outcome of breeding in successive generations based on mathematical probabilities. Mendel did not know about genes, chromosomes, and DNA, but from his experiments he deduced that two hereditary "units" (now known to be genes) determine a given trait, that one comes from each parent, and that one exhibits dominance over the other. What came to be called Mendel's law of segregation states that every individual carries pairs of factors for each trait and that members of the pair separate during gamete formation. Mendel also found that different traits were passed to offspring independently of one another, which led to the law of independent assortment. This law states that members of each pair of factors are distributed independently when gametes are formed.

Mendel published his manuscript, *Experiments with Plant Hybrids,* in the *Proceedings of the Brünn Natural History Society* in 1866, but it remained unappreciated during his lifetime. Mendel spent the remainder of his life devoted to the monastery, where he became abbot in 1868. It was not until the turn of the twentieth century that three scientists independently discovered the significance of Mendel's work as an explanation for how inheritance works. Since then, geneticists have shown that there are many exceptions to Mendel's rules. However, Mendel's discoveries hold true for the inheritance of many characteristics in plants and animals and provide the foundation for our current understanding of genetics.

LYNN M.L. LAUERMAN

Hugo von Mohl
1805-1872
German Botanist

German botanist Hugo von Mohl was one of the first to use the microscope to study the nature of plant cell structure and the physiology of higher plants. He is credited with recognizing and naming protoplasm and proposing that cells divide to form new cells.

Born in Stuttgart, Germany, on April 8, 1805, Mohl was part of a respected middle-class family who placed a great deal of value on scholarship. Brother Jules became a naturalized Frenchman and professor of Persian. Two other brothers were respected in the fields of economics and politics. However, from an early age Mohl's interest was science, especially botany and optics. He had a happy childhood and adolescence, and his university career and personal life were pleasant with few difficulties.

Mohl received a doctor of medicine at Tübingen in 1828. Interest in the puzzling mystery of climbing plants led to a doctoral dissertation on the stomata, or pores, in the leaves of plants. In 1832 he was given a post as professor of physiology at Bern, and in 1835 became a professor of botany at Tübingen, a position he held until his death.

The hobby of working with lenses and the microscope turned out to be a major focus of his vocation. He continued to develop and improve the microscope and wrote a how-to manual on the subject in 1846.

With the microscope, he determined that the basic structure of the cell included movement within, referred to as intracellular movement. He was the first to develop the idea that there is a nucleus within the granular flowing material that forms the main substance of the cell, naming this substance "protoplasm." This term, meaning "first form," was coined by Czech physiologist Jan Evangelista Purkinje (1787-1869) to refer to the embryonic material found in eggs. Mohl's application of the word was adopted into the knowledge of the time and survives as part of biology today. Protoplasm, a semi-fluid colloid, is the matter of life in all plant and animal cells.

Mohl's meticulous studies revealed the first description of the division of cells. Observing division of the alga *Conferva glomerata,* he recognized the tough secondary fibrous substance of the plant now known as the cell wall. In 1853 he published a very important work, *Die vegetabilishe Zelle* (*The Plant Cell*), which became a classic work in botany. The book presented a panorama of what was known about botany. He took the cell as the base, calling it "an elementary organ."

Mohl was very specific in his concrete descriptions and did not draw general conclusions from speculative thought. He appreciated the critique of his work by colleagues and contemporaries.

Among his varied scientific contributions were the discovery of the nature of small bodies within specialized cells, called plastids, and the process of osmosis, the passage of molecules through a membrane from a place of high concentration to low. He identified the stomata, the openings in leaves where carbon dioxide and water are taken in and oxygen is given off.

Mohl never married and did not consider it important to have a circle of friends or admiring students. Skill with his hands, meticulous work habits, and attention to detail brought him great satisfaction and recognition. He was one of the founders of *Botanishe Zeitung* (*Botany Times*), a prestigious journal that still exists today. His desire for outstanding science education for students led to the creation of the Faculty of Sciences at Tübingen, one of the first of its kind. He received many honors, including the Order of Wurtenberg in 1843 and an accompanying title of nobility. Mohl died in his sleep on Easter Sunday, April 1, 1872.

EVELYN B. KELLY

Lorenz Oken
1779-1851
German Natural Scientist

Lorenz Oken, a proponent of natural science and philosophy, asserted that there are fundamental units of life, which he called "infusoria." His general ideas about the elemental structures of living organisms, though specifically incorrect, anticipated the subsequent identification of the cell and development of cell theory. He was also a founder of scientific congresses or meetings.

Born at Bohlsbach bei Offenberg, Baden, Germany, in the Black Forest, Oken was the son of a peasant farmer. He attended the Universities of Freiburg, Wurzburg, and Göttingen. With a keen mind and great observation skills, he was interested in a wide variety of fields, including optics, minerals, and even military science. In 1803, at the early age of 24, he published a book entitled *Naturphilosophie*, translated as *Natural Philosophy*, in which he developed his ideas. The publication of this book solidly identified him with the school of thought called natural philosophy, or natural science, that had been founded by Friedrich Schelling (1777-1854), a German philosopher of romanticism and idealism.

Oken had a strong personality and presented his ideas with bombast and forcefulness. He was also involved in German revolutionary youth movements that the government strongly suppressed. Because of his strong and unpopular ideas, he frequently changed jobs. He graduated from Freiburg in 1804, then hopped from Göttingen to Jena, to Munich, and to Erlangen. He found a niche at the newly created University of Zurich, where he remained until his death in 1851.

The ideas of natural science and philosophy at the beginning of the nineteenth century were quite different from the scientific natural science of today. The concepts designated as part of the romantic movement were a revolt against the formal classicism in the past and dealt with primitive nature. Oken studied the bones, intestines, and umbilical cord, and established basic ideas about elementary units of living organisms, which he called the "infusoria." He believed all flesh is made of these infusoria and called them "animalcules" or little animals. He named these primal animals "Urthiere." The Urthiere is a romantic designation critical in natural thought, as the prefix "Ur" is used to describe primal animals or protozoans. This term is a key element of natural philosophy and the spirit of romanticism which sought to return to simple roots in nature.

Oken proposed that these primal animals are the basic materials of all beings, both plants and animals. The primal animals that make up the higher organisms are subject to that higher being. When the primal beings fuse together, they form the higher organism with each of the primals losing its own individual identity to be part of the whole. Actually, Oken was approaching the idea of the cell, but, as a romantic philosopher, always traced a thread of being from simpler to complex forms.

Oken promoted the idea of primal slime. The substance arose, it was thought, when conditions reached equilibrium, producing a sphere, a miniature of the planet. According to this idea, the primal mass is formed somewhere along the edge of the sea and land, and from this emerged the original infusorian.

Oken's proclamations fit into the fundamental concepts of nineteenth-century Romantic science or philosophy of the time. However, his enthusiasm and romantic ideas may be difficult for the modern reader to decipher.

One of his greatest contributions to science was the creation of scientific congresses outside the university setting. His first meeting was held in Leipzig in 1822. A prolific writer, Oken's books were printed and reprinted until the principles of cell theory, established by Theodor Schwann (1801-1882), superceded his work.

The strong opinions of romanticism were eventually overridden by the emergence of experimental natural science. However, in 1951, the centennial of Oken's death, popular articles were written about him and a scientific meeting was held in Freiburg in his honor.

EVELYN B. KELLY

Louis Pasteur
1822-1895
French Chemist and Microbiologist

Louis Pasteur began his scientific career as a chemist and made important contributions to that field. However, he is best remembered as the father of microbiology. His work toward identifying microorganisms and understanding how they cause disease was instrumental in the development of modern medicine. By discovering why food spoils and how to preserve it, Pasteur also helped to increase the safety and shelf life of the world's food supply.

Pasteur was born in 1822 in a small village near Dijon, France. As a youth he was remarkable mainly for his talent as a painter; he struggled with his schoolwork despite spending a great deal of time at it. One schoolmaster recognized that Pasteur's work was extremely methodical rather than simply slow, and he arranged for Pasteur to transfer to a private school in Paris. But the homesick country boy lasted only a few weeks. Back at home, Pasteur resumed his exacting study habits, attending a local college until returning to Paris a few years later. There he was finally admitted to the Ecole Normale, the national school for college professors.

While studying for his doctorate in chemistry, Pasteur began his research into the optical properties of crystals and made an important discovery concerning the polarization of light. Ordinary light waves can vibrate in any direction perpendicular to the direction in which the wave is moving. Polarized light consists of waves vibrating in only one direction. Some crystals have the ability to change the direction of polarization of a beam of light, because of the structure in which their molecules are arranged. Pasteur discovered materials with both "left-handed" and "right-handed" forms that

rotate the polarized light beam in different directions. This property had not been recognized before because when the two forms are mixed, the effect cancels out. With his work on crystals Pasteur founded the discipline of stereochemistry, the study of how particular arrangements of a substance's molecules affect its properties.

After obtaining his degree, Pasteur was offered an assistantship at the Strasbourg Academy. There he met his wife, Marie. The couple eventually had five children, of whom three died young. In Strasbourg Pasteur also encountered the project that was to direct his future path.

Pasteur was studying the effect of fermentation on some of his crystals. It seemed that the ones that turned polarized light to the left did not ferment, while the "right-handed" material did. Knowing he was working on fermentation, a businessman who distilled alcohol from beet juice approached Pasteur for help when his alcohol was going sour. Pasteur reluctantly left his laboratory and quickly discovered the problem at the distillery. Instead of alcohol, sometimes the fermentation would produce lactic acid, the same substance that develops in sour milk. Examining samples of properly fermenting and sour beet juice under the microscope, Pasteur discovered different microorganisms in them. Scientists had known about microorganisms, or microbes, since Anton van Leeuwenhoek (1632-1723) first identified them using a simple microscope in 1675. But there was no real understanding of what they did.

Eventually making his way back to Paris, for the next 14 years Pasteur studied the microbes responsible for the different types of fermentation. He discovered that some microorganisms were *aerobic*, thriving in the presence of oxygen, while others, called *anaerobic*, grew where oxygen was absent.

Winemaking was one of the most important French industries in Pasteur's time, as it still is today. The scientist was soon enlisted by Emperor Napoleon III in an effort to prevent wine from spoiling. But wine could be spoiled by any of several microorganisms. How could they all be destroyed? It was already known that if the wine was heated to high temperatures, it would keep indefinitely. Unfortunately, while the wine wasn't spoiled using this process, its taste was. Pasteur showed that the wine only needed to be briefly heated to between 131-158° F if it was then bottled without further contact with air. This method came to be called *pasteurization,* and it remains essential for food processing today, particularly for dairy products.

Pasteur's work on wine led him to believe that, contrary to the prevailing theory of the day, microbes did not appear by "spontaneous generation." He had shown that, if proper precautions are taken, they do not appear at all. This involved him in a great deal of controversy. The spontaneous appearance of life was in concert with the religious and philosophical outlook of the day, representing as it did a Creation in miniature. Pasteur was a religious man, but nonetheless he felt sure that microbial growth did not require divine intervention. He demonstrated that the germs were carried on the dust in the air. If these germs were prevented from getting to a medium, it remained unspoiled.

Pasteur recognized the implications of his discovery for the rather primitive medical treatment of the 1870s. Injured patients and those recovering from surgery often died of infections. If germs could be prevented from reaching the wounds, these patients could be saved. But French physicians disregarded what they considered to be the irrelevant opinions of a chemist. So Pasteur's work was first applied to the operating room by the English physician Joseph Lister (1827-1912), who originated antiseptic surgery.

Pasteur was subsequently pressed into service on a number of problems important to French industry and agriculture, such as beer brewing and silkworm disease. In each case his method was to identify the germ and proceed from there. Particularly important were his experiments on chicken cholera, in which he conferred immunity to the disease by using a weakened form of the microorganism. This provided a previously lacking scientific explanation for the smallpox vaccinations that had been developed by English physician Edward Jenner (1749-1823) in 1796.

In 1885 Pasteur was working on a vaccine for rabies. He had done some successful tests in animals when he was confronted by a dilemma in the form of Joseph Meister, a young boy brought to him after having been bitten by a rabid dog. While Pasteur did not consider that he was ready to try his vaccine on humans, the boy had no other hope for survival. With great trepidation, Pasteur administered his vaccine over the course of the next few weeks. To his great joy and relief, the boy recovered. Although Pasteur was not a physician, he had found a cure for a previously incurable disease.

Word soon spread of Pasteur's successful intervention and money began pouring in to establish the Pasteur Institute for rabies treatment, which opened in 1887. Eventually it developed into a research and teaching institution specializing in microbiology. Pasteur remained associated with his institute until his death in 1895, and became known as a great humanitarian.

SHERRI CHASIN CALVO

Jan Evangelista Purkinje
1787-1869
Czech Physiologist and Histologist

Jan Evangelista Purkinje made pioneering contributions to histology and physiology. His observations led to many important insights into the workings of the human body, especially various visual phenomena. He invented the microtome, an instrument that slices tissues into very thin samples, and demonstrated the importance of technical advances in microscopy for biological research. In addition to Purkinje cells, his name has been given to identify specific types of conduction, fibers, figures, layers, networks, phenomenon, systems, and Purkinje-Sanson images. Although he made some of the earliest observations of the cell nucleus, the characteristics of cells, and cell division, other scientists are generally credited with these discoveries. He introduced the term "protoplasm" to describe the contents of cells.

Purkinje, whose parents were Czech, was born in Libochowitz, Bohemia. He was educated by Piarist monks and intended to join the order, though went instead to Prague to study philosophy. This led him to a deep interest in science and medicine. His graduation thesis on subjective aspects of vision, completed at the University of Prague in 1818, brought him considerable attention. Purkinje observed that when the intensity of illumination decreases, objects of different color but equal brightness appear to be unequally bright; this is now known as Purkinje's phenomenon or shift. In other words, blue objects appear brighter than red objects in dimmer light. He also described the threefold images of an object that are seen by an observer in the eye of another person. These images are caused by an object reflecting from the cornea's surface and both sides of the lens. This phenomenon is now known as "Purkinje's images." Johann Wolfgang von Goethe (1749-1832) was impressed by Purkinje's research on vision and the two became

Jan Evangelista Purkinje. *(The Library of Congress. Reproduced by permission.)*

Purkinje emphasized the value of microscopy, especially the new achromatic microscopes, and developed several innovative teaching tools and a knife that was a precursor of the microtome. He also established the use of balsam sealed preparations and adapted Louis J. M. Daguerre's methods to produce the first photographs of microscopic materials. During the 1830s Purkinje conducted numerous studies of microscopic structures, including studies of the skin of various animals, ciliary motion, nerve cells, myelinated nerve fibers, the large flask-shaped cells with numerous dendrites in the cerebellar cortex (now known as Purkinje cells), the unusual muscle fibers found below the endocardium in a specific region of the ventricles of the heart (the fibers of Purkinje) which are important in the conduction of the cardiac impulse, and so forth. In 1850 he was elected a Foreign Member of the Royal Society.

LOIS N. MAGNER

Wilhelm Roux
1850-1924
German Biologist and Embryologist

Wilhelm Roux, the founder of experimental embryology, was primarily interested in the factors that governed the development of the embryo. Convinced that descriptive and comparative studies of embryonic development were inadequate, Roux demanded a new approach and saw himself as the founder of a new discipline, which he called developmental mechanics. To promote the advancement of this nascent field, he established a new journal called *Archive for Developmental Mechanics of Organisms* in 1894.

Roux, the son of a fencing master, was born in Jena. His life's work was profoundly influenced by two eminent teaches—Ernst Haeckel (1834-1919) at the University of Jena and Alexander Wilhelm Goette at Strassburg. Roux became a professor at Innsbruck, but soon moved to the University of Halle where he remained from 1895 to 1921. Throughout his life he remained interested in the great problems he took up as a student of Haeckel and Goette—phylogeny and the struggle for existence.

Roux believed that zoologists had focused too much attention on purely descriptive studies of the egg and embryo. He urged his contemporaries to begin experimental studies of the factors that allowed the egg to develop into an embryo. According to Roux, experimental

friends. In 1820 Purkinje studied the induction of vertigo that occurs when the body is rotated in an erect position, but he was unable to explain the phenomenon. In 1823 he was appointed to the Chair of Physiology at Breslau, where he founded an institute for histological research. Because university officials were unwilling to meet his demands for space and equipment, much of his research and teaching was carried out in his own home laboratory, which became known as the cradle of histology. Purkinje was highly respected as a teacher as well as a researcher. He accepted the Chair of Physiology at Prague in 1849, and established another institute for histological research. He directed research at the institute until his death in 1869.

In 1839 Purkinje was the first to use the term "protoplasm" in a scientific sense. Theologians had used the word protoplast for Adam, that is, "the first formed." As a physiologist, Purkinje used the term to refer to that which was first produced in the development of the individual plant or animal cell. Apparently, independent of Purkinje, Hugo von Mohl (1805-1872) used the word protoplasm to describe the part of the plant cell within the cell membrane. Although Purkinje's observations of the cell nucleus, or germinal vesicle, in the eggs of birds in 1825 preceded those of Robert Brown (1773-1858), he did not publish these studies until 1830.

embryology would proceed by breaking down complex developmental processes into simpler and simpler functional processes. Ultimately, these processes could be analyzed in physico-chemical terms. The influence of Charles Darwin (1809-1882) and Haeckel is obvious in Roux's 1881 paper "The Struggle of the Parts in the Organism: A Contribution to the Completion of a Mechanical Theory of Teleology."

Roux argued that embryologists must adopt experimental methods that would allow them to analyze the immediate causes of development and abandon useless philosophical debates about preformation and epigenesis. Roux attempted to address the key question of embryology by asking whether development occurred by means of "self-differentiation" or "correlative dependent differentiation." In part, Roux did this to avoid the old debates about preformation and epigenesis. Self-differentiation was defined as the capacity of the egg or parts of the embryo to undergo further differentiation independent of outside factors or of other parts of the embryo. Correlative dependent differentiation was defined as differentiation that was dependent on external stimuli or on other parts of the embryo. According to Roux, his definitions were valuable because they laid the framework for experimental testing. If each part of the embryo developed independently, like a mosaic, then the experiment demonstrated that the mechanism of development was self-differentiation. If interactions between groups of cells were necessary to development, then the mechanism was correlative dependent differentiation.

For theoretical reasons, Roux assumed that self-differentiation served as the mechanism of development. It seemed most likely that the fertilized egg was similar to a very complex machine. During development, parts of the machinery would be divided up and distributed the appropriate daughter cells. The mosaic model, therefore, predicted that external conditions should not affect the development of the embryo. Various experiments demonstrated to Roux's satisfaction that the development of embryos could be harmed by extreme changes in their environmental conditions, but that minor changes had little effect. Further experiments addressed the question of whether forces within the egg or embryo could change the pattern of development. First, Roux attempted to determine whether separate parts of the embryo could develop independently. As an experimental test, Roux destroyed one of the cells of a frog embryo at the two-cell stage. He pricked one of the cells with a hot needle and found that the other cell developed into a half-embryo. Roux concluded that his original prediction had been correct. According to his experimental demonstration, each cell develops independent of its neighbors. Under normal conditions, development was the result of the separate differentiation of each part. When other researchers performed similar experiments using embryos of different species, the results were quite different. Hans Adolf Eduard Driesch (1867-1941), another pioneer of experimental embryology, provided compelling evidence against Roux's model of mosaic development.

LOIS N. MAGNER

Matthias Jakob Schleiden
1804-1881
German Botanist

Matthias Jakob Schleiden and Theodor Schwann (1810-1882) are generally regarded as the first scientists to establish cell theory. Cell theory is a fundamental aspect of modern biology. This powerful generalization has played an essential role in explaining the basic unity of plant and animal life, the mechanism of inheritance, fertilization, development and differentiation, and evolutionary theory. Building upon the discovery of the cell nucleus by Robert Brown (1773-1858), Schleiden demonstrated that plants are composed of cells and cell products.

Schleiden studied law at the University of Heidelberg, but he was so unsuccessful in his attempts to establish a law practice in Hamburg that he was driven to suicide. Fortunately, his self-inflicted gunshot wound was not fatal. By the time he had recovered from his injury and depression, Schleiden decided to give up law and study natural science. He earned doctorates in medicine and philosophy and was appointed professor of botany at the University of Jena. Despite his success in research and teaching, he suffered from nervousness, fatigue, and depression. He resigned after 12 years and decided to rest his nerves and to travel. During a visit to Berlin he met with Schwann and described his ideas about plant cells.

Contemporaries generally described Schleiden as arrogant and unsympathetic towards rivals and predecessors. However, Schleiden did accord considerable respect to the work of Charles Brisseau-Mirbel (1776-1854), an eminent French botanist and microscopist. Brisseau-

Matthias Jakob Schleiden. *(The Library of Congress. Reproduced by permission.)*

Mirbel thought that cells were found in all parts of the plant. Schleiden generally agreed with Brisseau-Mirbel's suggestion that cells formed in some sort of primitive fermenting fluid.

Schleiden thought that most botanists were wasting their time arguing about old systems of taxonomy. He wanted to redefine botany as a new inductive science concerned with the forms and functions of the whole vegetable kingdom. He complained that botanists had discovered few facts and had established no new fundamental laws and principles. He believed that botanists should abandon systematic taxonomy and focus on the study of the chemistry, physiology, and microscopic structure of plants.

In 1838 Schleiden published his new ideas as "Contributions to Phytogenesis" in Müller's *Archives for Anatomy and Physiology.* Recognizing the importance of Robert Brown's discovery of the cell nucleus, Schleiden argued that the nucleus, which he renamed the cytoblast, was an essential component of all plant cells. He believed that all higher plants were aggregates of cells. The cells that made up the plant led a double life. In part they were independent entities, but they also served as integral parts of the plant. All aspects of plant physiology, therefore, resulted from the activity of the cells.

Although Schleiden described several possible methods of cell formation in "Contributions

to Phytogenesis" and later in his major treatise *Principles of Botany,* he generally supported the hypothesis known as "free-cell formation." That is, he thought that cell growth was rather like the process of crystallization. Presumably, granules in the cytoblastema, a fluid containing sugars and mucus, aggregated to form a nucleolus. More granules joined those that made up the nucleolus until the cytoblast (nucleus) formed around the nucleolus. Eventually, a young cell developed around the mature cytoblast and the rigid plant cell wall formed around the new cell. Schleiden though it was also possible for cells to form within cells in the growing plant. The contents of such cells would divide into two or more parts and a membrane would separate each part. He suggested that wood was formed when materials in plant juices were quickly aggregated. Although the mechanism by which cells multiplied was unclear, Schleiden was quite opposed to the doctrine of spontaneous generation. He was convinced that even the simplest plants, such as algae, lichens, and fungi, arose from parents of the same kind, not from spontaneous generation out of nonliving substances. Schleiden's work was confined to the plant world, but it was his work on cell theory that stimulated Schwann's study of the role of the cell in animals.

LOIS N. MAGNER

Max Johann Sigismund Schultze
1825-1874
German Zoologist and Cytologist

Max Johann Sigismund Schultze was a German zoologist and cytologist. Zoology is a part of biology that focuses its study on animal life and the animal kingdom in general. Cytology is the study of cells and how cells function.

Born in 1825 in Freiburg, Germany, Max Schultze attended school at the University of Greifswald in Germany, from which he graduated in 1849. He also studied at the University of Berlin. In 1859, following several years of teaching anatomy in Halle, he began teaching anatomy at the University of Bonn. In 1872 the University of Bonn offered Schultze a chair, a permanent position, and so he became the director of the Anatomical Institute.

Schultze's primary research focus was on a part of the cell he called protoplasm, which essentially includes all the material inside the cell membrane. As a cytologist Schultze's best-known work pertained to unicellular organisms.

In 1861 he was able to establish that cells of all organisms contain protoplasm. Today a more common term for protoplasm is cytoplasm.

Max Schultze's talents in the science arena were extensive. He was also a histologist, which is a part of biology that focuses on the structure and composition of animal and plant tissues as they pertain specifically to their functions. As a histologist he discovered and developed the stain used to aid in microscope viewing. The stain is called osmic acid and has the ability to allow the viewer to see the fine details of cell. His best use of this stain was to enhance the viewing of nervous tissues in the basilar membrane of the ear. His scientific writings covered other sensory organs: the internal ear (1858), the nose (1863), and the retina of the eye (1866), muscles, and nerve endings.

During his life Max Schultze was also able to show that the retinas of birds have rods and cones, two different sensory nerve endings that have separate functions. Birds and humans are the only organisms to have rods and cones. Light first passes through the liquid in the eyeball. The light then penetrates through several additional layers of cells before reaching the rods and cones, which are light sensitive. A rod is a specialized photoreceptor cell and is sensitive to light; however it is not sensitive to color. The rod responds in dim light is used primarily in peripheral vision. A cone is a color-sensitive photoreceptor cell and is associated with visual acuity. In 1866 Schultze proposed the duplicity theory of vision, which is that cones and rods are two different cells in the eye. His theory was a forerunner of modern theories of vision.

Max Schultze died in Bonn, Germany, in 1874. It is interesting to consider how Schultze was able to conduct his research. During his lifetime it was against the law to perform dissection on humans, and thus all of his research was performed on animals. His ability to directly correlate humans and animals was amazing. His observations have held up remarkably well in today's understanding of the senses.

BROOK HALL

Theodor Ambrose Hubert Schwann
1810-1882
German Biologist

Theodor Schwann was a German physiologist who is credited with publishing the most influential work on cell theory. He made significant findings in the study of digestion, fermentation, and tissues.

Schwann originally left his hometown of Neuss to study religion in Cologne. Instead, he gave up theology to enroll in premedical studies at the University of Bonn. Later, in Berlin, he prepared a dissertation under the guidance of Johannes Müller (1801-1858). Following his graduation, Schwann became Müller's assistant and devoted his time to research.

Having become interested in digestive processes, Schwann isolated from the lining of the stomach a chemical that was responsible for the digestion of protein. This chemical, which he called pepsin, was the first enzyme to be isolated from animal tissue.

Interested in disproving the theory of spontaneous generation (the production of living organisms from nonliving material), which was experiencing a resurgence in popularity among the German scientific community, Schwann began to research fermentation. He demonstrated that yeast consists of plant-like organisms and that the fermentation of sugar is the result of the life processes of yeast cells. Schwann later used the term "metabolism" to describe the chemical changes that occur in living tissue. His findings, however, received so much adverse criticism that he left Germany for Belgium, where he worked as a professor at the universities of Liege and Louvain. (His findings on fermentation would later be confirmed by Louis Pasteur.)

Previous to Schwann, no one had thought that an organism was composed solely of cells, although cells had been observed in both plants and animals under the microscope. While assisting Müller, he observed small units containing nuclei in animal tissue; he realized that these units were the animal equivalent to the plant cells studied by researcher Matthias Schleiden (1804-1881). Schwann also noticed that the fertilized egg from which an animal grows contains a nucleus and is surrounded by a membrane similar to the nucleus and membrane of animal tissue.

In 1839, one year after Schleiden presented his cell theory, Schwann published his own version of cell theory. While Schleiden's theory only applied to plants, Schwann offered a more precise theory that he extended to animals. This work, whose title in English is *Microscopical Researches on the Similarity in the Structure and Growth of Animals and Plants*, began with a presentation of Schwann's study of frog larvae. He

Theodor Schwann. *(The Library of Congress. Reproduced by permission.)*

asserted that the cells of the frog's notochord (the structure providing dorsal support) and cartilage are like plant cells; all contain a nucleus, membrane, and vacuoles. He then proceeded to argue that all organisms are made up entirely of cells or of the products of cells. The lives of individual cells, in turn, are subordinate to the life of the organism. Schwann also described the egg as a single cell that eventually develops into a complex organism. The cell, he maintained, is the basic unit of all life and the smallest unit capable of independent reproduction.

In his book Schwann wanted to supercede the religious view of living things being shaped by a divine plan with a scientific vision in which organisms are products of natural forces. Interestingly, Schwann eventually abandoned the pursuit of science for religious and mystical studies. Nevertheless, cell theory soon became widely accepted and is now considered one of the fundamental concepts of biology.

GARRET LEMOI

Nettie Maria Stevens
1861-1912
American Cytogeneticist

Nettie Stevens hypothesized that chromosomes in sperm determine the gender of embryos when fertilization of an egg occurs. Her research confirmed scientific speculation based on Gregor Mendel's nineteenth-century genetic experimentation showing that traits such as the sex of offspring are inherited from parents. Stevens proved her theory by researching insects and showing that chromosomes specified whether embryos were male or female. Although scientists at the time were skeptical of her findings, eventually her concept became a foundation of modern genetic knowledge.

Born in Cavendish, Vermont, Stevens was the daughter of Ephraim and Julia (Adams) Stevens. She grew up in Westford, Massachusetts, where she attended public schools and graduated from Westford Academy. Stevens taught high school classes in zoology, physiology, and Latin in New Hampshire before returning to Westford, where she was a teacher at her alma mater and a public librarian, earning enough money to enroll at Stanford University in September 1896. Stevens earned a physiology degree at Stanford in 1899 and completed a master's degree the next year. During the summers, she participated in biology research at the Hopkins Seaside Laboratory in Pacific Grove, California. Her article on protozoan parasites was published in the California Academy of Sciences' Proceedings.

Stevens accepted a research fellowship in biology at Bryn Mawr College where she began a doctoral program in 1900. She received funds to study at the Naples Zoological Station in Italy and the University of Wurzburg's Zoological Institute in Germany, where she worked with renowned biologist Theodor Boveri (1862-1915). Few American women studied in foreign laboratories at that time. Returning home, Stevens collaborated with professor Thomas Hunt Morgan (1866-1945) on regeneration. Using ultraviolet light to damage cells, Stevens proved that some organisms could not create new cells.

Stevens received a Ph.D. in 1903. She remained at Bryn Mawr as a postdoctoral researcher. Inspired by the regeneration examinations, Stevens studied chromosomes and heredity. Scientists were interested in Mendelian genetics, attempting to understand how traits were passed between generations. Many of Steven's scientific peers believed that nutrition and environment determined an embryo's gender during development. Stevens thought gender was established at conception. She successfully applied for a grant from the Carnegie Institution to focus on research into how chromosomes determine gender. Examining beetles, Stevens noted they produced sperm with X or Y chromosomes and that unfertilized eggs had two X chromosomes. She hypothesized that X chromosome sperm would fertilize an egg to produce female embryos and Y chromosome sperm would create a male offspring. The beetles reproduced as Stevens expected, and she replicated her experiment with other insects particularly aphids.

Steven's research was not accepted when she published her results in 1905. Edmund Wilson (1856-1939), a Columbia University zoologist with whom Stevens had worked on other projects, initially considered her idea implausible. He conducted similar investigations and later promoted chromosome determination of gender, receiving credit for this important discovery which became a basic premise of genetics. Morgan, who won the Nobel Prize in 1933 for his genetics work, is sometimes listed as the discoverer of Steven's chromosome results. Despite this professional dismissal and discouragement from her male colleagues, Stevens retained her enthusiasm for her work.

She devoted her life to her theoretical inquiries regarding chromosomes and published scholarly journal articles. In 1905 Stevens received the Ellen Richards Prize awarded to outstanding female scientific researchers. Developing breast cancer, Stevens went to Johns Hopkins University's hospital in Baltimore for treatment. She died in 1912, having attained the rank of associate professor at Bryn Mawr. The college had established a research professorship for Stevens, but she was too ill to begin work. By the late twentieth century, Stevens was identified as the scientist who recognized the fundamental genetic concept that chromosomes determine gender at conception.

ELIZABETH D. SCHAFER

Alfred Russel Wallace
1823-1913
English Naturalist, Explorer and Surveyor

Alfred Russel Wallace reached the conclusion that natural selection is the mechanism for evolution as did Charles Darwin, making him co-discoverer of the idea in the eyes of most historians. Although his work occurred twenty years later than did Darwin's, it propelled Darwin to publish his own theory of evolution—today considered one of the titanic achievements in the history of science. Wallace was the first naturalist to mount an expedition specifically to find proof of this theory. Collecting specimens in southeast Asia, he also noted a dividing line between animal species from Asia and Australia, still called the Wallace line.

Alfred Russel Wallace was born in Wales in 1823, the eighth of nine children. He went to school in Hertford, but formal education ended when he was sent to live with his brother William in London in 1836. He was an avid reader and never stopped learning.

Alfred made a small living apprenticed to his brother, a surveyor. He studied local plants, animals, and geology during surveying trips and collected fossils. When there was no more work, he was apprenticed to a watchmaker. Here he learned engraving and chemistry.

He then worked as a teacher, keeping just ahead of his students by reading. At this time he met Henry Bates, an entomologist interested in beetles and butterflies. They remained lifelong friends.

Alfred got a job surveying for a new railway, earning enough money to take an expedition to South America with Bates to study species in 1848. Wallace was 25 years old. Four years later, he returned home, but on the way his ship was destroyed by fire. He lost many of the specimens he had collected. Fortunately, he had sent some specimens home, which he then sold to museums and collectors. He met members of the Zoological and Entomological Society of London and was aided by Thomas Huxley (1825-1895), a friend of Darwin, to publish his first article in 1853.

Later, Wallace embarked alone on an expedition to the Malay Archipelago, where he remained for eight years. It was the most significant period of his life, during which time he developed his theory of evolution and natural selection, observed differences between animals of same species in different locations, and wrote many articles. One—"On the Tendency of Varieties to Depart Indefinitely from the Original Type"—he sent to Charles Darwin (1809-1882). When Darwin received it, he was shocked to know someone else had reached the same conclusions he had. On July 1, 1858, a short treatise by Darwin and Wallace's paper were read jointly to a meeting of the Linnean Society. The occasion established Darwin's priority on the subject but showed that Wallace was co-discoverer of these evolutionary ideas. Most of Darwin's landmark book, *Origin of the Species,* was written by that point, and it was much more detailed than

Alfred Russel Wallace. *(The Library of Congress. Reproduced by permission.)*

Wallace's own writings. Wallace acknowledged this in a letter to Darwin. They differed on many aspects of the subject and engaged in a lively correspondence for years.

Wallace returned to England in 1862, sold most of his specimens, invested the money in railroad stocks, and settled down in Dorset. With a steady income, he gave lectures and wrote about his ideas and travels. In the next 50 years, he wrote 24 books, 240 articles, 100 reviews, and countless letters. When he was 72, he went on a natural history collecting expedition to Switzerland. He received prizes and medals, was elected to the Royal Society, and was generally respected for his work. His last book was published just before he died in 1913.

LYNDALL BAKER LANDAUER

August Friedrich Leopold Weismann
1834-1914
German Biologist

August Weismann, an early adherent to Darwin's theory of evolution, became famous for his studies on heredity. He denied that organisms could inherit acquired characteristics, touching off an important debate in the late nineteenth century between his own school, neo-Darwinism, and the opposing neo-Lamarckians who believed organisms could inherit acquired characteristics.

Weismann was born in Frankfurt-am-Main, Germany, and attended Göttingen University to study medicine. Beginning in 1866, he taught at the University of Freiburg, where he held a position as professor of Zoology until 1912. During his tenure, he convinced the university to establish a zoological institute and museum, and served as its director. As his vision began to deteriorate, Weismann turned his attention from microscopic studies to theoretical speculation concerning evolution and the mechanisms of hereditary transmission.

In the 1880s Weismann began to publish his views on evolution and heredity. His main works were *Studies in the Theory of Descent* (2 vols., 1882), *The Germ-Plasm—a Theory of Heredity* (1893), and *The Evolution Theory* (2 vols., 1904). Weismann believed that in order to understand heredity, one must distinguish between reproductive cells (containing hereditary material that Weismann called "germ plasm," but today would be called genes) and all other cells in an organism (somatoplasm). He believed that heredity was influenced solely by the reproductive cells without any influence from the somatic cells.

This view is often referred to as "hard heredity." When the function of the chromosome was discovered in the late nineteenth century, Weismann recognized that it might be the physical unit containing germ plasm, a hunch that was later confirmed.

Weismann rejected the Lamarckian theory—a view still popular in his day—that evolution was driven by the inheritance of acquired characteristics (such as the use or disuse of organs), also known as "soft heredity." For example, if called on to explain the evolutionary development of blind cave fish, Weismann would claim that it was not the disuse of the organs that directly caused the atrophy of their eyes, but rather natural selection favored the fish without eyes. Weismann believed that selection pressure—in this case, light to give the fish with eyes an advantage—was necessary to keep organisms from degenerating.

Weismann was not claiming that the environment has absolutely no influence on heredity. He acknowledged that alcohol, poisons, or other substances taken into the body might have deleterious effects on the germ plasm and result in degeneration. But he insisted that it was only effects on the germ plasm, not effects on the body, that could be inherited. As with Charles Darwin (1809-1882), Weismann never discovered how variations arose to produce evolutionary novelties, but he showed that they had to occur in the germ plasm or genetic material.

Weismann conducted a famous experiment to try to disprove the inheritance of acquired characteristics. He cut the tails off of several generations of mice and measured the tails in their offspring. There was no decrease in length, leading Weismann to conclude that the information passed on from an organism to its offspring is independent of the influence of the body cells.

Weismann's views on "hard heredity" spread rapidly, especially in Germany, England, and the United States. Many biologists, however, were not convinced by Weismann's theory of hard heredity. Ernst Haeckel (1834-1919), the most famous Darwinian biologist in Germany in the late nineteenth century, continued to believe that the inheritance of acquired characteristics produced hereditary change, though he, like Darwin, also insisted on the efficacy of natural selection. In the 1890s neo-Lamarckians vociferously opposed Weismann, since they did not think he solved the problem of the source of evolutionary variation. Around 1900 scientists in France, Spain, Latin America, and many other parts of the world leaned more toward Lamarck and soft heredity than to Weismann's hard heredity.

However, Weismann's views prevailed in the scientific world through the discovery of the chromosome, the rediscovery of Mendelian genetics around 1900, and, after his death in 1914, the subsequent neo-Darwinian synthesis of the 1930s and 1940s.

RICHARD WEIKART

Friedrich Wöhler
1800-1882
German Chemist

Friedrich Wöhler, along with Jöns Jacob Berzelius (1779-1848) and Justus von Liebig (1803-1873), were pioneers in applying the techniques of organic chemistry to the parts and products of living things. Their work, originally known as "animal chemistry," established the foundations of modern biochemistry. Wöhler and Liebig are important in the history of chemistry for their recognition of the benzoyl radical. Wöhler is best known for demonstrating that the isomerization of ammonium cyanate produced urea, thus making him the first to synthesize an organic compound from an inorganic compound.

Wöhler, the son of a schoolmaster, was born in Eschersheim, near Frankfurt-am-Main. Although Wöhler studied medicine at Marburg and Heidelberg and earned the M.D., he came under the influence of Leopold Gmelin (1788-1853) and decided to make chemistry his career. He never practiced medicine. Wöhler was not a distinguished student, but flourished at Heidelberg when he was given permission to abandon routine coursework and devote himself to research. A year of work with Berzelius in Sweden (1823-24) established the future course of his research career. He taught at technical schools in Berlin (1825) and Cassel (1831) before being appointed Professor of Chemistry in Göttingen in 1835. He remained at that university until his death in 1882. Wöhler was greatly respected as a teacher and exerted tremendous influence on the development of organic chemistry though his numerous students.

As a medical student, Wöhler initiated an investigation of the derivatives of cyanogen and subsequently discovered cyanic acid. In 1828 he carried out his most famous experiment, in which he demonstrated that ammonium cyanate, the ammonium salt of cyanic acid, could be isomerized (transformed) into urea. This was the first time that a chemical normally produced only by living beings was synthesized from materials that could be obtained, at least in theory, from nonliving matter. In a famous letter to a colleague Wöhler wrote that he could "make urea without kidney of man or dog." Although Wöhler's preparation of urea is often seen as proof that inorganic and organic chemicals are theoretically equivalent, many other demonstrations were needed to abolish the ancient belief that "animal chemistry" was fundamentally different from inorganic chemistry. Indeed, Wöhler never claimed that his synthesis of urea signaled the death of vitalism. The synthesis of acetic acid by Hermann Kolbe (1818-1884) is generally considered the first complete *in vitro* synthesis of an organic compound.

The study of cyanogen compounds also led to the isolation of benzaldehyde from pure oil of bitter almonds. In 1832 Wöhler and Liebig found that when they put benzaldehyde through a variety of reactions, a cluster of atoms seemed to act like a chemical element. That is, it apparently remained unchanged throughout all the reactions of benzaldehyde that they were able to test. They named this the group of atoms benzoyl. This concept brought order and direction to further work in organic chemistry. Eventually, other examples of such "compound radicals" were discovered. Wöhler published a landmark article on the chemistry of metabolism in 1842. He demonstrated that when benzoic acid was consumed with food, it was excreted in the urine as hippuric acid. Other investigations involved the quinones, alkaloids, and derivatives of uric acid.

Although Wöhler's most famous achievements were in organic chemistry, he also remained interested in inorganic chemistry. His many admirers said that there were hardly any known elements that he had not studied. Indeed, his work included the isolation of aluminum, beryllium, silicon hydrides, analyses of various minerals, and studies of meteorites.

LOIS N. MAGNER

Biographical Mentions

~

Louis Agassiz
1807-1873

Swiss geologist and naturalist who first popularized the idea of an ice age. Agassiz was a renowned naturalist who, as a professor of natural history at Neuchâtel, determined that glaciers had once covered Europe. In 1847 he moved to the United States to become a professor at Harvard University. While there he discovered evidence that the Ice Age had affected North America too. His ice age theories caught the public fancy. A committed Creationist, Agassiz never accepted the evolution of species, despite his acceptance of an ancient Earth.

Henry Walter Bates
1825-1892

English naturalist whose theory of mimicry (now called Batesian mimicry) explained that color patterns of different species can be similar because nonpoisonous species mimic the bright warning patterns of poisonous species. His hypothesis supported the theory of evolution by natural selection proposed by his contemporaries Charles Darwin and Alfred Russel Wallace. Bates also collected almost 15,000 animal species during 11 years of field work in the

Amazon Basin and described more than 700 scarab species new to science.

Martinus Willem Beijerinck
1851-1931

Swiss botanist who discovered a new form of life even smaller than bacteria (now known to be viruses) while studying a contagious disease that affects tobacco plants (the tobacco mosaic virus). He described what remains the most distinctive feature of viruses: to reproduce, a virus must incorporate itself into the living cellular machinery of its host. Beijerinck also made many significant contributions to agriculture through his research in the field of soil microbiology.

Edouard van Beneden
1846-1910

Belgian cytologist whose research helped explain how cells divide and distribute chromosomes equally to daughter cells. He was one of several scientists who determined that chromosomes duplicate by longitudinal splitting during the process of cell division. Van Beneden discovered that all individuals within a given species have a fixed number of chromosomes. He also demonstrated that egg and sperm cells are haploid (they contain half the number of chromosomes as body cells) and that fertilization reinstates the diploid state.

George Bentham
1800-1884

English botanist who was one of the most prolific botanists of his time. Among his many written contributions was *Genera Plantarum,* in which he and Joseph Dalton Hooker described and classified the flowering plants and gymnosperms. Although their classification scheme was replaced by one that recognized evolution, their generic descriptions were excellent and many of their familial and generic categories remain valid today. During his lifetime, Bentham donated as many as 100,000 botanical specimens to the Royal Botanic Garden.

Theodor Ludwig Wilhelm von Bischoff
1807-1882

German physician recognized for his work in comparative embryology. Born in Hanover, Germany, he received both M.D. and Ph.D. degrees and was a professor at Bonn, Heidelberg, Giessen, and Munich universities. He first described the maturation and release of eggs from the ovary, although he did not recognize the role of sperm in fertilization. He compared the embryological development of many animals including rabbits, guinea pigs, deer, and dogs, as well as humans. He wrote on the fetal development of the human brain and compared human embryology to the development in apes.

Johann Friedrich Blumenbach
1752-1840

German physiologist known as the founder of physical anthropology. Born in Gotha, he received the doctor of medicine from the University of Göttingen in 1775 and remained there as professor until his death. He was the first to classify races into five groups: Caucasian, Mongoloid, Ethiopian, American, and Malay. The term Caucasian, referring to white people or Europeans, originated from his belief that the race developed in the Caucasus Mountains. He recognized the Caucasian or white race as the basic type, with other races evolving because of the demands of the environment. He measured skulls and collected many samples for his studies of comparative anatomy.

Jacques Boucher de Crevecoeur de Perthes
1788-1868

French archaeologist who determined that humans had existed millions of years ago, after finding flint axes and other tools in deposits with the bones of extinct animals. While his theory met with disbelief in his day, it advanced the idea of human antiquity, set the precedent for careful stratigraphic dating, and inspired the work of other notable scientists, including Charles Lyell.

Theodor Boveri
1862-1915

German cytologist who proved that chromosomes are independent entities. Emphasizing that chromosomes are organized structures, he showed that each chromosome is responsible for certain hereditary characteristics. Boveri and Edouard van Beneden independently discovered a structure connecting the chromosomes during cell division. Calling it the centrosome, Boveri demonstrated that it provides the division center for the cell. Boveri's work has greatly influenced subsequent cytological interpretation of genetic phenomena.

Henri Braconnot
1780-1855

French botanist Henri Braconnot is credited as the first person to isolate and identify chitin, a substance commonly found in the exoskeletal shells of invertebrates, which he located in the cell walls of fungi. Additionally, his experiments using plant and animal components lead him to discover other common biological molecules and substances such as glycine, leucine, and nitrocellulose.

Alexandre Brongniart
1770-1847

French mineralogist whose research helped to demonstrate how fossils could be used to trace geologic changes over time in a particular area, thus introducing the practice of geologic dating. Brongniart was the first to distinguish minerals from rocks, and he shares the credit for naming the period of the Mesozoic era known as Jurassic. Later in his career he turned to ceramic technology and worked to improve the art of enameling in France.

Robert Brown
1773-1858

Scottish botanist Robert Brown revolutionized cell theory with his discovery of the nuclei in many different types of plant tissue. He observed one "nucleus" in each cell. He also made substantial contributions to the field of plant taxonomy, the classification of plants based on physical characteristics. Brown's description of the lack of ovary around the ovule in conifers and other similar plants provided a fundamental distinction between gymnosperms and angiosperms. Brown, who shared his fascination of plants with the British public, was the first to secure a nationally owned botanical collection available for public viewing.

William Buckland
1784-1856

English geologist who attempted to reconcile the conflict between science and religion. Publishing works on the world's creation and the great flood referred to in the Bible, Buckland displayed a keen interest in using geologic evidence to support Christian beliefs. In addition, Buckland compared certain gouged rocks in England to similar ones in Switzerland to show that glaciers had at one time covered Scotland and England.

Alphonse de Candolle
1806-1893

Swiss botanist who took on the task of completing his father Augustin de Candolle's multi-volume work on plant classification called the *Prodromus*. Alphonse Candolle was president of the International Botanical Congress in 1866 and in this capacity was instrumental in the publication of the first of the international rules for the naming of plants. His other significant project was a two-volume work on plant geography, on how climate and terrain affect the distribution of plants.

Edward Drinker Cope
1840-1897

American paleontologist and zoologist who was among the first to discover complete dinosaur skeletons. His conflicts with rival paleontologist Othniel C. Marsh and their competing fossil-hunting expeditions in the American West increased public interest in paleontology. Cope was a proponent of the Neo-Lamarckian view of evolution. This theory regarded attributes developed through use, rather than natural selection, to be the mechanism of evolutionary change.

Elliott Coues
1842-1899

American naturalist, bibliographer, anatomist, historian, and surgeon. Coues was an exceptional field naturalist and the leading American ornithologist of the late nineteenth century. An eloquent and prolific author, he wrote many descriptive works on North American birds and mammals and compiled bibliographies of American and British ornithology. In addition, he edited fifteen volumes of travel journals of western explorers. He served as Assistant Surgeon, U.S. Army, from 1864-81 and traveled extensively in the American West, collecting animal specimens. He was elected to the National Academy of Sciences in 1877. In 1883 he co-founded the American Ornithologists' Union, and he later (1892-5) served as its president.

Anton Dohrn
1840-1909

German marine biologist who founded the Naples Zoological Station, a research facility where many biologists, especially Germans, studied marine biology. Dohrn was a student of Ernst Haeckel (1834-1919), under whom he became a committed Darwinist. Dohrn was remarkably successful in raising funds for his station, procuring funds from the German government as well as individuals such as Charles Darwin (1809-1882).

Marie Eugène François Thomas Dubois
1858-1940

Dutch anatomist and paleoanthropologist who discovered the first fossils belonging to a direct ancestor of modern humans. Dubois was seeking Darwin's "missing link" on the island of Java, hence "Java man," the common name of his find. Dubois named his discovery *Pithecanthropus erectus*, which was later reclassified as *Homo erectus*. Dubois's work reflected two contemporary theories, that upright posture was of primary importance in human evolution and that the East Indies, not Africa, was the site of humankind's origins.

Emil Heinrich Du Bois-Reymond
1818-1896

German physiologist who conducted ground-breaking research into nerve and muscle stimulation. He found that the stimulation of a nerve membrane results in a wave, the electrical impulse, which travels along the nerve. Du Bois-Reymond also observed the same phenomenon occurring during muscular contraction. His work founded the field of electrophysiology. He emphasized the importance of applied physics and chemistry to physiology, helping to dispel the idea that organic matter contained a "life force."

Jean Baptiste André Dumas
1800-1884

French chemist who conducted pioneering work in organic chemistry. He developed a method for determining the nitrogen content of organic compounds and demonstrated that in organic compounds halogens could replace hydrogen. Dumas collaborated with Eugene-Melchior Peligot to isolate methyl alcohol and establish the alcoholic series. He also worked with Justus von Liebig to study organic chemical reactions. Dumas's work on atomic weights helped to supercede the theories of Jons Berzelius.

André Marie Constant Dumeril
1774-1860

French zoologist who led a scientific mission to Central America, including the vast area between Panama and the southwest United States. As a result he co-authored the first comprehensive books on reptiles of these geographical areas. Today Dumeril's name lives on as herpetologists, people who study reptiles and amphibians, continue to name new species after him. Dumeril's ground boa (*Acrantophis dumerili*) and Dumeril's monitor (*Varanus dumerilii*), a semiaquatic lizard that can remain submerged for up to 75 minutes, are both found in Indonesia and Madagascar.

(Rene Joachim) Henri Dutrochet
1776-1847

French physiologist who discovered osmosis. He observed the diffusion of a solvent through a semipermeable membrane, calling the process osmosis. In the area of plant physiology, Dutrochet recognized the significance of green pigment in the use of carbon dioxide, investigated the mechanisms of respiration, light sensitivity, and geotropism, and showed that internal plant transport involves osmosis. He demonstrated that mushrooms are the reproductive bodies of the mycelium. Dutrochet was among the first to recognize the importance of an organism's individual cells.

Christian Gottfried Ehrenberg
1795-1876

German biologist who discovered microorganisms in fossils. Ehrenberg discovered microscopic fossil organisms in different geologic formations, in the process founding the field of micropaleontology. He demonstrated that fungi come from spores and showed that mushrooms and molds reproduce sexually. Ehrenberg was the first to study coral in detail and he showed that phosphorescence in the sea is caused by planktonic microorganisms. While correctly opposing the theory of spontaneous generation, he mistakenly believed that all organisms possess complete organ systems.

George Engelmann
1809-1884

German-American botanist, physician, and meteorologist best known for his botanical monographs. Engelmann earned his M.S. degree from the University of Würzburg, Germany, in 1813. His most important study was of the *Morphology of Monstrosities*. He immigrated to the United States in 1833. A pioneer in his field, his efforts included the discovery of immunity in the North American grape to the plant lice *Phyllaxera*.

Walther Flemming
1843-1905

German anatomist (1843-1905) who first observed and identified the stages of mitosis. Using new synthetic dyes, Flemming found material, which he named chromatin, within the cell nucleus. Observing the chromatin at different phases, he traced the process of cell division, calling it mitosis. Although Flemming himself was not aware of the relation of his findings to genetics, his work provided the underlying physical basis for inheritance when Gregor Mendel's theories were rediscovered.

Marie Jean Pierre Flourens
1794-1867

French physiologist who determined the functions of the major parts of the vertebrate brain. Using pigeons, he observed the physiological changes that occurred when certain portions of their brains were removed. From his studies, he demonstrated that higher intellectual abilities are found in the cerebral hemispheres, that movement is regulated by the cerebellum, and that vital bodily functions are controlled by the medulla oblongata. He also linked the semicircular canals of the inner ear to the body's equilibrium.

Hermann Fol
1845-1892

Swiss physician and zoologist who in 1877 first observed the penetration of an egg by a sperm cell while studying sea urchins. He showed that the egg nucleus is part of the ovum, and provided evidence supporting the claims of Oskar Hertwig that the nuclei of the cells in an embryo descend from the original egg and sperm pair. Fol was lost at sea during an expedition to study sponges off the coast of Tunisia.

Edward Forbes
1815-1854

English naturalist who co-founded the science of oceanography and who was a pioneer in biogeology. Forbes was one of the first to divide and study ocean zones in scientific terms. He also detailed how most of the plants and animals of England migrated there from the European continent during distinct episodes of the glacial epoch. His scientific contributions fostered several areas within natural history and made the study of them systematic.

Karl Friedrich von Gaertner
1772-1850

German physician and botanist Karl Gaertner was instrumental in unraveling the mystery surrounding plant reproduction. He established that plants are sexually reproducing organisms through his experimentation and careful observation of the physiologic changes that take place in flower parts during fertilization. His research on hybrids, offspring that result from a cross between members of two different species or subspecies, provided the groundwork for Gregor Mendel (1822-1884), founder of the laws of inheritance.

Sir Francis Galton
1822-1911

British gentleman scientist and cousin of Charles Darwin (1809-1882) who studied meteorology and anthropology, but is best known for founding the eugenics movement. Eugenics, a term coined by Galton, refers to the attempt to improve human heredity by rationally influencing or controlling human reproduction. His ideas were largely influenced by Darwin's *Origin of Species* (1859) and the concept of natural selection. Galton advanced his eugenic views, which he and many proponents considered applied science, in *Hereditary Genius* (1869) and other works.

Hans Christian Joachim Gram
1853-1938

Danish physician and pharmacologist who developed the "Gram staining technique," still used today to identify bacteria as "gram positive" or "gram negative". By modifying an existing method, Gram was able to differentiate bacteria based on their cell wall structure. He taught and practiced internal medicine, and his publications advocated rational pharmacotherapy and a clinical science approach to illnesses and disease. Gram was able to end the use of many ineffective and obsolete therapeutic methodologies, and helped reinvigorate and modernize pharmacology.

Asa Gray
1810-1888

American botanist who was the leading authority on the subject in the United States for four decades. Educated as a physician, Gray became the first professor of botany at Harvard University, where he also supervised the botanical garden and established the herbarium of dried plant specimens. He traveled widely on two continents but collected little himself, preferring to classify plants gathered by others, many of whom he trained. His classic *Manual of Botany* (1848) was often revised by his successors and remained in print long after his death. He was the foremost American advocate for Darwin's theory of evolution.

Oskar Wilhelm August Hertwig
1849-1922

German zoologist and embryologist who in 1876 postulated that fertilization is a result of the fusion of germ cell nuclei from the male and female parents. This was a departure from the traditional view that the sperm induced a physical vibration or chemical change in the egg that caused it to develop. He observed that only one spermatozoon is required to fertilize the egg, and that after this occurs, additional sperm cells are prevented from entering.

Wilhelm His
1831-1904

Swiss anatomist who studied the cornea, lymph vessels and glands, the thymus, the central nervous system, and the heart. His recognized the differences between true body cavities and linings, such as the endothelia of blood vessels, and he delineated the embryonic development of individual organs and organs systems. He constructed the first microtome to section tissue, and advanced the use of photography, wax model embryos, standardized anatomical nomenclature, and embryology charts as teaching and research tools.

Wilhelm Friedrich Benedikt Hofmeister
1824-1877

Self-taught German botanist who published two groundbreaking textbooks in his field. His research distinguished flowering from nonflowering plants and demonstrated how plant generations alternate between sexual and nonsexual generations. His first published handbook was on plant physiology. His other textbook became the first general work describing how plant tissues and organs are formed and differentiated. He is considered a pioneer of comparative plant morphology.

Baron Friedrich Heinrich Alexander von Humboldt
1769-1859

German-born and internationally renowned natural historian, traveler, and diplomat. Humboldt was a student at several German universities. From 1799-1804, he visited Latin America, studying natural history and geology (with botanist Aimé J. A. Bonpland). He demonstrated the connection between the Orinoco and Amazon rivers. He also accomplished pioneering work in plant geography, and he established a system of meteorological observation stations in Eurasia and the British colonies. In 1845 the first volume of *Kosmos,* a summary of his views concerning the physical history of the world and the complexity and unity of nature, appeared; five volumes in all were published. See long biography on p. 77.

Karl Martin Leonhard Albrecht Kossel
1853-1927

German biochemist who, for his work in cell and protein chemistry, won the Nobel Prize for physiology or medicine. His research demonstrated that the cell substance nuclein (now called nucleoprotein) contains both protein and nonprotein (nucleic acid) parts. He also had great success in investigating the composition of proteins. Correctly concluding that nuclein, or nucleoprotein, is related to the formation of flesh tissue, his work foresaw modern investigations of acids as the bearers of genetic information.

Edouard Amant Isidore Hippolyte Lartet
1801-1871

French paleontologist and archeologist credited with the discovery of more than 90 genera and species of fossil mammals and reptiles recovered from the tertiary terrain of Gers, France. His excavations at cave sites in central France provided definitive proof that human beings co-existed with extinct prehistoric animals The products of these excavations served as the foundation for the science of prehistory, supporting the evolutionary account of human development.

Pierre André Latreille
1762-1833

French entomologist best known for his classifications of crustaceans, arachnids, and insects. An ordained priest and professor of natural history at the Museum of Natural History in Paris, Latreille earned distinction as a founder of modern entomology.

Gideon Algernon Mantell
1790-1852

British surgeon and amateur naturalist who is credited with first discovering dinosaur fossils in 1822. According to the commonly held story, Mantell's wife, Mary Ann, discovered a fossil tooth in a rubble pile where a road crew was working near the Ouse River in England. Mantell later discovered similar teeth and bones in Sussex and came to the conclusion, despite the skepticism of the scientific establishment, that they were from a large ancient reptile. At the Hunterian Museum in London, medical student and iguana researcher Samuel Stutchbury told Mantell that the teeth reminded him of those of a marine iguana. Mantell thus named his fossil reptile *Iguanodon*, or iguana tooth. The group that the *Iguanodon* was part of was later christened the Dinosauria (terrible reptiles) by Richard Owen in 1841.

Othniel Charles Marsh
1831-1899

American vertebrate paleontologist who helped establish the science of vertebrate paleontology in the United States. A Yale graduate and faculty member, Marsh later worked for the U.S. Geological Survey and was president of the National Academy of Sciences. He created a large collection of specimens at Yale based upon his own expeditions and on the work of hired collectors. He described the earliest fossil mammals then known, and he worked on the evolution of North American horses, the fossil reptiles of the West, and the reptilian origins of birds. He was a great rival of paleontologist Edward Drinker Cope (1840-1897).

William Martin
1767-1810

English geologist who proposed that fossil evidence be used as the basis for establishing a science devoted exclusively to the study of ancient plants and animals.

Johann Friedrich Miescher
1844-1895

Swiss scientist who discovered nucleic acids. Miescher found that the nuclei of the white blood cells found in pus contained a substance containing phosphorous and nitrogen. The substance was first called nuclein, but after Miescher separated it into protein and an acid molecule, nuclein became known as nucleic acid. The name changed again to its current one, deoxyribonucleic acid (DNA). He also discovered that the concentration of carbon dioxide, not that of oxygen, in the blood regulates breathing.

Conwy Lloyd Morgan
1852-1936

British zoologist and psychologist who has been called the founder of comparative, or animal, psychology. Morgan's work in comparative psychology emphasized the importance of objectively describing animal behavior without resorting to anthropomorphism. He studied animal behavior independently of human mental evolution. Morgan argued that no action could be attributed to a higher mental faculty if it could be linked to a lower one. This idea has since become known as the principle of parsimony.

(Louis-Laurent-Marie) Gabriel de Mortillet
1821-1898

French archeologist who created the first classification system dividing man's prehistoric cultural development into chronological epochs. His ordering of the Paleolithic, or Stone Age, epoch carried on into the twentieth century as the basis for anthropological classification. Mortillet also studied the geology and paleontology of the Alps.

Karl Wilhelm von Nägeli
1817-1891

Swiss botanist who studied plant cells. Nägeli identified structures later to be called chromosomes. He observed cell division and osmosis in algae. Nägeli introduced the idea of a group of plant cells, the meristem, always capable of division. He and Hugo von Mohl distinguished the wall of a plant cell from the interior. Nägeli unfortunately rejected a paper sent to him by Gregor Mendel that would later become the material for Mendel's laws of inheritance.

Richard Owen
1804-1892

British paleontologist who studied many important fossils, including Charles Darwin's South American collection and the Jurassic age *Archaeopteryx*. Owen was passionate about evolution and argued against natural selection, insisting instead that the Creator endowed embryos with innate abilities to deviate from the parental type and form new species. His research produced a theoretical hierarchy that elevated humans above all other anthropoid apes, a distinction based on the unique mental capacities derived from the enlarged human cerebrum.

Christian Pander
1794-1865

Russian zoologist who discovered and introduced the concept of different embryonic tissue layers. He received his training in embryology while living in Germany. Embryology is the study of how individual cells develop from a fertilized egg into a new organism. In 1817 Pander began his observations on the embryos of chicken eggs. This study helped him discover embryonic tissue layers, which he called the primordial or germ layers. Today we know these layers as the ectoderm, mesoderm, and endoderm. It is interesting to note that Pander began this study because he was one of the few people who could afford to purchase the eggs and pay an attendant to watch the incubators.

Anselme Payen
1795-1871

French chemist who made advances in the study of carbohydrates and enzymes. Payen developed a charcoal filter used to decolor sugar. He discovered diastase, an organic catalyst that converts starch to the sugar maltose and the first enzyme produced in concentrated form. From the cell walls of plants he isolated cellulose, which he found to be similar to starch. Following Payen, the names of enzymes would come to end in *ase* and carbohydrates in *ose*.

William Dandridge Peck
1763-1822

American naturalist who was among the first to study etymology in the United States. Graduating from Harvard College, Peck farmed in Maine where he collected specimens of indigenous animals and plants. In a 1794 paper he classified fish he found in New Hampshire, considered one of the first American scientific zoological publications. Peck focused on insects detrimental to agriculture in New England and their economic impact. Selected chair of natural history at Harvard, Peck taught applied entomology classes and established a botanical garden.

Pierre-Joseph Pelletier
1788-1842

French chemist whose work in alkaloid chemistry contributed much to the field of medicine. Working with Joseph-Bienaime Caventou, Pelletier demonstrated that alkaloids, a group of organic compounds derived from plants, contain oxygen, hydrogen, carbon, and, particularly, nitrogen. Together, they isolated chlorophyll, strychnine, brucine, quinine, and caffeine. From their work, chemists have been able to isolate and produce alkaloids for medicinal uses, including quinine, which is used to treat malaria.

Wilhelm (Friedrich Philipp) Pfeffer
1845-1920

German botanist who studied osmotic pressure. While working on cell metabolism, Pfeffer developed a semipermeable membrane with which he studied osmosis. In addition to creating a method for measuring osmotic pressure, he proved that osmotic pressure depends on the size of those molecules too large to pass through the membrane. In so doing, he was able to measure the size of giant molecules.

Nathanael Pringsheim
1823-1894

German botanist who investigated reproduction in plants. Pringsheim was among the first to observe sexual reproduction in algae. He showed that these tiny organisms release sperm and egg cells into the water, where they combine. He also described the alternation of generations, or reproduction by spores, in mosses. Pringsheim was independently wealthy and conducted private research, as well as established the German Botanical Society and the journal *Jahrbucher fur wissenschaftliche Botanik.*

Frederick Pursh
1774-1820

German-American botanical traveler whose *Flora Americae Septentrionalis,* a book published in England in 1814, was the earliest attempt to describe the complete flora of North America. Pursh eked out a living working as a gardener for others; he traveled and collected plant specimens throughout the eastern United States, and also studied plants collected in the West by Lewis and Clark. He was making a collection of Canadian plants when he died a pauper in Montreal.

(Ferndinand Gustav) Julius von Sachs
1832-1897

German botanist who contributed much to the field of plant physiology. Sachs described how plant root hairs function and stated that absorbed water moves through tubes in plant walls without the aid of living cells. He demonstrated that chlorophyll is not diffused in plant tissue but contained in special cellular bodies (chloroplasts). He showed that starch results from the absorption of carbon dioxide and that starch is the first visible product of photosynthesis.

Nicolas Theodore de Saussure
1767-1845

Swiss chemist and plant physiologist whose work provided the basis for phytochemistry. After studying the formation of carbonic acid in plant tissues, he proved Steven Hales's theory that plants increase in weight after absorbing water and carbon dioxide in sunlight. Saussure showed the dependence of plants on the absorption of nitrogen from the soil. He also conducted important analyses of biochemical reactions in plant cells.

Carl Theodor Ernst von Siebold
1804-1885

German zoologist who studied invertebrates and contributed much to the field of parasitology. One of the first important texts in comparative anatomy involved a collaboration between Siebold, who contributed the work on invertebrates, and Friedrich Hermann Stannius, who was responsible for the information on verte-

brates. Based on factual observations, the text departed from earlier philosophical approaches. Siebold's work in parasitology involved the new idea that the stages in a parasite's life cycle alternate between hosts.

Herbert Spencer
1820-1903

English philosopher, psychologist, and sociologist who synthesized the social and natural sciences into the framework of nineteenth-century evolutionary thought. By asserting the relationship between progress and science, Spencer posited the progressive aspects of evolution and coined the term "survival of the fittest." He had a low standing among scientific specialists, but was well respected by scientific innovators such as Charles Darwin, Francis Galton, and Alfred Wallace. Spencer's work had only moderate influence in Europe, but it was widely accepted in the United States.

Eduard Adolf Strasburger
1844-1912

German plant cytologist who studied nuclear division in plants. Strasburger accurately described the embryonic sac in gymnosperms (conifers and others) and angiosperms (flowering plants). He explained the basic principles of mitosis and declared that new nuclei can arise only from the division of other nuclei, a modern law of plant cytology. Strasburger invented the terms cytoplasm and nucleoplasm. He showed that in angiosperms the primary structure involved in heredity is the nucleus and that the nuclei of angiosperm germ cells undergo meiosis.

Sir Charles Wyville Thomson
1830-1882

Scottish naturalist and marine biologist who was one of the first to describe deep-sea life. On deep-sea dredging expeditions to the north of Scotland, Thomson discovered a diversity of marine invertebrates, many of which had been believed to be extinct. His finding that deep-sea temperatures are not as constant as had been thought indicated the presence of oceanic circulation. On another of his expeditions, the crew conducted observations of the three great ocean basins.

Gustave Thuret
1817-1875

French botanist whose greatest interest was with the group of algae called the rhodophyta, or red algae. In addition to his discovery of new species, he examined how algae reproduced. The reproductive success of algae was not well understood until Thuret described his findings. Travels to Egypt and Syria contributed to his knowledge of seaweed. One of his works was the establishment of a great botanical garden at Antibes near the Mediterranean Sea.

Sir Edward Burnett Tylor
1832-1917

English anthropologist who is considered the founder of cultural anthropology. Tylor was influenced by Charles Darwin's work and theorized that there is an evolutionary relationship between primitive and modern cultures. He saw a progressive development from a primitive to a civilized state. Tylor believed that culture should be studied at all stages of man's development. During the late nineteenth-century controversy over the races of mankind, Tylor argued for man's physical and mental unity.

Rudolf Ludwig Carl Virchow
1821-1902

German physician, pathologist, and physical anthropologist who made major contributions to cell theory and established the doctrine of cellular pathology. Virchow established the modern concept of cell theory, which maintains that all cells come from cells. He published pioneering studies of "white blood" (leukemia), phlebitis, embolism, pyaemia, and inflammation. A statesman as well as a pathologist, Virchow established many improvements in the sanitary and medical conditions in Berlin. He was a founder of the Berlin Society of Anthropology, Ethnology and Prehistory.

Carl Vogt
1817-1895

French zoologist who helped pioneer the use of photography in scientific publication. Vogt's photographic image of the Berlin *Archaeopteryx macrura* originally appeared in the *Revue Scientifique* in 1879. *Archaeopteryx* at the time was believed to be an intermediate form between birds and reptiles. His illustration of the actual slab containing the fossil of this animal was a photograph, not a drawing as was the usual convention, and he received a great deal of criticism because the photograph did not provide as much detail as a drawing. Today, however, photographs are included in scientific work all the time.

Edmund Wilson
1856-1939

American cytologist who helped discover the existence and nature of sex chromosomes. Wilson's "The Cell in Heredity and Development" (1928) integrated cell structure and function with heredity, adaptation, and evolution, and helped advance Mendelian genetics. Wilson identified the spiral cleavage of annelids, arthropods, and mollusks, and the radial cleavage of echinoderms, chordates, and vertebrates as the two major patterns of embryo development. Wilson insisted that the scientific method of careful observation, testable hypothesis formulation, and precise experimentation be used throughout biological research.

Bibliography of
Primary Sources

Books

Agassiz, Jean Louis Rodolphe. *Studies on Glaciers*. 1840. In this work Agassiz first introduced the concept of the Ice Age and the mobility of glaciers. According to Agassiz, the Ice Age caused the mass extinction of plant and animal life, accounting for differences among past and living species.

Audubon, John James. *Birds of America*. 4 vols. 1827-1838. Audubon's most important publication includes 435 hand-colored illustrations. Two companion works were also published: the five-volume *Ornithological Biography* (with William MacGillivray; 1831-1839) includes the text to accompany the illustrations, while *A Synopsis of the Birds of America* (1839) serves as a index volume.

Audubon, John James. *Viviparous Quadrupeds of North America*. 2 vols. 1845-46. Audubon's last project is a study of American land mammals, with the 150 paintings accompanied by a three-volume text by North Carolina minister and mammalogist John Bachman (1846-54).

Baer, Karl Ernst von. *History of the Development of Animals*. Vol. 1, 1828; Vol. 2, 1837. Here biologist Karl von Baer demolished the notion that "ontogeny recapitulates (repeats) phylogeny"—that at various stages of development the embryos of any animal species, including man, resembled a primitive form of the adult of the species. Baer established that man did not go through primitive evolutionary stages and that, in fact, there was no progression through any species hierarchy of human development. In essence Baer's work established that there were no "less evolved races" of humans.

Candolle, Augustin de. *Théorie élémentaire de la botanique*. 1813. In this work, considered Candolle's most important, the Swiss botanist argued that plant anatomy, not physiology, should be the sole basis of classification. Candolle also coined the term taxonomy to describe his system of classification.

Candolle, Augustin de, and Alphonse de Candolle. *Prodromus Systematis Naturalis Regni Vegetabilis*. 17 vols. 1824-73. In this work Augustin de Candolle provided a description of the whole range of plants and attempted to include all known species. He presented information not only on classification but on ecology, agriculture, and phytogeography (the study of how climate and terrain influence the distribution of plant species). Several of the volumes were written and published after Candolle's death by his son Alphonse.

Cuvier, Georges. *The Animal Kingdom.* 1817. In this work Cuvier perfected the classification system of Carolus Linnaeus, adding the phylum category to show relationships between broader groups of species. Cuvier also compared the fossil remains of various animals to apparently related, living species, and included these fossil species in his classification scheme.

Cuvier, Georges. *Researches on the Fossil Bones of Quadrupeds.* 1812. In this work Cuvier, founder of paleontology, argued that the earth had experienced a series of massive geologic upheavals that occurred on a global scale and rearranged the face of the earth. According to Cuvier, these upheavals also caused the extinction of many animal species, which he demonstrated by studying and reconstructing the fossil bones of various species.

Darwin, Charles. *The Descent of Man and Selection in Relation to Sex.* 1871. In this work Darwin further elaborated his theory of evolution by natural selection, focusing on the topic of sexual selection, and discussed the place of humans along the evolutionary path.

Darwin, Charles. *On the Origin of Species by Means of Natural Selection, or the Preservation of Favoured Races in the Struggle for Life.* 1859. In this landmark work Darwin established his theory of evolutionary development by natural selection. According to Darwin, current species evolve from past species through hereditary genetic variations, a process of "natural selection" whereby favorable traits are passed on to younger generations of a given species to provide for the progressive adaptation and survival of that species. Darwin's theory, especially its application to the evolution of man, had a wide-reaching and controversial impact on subsequent scientific, philosophical, religious, and social thought. It also became a cornerstone of modern biological study.

Davy, Humphry. *Elements of Agricultural Chemistry.* 1813. Britain's Board of Agriculture commissioned Davy to give lectures on connections between chemistry and agriculture; the result was this influential work.

Galton, Francis. *Hereditary Genius.* 1869. In this work Galton proposed that human intelligence, specifically genius, is a hereditary trait. By studying the patterns of intelligence among distinguished families, Galton deduced that parents who reproduce with a person of higher or lower intelligence have children of corresponding mental abilities.

Henle, Friedrich Gustav Jakob. *Allgemeine Anatomie.* 1841. This work represents the first systematic treatise on histology, the study of minute cell and tissue structures. Henle described the microscopic structure of plant and animal cells, and proposed the modern germ theory of communicable diseases, whereby germs are recognized as living, parasitic organisms.

Hofmeister, Wilhelm. *Handbuch der Physiologischen Botanik.* 1865. In this work Hofmeister, considered a founding figure of modern botany, provided the first major textbook on plant forms and structures.

Lamarck, Jean Baptiste de. *Zoological Philosophy.* 1809. In this work, Lamarck put forth the principles of his theory of evolution, known as "Lamarckism." According to Lamarck, living organisms are compelled to adapt for survival and their acquired characteristics, or those characteristics accumulated during their lifetime, are inherited by their immediate descendants. Lamarck also introduced the "use and disuse" hypothesis, which posited that certain body parts mutate and evolve because they are necessary or useful, while others disappear because they are not, providing an explanation for the physical evolution of a species over time.

Liebig, Justus von. *Organic Chemistry and Applications to Agriculture and Physiology.* 1840. This book is generally considered the central work in nineteenth-century agricultural chemistry. In the work Liebig explained that the exact requirements for plant growth could be measured through laboratory combustion analyses, so that scientists could tell farmers precisely what they needed to maintain a balance of chemical inputs and outputs.

Malthus, Thomas R. *An Essay on the Principles of Population.* 1798. In this work economist Thomas Malthus presented his dire formula for population growth and human suffering. As Malthus observed, population increases at a geometric rate while food resources to sustain that population increase at only an arithmetic rate. Malthus suggested that war, starvation, poverty, and disease serve as natural, and inevitable, deterrents to population growth. This work influenced Charles Darwin's theory of evolution by natural selection and had a important impact on subsequent social, moral, and political thought.

Mohl, Hugo von. *Die Vegetabilische Zelle.* 1853. In this work, considered a classic botanical text, Hugo von Mohl provided a comprehensive summary of plant research and related cell theory to that date.

Morgan, Conwy Lloyd. *Animal Life and Intelligence.* 1890. In this work Morgan founded the study of comparative, or animal, psychology. Morgan drew attention to the importance of scientific observation of animal behavior and the false tendency of researchers to attribute human characteristics to their animal subjects.

Morgan, Conwy Lloyd. *An Introduction to Comparative Psychology.* 1894. In this work Morgan established the principles of comparative psychology and first identified trial-and-error learning among animals. He also introduced the principle of parsimony, based on Occam's razor, which asserts that no psychological activity should be attributed to a higher mental faculty if it can be linked to a lower one.

Pfeffer, Wilhelm. *Handbuch der Pflanzenphysiologie.* 1881. In this work Pfeffer, a pioneering investigator of osmosis and semi-permeable membranes, provided an influential delineation of plant physiology.

Schwann, Theodor. *Microscopical Researches into the Accordance in the Structure and Growth of Animals and Plants.* 1839. In this work Schwann established the basic principles of cell theory, including the idea that all plant and animal tissues are composed of cells, which by definition consist of a nucleus, and cell products.

Virchow, Rudolf. *Cellular Pathology.* 1858. In this work Virchow established the fundamental principles of modern cell theory, revolutionizing the study of biology and medicine. He correctly asserted that the cell is the primary unit of life, that every cell is the product of a pre-existing cell, and that diseased cells and tissues are the descendants of normal, healthy cells.

Mathematics

Chronology

1801 French mathematician Louis François Antoine Arbogast develops what comes to be known as the Arbogast method for deriving coefficients.

1801 German mathematician Johann Karl Friedrich Gauss publishes *Disquisitiones Arithmeticae,* a seminal work of number theory.

1807 French mathematician Jean Baptiste Joseph Fourier announces his famous theorem concerning periodic oscillation, which will prove invaluable to the study of wave phenomena.

1812 French astronomer and mathematician Pierre Simon Laplace establishes the modern form of probability theory in *Théorie Analytique des Probabilités.*

1822 Jean-Victor Poncelet, a French mathematician, lays the foundations for projective geometry as a separate branch of mathematics with the publication of *Traité des Proprietés Projectives des Figures.*

1827 German mathematician August Ferdinand Möbius introduces the concept of analytical geometry in *Der Barycentrische Calcul.*

1829 Russian mathematician Nicolai Ivanovich Lobachevski discovers non-Euclidean geometry, paving the way for the mathematics of curved surfaces.

1847 English mathematician and logician George Boole becomes the first to treat mathematical symbols primarily as forms rather than as units of value; seven years later, he introduces Boolean algebra.

1854 Georg Friedrich Bernhard Riemann, a German mathematician, introduces a theory of multi-dimensional space which provides a geometric foundation for modern physical theory.

1857 English mathematician Arthur Cayley pioneers the ideas of groups and matrices.

1874 With his revolutionary work on set theory and the theory of the infinite, German mathematician Georg Ferdinand Ludwig Philipp Cantor opens new fields for mathematical research.

1880 The topology studies of French mathematician Jules Henri Poincaré provide the foundation for twentieth-century algebraic or combinational topology.

1894 Building on the work of German mathematicians Friedrich Ludwig Gottlob Frege and Julius Wilhelm Richard Dedekind, Italian mathematician and logician Giuseppe Peano provides the framework for symbolic logic in *Formulaire de Mathématiques.*

Overview:
Mathematics 1800-1899

As the eighteenth century drew to a close, mathematics was in a state of rapid change. New areas of mathematics remained wide open to research, while older, established areas of mathematics were finding new applications. Advances in analytic geometry, differential geometry, and algebra all played important roles in the development of mathematics in the eighteenth century. It was calculus, however, which commanded most of the attention of eighteenth-century mathematicians. Discovered by Isaac Newton (1642-1727) and Gottfried Leibniz (1646-1716) late in the seventeenth century, the theory and applications of calculus dominated the mathematical scene throughout the eighteenth century. New methods in calculus were developed by some of the greatest mathematicians in history: Newton, Leibniz, brothers Jakob Bernoulli (1654-1705) and Johann Bernoulli (1667-1748), Leonhard Euler (1707-1783), Joseph Louis Lagrange (1736-1813), and Pierre Simon Laplace (1749-1827), to name a few. However, as these techniques and applications of calculus were developed, the absence of rigor slowly began to become a more important question. Calculus worked: that much could not be argued. But what was the logical basis for the new techniques? Many mathematicians and philosophers of the eighteenth century addressed this question, but it was not finally answered until the nineteenth century. The development of calculus led to huge breakthroughs in the application of mathematics to the sciences and set the stage for much of the mathematical work of the nineteenth century.

Building upon the mathematical work of the eighteenth century and developing new areas of mathematics previously unknown, the nineteenth century witnessed some of the greatest discoveries in mathematics since Euclid (330?-260? B.C.) and the ancient Greeks. New areas of mathematics were discovered, important new applications of established mathematics were found, and advances in the understanding of the logical foundations of mathematics were made. Amidst these mathematical developments in the nineteenth century, the foundations of a professional mathematics community were being built.

Many new fields developed within mathematics in the nineteenth century. These emerging areas of mathematics had important ramifications for not only mathematics, but also for how man understood the world around him. For many centuries, since Euclid wrote *The Elements* around 300 B.C., Euclidean geometry was thought to be the only possible form of geometry. The development of non-Euclidean geometry by Wolfgang Bolyai (1775-1856), Nicolai Lobachevski (1793-1856), and Johann Gauss (1777-1855) radically changed how mathematics was viewed. Broken from the constraints of Euclid's ancient axioms, mathematicians began to think about how basic assumptions affected their work. In addition, the existence of non-Euclidean geometry caused doubts to arise concerning the very nature of human knowledge. If such an ancient and universal system of thought had alternative meanings, what other shrines of the human intellect would also be challenged?

The work of Georg Cantor (1845-1918) and others in the emerging field of set theory was another important development in nineteenth-century mathematics. This work led to new conceptions concerning the foundations of mathematics and new ideas about infinity, including the stunning conclusion that different sizes of infinity existed. Another development in mathematics that eventually changed our world was a new algebra invented by the English mathematician George Boole (1815-1864). Boolean algebra, developed by Boole as a logical symbolic language, became one of the foundations upon which computer science was built in the twentieth century. Other new developments in mathematics include the introduction of the Fourier series and the development of hypercomplex numbers. Also important was the work by Gauss, Georg Riemann (1826-1866), August Möbius (1790-1868) and others, which would form the basis of topology.

Many areas of mathematics, although not entirely new, received much attention during the nineteenth century. Euler had revived interest in number theory in the eighteenth century. New work in number theory by Gauss, Peter Dirichlet (1805-1859), Ernst Kummer (1810-1893), Riemann, and others in the nineteenth century led to renewed attempts to prove Pierre de Fermat's Last Theorem and the Prime Number Theorem. New concepts in probability and statistics revolutionized ideas concerning determinism in nature. Applications of these concepts led to revolutionary work by James Clerk Maxwell (1831-1879) and Laplace in the physical sciences and by Lambert Quetelet (1796-1874) and others in the social sciences. Calculus, which had been an indispensable tool for scientists for over a century, was finally given a rigorous foundation by Augustin Louis Cauchy (1789-1857), Riemann, and Karl Weierstrass (1815-1897), among others.

Other important advances in mathematics in the nineteenth century included work on descriptive geometry by Jean-Victor Poncelet (1788-1867), Gauss's proof of the Fundamental Theorem of Algebra, the work of Niels Abel (1802-1829) and Karl Jacobi (1804-1851) on elliptical functions, the codification of mathematical induction, advances in the theory of complex numbers, and the work of Jules Henri Poincaré (1854-1912), which became the seed for the twentieth-century discovery of chaos theory.

Without the developments in mathematics in the nineteenth century, many of the breathtaking advances in science would have been impossible. The work of George Green (1793-1841) involving electricity and magnetism, Maxwell's statistical theory of gases, and Laplace's extraordinary work in celestial mechanics all involved difficult mathematical concepts not available to scientists only a century earlier. Imaginary numbers, not even accepted by most mathematicians until the previous century, found application in electrical engineering, wave mechanics, and other areas of physical science.

Finally, the organization of the mathematical community underwent drastic changes in the nineteenth century. Prior to he nineteenth century, there was no "mathematical profession." Mathematics did not exist, or existed very tenuously at best, outside of its scientific applications. The nineteenth century, however, saw several changes that allowed for the development of a separate mathematics profession. Early in the

century, a distinction began to appear between applied mathematics and pure mathematics. Mathematical research was accepted for its own sake, independent of its applications. This led to the establishment of new professorships in mathematics at universities throughout Europe, to new periodicals dedicated to pure mathematics, and to professional societies for mathematicians. All of these things were virtually nonexistent only a century before.

The new discoveries in mathematics, the new applications of mathematics to the sciences, and the professionalization of the field of mathematics also led to new ideas in mathematics education. Mathematics became required knowledge for the educated person who was interested in keeping up with the flurry of changes in the world. All in all, the developments of the nineteenth century had profound affects upon how the nature of mathematics was viewed, how mathematics was applied to science, and even how man viewed the world in which he lived. These developments impacted society in ways that continue to be studied by historians today.

The mathematical discoveries of the nineteenth century led to more breakthroughs in the twentieth century. New fields of mathematics such as abstract algebra and topology were developed. New applications of mathematics in such diverse areas as chaos theory and cryptography were found. The entire foundation of mathematics has come into question with work on set theory, logic, and with Kurt Gödel's incompleteness theorems. Finally, long-standing problems in mathematics have been solved. One of these problems, the four-color problem, introduced computers into mathematical proofs. Another, the proof of Fermat's Last Theorem, solved the most famous problem in modern mathematics. Each of these developments in mathematics in the twentieth century found their roots in the mathematics of the nineteenth century.

TODD TIMMONS

Fourier Analysis and Its Impact

Overview

Jean-Baptiste Joseph Fourier, in studying the conduction of heat in solid bodies, devised a way to analyze it using an infinite series of

trigonometric terms. Similar mathematical problems arise in almost every branch of physics, and Fourier's methods have been applied in many fields of science and engineering.

Background

The French Revolution (1789-99) was a dangerous time to be an intellectual. Jean-Baptiste Joseph Fourier (1768-1830) was the mathematically gifted son of a tailor. The idealistic young man had considered studying for the priesthood until he encountered the Revolution's promises of equality and the rights of man, with freedom from both the monarchy and the Church. But he was horrified by the Reign of Terror that followed and became embroiled in disputes between factions, barely escaping with his head.

He was, however, in the right place at the right time when Napoleon Bonaparte took over after the Revolution. It was easy to see the problems inherent in government by an unruly mob, so Napoleon set out to banish ignorance by establishing schools. Since the guillotine—the infamous execution device used heavily during the Revolution—had drastically reduced the supply of teachers, Napoleon founded the *Ecole Normale* in Paris to train new ones. Fourier was among the first students there and was on the faculty within a year after graduating. He later went with Napoleon to Egypt, becoming an expert on its antiquities. In 1802 he was appointed prefect of the French region of Isere, with headquarters in Grenoble. There he proved an able administrator, all the while continuing his studies of mathematics and Egyptology.

Napoleon fell from power in 1815, and Fourier was re-assigned to a quiet post in Paris that gave him the freedom to enjoy a scholarly life. He was elected to the Academie des Sciences, and became its permanent secretary. Most importantly, in 1822 he finally had the time to finish the work on the mathematics of heat conduction that he had begun 15 years before in Grenoble.

Fourier's "Theorie analytique de la chaleur" ("The Analytical Theory of Heat") dealt with problems such as finding the temperature in a conducting plate if the initial temperatures at the edges of the plate are known. *Boundary-value problems* of this type are among the most common in physics. They describe situations in which the known quantities are initial or final states, or conditions at the physical edges or boundaries. The goal is to figure out what happens in between.

To obtain a solution to the problem of heat conduction, Fourier expressed it as the sum of an infinite mathematical series with sines and cosines as terms.

Jean-Baptiste Joseph Fourier. *(The Library of Congress. Reproduced by permission.)*

$$y = \tfrac{1}{2}a_0 + (a_1 \cos x + b_1 \sin x)$$
$$+ (a_2 \cos 2x + b_2 \sin 2x) + \ldots$$

These trigonometric functions can be plotted as smooth, repeating wave-like curves. Fourier's methodology had been tried during the previous century, when Leonhard Euler (1707-1783) and Daniel Bernoulli (1700-1782) had studied vibrating strings fixed at their ends. However, the eighteenth-century mathematicians had distrusted the validity of using infinite series in their solution. It was Fourier who brought this method into mathematical physics, and today it is called "Fourier analysis."

Impact

One of the reasons that infinite series solutions had disturbed previous generations of mathematicians was that it was unclear whether the sum of the series was finite or not. Even though the series had an infinite number of trigonometric terms, their *coefficients*, or multipliers, could be either positive or negative, causing terms to cancel out. Fourier devised rules for obtaining these coefficients so that the series would be finite, and thus *converge* to a useful solution.

Fourier series can be used to approximate any waveform. The more terms of the series that are used, the closer the approximation will be. Sines and cosines are "well-behaved" functions, meaning that the techniques of algebra and calculus can be easily applied to them. So a series of these terms can be used instead of a waveform that would not be convenient to work with otherwise.

For example, consider the sharp-edged "sawtooth" wave. In calculus we refer to the slope or rate of change of a function at a particular point as its *derivative* at that point. But a derivative requires a smooth curve. At sharp points like those on the sawtooth wave, we can define no derivative. However, if we express the sawtooth wave as the sum of a series of trigonometric terms, we can take the derivative of each term individually.

Another key point in Fourier analysis is the *periodicity,* or repeating nature, of the trigonometric terms in the series. There are many periodic phenomena in nature. Some, like a radio wave of a specific frequency, can be represented by a single sine or cosine wave. Others have more complex patterns.

Imagine playing a note on the piano while your friend plays the same note on the violin. The pitch is the same, and if you work at it you could get the volume to be the same, but nevertheless the sounds would be different. Different instruments playing the same note produce the same *fundamental* tone, a simple trigonometric waveform. But the specific sound each makes is a result of its pattern of *harmonics,* or overtones of frequencies that are multiples of the fundamental. The sum of the fundamental and its harmonics produce a complex waveform that is unique to the particular instrument.

Using Fourier analysis, it is possible to analyze a sound or other waveform in terms of its constituent harmonics; in fact, Fourier analysis is sometimes called "harmonic analysis." Each term of a Fourier series represents a different frequency. For example, a term in *sin 2x* represents a sine wave with a frequency twice that of a term in *sin x*. By building up these terms, the desired complex waveform can be produced. Generally it only requires the fundamental plus a few harmonics to achieve a reasonable facsimile of the original. For example, an electronic synthesizer can imitate the sound of many musical instruments by using its tone generators to produce the appropriate harmonics.

Fourier analysis also provides important tools for handling scientific data. For example, a common scientific instrument is the spectrometer. It measures the different levels of energy an object gives off over a range of frequencies. Spectrometry is a central technique of astrophysics, because the frequency distribution of the light from, say, a star, can tell scientists about its temperature and other properties.

A method of Fourier analysis called the "Fourier transform" provides a way to go back and forth between data given in terms of frequency and time. So, for example, a frequency peak in spectrometer data from a distant star or galaxy can be interpreted as a periodic fluctuation of light or energy over time. The rotating disk of matter around a black hole and the energy bursts emitted from pulsars are examples of phenomena that can be investigated in this way.

Today Fourier analysis is generally done with the aid of computers. The "fast Fourier transform" is a technique that allows drastically reducing the number of numerical operations required, making the analysis much faster and cheaper. When it was introduced in the 1960s, it revolutionized the digital processing of waveforms.

SHERRI CHASIN CALVO

Further Reading

Butzer, Paul Leo. *Fourier Analysis and Approximation.* New York: Academic Press, 1971.

Cartwright, Mark. *Fourier Methods for Mathematicians, Scientists and Engineers.* New York: Ellis Horwood, 1990.

Edwards, R. E. *Fourier Series: A Modern Introduction.* New York: Springer-Verlag, 1979.

Folland, Gerald B. *Fourier Analysis and Its Applications.* Pacific Grove, CA: Wadsworth & Brooks, 1992.

The Development of Number Theory during the Nineteenth Century

Overview

Number theory—the study of properties of the positive integers—is one of the oldest branches of mathematics. It has fascinated both amateurs and mathematicians throughout the ages. The subject is tangible, and a great many of its problems are simple to state yet very difficult to solve. "It is just this," said the great nineteenth-century mathematician Carl Friedrich Gauss (1777-1855), "which gives number theory that magical charm which has made it the favorite science of the greatest mathematicians." Indeed, Gauss himself made seminal contributions to the subject, as did such other nineteenth-century greats as Lejeune Dirichlet (1805-1859), Ernst Kummer (1810-1893), Richard Dedekind (1831-1916), Bernhard Riemann (1826-1866), and Leopold Kronecker (1823-1891). Moreover, since the number-theoretic problems they tackled *were* very difficult, they often had to resort to "nonelementary" means—mainly algebraic and analytic—to deal with them. ("Elementary" methods are not necessarily simple; rather, they are merely methods that do not use advanced mathematics.)

Background

A supreme masterpiece about number theory that set the stage for the century's advances was Gauss's *Disquisitiones Arithmeticae* ("Arithmetical Investigations"), published in 1801 but completed in 1798—when Gauss was only 21! The title of his book refers to the fact that in previous centuries "number theory" was called "arithmetic." Pre-nineteenth-century number theory contained many brilliant results but often lacked thematic unity and general methodology. In the *Disquisitiones* Gauss supplied both. He systematized the subject, provided it with deep and rigorous methods, solved important new problems, and furnished mathematicians with new ideas to guide their researches for much of the nineteenth century.

The fundamental theorem of arithmetic, a cornerstone of the subject, states that every integer greater than 1 is a *unique* product of primes. Put another way, the primes are the (multiplicative) "building blocks" of the integers: products of primes will generate (uniquely) all the integers. This result was undoubtedly known to mathe-

maticians of past centuries, but Gauss, in the *Disquisitiones,* was the first to state it formally and give a rigorous proof. Here also appears the first formal definition of the notion of congruence.

Since the primes are the "atoms" that make up the integers, to understand the latter it is imperative to understand the former. In 300 B.C. Euclid (c. 330-260 B.C.) proved that there are infinitely many primes. But how are they distributed among the integers? Do they follow a pattern? (The first 20 primes are 2, 3, 5, 7, 11, 13, 17, 19, 23, 31, 37, 41, 43, 47, 53, 61, 67, 71, 73, 79.) This question baffled mathematicians for centuries. Numerical evidence showed that the primes are spread out irregularly among the integers, in particular that they become scarcer—but not uniformly—as the integers increase in size. For example, there are 8 primes between 9991 and 10090 and 12 primes between 67471 and 67570. Furthermore, arbitrarily large gaps exist between primes: it is easy to produce a sequence of a billion *consecutive* nonprime integers. On the other hand, considerable evidence suggests that there are infinitely many pairs of primes as close together as can be, namely primes p and q for which q – p = 2 (they are called "twin primes"). This apparent irregularity in the distribution of primes prompted Leonhard Euler (1707-1783) in the eighteenth century to say: "Mathematicians have tried in vain to this day to discover some order in the sequence of prime numbers, and we have reason to believe that it is a mystery which the human mind will never penetrate."

Euler's pessimism was, in an important sense, unjustified. It is true that there is no regularity in the distribution of primes considered *individually;* in particular, it is most unlikely that we could find a formula that will produce *all* the primes and *only* primes. But there *is* regularity in the distribution of the primes considered *collectively*. In fact, such regularity was later conjectured by Euler himself, and subsequently by Adrien-Marie Legendre (1752-1833) and Gauss.

In mathematics one must be able not only to give the right answers but to ask the right questions. Instead of looking for a rule that will generate successive primes, one might ask for a description of the number of primes in a given interval. Put differently, one might try to describe not "how" but "how often" the primes occur among the integers. Gauss and others made a conjecture about this issue, but it took close to a century to give a proof, mainly because basic tools were lacking that were developed during the nineteenth century. A major step toward the proof was taken in mid-century by Riemann, who introduced for this purpose what came to be known as the *Riemann zeta function.* While working on this problem, Riemann introduced a conjecture, still open 150 years later, that came to be known as the *Riemann hypothesis.* It is arguably the most celebrated unsolved problem in mathematics.

Building on Riemann's work, a proof of Gauss's conjecture was finally given at the century's end, independently by Jacques Hadamard (1865-1963) and Charles Jean de la Vallée-Poussin. The result is now known as the *prime number theorem,* a central result in number theory. It says that the number of primes less than or equal to x (x being a real number) is approximately equal (or asymptotic) to x/log x. In order to arrive at their proof Hadamard and de la Vallée-Poussin had to introduce important new ideas in complex analysis (the calculus of complex functions).

Two significant observations derive from these considerations. First, that it is often specific problems that motivate the development of theoretical results (in this case, it was attempts to prove the prime number theorem that motivated the introduction of important ideas in complex analysis). Second, that analysis—the study of the *continuous*—enters to resolve problems in number theory—the study of the *discrete.* This is surely a surprising phenomenon. In fact, several number-theoretic problems led in the nineteenth century to the founding of a new field, analytic number theory, which is to this day of great importance. (In a most surprising development, Paul Erdös and Atle Selberg proved the prime number theorem in the 1940s by "elementary" methods, without using complex analysis; the proof, however, was far from simple.)

As we mentioned, Euclid proved that there are infinitely many primes. Since 2 is the only even prime, this result can be rephrased to say that there are infinitely many primes in the arithmetic sequence 2n + 1 (n = 0, 1, 2, 3,...) consisting of the odd integers. In the 1830s Dirichlet proved a grand generalization of this result by showing that *any* arithmetic sequence an + b (n = 0, 1, 2, 3,...), namely b, b + a, b + 2a, b + 3a,...,

contains infinitely many primes, with the obvious exclusion of the situation in which a and b have a common factor greater than 1 (in which case, of course, none of an + b is prime). To prove this result Dirichlet introduced important and far-reaching ideas from analysis. Here was another celebrated example of analytic number theory.

Despite these triumphs, open problems abound in the distribution of primes. For example, is every even number greater than 2 a sum of two primes (as the evidence suggests)? Is there a prime between n^2 and $(n + 1)^2$ for every positive integer n? Undoubtedly, Euler's statement we quoted earlier about the mysterious nature of the primes has considerable merit.

Impact

Many other important concepts were introduced during the century as a result of work on number theory, and some of these influenced and/or gave rise to other branches of mathematics. For instance, the notions of integral domain, unique factorization domain, Dedekind domain, and ideal—adumbrated or introduced by Gauss, Kummer, Dedekind, and Kronecker—are important concepts in algebra. Their development led to yet another branch of number theory, algebraic number theory, in which the tools of algebra are brought to bear on the study of the integers. In fact, by the end of the century, the very term "integer" could no longer be used with impunity: there were now many types of integers—"ordinary" integers, Gaussian integers, and cyclotomic integers, to name but a few. From these and other developments, we can see that the scope and methods of number theory were enormously enlarged in the nineteenth century.

ISRAEL KLEINER

Further Reading

Adams, William and Larry Goldstein. *Introduction to Number Theory.* Englewood Cliffs, NJ: Prentice-Hall, 1976.

Apostol, Tom. *Introduction to Analytic Number Theory.* New York: Springer-Verlag, 1976.

Edwards, Harold. *Fermat's Last Theorem: A Genetic Introduction to Algebraic Number Theory.* New York: Springer-Verlag, 1977.

Frei, Günther. "Number Theory." In *Companion Encyclopedia of the History and Philosophy of the Mathematical Sciences,* edited by I. Grattan-Guinness. Volume 1. New York: Routledge, 1994.

Goldman, Jay. *The Queen of Mathematics: A Historically Motivated Guide to Number Theory.* Wellesley, MA: A. K. Peters, 1998.

Hardy, Godfrey and E. Wright. *An Introduction to Number Theory.* New York: Oxford University Press, 1962.

Ireland, Kenneth and Michael Rosen. *A Classical Introduction to Modern Number Theory.* New York: Springer-Verlag, 1982.

Kline, Morris. *Mathematical Thought from Ancient to Modern Times.* New York: Oxford University Press, 1972.

Ore, Oystein. *Number Theory and its History.* New York: McGraw-Hill, 1948.

Pollard, Harry and Harold Diamond. *The Theory of Algebraic Numbers.* Washington, DC: The Mathematical Association of America, 1975.

Ribenboim, Paulo. *The Book of Prime Number Records.* 2nd ed. New York: Springer-Verlag, 1989.

Scharlau, Winfried and Hans Opolka. *From Fermat to Minkowski: Lectures on the Theory of Numbers and its Historical Development.* New York: Springer-Verlag, 1985.

Projective Geometry Leads to the Unification of All Geometries

Overview

The nineteenth century witnessed a great change in the nature of geometry. From beginnings in perspective drawing of artists in the eighteenth century, mathematicians developed projective geometry, and with the work of Jean-Victor Poncelet geometry became the study of properties of figures that remain unchanged under families of transformations. These transformations now became the objects of study, and different geometries would eventually be seen as parts of one unified whole.

Background

Since ancient times, geometry has always been at the heart of mathematics; mathematicians in ancient Greece were referred to as "geometers." During the Renaissance, when Europe had recovered from the Black Death (bubonic plague), dynastic rule became the standard form of government, and economies were again growing, there was a renewed interest in the classical works of the ancient Greeks. Thus, geometry returned to take its place in the curriculum of the newly established European universities. Classical works in geometry from Euclid, Archimedes, Apollonius, and Pappus were returned to Europe via Byzantium and the Arab states, and these works were restored, translated, and studied. This set the stage for the later development of new branches of geometry.

Also contributing to the rebirth of geometry was royal patronage of mathematics and mathematicians. Of utmost importance to sovereigns were map making, shipping, trade, and war. It was imperative, then, for these rulers to lend royal support to the study of cartography,

Jean-Victor Poncelet. *(Corbis-Bettmann. Reproduced by permission.)*

hydraulics and hydrostatics, astronomy, and ballistics. Royal scientific societies were chartered, and mathematicians had the means to devote themselves to the development of the discipline. The arts also enjoyed royal patronage, and artists now had the means to work.

The growth of commerce would lead to the emergence of practical mathematics outside the university curriculum, such as reckoning with the Hindu-Arabic place value system and algebra. Geometry itself would be put to practical use as well. Our story begins with the use of geometry in art.

Perhaps for the first time, Renaissance painters began to try to give the appearance of depth on a flat canvas. What size should a given object be in an attempt to make the picture seem more realistic? In mathematical terms, how can the artist project a three-dimensional object on a two-dimensional canvas? The answers came from the geometry of perspective. We have all seen pictures of a road vanishing in the distance or perhaps of a body falling from a great height. Objects further away must be drawn smaller. The Italian Filippo Brunelleschi (1377-1446) studied the theory of perspective, and later Leon Battista Alberti (1404-1472) wrote his text on the subject, *Della Pittura* ("On Painting"; 1435). In it he describes the ideas of vanishing point and vanishing or horizon line. All groups of parallel lines intersect at a point on the horizon line. The set of parallel lines that are meant to be seen as perpendicular to the plane of the picture intersect at the vanishing point. The notions of "ideal" points where parallel lines "intersect" and project a three-dimensional object onto a two-dimensional canvas would be exploited further in the nineteenth-century works of Jean-Victor Poncelet (1788-1867) and others.

Girard Desargues (1591-1661) was a French engineer with a professional interest in perspective as used by artists and architects. Desargues studied the works of the classical Greeks and wanted to include the theory of perspective into the larger framework of geometry. His innovation was to study the properties of figures that remain invariant (constant or unchanging) under projection. Since the property of being a conic section is such an invariant, Desargues could attempt to unify the study of conics. He introduced the notion of points at infinity where parallel lines meet. His new terminology included "ordinance" for a collection of parallel lines or of a collection of lines intersecting at the same point (what today is called a pencil of lines) and "butt" for the point of intersection of the lines in the ordinance. Desargues had the unfortunate luck to be working in synthetic geometry (geometry of figures without recourse to formulas) just as analytic geometry was being born. Analytic geometry flourished, while Desargues' work was largely ignored. It was forgotten until being rediscovered by French geometer Michel Chasles (1793-1880) in 1845.

The French Revolution (1789-1799) helped make possible the further development of geometry. As mentioned earlier, royal patronage and the establishment of universities helped revive the study of classical geometry. In France military schools were also established to provide a scientific education for military engineers. Naturally, these schools supported the monarchy during the Revolution, and they were subsequently closed. The revolutionary government deemed it necessary to create its own schools, and it did so, creating both an engineering school (the École Polytechnique in 1794) and a school for teachers (the École Normale Superieure in 1795). In keeping with the romanticism of the time, these schools led to the expanding of literacy and opportunity for many, and this in turn led to the successful mathematical careers of many men who may not have had the same opportunity under the old regime. This period is also marked by the rejection of the Age of Enlightenment's emphasis on reason and the coming of the Age of Romanticism, which favored the use of intuition and imagination, two qualities necessary to challenge the dominance of Euclidean geometry.

Gaspard Monge (1746-1818) was instrumental in organizing these new schools, which had as their instructors such prominent mathematicians as Joseph Louis Lagrange (1736-1813), Pierre Simon Laplace (1749-1827), and Monge himself. When Napoleon took over France, many mathematicians, including Monge and Poncelet, accompanied him on his excursions. As we shall see, Poncelet's trip with Napoleon was to be particularly eventful.

Impact

In 1822 Poncelet published *Traité des propriétés projectives des figures* ("Treatise on the Projective Properties of Figures"). This work set the stage for a reformulation and unification of geometry that was to culminate in the work of Felix Klein (1849-1925) in 1872. Poncelet was in Napoleon's army as an engineer and accompanied him during the Russian campaign, when he was taken prisoner. During his imprisonment Poncelet began reworking geometry for himself (having no books in prison), and the result is his *Traité*. One of the ideas discussed in this work is the theory of polars in conic sections. However, the most important study that Poncelet undertook in *Traité* was the study of central projections. Two figures in two distinct planes are related by a central projection if corresponding points in the two figures can be joined by concurrent lines all passing through a fixed point (much like the vanishing point in perspective drawing or the eye when looking at an object). While this sounds complicated, if you have seen a light cast a shadow, you have seen a central projection. You can make a shadowbox to see the idea. Hang a light in a thin, transparent "box" and draw a picture on the side of the box. The figure on the box and the shadow on the table are related by a central projection.

Poncelet's work caused many mathematicians to take a look at projective geometry. The analytic geometry of Descartes and Fermat had led the way towards calculus, but now analytic geometry would return to its roots as mathematicians employed analytic methods in the study of projective geometry. The non-Euclidean geometry of Nikolai Lobachevsky (1792-1856) was published in the early nineteenth century (1829), and it too would eventually find its way back to converge with Poncelet's study of invariants.

In *Traité* Poncelet arrived at a notion of duality using poles and polars. Joseph-Diaz Gergonne (1771-1859) is in fact responsible for the term "polar," and the principle of duality was first explicitly stated (without need of reference to polars) by Gergonne in 1826, four years after Poncelet's groundbreaking work. In the principle of duality a true proposition about "points" and "lines" remains true of the words "points" and "lines" are interchanged. For example the statement "Not all points are on the same line" can be paired with a dual statement that "Not all lines contain the same point." This notion of duality is a valuable tool in projective geometry for discovering new properties. Gergonne would go on to use duality in investigating algebraic curves. He studied the relationship between points lying on such a curve and the dual notion of tangents to such a curve. Although he made some missteps (not allowing for vertical tangents, for example), Gergonne did further the use of duality and projective techniques beyond elementary geometry and conic sections.

A final note on the impact of the French work on projective geometry was the founding of the journal *Annales de Mathématiques* by Gergonne in 1810 to publish the works of former students of the École Polytechnique. One cannot underestimate the importance of the emergence of mathematical journals in contributing to the growth of mathematics in this century.

Perhaps the last of the great French geometers of this period was Michel Chasles. Recall that it was Chasles who in 1845 found a copy of Desargues' work and rescued it from obscurity. Chasles was also the co-discoverer (with August Möbius) of the most important projective invariant, the cross ratio of four points on a line. With

the notable exception of the work of Jacob Steiner (1796-1863), analytic methods would come to dominate future work in projective geometry. Chasles himself was forced to admit that leadership in projective geometry had passed to Germany by 1837, with the majority of German geometers favoring tools from analytic geometry. Not coincidentally, the passing of the torch to Germany coincides with the fall of Napoleon (and his supporters), the restoration of the monarchy and the conservative order in France, and the growth in power of the German states.

An axiomatic system for projective geometry was established in 1847 by Karl Georg Christian von Staudt (1798-1867) in his work *Geometrie der lage* ("Geometry of Position"). This axiomatic system did not make use of distance; instead, von Staudt was able to create within the axiomatic system a notion of coordinates intrinsic to geometry, as opposed to the extrinsic notion of coordinates based on distance. This allowed the notions of addition and multiplication to be defined via geometrical constructions; also, this intrinsic introduction of coordinates would eventually lead to generalized projective geometry using abstract algebraic systems as coordinates instead of the real numbers. A simple example of such an abstract algebraic system would be the set {0, 1, 2, 3, 4} with arithmetic module 5 (for example 3 + 4 = 2 and 2 x 3 = 1). Using this system as coordinates leads to finite geometries, combinatorics, and the theory of codes. Von Staudt's axiomatization of projective geometry also paved the way for new axiomatizations of Euclidean and non-Euclidean geometries.

Side by side with the synthetic projective geometry of Steiner and von Staudt, analytic projective geometry also flourished in the nineteenth century, with Augustus Ferdinand Möbius (1790-1868) and Julius Plücker (1801-1868) as its main adherents. These men independently invented what are now known as homogeneous coordinates for the projective plane, that is, the Euclidean plane augmented by points at infinity. The use of homogeneous coordinates helped to crystallize the relationship between geometry and algebra. Transformations between geometric objects can now be seen as substitutions. From these very types of transformations the theory of matrices was developed by James Joseph Sylvester (1814-1897) and Arthur Cayley (1821-1895).

With all the pieces falling into place, Klein could unify and classify the geometrical studies of the nineteenth century. He did so in his famous *Erlanger Programm* of 1872. As was the custom in

German universities, a professor would give an inaugural lecture upon his appointment. Klein's appointment to the University of Erlangen was the occasion for his program for the unification of geometry. In this lecture Klein defined what a "geometry" meant, noting essentially that new geometries could be defined by starting with different groups of transformations; thus, he showed that projective geometry does indeed serve as the all encompassing geometry.

The trip from Poncelet's question about projective invariants to a new definition of geometry took most of the nineteenth century. Our fellow travelers' influence continues into the twentieth century as well. Higher dimensional geometry grew by taking the idea of coordinates beyond three dimensions. This higher dimensional geometry, combined with the study of electricity and magnetism, led to the development of vector analysis. Loosening the notion of transformation even further to include stretching, shrinking, and bending (but not tearing or puncturing) leads to topology, so-called "rubber sheet geometry." If one considers looking beyond the family of conic sections (which are invariant under projections) to other higher degree curves, one enters the realm of algebraic geometry, which developed around the attempt to classify curves of a certain degree and the search for invariants under transformations. Projective methods are very useful in algebraic geometry, where curves are simplified when viewed projectively. Also,

the behavior of a curve at infinity can be studied in the same way as the behavior of the curve at an ordinary point. Plücker and Cayley were very much involved in this study. One should also note that algebraic geometry was an important tool in Andrew Wiles's (b. 1953) proof of Fermat's last theorem in 1995.

Euclidean geometry had been considered by philosophers such as Descartes, Hume, Spinoza, and Kant to be the paradigm of logical thought and certainty. With the discovery of non-Euclidean geometries and the reduction of Euclidean geometry to a sub-discipline of other wide-ranging mathematical theories, this paradigm was shattered. Anyone who looked for absolute right or wrong was forced to look elsewhere. This had a profound impact on philosophy and what people think about human knowledge. If the geometric nature of the universe is open to experiment, there are no necessary truths about space and time.

GARY S. STOUDT

Further Reading

Coxeter, H. S. M. *The Real Projective Plane*. New York: Springer-Verlag, 1993.

Fauvel, John and Jeremy Gray, eds. *The History of Mathematics: A Reader*. London: The Open University, 1987.

Kline, Morris. *Mathematics in Western Culture*. New York: Oxford University Press, 1974.

Wallace, Edward C. and Stephen F. West. *Roads to Geometry*. Upper Saddle River, NJ: Prentice Hall, 1998.

The Shape of Space: The Beginning of Non-Euclidean Geometry

Overview

During the nineteenth century several mathematicians realized, at about the same time, that a geometry based on Euclid's classic system was not the only possibility. A non-Euclidean universe was too strange for many to accept at first. Yet it was non-Euclidean geometry that paved the way for Albert Einstein's theory of general relativity in the early 1900s and the modern understanding of space-time.

Background

Geometry is one of the oldest branches of mathematics. Because it deals with the properties of

objects and space, such as lengths, areas, volumes, and angles, it was of immediate practical significance. Ancient artists, artisans, engineers, architects, warriors, navigators, and astronomers all had objects to measure and build, courses to plot, or trajectories to predict.

The familiar geometry of planes and solids that most modern students learn in high school was first formalized in Euclid's (c. 330-260 B.C.) *Elements* in about 300 B.C. The *Elements* were based on ten statements called axioms, or postulates. These assertions were used as starting points and accepted as given. From the ten postulates, several hundred theorems were then proven by deductive logic. This axiomatic-deductive method has become one of the standard ways of doing mathematics.

Euclid's fifth postulate is also called the *parallel postulate*. It states that one and only one line can be drawn parallel to a given line through a point not on the line. This is a more complex statement than is usual for a postulate, and in fact Euclid himself seemed rather reluctant to introduce it as such. He proved 28 theorems from the first 4 postulates alone, a group that has since become known as *absolute geometry*. As other mathematicians worked with the *Elements,* the number of theorems in this group continued to grow. Eventually, many assumed, the fifth postulate would be derived as a theorem from the first four.

After Euclid's time, many Greek, Islamic, and later European mathematicians tried to accomplish this feat. None ever succeeded, although some erroneously convinced themselves that they had. In most cases what they had actually done was to replace the parallel postulate with an equivalent but differently worded statement that they regarded as self-evident, such as "two parallel lines are equidistant." However, the mathematicians were correct to recognize the significance of postulate five. It was the point at which two great branches of geometry were to diverge.

Impact

Like many mathematicians before him, the Hungarian Farkas Bolyai (1775-1856) was obsessed with trying to prove the parallel postulate, and he passed this preoccupation along to his son János (1802-1860). The younger man eventually concluded that a proof was impossible and recognized that absolute geometry branched into two groups of cases: one group in which the parallel postulate was accepted, and one group in which it was not. He developed a geometry without it.

János Bolyai's creation was a form of hyperbolic geometry, in which the fifth postulate was replaced by one allowing more than one parallel line through the fixed point. In 1823 he described his work in a paper he called "Appendix Scientiam Spatii Absolute Veram Exhibens" ("Appendix Explaining the Absolutely True Science of Space").

Farkas Bolyai was an old friend of Carl Friedrich Gauss (1777-1855), and he proudly sent a draft of his son's paper to the eminent mathematician. Gauss replied that he had in fact worked such a system out himself decades before. Although he had never published his work, and therefore had no claim to precedence, his response stunned János Bolyai. The paper was published in 1832 as an appendix to a textbook authored by his father, but Bolyai himself faded into relative obscurity.

Nikolai Lobachevsky (1792-1856), a professor at the University of Kazan on the fringes of Siberia, was not to be so easily discouraged. His non-Euclidean geometry was the first in print, appearing in the *Kazan Messenger* in 1829. It was very similar to that of Bolyai, who was completely unknown to Lobachevsky. What Gauss might have thought of decades before but declined to publish did not concern him. Lobachevsky wrote several expositions of his system, culminating in the book *Pangeometrie* in 1855. However, his fame also grew mainly after his death, when the implications of non-Euclidean geometry were better understood.

It was Georg Friedrich Bernhard Riemann (1826-1866) who first began shedding light on those implications. Riemann's classic paper was written to secure his admission to the University of Göttingen as a *Privatdozent,* or unpaid lecturer dependent on student fees. It was entitled "Über die Hypothesen, welche der Geometrie zu Grunde Liegen" ("On the Hypotheses that Form the Foundations of Geometry"), and included another formulation of a non-Euclidean geometry.

Riemann's system differed not only from Euclid's but from those of Bolyai and Lobachevsky as well. Euclid had postulated only one parallel line through a point not on the line. Bolyai and Lobachevsky had allowed more than one. Riemann's elliptical geometry was built on the postulate that there are no lines parallel to a given line through a point not on it. Furthermore, all lines are of finite length. This can be visualized by considering a globe (since a sphere is just an especially symmetrical ellipsoid), on which all the meridians, or lines of longitude, meet at the poles. In his paper Riemann posed questions about what type of geometry represented that of real space. Thus began the idea that non-Euclidean geometry might have physical meaning.

In 1872 Felix Klein (1849-1925) published two papers entitled "On the So-called non-Euclidean Geometry." Klein's major contribution to this field was the idea that both Euclidean geometry and the non-Euclidean geometries of Lobachevsky and Riemann are special cases of a more general discipline called projective geometry.

Geometries may be classified by the type of *transformations,* or mathematical manipulations, that can be performed without contradicting any of their theorems. The more transformations under which the geometry is invariant (unchanging), the more general it is. Projective geometry is concerned mainly with properties such as when points lie on the same plane and when a set of lines meet in a single point. These properties are invariant under a larger group of transformations than the congruence, or equality of lengths, angles, and areas central to Euclidean geometry.

Hermann Minkowski (1864-1909) developed a geometry encompassing the usual three dimensions of space and adding time as a fourth dimension. This geometrical system has since come to be called *Minkowski space.* It was in the language of non-Euclidean geometries like Minkowski space that Albert Einstein (1879-1955) was able to frame his general theory of relativity. In general relativity gravity is described in terms of the curvature of space-time. For example, imagine a sheet of rubber with grid lines like graph paper, suspended horizontally so that it forms a flat surface. With no weight on it, the grid has straight lines and right angles, corresponding to the "flat space" of Euclidean geometry.

If you place a ball on the surface, the rubber sheet stretches around it. The curvature of the grid increases as it gets closer to the ball. This corresponds to the curvature of space-time near a massive object. The description of this curved space was made possible by a few nineteenth-century mathematicians, who were willing to consider the idea that there were geometries beyond that of Euclid.

SHERRI CHASIN CALVO

Further Reading

Boi, L., D. Flament, and J.-M. Salanskis, eds. *1830-1930: A Century of Geometry, Epistemology, History, and Mathematics.* New York: Springer-Verlag, 1992.

Gray, Jeremy. *Ideas of Space: Euclidean, Non-Euclidean, and Relativistic.* New York: Oxford University Press, 1989.

Greenberg, Marvin Jay. *Euclidean and Non-Euclidean Geometries: Development and History.* New York: W.H. Freeman, 1993.

Kolmogorov, A. N., ed. *Mathematics of the 19th Century: Geometry, Analytic Function Theory.* Translated from the Russian by Roger Cooke. Boston: Birkhäuser Verlag, 1996.

Majer, U. and H.-J. Schmidt, eds. *Reflections on Spacetime: Foundations, Philosophy, History.* Boston: Kluwer Academic Publishers, 1995.

Topology:
The Mathematics of Form

Overview

Mathematics began in earliest times as a collection of practical methods for counting and measuring. In the nineteenth century a branch of geometry arose in which sizes and distances were irrelevant. *Topology,* sometimes called "rubber sheet geometry," deals with properties that remain the same when an object is bent or stretched. These properties include the number of times a curve intersects itself and whether a surface is open or closed.

Background

During the lifetime of the famous Swiss mathematician Leonhard Euler (1707-1783), there were seven bridges in the Prussian town of Königsberg, where two branches of the River Pregel flowed around an island. The townspeople amused themselves by considering whether it was possible to cross all seven bridges in one continuous trip without re-crossing any of them. Eventually the so-called Königsberg Bridge Problem became a well-known puzzle, but no one was able to find a solution.

Euler approached this problem by replacing the land areas by points, or *vertices,* and the bridges by lines connecting the vertices. An *odd vertex* is one in which the number of lines connected to it is odd, and an *even vertex* has an even number of lines connected to it. The problem may then be viewed as a graph, and the object is to traverse the graph without backtracking or lifting the pencil from the paper.

The solution Euler published, in a paper that is considered to be the first publication in the field of topology, was that a graph of this type can be traversed if it has only even vertices. If it has one or two odd vertices, it can be traversed in one trip, but not with a return to the starting point. In general, a graph with $2n$ odd vertices, where n is any integer, will require n distinct trips. In the Königsberg problem, all four vertices are odd. Therefore, its traversal could not be accomplished in a single trip.

Another contribution Euler made to topology was his famous formula for a *polyhedron*, or many-sided solid:

$$v-e+f = 2$$

where v is the number of vertices of the polyhedron, e is the number of edges, and f is the number of faces.

Euler published his formula in 1752. In 1813 Antoine-Jean Lhuilier (1750-1840) literally found a hole in it. That is, the formula does not work if there are any holes in the solid. The general formula is

$$v-e+f = 2 - 2g$$

where g is the number of holes. This distinction among objects with different numbers of holes was key to the further development of topology.

Impact

Topology holds all objects that can be continuously deformed into one another to be equivalent, regardless of their original shape. To understand distinctness in topology, it is helpful to think of a ball of clay. The rules are that you can stretch or mold it, but you are not allowed to push a hole into it, tear it, or stick unattached edges of it together. You can mold the ball into a cube, and in fact in topology there is no difference between a cube and a sphere.

Suppose you want to make a *torus*, or doughnut shape, out of your ball of clay. One way might be to flatten it out and push a hole through the center with your thumb. But according to the rules of topology, that isn't a valid deformation. So you try molding your sphere into a cigar shape. So far, so good. Now all you need to do to make a torus is to wrap your cigar shape around and stick the two ends together... but that's not allowed either. A torus is fundamentally different from a sphere.

The term *topology*, the study of shape, was first used by Johann Benedict Listing (1802-1882). Having introduced the term in correspondence beginning about 1837, he published *Vorstudien zur Topologie* ten years later. Listing's ideas on topology owed much to Carl Friedrich Gauss (1777-1855). Among Gauss's innovations was the linking number, which determines whether circles are linked and is invariant even when the circles are continously deformed.

Around 1850 Georg Friedrich Bernhard Riemann (1826-1866) was at the University of Göttingen studying for his Ph.D. under Gauss's supervision. Listing was also a professor there. In Riemann's thesis, which received an uncharacteristically enthusiastic response from the rather haughty Gauss, he developed the concept that was later to be called the *Riemann surface*. The importance of this work was that it introduced topological methods into the theory of complex variables. Complex numbers are those of the form $a + bi$, where i is the imaginary square root of -1. The Riemann surface is multilayered, allowing a multi-valued function of a complex variable to be interpreted as a single-valued function.

One of the most famous constructions in topology is the Möbius strip. This was discovered independently by August Frederick Möbius (1790-1868) and Listing in the early 1860s. The Möbius strip is intriguing because at first glance it appears to be a simple ring with inside and outside surfaces. However, it is actually a single-sided surface.

You can construct a Möbius strip yourself by taking a strip of paper and giving it a half-twist before taping the two ends together. Then take a pencil and, beginning anywhere along the strip, follow it around its length, keeping always to the middle of the strip. You will find that, because of the half-twist, you will continuously traverse what appears to be both "sides" of the paper until your pencil arrives back it its starting point, demonstrating that the construction really only has one side. If you cut along the line you have made, you don't get two narrow rings as you might expect; instead, the strip stays in one piece. Repeat the procedure and you get two separate, intertwined strips.

The Klein bottle is analogous to the Möbius strip but in three dimensions. It is a closed surface, but it has no inside or outside. Unfortunately, the Klein bottle does not lend itself to demonstration. Just as the two-dimensional Möbius strip is constructed with a half-twist through the third dimension, constructing a Klein bottle would require manipulation in a fourth spatial dimension.

The topological concept illustrated by the Möbius strip and Klein bottle is that of connectivity. Connectivity was among the ideas formalized by the French mathematician Jules-Henri Poincaré (1854-1912) in 1895 in a series of papers called *Analysis situs* ("Positional Analysis"). Poincaré was among the first to use the tools of algebra in topology.

One familiar topological problem first posed in the middle of the nineteenth century was not solved until more than 100 years later. Maps are generally drawn so that no two regions sharing a boundary are the same color. Mathematicians were interested in determining the minimum number of colors required to color any map in this way. Three colors were clearly not enough; a map of four regions could easily be drawn in which each region shared a boundary with three others, and that required four colors.

In 1890 it was proven mathematically that five colors would always be sufficient. A map had never been found in which more than four colors were required. Individual proofs had been constructed for four-color mapping of up to about 40 regions by the middle of the twentieth century. Still, a general proof that four colors were always enough remained elusive. The problem was finally solved in 1976 by mathematicians at the University of Illinois, with the aid of more than 1,000 hours of computerized number-crunching.

SHERRI CHASIN CALVO

Further Reading

Aull, C. E. and R. Lowen, eds. *Handbook of the History of General Topology.* Boston: Kluwer Academic Publishers, 1997.

James, I. M., ed. *History of Topology.* New York: Elsevier Science, 1999.

Manheim, Jerome H. *The Genesis of Point Set Topology.* New York: Macmillan, 1964.

The Rise of Probabilistic and Statistical Thinking

Overview

The rise of probabilistic and statistical thinking is considered one of the most important accomplishments of nineteenth-century science. Probability and statistics played a major role as science changed the way our world was understood. At the beginning of the nineteenth century, science involved the search for universal and absolute truths. By the end of the century, scientists were dealing with the notion that some things could only be known to varying degrees of certainty. This degree of certainty was measured by probability.

Background

Although the roots of probability extend far back into history, the beginning of mathematical probability is usually traced to a series of letters exchanged between the French mathematicians Pierre de Fermat (1601-1665) and Blaise Pascal (1623-1662) in the seventeenth century. This correspondence was based on the problem of how to distribute an amount of money wagered on a game if that game is inter-

Pierre Laplace. *(The Library of Congress. Reproduced by permission.)*

rupted before its conclusion. The correspondence between Pascal and Fermat influenced Christiaan Huygens (1629-1695), who published the first text on probability in 1657. Much of the writing on mathematical probability in the next 100 years was presented in gambling terms and examples.

The changes brought about by the development of probability in the nineteenth century are best understood in the context of developments in the previous century. The eighteenth century was a time of great expectations for science. Using the incredible work of Isaac Newton (1642-1727) as a starting point, scientists were making important discoveries in all parts of the natural world. The success of Newtonian principles in the physical sciences led others to apply similar principles to their own fields of study. Specialists in the biological sciences, earth sciences, and even social sciences had hopes of quantifying their studies following Newtonian principles. During this flurry of quantification, the field of probability began to take shape.

Other mathematicians contributed to the theory of probability before the nineteenth century. James Bernoulli's *The Art of Conjecturing,* published posthumously in 1713, addressed problems in probability related to gambling. Abraham De Moivre's *Doctrine of Chances* became the first English-language work on probability in 1718. In this work Moivre (1667-1754) introduced the famous normal curve, usually referred to as the bell curve today. Despite these early discussions of probability, it was not until the latter part of the eighteenth and early part of the nineteenth centuries that probability and statistics began to seriously influence science and society as a whole.

In addition to questions addressing gambling, the emerging business of insurance required an ever-increasing understanding of probability and statistics. Insurance rates had been based on guesswork and myth. Eventually, actuaries (mathematicians who worked with insurance statistics) began to pay increased attention to such things as mortality tables. A mortality table provided data on life expectancies based on the current age of the insured party. Thanks to work done by mathematicians such as Charles Babbage (1792-1871) in England, statistics and probability gradually became an important part of the insurance business.

Impact

One of the leaders in the development of the theory of probability in the nineteenth century was French mathematician Pierre-Simon Laplace (1749-1827). In 1812 Laplace published a very influential work called *Analytical Theory of Probability.* In this and other works on probability, Laplace described the method of inverse probability and gave a formal proof of the least squares rule. The least squares rule is a method for fitting data to a curve. This method first appeared in a work by Adrien-Marie Legendre (1752-1833) in 1805, although Carl Friedrich Gauss (1777-1855) later claimed that he had been using the method since 1795. Regardless of who invented the method of least squares, it became an important tool in probability. Although initially applied to errors occurring in astronomical data, the use of least squares quickly spread to many other fields. Mathematical advances such as the method of least squares made probability and statistics applicable to a diverse collection of problems.

Laplace, like other mathematicians working on probability, developed his techniques for specific applications. In Laplace's case his work in probability was used in astronomy. Laplace noticed that the errors involved in astronomical observations followed the normal distribution introduced by De Moivre almost a century earlier. This normal distribution was to become a foundation for many of the applications of statistics and probability in the nineteenth century.

Laplace also influenced the way probability was perceived in the nineteenth century. Historically, there had been considerable resistance to the incorporation of probability into scientific disciplines. The perception was that probabilistic knowledge was imperfect knowledge. Laplace, and others, applied probability when absolute knowledge was not possible. Laplace believed that probability addressed "the important questions of life" for which complete knowledge was "problematical." This philosophy, known as classical probability, maintained that probability was a measure of the degree of certainty or rational belief. In this philosophy, probability was a measure of man's ignorance. However, probability was also a guide that would help a rational man plan his actions. The acceptance of probabilistic knowledge in place of absolute knowledge in science represented a fundamental change in Western culture.

The classical conception of probability was applied to a variety of life's problems. Its uses ranged from fairly obvious applications such as life insurance to less apparent applications such as deciding the credibility of a trial witness or the rationality of administering the smallpox vaccine. Probability affected the lives of nineteenth-century humans in many ways. It was not until late in the nineteenth century that the classical interpretation of probability gave way to the frequentist view. This view maintained that probabilistic events were not simply a result of man's lack of certain knowledge. Instead, it was believed that probability was an inherent part of the world's structure. In other words, there were some things that could not be known absolutely but only probabilistically.

Ever since Newton quantified astronomy and mechanics, other sciences had attempted to follow a similar pattern of quantification. Even in areas traditionally qualitative, such as the study of man and society, the Newtonian system was looked upon as a model to emulate. In the nineteenth century, this model was applied to statistics describing man and society.

Previous attempts had been made to apply probability to the statistics of society. However, it was Belgian astronomer Adolphe Quetelet (1796-1874) who developed extensive applications for probability in analyzing social statistics. Quetelet noticed that statistics as varied as the height and weight of individuals and the number of crimes and suicides that occurred in a society fit a normal curve. This normal curve, which had already been applied to astronomy, was used by Quetelet to develop what he called a "social physics." Quetelet's social physics was an attempt to quantify the social sciences in the same way that Newton had quantified the physical sciences. He was interested in subjecting the seemingly chaotic data of society to statistical laws. Quetelet looked for these laws using the same techniques as astronomers used in analyzing errors in data. By accumulating statistics related to man's physical dimensions and his personal characteristics, Quetelet believed he could describe the "average man." The average man represented the true type of the human species. Any deviations from the average were considered "errors." These terms were chosen by Quetelet to emphasize the similarity to statistical work in astronomy.

Quetelet showed that births, deaths, marriages, crimes, and suicides were almost constant in any given country, independent of individual actions. This seemed to Quetelet to confirm that stability might be found in any data

given a large enough set of statistical information. This stability represented an example of the law of large numbers applied to social conditions. The phrase "law of large numbers" was coined by Simeon-Denis Poisson in 1835. It stated that over a long period of time, the frequency of an event must become ever closer to the probability of its outcome. For example, the probability that a coin flip will result in heads is one half. If a coin is flipped repeatedly, the law of large numbers says that the frequency of heads will approach fifty percent. Quetelet held that important conclusions might be drawn from the statistics of society using the law of large numbers. He believed that the conclusions of social physics could be used as a moral standard and that society might be improved through an understanding of sickness, crime, and other societal ills.

Quetelet is also important because of his influence on other scientists who used probability to revolutionize their fields. Three such men were Francis Edgeworth, Francis Galton (1822-1911), and Karl Pearson (1857-1936). Edgeworth applied error analysis and the normal curve to topics in economics, and Galton applied the same concepts to the study of heredity. Galton also developed an important method in statistics called correlation. Correlation measures the tendency of one variable to change as a related variable changes. For instance, correlation would measure the tendency of lung cancer cases to increase as the number of smokers in a population increases. Pearson further developed the work of Galton and in the process created the field of biometrics, which is the application of statistics to the biological sciences.

Quetelet's work also exerted a strong influence over physicist James Clerk Maxwell (1831-1879). Maxwell, along with Ludwig Boltzmann (1844-1906), developed a statistical theory of gases that revolutionized physics and created the field of statistical mechanics. Maxwell's theory of gases relied upon the idea that something that cannot be understood individually (the motion of a single molecule) may be understood as an aggregate (the motion of an accumulation of molecules). This idea mirrored the way probability was used in Quetelet's social physics.

These examples are by no means the only areas influenced by probability and statistics. Psychology, agronomy, medicine, and numerous other fields of study adopted probability as a key tool for evaluating data. Most importantly, though, the rise of probabilistic and statistical thinking forever changed the way humankind viewed knowledge.

TODD TIMMONS

Further Reading

Daston, Lorraine J. *Classical Probability in the Enlightenment.* Princeton, NJ: Princeton University Press, 1988.

Hacking, Ian. *The Emergence of Probability.* Cambridge: Cambridge University Press, 1975.

Krüger, Lorenz, Lorraine J. Daston, and Michael Heidelberger, eds. *The Probabilistic Revolution.* Cambridge, MA: MIT Press, 1987.

Porter, Theodore M. *The Rise of Probabilistic Thinking, 1820-1900.* Princeton, NJ: Princeton University Press, 1986.

Stigler, Stephen M. *The History of Statistics: The Measure of Uncertainty Before 1900.* Cambridge, MA: Cambridge University Press, 1986.

Solving Quintic Equations

Overview

By the nineteenth century, mathematicians had long been interested in solving equations called polynomials. However, Paolo Ruffini (1765-1822) and Niels Abel (1802-1829) proved that some polynomials could not be solved by previously known methods. Partly in response, Evariste Galois (1811-1832) developed a new way of analyzing and working with these types of equations. This method is called group theory, and it was to have implications in other scientific fields, such as mineralogy, physics, and chemistry.

Background

Polynomial equations are used in almost every branch of mathematics and science. An example of a polynomial equation is $3x^2 + 4x + 5 = 0$. This equation is called a second degree polynomial because the highest power of x it contains is 2. The degree of a polynomial indicates the number of solutions it has. A number is said to be a *solution* of a polynomial equation if substituting it into the equation makes the equation true. For instance, the number 7 is a solution of the equation $x + 5 = 12$. By the nineteenth century, mathematicians had already discovered ways to solve second, third, and fourth degree polynomial equations. They next turned their attention to solving fifth degree, or quintic, equations. (An example of a quintic equation is $6x^5 + 3x^4 + 3x^2 + 5x + 6 = 0$.)

The fundamental theorem of algebra would come to be important in finding solutions to quintic equations. Carl Gauss (1777-1855), who is sometimes referred to as the founder of modern mathematics, proved this theorem in 1801. Gauss's theorem deals with the relationship between the coefficients of a polynomial equation and its solutions. (For the polynomial $4x^2 + 7x = 0$, the *coefficients* are the numbers 4 and 7.) Specifically, the fundamental theorem of algebra concerns polynomials with *complex* coefficients. Complex numbers consist of two parts: a *real* part and an *imaginary* part. The real numbers are the positive numbers, negative numbers, and zero. Imaginary numbers are the product of a real number and *i*. (*i* represents the square root of -1.) The number $5 + 3i$ is an example of a complex number; 5 is the real part, and $3i$ is the imaginary part. The number 7 is also a complex number; 7 is equal to $7 + 0i$. Gauss's fundamental theorem of algebra states that every polynomial equation with complex coefficients has at least one complex solution.

Polynomial equations of degree less than five are said to be solvable by radicals. This means that they can be solved by a combination of addition, subtraction, multiplication, division, and the taking of roots. In 1796 Paolo Ruffini, an Italian mathematician, proposed that quintic equations could not be solved by radicals. However, because of the fundamental theorem of algebra, Ruffini knew that solutions to quintics must exist. Therefore, his hypothesis suggested that much more complicated mathematics would be necessary to solve these equations. Ruffini attempted a proof of his proposal in 1799, but he was not entirely successful. In addition, much of his work was so complicated that even leading mathematicians could not understand it. Ruffini published additional proofs in attempts to convince others of his discovery. Yet, he continued to have great difficulty in achieving recognition for his work.

In 1824 Norwegian mathematician Niels Abel also presented a proof that, in general, quintic equations could not be solved by radicals. Abel sent a copy of his paper to Carl Gauss, who, like the mathematicians who had seen Ruffini's proofs, failed to realize its importance. Within five years, however, Abel's work became known and accepted throughout the European mathematical community. Abel also showed that in some special cases, quintic equations could be solved by radicals. These equations are now called Abelian equations. For instance, it is easy to see that the quintic equation $x^5 - 1 = 0$ is true when $x = 1$. This fact made Abel wonder whether if there is a way to determine whether a quintic equation is easily solvable. However, he died at the age of 26 before he could begin to investigate this question.

The French mathematician Evariste Galois took up Abel's work. In particular, Galois studied Abel's ideas concerning groups. A *group* is a set of numbers that can be combined in pairs so that the resulting numbers are also in the set. (For example, integers form a group for the operation of addition. Whenever two integers are added, the result is always another integer.) Galois developed what became known as the group theory of algebra. Group theory is a branch of mathematics concerned with identifying groups and studying their properties. His work on group theory was not widely recognized until it was published in 1846, 14 years after his death.

Impact

Group theory led to an entirely different way of searching for and analyzing the solutions of polynomial equations. It involved examining the *permutations* of the solutions of an equation. A permutation is a combination of a group of objects in which the order of the objects is important. (For example, the permutations of the letters A and B are AB and BA.) All of the permutations of the solutions of an equation form a group. These permutations can then be combined in different ways to form subgroups. By analyzing the ways in which these subgroups are related, Gauss could determine whether or not a polynomial equation was solvable by radicals.

Galois's group theory is one indication of a major transition that occurred in the field of mathematics during the nineteenth century. Rather than performing calculations to solve a particular problem (such as finding the two solutions of the equation $x^2 + 5x - 6 = 0$), mathematicians began to work with extremely complicated analyses of general problems (such as determining which types of polynomials are

solvable by radicals). Galois realized that his method of using groups was extremely theoretical, and he did not intend for it to be a practical method of solving equations. In fact, Galois's analyses of the permutations of the solutions of equations was performed without actually knowing the numerical values of the solutions themselves.

In 1870 French mathematician Camille Jordan (1838-1922) published an edited version of Galois's theory in his book *Treatise on Substitutions and Algebraic Equations*. Many modern concepts of group theory first appeared in Jordan's work. For instance, he defined a solvable group as a group that belonged to an equation solvable by radicals. Jordan also answered the question posed by Abel, that of determining which quintic equations are solvable by radicals. Jordan concluded that a quintic equation is solvable by radicals if its solutions form a solvable group.

Charles Hermite (1822-1901), a French mathematician, published a solution to quintic equations in 1858. His solution involved the use of elliptic functions. A *function* defines a relationship between an independent variable and a dependent variable. For example, $y = 3x + 5$ is a function in which x is the independent variable and y is the dependent variable. An elliptic function can be used to calculate the perimeter of an ellipse. The dependent variable of an elliptic function is complex (in other words, it has a real part and an imaginary part). Therefore, Hermite showed that complex functions could be used to solve quintic equations.

In 1878 Ludwig Kiepert, a German mathematician, wrote an article describing a systematic procedure that could be used to solve quintic equations based upon Galois's group theory and Hermite's use of elliptic functions. However, Kiepert's procedure, as well as those of other mathematicians, were largely impractical because the mathematics involved were simply too complex to be performed at that time. Not until the late twentieth century and the development of computers could such procedures be put to practical use.

Another application of Galois's group theory is that it can be used to study the symmetry of physical objects. An object has *symmetry* if a change in its position in space seems to leave it unmoved. For example, when a square is rotated 180°, it seems not to have moved at all. A rotation of 180° is therefore an element of symmetry for a square. Elements of symmetry may include rotation, reflection, and translation.

Felix Klein (1849-1925) was one of the first mathematicians to use group theory and the solutions of polynomial equations to study the symmetry of physical objects. Specifically, he studied quintic equations. A quintic polynomial has five solutions. The group composed of the permutations of these five solutions consists of 120 elements. (There are 5! or 120 permutations of five objects taken five at a time.) An icosahedron is a three-dimensional figure with 20 faces and 120 elements of symmetry. These elements of symmetry form a group. Therefore, the group of a quintic equation can be used to study the symmetry group of an icosahedron, and vice versa. Klein first described this relationship in 1884.

Quintic equations are not the only polynomials that can be used to study symmetry. For example, quartic, or fourth degree, polynomials can be used to analyze the symmetry of tetrahedrons. A tetrahedron has four faces, each of which is an equilateral triangle. The permutation group of quartic polynomials has 4! or 24 elements, and the symmetry group of a tetrahedron also has 24 elements.

Auguste Bravais (1811-1863), a French physicist and mineralogist, used group theory as it related to symmetry to determine the structure of crystals. The atoms in a crystal are arranged in a definite pattern. These arrangements exhibit elements of symmetry. Bravais analyzed the permutations of the solutions of polynomials in order to study this symmetry. In 1849 he published a paper in which he proposed the 32 classes of molecular structures found in crystals.

In the twentieth century, physicists used group theory to study the interactions of subatomic particles. By analyzing symmetry, they could determine which interactions between particles were possible and which were not. Chemists also studied symmetry to determine the arrangement of atoms in molecules. For example, they determined that a molecule of methane has the shape of a tetrahedron. A methane molecule consists of one atom of carbon and four atoms of hydrogen. The carbon atom is located at the center, and the hydrogen atoms are positioned around it at the four corners of a tetrahedron. Therefore, the symmetry of a molecule of methane can be described by a quartic polynomial equation. There are also molecules that can be studied with quintic equations. For example, a chemical called *o*-carborane consists of two atoms of carbon, 10 atoms of boron, and 12 atoms of hydrogen. The two carbon atoms are at the center of the molecule, and the boron and hydrogen atoms form the corners of an icosahedron. Today, scientists and mathematicians in many fields continue to find new applications of group theory.

STACEY R. MURRAY

Further Reading

Books

Amdahl, Kevin, and Jim Loats. *Algebra Unplugged.* New York: Clearwater Publishing Company, 1996.

Bell, Eric Temple. *Men of Mathematics.* New York: Touchstone Books, 1986.

Other

The MacTutor History of Mathematics Archive. University of St. Andrews, 1999. http://www-groups.dcs.st–and.ac.uk/~history

"A Short History," in *Solving the Quintic.* Wolfram Research Resource Library. 2000. http://library.wolfram.com/examples/quintic/timeline.html

Advances in Logic during the Nineteenth Century

Overview

The nineteenth century witnessed the formalization of traditional logic along symbolic lines, followed by an attempt to recast the foundations of mathematics in rigorous logical form. The extensive development of mathematical logic was motivated in part by the discoveries of new geometries and new number systems that caused many mathematicians to question the logical soundness of traditional mathematical ideas.

The attempt to impose the strictest rigor on arithmetic, the most fundamental area of mathematics, would eventually lead to a number of surprising results, results that caused many philosophers and mathematicians to modify their views of the very nature of mathematics. The same developments would nonetheless provide techniques essential to the development of digital computers, artificial intelligence, and modern theories of language.

Background

Logic is the subject that deals with the drawing of correct conclusions from statements assumed to be true. It was first formalized by Greek philosopher Aristotle (384-322 B.C.) in terms of verbal examples that assumed the grammatical structure of the Greek language. The methods of deductive logic are readily applicable to mathematical reasoning, and the proofs of Euclidean geometry are traditionally presented as a matter of logical deduction. The high esteem in which Aristotle was held during the Middle Ages insured that his system would be studied and elaborated upon by generations of philosophers and theologians.

German philosopher and mathematician Gottfried Wilhelm Leibniz (1646-1716) was the first modern figure to suggest that the essence of logic could be captured in a set of rules for the manipulation of symbols. In fact, Leibniz had gone a bit further to envision a "universal language," a language so precise that one could not draw incorrect conclusions from it. Although Leibniz worked towards his "calculus of reasoning" as well as on many other scientific problems for a 35-year period, his logical work did not attract much immediate interest. The first sustained treatment of symbolic logic was provided in 1847 by George Boole (1815-1864), who followed with other publications and a book, *An Investigation into the Laws of Thought,* in 1854. Boole provided a formalism in which the traditional laws of reasoning could be expressed in purely symbolic terms. He noted that statements could be assigned numerical values of 1 if true and 0 if false, so that the truth values of the various possible logical combinations could be computed in an algebraic way. Boole also noted the connection between logic and the theory of sets. In 1847 Augustus DeMorgan (1806-1871) published a book, *Formal Logic,* that complemented Boole's ideas and indicated how logical combinations, written in symbolic form, cold be transformed into equivalent forms.

In 1879 German mathematician and philosopher Gottlob Frege (1848-1925) published an 88-page booklet entitled *Begriffsschrift* ("Concept Language"), which might be regarded as a fresh start towards Leibniz's "calculus of reasoning." In this work he presented an entirely symbolic language that could express the logical possibilities in any situation. Frege made a number of important innovations. In particular, Frege's formalism allowed for statements that include quantifiers, such as "for all" or "for some." Frege also provided for statements that included variables, that is, symbols that could stand for other concepts. Unfortunately, Frege's notation, which made use of vertical and horizontal lines, proved quite unwieldy and not well suited for publication.

Frege's motivation was, in part, to establish the laws of arithmetic on a firm logical foundation. In 1884, in a publication entitled *Foundations of Arithmetic,* Frege attempted to prove that all arithmetic reasoning could be established on the basis of logic alone. Frege defined the equality of numbers in terms of the establishment of a one-to-one correspondence between elements of two sets. He also proposed that each of the individual integers might be interpreted as the class or infinite set of sets containing all sets with the same number of members.

The next major mathematician to be concerned with symbolic logic was Giuseppe Peano (1858-1932). Peano was unaware of the work of Frege when he published *The Principles of Arithmetic* in 1889, but he acknowledged Frege in subsequent publications. Peano introduced symbol manipulation rules equivalent to those of Frege but in far more tractable notation, and in particular presented a set of axioms for the natural numbers that conclude with a statement of the principle of induction. Peano's axioms, though largely a restatement of the ideas of Richard Dedekind (1831-1916), are perhaps the most precise formulation of the properties of integers to emerge in the nineteenth century. Unlike Frege, Peano was able to assemble a group of colleagues who would apply his approach in different areas of mathematics, resulting in the publication over the years 1895-1905 of a 5-volume *Formulary of Mathematics,* which made it credible that his approach could form the basis for a complete foundations of mathematics.

Impact

For mathematics the nineteenth century began and ended in an atmosphere of confidence and achievement but was interrupted by periods of doubt and controversy. In 1800 most mathematicians could accept the notion that mathematics was a study of the self-evident properties of space and number, a study that had proven its value to science and engineering. The discovery of non-Euclidean geometries and of number-like objects such as quaternions, matrices, and even Boole's truth-values suggested that the principles of mathematics were not as self evident as had been supposed, and thus there was a new emphasis on rigorous argument. By the close of the century there was confidence again that mathematics was consistent if arithmetic reasoning was consistent, and it seemed that the followers of Frege and Peano would soon be able to derive the principles of arithmetic from logic itself. Some mathematicians even imagined formulating a set of axioms and transformation rules for all of mathematic in a symbolic language such that any meaningful combination of symbols could be determined to be true or false by an automatic procedure. German mathematician David Hilbert (1862-1943) summarized this expectation in 1900 in an address to the Second International Congress of Mathematics, when he predicted that proofs of the decidability and completeness of mathematics would be found. As will be seen the exact opposite turned out to be the case.

The nineteenth century had left some unfinished business. The set theory of Boole, DeMorgan, and German Georg Cantor (1845-1918) allowed for sets of different infinite sizes and allowed the possibility of sets containing themselves as members. In a letter written to Frege in 1901, English philosopher Bertrand Russell (1872-1970) stated what has come to be known as Russell's paradox, that any set theory that allows for a set to have membership in itself (as Frege's theory of numbers) did would allow for a set of all sets not members of themselves, and that it is impossible to classify this set as either a member or not a member of itself.

An attempt to eliminate the paradoxes of set theory from symbolic logic led to an elaborate "theory of types" developed by Russell and philosopher Alfred North Whitehead (1861-1947) in the three-volume *Principia Mathematica,* first published over the years 1910-13. The *Principia* also introduced the formal rules of inference lacking in Peano's original work and attempted to eliminate the principle of induction by a more general principle of "equivalence."

The possibility that a single set of axioms and rules of inference, such as those in *Principia Mathematica,*, could form a firm basis for all of mathematics was ruled out permanently in a remarkable paper published in 1931 by Austrian mathematician Kurt Gödel (1906-1978). Gödel showed that any formal system that included the usual rules for the multiplication of integers would necessarily allow the existence of meaningful statements that could neither be proven nor disproved within the system. Gödel's result would have a profound effect on the philosophy of mathematics, discouraging attempts to find a single axiomatic basis and stimulating a debate between "formalists" who believed that the content of the subject was only

what the manipulations revealed, and "intuitionists" who believed that mathematical objects, no matter how "idealized," were nonetheless real and could be thought about.

The possibility that the truth of mathematical statements could be decided by a mechanical procedure was ruled out with the publication of a paper by British mathematician Alan Turing (1912-1954) in 1937. After formalizing the notion of mechanical procedure for manipulating symbols, Turing was able to demonstrate that there was no way to tell if such a mechanical procedure applied to an arbitrary string of mathematical symbols would give a result in a finite number of steps. Turing's realization was disappointing from the standpoint of Hilbert's hope that all mathematical statements would be decidable by a single method. However, in his analysis of mechanical procedure Turing conceived the modern notion of a digital computer as an information processing machine that could be programmed to carry out tasks expressed in symbolic form.

The computer would, of course, turn out to profoundly affect almost every area of human activity, and it is not surprising that by the 1950s a serious interest in the development of machine or "artificial" intelligence would begin to occupy philosophers, psychologists, and engineers. These disciplines would begin to concern themselves with how knowledge was representable in computer memory, which brought up further questions about the representation of knowledge in the human brain. While the issues are far from resolved, much artificial intelligence research is based on the first order predicate calculus, a clear descendent of Frege's *Begriffsschrit.* Indeed, one of the first successes of artificial intelligence was a program named "Logic Theorist," which could prove theorems in the system of *Principia Mathematica.*

DONALD R. FRANCESCHETTI

Further Reading

Bell, Eric Temple. *The Development of Mathematics.* New York: McGraw-Hill, 1945.

Boyer, Carl B. *A History of Mathematics.* New York: Wiley, 1968.

Kline, Morris. *Mathematical Thought from Ancient to Modern Times.* New York: Oxford University Press, 1972.

Kline, Morris. *Mathematics: The Loss of Certainty.* New York: Oxford University Press, 1980.

Taton, Rene. *History of Science: Science in the Nineteenth Century.* New York: Basic Books, 1965.

Van Heijenoort, Jean. *From Frege to Gödel: A Source Book in Mathematical Logic, 1879-1931.* Cambridge, MA: Harvard University Press, 1967.

Set Theory and the Sizes of Infinity

Overview

Set theory, and its transformation of mathematician's ideas of infinity, was mainly the work of one man, the nineteenth-century German mathematician Georg Cantor (1845-1918). Cantor found ways to work with infinite sets, which many believed could not exist. He further alarmed his contemporaries by demonstrating that while all infinite sets are indefinitely large, some are nonetheless larger than others.

Background

One of the earliest philosophers of infinity was Zeno of Elea (495-435 B.C.). His ponderings led him to paradoxes such as one in which Achilles running to overtake a crawling tortoise could never accomplish the feat. First, you see, he must reach the place where the tortoise started, and by that time the tortoise is no longer there. In fact, Zeno "proved" that the entire idea of motion was absurd. Finally, the local authorities lost patience with him and had him executed for treason.

Yet Zeno had been grasping at some ideas that were very profound indeed—but without the mathematical language to handle them. He realized that to frame problems of space and time accurately, they should be broken up into an infinite number of points or instants. Being able to do so awaited understanding of how to work with such infinitesimals using the tools of calculus.

The fourteenth-century scholar Albert of Saxony (c. 1316-1390) was among the first to put forth the idea of comparing infinite quantities. He proved that a beam of infinite length has the same volume as space itself. He did this by conducting a thought experiment in which he sawed the beam into an infinite number of imaginary pieces to build an infinite number of concentric shells in space.

Mathematicians generally regarded *infinity* as a way of speaking about a limit that could never be reached. They balked at the idea of considering it in more concrete terms, such as an infinitely large group of items. Carl Friedrich Gauss (1777-1855) wrote of his "horror of the actual infinite." Bernhard Bolzano (1781-1848), an Austrian priest, defended the concept of an infinite set in 1847. However, it was the work of Georg Cantor in the 1870s that put the theory of

sets, both finite and infinite, in formal mathematical terms.

A *set* is a grouping of numbers, objects, or ideas of any kind, grouped so that it can be considered as a single entity. For example, a class may be described as a set of students. An individual item within a set (an individual student in our example) is a called an *element* of the set.

The *union* of two sets contains all the elements of both sets. If *A* is the set of all American League baseball teams, and *N* is the set of all National League baseball teams, then the union of *A* and *N* is the set of all major league baseball teams. The *intersection* of two sets is the set of elements that appear in both sets. Since no team belongs to both the American League and the National League, here the intersection of *A* and *N* is a set with no elements, called the *null set*.

Two sets *A* and *B* are equal if all the elements of *A* are in *B*, and vice versa. Suppose a clothing store sells jeans. The set of all the jeans it has for sale is called *J*. The set of blue jeans it sells is called *B*. If the store only sells blue jeans, then *B = J*. However, if the store sells several colors, the set *B* of blue jeans is a *subset* of the set *J* of jeans. A *proper subset* is one that is not the same as the original set, which technically may be considered a subset of itself.

For finite sets, a proper subset is smaller than the original set. That is, suppose we have a set *A* of integers between 1 and 10 inclusive. Sets are often written in brackets, so here *A* = {1, 2, 3, 4, 5, 6, 7, 8, 9, 10}. Now let's define a set *B* consisting of only the even numbers between 1 and 10 inclusive. So *B* = {2, 4, 6, 8, 10}, and *B* is a subset of *A*.

In this example, we can see that *B* is a smaller set than *A*. If we try to put the elements of *A* and *B* into a *one-to-one correspondence,* we can't do it. We can start out, matching up the 1 from *A* with the 2 from *B*, then the 2 from *A* with the 4 from *B*, and so on. But clearly we are going to run out of elements in *B* before we get through with *A*.

How about one-to-one correspondences for infinite sets? It is easy to see intuitively, for example, that the set of odd integers and the set of even integers can be paired off as far out as you care to count. These infinite sets are the same "size"; in set theory, we say that they have the same *cardinality*. For finite sets, the cardinality is the number of elements. One-to-one correspondences are used to determine the cardinality of infinite sets.

Bolzano first gave examples of infinite sets that, unlike finite sets, could be paired in a one-to-one correspondence with one of their own proper subsets. For example, the set of all integers can be put into a one-to-one correspondence with the set of odd integers. Since both sets are infinite, we don't have to worry about running out, like we did when we limited the set to integers between 1 and 10. The key issue is that the integers are countable. Once we've matched up a pair and moved on, we don't have to worry about finding more elements between those we've already counted.

Bolzano's examples lent credence to the popular idea that all infinities were the same size. Cantor then came along to prove that this was not the case. He called the cardinality of such countable infinite sets as we have described $\aleph_0$. Then he gave examples of infinite sets that are not countable in the same way.

Consider the real numbers; that is, not only the integers but everything in between, including irrational numbers with decimals that go on forever. You can start matching these up with the set of integers, but at any time you could go back and discover additional irrational numbers between those you'd already counted. The real numbers are uncountable, infinitely more numerous than integers. Therefore, the cardinality of the set of real numbers is greater than that of integers. Some infinities, in other words, are bigger than others; their cardinalities are denoted $\aleph_1$, $\aleph_2$, etc.

Impact

Cantor's view of infinites as having an actual existence quickly became controversial. In particular, he was opposed by the influential mathematician Leopold Kronecker (1823-1891), who believed that "God made the integers, and all the rest is the work of man." For the rest of Kronecker's life he bedeviled Cantor, getting his papers rejected from journals and blocking his appointment to teach at the University of Berlin.

Around the turn of the century, mathematicians including Bertrand Russell (1872-1970) discovered a few disturbing paradoxes in Cantor's formulation. Yet at the same time set theory was becoming recognized as fundamental to topology and the analysis of functions. In addition, basic set theory has become a standard part of the modern elementary curriculum introduced in the 1960s as the "new mathematics."

SHERRI CHASIN CALVO

Further Reading

Dauben, Joseph Warren. *Georg Cantor: His Mathematics and Philosophy of the Infinite*. Princeton, NJ: Princeton University Press, 1990.

Ferreirós, José. *Labyrinth of Thought: A History of Set Theory and Its Role in Modern Mathematics*. Boston: Birkhäuser Verlag, 1999.

Johnson, Phillip E. *A History of Set Theory*. Boston: Prindle, Weber & Schmidt, 1972.

Zippen, Leo. *Uses of Infinity*. New York: Random House, 1962.

Development of Higher-Dimensional Algebraic Concepts

Overview

During the nineteenth century several attempts were made to generalize the algebra of complex numbers, which provided an adequate description of displacements and rotations in the plane, to describe objects in three dimensions. A number of generalized or hypercomplex systems were found, but none that preserved the properties of multiplication that were common to the real and complex numbers. Over time the form of vector algebra developed and popularized by the American J. Willard Gibbs came to be accepted as standard in scientific use, although some of the issues raised by the earlier work have left a lasting impact on pure and applied mathematics.

Background

At the start of the eighteenth century, mathematicians were generally familiar with the appearance of the square root of negative integers in the solution of algebraic equations, but were distrustful of such results, considering them spurious or fictitious. By the middle of that century the great Swiss mathematician Leonhard Euler 1707-1783) had shown by manipulating infinite series that one could make the identification:

$$e^{i\theta} = \cos \theta + i \sin \theta$$

where θ is an angle measured in radians, $e = 2.718...$ is the basis of natural logarithms, and i denotes the square root of -1. A more complete interpretation was provided in 1797 by Caspar Wessel (1745-1818), a Norwegian surveyor whose paper was published in a journal little read by mathematicians, so that credit is generally given to the French mathematician J. R. Argand (1768-1822) who independently obtained the same result in 1806. Wessel and Argand showed that every point in the plane could be represented by a combination of the form x+iy, or equivalently, by a line connecting the origin of coordinates with the point (x,y). Euler's result then represented a line of unit length. The addition of complex numbers corresponded to what we now call vector addition in two dimensions. The multiplication of the two complex numbers (a+ib) and (c+id) yields (ac-bd) + i(ad+bc) or a line of length numerically equal to products of the lengths corresponding to the two numbers multiplied and making an angle equal to the sum of the angles made by the two lines with the x-axis. With Argand's interpretation it was easy to interpret the subtraction and division of two complex numbers so that complex numbers possessed all the arithmetic properties of real numbers.

Given the success of mathematics in defining complex numbers that obeyed ordinary arithmetic rules, it seemed plausible that it would be possible to find a similar interpretation of points in three-dimensional space. The Irish astronomer Sir William Rowan Hamilton (1788-1856) reported the first significant success in this direction in 1843. Hamilton was, however, unable to find any procedure involving triples of numbers that would obey the right arithmetic rules. Instead, he introduced a system of quaternions, a set of numbers of the form w+ix+jy+kz, where the new units i, j, and k obey the multiplication rules

$$i^2 = j^2 = k^2 = ijk = -1$$
$$ij = k = -ji \; jk = i = -kj \; ki = j = -ik$$

Multiplication of quaternions was unlike the multiplication of real or complex numbers in that it was not commutative, that is, the product of two quaternions was different when the order is interchanged, so that *ab* was not necessarily equal to *ba*.

The quaternion system clearly included more information than was needed to describe displacements in three-dimensional space. Hamilton suggested calling the first term the scalar term and the remainder the vector term. When two quaternions with zero scalar terms are multiplied, one obtains a product with a scalar term equal to the modern scalar product of the two vectors and a vector term equal to the modern vector product.

Hamilton considered the discovery of quaternions to be his greatest achievement, devoting most of his attention to them for the remainder of his life. One of his corespondents, Peter Guthrie Tait, assumed leadership of the field on Hamilton's death in 1865, producing eight books on the subject with various collaborators Tait had assumed the chair of Natural Philosophy at Edinburgh University in 1960, becoming one of the collaborators of William Thomson (1824-1907), who would later become Lord Kelvin and a leader of British physics. Thomson considered quaternions unnecessarily elaborate, and placed Tait in the curious position of co-authoring several books making no mention of quaternions while promoting their use in his other works. Some of Tait's methods were adopted, however, by the Scottish mathematical physicist James Clerk Maxwell (1831-1879), who had been his childhood friend, in his analysis of electromagnetic phenomena.

The modern system of vector analysis, which has come to supplant quaternions, is generally attributed to the American physicist Josiah Willard Gibbs (1839-1903), a professor of mathematical physics, and the Englishman Oliver Heaviside (1850-1925), a self-taught scientist interested in the new electrical technology, who worked independently. Both individuals traced their interest in vector quantities to Maxwell's *Treatise on Electricity and Magnetism*, published in 1873, which had employed quaternion methods. Gibbs and Heaviside found the combination of vector and scalar products implicit in quaternions to be artificial and produced a streamlined version with the scalar and vector products considered as separate entities. This resulted in a simpler formalism that met the needs of electromagnetic theory and physics and engineering in general. Gibbs wrote the first textbook of vector analysis and taught the first university courses in the subject. In twentieth-century physics texts, quaternions are mentioned as a historical footnote, if at all.

Impact

Over the course of the history, the concept of number has had to be broadened numerous times. To the natural or counting numbers, 1,2,3, ..., had to be added a zero, then negative numbers, then rational numbers, then irrational and complex numbers. In each case a guiding principle had been that the basic properties of the fundamental arithmetic operations, addition and multiplication, would still apply to the broadened set of numbers. This expectation was even formalized by the English mathematician George Peacock in 1830 as a "principle of permanence of formal operations."

The fact that a consistent algebra in which numbers do not commute under multiplication could exist and be useful in the description of nature came as something of a shock to mathematicians, who tended to regard mathematics as a set of self-evident truths about the universe. This shock followed on the discovery that alternative geometries exist, geometries in which the sum of the angles of a triangle could exceed or be less than the sum of two right angles. This loss of certainty about what were once considered self-apparent rules of arithmetic and geometry stimulated an increased attention to the foundations of mathematics, an investigation which was to occupy many of the major figures in mathematics and mathematical logic in the twentieth century.

Among the sets of noncommuting mathematical objects to be introduced into mathematics are the matrices introduced by the English mathematician Arthur Cayley (1821-1895) in 1858. These are arrays of numbers, real or complex, which obey a multiplication rule:

> If A and B are matrices then in general AB is a different matrix from BA so the multiplication of matrices, like that of quaternions does not obey the commutative law. In fact sets of 3x3 matrices of real numbers exist which obey the same multiplication rules as Hamilton's quaternions, provided the number "1" is interpreted as the matrix with each diagonal element equal to l and 0s in all the other positions.

Matrices are used in many branches of higher mathematics and in all areas of physics. One very important use is the representation or rotations of the xyz coordinate system by 3x3 matrices of real elements. If the components of a vector are listed in a column, then placing the vector on the right-hand side of such a matrix and performing the matrix multiplication operations yields a column with the vectors x-, y-, and z-components in the rotated coordinate system. The effect of two subsequent rotations about two different axes is then represented by the multiplication of the two rotation matrices. Since the result of two rotations in three-dimensional spaces depends on the order in which they are performed, it is not surprising that matrix multiplication is not commutative. The difference in the products of two matrices is called the commutator or commutation relation, denoted [A,B]=AB-BA.

In modern physics, the Heisenberg uncertainty principle states that it is generally not possible to measure two properties of a system at the same time, since measurements generally affect the states of the systems. Thus the outcome of two measurements performed in sequence will depend on the order in which they are performed. Heisenberg used this line of reasoning as a basis for developing a matrix mechanics to describe subatomic states. In this matrix approach, which is still used in quantum mechanics and elementary particle physics, it is the commutation relations which are used to calculate all the measurable properties of the systems.

It is noteworthy that almost all the individuals involved in the development of the theory of vectors were employed as physical scientists. At the start of the nineteenth century almost all mathematics was done by people concerned with the applications of mathematics to physical problems or engineering. By the end of the century a new group of professional mathematicians, with their own journals and academic posts and societies, had come into being, in part because of the demand for teachers of mathematics for students of the new technical fields. The pragmatic vector analysis of Gibbs met the needs of science for a time, leaving the new mathematicians to contemplate the significance and potential of noncommuting algebras, until they too were needed by the physicists about a quarter century later.

DONALD R. FRANCESCHETTI

Further Reading

Bell, Eric Temple. *The Development of Mathematics.* New York: McGraw-Hill, 1945.

Boyer, Carl B. *A History of Mathematics.* New York: Wiley, 1968.

Crowe, Michael J. *A History of Vector Analysis.* New York: Dover, 1985.

Kline, Morris. *Mathematical Thought from Ancient to Modern Times.* New York: Oxford University Press, 1972.

Kline, Morris. *Mathematics, The Loss of Certainty.* New York: Oxford University Press, 1980.

George Green Makes the First Attempt to Formulate a Mathematical Theory of Electricity and Magnetism (1828)

Overview

Over the course of the nineteenth century the science of electricity and magnetism advanced from laboratory curiosity to a fully developed theory that would provide the basis for several major technologies. Essential to this development was the development of a mathematical apparatus to describe the behavior of fields, physical states characterized by a vector or scalar at every point in space. A critical initial step was provided in 1838 by George Green (1793-1841), then a self-taught amateur mathematician. A complete formulation of the behavior of electromagnetic fields was achieved over the next 35 years.

Background

While the ancient Greeks were familiar with both static electricity and permanent magnets, the nature of the two phenomena remained a subject of speculation until the beginning of the nineteenth century. In 1800 the Italian physicist Count Alessandro Volta (1745-1827) created the Voltaic "pile," a dependable source of electric current, and vast new experimental possibilities arose. In 1820, the Danish physicist Hans Christian Oersted (1777-1851) reported that a current carrying wire had an effect on compass needles placed around it, a report that quickly brought new investigators into the field. By 1821 Oersted's experiments were being reproduced and expanded upon by two men who would play a major role in the development of the new science of electromagnetism—André Marie Ampère (1775-1836) in France and Michael Faraday (1791-1867) in England.

It was recognized that the electric and magnetic forces had some of the same characteristics as the gravitational force, but were somewhat more complex in character. It was easier, particularly in the case of magnetism, to think of each charged or magnetic object as setting up a disturbance in the space around it which would determine the force that would act on a charged or magnetic object placed at that point. The electric and magnetic fields each assigned a vector quantity to each point in space. As one went from any point to neighboring points, the magnitude and direction of the field would change—the rate of change being determined by the material present.

The analogous problem in fluid flow had been treated by Swiss mathematician Daniel Bernoulli (1700-1782) in a 1789 book on hydrodynamics. In a paper on fluids in 1752, the prolific Swiss mathematician Leonhard Euler (1707-1783) showed that the potential function satisfied a very simple equation involving second partial derivatives, a equation now generally known as Laplace's equation after the French mathematician Pierre Simon Laplace (1749-1827).

In March of 1828, George Green, a self-taught English amateur mathematician, published a work entitled "An Essay on the Application of Mathematical Analysis to the Theories of Electricity and Magnetism." In this work Green introduced the notion of potential functions for the electric and magnetic field and showed how to construct the function by adding contributions from each charge. This essay included a very important formula, now known as Green's theorem.

Green's work, distributed to only the 52 individuals who had financially supported the work, might have been lost had it not been for Sir William Thomson (aka Lord Kelvin; 1824-1907), who had it reprinted in a German mathematics journal in 1850. A few years earlier Thomson had noted that the solution to Laplace's equation that took on a defined set of values on the boundary of a region of space would be, of all sufficiently "smooth" functions that satisfied the boundary conditions, that function which minimizes an integral commonly known as the Dirichlet integral after the German Mathematician Lejeune Dirichlet (1805-1859). Green and Thomson's results together make it possible to determine the potential over a region of space, given either its values or that of its derivatives (the field) over the boundary.

Over the course of the nineteenth century, considerable effort was devoted to elucidating the nature of light. This was motivated by the discovery of the polarization of light on reflection by Etienne Louis Malus (1775-1812) in 1808 and the strange "double refraction" of light into polarized beams in crystalline materials such Iceland spar, a form of calcium carbonate. Researcher in optics at the time considered light to be a vibration in a medium, the "luminiferous aether" which filled all space and somehow interacted with matter, but did not slow the motions of bodies moving through it. The Irish mathematician and astronomer Sir William Rowan Hamilton (1805-1865) devoted several years of effort to this problem, not leaving any results of lasting value to optics, but developing the mathematical techniques later to be successfully applied by himself and others, including the German Karl Gustav Jacobi (1804-1851), to problems in mechanics.

The true nature of light became apparent when the English mathematical physicist James Clerk Maxwell (1831-1879) formulated the set of four Maxwell equations describing the behavior of the electric and magnetic fields in space. Applying the mathematical techniques of Green and others to the experimental observations of Ampère and Faraday, Maxwell derived in 1864 a set of four coupled partial differential equations which in empty space could take on the form of wave equations for the components of the electric and magnetic fields. The velocity of the waves was given in the equations in terms of the fundamental force constants of the electric and magnetic force, and turned out to be 300,000 km per second, exactly the measured speed of light in vacuum. There could then be little doubt that light was a form of electromagnetic radiation and that the effects of matter on light, reflection, refraction, and polarization could be calculated from the interactions of the electromagnetic field with the charged particles making up the matter in question.

Impact

It is interesting to note the unconventional educational backgrounds of some of the pioneers of electromagnetism. Ampère was born into an upper middle-class family and would most likely have prepared for legal practice or the church were it not for the excesses of the French Revolution, which lead to the execution of his father and the loss of the family fortune. Although he never received a degree, he would be appointed to numerous important university posts in post-revolutionary France. Faraday was born into a working-class family and received an education working in an institution originally founded to provide scientific instruction to the working class. Green was a baker's son who left school early but was nonetheless able to educate himself through independent reading. After making his important contributions Green was admitted to Caius College as a scholar at the age of 40. The science of electromagnetism was thus developed very rapidly during a time of rapid social and economic change by men who would not have been considered educated by traditional standards.

It would be difficult to overstate the impact of the development of electromagnetic theory on the conditions of life in the industrialized world. The principle of electromagnetic induction, discovered by Faraday and incorporated into Maxwell's equations, made possible the design of electric generators and motors, which in turn made it possible to separate the production of electrical energy from its use in industrial production. Maxwell's identification of light as an electromagnetic wave led directly to the discovery of radio waves and the revolution in communications and mass culture that followed. One of the more astonishing conclusions of consequences of Maxwell's theory was that the speed of light would have the same velocity regardless of the relative velocities of the source and the observer. This conclusion led Albert Einstein (1879-1955) to recast the principles of mechanics in the special theory of relativity. It also abolished any need for a "luminiferous aether" in physics. Over the next century, understanding the atomic structure of matter and its interaction with radiation have created a world of lasers and optical fiber communication, none of which would be possible without the mathematical techniques developed by Green and his contemporaries.

DONALD R. FRANCESCHETTI

Further Reading

Bell, Eric Temple. *The Development of Mathematics.* New York: McGraw-Hill, 1945.

Boyer, Carl B. *A History of Mathematics.* New York: Wiley, 1968.

Crowe, Michael J. *A History of Vector Analysis*. New York: Dover, 1985.

Kline, Morris. *Mathematical Thought from Ancient to Modern Times*. New York: Oxford University Press, 1972.

Whittaker, Sir Edmund T. *A History of the Theories of Aether and Electricity*. Vol. 1. New York: American Institute of Physics, 1987.

Advances in Understanding Celestial Mechanics

Overview

Modern celestial mechanics began with the application of Isaac Newton's laws of motion to the observations of astronomy. Mathematicians of the eighteenth and nineteenth centuries worked to understand the workings of the solar system in terms of all its gravitational forces. While the gravitational pull of the Sun keeps the planets in their orbits, each planet's path is slightly disturbed by the presence of the others.

Background

Early astronomy was primarily concerned with celestial mechanics in a broad sense; that is, understanding the apparent motion of the stars and planets. Without telescopes, the Moon was the only celestial body upon which details could be observed. Everything else was too distant, except for the Sun, which was too bright. So theories could be devised about the nature of the other bodies, but the only property that could actually be measured was their motion.

The word *planet* comes from the Greek word for wanderer. As seen from Earth, the stars all march across the sky at the same steady pace, but the planets move at a variable rate, sometimes even changing direction. We now know that we see the planets in this way because they, like Earth, are moving in their separate orbits around the Sun. The apparent regular motion of the distant stars arises from our changing point of view as Earth rotates on its axis and revolves around the Sun.

Of course, none of this was obvious to the ancients, who assumed that the entire cosmos revolved around Earth. The most influential ancient astronomer, Ptolemy, lived in Alexandria around 150 A.D.. He explained the variable motion of the planets by assuming that they revolved in small orbits called *epicycles*, on a larger circle called a *deferent* that orbited Earth. Ptolemy adjusted the speeds and distances in his system until he was able to make accurate predictions of planetary positions. His system, preserved in a work his successors called the *Almagest*, or the "Greatest," was used until the Renaissance.

The Polish astronomer Nicolaus Copernicus (1473-1543) is generally considered the father of modern-day astronomy. He realized that Earth orbits the Sun like the other planets and that our view of the sky is affected by Earth's motion. He put forth his *heliocentric*, or sun-centered, view of the cosmos in his great work *Concerning the Revolutions of the Celestial Sphere* (1543). Since he still assumed, incorrectly, that the orbits of the planets were circular, he had to retain some of Ptolemy's epicycles to accurately match their positions.

Johannes Kepler (1571-1630), using twenty years of precise measurements made by his mentor Tycho Brahe (1546-1601), developed his famous laws of planetary motion, the first of which states that the orbits are elliptical. Kepler's laws were accurate, but empirical; that is, they accounted for the observed data, but had no explanation for why the orbits should be shaped as they are. It was Isaac Newton (1642-1727) who provided that explanation. Newton derived three general laws of motion, and concluded that the gravitational force between two bodies was proportional to the product of their masses and the inverse square of the distance between them.

With the new understanding of gravity and the laws of motion, it was now possible to consider the planets as an *n-body problem*; that is, a theoretical problem of predicting the behavior of a fixed number of masses interacting by means of their gravitational fields. Such problems are complex because every object affects every other. So, for example, in our Solar System, the main determinant of the planetary orbits is the gravitational field of the Sun. However, each planet experiences perturbations in its own orbit because of all the others. Once the basic mechanism of orbits was understood, the discipline of celestial mechanics began to concentrate on understanding the perturbations.

Impact

Planetary motion is described in terms of differential equations. Differential equations involve derivatives, mathematical expressions for the rate of change of one quantity with respect to another. For example, the derivative, or rate of change, of distance with respect to time is velocity. The rate of change of velocity with respect to time is acceleration, and acceleration is determined by the gravitational or other force on an object. The equations describing position, velocity, and force necessarily involve derivatives.

Differential equations are themselves an important field of study in mathematics. Many are extremely difficult to solve. While some have exact solutions, the solutions to others must be approximated, and many techniques have been developed to do so. Today differential equations are sometimes solved using numerical computing techniques.

In celestial mechanics, a "two-body problem" such as the rotation of a single planet around the Sun without taking any other masses into account is described by a relatively simple differential equation. It can be solved exactly, and the solution reproduces Kepler's laws. Add in the masses of the other planets, however, and the situation becomes much more complicated. Even the great Newton essentially threw up his hands and attributed the stability of the Solar System to occasional divine intervention in which everything was nudged back into place.

The stability problem was taken up by the French astronomer Pierre-Simon Laplace (1749-1827). Observations seemed to indicate that the orbit of Jupiter was continuously shrinking while that of Saturn was expanding. In 1786 Laplace showed that the eccentricities of the planetary orbits and the angles at which they are inclined with respect to one another will remain small and self-correcting. Because the perturbations in the motion are periodic, they do not accumulate and disrupt the stability of the solar system. In the case of Jupiter and Saturn, the effect being observed had a period of 929 years.

Between 1798 and 1827 Laplace's five-volume work on planetary motion, *Traité de mecanique céleste* ("Treatise on Celestial Mechanics"), was published. In it he provided a complete method for calculating the movements of the planets and their moons, including gravita-

tional perturbations and the tidal disturbances of the bodies' shapes. The book quickly became a classic. It was updated and enlarged by both the American mathematician Nathaniel Bowditch (1773-1838) and the French astronomer Félix Tisserand (1845-1896). Tisserand's version, four volumes published between 1889 and 1896, remains an important reference in the field.

For any given planet, it is generally convenient to consider the sum of the forces from all the others as a net perturbation of the orbital ellipse. The French-Italian mathematician Joseph-Louis Lagrange (1736-1813) expressed this in a set of differential equations sometimes called the *Lagrange planetary equations*. They must be solved numerically or by means of successive approximations using a series of mathematical terms to increase the range over which a solution is a good fit to the actual motion. Lagrange used his equations to explain the libration of the Moon; that is, the oscillations seen as a slight change in the position of the visible lunar features. The techniques also resulted in the discovery of the planet Neptune in 1846, after its position was predicted from perturbations of the orbit of Uranus.

In 1889 the French savant Henri Poincaré (1854-1912) won a prize offered by King Oscar II of Sweden for a contribution to the n-body problem. Poincaré applied to the problem several of the advances in mathematical analysis that had been developed since the time of Laplace and Lagrange and added a few important techniques of his own invention. His work was published between 1892 and 1899 as *Les méthodes nouvelles de la méchanique céleste* ("The New Methods of Celestial Mechanics").

Poincaré was more concerned with the mathematics of celestial mechanics than with obtaining precisely accurate predictions of planetary motion. He studied the series approximations used with the differential equations of motion, to understand when they would converge to a useful solution. In cases where they did not converge to a general solution, he showed under what conditions they could be used to approximate the motion for a significant period of time.

The advances made by nineteenth-century mathematicians and astronomers with respect to celestial mechanics set the stage for further developments in the twentieth century. Poincaré's work, for instance, anticipated the modern idea of chaotic motion—or chaos theory—in which for some initial conditions the future state of a system becomes unpredictable within an allowable range.

SHERRI CHASIN CALVO

Further Reading

Arnold, V. I., ed. *Mathematical Aspects of Classical and Celestial Mechanics.* New York: Springer-Verlag, 1993.

Collins, George W., II. *The Foundations of Celestial Mechanics.* Tucson, AZ: Pachart Pub. House, 1989.

Poincaré, Henri. *New Methods of Celestial Mechanics.* Edited and introduced by Daniel L. Goroff. Woodbury, NY: American Institute of Physics, 1993.

Roy, Archie E. and Bonnie A. Steves, eds. *From Newton to Chaos: Modern Techniques for Understanding and Coping with Chaos in n-body Dynamical Systems.* New York: Plenum Press, 1995.

Sternberg, Shlomo. *Celestial Mechanics.* New York: W. A. Benjamin, 1969.

A New Realm of Numbers

Overview

Great strides were made in the 1800s toward moving back to a rigorous, logical base for mathematics. Essential to this effort was progress in number theory. Joseph Liouville (1809-1882) expanded the understanding of real numbers when he proved the existence of transcendental numbers. Later, Charles Hermite (1822-1901) demonstrated that *e*, the natural logarithmic constant, was a transcendental number. In 1882, Ferdinand Lindemann (1852-1939) answered "no" to the classic challenge, "Can the circle be squared?" when he proved that pi (π), the ratio of the circumference of any circle to its diameter, was also a transcendental number. Julius Wilhelm Richard Dedekind (1831-1916) completed the view of real numbers by explaining them in terms of irrational and irrational numbers. The establishment and characterization of real numbers extended the rigor of mathematics, improving the quality of proofs. It affirmed the concept of limits and allowed the rigorous development of analysis, which is essential to solving many difficult problems of engineering and science.

Background

During the 1800s, mathematicians took on the challenge of making their discipline more rigorous. While the Greeks carefully defined terms and worked out proofs based on logic and consistency, their successors took a more practical approach. For over 1,000 years, most mathematicians relied primarily on intuition and geometry to expand their understanding of the quantitative world and to solve practical problems in science and engineering. Galileo (1564-1642) used geometry to formulate and express his understanding of the motion of bodies. René Descartes (1596-1650) developed graphs to synthesize algebra and geometry, creating analytical geometry. Even calculus failed the test of rigor. In 1734, the philosopher George Berkeley (1685-1753) criticized the use of limits by Isaac Newton (1642-1727) and Gottfried Leibniz (1646-1716) because they did not have an adequate logical basis. Mathematics had had a long run of successes based on visual and intuitive approaches to mathematics, but, by the eighteenth century, the rigorous base provided by the Greeks was insufficient to the task supporting detailed analysis of data in maturing sciences. A return to rigor was needed if progress was to continue.

In 1821, Augustin Cauchy (1789-1857) started the move toward making mathematics more rigorous with his work *Cours d'analyse de l'École Royale Polytechnique.* But his work, and the work of his colleagues, presupposed a number system which did not yet exist. Calculus had demonstrated that there were precise results that were not solvable by rational numbers (integers and fractions). It relied on the notion of limits, but no one had created a number system in which convergent sequences had limits. What mathematicians needed to develop was a deep understanding of real numbers, one which divided these numbers into algebraic and transcendental, and, alternatively, into rational and irrational. This would provide what was needed to support rigorous proofs in analysis.

Paris was a hotbed of scientific and mathematical thought in the nineteenth century. In addition to Cauchy, mathematicians like André Ampère (1775-1836) and Pierre Laplace (1749-1827) and physicists like Jean Baptiste Biot (1774-1862) and Dominique Arago (1786-1853) collaborated, and often competed, for recognition, funding, and public attention. Joseph Liouville (1809-1882) rose to become one of the leading mathematicians of this period. Liouville was much influenced by Cauchy and was a student of Ampère. He moved to the center of activity in mathematics as a scholar, teacher, and publisher.

When Liouville read the correspondence of Christian Goldbach (1690-1764) and Daniel Bernoulli (1700-1782) from the previous century, he became intrigued by their speculation on existence of transcendental numbers. A transcendental number is a real number, that is, it is the limit of a convergent series of rationals. But unlike algebraic numbers, which are real numbers that can serve as solutions for polynomials with rational (integer or fractional) coefficients, transcendental numbers can never serve as solutions for these equations. Liouville decided he would prove that the natural logarithmic constant e was a transcendental number.

This was an interesting target. The natural logarithmic constant e is critical to calculations of compound interest and to the description of natural phenomena like radioactive decay and population growth. It is also a number that intrigues mathematicians because the derivative of e^x is e^x, and is part of $e^{i\pi} + 1 = 0$, an equation that links the five most important numbers in mathematics—e, i, π, 1, and 0. Johann Lambert, the man who proved π was irrational, had speculated that both π and e were transcendental numbers.

Though Liouville never succeeded in proving e was transcendent, he did succeed in 1844 in becoming the first person to prove that transcendental numbers existed at all. He did this by showing that in some instances where the number of algebraic solutions must necessarily be finite, you can still find cases where the number of solutions is infinite. These numbers, which clearly are both non-algebraic and real, are transcendental. Liouville identified an infinite class of such numbers. He also took the first steps toward proving that e was transcendental, but he was not able to complete the job. This work was to be accomplished by his student, Charles Hermite (1822-1901).

Hermite was one of the leaders in adding rigor to mathematics in the nineteenth century. He provided the first solution to the general equation in the fifth degree, and he was a key contributor to the development of the theory of algebraic forms. Taking up where Liouville left off, he set out to prove that e was transcendental.

Essentially, while Liouville had shown that e could not solve some polynomial equations, Hermite showed in 1873 that e could not solve *any* polynomial equations. It was a transcendental number.

Just a year earlier, another important view of real numbers had been established: rational and irrational. Pythagoras (c. 580-500 B.C.) had discovered the first irrational number (the square root of two) over two millennia before. Legend has it that his followers executed the man who let this secret out, and irrationals had troubled mathematicians thereafter. But in 1872, the German mathematician Julius Dedekind (1831-1916) looked at irrationals and rationals as cuts in a line. This concept allowed manipulation and use of both irrationals and rationals in a consistent way. The irrationals and rationals were thereby brought together to comprise all the real numbers.

In addition to being transcendental, e was known to be irrational. Was the same true for π? The answer would prove to be of both mathematical and cultural importance. Arguably, π is the number that has most fascinated both amateur and professional mathematicians throughout history. Famous for the expression πr^2, π is the ratio of the circumference of any circle to its diameter. It is also central to one of the classic challenges of classical Greece (included in Euclid's *Elements*), popularly known as "squaring the circle." Using only an unmarked ruler and a compass, is it possible to construct a square with the same area as a circle? In 414 B.C., Aristophanes mentioned squaring the circle in his play, *The Birds*. Since that time, squaring the circle has been used as an expression for doing the impossible. But great minds were not discouraged. Hippocrates (c. 460-377 B.C.), Archimedes (c. 287-212 B.C.), and even Leonardo da Vinci (1452-1519) had taken on the challenge. For centuries, squaring the circle had provided an intellectual stimulus and led mathematicians to a deeper understanding of their discipline. In the seventeenth century, François Viète (1540-1603) used a polygon with 393,216 sides to approximate squaring the circle and determined π to 10 decimal places, the most accurate measure of π to that date. Proving that π was transcendental would prove that squaring the circle could not be done, since then, no line could be π (or its multiple) units in length.

By 1873, π was known correctly to the 527th digit, and no pattern was evident in the decimals. This is what would be expected from a transcendental number, but could it be proven? Ferdinand Lindemann (1852-1939) decided to find out. Using the methods of Hermite, he succeeded in doing so in 1882. Lindemann had shown Aristophanes was right. For all the effort that had been put into squaring the circle, it could not be done. The answer to Euclid's challenge was "no."

Impact

From a popular standpoint, the biggest impact of nineteenth century was the end to the pursuit of squaring the circle for most reasonable people. Like the building of perpetual motion machines, squaring the circle is one of those challenges that delights amateurs and annoys professionals. Even before Lindemann's proof of the impossibility of the challenge, both the French Academy (a century earlier, in 1775) and the Royal Society in London had declared that they would no longer review "solutions." Although there will probably always be a cadre of people who will continue to try, Lindemann forever shut the door on serious attempts to square the circle. (Approximations of the solution, however, are still fair game. The brilliant mathematician Srinivasa Ramanujan (1887-1920) provided a solution in 1914 that would correlate to less than an inch error in each side of a square constructed from a circle of 8,000 miles [12,872 km] in diameter.)

The work of Liouville, Hermite, Dedekind, and Lindemann helped make mathematics a more rigorous discipline. As Gödel was to prove, every mathematical system contains unresolvable paradoxes. One expression of this is in number theory. Natural numbers become incomplete in the face of subtraction—negative numbers must be postulated. Integers (both positive and negative) are insufficient in the face of division. Rational numbers must be postulated, which include fractions and take everything into account except division by zero.

Real numbers, including transcendental numbers, moved the boundary forward, allowing for rigor and consistency in analysis (and, of course, moving mathematics forward to new paradoxes and inconsistencies). Working with transcendental numbers introduced new approaches to approximation and made limits an accepted concept in mathematics. This opened the door for the solution of many important problems in science and engineering and brought the mathematical discipline of analysis into its own. However, the practical utility of real numbers has been limited. Using real numbers to calculate answers to problems in applied mathematics is very difficult. For instance, while real numbers help provide a theoretical basis for computers, they cannot be used by a computer. Exact real numbers are infinite, so they can't be stored or manipulated. Computer scientists are forced to use floating point numbers to approximate answers, something which can lead to errors (many of which are not easily anticipated). This lack of precision is part of our heritage in the mismatch between the number system and the tools we inherited from the nineteenth century. It is an area of active investigation.

Perhaps the most far-reaching effect of nineteenth-century work in number theory was the forceful shift of science and mathematics from the visual to the textual. The rigor of exact equations came to be a prerequisite for any theory or proof to gain acceptance by professionals in mathematics, science, and engineering. It is only as the visual capability of computers has risen to a high level that confidence in visual demonstrations has begun to reemerge.

PETER J. ANDREWS

Further Reading

Books

Asimov, Isaac. *Isaac Asimov's Biographical Encyclopedia of Science and Technology.* New York: Doubleday and Co., 1976.

Maor, Eli. *E: The Story of a Number.* Princeton, NJ: Princeton University Press, 1998.

Pickover, Clifford A. *Chaos in Wonderland.* New York: St. Martins Press, 1995.

Pickover, Clifford A. *Keys to Infinity.* New York: John Wiley and Sons, Inc., 1995.

Other

Jones, R. B. "Numbers—A Logical Development." http://www.rbjones.com/rbjpub/maths/math007.htm

George Boole and the Algebra of Logic

Overview

Until the English mathematician George Boole (1815-1864) came along in the nineteenth century, logic was regarded as a branch of philosophy. By codifying it in algebraic form, Boole brought it into the realm of mathematics. His work formed the basis for the digital logic that drives today's computer and communications technologies.

Background

Logic deals with rules of correct reasoning. There are two types of logical arguments. *Inductive* reasoning generalizes from previous experience. Therefore it provides us with probabilities rather than certainties. Inductive reasoning is the basis of what we often call "common sense." For example, if you are eating a bunch of grapes, and every grape you have eaten so far is sweet, you expect the next one to be sweet as well. It probably will be, but there are no guarantees.

Deductive logic does not depend upon experience. Rather, it applies rules to determine whether a statement is *valid* based on the *premises* from which it is derived. It is important to note that "valid" does not necessarily mean "true." If the premises are true, and deductive logic is applied correctly, then the conclusion will be true. However, if any of the premises are false statements, then a perfectly valid chain of reasoning can lead to a nonsensical result.

Suppose you begin with the premise "All mammals are blue." Your second premise is "All cows are mammals." Therefore, you deduce that all cows are blue. This argument is valid given the premises you used. However, the first premise is false, so your conclusion is nonsense. Among computer programmers, this pitfall is expressed as "garbage in, garbage out."

A set of statements like the example just given, consisting of two premises and a conclusion, is called a *syllogism*. The ancient Greek philosopher Aristotle (384-322 B.C.) was among the first scholars to conduct a systematic study of the rules of deductive logic. His works on the subject are collected in a set of books called the *Organon*, or "instrument." The name is derived from the exploration of thought as the instrument by which knowledge is derived.

Aristotle was among the most influential thinkers of Western civilization. He planted logic firmly in the discipline of philosophy, and there it stayed for more than 2,000 years, until George Boole began looking for new applications for algebra.

Boole came from a background that did not augur well for him entering the mainstream of mathematics. First of all, he was English, at a time when the field was centered in Germany and France. Even in his native land, many would not look twice at him. England was a class-conscious society, and Boole was the son of a poor tradesman.

Boole was a youth of scholarly ambitions but limited economic prospects. By his mid-teens, after a few years in a mediocre working-class school and some tutoring in mathematics by his father, he found himself in need of a job to help support his family. Making the best of the situation, he became a schoolteacher, and he continued to educate himself in mathematics from books and periodicals. When he was 24 he began submitting papers to the *Cambridge Mathematical Journal*. The able editor of this publication, a Scotsman named D. F. Gregory, fortunately took an interest in the originality of Boole's work and overlooked the obscurity of the author.

Before he was 30 Boole was awarded a medal by the Royal Society for his contributions to the study of algebra. His work involved applying it to difficult problems in calculus and differential equations. Soon he realized it could be applied to logic as well. The algebraic symbols could be used to express logical operations and relationships.

In 1847 Boole published a pamphlet entitled "Mathematical Analysis of Logic." In it he first argued that logic should be considered a branch of mathematics rather than philosophy. On the strength of this work and his other publications, he was appointed to the faculty of Queen's College in Cork, Ireland, despite his lack of a university degree. He taught there for the rest of his life. Boole published a full exposition of his algebra of logic in 1854, in a book called *An Investigation into the Laws of Thought, on which Are Founded the Mathematical Theories of Logic and Probabilities.*

Impact

Boole's algebra of logic has had a profound influence on society, especially beginning with the computer revolution of the 1950s. All computers use Boolean operations to function. Since computers are at the heart of the world economy at the turn of the twenty-first century, the legacy of Boole's contributions is indeed enormous.

The symbolic method of logical inference that has come to be called *Boolean algebra* is based upon a two-valued, or *binary* scheme. The values may be expressed in many ways, such as true or false, one or zero, or on and off. It is this property that makes it so useful for implementing logic through electronic circuits. For example, "on" and "off" might be implemented as two different voltage levels in a circuit, or the presence or absence of current flow through a switch.

Boolean operations were once known mainly among the ranks of engineers and computer programmers. However, they have become more familiar to the general public as applied to searches in databases and on the Internet. While interfaces sometimes employ "natural language" and other mechanisms to assist users in constructing a search, the Boolean operations remain an efficient way to clearly express the desired action. When searching, these operations are applied to *sets*, or groups of objects or ideas.

The Boolean operation AND, for example, means that two or more conditions must be true. Suppose you are searching a used-car database for a blue Corvette. In order for a car to meet your criteria, both conditions must be met. Your search might read "color = blue AND model = Corvette."

The Boolean OR expresses a situation in which the criteria are met if any of the conditions are true. If you want a sports car, but would look at either Corvettes or Mustangs, you could enter a search such as "model = Corvette OR model = Mustang."

What would happen if you said "model = Corvette AND model = Mustang"? No car is both a Mustang and a Corvette, so you would get no car listings back. You have constructed a condition whose solution is the *null set*.

The OR operation will return objects in both sets referred to by its conditions. Often these conditions overlap. For example, suppose you are searching the library catalog for something to read. You enjoy reading science fiction. Two of your favorite authors are Jerry Pournelle and Larry Niven, who have collaborated on many books. You've read all their collaborations, though, and you'd like to know what else they've written. If you search for "Niven OR Pournelle," you will get all the books written by the two authors, including their collaborations. For the list you want, you might use the *exclusive or,* or NOR operation. This would give you books that are in the "Niven" set or the "Pournelle" set, but not those that are in both.

The *not-and* or NAND operation is the opposite of AND. It returns those items that meet neither condition. Finally, the NOT operator can be used alone, to express the set of objects that does not meet a given condition. It is called a *unary operator* because it refers to a single condition. The AND, OR, NAND, and NOR are *binary operators* because they refer to the relationship between two conditions.

In digital electronics the Boolean operators are implemented using *logic gates,* which are often etched onto integrated circuit chips. These process signals in accordance with their function. For example, the unary NOT gate has one input and one output. If the input is "0," the output is "1," and vice versa. The binary operations are implemented using logic gates with two inputs and one output. An OR gate has a "1" on the output if either input is "1." An AND gate requires both inputs to be "1" in order to have an output of "1."

These simple logic gates are used to design digital circuits of many kinds, found in devices ranging from coffeepots and calculators to theater-quality audio and video systems and the most powerful computers. Many of these devices would have been beyond Boole's wildest imaginings, but they owe their existence to the logical algebra he devised.

SHERRI CHASIN CALVO

Further Reading

Barry, Patrick D., ed. *George Boole; A Miscellany.* Cork, Ireland: Cork University Press, 1969.

Brown, Frank Markham. *Boolean Reasoning: The Logic of Boolean Equations.* Boston: Kluwer Academic Publishers, 1990.

Gregg, John. *Ones and Zeros: Understanding Boolean Algebra, Digital Circuits, and the Logic of Sets.* New York: IEEE Press, 1998.

Houghton, M. Janaye and Robert S. Houghton. *Decision Points: Boolean Logic for Computer Users and Beginning Online Searchers.* Englewood, CO: Libraries Unlimited, 1999.

Solomon, Alan David. *The Essentials of Boolean Algebra.* Piscataway, NJ: Research and Education Association, 1990.

The Promotion of Mathematical Research

Overview

The makers of mathematics and their modes of practice were transformed during the nineteenth century. New mathematical ideas were publicized in mathematical journals. Professional mathematicians began to meet together to share their research in local societies, which grew into national organizations. At the same time, these mathematicians created opportunities to complete their research as university professors and to train mathematics professionals. Toward the end of the century, an international congress was established. Indeed, by 1900 a growing community of specialized mathematicians cooperated around the world.

Background

Before 1800, mathematicians were generally isolated in their own nations. Mathematicians supported themselves through patronage from nobles or the state, or with their own personal wealth; even Isaac Newton (1642-1727) served as Master of the Mint in London. They generally exchanged ideas sporadically, in private letters or by publishing books containing their life's work. Although it remained possible for one person to master all existing mathematical knowledge, the body of mathematics was increasing. Still, amateurs and individuals outside institutions continued to dominate the exploration of mathematics since mathematics teachers were usually too busy to do research or to learn the new results on their own.

Around the turn of the nineteenth century, however, mathematicians increasingly gained awareness of mathematical improvements in other nations. One of the most striking examples of this new awareness was in Great Britain. Because of the bitter priority controversy between Newton and Gottfried Wilhelm Leibniz (1646-1716) over the invention of the calculus, British mathematicians were especially disconnected from mathematical developments on the European continent during the eighteenth century. By the 1810s, though, professors such as John Playfair (1748-1819) and Robert Woodhouse noted the superiority of Continental analysis. Then, three students at Cambridge University, Charles Babbage (1792-1871), John Herschel (1792-1871), and George Peacock (1791-1858), founded an "Analytical Society" in 1813 to urge the university to use Leibniz's differential notation, rather than Newton's fluxions, on the annual tripos examination. They also argued for a translation of a calculus textbook by Silvestre François Lacroix, which was printed in 1816. The achievements of the society were limited overall, but the group served as a harbinger of a new, modern attitude toward mathematics in Great Britain.

Communication between mathematicians was further facilitated by the publication of journals. While earlier periodicals such as the English *Ladies' Diary* popularized mathematics and problem-solving, and encyclopedias informed a larger public of relatively recent results, French mathematicians started to directly report their research in publications such as *Journal de l'École Polytechnique*. In 1810 Joseph Diaz Gergonne (1771-1859) established the first private journal wholly devoted to mathematics—*Annales de Mathématiques Pures et Appliquées*. Although Gergonne's original intent was to provide material for teachers, he mainly published new results. By 1832, when the journal ceased publication, Gergonne had publicized areas of mathematics often neglected in France and influenced the direction of future research.

After the middle of the century, mathematicians who sought places where they could discuss mathematical journals and research formed societies devoted specifically to mathematics. These societies eventually grew beyond local boundaries. The first to flourish, founded in 1865, was the London Mathematical Society, originating as a college club at University College. Because the group was supported by Augustus De Morgan (1806-1871), one of Britain's leading mathematicians, other major British mathematicians soon joined the society. The group began publishing original papers immediately, with the first by James Joseph Sylvester (1814-1897). By the end of 1866, the society was national in membership and activity.

Mathematical meetings turned fully international in scope with the congress held in conjunction with the 1893 World's Fair in Chicago. The New York Mathematical Society (renamed the American Mathematical Society the following year) issued invitations to prominent European mathematicians. Felix Klein (1849-1925), a German mathematician additionally involved in educational reforms and the application of mathematical ideas, accepted. From August 21-26, 1893, he gave talks surveying all of mathematics. Afterwards, Klein stayed an additional two weeks at Northwestern University for a colloquium during which he presented 12 lectures exposing his audience—from Austria, France, Germany, Italy, Russia, Switzerland, and the United States—to current areas of mathematical research. He also met personally with the participants. The New York Mathematical Society later issued these lectures in book form. The congress inspired further meetings both in the United States and as a series of international mathematics congresses held every four years, beginning in 1897 in Zurich.

Impact

These developments both influenced and were influenced by changes in the practice of mathematics. For example, the technical standards of mathematics were elevated as mathematicians such as Augustin-Louis Cauchy (1789-1857) strove for greater rigor in proof. Furthermore, as mathematicians expanded their areas of study, mathematics as a whole became a more specialized endeavor. It was no longer possible for one person to master all mathematical knowledge. New fields ranged from the mathematical theory of heat studied by Joseph Fourier (1768-1830), to research in elliptical functions by Carl Gustav Jacobi (1804-1851), to the invariants studied by Arthur Cayley (1821-1895) and Sylvester, all mathematicians who reported their results to mathematical societies and in journals.

Mathematicians who studied these specialized subfields needed certain qualifications, most notably a doctoral degree. Traditional higher education policies emphasizing Euclid's *Elements,* the value of mathematics in training the mind to reason properly, and memorization were increasingly considered unsatisfactory. Rather, military academies often led the way early in the century by providing specialized and technical training. Then, colleges and universities in Europe and the United States gradually focused on advanced mathematics. German institutions established the model for the modern graduate school, where students met in seminars with professors and were awarded doctorates in mathematics departments upon completing dissertations on original research.

These trained professionals were then recruited back to universities to conduct research. University leaders had recognized the prestige mathematical research could bring to their institutions. For example, after Daniel Coit Gilman (1831-1908) established Johns Hopkins University in Baltimore as primarily a graduate school, he hired Sylvester in 1876 to lead the mathematics department. Sylvester relished the opportunity to escape teaching introductory mathematics courses. He carried on his research into the properties of invariants, among other interests, and founded the *American Journal of Mathematics* in 1878, securing contributions from prominent European mathematicians. In turn, with the university's new mission to both transmit and generate mathematical ideas, professors such as Sylvester trained advanced students who earned doctorates and generally embarked on careers as mathematical researchers themselves.

The success of Gergonne's *Annales* opened the way for additional mathematical journals. For instance, the *Journal für die reine und angewandte Mathematik,* founded by August Leopold Crelle (1780-1855) in 1826, took the European lead as a periodical of advanced content after *Annales* folded. "Crelle's journal" accepted contributions from nations outside the German lands and promoted travel and research by mathematicians. Meanwhile, by mid-century, French mathematicians published their work in more than 30 journals, magazines, and newspapers, including the *Journal de Mathématiques Pures et Appliquées,* founded by Joseph Liouville (1809-1882) in 1836. In the last quarter of the century, Italian journals promoted the dissemination of mathematical research and bridged the gulf between French and German mathematicians caused by the Franco-Prussian War, while *Acta Mathematica,* based in Scandinavia, successfully drew international contributions on the largest scale for that time. These periodicals also started to include material on the history of mathematics. The major journals in the United States were Sylvester's *American Journal of Mathematics* and *Annals of Mathematics,* founded in 1884.

Similarly, the London Mathematical Society's example in facilitating communication between mathematicians, publicizing international developments, and holding talks given by and to mathematics researchers was followed by mathematical societies established around the world in the last quarter of the nineteenth century. These organizations included the Moscow Mathematical Society, founded in 1867, the Société Mathématique de France, founded in 1872, the Edinburgh Mathematical Society, founded in 1883, and the Circolo Matematico di

Palermo, founded in 1884. These societies often founded journals to publicize research results or published the proceedings of their meetings. Sometimes they presented awards to outstanding mathematicians.

In addition, new centers of mathematics rose in importance. While mathematicians in Paris dominated mathematical achievement into the 1830s, German mathematicians were then generally most productive for at least the next third of the century. London and Italy were also home to accomplished mathematicians. The United States as a whole rapidly gained prominence late in the nineteenth century, with the success of the 1893 congress bringing specific attention to the University of Chicago as an educational institution offering advanced mathematical training comparable to that available in Europe.

Mathematicians came to view mathematics as an enterprise dependent upon international cooperation. National societies brought in distinguished mathematicians by electing foreign members. For instance, just two years after the London Mathematical Society was founded, French geometer Michel Chasles (1793-1880) was named to an honorary position. In the last decade of the nineteenth century, in addition to the congresses that followed the 1893 meeting, the International Mathematical Union was founded. Although the upheavals of the two world wars would disrupt its activities in the twentieth century, the Union would foster communication and interaction between mathematicians on an unprecedented level. It then became global as mathematicians in such areas as eastern Asia and South America professionalized and communicated with other mathematicians.

None of these developments could have taken place without the new technologies of the nineteenth century. Improvements in transportation, such as the railroad, made it easier for mathematicians to visit top researchers. Shorter travel times also meant books and journals reached distant mathematicians soon after they were printed, while changes in printing materials and techniques enabled mathematical research to be published more quickly and cheaply.

In summary, the promotion of mathematical research in the nineteenth century coincided with the evolution of a new definition for "mathematician." A professional mathematician had earned a doctoral degree, produced research, and shared the results in talks and papers with other mathematicians, a community of scholars with a sense of identity. Nineteenth-century mathematicians promoted professionalization by establishing journals and forming mathematical societies. On the other hand, the dissemination of research made it possible for more people to engage in advanced mathematics and for mathematical productivity and diversity to increase even more dramatically. As these various factors acted upon each other, the result was an attitude among mathematicians that centered upon the promotion of mathematical research.

AMY ACKERBERG-HASTINGS

Further Reading

Books

Ausejo, Elena, and Mariano Hormigón, eds. *Messengers of Mathematics: European Mathematical Journals (1800-1946)*. Madrid: Siglo XXI de España Editores, S. A., 1993.

Grattan-Guinness, Ivor. *The Norton History of the Mathematical Sciences: The Rainbow of Mathematics*. New York and London: W. W. Norton, 1997.

Lehto, Olli. *Mathematics Without Borders: A History of the International Mathematical Union*. New York: Springer, 1998.

Parshall, Karen Hunger, and David E. Rowe. *The Emergence of the American Mathematical Research Community, 1876-1900: J. J. Sylvester, Felix Klein, and E. H. Moore*. Providence, RI: American Mathematical Society, 1994.

Tobies, Renate. "On the Contribution of Mathematical Societies to Promoting Applications of Mathematics in Germany." In *The History of Modern Mathematics*, edited by David E. Rowe and John McCleary, Vol. 2. San Diego: Academic Press, 1989.

Periodicals

Enros, Philip C. "The Analytical Society (1812-1813): Precursor of the Renewal of Cambridge Mathematics." *Historia Mathematica* 10 (1983): 24-47.

Rice, Adrian C., and Robin J. Wilson. "From National to International Society: The London Mathematical Society, 1867-1900." *Historia Mathematica* 25 (1998): 185-217.

Nineteenth-Century Efforts to Promote Mathematics Education from Grade School to the University Level

Overview

Economic and political factors combined to foster an increased emphasis on mathematics education over the course of the nineteenth century. The demand for technical workers and for military officers who could understand the complexities of more powerful weapons resulted in new polytechnic schools and military academies throughout northern Europe and the United States. The increased need for teachers of mathematics provided employment for a vastly increased number of mathematicians. In addition, newly formed mathematical societies took an interest in the mathematics curriculum and teaching methods at the elementary level.

Background

Ever since Plato refused some 2,400 years ago to admit to his academy students who were ignorant of mathematics, there has been a connection between mathematics and education. The inclusion of three fields of mathematics—arithmetic, geometry, and logic—among the seven traditional liberal arts assured some emphasis on mathematical topics in the schools of the classical period and the middle ages. Printing with movable type became possible in 1454, and some of the first books printed were manuals of mathematics. The earliest arithmetic text was published in Trevisi, Italy, in 1478. The first printed edition of famed Greek mathematician Euclid (c. 330-260 B.C.) appeared in Venice in 1482. Prior to the nineteenth century, however, most mathematical research was conducted outside the universities. Mathematicians were supported by patrons and then by the scientific academies, which also produced the first scientific journals.

Expanded trade and new technologies called for increased emphasis on the mathematics of the marketplace: counting, weights and measures, currency exchange, and the calculation of integers—topics not normally included in the training of young gentlemen or future clergy. The industrial revolution and a demand for faster water and land transport called for the solutions of the mathematical problems of fluid motion, bridge building, and heat transfer—the bases of modern civil and naval engineering.

At the university level, revolutionary France led the way. The government in 1794 authorized the establishment of the schools that were to become the École Polytechnique, later converted into a military school by Napoleon, and the École Normal Superiore, to train future teachers. The École Normal would soon boast Joseph Lagrange (1736-1813), Adrien-Marie Legendre (1752-1833), and Pierre Simon Laplace (1749-1827) on its faculty. The École Polytechnique published the first journals devoted to mathematical research, when in 1810 Joseph-Diaz Gergonne (1771-1859) inaugurated the *Annals of Pure and Applied Mathematics* in French. German and English journals would soon follow. In 1826 August Leopold Crelle (1780-1855) inaugurated the *Journal for Pure and Applied Mathematics,* published in German. In 1841 the Berlin Mathematical Society began publishing *The Archives of Mathematics and Physics with Special Consideration of the Needs of Teachers* as a supplement to its *Proceedings*. Under the editorship of G. Grinert, this was the first journal to deal with mathematics teaching, as opposed to new research. In 1865 the London Mathematical Society was formed and began to publish its *Proceedings*.

The situation of geometry in English universities and preparatory schools was a special case. An English edition of Euclid's *Elements* had been published in 1758 by Robert Simson, a professor of mathematics at the University of Glasgow, and was to provide the basis for geometry instruction until late in the nineteenth Century, despite the fact that the inadequacies of many of Euclid's demonstrations were well known to mathematicians. In 1870 the Association for the Improvement of Geometrical Thinking was formed in England. In 1875 the Association published a *Syllabus of Plane Geometry,* which was based on Euclid's first six books but which was intended to remedy many minor deficiencies. The *Syllabus* gained the approval of the British Association for the Advancement of Science in 1877. In 1897 the Association changed its name to the "The Mathematical Association."

The development of university education in the United States proceeded in a very uneven manner. None of the colleges founded in the colonial period required extensive preparation or study of mathematics. A text, "A New and Complete System of Arithmetic Composed for the Use of Citizens of the United States," published in 1872 was adopted by Harvard, Yale, and Dartmouth colleges. It provided an overview of commercial mathematics, including currency conversions, weights and measures, the calendar, and the calculation of interest. Geometry was not required until the Civil War (1861-65). The first technical school to be established was the United States Military Academy at West Point, New York, in 1802. By the end of the century, the establishment of state universities and technical schools where the new engineering disciplines could be studied provided a basis for a growing American mathematical community.

Perhaps the most important influence on the teaching of mathematics in the modern elementary school was the work of Swiss educational reformer Johann Heinrich Pestalozzi (1746-1827), a disciple of the philosopher Jean Jacques Rousseau. Pestalozzi wrote that instruction should proceed from the familiar to the new, from the concrete to the abstract, and that the student of mathematics should retrace the path to mathematical or scientific knowledge that the original investigators trod in the first place. Pestalozzi was allowed to test his theories when the revolutionary French military forces gained control of the Swiss government for a time. Pestalozzi's influence was transplanted to the United states when William Maclure (1763-1840) published his *First Lessons in Arithmetic on the Plan of Pestalozzi, with Some Improvements* in Boston in 1921. In use for nearly a century, it has been described as the most popular arithmetic text ever.

Impact

It is perhaps ironic that the development of mathematics, perhaps the most austere and purely academic discipline of all, has been influenced in its development so strongly by political, military, and social factors. Throughout the twentieth century mathematics and mathematics education have been understood as components of national and economic power. Mathematicians played a prominent part in the Second World War in the development of atomic weapons and of electronic computers and computer languages. When the former Soviet Union appeared to have gained a temporary advantage in a "space race" with the United States by launching Sputnik, the first artificial Earth satellite, part of the American response was to develop new curricula for science and mathematics

instruction. One component of the new curriculum was the "new math," an emphasis on abstract mathematical principles in place of rote learning of mathematical facts. The "new math" was at best a partial success. Ignoring the tradition of Pestalozzi and his followers, it expected students to deal with abstract concepts without having had the opportunity to become familiar with the facts and techniques of ordinary addition and multiplication at the concrete and operational level.

In hindsight, Pestalozzi appears to have been part of an educational tradition that includes Maria Montessori, John Dewey, and Jean Piaget. The research of these educators and psychologists has focused on the mental development of children and their growth through gradual progression through stages of concrete thinking, operationalism, and later abstract thinking, with the abstract conceptual stage considered the highest or most mature stage of development. This position is not without controversy. Any alternative school of thought holds that individuals differ in "learning style," and that individuals not comfortable with abstract relationships may function as well or better in all important human activities than "abstract thinkers" do. This group includes teachers who consider themselves "ethnomathematicians" and feel that traditional mathematics instruction depends too heavily on the assumptions of white European culture, and that more concrete ways of thinking also provide a valid approach to mathematical truth.

A number of important associations of mathematicians and mathematics teachers are now involved in decisions affecting mathematics education in the United States. The American Mathematical Society was originally founded as the New York Mathematical Society in 1888 and, adopting its present name in 1894, is devoted primarily to original research in mathematics. The Mathematical Association of America was organized 1n 1914 for individuals primarily interested in the teaching of mathematics on the college level. Concern with the teaching of mathematics at the primary and secondary levels would rest with the National Education Association , originally organized as the National Teacher's Association in 1875. This organization appointed a number of committees over the period 1892-1912 to make recommendations about the content of secondary school mathematics and the mathematics to be required for college admission. In 1920 the

National Council of Teachers of Mathematics was formed by the National Education Association in order to bridge the gap between the more general concerns of the National Education Association and the Mathematical Association of America.

It is likely that debate about the proper way or ways to teach mathematics will continue for some time. In recent years the relatively poor showing of American students compared to students from other nations has served to renew debate on the best ways to tech mathematics in a culturally diverse society. With the number of employment opportunities for the mathematically literate increasing rapidly, means will need to be found, either to involve more professional mathematicians in the educational process or to provide the future mathematics teacher with a better understanding of mathematics and the methods of teaching it to contemporary students.

DONALD R. FRANCESCHETTI

Further Reading

Bell, Eric Temple. *The Development of Mathematics.* New York: McGraw-Hill, 1945.

Bidwell, James K. and Robert G. Clason. *Readings in the History of Mathematics Education.* Washington, DC: National Council of Teachers of Mathematics, 1970.

Boyer, Carl B. *A History of Mathematics.* New York: Wiley, 1968.

Cajori, Florian. *A History of Elementary Mathematics.* New York: Macmillan, 1930.

Kline, Morris. *Mathematical Thought from Ancient to Modern Times.* New York: Oxford University Press, 1972.

National Council of Teachers of Mathematics. *A History of Mathematics Education in the United States and Canada.* Washington, DC: National Council of Teachers of Mathematics, 1970.

The Return of Rigor to Mathematics

Overview

Rigor was a characteristic of mathematics going back to Greek times. For much of the Renaissance and the Enlightenment mathematics in general and calculus in particular were more about problem solving than about proving with logical exactness the correctness of theorems. The nineteenth century saw the return of logical rigor to mathematics, accompanied by a more thorough investigation of the foundations of mathematics than had been attempted previously, thanks largely to progress in mathematical logic.

Background

One of the great works of Greek mathematics to come down to the world of scholarship after the Middle Ages was the long-awaited translation of the *Elements* of Euclid of Alexandria (fl. c. 300 B.C.), which arranged the results of geometry in a sequence of definitions, axioms, postulates, and theorems, where each of the theorems was proved using the earlier theorems. Thanks to the arrangements of the results in the *Elements,* students could have complete confidence in geometrical theorems, although they had to worry about how to apply them to the world of concrete objects. Other disciplines sought to emulate the form of Euclid's work in an effort to cap-

ture the same aspect of certainty, but the attempt was never carried out with equal completeness.

Mathematicians of the Renaissance, however, found themselves more occupied with solving problems than with trying to prove the results. As a result, the confidence in the methods being used came from giving correct results rather than from basing the methods on theorems proved earlier. Perhaps the culmination of this approach was the invention of the calculus in the seventeenth century by Sir Isaac Newton (1642-1727) and Gottfried Wilhelm Leibniz (1646-1716). The calculus, which was devoted to the study of change and the finding of volumes, proved to be an invaluable tool for settling all sorts of questions that the geometry of Euclid was unable to solve. What was disturbing was that neither Newton nor Leibniz nor any of their followers was able to come up with an explanation for why the methods of the calculus worked. They referred to "infinitely small" quantities without explaining what they were. The attempts to lay a foundation for the calculus always involved an appeal to intuition, whether physical or mathematical, rather than a sequence of persuasive arguments.

In the next century Leonhard Euler (1707-1783) extended many of the results that Newton and Leibniz had obtained, but he was no more successful in providing a convincing explanation for why the impressive edifice of calculus worked. One of Euler's chief subjects of research was the use of series to represent functions like the trigonometric and exponential. Others who tried to follow Euler's work found themselves writing down meaningless statements in the absence of any basic principles to serve as a guide for the study of such series. Some of Euler's work had looked puzzling, but it turned out to be useful nonetheless. The mathematicians who tried to continue his work could not figure out what distinguished sense from nonsense in the foundations of the calculus.

By the nineteenth century there was a widespread conviction that mathematics in general needed more secure foundations. Those responsible for writing textbooks for students of the calculus felt an obligation to make the material convincing as well as comprehensible. It was hard to require students to avoid making nonsensical statements when mathematicians doing research were unable to avoid the same trap. The physics community was not likely to have confidence in the statements coming from mathematics if the statements looked irrelevant to the physical objects being studied.

Impact

One of the first breakthroughs in restoring the level of rigor that had long been lost to mathematics was a paper written by Carl Friedrich Gauss (1777-1855), perhaps the greatest mathematician of the century. Gauss's paper dealt with the hypergeometric series, a particular sort of representation of a function, but what made it so important was his great attention to the convergence of the series he discussed. A series is convergent if it can be summed to a finite value, while a divergent series goes off to infinity. All sorts of problems arose from the appearance of infinity in mathematics, whether the infinitely large or the infinitely small Gauss made sure that the infinitely large would not be a problem for the sort of series that he was investigating.

The individual who did the most to put mathematics back on a rigorous footing was Augustin-Louis Cauchy (1789-1857), a writer who contributed an immense amount to the teaching of mathematics as well as to mathematical research. Cauchy was the first to take the basic objects of the calculus—the derivative (a way of expressing the rate of change of a function) and the integral (an expression for the area

under the graph of a function)—and present them in a way that did not require the sort of appeal to the infinite that had been present since the time of Newton. What Cauchy did was to make the idea of a limit central to the definitions of derivative and integral and to base limits on the way real numbers behaved. Although he did not have an explicit picture of the basis for the real numbers, this was already a great improvement on the kind of slipshod argument prevalent previously.

On the strength of the notion of limit, Cauchy was able to make precise what it meant for a function to be continuous. From an intuitive point of view, a function is continuous if its graph can be drawn without lifting one's pencil from the paper. This is not helpful in trying to characterize which functions are continuous, and the class of continuous functions was of central importance in applying the calculus. Many of the fundamental theorems of calculus only apply to continuous functions, so Cauchy's definitions proved to be essential in identifying them. In the same way, Cauchy was able to make the idea of a convergent sequence more explicit than it had been in the work of earlier writers. The work of Newton and Leibniz was justified by Cauchy's efforts, as he had finally succeeded in giving a basis for their use of calculus.

After Cauchy, work on making mathematics even more rigorous proceeded in two directions. One of these was led by Karl Theodor Wilhelm Weierstrass (1815-1897), who recognized that there were important functions that did not behave as well as some of those Cauchy studied. It had been typical to think of a function as represented by a smooth curve, but physicists as well as mathematicians recognized the benefits of studying curves that had jumps or sharp corners. Weierstrass took Cauchy's definitions and made them even more straightforward to apply, removing appeal to anything beyond arithmetic. As an influential teacher, Weierstrass was able to bring this arithmetization of calculus to a generation of students and created the approach to the objects of calculus like limits, derivatives, and integrals used today.

The other direction for researchers was to understand the real numbers themselves better. One of the leaders of this movement was Richard Dedekind (1831-1916), author of the book *What Are and What Should Numbers Be?* Dedekind was aware that even Cauchy's careful study of the foundations of calculus depended on the underlying numbers, which had received less than rigorous attention even from the Greeks. The Greek mathematicians had put geometrical objects at the center of mathematics and used lengths to represent numbers. Dedekind developed a sophisticated account of ordinary real numbers by starting with fractions and regarding the irrational (nonfractional) numbers as the dividing lines between sets of fractions. By use of this method of cuts, Dedekind was able to prove results about the numbers that had been only presented by intuition before.

The whole theory of sets was given a new foundation by Georg Cantor (1845-1918), the mathematician who did the most to make the infinite respectable again after all the years of problems it had created. Cantor presented a theory of infinite numbers with an explanation of when two infinite sets had the same number of members. This also allowed him to talk about two infinite sets being different in size, which had been perplexing to previous students of the infinite. Cantor's work was not uniformly popular with those who felt that the infinite had no place in mathematics, but it has proved central in going beyond the limits of the functions known and studied before the infinite was tamed.

The work of Dedekind and Cantor formed part of the new discipline that was to be given the name of mathematical logic. It received that name for two reasons: it was the study of logic as used in laying the foundations of mathematics, and it involved mathematics in studying logic itself. Well into the nineteenth century the study of logic was restricted to the work of Aristotle (384-322 *B.C.*), or at best the commentaries it continued to received. The philosophical sophistication of Aristotelian logic was substantial, but it was not devised particularly with mathematics in mind. English mathematician George Boole (1815-1864) undertook a mathematical analysis of logic that went beyond the limits of Aristotelian logic and drew on some of the ideas current in the study of algebra. These new techniques gave rise to the capacity to study the foundations of mathematics and proofs themselves in detail. By the end of the century there were several ambitious attempts to provide a rigorous foundation for all of mathematics, although subsequent work was to point out why they were doomed to fall short.

The revolution in rigor that took place in mathematics in the nineteenth century was a response to the problems that had been encountered in teaching and applying calculus for more than a hundred years. It was essential to broaden the range of objects to which calculus could be applied, but that was scarcely possible if the calculus itself were so little understood. The movement in the direction of greater rigor enabled mathematics to have more to say about physics in the twentieth century than it ever had before. In addition, the detailed study of mathematical objects and proofs led to the development of mathematical logic, an essential tool on the road to computer science. Even those who had only wanted a return to Greek rigor as practiced by Euclid were the beneficiaries of the techniques that went well beyond the elementary.

THOMAS DRUCKER

Further Reading

Bottazzini, U. *The Higher Calculus: A History of Real and Complex Analysis from Euler to Weierstrass.* New York: Springer-Verlag, 1986.

Grabiner, Judith V. *The Origins of Cauchy's Rigorous Calculus.* Cambridge, MA: MIT Press, 1981.

Grattan-Guinness, Ivor. *The Development of the Foundations of Mathematical Analysis from Euler to Riemann.* Cambridge, MA: MIT Press, 1970.

Hawkins, Thomas W., Jr. *Lebesgue's Theory of Integration: Its Origin and Development.* Madison, WI: University of Wisconsin Press, 1970.

Kitcher, Philip. *The Nature of Mathematical Knowledge.* Oxford: Oxford University Press, 1983.

Tiles, Mary. *Mathematics and the Image of Reason.* London: Routledge, 1991.

The Specialization of Mathematics and the Rise of Formalism

Overview

Mathematics is the study of the relationships among, and operations performed on, both tangible and abstract quantities. In its ancient origins, mathematics was concerned with magnitudes, geometries, and other practical and measurable phenomena. During the nineteenth century, mathematics, and an increasing number of mathematicians, became enticed with relationships based on pure reason and upon the abstract ideas and deductions properly drawn from those relationships. In addition to advancing mathematical methods related to applications useful to science, engineering, or economics (hence the term "applied mathematics"), the rise of the formalization of symbolic logic and abstract reasoning during the nineteenth century allowed mathematicians to develop the definitions, complex relations, and theorems of pure mathematics. Within both pure and applied mathematics, nineteenth-century mathematicians took on increasingly specialized roles corresponding to the rapid compartmentalization and specialization of mathematics in general.

Background

Well into the nineteenth century mathematicians continued to scramble to invent and refine analytical methods that would be of use in solving the seemingly endless list of questions and problems being raised by the emerging European industrial revolution's demand for increased experimentation in physics, astronomy, and engineering. By the middle of the century, however, attention began to shift toward the operations of mathematical logic, and, as a consequence, there was an increased emphasis on the relationships and rules for evaluating axioms and postulates.

Building upon the calculus of the seventeenth-century giants Sir Isaac Newton (1643-1727) and German mathematician Gottfried Wilhelm von Leibniz (1646-1716), nineteenth-century mathematicians extended the accuracy and precision of mathematical calculations. In particular, nineteenth-century mathematicians built applied theory upon foundations laid by Swiss mathematician Leonhard Euler's (1707-1783) work in mechanics, differential and integral calculus, geometry, and algebra.

The large strides being taken in applied mathematics, however, left much theoretical and logical ground untouched. After centuries of emphasis on practical applications, the nineteenth century also proved ripe for the development of pure mathematics.

Impact

The ultimate end of the increasing specialization of mathematics was the schism between pure and applied mathematics. A simplest view of this division rests on the definition of pure

mathematics as composed mathematics advanced for theoretical interests while applied mathematics develops tools and techniques used to solve problems in science, engineering, and economics. Such a simplistic definition, however, denies the common history and crossovers between the two divisions.

Starting in the nineteenth century there evolved a profound difference in methodology of pure and applied mathematics. Within pure mathematics, deductions are valid if properly derived from a given hypotheses. In the reasoning of applied mathematics, deductions are grounded in experimental evidence, and the goal is to correctly identify the hypotheses from which the deductions can be properly deduced. In this sense, pure mathematics is a matter of correctly following the laws of reasoning—and applied mathematics is a matter of identifying the hypothesis (e.g., scientific law) and the correspondence of results properly deduced from those hypotheses. Another view of the division between pure and applied mathematics allows pure mathematicians the ability to substitute formal calculation with conceptual analysis of a problem.

During the nineteenth century pure mathematics and its abstract calculations became a method to explore the methodology of deduction itself. In contrast, applied mathematics became the methodology used to understand or predict results, particularly the results of scientific and technological experimentation, produced by objective procedures in terms familiar to already established and agreed upon deductive systems.

At the beginning of the nineteenth century the work of German mathematician and physicist Johann Carl Friedrich Gauss (1777-1855) embraced and embodied both pure and applied mathematical concepts. Gauss's practical mathematical applications included advances in the study of the shape of the Earth (geodesy), planetary orbits, and statistical methodology (i.e., least-square methodology). In addition to this practical work, Gauss established himself as true polymath by also advancing the cause of pure mathematics through seminal works in number theory, representations of complex numbers, quadratic reciprocity, and proof of the fundamental theorem of algebra.

In 1847 English mathematician George Boole (1815-1864) published his *Mathematical Analysis of Logic* and followed it in 1854 with another important publication, *The Laws of Thought*, in which he asserted that the development of symbolic logic and pure mathematical reasoning was being retarded by an overdependence on applied mathematics. Boole advocated the expression of logical propositions in symbols to represent logical theorems that could be proven just like any other mathematical theorem. Boole championed pure mathematical reasoning by attempting to dissociate and abstract mathematical operations without regard to their underlying applications.

Boole's publication was akin to drawing a line in the sand for mathematicians and demarked a trend away from Gaussian-like mathematical universalism toward an increased specialization within mathematics community. In particular there became an increasing divergence between pure and applied mathematics. As a consequence of the popularity of formalism, there also resulted an increasing number of mathematicians dedicated to pure mathematics of minimal consequence to applications to science and technology.

This divergence was not always viewed with favor. With regard to mathematical logic, although initially created as a sub-discipline of mathematics, the field was widely ignored or held in disdain by many mathematicians. By the end of the century, however, symbolic logic, progressed from academic obscurity to become popular entertainment. Lewis Carrol's books on logic, *The Game of Logic* (1887) and *Symbolic Logic* (1896), became popular topics of conversation and entertainment both for scholars and laymen in Victorian England. Despite initial resistance, mathematical logic made strong inroads into philosophy after William Stanley Jevons heralded its use in his widely read 1874 publication, *Principles of Science*. Jevon argued that symbolic logic was of importance to both philosophy and mathematics.

The increasing volume of work relating to number theory also lead to hierarchies within the emerging specialties. Fueled by Gauss's work on the theory of numbers, algebraic theories of numbers took on a preeminent position in pure mathematics.

Some initially pure mathematical theory was met with outright derision and scorn. Among the most controversial of advances in mid-nineteenth-century mathematics was the publication of non-Euclidean geometries championed by German mathematician Georg Friedrich Bernhard Riemann (1826-1866). Riemann asserted that Euclidian geometry was but a subset of all possible geometries. Riemann's expanded concepts of geometry treated the properties of curved space and seemed abstract and useless to nineteenth-century Newtonian physics. In addition, many mathematicians thought Riemann's conceptualizations bizarre. Regardless, Riemann's theories proved of enormous consequence and value to the expansion of concepts of gravity and electromagnetism—and of fundamental importance to the twentieth-century theoretical work of Albert Einstein (1879-1955) regarding a formulation of theories describing the interactions of light and gravity. Topology as a specialization of mathematics was born in the advances of nineteenth-century geometry.

Non-Euclidean geometry was so controversial that it may have prevented Gauss from advancing the concept. Eventually found among Gauss's unpublished works were the essential elements ultimately formulated by Riemann.

Later in the nineteenth century when German mathematician Georg Ferdinand Ludwig Philipp Cantor (1845-1918) proposed his transfinite set theory, many thought it the height of abstraction. Advances in twentieth-century physics, however, have also found use of Cantor's theories.

Not all developments were polarized into the pure and applied camps. Early in the nineteenth century, French mathematician Jean Baptiste Joseph Fourier's (1768-1830) work with mathematical analysis allowed him to establish what is known to modern mathematicians as the Fourier series, which is central to Fourier analysis and an important tool for both pure and applied mathematicians. Late in the nineteenth century, group theory made possible the unification of many aspects of geometric and algebraic analysis.

Although there was an increasing trend toward specialization throughout the nineteenth century, near the end of the century French mathematician Jules Henri Poincaré (1854-1912) embodied Gauss's universalist spirit. Poincaré's work touched on almost all fields of mathematics. His insights provided significant advances in applied mathematics, physics, analysis, functions, differential equations, probability theory, topology, and the philosophical foundations of mathematics. Poincaré's studies of the chaotic behavior of systems subsequently provided the theoretical base for the continually evolving chaos theory of late twentieth-century mathematics.

Mathematical rigor, defined in the early part of the century by the applications of calculus, was broadened near the end of the century by German mathematician Karl Theodor Wilhelm Weierstrass (1815-1897) into the types of analysis familiar to modern mathematicians.

The advancement of elliptic functions principally through the work of Norwegian mathematician Neils Henrich Abel (1802-1829) and Prussian mathematician Karl Gustav Jacob Jacobi (1804-1851) provided mathematical precision in calculations required for discoveries in astronomy, physics, algebraic geometry, and topology. In addition to their use in applied mathematics, however, the development of the theory of elliptic functions also spurred the study of functions of complex variables and provided a bridge between the widening chasm between pure and applied mathematics.

Although there were subtle divisions of mathematics at the beginning of the nineteenth century, by the end of the century there were full and formal divisions of pure and applied mathematics. University appointments and coursework reflected these divisions, and an increasing number of professorial positions were designated for pure and applied mathematicians.

K. LEE LERNER

Further Reading

Boyer, C. B. *A History of Mathematics*. Princeton, NJ: Princeton University Press, 1985.

Brooke, C., ed. *A History of the University of Cambridge*. Cambridge: University of Cambridge Press, 1988.

Carroll, L. *The Game of Logic*. London: Macmillan, 1887. Reprinted in 1958.

Dauben, J. W., ed. *The History of Mathematics from Antiquity to the Present*. Garland Press, 1985.

Kline, M. *Mathematical Thought from Ancient to Modern Times.*. Oxford: Oxford University Press, 1972.

Codification and Employment of the Principle of Mathematical Induction

Overview

Over the course of the nineteenth century, mathematicians and philosophers were forced to change their view of the subject matter of mathematics. At the start of the century many scholars considered it to be a collection of self-evident truths about nature. By 1900 many were convinced that mathematics was principally a creation of the human imagination and as such subject to error. This shift in view resulted from the discovery of new geometries and number systems calling into question whether the traditional mathematical description of nature using Euclidean geometry and real numbers was necessarily correct and free of potential contradictions. The principle of induction was formalized during this period in an attempt to put arithmetic on a firm logical foundation.

Background

The principle of induction is simply stated: If P(n) denotes a statement in which n can be any of the integers and P(n) is true when n=1 and it can further be proven that P(n+1) is true whenever P(n) is true, then P(n) is true for all the positive integers n. As a simple example we let P(n) be the statement: "The sum of the first n integers is n(n+1)/2." This statement is clearly true when n=1. With simple algebra we can show that adding n+1 to n(n+1)/2 yields (n+1)((n+1)+1)/2, so the statement must be true for all integers n, no matter how large. The principle of induction has been used implicitly many times throughout the history of mathematics. Euclid (c. 300-260 B.C.) assumed it in his famous proof that there is no largest prime number. It was used explicitly by the mathematician Mauroly in an arithmetic text published in 1575, and by philosopher and mathematician Blaise Pascal (1623-1662) in 1665 in the derivation of what we now call Pascal's triangle. It was not given a name, however, until 1838 when an article entitled "Induction (Mathematics)" by Augustus De Morgan (1806-1871) appeared in the *Penny Cyclopedia*. By then it was being used, or at least assumed, by the growing number of mathematicians and scientists using infinite series to solve problems in applied mathematics. By the end of the century it was nonetheless to become a point of considerable controversy in the attempt to put mathematics on a fully rigorous basis.

By the beginning of the nineteenth century mathematical proof had come to be regarded as a model of clear reasoning. The importance of proof varied, however, from one area of mathematics to another. In geometry, the enduring influence of Euclid's classic text, with its emphasis on proofs, had set the tone for future work. In arithmetic and the theory of numbers, the properties of the integers were considered sufficiently well known that there was less attention to deriving results from axioms and postulates. The situation changed, however, as mathematicians discovered that the famous parallel postulate of Euclid could be replaced by other postulates and still leave a consistent geometry. One could not then be certain that the postulates of Euclid were self-evident truths about space, or even, as the philosopher Kant had maintained, inherent in the way in which the human mind thought about space.

Considerable attention was then devoted to the foundations of arithmetic and the number system. One major thrust was the development of the theory of sets by the German mathematician Georg Cantor (1845-1918), beginning in 1873. At first set theory appeared to offer a true means of developing all the important ideas of mathematics within a common framework. The number three, for instance, could be thought of as the set with all the possible sets of three elements as members. Set theory also allowed an approach to discussing infinite quantities. A set was said to contain an infinite number of elements if its elements could be set into a one-to-one correspondence with the elements. Thus, since the set of integers {n} could be put into a one-to one correspondence with the set of even integers {2n} it is an infinite set. Unfortunately, paradoxes soon became apparent. One could conceive of a set of all sets. This set would be a member of itself, that is, of the set of all sets. One could as easily consider a set of all sets not members of themselves. If this set were a member of itself it would not meet the requirement for membership in itself. If it were a member of itself it would clearly not qualify for membership in itself. Serious paradoxes were inevitable if sets were allowed to contain other sets as members.

An alternative approach to the foundations of arithmetic was provided in 1889 by Giuseppe Peano (1858-1932), a professor of mathematics at the University of Turin in Italy, who developed a set of five axioms to describe the set of natural numbers, {1, 2, 3, ...}, based on the idea of succession. Peano's axioms can be stated as:

1. 1 is a natural number.

2. The successor of any natural number is a natural number.

3. If two natural numbers are equal, their successors are equal.

4. 1 is not the successor of any natural number.

5. If S is any set, and 1 is an element of S, and if the fact that any natural number belongs to S implies that its successor belongs to S, then S contains all the natural numbers.

Peano proved that each of his axioms was independent of the others. The fifth axiom is a rigorous statement of the principle of mathematical induction

Impact

Increased concern with the foundations of mathematics was perhaps an inevitable consequence of the growth of professional mathematics that has continued steadily since the seventeenth century. Initially mathematicians communicated through correspondence and the occasional publication of books. In the eighteenth century there were a few paid positions in national academies. By the nineteenth century the need for advanced technical education was appreciated by national leaders and a growing industrialist class in France, Germany, the Netherlands, the United Kingdom, and, somewhat later, the United States. Technical institutes and universities were established in these countries, creating a need for full-time instructors in mathematics. Newly established journals were devoted to pure and applied mathematics as distinct from the physical sciences. Mathematicians developed a separate identity as mathematicians, and there was both the manpower and a forum for the close examination of mathematical ideas.

Traditionally, logic has been divided into two parts: deductive logic, which allows us to deduce the consequences of a given set of statements assumed to be true, and inductive logic, which permits us to infer general truths from specific examples. Euclidean geometry is often taken to be an excellent example of deductive logic, while mathematical induction is a form of inductive logic, since it allows conclusions to be drawn about all the integers based on a relation between successive integers.

A desire to avoid the paradoxes to which set theory might lead led to the emergence of the logistic school of mathematics, the group with which Peano himself identified. The goal of this group was to establish the principles of mathematics on the basis of deductive logic alone, without any dependence on the properties of numbers. The most impressive product of this school is the three-volume *Principia Mathematica* first published over the years 1910 to 1913 by the English mathematicians and philosophers Bertrand Russell (1872-1970) and Alfred North Whitehead (1861-1947). Over a thousand pages long with very sparse text interrupting many thousands of lines of symbolic manipulation, this work progresses so carefully that the number "one" is not defined until over a hundred pages of preliminary results. Russell and Whitehead attempted to avoid the difficulties associated with sets containing other sets as members by developing an elaborate "theory of types," in effect eliminating the possibility that sets could be considered to belong or not belong to themselves. They also resisted assuming the principle of induction, but in the end were forced to adopt an "axiom of reducibility," which most mathematicians found even less acceptable.

The possibility of the logistic approach providing a firm basis for all of mathematics was ruled out permanently in a remarkable paper published in 1931 by the Austrian mathematician Kurt Gödel (1906-1978). Gödel's proof, as it has come to be known, showed that any formal system which could lead to the usual rules for the multiplication of integers would necessarily allow the existence of meaningful statements which could neither be proven nor disproven within the system. The proof was based on a clever coding scheme, which made it possible to assign a unique integer to the assertion that a given string of symbols constituted a proof of a particular assertion. Gödel's scheme, which made use of the availability of an infinite number of distinct prime numbers (as established by Euclid using implicit induction), made it possible to then construct statements which asserted their own non-provability, in any such logical system.

The full impact of Gödel's proof is still being examined by mathematicians and philosophers. Many would agree that it has eliminated any possibility of mathematics becoming a "closed book," with all the important results to be established once and for all. Instead, many mathematicians regard it to show that mathematics is in its essence an arena for human creativity, not unlike music or poetry, with an endless capacity for invention and discovery.

DONALD R. FRANCESCHETTI

Further Reading

Boyer, Carl B. *A History of Mathematics*. New York: Wiley, 1968.

Kline, Morris. *Mathematical Thought from Ancient to Modern Times*. New York: Oxford University Press, 1972.

Kline, Morris. *Mathematics, The Loss of Certainty*. New York: Oxford University Press, 1980.

Elliptic Functions Lay the Foundations for Modern Physics

Overview

Elliptic functions are considered a special class of analytic mathematical functions that are used to analyze and solve problems in physics, astronomy, chemistry, and engineering. More specifically, elliptic functions (known to modern mathematicians as elliptic integrals) are a large class of integrals related to, and containing among them, the expression for the arc of an ellipse. The advancement of elliptic functions during the nineteenth century provided mathematical precision in calculations required for discoveries in astronomy, physics, algebraic geometry, and topology. In addition to their use in applied mathematics, the development of the theory of elliptic functions also spurred the study of functions of complex variables and provided a bridge between pure and applied mathematics.

Background

Although not called elliptical functions until the nineteenth century, modern study of elliptical functions began in the middle of the seventeenth century. Several mathematicians published works examining the arc length of an elliptical path and Sir Isaac Newton (1642-1727) published works regarding the mathematics of elliptical orbits. As a particular consequence of Newton's work, the mathematical development of the elliptic functions and integrals emerged from centuries of struggle to accurately (i.e., mathematically) explain the mechanics of motion, including the motions of the Sun and planets.

In addition to descriptions of the elliptical paths and orbits of moving bodies, in 1679 Swiss mathematician Johann Bernoulli (1667-1748), while attempting to mathematically describe a spiral path, found that the true description for the deformation of an compressed elastic rod was elliptical. Bernoulli determined that the resulting curve describing the deformation was, by definition, what would later be known as an elliptical function.

During the later years of the eighteenth century, French mathematician and astronomer Pierre Simon Marquis de Laplace (1749-1827), the subsequent discoverer of the Laplace theorem that bears his name, and others studied unexplained variations in the orbits of Jupiter and Saturn. These enigmatic changes (a contraction in Jupiter's orbit and an expansion in Saturn's orbit) seemed an important discrepancy to the Newtonian-based cosmological models of eighteenth- and nineteenth-century science and philosophy that assumed a static and unchanging universe. Although Laplace eventually succeeded in accounting for the orbital variations with a periodicity predicted by Newtonian gravity (and hence compatible with existing cosmological models), his work stirred a greater scrutiny of celestial mechanics and set the stage for the advancement of elliptical functions as mechanisms to calculate and predict celestial movements.

Laplace also published works that dealt with the precise motion of the Moon around Earth, a problem that had frustrated earlier generations of mathematicians. Laplace was the first to account for the influence of the Sun on the lunar orbit (i.e., the influence of the Sun's gravity on the two body Earth-Moon system) and of the Sun and planets on ocean tides . Laplace's work, made popular with his 1796 publication of *Exposition du Système du Monde*, stirred interest in developing refined mathematical tools that would enable astronomers to more easily and fully study perturbations of celestial movement.

This heightened interest set the stage for the ideas and terminology embodied by elliptical functions that were to evolve from Swiss mathematician Leonhard Euler's (1707-1783) elegant explanations of mechanics using differential equations, German Johann Carl Friedrich Gauss's (1777-1855) work that anticipated many properties of elliptical functions, and of the Italian-born Joseph-Louis Lagrange's (1736-1813) theories of functions related to celestial mechanics.

Impact

In the nineteenth century early understanding and application of elliptic functions was principally nourished by the work of the great Norwegian mathematician Neils Henrich Abel (1802-1829) and Prussian mathematician Karl Gustav Jacob Jacobi (1804-1851). Although Abel and Jacobi were both pioneers of modern mathematics, they came from vastly different circumstances, drawn together in history only through their competitive quest to understand and describe the properties of elliptical functions.

Abel's brief but brilliant career was on constant battle with destitution. When Abel was 18, the death of his father put Abel's educational prospects in doubt. With help from professors who recognized his mathematical talent, Abel entered the University of Oslo in 1821. By 1923, despite his modest circumstances, Abel paid for the publication of solutions to mathematical problems that had previously vexed mathematicians for hundreds of years. These solutions also promoted the work of others, including Jacobi's development of integral equations. Despite his brilliance, Abel wandered through European academia, unable to find a permanent position. He lived hand-to-mouth by tutoring and substitute teaching. Eventually, while working in Berlin, Abel began to dedicate his energies to the study of elliptical functions.

After reading Jacobi's publications regarding transformations of elliptic integrals, Abel realized that Jacobi's work was based, in part, on Abel's own unpublished insights, Able scrambled to compete with Jacobi and managed to swiftly publish several papers on elliptical functions, including a 1827 work titled *Recherches sur les fonctions elliptiques*. Abel's discovery that elliptical functions were actually the inverse of elliptical integrals brought him world-wide acclaim and fame. Before he could reap the rewards of his success, however, and just as he was about to be appointed to a professorship, Abel succumbed to tuberculosis contracted during his travels.

In contrast, Jacobi (1804-1851) enjoyed a stable academic career. First at the University of Königsberg and then, 15 years after the tragic death of Abel, in Berlin. Jacobi worked on the description and application of elliptical functions throughout his career and, in 1829, set forth a well-regarded description of the functions in his *Fundamenta Nova Theoria Functionum Ellipticarum*.

In 1830, Paris Academy awarded the Academy's Grand Prix to Abel (posthumously) and Jacobi for their outstanding work.

French mathematician Adrien Legendre (1752-1833) also made substantial contributions to the study of elliptical integrals. In fact, the term elliptic function arguably first appeared in Lengendre's 1811 publication *Exercises du Calcul Intégral.*

Although elliptic functions were simple in form, defined as r(x, p(x))dx where r(x,y) is a rational function in two variables and p(x) is a 3rd or 4th degree polynomial without repeated roots, the development of elliptical functions had profound consequences on the analysis of the mechanics of motion. One problem, for example, that had confounded engineers and scientists was the ability to accurately and quickly determine the perimeter of an elliptical oval or the swing of a pendulum. Because of non-linear elements in these problems, the problems could not be easily solved using standard elementary functions, and application of elliptic functions as an analytical tool was required in order to make accurate descriptions and predictions of pendular movement.

Elliptical functions also allowed more accurate descriptions and predictions of the celestial mechanics of Keplerian orbital motions (i.e., gravitationally bound two-body systems or the bound motion of a one body system interacting with an attractive central force). The Sun and planets comprise such a two-body system that can be described by elliptical functions and, in accord with Kepler's laws of planetary motion, the application of elliptic functions provided very precise descriptions of planetary motions that allowed the subsequent calculation of perturbations in those orbits caused by other bodies in the solar system.

The great power of elliptical functions was their ability to elegantly and accurately describe rotational dynamics, in particular, the highly complicated motions of celestial bodies that consumed nineteenth-century astronomers. Accordingly, the theory of perturbations and the understanding of planetary orbits owed much to the development of elliptic functions because they allowed astronomers to plot graphs of the orbits and distances traveled by a planet or comet along an orbital path as a function of time. More than a century after their articulation, elliptic integrals are still used to calculate spacecraft trajectories—especially those sent out on interplanetary missions.

In the middle of the nineteenth century, irregularities in the orbit of Uranus prompted astronomers and mathematicians to seek the cause of such perturbations. Precise calculations of Uranus' orbit using elliptical integrals showed that the perturbations could be explained by the presence of an undiscovered planet. Although some of these calculations were completed as early as 1845 by the British astronomer John Couch Adams (1819-1892) of Cambridge University, credit for the discovery of the planet, eventually named Neptune, was given to the brilliant French mathematician and astronomer Urbain Jean Joseph Le Verrier (1811-1877). In addition, Le Verrier's use of elliptical functions to describe a discrepancy in the orbital motion of Mercury (e.g., the advance of the perihelion of Mercury) became an important stimulus to the subsequent formation and proof of Albert Einstein's general theory of relativity.

French mathematician Jules Henri Poincaré's (1854-1912) use and advancement of elliptical functions in the fields of celestial mechanics (three-body problems) and in the emerging theories of light and electromagnetic waves provided significant contributions to the further development of Scottish physicist James Clerk Maxwell's (1831-1879) profoundly important equations of the electromagnetic field. Maxwell's equations describing the propagation of electromagnetic waves were derived from both hyperbolic and elliptical components (i.e., the unified equations make an important synthesis of hyperbolic and elliptic functions). Maxwell's equations became the essential key to understanding the scope of the electromagnetic spectrum and laid the essential foundations for the formation of twentieth-century quantum and relativity theory.

Beside wide-ranging use in number theory and celestial mechanics, elliptical functions continue to be used in many engineering applications and solving many problems in electromagnetism and gravitation.

K. LEE LERNER

Further Reading

Books

Grattan-Guinness, I., ed. *Companion Encyclopedia of the History and Philosophy of the Mathematical Sciences.* New York: Routledge, 1994.

Stillwell, J. *Mathematics and History.* New York: Springer Verlag, 1989.

Periodicals
Stander, D. "Makers of Modern Mathematics: Carl Gustave Jacob Jacobi." *Bull. Inst. Math. Appl.* 24 (1988): 27-28.

Biographical Sketches

Niels Abel
1802-1829
Norwegian Mathematician

Niels Abel, in his tragically short life, made fundamental contributions to the study of mathematics. He may be best known for his work on elliptic functions and definite integrals, but he also proved the insolvability of quintic equations (equations in which a factor is raised to the fifth power) and a class of functions named for him, the Abelian functions. He also established new standards for mathematical rigor in his work and developed a general proof for Leonhard Euler's binomial theorem.

Abel was born the son of a Lutheran minister in a small town in Norway. He did not attend formal schooling until he was 13, and that school was inadequately staffed, including the Mathematics department. However, shortly after starting, the mathematics teacher was dismissed and replaced by a junior professor from a nearby university, Bernt Holmboe. Under Holmboe's tutelage and encouragement Abel's mathematical skills blossomed. Before long, Abel's skills overtook his teacher's.

Before entering university studies, Abel made his first major contribution to mathematics, trying to develop a general solution to the quintic equation, a problem that had vexed mathematicians for over 200 years. Thinking he had found a solution, he wrote a proof and sent it to the Danish mathematician Degen. Quickly finding a flaw in his work, Abel continued to work and eventually was able to prove that there was no general solution for problems of this sort. He sent this off to Degen, too (there were no Norwegian mathematicians at that time who were qualified to critique his work). Degen was impressed and suggested that Abel study elliptic integrals, another particularly difficult problem for mathematicians.

Abel's father died in 1821, just as Abel was beginning his studies. Abel supported himself and helped support his family by tutoring and with grants from the university and his professors. Nevertheless, he completed his graduation requirements in one year and continued to work on his research at the same time. In 1823 he published his next major work, which included the first solution to an integral equation. This

Niels Abel. *(The Library of Congress. Reproduced by permission.)*

was followed by an important work on the integration of functions.

Unfortunately, Abel's work was published in Norwegian, a language not read by most mathematicians, who worked in French or German. Because of this, the general mathematical community ignored his early papers and he was not offered the professorship his work had already earned him. In fact, papers he sent to Carl Gauss (1777-1855), Adrien-Marie Legendre (1752-1833), and Augustin Cauchy (1789-1857) were completely ignored.

In 1825 the Norwegian government gave Abel a grant to visit mathematicians elsewhere in Europe. Unfortunately, Degen, his first stop, had died and he managed to visit Paris when virtually all mathematicians were on vacation. He did manage to meet with August Crelle (1780-1855) who, in spite of language difficulties, recognized Abel's genius and later published some of his work.

Returning to Norway, Abel was dejected. He had received virtually no recognition for his work, had not been offered a professorship that he desperately needed, and had contracted tuberculosis during his trip. On top of that, upon his return he found that a Norwegian teaching position had been offered to someone else, leaving him without a job. As before, he survived on grants and generosity while continuing to work on his mathematics.

During the next few years, understanding that he was ill, Abel worked at a feverish pace. He finally began to receive some degree of recognition from mainstream European mathematicians, including Legendre and Karl Jacobi (1804-1851). He was finally offered a much-deserved professorship at the University of Berlin in 1829. Unfortunately, he died shortly before the offer was sent. After Abel's death, a colleague commented, "he has left mathematicians something to keep them busy for five hundred years."

P. ANDREW KARAM

Charles Babbage
1791-1871
English Mathematician and Inventor

The development of the modern computer owes a debt to Englishman Charles Babbage, who developed plans for an analytical engine (the forerunner of today's digital computer) as early as the mid-1830s.

Born the day after Christmas in 1791 in London, England, Babbage entered the University of Cambridge around 1809. In 1812, he helped to organize Analytic Society as a means of introducing new theories emerging from other European countries into the British field of mathematics. He became well known for his creative organizational ideas and, in 1816, was elected a fellow of the Royal Society of London. He also founded Royal Astronomical (1820) and Statistical (1834) societies.

Around the time he was organizing the Analytic Society, Babbage came up with a concept far ahead of his time: a machine that could calculate various mathematical functions with an accuracy of upto eight decimals. Later, in 1823, he was able to interest the British government in supporting his design of a machine accurate to 20 decimals. Naturally, this new machine was dependent on an entirely new set of mechanical engineering functions, which Babbage had to develop on his own.

At this time, he was also actively engaged in his scholastic career and served as Lucasian professor of mathematics at the University of Cambridge (1823-1839).

Charles Babbage. *(The Library of Congress. Reproduced by permission.)*

As he began serious work on his so-called analytical engine, Babbage began to turn his ideas into reality. The possibilities he saw were enormous: using punched cards as the basis for commands to the machine to perform any arithmetical operation: addition, subtraction, division, percentage, etc., plus a storage or memory capability where numbers, sequences, and other data could be stored and retrieved.

As we consider the remarkable work of Babbage, it becomes more and more amazing. Today, the word computer is not limited to a purely electronic device, but encompasses three generic types: analog, digital and hybrid. They can also differ widely in size—some as small as a wristwatch, while others take up large rooms.

However, in the mid-1800s, the idea of a machine replacing human calculators must have seemed as unreal as our present, taken-for-granted keyboards and monitors. Unfortunately, Babbage was never able to complete his analytical engine, and the concept was shelved and forgotten until 1937 when many of his unpublished notebooks were discovered. Finally, in 1991, British scientists got around to constructing a machine called Difference Engine No. 2 (accurate to 31 digits). It was built to Babbage's specifications.

In today's world, all industrialized nations depend heavily on computers to perform tasks that formerly required thousands of man-hours with a large margin of error. The intricate programs which control all forms of manufacturing, travel and shipping records, inventories, bank records, scientific research, and so on, are seldom used or understood by most of us. Instead, even people with the least possible knowledge of technical processing use computers in their homes, offices, schoolrooms, and on their laps in the most improbable places.

As for Babbage, he didn't stop with his ideas for mechanical calculators. He gave his country a technical jump on the rest of the world by assisting in the development of the modern postal system (still in use), and used his knowledge of sequential numbering to develop actuarial tables. In his spare time, he invented a type of speedometer and the cowcatcher used on locomotives well into the twentieth century. Babbage died in London on October 18, 1871.

GERALD F. HALL

Janos Bolyai
1802-1860
Hungarian Mathematician

Janos Bolyai was among the founders of non-Euclidean geometry. Non-Euclidean geometry concerns itself with internally consistent mathematical systems in which Euclid's parallel axiom does not apply. The parallel axiom states that only one line can be drawn parallel to a given line through a point not on the line. Non-Euclidean geometry later gained importance as the mathematical foundation of the general theory of relativity.

Bolyai was born in Kolozsvar, Hungary (now Cluj, Romania), on December 15, 1802. His father, Farkas, also called Wolfgang, was a mathematician and a lifelong friend of Carl Friedrich Gauss (1777-1855). By the time he was 13 years old, Janos Bolyai had been taught geometry and calculus by his father. He was also a gifted violinist, and later became an accomplished swordsman. Bolyai continued his education at the Royal Engineering College in Vienna from 1818 through 1822, and served as an officer in the engineering corps of the Austrian army.

Farkas Bolyai, inspired by his experience of tutoring his son, published his principal work, *Tentamen Juventutem Studiosum in Elementa Matheseos Purae Introducendi* ("An Attempt to Introduce Studious Youth to the Elements of Pure Mathematics") in 1832. This treatise on the foundations of mathematics included a rigorous presentation of geometry.

The elder Bolyai had long been almost obsessed with proving the parallel axiom. His son became absorbed in this quest as well, in spite of his father's warnings. "I entreat you, leave the science of parallels alone," wrote Farkas Bolyai in a letter to Janos. "I have traveled past all reefs of the infernal Dead Sea and have always come back with a broken mast and torn sail."

In 1820, Janos Bolyai convinced himself that a proof of the parallel axiom was impossible. This led him to try to construct a complete and consistent geometry that did not depend on it. In 1823, he drafted a presentation of such a system, and sent it to his father, exclaiming in his youthful exuberance: "I have created a new universe from nothing!" The 24-page work was published as an appendix to the *Tentamen*, entitled "Appendix Spatii Absolute Veram Exhibens," or "Appendix Explaining the Absolutely True Science of Space." This appendix was the only work of Janos Bolyai that was published during his lifetime, although he also applied himself to the field of complex variables.

When Farkas Bolyai sent the *Tentamen* to his friend Gauss, proudly pointing to the appendix written by his son, Gauss replied that he himself had had the same ideas decades earlier. In fact this assertion is confirmed in his private papers; he had even written of his thoughts to Farkas Bolyai. Gauss, fearing ridicule, had never published his revolutionary ideas on the subject, so he had no valid claim of precedence, nor had he systematized them to the extent that Janos Bolyai had.

Farkas Bolyai interpreted Gauss's remark that his son's ideas "coincide almost entirely with my meditations which occupied my mind for the last thirty or thirty-five years" as high praise. Yet Gauss's claim of precedence was a profound shock to Janos Bolyai, one he never truly overcame. In a further blow, he later discovered that the Russian mathematician Nikolai Lobachevsky (1792-1856) had also been working independently on non-Euclidean geometry, and had published a paper very similar to his three years earlier.

The appendix to the *Tentamen* gained little attention, and the unhappy Bolyai remained largely unknown during his lifetime. Later, however, George Bruce Halsted called his work "the most extraordinary two dozen pages in the whole history of thought."

Gauss was acquainted with Lobachevsky as well as the Bolyais, but never bothered to introduce the two younger mathematicians, and there is no evidence that Lobachevsky knew of Bolyai at all. Bolyai, although too retiring to make Lobachevsky aware of his own work, did read a German translation of one of Lobachevsky's books in 1848, and praised it warmly. Janos Bolyai died on January 27, 1860, in Marosvasarhely, Hungary (now Targu Mures, Romania).

SHERRI CHASIN CALVO

George Boole
1815-1864
English Mathematician

Geroge Boole was an English mathematician who was among the founders of modern symbolic logic. His algebra of logic, which has come to be called *Boolean algebra*, is basic to the design of digital computers.

Boole was born on November 2, 1815, in Lincoln, England. His father was a tradesman, and taught him how to make optical instruments, as well as the rudiments of mathematics. A local bookseller taught him Latin. Aside from this early training and a few years at local schools, Boole was self-educated. He had to earn an income to help support his family, and by the age of 20 he was an experienced schoolteacher with his own school in Lincoln. In his limited free time, he studied mathematics, including such classic works as Isaac Newton's (1642-1727) *Principia*, Pierre Simon de Laplace's (1749-1827) *Traité de mécanique céleste*, and Joseph Louis Lagrange's (1736-1813) *Mecanique analytique*.

Beginning in 1839, Boole submitted a number of papers to the *Cambridge Mathematical Journal*. This new journal was edited by Duncan Gregory, who acted as Boole's mentor. In 1844, Boole published an important paper in the *Philosophical Transactions of the Royal Society*, on using algebraic methods to solve differential equations. The paper resulted in his being awarded the Society's Royal Medal.

Boole soon realized that algebra could be applied to logic as well. In 1847, he published a pamphlet entitled *Mathematical Analysis of Logic*, arguing that logic should be considered a branch of mathematics rather than a branch of philosophy. A more complete treatment of his symbolic logic was published in 1854 as *An Investigation into the Laws of Thought, on Which are Founded the Mathematical Theories of Logic and Probabilities.*

George Boole. *(The Library of Congress. Reproduced by permission.)*

The key to Boole's system of symbolic logic, which came to be known as *Boolean algebra*, was the separation of the symbols of operation from those of quantity. The operational symbols, such as + , can be used to represent logical relationships and operations. Symbols expressing quantity are not used in Boolean logic because it is a two-valued system referring only to the presence or absence of some attribute. This is generally expressed as one and zero, on and off, or true and false.

Despite his lack of a university degree, Boole was appointed a professor of mathematics at Queen's College in Ireland in 1849 on the strength of his published work. Later he was elected a Fellow of the Royal Society. In 1855, he married Mary Everest, the niece of mountaineer George Everest. They had five daughters.

In addition to his work on symbolic logic, Boole wrote two influential books, the *Treatise on Differential Equations* (1859) and the *Treatise on the Calculus of Finite Differences* (1860). These were used as university texts for many years.

Boole remained at Queen's College for the rest of his life, and enjoyed a reputation as an excellent teacher. Unfortunately, his career started late and ended early, when he died at the age of 49, on December 8, 1864, of a feverish cold that progressed to pneumonia. However, his work is the foundation upon which the Information Age was built.

SHERRI CHASIN CALVO

Georg Ferdinand Ludwig Philipp Cantor
1845-1918
German Mathematician

Georg Cantor is regarded as the founder of set theory. He also introduced the concept of *transfinite* numbers. This is the idea that there is not just one quantity of "infinity," but many quantities that are distinct from one another but all indefinitely large.

Cantor was born in St. Petersburg, Russia, on March 3, 1845. His father was a well-to-do merchant, and his mother was an artistic woman from a family of musicians. In 1856, the family moved to Germany. Cantor's mathematical talents soon became apparent at *gymnasien*, or secondary schools, in Darmstadt and Wiesbaden.

As an undergraduate student at the University of Berlin, Cantor specialized in mathematics, physics, and philosophy. There, he first met the mathematician Leopold Kronecker (1823-1891), who was later to become his archrival. In 1867, Cantor received his doctorate from the University of Göttingen, with a thesis entitled *In re mathematica ars propendi pluris facienda est quam solvendi,* or, "In mathematics the art of asking questions is more valuable than solving problems." The thesis addressed an unsettled question from *Disquisitiones Arithmeticae*, an 1801 work by Carl Friedrich Gauss (1777-1855).

Cantor spent a short time teaching at a girls' school in Berlin, and then joined the faculty at the University of Halle. He remained there for the rest of his life. Between 1869 and 1873, he published a series of papers on number theory and trigonometric series. It was in the 1870s that he began working on set theory, which led him to the concept of transfinite numbers.

A *set* is a group of objects or numbers that retain their individuality while having some property in common. The group can be either finite or infinite. For example, the set of students in a class is finite. The set of integers is infinite. Cantor concerned himself in particular with one-to-one correspondences. That is, the sets {*a,b,c*} and {*1,2,3*} are in a one-to-one correspondence because *a* can be paired with *1*, *b* can be

Georg Cantor. *(The Library of Congress. Reproduced by permission.)*

paired with 2, and *c* can be paired with 3. However, for infinite sets, one cannot simply compare the number of members, since the sets are indefinitely large.

Cantor developed a theory of countability based on one-to-one correspondences between infinite sets. Rational numbers, for example, are countable even though infinite. They can be placed in a one-to-one correspondence with integers. The real numbers, consisting of the set of rational and irrational numbers taken together, are uncountable. This led him to the realization that some infinite sets are larger than others, and thus to the idea of transfinite numbers.

Cantor's first paper on set theory was at first refused for publication by *Crelle's Journal* due to the vehement opposition of Leopold Kronecker. Most scholarly journals ask professionals in the field to *referee* papers before publishing, and Kronecker was serving in that capacity. The paper was published the next year, but the antagonism between the two mathematicians remained. Its basis was essentially philosophical. Cantor drew on his childhood religious training and Platonic metaphysics to come to the conclusion that infinite numbers had an actual existence. Kronecker was more limited in his view. "God made the integers," he asserted, "and all the rest is the work of man."

As for the work of Cantor, it became the basis for an entire field of study, the mathematics of the infinite. It was fundamental to the development of function theory, analysis, and topology. It also changed educators' ideas about the foundations of mathematical thought. Basic set theory and functions are part of today's elementary mathematics curriculum, first introduced in the 1960s as part of the "new math."

In 1874, Cantor married Vally Guttman. They had five children. The mathematician suffered from bouts of depression from about 1884, but continued to work as his health permitted. In 1897 he was involved in organizing the first international mathematical congress, which took place in Zurich. He died in Halle on January 6, 1918.

SHERRI CHASIN CALVO

Augustin-Louis Cauchy
1789-1857
French Mathematician

During his impressive career Augustin-Louis Cauchy published 789 scientific papers, more than almost any other scientist in the history of science. These papers spanned all of the fields of mathematics of his day and helped to place calculus on a firm theoretical footing, expanding its utility in the study of physics and astronomy. Cauchy held very conservative political views, which caused him a great deal of trouble politically and professionally, but he adamantly refused to act in a way contrary to his beliefs at any time in his life.

Cauchy was born in Paris a month after the Bastille was stormed in the first of the French revolutions. His father, a government official, supported the King, which made life difficult for Cauchy and his family during his youth. Cauchy was educated at home, where Pierre Laplace (1749-1827) and Joseph Lagrange (1736-1813) were frequent visitors. Both were impressed by the young Cauchy's abilities and took an interest in his mathematical education. They encouraged his study of classical languages and mathematics at the Ecole Polytechnique, where he graduated in 1805. After graduation he was put to work as a military engineer for a time, until turning his attention entirely to mathematics in his mid-twenties.

Cauchy applied for several academic positions over the next few years without success. Finally, in 1815 he earned an appointment as an

Augustin-Louis Cauchy. *(The Library of Congress. Reproduced by permission.)*

assistant professor of analysis at the Ecole Polytechnique and, at age 27, he became the youngest person to be elected to the Academie of Sciences, based on his work the theories of definite integrals and complex functions.

One of Cauchy's most significant accomplishments involved establishing conditions under which an infinite series will converge on a solution. He also established the first mathematically rigorous definition of an integral, a concept of fundamental importance to calculus. However, perhaps his greatest achievement was in developing the basic concepts of calculus in a rigorous manner. This placed calculus on a firm mathematical footing, helping to ensure its viability as a useful branch of mathematics. In particular, his theory of the continuity of functions and his work on limits not only helped to erect a logical framework for differential calculus, but are still taught today as essential for a full understanding of some of the most basic concepts of calculus.

During this period of his life, Cauchy earned a reputation for treating fellow mathematicians less than favorably. He refused to read some of the early work of Niels Abel (1802-1829), causing Abel to write that "Cauchy is mad and there is nothing that can be done about him, although, right now, he is the only one who knows how mathematics should be done." His treatment of Evariste Galois (1811-1832) was poor, too, and he was so harsh with Jean-Victor Poncelet (1788-1867) that Poncelet gave up trying to earn any regard or scientific respect at all.

Cauchy's religious and political beliefs kept him in trouble through much of his professional life. His refusal to take an oath of allegiance to the new French government in 1831 resulted in his losing his position at the university. He left France for several years, exiling himself to Italy where he continued to teach, at one point acting as tutor to the son of Charles X when the royal family was in exile, too. Cauchy eventually returned to France in 1838, regaining his position with the Academy. Later, after another change in government, the issue of loyalty oaths arose again. This time, however, Cauchy and Dominique Arago (1786-1853) were exempted from this requirement because of their importance to French science.

The final years of Cauchy's life were spent in more disputes, both scientific and personal. This led to some degree of bitterness and sadness that was apparent to his children and friends, if not to his colleagues and enemies. Cauchy died in 1857 at the age of 67, leaving behind a great legacy of work touching on almost every part of mathematics.

P. ANDREW KARAM

Julius Wilhelm Richard Dedekind
1831-1916
German Mathematician

A brilliant mathematician who made significant contributions in set theory, number theory, and mathematical induction, Julius Dedekind was also well known as the editor of the collected works of Peter Dirichlet (1805-1859), Carl Gauss (1777-1855), and Georg Riemann (1826-1866). Dedekind is the father of the "Dedekind cut," a method of representing rational numbers in terms of division into two sets by "cutting" the set of rational numbers with a single real number of value "r."

Dedekind was born in the duchy of Braunschweig, now part of Germany. He was the son of a professor at the Collegium Carolinum, and his mother was the daughter of another professor at Carolinum.

Dedekind's first interest in school was physics, not mathematics. However, the relative lack of rigor and logical structure in physics left him unsatisfied and he quickly turned to mathematics. He entered the Collegium Carolinum at age 16, receiving a solid grounding in basic mathematics, including calculus and analytic geometry. At that time, Carolinum was more an intermediate step between high school and university, so Dedekind left for more advanced studies at the University of Göttingen in 1850.

In spite of the presence of Gauss as a professor, Göttingen was not a hotbed of mathematics at that time. Dedekind took his first advanced course from Gauss in the autumn of 1850, a course in least squares. According to a contemporary, "fifty years later Dedekind remembered the lectures as the most beautiful he had ever heard." As Gauss's last doctoral student, Dedekind finished his dissertation in 1852. However, he realized he still had deficiencies in advanced mathematics, deficiencies he determined to rectify by further study.

In 1854, after two years of study, Dedekind was fully qualified to teach at the university level. He began teaching probability and geometry at Göttingen and, upon Gauss's death the following year, assumed Gauss's chair in the mathematics department. During this time he continued to take classes, mostly from Dirichlet, who quickly became a friend and research collaborator. Later, in 1858, Dedekind was offered a position at the Polytechnikum in Zurich, which he accepted immediately. By 1862 the Collegium Carolinum had been upgraded to university status as the Brunswick Polytechnikum and offered Dedekind a position. Dedekind accepted and moved back to his hometown, never to leave again.

Dedekind continued teaching until his retirement in 1894, receiving high praise for his teaching skills as well as continuing research in mathematics. It was during this time that he edited volumes on the lectures and papers of Riemann, Dirichlet, and Gauss. During this period, too, he developed concepts that helped advance the field of number theory and ring theory.

In addition to these contributions, Dedekind also earned renown for his explanatory style. He was able to express his ideas clearly and understandably so that most mathematicians were able to follow his arguments. This led to faster and more complete acceptance of his ideas than would have been the case otherwise. This, in turn, left a lasting mark on all of mathematics. According to another contemporary, Dedekind contributed "not only important theorems, examples, and concepts, but a whole style of mathematics that has been an inspiration to each succeeding generation."

By the time of his death, Dedekind had received numerous honors for his work, although he retained a deep sense of modesty throughout. He was elected to the Berlin Academy, the Rome Academy, the Académie des Sciences in Paris, and other similar organizations during his life, and received a number of honorary doctorates. Dedekind never married, living with relatives for most of his life. He died at the age of 84 in his native Brunswick.

P. ANDREW KARAM

Johann Peter Gustave Lejeune Dirichlet
1805-1859
German Mathematician

Lejeune Dirichlet. *(Archive Photos. Reproduced by permission.)*

Lejeune Dirichlet was a professor at the University of Berlin prior to accepting a chair—previously held by Carl Gauss (1777-1855)—at the University of Göttingen. Perhaps Dirichlet's most valuable contribution to science was in the theory of the Fourier series, which he originated. He also made important contributions to algebraic number theory and harmonic functions.

Dirichlet was born in Düren, then part of the French Empire and now part of Germany. The first mention of his work appears in 1826, when he proved one of Gauss's conjectures about prime numbers. The next year he was awarded a teaching position at the University of Breslau, after which he moved to the University of Berlin in 1828 and, in 1855, to Göttingen.

It was during his time at Berlin that Dirichlet first proposed what is the currently accepted definition of a function: If a variable y is so related to a variable x that whenever a numerical value is assigned to x, there is a rule according to which a unique value of y is determined, then y is said to be a function of the independent variable x.

In this definition, the key concept is that, for each number x, there is a unique value for y, something that had not previously been stated so simply. Or, put another way, any line drawn that is parallel to the y-axis will pass through the function no more than one time.

Dirichlet is best known, however, for his papers on the convergence of trigonometric series toward a solution, and in the use of such series to represent arbitrary functions. This work applied especially well to the Fourier series. The previous work of Simeon Poisson (1781-1840) on the Fourier series was shown by Augustin Cauchy (1789-1857) to be inadequate, but Dirichlet showed flaws in Cauchy's work. This led to Dirichlet's earning the moniker "founder of the theory of Fourier series."

Dirichlet, during his tenure at Göttingen, worked with Dedekind as both teacher and collaborator. At one time a student remarked that "he only knew Dedekind by sight because Dedekind always arrived and left with Dirichlet and was completely eclipsed by him." Dedekind himself wrote that "What is most useful to me is the almost daily association with Dirichlet, with whom I am for the first time beginning to learn properly.... I thank him already for infinitely many things, and no doubt there will be many more."

However, in spite of his accomplishments as a mathematician and mentor, Dirichlet paid little attention to himself. This led one colleague to note, "He is a tall, lanky-looking man, with moustache and beard about to turn grey with a somewhat harsh voice and rather deaf. He was unwashed, with his cup of coffee and cigar. One of his failings is forgetting time, he pulls his watch out, finds it half past three, and runs out without even finishing the sentence."

In addition to his work in mathematics, Dirichlet made some contributions to mathematical physics. In particular, he studied potential theory and the mechanics of systems in equilibrium. Dirichlet was elected a Fellow of the Royal Society in 1855 and is remembered by Crater Dirichlet on the Moon. He died in 1859 at age 54.

P. ANDREW KARAM

Jean-Baptiste Joseph Fourier
1768-1830
French Mathematician, Egyptologist and Administrator

Jean-Baptiste Joseph Fourier was a French mathematician, Egyptologist, and administrator with the French government. He is best remembered for his work on the analysis of functions in terms of a type of infinite mathematical series now known as *Fourier series*. This type of analysis has many important applications in the physical sciences.

Fourier was born on March 21, 1768, in Auxerre, France, the ninth child of a tailor and his second wife. Both his parents were dead by the time he was 10, and he was sent to a local military school run by Benedictine monks. He exhibited a great proficiency in mathematics, and eventually became a teacher there.

Fourier considered training for the priesthood, but became embroiled in politics in the wake of the French Revolution. He joined the local Revolutionary Committee, and became enamored of "the sublime hope of establishing among us a free government exempt from kings and priests." Although he deplored the subsequent Reign of Terror and tried, unsuccessfully, to resign from the committee in protest, he was at this point hopelessly entangled. As various factions rose and fell, fortunes changed. Fourier was twice jailed, fearing the guillotine. Each time, fortunately, he was released. Between prison stays, Fourier was associated with newly founded schools in Paris, the *Ecole Normale* and the *Ecole Polytechnique*.

In 1798, Fourier joined Napoleon on his expedition to Egypt. For three years, he conducted research on Egyptian antiquities, consulted on engineering projects and diplomatic undertakings, and served as the secretary of the Institut d'Egypte. After his return to France, he was responsible for publishing the massive body of scientific and literary discoveries from Egypt, a project that became the *Description d'Egypte*. This work includes Fourier's significant historical preface on Egypt's ancient civilization.

Between 1802 and 1814, Fourier was based in Grenoble as prefect for the French region or *departement*, of Isère. While supervising such official projects as the draining of swamps, he also continued his work in mathematics and Egyptology. Napoleon rewarded him in 1809 with the title of baron, but fell from power in 1815. This resulted in Fourier being transferred to head the Statistical Bureau of the Seine, allowing him to resume his academic life in the city of Paris. He was elected to the Académie des Sciences in 1817, and the Académie Française in 1826.

Fourier series arose from work that the mathematician began in Grenoble and finished in Paris in 1822. In his *Théorie analytique de la chaleur* ("Analytical Theory of Heat"), Fourier showed how the conduction of heat in solid bodies could be analyzed in terms of series with sines and cosines as terms. The implications of this method were much wider-ranging than heat conduction, affecting data analysis techniques in many scientific and engineering fields. It also had a great influence on the theory of functions of a real variable.

During his last years, Fourier published a number of papers in both pure and applied mathematics. He died in Paris on May 16, 1830.

SHERRI CHASIN CALVO

Friedrich Ludwig Gottlob Frege
1848-1925
German Mmathematician and Philosopher

Gottlob Frege, a philosopher of mathematics and a mathematician in the realm of philosophy, was the founder of modern mathematical logic and one of the most important figures in the histories of logic, mathematics, and the philosophy of mathematics. Frege also helped to shape the discipline of the philosophy of language and influenced, among others, Ludwig Wittgenstein (1889-1951) and J. L. Austin.

Frege was born November 8, 1848, in Wismar, Mecklenburg-Schwerin, in what is now eastern Germany. His father, Alexander Frege, a high school principal in Wismar, died when Gottlob was 18. Three years later, in 1869, Frege became a student at the University of Jena, but after two years went on to the University of Göttingen, where he studied mathematics, philosophy, and the physical sciences. In 1871 he returned to Jena to become a lecturer and was promoted first to associate professor in 1879, then to full professor of mathematics in 1896.

Though a stunning intellect, Frege's political and personal beliefs reflected a life of bitterness. He resented the outcome of the Treaty of Versailles—the peace treaty signed between the Allies and Germany after the First World War—and thought the terms imposed by the Allies crippling to his native Germany. As a result he came to despise democracy, socialism, Catholics, Jews, and the French—blaming all of them for Germany's post-war decline. The aloof professor also felt, well before the war, as if his intellectual peers overlooked the importance of his work, which he not immodestly considered to be both vital and innovative. He retired from teaching in 1917 not long after the death of his wife and died July 26, 1925, in Bad Kleinen, Germany.

Frege's major contribution to mathematical logic was made in *Concept-Script* (1879). In this work, he first came to the idea of a formal system, that is, a system that uses a quantifier-variable notation to make the language of proof regulated and unambiguous. Such a system, he thought, would render the logic behind proofs clearer than it appears in ordinary language and provide philosophers with an easily understood means for representing their arguments.

Arguing against Kant and Mill, Frege next tried to define the basic principles of arithmetic in purely logical terms, a project which he undertook in his deceptively simple work, *The Foundations of Arithmetic* (1884). Here he reasoned that if he could explain arithmetic without recourse to non-logical concepts, he would disprove Kant's notion that arithmetic is "synthetic;" put slightly differently, Frege wanted to show that the laws of arithmetic are a priori, meaning not subject to experience but derivable from deduction alone.

In *The Foundations of Arithmetic,* Frege also developed the idea that words only gain meaning when embedded in a context. This led in 1892 to his famous distinction between "sense" and "reference," the latter term representing the actual object about which a speaker is talking, the former the various names given to that object. The important point of this distinction is that sense and reference can be at variance with one another and can thus lead to confusion in statements about identity. For instance, the planet Venus was for centuries known as the "Morning Star" in the eastern sky but as the "Evening Star" in the western sky without any knowledge on the part of speakers that both words denoted the same celestial body. Ever since its articulation, this distinction between sense and reference has been a source of ongoing debate among philosophers, yet, as with many of Frege's ideas, its subsequently revealed imperfections have not necessarily lessened its lasting importance.

MATT KADANE

Evariste Galois
1811-1832
French Mathematician

Evariste Galois was born on October 25, 1811, in Bourg-la-Reine near Paris during one of France's hectic political periods. His father, Nicholas-Gabriel Galois, an active and well-known citizen, was elected mayor in 1815 during the Hundred Days regime that followed Napoleon's escape from Elba.

In addition to his father's prominent position, his mother, Adelade-Marie Demante, was a member of a distinguished family, all of whom believed strongly in education for both men and women. She personally educated Galois at home until 1823, when she decided to place him in the College Royal de Louis-le-Grand. This proved to be a mistake, as he was left to the mercy of mediocre teachers who lacked knowledge and the ability to inspire.

Fortunately, one of his teachers, Louis Richard, recognized Galois's unusual ability in mathematics and encouraged him to take up a more intensive study of algebra. When Galois was about 16, he embarked on a mathematical journey that would eventually make him world famous as a contributor to the study of higher algebra known as group theory. His "Galois theory" involved the solving of a long-standing group of mathematical puzzles, such as the impossibility of trisecting the angle and squaring the circle.

While Galois continued his brilliant and imaginative studies, his personal life became both disappointing and tragic. Months of work were lost when three papers submitted to mathematicians at the Academy of Sciences were lost or rejected by its reviewers. Galois then made two attempts to enter the Ecole Polytechnique, the leading school of French mathematics. On both occasions he had traumatic encounters

Evariste Galois. *(Corbis-Bettmann. Reproduced by permission.)*

with one or more of the oral examiners and was rejected as a future student.

After his second refusal, he suffered another blow when his father, following bitter clashes with politicians in his hometown, committed suicide in 1829. That same year, aware that his career as a professional mathematician was unlikely, Galois turned to the less prestigious Ecole Normale Superieure and devoted much of his time to political activism.

Galois continued his research independently and, in 1830, submitted another paper on algebraic functions to the Academy of Sciences. Again, the material was "lost," this time by Jean Baptiste Joseph Fourier (1768-1830).

In this same year, the citizens' revolution sent Charles X into exile and placed Louis-Philippe upon the throne. Galois published a vigorous condemnation of the proceedings and was promptly expelled from the Ecole Normale Superieure. Following this action, he was arrested on two separate occasions for his republican activities and, although he was acquitted the first time, he spent six months behind bars for the second offense.

A cloud of uncertainty continues to shroud the circumstances of Galois's death in Paris on May 31, 1832. There are three possible explanations offered for the cause of the duel that brought about his demise: a quarrel over a woman; challenges by royalists who resented and decried his republican views; or the possibility that an agent provocateur of the French police was the assassin. When French author Alexandre Dumas published his memoirs in 1865, he named Pecheux d'Herbinville as the man who shot Galois.

After years of arguments and procrastination, Galois's manuscripts, with annotations by Joseph Liouville (1809-1882), were finally published in 1846 in the *Journal de Mathematiques Pures et Appliquees*. In 1870 a lengthy treatment of Galois's theory, *Traité des Substitutions,* was published by French mathematician Camille Jordan (1838-1922). These subsequent validations made Galois's discoveries accessible to the mathematical community and secured his reputation in the history of mathematics.

On June 13, 1909, Galois was posthumously honored with a plaque at his birthplace in Bourg-la-Reine. Jules Tannery, a French mathematician and brother of Paul Tannery (1843-1904), made an eloquent speech of dedication, which was published the same year in the *Bulletin des Sciences Mathematiques*.

BROOK HALL

Carl Friedrich Gauss
1777-1855
German Mathematician, Astronomer and Physicist

Carl Friedrich Gauss is considered to be one of the greatest mathematicians of all time, mentioned in the same breath as Archimedes (c. 287-212 B.C.) and Isaac Newton (1642-1727). He made revolutionary strides in pure mathematics, especially the theory of complex numbers, as well as important contributions to the fields of astronomy, geodesy, and electromagnetism. He showed how probability can be represented by a bell-shaped curve, a concept with major implications in understanding statistics. He was among the first to consider the possibility of a non-Euclidean geometry.

Gauss was born in Brunswick, Germany, the only son of an impoverished bricklayer and his wife. His gifts for mathematics and languages were apparent at an early age. Persistent efforts by impressed teachers and his devoted mother brought him to the attention of the Duke of Brunswick, who arranged financing for his continuation in secondary school and attendance at the University of Göttingen.

In his doctoral dissertation, obtained in absentia from the University of Helmstedt, Gauss completed the proof of the Fundamental Theorem of Algebra. The Fundamental Theorem of Algebra states that any polynomial equation has solutions. It had been partially proven before Gauss, but he was able to

YOU, TOO, CAN BE A MATHEMATICAL GENIUS

~

The following story has often been told about Carl Gauss, though it may or may not actually be true. The math involved, however, is real. When Gauss was nine years old, his teacher gave his class the problem of adding the integers from 1 to 100. Gauss responded with the correct answer within seconds, astonishing his teacher. Gauss solved the problem by noticing a pattern: When the greatest number in the series (100) is added to the least number (1), the result is 101; when the second highest number (99) is added to the second lowest number (2), the result is 101; when the third highest number (98) is added to the third lowest number (3), the result is 101; and so on. Because there are 50 pairs of integers between 1 and 100, the sum of these integers is 101 x 50 or 5,050. The pattern Gauss noticed can be used to find the sum of the integers from 1 to any even number—from 1 to 200, for example.

Step 1: Add 1 to the even number. 1 + 200 = 201

Step 2: Divide the even number by 2. 200 ÷ 2 = 100

Step 3: Multiply Step 1 by Step 2. 201 x 100 = 20,100

See if you can use these steps to find the sum of the integers from 1 to 1,000,000.

demonstrate that every algebraic equation with complex coefficients (that is, coefficients given in terms of the imaginary square root of -1) has complex solutions. He managed to find a way to finish this proof without the actual manipulation of complex numbers, as their theory was yet to be well developed. In fact that task would soon be accomplished by Gauss himself.

At the age of 24, Gauss published his first great work, *Disquisitiones Arithmeticae*. This far-ranging treatise on number theory included the formulation of complex numbers as $a + b\sqrt{-1}$ where a and b are real numbers. Although it may seem that arithmetic involving imaginary numbers would not have many practical applications, in fact this mathematical construct has been instrumental in expressing and solving many physical problems. Later Gauss took a vital step in this direction by showing how complex numbers could be represented on an (x,y) plane.

Gauss also developed a way to use number theory to determine whether a regular polygon with a given number of sides could be geometrically constructed using a compass and ruler. Finding that a 17-sided polygon could be constructed, he developed a method to do so. His achievement was the first progress in this area since Euclid (c. 330-260 B.C.), who lived 2,000 years earlier.

In 1801, the asteroid Ceres was discovered, but astronomers were unable to calculate its orbit. Gauss developed a technique that predicted the orbital track with enough precision for the asteroid to be observed repeatedly as it continued on its path. He applied his technique again to the asteroid Pallas, refining it to account for the effects of the planets. Gauss's methods are still in use today, implemented by modern computers. In 1807 he became a professor at the University of Göttingen, and director of its observatory. He would remain there for the rest of his life.

Gauss started working in the field of geodesy, the measurement of the Earth's surface, in about 1820. He invented the *heliotrope*, a more accurate surveying instrument than was previously in use. It was in attempting to understand the distribution of repeated measurements that he introduced the famous bell curve, or *normal distribution of variation*.

Thinking about the curved surface of the Earth led Gauss back to pure mathematics, and he developed a way to determine characteristics of a surface by measuring the lengths of curves that lie upon it. This work would inspire his student, Bernhard Riemann (1826-1866), to embark upon the geometrical theories that would lay the foundation for general relativity. Gauss himself came to the conclusion that it was possible to have a self-consistent geometry in which Euclid's parallel axiom (that only one line can be drawn parallel to another line and through a point not on it) did not apply. But the conservative mathe-

Carl Friedrich Gauss. *(The Library of Congress. Reproduced by permission.)*

matician was reluctant to proclaim such an idea until it was independently published by Janos Bolyai (1802-1860) and Nicolai Lobachevsky (1792-1856) three decades later.

Gauss's work in mathematical physics contributed to potential theory and the development of the principle of conservation of energy. He calculated that the net electric flux through a surface, regardless of its shape, equals a constant times the net charge inside the surface, a result that has become known as Gauss' Law. He worked closely with the physicist Wilhelm Weber (1804-1891) on experiments in electromagnetism that would be instrumental in the development of telegraphy.

SHERRI CHASIN CALVO

George Green
1793-1841
English Mathematician

During his career in mathematics, George Green developed some of the most important ideas in the area of mathematical physics. Chief among these was the development of potential function, subsequently used to describe electrical and magnetic fields as well as the energy present in some mechanical systems. His contributions are even more remarkable

because he had virtually no formal schooling, rarely had time to devote to research, and was largely self-taught in mathematics.

Green was born to an English baker, also named George Green. His formal schooling was minimal, consisting of only a single year at age eight. However, it is thought that a mathematics student at nearby Cambridge, John Coplis, must have tutored Green in not only mathematics, but French as well. French was important because of the prominence held by French mathematicians at that time; many important papers were written in French and the new "French style" of mathematics was considered an important advance.

For the next several years, Green studied math on his own in the top floor of his father's grain mill, where he worked full-time. In addition to his duties in the family mill, Green was also the father of seven children, although he never formally married their mother. During this time, too, Green's parents died, leaving him the mill. In spite of his lack of formal education and his sketchy background in mathematics, Green published what was to become one of the most important mathematical works of his century and, arguably, of any century.

Green's *Essay on the Application of Mathematical Analysis to the Theories of Electricity and Magnetism* was published by subscription in 1828. By chance, one of the subscribers understood the importance of Green's work, though he did not understand the mathematics itself. Sir Edward Bromhead offered to send future papers of Green's to any of a number of scientific bodies for publication, an offer that Green eventually accepted. Between 1830, when they first met in person, and 1834, Green and Bromhead met regularly and Green wrote three major papers, two on electricity and one on hydrodynamics. All three were published, though received little attention.

During this time Green consistently underrated his abilities because of his lack of formal education. He simply did not realize the importance of his work. Finally, at age 40, Green enrolled in Cambridge as an undergraduate, completing his degree in 1837. Following his graduation, Green continued his work on hydrodynamics, writing two papers that were published in 1838 and 1839. Around this time he was awarded the Perse fellowship at Cambridge. Unfortunately, his tenure was to be short as he fell ill in 1840, dying just a year later of an unknown illness. His obituary, printed in the local newspaper, stated, "Had his life been prolonged, he might have stood eminently high as a mathematician." Ironically, at the time of his death even Green did not apprehend the importance of his work, or the recognition that would later come to him.

Green's work was not fully appreciated during his life. However, shortly after his death William Thomson (1824-1907), later Lord Kelvin, came across a reference to one of Green's papers. Tracking this down four years later, Thomson realized the importance of the work. He shared copies of the paper with Joseph Liouville (1809-1882) and Jacques Sturm (1803-1855), both of whom shared his excitement over the work. Thomson republished Green's paper, making it accessible to James Clerk Maxwell (1831-1879) and others who could use and build upon the work of this unknown, largely self-educated son of a miller.

P. ANDREW KARAM

Sir William Rowan Hamilton
1805-1865
Irish Mathematician

Sir William Hamilton is perhaps best known for developing the theory of quaternions, the first non-cummutative form of algebra. However, he also made significant contributions to the theories of optics and other areas in mathematical physics.

Hamilton was born in Dublin, the son of an attorney. He showed signs of genius early, learning Latin, Greek, and Hebrew by the age of five. At age 13 he was introduced to mathematics and, at age 17, he brought an error in Pierre Laplace's book on celestial mechanics to the attention of John Brinkley, Ireland's Astronomer Royal. After this first meeting, Brinkley commented, "This young man, I do not say will be, but is, the first mathematician of his age."

During his undergraduate years, Hamilton was appointed professor of astronomy at Trinity College. This distinction was followed by his appointment as Astronomer Royal of Ireland. It turned out that both of these appointments were unsatisfying, as Hamilton's primary love was mathematics, not astronomy, and it was in mathematics that he did his most important work.

One of Hamilton's first major accomplishments was a theory of conical refraction, of interest to opticians and verified experimentally within months. He then began work in complex

Sir William Rowan Hamilton. *(The Library of Congress. Reproduced by permission.)*

numbers, systems of numbers in which one term is the square root of -1. Hamilton was among the first to treat complex numbers as points on a plane in which one coordinate is a real number and the other, an imaginary number. (An imaginary number is a number multiplied by i, the square root of -1, and is usually expressed as $2i$, for example.)

He also introduced the concept of the Hamiltonian, a way of describing the total energy of a system. In this system, the total energy H of a body is constant over any path it can take, making the dynamical properties of the body more easily and accurately described. In this same area of mathematical physics, Hamilton also described Hamilton's principle, worked on the Hamilton-Jacobi equations, and developed the Hamiltonian operator. All of these are considered developments of fundamental importance to the study of dynamical systems, and all are still taught to physicists today.

Hamilton spent many years working on quaternion theory in one form or another. This work was started by realizing that a description of the motion of an object in three dimensions requires a mathematical description in four dimensions. Such problems could not be handled by the mathematics of the day. After nearly a decade of work, Hamilton finally understood the solution to this problem in the form of quaternions, which could be solved by using the first non-commutative algebra. An algebra is commutative if changing the order of the terms has no effect on the final solution. For example, 2 x 3 gives the same answer as 3 x 2. However, quaternions do not behave in this manner—they produce a different answer if the order of the terms is changed. Quaternions were not of much immediate use and, in fact, were not nearly as important as Hamilton had hoped. William Thomson (1824-1907), Lord Kelvin, wrote, "Quaternions came from Hamilton after his really good work had been done and, though beautifully ingenious, have been an unmixed evil to those who have touched them in any way." However, several decades later, Josiah Gibbs (1839-1903) was to find them exceptionally useful in developing vector calculus which, in turn, is widely used in physics today.

In marked contrast to his professional successes, Hamilton's personal life was a disappointment to him. In his early twenties he fell in love with a woman, but lost her to a more financially successful rival. After that, he seemed not to care much for any other women, and eventually married for the sake of marrying. He also had periodic problems with alcohol abuse, alternately controlling and losing control to his cravings. He died of a severe attack of gout at the age of 60, shortly after his election as the first foreign member of the National Academy of Sciences in the United States.

P. ANDREW KARAM

Karl Gustav Jacob Jacobi
1804-1851
German Mathematician

Karl Jacobi made his most notable contributions to mathematics in the area of elliptic functions. His book *Concerning the Structure and Properties of Determinants* was an important work in that branch of mathematics, and his work on partial differential equations proved important in the formulation of quantum mechanics.

Jacobi was born into a relatively prosperous family in Potsdam, Germany, in 1804. His father was a banker, assuring Jacobi a good education as a child and, later, at the University of Berlin. He completed his Ph.D. at Berlin in 1825, then taught mathematics at the University of Königsberg from 1826 until 1844.

Jacobi's main area of interest was in the branch of mathematics that dealt with elliptic

Karl Gustav Jacobi. *(The Granger Collection. Reproduced by permission.)*

functions. These functions were first studied in the mid-seventeenth century when mathematicians began investigating ways to determine the length of an arc of arbitrary length and position in an ellipse. Since the curvature of an ellipse varies along its circumference, this can be a difficult problem. Such curves that vary with two levels of periodicity are called "doubly periodic" functions. Elliptic functions are important in dealing with problems in physics that examine the shape of bars under stress, the effects of stress on rods and bars, and other, similar problems that are frequently seen in engineering fields today. The problem of dealing with elliptic functions was addressed with varying degrees of success by such mathematicians as John Wallis (1616-1703), Isaac Newton (1642-1727), and Jakob Bernoulli (1654-1705).

In addition to his work on elliptic functions, Jacobi carried out very significant research into partial differential equations and their application to problems in dynamics (problems involving moving bodies). Some of this research was carried out in collaboration with the Irish mathematician William Hamilton (1805-1865), resulting in the Hamilton-Jacobi equation, which was to play a great role in the formulation of quantum mechanics in the early twentieth century.

Jacobi's other significant contribution to mathematics was in the area of determinants. A determinant is the result of a series of mathematical operations performed on a matrix. In this case, the matrix would be set up to help solve a set of equations called linear equations, and each line in the matrix would represent the numerical coefficients in a single mathematical equation. Jacobi was able to show that if functions with the same number of variables are related to each other, then the Jacobian determinant is equal to zero. Any non-zero value shows that these functions are not related.

In addition to his mathematical skills, Jacobi had a reputation as an excellent professor. Perhaps his most important innovation was the introduction of the seminar style of teaching, in which the students play an active role in both teaching and learning. Jacobi used seminars extensively to present the most advanced topics in mathematics to his students in a less formal setting. In fact, it was not uncommon for other respected mathematicians to attend Jacobi's seminars in order to learn better what their counterparts elsewhere were doing.

Jacobi once commented on the significance of mathematics: "It is true that Fourier had the opinion that the principal aim of mathematics was public utility and explanation of natural phenomena; but a philosopher like him should have known that the sole end of science is the honor of the human mind, and that under this title a question about numbers is worth as much as a question about the system of the world."

P. ANDREW KARAM

Felix Klein
1849-1925
German Mathematician

Felix Klein's work had a profound effect on mathematical thought. He unified Euclidean geometry with the non-Euclidean geometries of Nikolai Lobachevsky (1792-1856) and Georg Riemann (1826-1866) by showing that they all could be derived as special cases of a larger system called *projective geometry*.

Projective geometry is more fundamental than Euclidean geometry because it deals with properties such as when points lie on the same line and when a set of lines meet in one point. These properties are *invariant* under a larger group of transformations, or mathematical manipulations, than the *congruence*, or equality of lengths, angles, and areas with which Euclidean geometry mainly concerns itself. "Projective

geometry has opened up for us with the greatest facility new territories in our science," Klein wrote, "and has rightly been called a royal road to its own particular field of knowledge."

Klein, whose full given name was Christian Felix, was born on April 25, 1849, in Düsseldorf, Germany, and attended the *gymnasium*, or secondary school there. Later he attended the University of Bonn. As a postgraduate student sojourning in Paris, he worked on group theory with the Norwegian Sophus Lie (1842-1899), and became interested in its possibilities for unification of disparate mathematical situations.

In 1872, he became a professor at the University of Erlangen. It was there, in a lecture written to inaugurate his appointment, that he set forth his views on geometry. His system thus became known as the *Erlanger Programm*. It was not until 1916 that new geometries were discovered that did not fit the *Erlanger Programm*, thus requiring a wider-ranging synthesis. However, for the classes of geometries that it covers, Klein's analysis is still useful. It was especially influential in the United States for 50 years after its publication.

Klein taught at the Technical Institute in Munich from 1875 to1880. There he married Anne Hegel, the granddaughter of the great German philosopher Georg Wilhelm Friedrich Hegel. Later Klein moved to the universities of Leipzig (1880-1886) and Göttingen (1886-1913). Beginning in 1872, he edited the *Mathematische Annalen* of Göttingen. He began work on a major mathematical encyclopedia, the *Encyklopädie der Mathematischen Wissenschaften* ("Encyclopedia of the Mathematical Sciences") in 1895, and supervised it for the rest of his life.

The eminent mathematician was the author of many popular books on the theory and history of mathematics and mathematical education, including *Elementare Mathematik von höheren standpunkte aus* ("Elementary Mathematics from an Advanced Standpoint," 1908). His view of mathematics was that it was "the science of self-evident things." Once a proof or calculation had been finished, no further marshaling of facts or opinions was necessary.

He was a leader in educational reform, and spearheaded the movement encouraging *functional thinking*. This advanced the idea that as part of a general education, students should learn to think in terms of *variables* and the *functions* that describe how one variable depends upon others. The importance of this concept is that it is used in every field of mathematics and science as well as in daily life. For example, a function can be used to describe how the funds in a bank account grow over time.

Among Klein's other important works were *Vorlesungen über das Ikosaeder* ("Lectures on the Icosahedron," 1884), in which he showed how the rotation of regular solids such as the 20-sided icosahedron could be applied to solving difficult problems in algebra.

Always of delicate health, Klein tended to overwork. His physical and mental condition collapsed in 1882, and he suffered from depression for the next few years. This period essentially marked the end of his own career in research. However, during the second half of his professional life he established one of the world's finest mathematical research centers at Göttingen. He also concentrated on teaching and his efforts in improving mathematical education. He retired in 1913, but tutored students in his home during the First World War. He died on June 22, 1925, in Göttingen.

SHERRI CHASIN CALVO

Pierre-Simon Laplace
1749-1827
French Mathematician and Astronomer

Pierre-Simon Laplace, sometimes called "the Newton of France," was a mathematician and astronomer who made many important contributions to the fields of mathematical astronomy and probability. Laplace's *Mécanique Céleste* (*Celestial Mechanics*) was the most important work in mathematical astronomy since Isaac Newton. His *Théorie Analytique des Probabilités* (*Analytical Theory of Probability*) influenced work on statistical probability for most of the nineteenth century. These two works alone guaranteed Laplace's place among the great scientists of the age.

Born to a middle-class family in Normandy, Laplace attended a Benedictine school between the ages of 7 and 16, thanks to wealthy neighbors who noticed Laplace's abilities at an early age. At the age of 16 Laplace entered Caen University to study theology. Laplace soon found that his talents and interests lay in mathematics and began a long and successful career pursuing those interests.

At the age of 19 Laplace was appointed to a chair of mathematics at the Military Academy of Paris. After becoming an associate member of the Paris Academy of Sciences in 1773, Laplace read a paper in which he showed that the planetary motions were stable. The question of the stability of the solar system had been a point of contention between scientists since Isaac Newton (1642-1727) had speculated that God's intervention might be occasionally required to keep the planets in their respective orbits. Laplace became a full member of the Academy of Sciences in 1785.

In 1796 Laplace published his famous nebular hypothesis. This hypothesis states that the solar system evolved from a mass of rotating gases which, as it cooled, had rings break away from its outer edges. These rings cooled further and condensed to form the planets. The sun is the remaining central core of the original gases.

In *Mécanique Céleste,* published in five volumes from 1799 to 1825, Laplace expanded his previous work on the stability of the solar system. In this monumental work, Laplace accounted for the effects of gravitation on all the bodies in the solar system. When presented a copy of *Mécanique Céleste,* Napoleon asked Laplace why God was never mentioned in his book. Laplace replied, "I had no need of that hypothesis."

The second branch of mathematics that occupied much of Laplace's thought was probability. Laplace's 1812 classic, *Théorie Analytique des Probabilités,* became the model of classic probability. Laplace's contributions to mathematical probability included a formal proof of the least squares rule, the method of inverse probability, and the first statement of the central-limit theorem. These and other discoveries in probability by Laplace were made primarily for use in his mathematical astronomy.

In addition to his mathematical contributions, Laplace's philosophy of probability influenced the way scientists thought in the nineteenth century. Laplace made his philosophy known in *Essais philosophique sur les probabilités* (1814), a popularized account of *Théorie Analytique des Probabilités.* Laplace defined probability as the ratio of the number of particular cases in question to the total of all the cases possible. Laplace believed that probability was a measure of the degree of certainty or a rational belief that an event would occur. This philosophy, called classical probability, dominated scientific thought on probability through much of the nineteenth century.

Laplace wrote that probability addresses "the important questions of life" for which complete knowledge was "problematical." One of the questions addressed by probability concerned legal testimony. Laplace wrote that the reliability of testimony depends on the reputation of the witnesses, the number of witnesses, whether there were conflicting testimonies, and whether the testimony was an eyewitness account or passed through other people. For Laplace, probability was a guide for the actions of a rational man.

In addition to his work in mathematics, Laplace was politically active through a very tumultuous time in French history. Laplace was involved in the commission that designed the metric system in 1790. He became a member, then chancellor, of the Senate under Napoleon.

"THUS IT PLAINLY APPEARS"

~

In *Mécanique Céleste,* Laplace did not generally acknowledge the work of his predecessors, implying that the results were his even when they were not. In addition, Laplace's text was very dense and difficult to follow, even for accomplished mathematicians. He often gave results with little or no details concerning their derivation, prefacing many difficult results with comments such as "It is easy to see...." Nathaniel Bowditch (1773-1838), the American mathematician who translated and commented on the first four volumes of *Mécanique Céleste,* stated that "Whenever I meet in LaPlace with the words 'Thus it plainly appears,' I am sure that hours, and perhaps days, of hard study will alone enable me to discover *how* it plainly appears."

Laplace received the Legion of Honor in 1805, became Count of the Empire in 1806, and was named a Marquis in 1817.

TODD TIMMONS

Joseph Liouville
1809-1882
French Mathematician

French mathematician Joseph Liouville was an accomplished teacher and a gifted researcher. His work in mathematical physics influenced the study of electrodynamics, heat flow, and addressed problems in astronomy. In addition, his purely mathematical contributions included the integration of certain algebraic functions, transcendental numbers, and examination of boundary values in differential equations. During his incredibly productive career, he taught almost continuously and published over 400 scientific papers.

Liouville was the son of a captain in Napoleon's army, causing him to live with his uncle until his father's return from the Napoleonic wars. After the wars, Liouville attended school, eventually attending the Collège St. Louis in Paris. It was there that he first studied high-level mathematics and began writing papers (though none were published). In 1825 Liouville entered the Ecole Polytechnique, taking classes from André Ampère (1775-1836), Dominique Arago (1786-1853), and Simeon Poisson (1781-1840).

In 1831 Liouville took his first academic position. For the next several years, he taught an average of 35-40 hours weekly at several different schools in the Paris area. In spite of this, he continued his research, but sometimes at the expense of his students. In 1836, unhappy with the low quality of mathematics journals in France, he founded a journal devoted to pure and applied mathematics, which became known as the *Journal de Liouville.*

Liouville developed an international reputation through his publications in August Crelle's journal and, based in part on this, he was elected to the Astronomy Section of the Acadèmie des Sciences in 1839. The following year, he was elected to the Bureau des Longitudes to fill a vacancy left by the death of Poisson. This was a significant event because it secured Liouville's career, allowing him to help other mathematicians with theirs. For the next 20 years he worked extensively toward helping younger mathematicians develop and promote their own ideas and work.

During this time, Liouville suffered a painful indignity when he was passed over for a chair in the Collège de France. He was so upset at the appointment of a professional rival over him that he immediately resigned. Expressing his outrage, he stated, "I am profoundly humiliated as a person and as a geometer by the events that took place yesterday at the Collége de France. From this moment on it is impossible for me to lecture at this institution."

Liouville, encouraged by Arago, also made a successful bid for political office. However, his failure to win reelection seemed to have a profound negative impact on him. Prior to this defeat, his friends noted that he never failed to stand up for his beliefs, while afterwards he simply became bitter and was beset by melancholic thoughts that interrupted even his mathematical notes and research.

Liouville reached his most productive years in 1856 and 1857, prior to becoming bogged down with excessive teaching duties again. After this time, his papers, while still numerous, were less detailed and not as polished, often leaving proofs to others rather than completing them himself.

Liouville's most significant contribution was developed in the 1830s, when he and Jacques Sturm (1803-1855) worked on boundary value problems in differential equations. By helping to solve these problems, Liouville contributed immeasurably to mathematical physics and to the solution of integral equations. His work on some aspects of Hamiltonian dynamics produced results that were—and continue to be—of major importance in statistical mechanics.

In spite of his personal and professional disappointments, Liouville won esteem during his life and remains highly regarded as a mathematician. Any single one of his major contributions would have been sufficient to secure his reputation as a mathematician of the first rank; his ability to contribute so much in so many areas marks him indelibly as a genius.

P. ANDREW KARAM

Nikolai Ivanovich Lobachevsky
1792-1856
Russian Mathematician

The Russian mathematician Nikolai Lobachevsky, along with the Hungarian Janos Bolyai (1802-1860), is considered to be the founder of non-Euclidean geometry. Neither man's contribution was fully recognized until after their deaths, despite Lobachevsky's perseverance in publishing in French and German as well as Russian. Non-Euclidean geometry was later to become an essential building block for Albert Einstein's (1879-1955) theory of general relativity.

Lobachevsky was born on December 1, 1792, in Nizhny Novgorod, Russia, into very limited economic means. His father was a low-ranking government official, and died when young Nikolai was only seven years old. The family then moved to Kazan, at the edge of Siberia. Lobachevsky received a public scholarship to the university in Kazan at the age of 14, intending to study medicine. Under the influence of a skilled mathematics teacher, Johann Bartels, who had taught the eminent Carl Friedrich Gauss (1777-1855), he soon switched to that field. Lobachevsky remained at Kazan for the rest of his life. He became a professor in 1816.

In 1826, Lobachevsky first announced his challenge of Euclid's fifth postulate, that one and only one line parallel to a given line can be drawn through a fixed point external to it. A successful challenge required developing a system of geometry without this postulate and which did not include any internal contradictions. Lobachevsky and Bolyai each accomplished this task independently at about the same time. However, Bolyai essentially withdrew even before becoming aware of Lobachevsky's work, when Gauss, a good friend of his father's, claimed to have had the idea decades before. Lobachevsky, probably realizing that an idea without follow-up or publication was not a valid claim of precedence, was not bothered by it. Although Gauss was acquainted with both mathematicians and knew they were working in the same field, he never introduced them. In fact, there is no evidence that Lobachevsky knew of Bolyai at all.

Lobachevsky published the first account of his non-Euclidean geometry in 1829, in the *Kazan Messenger*. He later wrote several more complete expositions of his work, including *Geometrical Researches on the Theory of Parallels* (1840), which garnered a favorable reaction from Gauss, and *Pangeometrie* (1855). He also worked in the areas of infinite series, integral calculus, and probability.

In addition to his mathematical research, Lobachevsky was a skilled university administrator, serving as dean of mathematics and physics, librarian, and rector as well as maintaining his professorship. The university had suffered under the reign of Tsar Alexander I. The tsar distrusted modern science and philosophy, regarding them as aberrant products of the French Revolution and a threat to orthodox religion. The result was factionalism and lowered standards in academic life, and the departure from Kazan of some of the best professors, including Lobachevsky's old teacher, Bartels.

When Tsar Nicholas I succeeded Alexander in 1826, a more tolerant period began. Lobachevsky was able to turn the university around, reestablishing it as a place of high academic standards and a collegial environment. He enforced sanitary precautions to limit the university's suffering from the cholera epidemic of 1830, rebuilt several buildings after a catastrophic 1842 fire, and improved primary and secondary education in the surrounding area. Despite all his efforts and accomplishments, in 1846 the capricious government abruptly relieved him of his university posts without explanation and despite the protests of all his colleagues. He died in 1856 in Kazan.

SHERRI CHASIN CALVO

Guiseppe Peano
1858-1932
Italian Mathematician

Guiseppe Peano was one of the outstanding mathematicians of the nineteenth century. He produced original and important work in the areas of calculus and set theory, and his work on solving systems of linear equations developed simultaneously with that of Emile Picard (1856-1941) and Hermann Schwarz (1843-1921). He spent the last part of his career, however, working on large projects of minor importance to mathematics.

Peano was born into a farming family in Italy. Recognizing his talents, his uncle, a priest and lawyer, took him to Turin where he attended secondary school. He was admitted to the University of Turin at the age of 18. There, he studied under some of the premier Italian mathematicians of the day.

Although initially planning on pursuing studies in engineering, by his third year, Peano had decided on a career in mathematics. He graduated with a doctorate in mathematics in 1880, at the age of 22, and began teaching at Turin that same year.

Peano's first work of importance was editing a textbook written by his mentor, Angelo Gennocchi in 1884. That same year he was made a professor of mathematics at Turin, taking over much of Gennocchi's teaching load. Peano's first important contribution to mathematics came in 1886 when he expanded on some work by Augustin Louis Cauchy (1789-1857) and Rudolf O. S. Lipschitz (1832-1903) in differential equations, later expanding this work even further. This was quickly followed by his work on solving systems of linear equations by the method of

successive approximations, although this work turned out to have been anticipated by others. His next major accomplishment was the publishing of his book, *Geometrical Calculus*, in 1888, in which he introduced Hermann Grassmann's (1809-1877) important work in a much more approachable manner than Grassmann had done. This work also marked the beginning of Peano's work in mathematical logic, an area in which he was to make many more important contributions.

Yet another achievement of this early period was Peano's work in set theory. This began in 1889 with his publications of the Peano axioms, which defined the natural numbers in terms of sets. A year later, he introduced the concept of "space-filling curves," previously thought not to exist and whose proof was called one of the most remarkable facts in set theory.

Peano was well-known for his clarity of thought and his ability to spot logical flaws in arguments presented by others. This gave him an almost unique ability to find exceptions to mathematical proofs or theorems presented to him as well as to find small flaws in arguments. This trait also served to irritate many of his colleagues. Of this ability, Bertrand Russell commented: "In discussions...I observed that he was always more precise than anyone else, and that he invariably got the better of any argument on which he embarked. As the days went by, I decided that this must be owing to his mathematical logic."

In 1896, at the age of 38, Peano effectively ended the outstandingly creative period of his career to pursue work on a new project, the *Formulario Mathematico*. This was to be a compendium of all known mathematical theorems, organized by subject area and described solely with mathematical notation. In fact, after about 1900 Peano made virtually no additional discoveries of importance in mathematics, although work on the *Formulario* continued.

In addition to working on the *Formulario*, Peano began work in 1903 on a universal language that would be based on Latin but from which all grammar would be removed. This language, which he called *Latino sine flexione* (later called Interlingua), was to incorporate vocabulary from many of the world's languages, and Peano hoped it would eventually become widespread. Peano continued work on these and similar projects until his death in 1932.

P. ANDREW KARAM

Jules-Henri Poincaré
1854-1912
French Mathematician, Scientist and Philosopher of Science

Jules-Henri Poincaré was one of the greatest mathematicians of modern times. In an age of increasing specialization, he was among the last to see all mathematics, pure and applied, as his domain. He used mathematics to investigate many scientific problems, including the motion of the tides and planets, and theories of electricity and light. He was particularly skilled at developing general methods superseding whole collections of mathematical "tricks" used in specific situations. His literary abilities also contributed to the influence of his ideas. He published several widely read books on the philosophy of science, and was elected to the Académie Française on the strength of his writing.

Poincaré was born in the town of Nancy, France, on April 29, 1854, into an accomplished family. His father was a distinguished physician, and his mother was a gifted woman who was instrumental in the education of her rather sickly son. His first cousin, Raymond Poincaré, later served as president of the French Republic during World War I.

Between 1872 and 1875, Poincaré attended the Ecole Polytechnique in Paris, and won many prizes in mathematics there. He had a very good memory, could visualize and solve complex equations in his head, and could quickly write papers requiring little revision. In 1879, Poincaré received his doctorate from the Ecole Nationale Supérieure des Mines, an engineering school. In 1881, he accepted an appointment at the Université de Paris, where he was to teach for the rest of his life.

Poincaré developed the theory of the *automorphic function*, one that doesn't change when put through certain types of transformations. He demonstrated how automorphic functions could be used to integrate linear differential equations and to express the coordinates of any point on an algebraic curve in terms of a single variable. The entire class of elliptic functions, of which the trigonometric functions are a subclass, is included in this theory. Some automorphic functions were associated with transformations that arose in non-Euclidean geometry. Poincaré was elected to the Academie des Sciences in 1887 in recognition of his mathematical achievements.

In 1887, the Swedish King Oscar II offered a prize for a solution to the "n-body problem," that is, the gravitational interactions and movements of a system containing an arbitrary number of masses. This problem is important for the theory of orbits. For example, the interactions of the Sun, Earth, and Moon constitute a three-body problem. Poincaré's solution, although incomplete, included important new work in mathematics, including the behavior of integral curves of differential equations near *singularities*, points where the derivative of a function ceases to exist. Not only was he awarded the prize, he was made a knight of the French Legion of Honor as well.

Poincaré also contributed to the fields of topology, number theory, and the dynamics of rotating fluid masses, applying the latter to the evolution of the planets and stars. He studied the dynamics of the electron and independently obtained some of the same results as Albert Einstein (1879-1955) did in developing the special theory of relativity. However, Einstein's formulation gained the upper hand because it was based on fundamental considerations, while Poincaré's was predicated on the later-discredited ether theory of light transmission.

Later in his career, Poincaré turned his gifts to explaining the meaning of science and mathematics to the general public. His works, including *La Science et l'Hypothèse* (*Science and Hypothesis*, 1903), *La Valeur de la Science* (*The Value of Science*, 1905) and *Science et Méthode* (*Science and Method*, 1908) were translated into several languages. His view of mathematical creativity was *intuitionist*. That is, he believed sudden insights to be a product of previous work done in the subconscious mind. He also emphasized the role of *convention*, the arbitrary choice of concepts or expression based on a group consensus, in the scientific method.

In 1906, Poincaré was elected president of the Académie des Sciences. Two years later, he was elected to membership in the Academie Française, the highest honor in the French literary world. He died suddenly on July 17, 1912, in Paris.

SHERRI CHASIN CALVO

Jean-Victor Poncelet
1788-1867
French Mathematician

Jean-Victor Poncelet was one of the founders of projective geometry. He was born on July 1, 1788, in Metz, France, and grew up to become an officer in Napoleon's army. As a lieutenant of engineers in 1812, he was badly wounded during the Russian campaign and left behind at Krasnoy, where he was thought to be dead. Instead, he was taken prisoner and remained at a camp in Saratov until 1814, when he returned to France. During his two years in prison, Poncelet busied himself with an intensive study of projective geometry—a facet of mathematics dealing with relationships between geometric figures and the projected images (or mappings).

Poncelet went on to study military engineering at Metz (1815-25) and soon became a professor of mechanics at the Ecole d'Application there (1825-35). In 1826, by applying mathematics to working functions of turbines and water wheels, he proposed a design for the first inward-flow turbine, which was not actually built until 1838. In that year, Poncelet became a professor to the faculty of sciences in Paris and held this post for the next 10 years, when he was appointed commandant of the Ecole Polytechnique with the rank of General.

Prior to Poncelet's remarkable theories, there had been only two momentous geometrical developments. They had taken place during the Greek period and at the end of the eighteenth century. Both concerned the eventual subject of projective geometry. One was a theorem discovered and proved by Gerard Desargues (a French mathematician) in 1639; the second was a significant broadening of an earlier theorem (credited to Pappus of Alexandria in the fourth century) by Desargues's countryman, Blaise Pascal, in 1640.

Poncelet's years of study led to his personal belief that geometry could be founded on a series of fundamental principles as general as those on which algebra was based. He carried this belief forward by suggesting that since every straight line and plane extend to a point of infinity, then any new points on a line would be the same for specific parallel lines (or planes). At this time in mathematical history, it was accepted that all infinite elements of space were supposed to lie on the infinite plane of space, known as the projective plane.

Early on, Poncelet's colleagues were extremely reluctant to accept his ideas and resulting theories. However, as time went on, several prominent German mathematicians not only accepted them but contributed to the emerging new science. Among those who recognized Poncelet's breakthrough in the field were Karl Georg Christian von Staudt, Felix Klein, Georg Cantor, Richard

Dedekind, and Moritz Pasch. They were later joined by Otto Stolz of Austria who made his own contributions to the field.

In spite of these isolated encouragements, Poncelet finally decided that he could be more effective in his chosen field by returning to his original work and applying mathematics to the design of machines and other technology.

He died on December 22, 1867, in Paris without the recognition he deserved for his brilliant contributions to the world of numbers. His *Treatise on the Projective Properties of Figures* (1822) is still regarded as the pioneer work in the field.

BROOK HALL

Lambert-Adolphe-Jacques Quetelet
1796-1874
Belgian Statistician, Astronomer and Meteorologist

Adolphe Quetelet envisioned a new scientific discipline that he called "social physics." It is one of the precursors of sociology. Social physics combines statistical enumeration with the analytical tools of probability theory. He coined the term "average man," a fictitious person with average build and average mental characteristics. To coordinate the collection of statistical data Quetelet organized international statistical congresses.

Lambert-Adolphe-Jacques Quetelet was born in 1796 in Ghent, in the French speaking part of Belgium. Just before his birth Belgium had been annexed by France, and when French leader Napoleon Bonaparte was defeated in 1816 the Allies joined Belgium with the Netherlands to form a stronger buffer against future French ambitions. All during his early life Quetelet would strive for Belgium to become an autonomous nation.

Quetelet's dissertation was in geometry. In 1823 he traveled to Paris to seek advice for his plan to build an observatory for astronomy, meteorology, and geomagnetism in Brussels. There he met famous scientists such as Pierre Laplace (1749-1827) and Siméon Poisson 1781-1840(). Through these contacts Quetelet became interested in statistics and probability theory.

In 1826 he published his first statistical study, a birth- and death-statistics for the city of Brussels to form a basis for a reliable life insurance. In 1831, in a study on crime statistics, he

Lambert-Adolphe-Jacques Quetelet. *(The Library of Congress. Reproduced by permission.)*

heated discussions were held to decide whether regular crime statistics are compatible with the assumption that people have a free will.

For the remainder of his scientific career Quetelet became the untiring promoter of international cooperation in the collection of statistical data. He was one of the first foreign members of the American Statistical Association founded in 1839, and in 1853 Quetelet organized the International Congress of Statistics. Its goal was to introduce unity in the official statistical investigations of the participating countries to make the results comparable. Eight more of these congresses were to follow. Looking back, Quetelet claimed later that statistics had been as equally important for the nineteenth century as the metric system, the steam engine, and photography.

Quetelet's funeral was a gathering of scientists and politicians from all over the world. A statue of him was erected in 1880. It shows him seated in an armchair, the fingers of his left hand spread out on a nearby globe, and his head raised as he peers into the secrets of space.

ZENO G. SWIJTINK

introduced his famous concept of "average man." The "average man" is a fictitious being for whom everything happens according to the average results obtained for the whole society. Just as stable mortality statistics proved that the average man had a stable probability to die, Quetelet argued, he had also a propensity to commit a crime: "Thus we pass from one year to another with the sad perspective of seeing the same crimes reproduced in the same order and calling down the same punishments in the same proportions.... There is a budget which we pay with a frightful regularity; it is that of prisons, chains, and the scaffold."

These ideas obtained wide circulation in his famous treatise of 1835, "A Treatise on Man and the Development of his Faculties," published in Paris. This work was translated into many languages. It is subtitled "An Essay on Social Physics"; in it the average man is on the one hand an analog of averaging large numbers of observations to find planetary positions and on the other an expression of societal forces. According to Quetelet "society includes within itself the germs of all the crimes committed.... It is the social state, which prepares these crimes, and the criminal is merely the instrument to execute them."

These ideas led to accusations of fatalism and materialism, and all over the Western world

Georg Friedrich Bernhard Riemann
1826-1866
German Mathematician

Georg Friedrich Bernhard Riemann was a brilliant German mathematician who recognized the application of his work in non-Euclidean geometry to physics, including the shape of space itself. This mathematical advance made an immense contribution to modern theoretical physics, including laying the foundation for Albert Einstein's (1879-1955) theory of general relativity.

Riemann was born on September 17, 1826, in Breselenz, Germany, the second child of a Lutheran pastor. His family life was happy, and he progressed from his father's early tutelage to the local gymnasium, or secondary school, where his mathematical abilities quickly outstripped those of his teachers. The director of the gymnasium made mathematical texts available to him and allowed him to study them on his own. Otherwise, his education in the standard classical curriculum progressed normally, and he went on to the universities of Gottingen and Berlin, returning to Gottingen for his graduate work. There, after obtaining his father's per-

Georg Riemann. *(Corbis-Bettmann. Reproduced by permission.)*

mission to transfer from the department of theology, he studied both mathematics and physics, including *Naturphilosophie*, the quest to derive universal principles from natural phenomena.

Riemann obtained his doctorate at Göttingen in 1851, with a thesis entitled "Foundations for a General Theory of Functions of a Complex Variable." Complex numbers are those expressed in terms of a+bi where i is the square root of -1, an imaginary number. Despite the imaginary nature of i, complex variables have many applications in the physical sciences. Riemann's work led to the idea of a multi-layered surface, later called a *Riemann surface*, allowing a multi-valued function of a complex variable to be interpreted as a single-valued function. The thesis won coveted praise from the eminent mathematician Carl Friedrich Gauss (1777-1855).

Riemann's next major body of work was prepared to meet the requirements for admission to the university as a lecturer. This effort led him to an independent formulation of a non-Euclidean geometry. He was apparently unaware that Janos Bolyai (1802-1860) of Hungary and the Russian mathematician Nikolai Lobachevsky (1792-1856) had already developed geometries without Euclid's parallel postulate that states that one and only one line can be drawn parallel to a given line through a point not on the line.

Riemann's geometry differed in that it postulated that there are no lines parallel to another line through a point not on it. He referred to the physical example of two ships traveling on the Earth's curved surface along meridians of longitude and meeting at a pole. It was the idea of space itself having a curved shape, such that it must be described using non-Euclidean geometry, which Einstein drew upon in developing his model of space-time. Riemann's work, which was destined to become a classic in the mathematical field, sufficed to secure his appointment as a *Privatdozent*, an unpaid lecturer dependent upon student fees.

Riemann was soon appointed professor, but suffered from deaths in his family, his own poor health, and the effects of overwork. Still, he was appointed to the Berlin Academy of Sciences in 1859, and continued to publish influential papers. Many important mathematical methods, theorems, and concepts still bear his name. In 1862, he married Elise Koch, a friend of his sister's, and they had one daughter. He died before his 40th birthday, on July 20, 1866, while attempting to recuperate from pleurisy and tuberculosis in Selasca, Italy.

SHERRI CHASIN CALVO

Karl Theodor Wilhelm Weierstrass
1815-1897
German Mathematician

An exceptional teacher and researcher, Karl Weierstrass was one of the most influential mathematicians of the nineteenth century. He taught and inspired some of Europe's most gifted mathematicians and made important contributions in elliptic functions, the calculus of variations, and other areas of mathematics.

Weierstrass was born in Prussia, now part of Germany, to a mid-level bureaucrat. His father, a well-educated and intelligent man, was seemingly content working at a lower level than his abilities permitted. His father's position with the main tax office convinced him that his son should study accounting and law, though Weierstrass's interests lay elsewhere, in mathematics. This dilemma played itself out internally, with Weierstrass paying little attention to his studies at all for a time.

Eventually he decided to pursue mathematics, showing an immediate aptitude for the field.

Karl Weierstrass. *(Corbis-Bettmann. Reproduced by permission.)*

He changed universities before taking his graduation exams, leaving the University of Bonn to attend the Theological and Philosophical Academy of Münster. There he studied under the mathematician Gudermann, whose reputation lured him to Münster.

Graduating from Münster in 1842, Weierstrass took a teaching position at the Gymnasium at Münster where, in addition to mathematics, he taught many other subjects. This position, like those his father held, was far below his abilities and he soon tired of it. At the same time, he continued working on his research, publishing some papers on elliptic and complex functions in relatively obscure publications. Unfortunately, the stress of his unloved teaching job and his research began to take a toll on Weierstrass's health, leaving him with frequent attacks of dizziness and nausea. These spells were to recur frequently for much of the rest of his life.

Weierstrass finally gained recognition with the publication of a paper on the theory of Abelian functions in August Crelle's journal in 1854. In fact, based solely on this paper, he was given an honorary doctorate from the University of Königsberg that same year.

Over the next few years a number of European universities fought to attract Weierstrass to join their faculty. He finally accepted an offer from the University of Berlin, his original university of choice.

As a lecturer, Weierstrass excelled, attracting students from all over the world. Among those influenced by him were many of the best mathematicians of his time, including Georg Cantor (1845-1918), Ferdinand Frobenius (1849-1917), Felix Klein (1849-1915), Hermann Schwarz (1843-1921), Gösta Mittag-Leffler (1846-1927), Sophus Lie (1842-1899), and Sonya Kovalevskaya (1850-1891). Weierstrass was especially taken with Kovalevskaya, helping her to receive an honorary doctorate from Göttingen and to receive a position in Stockholm later. Her early death was a severe blow to him.

Although Weierstrass published very little during his career, many of his findings were announced in his lectures, which were collected and published during his later years and after his death. A colleague, who noted Weierstrass's reputation as "the father of modern analysis," summarized his achievements: "Weierstrass devised tests for the convergence of series and contributed to the theory of periodic functions, functions of real variables, elliptic functions, Abelian functions, converging infinite products, and the calculus of variations. He also advanced the theory of bilinear and quadratic forms." All in all, an impressive professional legacy.

Weierstrass died at the age of 81, spending the last three years of his life confined to a wheelchair. The first two volumes of his collected works were published before his death, and another five were published posthumously.

P. ANDREW KARAM

Biographical Mentions

Louis François Antoine Arbogast
1759-1803

French mathematician who introduced discontinuous functions and first considered the concept of calculus as operational symbols (operators). Arbogast was interested in the philosophy of mathematics as well as its practice. He started and maintained a classification system for some of the outstanding papers of his time, forming an exceptionally important collection that is still widely used. His introduction of operators in calculus helped to standardize the field, making many discoveries more readily accessible to other mathematicians.

Jean-Robert Argand
1768-1822

French-Swiss mathematician who made important contributions in the area of complex numbers. Although Argand was, by profession, an accountant, he performed some high-quality work in mathematics. Perhaps most important was his geometrical interpretation of complex numbers in which *i* (the imaginary component) is plotted versus the real component of complex numbers on a plane. Other works included introducing the concept of the modulus of complex numbers and a proof of the fundamental theorem of algebra.

Eugenio Beltrami
1835-1900

Italian mathematician who showed that the non-Euclidean geometry of Nikolai Lobachevski is consistent by mapping it onto Euclidean geometry—he identified the straight lines of this non-Euclidean geometry with geodesics on surfaces of constant curvature—and so opened a new way of thinking about mathematical structures. In his later work in applied mathematics, on fluid theory and elasticity, Beltrami remained interested in the underlying conceptions of space in these theories.

Friedrich Wilhelm Bessel
1784-1846

German astronomer and mathematician whose work in astronomy, celestial mechanics, and mathematics led him at age 26 to the Königsberg Observatory directorship, a position he held all his life. He is best remembered for a mathematical function that carries his name, one that he developed to solve Kepler's problem of a three-body system under mutual gravitation. Together with Jacobi and Franz Neumann, Bessel is usually associated with teaching reform in German universities.

Enrico Betti
1823-1892

Italian mathematician who made major contributions in algebra and topology (the study of surfaces). Betti developed proofs for many of the algebraic concepts of his colleague, Evariste Galois. Betti was also the first to show that the quintic function—a function in which a variable is raised to the fifth power—can be solved using integrals of elliptic functions. In addition to his mathematical work, Betti served in several political positions during his life.

Luigi Bianchi
1856-1928

Italian mathematician whose work on non-Euclidean geometries (i.e., geometries that are not based on figures on flat surfaces) was used by Albert Einstein in his general relativity theory. Bianchi's major work involved the theory of surfaces and differential geometry, a topic upon which he wrote several influential papers. In addition to his work in geometry, he also made fundamental contributions to the theory of groups and theories of complex variables.

Bernhard Placidus Johann Nepomuk Bolzano
1781-1848

Czech philosopher, mathematician, and theologian who worked to develop theories of mathematical infinities. Bolzano was an ordained Roman Catholic priest as well as a mathematician. In addition to his work in mathematics, he taught and wrote philosophical and political works, many of them banned by his government. He independently defined what is now called the Cauchy sequence, while his other work anticipated Georg Cantor's theory of infinite sets. Unfortunately, much of his work remained unpublished until after his death and went unrecognized during his lifetime.

Mary Everest Boole
1832-1916

English mathematics educator who invented the use of "string geometry" to illustrate the principles of angles and shapes. Mary Everest Boole worked as a librarian at Queens College, which was the first women's college in England but where women were not permitted to teach. She began to tutor students informally, and was eventually widely recognized for her teaching skills. Boole became interested in the psychology of the learning process, and her book *Preparing the Child for Science* (1904) had a major impact on progressive education in England and the United States.

(Augusta) Ada Byron, Countess of Lovelace
1815-1852

English mathematician and pioneer of computer science. The daughter of poet Lord Byron, Ada taught herself geometry and, encouraged by her mother, was trained in astronomy and mathematics. She was later helped with advanced studies by Augustus De Morgan, first professor of mathematics at the University of London. Byron is best known for her work with mathematician Charles Babbage, whose work she translated and published along with her own annotations and important comments. Though underappreciated during her life, she is now highly regarded as an early computer program writer. ADA, the high-level computer programming language, is named in her honor. Byron became Countess of Lovelace in 1838 through marriage to William King, an earl.

Moritz Benedikt Cantor
1829-1920

German historian of mathematics whose four-volume *Vorlesungen über Geschichte der Mathematik* (*Lectures on the History of Mathematics*, published 1880-1908) is still considered one of the finest works in the field. It covers mathematical developments from the earliest times through the eighteenth century. Cantor spent his career as a professor at the University of Heidelberg. He was an editor of the journal *Zeitschrift für Mathematik und Physik*, in which several of his papers were published.

Lazare Nicolas Marguérite Carnot
1753-1823

French mathematician and military engineer who headed the Army of the North during the French Revolution, thus acquiring the nickname "organizer of victory." His major scientific contributions are often overshadowed by two events : first, his appointment, along with Gaspard Monge, in 1794 to take charge of what is known today as the Ecole Polytechnique; and second, the birth of his son, Sadi Carnot (1796-1832), the author of a masterpiece on the thermodynamics of the steam engine.

Arthur Cayley
1821-1895

English mathematician who created several branches of abstract algebra and restored the international prestige of British mathematics. A lawyer by profession, he was a prolific writer and contributed to the foundations of linear algebra and invariant theory. His work on matrices played a part in mathematical physics in the twentieth century as did his approach to geometries more general than the traditional Euclidean version.

Michel Chasles
1793-1880

French mathematician known for his contributions to algebraic and projective geometry.

Chasles wrote an important historical reference, still in use today, on the development of methods used in geometry. He also developed the theory of cross ratios independently of the famous mathematician August Möbius. Chasles also won unfortunate distinction as the victim of a fraud; he paid over $30,000 for forged letters from famous scientists. He was elected to the Royal Society and awarded the Copley Medal for his scientific contributions.

Pafnuty Lvovich Chebyshev
1821-1894

Russian mathematician whose best known papers deal with prime numbers. Appointed professor of mathematics at the University of St. Petersburg in 1847, Chebyshev also studied problems involving probability theory, quadratic forms, orthogonal functions, and the theory of integrals. In mechanics, he worked on solutions to convert, by mechanical coupling, rotary motion into rectilinear motion, resulting in what is called Chebyshev parallel motion. Later in life, he became a foreign member of the Institut de France and the Royal Society of London.

Elwin Bruno Christoffel
1829-1900

German mathematician who made major contributions in a number of mathematical disciplines, including tensor analysis, differential equations, and several areas of mathematical physics. Christoffel was considered one of the best professors and practitioners of mathematics during his era. His work on tensor analysis made possible the later work of Tullio Levi-Civita and Gregorio Ricci-Curbastro, leading, in turn, to the development of relativity theory.

Rudolf Friedrich Alfred Clebsch
1833-1872

German mathematician whose work on algebraic geometry was considered the first and most fundamental work in that branch of mathematics. Clebsch also developed a brilliant interpretation of some of Georg Riemann's work on function theory and investigated some aspects of algebraic geometry using approaches that differed from convention, focusing on algebraic interpretations rather than geometrical ones. Clebsch also founded an important journal, *Mathematische Annalen* before his premature death at the age of 39.

William Kingdon Clifford
1845-1879

English mathematician best known for his work in the areas of non-Euclidean geometry and topology (the study of surfaces). Clifford was influenced by Georg Riemann and Nikolai Lobachevsky, publishing research in 1870 on properties of matter, energy, and space that were fundamental to Albert Einstein's development of general relativity theory. He also made significant contributions to William Rowan Hamilton's quaternion theory, using them to describe types of motion in what are now known as "Clifford-Klein spaces." Clifford died at the age of 35, apparently from stress induced by severe overwork.

Antoine Augustin Cournot
1801-1877

French mathematician who was among the first to apply mathematical principles and techniques to the study of economics. Cournot's early work in mechanics (a branch of physics) won him plaudits from Siméon Poisson, a prominent French mathematician of the time. With Poisson's recommendation, Cournot became a professor at Lyon and, later, in Grenoble, where he changed his interest to mathematical economics. Cournot's definition of the basic term "market" is still in use today, as are many of the tools he developed.

August Leopold Crelle
1780-1855

German mathematician who in 1826 founded the first journal devoted entirely to mathematics, known today as *Crelle's Journal*. Crelle, a civil engineer who worked for the Prussian government, also published a number of mathematical textbooks and many editions of multiplication tables. He early realized the importance of work by Niels Abel, including his proof that quintic equations could not be solved using radicals. Abel and Swiss mathematician Jakob Steiner were major contributors to the premier issue of Crelle's periodical.

Antonio Luigi Gaudenzio Guiseppe Cremona
1830-1903

Italian mathematician who developed the "Cremona transformations" for studying surfaces and singularities (places where curves are not amenable to solution, such as at points). Cremona's early attempts to lead an academic life were hampered by his role as a soldier in Italy's failed revolution against Austrian rule. He was awarded a teaching position following Northern Italy's independence in 1858. Cremona became a senator in the new Italian republic in 1879, effectively ending his mathematical career.

Jean Gaston Darboux
1842-1917

French mathematician who made important contributions to differential geometry and analysis. Darboux made important discoveries in the studies of one surface rolling over another, a difficult process to describe mathematically. He also had a rare ability to combine both algebraic and geometrical approaches to some problems, arriving at solutions that were both elegant and brilliant. Darboux was considered one of the finest teachers of mathematics and he received many honors as both a teacher and a mathematician.

Augustus De Morgan
1806-1871

English mathematician whose work in logic helped spur the development of mathematical logic as a separate discipline. He was instrumental in making mathematics a profession in England and in taking it out of the narrow confines of the traditional universities of Oxford and Cambridge. A man of principle and great learning, he ensured that mathematics was part of the training for students planning to pursue technical careers.

Ulisse Dini
1845-1918

Italian mathematician whose most important work was on the theory of functions of real variables. He also conducted studies on surfaces and on some of the work performed by Joseph Liouville and Eugenio Beltrami, solving a difficult problem posed by the latter in one of his publications. Dini is also known for his discovery of the Dini condition, which ensures the convergence of a Fourier series in terms of the convergence of a definite integral.

Philbert Maurice D'Ocagne
1862-1938

French mathematician who created the field of nomography—the study of nomograms being used by engineers and others for graphical calculation of equations. Many of the equations can be represented graphically by three curves, often straight lines, in a plane such that the relationship among three variables is represented by the alignment of three points, one on each curve. D'Ocagne's work was translated into more than twelve languages.

Ferdinand Gotthold Max Eisenstein
1823-1852

German mathematician who in his short life published a tremendous amount of research in the areas of quadratic and cubic forms and elliptic functions. Eisenstein was inspired to study mathematics after meeting William Rowan Hamilton, who gave him a copy of a paper he had recently written on a difficult problem in mathematics. Eisenstein pursued his research in spite of chronically poor health, writing papers even while bedridden. He died of tuberculosis at the age of 29 in his native Berlin.

Karl Wilhelm Feuerbach
1800-1834

German mathematician who discovered the Euler circle, also called the "nine point circle of a triangle." Feuerbach's other significant contribution to mathematics was his work on homogeneous coordinates, which he developed simultaneously with (and independently of) August Möbius in 1827. A brilliant student with great potential, Feuerbach was ill for much of his life and practiced mathematics for only six years.

Ferdinand Georg Frobenius
1849-1917

German mathematician who established the mathematical framework for the study of abstract groups. In particular, his paper on group characters was of fundamental importance to this field. In 1897, learning of work on matrices by Molien, Frobenius successfully reformulated much of his own work. Frobenius's work on character theory was used by William Burnside and received favorable mention in Burnside's *Theory of Groups of Finite Order.* Frobenius's theory on finite groups provided important mathematical tools to the field of quantum mechanics.

Joseph-Diaz Gergonne
1771-1859

French mathematician who introduced the term *duality* into mathematics and founded the influential journal *Annales de Mathematiques Pures et Appliquées.* A duality is a symmetry in a mathematical theory such that if certain objects or relations are interchanged—such as line and point in projective geometry—all theorems remain valid. Gergonne's work had the philosophical bent of conceptual analysis, as in his work on definition

theory, in which he made the distinction between implicit and explicit definitions.

Sophie Germain
1776-1831

French mathematician who contributed to the studies of acoustics, elasticity, and the theory of numbers. Born in a time when women were not permitted to attend schools of higher learning, Germain advanced her education early on by reading whatever she could in her father's library. She mastered calculus in this manner, then took correspondence courses (under the pseudonym of M. Leblanc) from the Ecole Polytechnique in Paris, where women were not accepted. Eventually, she became so highly regarded by prominent mathematicians of her era that she was recommended for an honorary Ph.D from the University of Göttingen. However, she died before her degree was presented.

Paul Albert Gordan
1837-1912

German mathematician who was a major contributor to the field of invariant theory. He collaborated with Rudolf Clebsch on both invariant theory and algebraic geometry, and also developed proofs demonstrating that the numbers e and π are transcendent numbers, numbers that are not the root of any algebraic equation with rational coefficients. A strict mathematical logician, Gordan commented that David Hilbert's less formal approach to mathematics was "not mathematics, it is theology."

Hermann Günter Grassmann
1809-1877

German-Polish mathematician best known for his work on the calculus of vectors. Grassmann also developed a geometrical algebra in which symbols represented geometrical entities instead of numbers. In addition, his development of what he called "external algebra" was adapted by William Clifford to form what is now the mathematical basis for relativistic quantum mechanics. At 53, upset at the lack of recognition his mathematical research had generated, Grassmann left mathematics for work in Sanskrit studies, writing a Sanskrit dictionary that is still respected.

Jacques Salomon Hadamard
1865-1963

French mathematician who made important contributions in prime number theory, integral functions, and singularities of functions. He also made significant contributions in the partial differential equations used extensively in mathematical physics and in boundary value problems. Hadamard was active in politics, especially in the controversial Dreyfus Affair, which involved his relative Alfred Dreyfus. Hadamard's left-leaning politics and campaigns for peace following World War II made him politically unwelcome in the United States, though the support of his colleagues never wavered.

Hermann Hankel
1839-1873

German mathematician who developed the Hankel functions, variations on the Bessel function. Hankel also worked on complex numbers, the history of mathematics, and function theory during his mathematical career, but his most important work was his theory of complex number systems in which he introduced and expanded on Hermann Grassmann's ideas in these areas. The Hankel transformation is also widely used in the study of functions.

Heinrich Eduard Heine
1821-1881

German mathematician who made important contributions to the field of mathematical analysis. He is best known for formulating Heine's theorem, also known as the Heine-Borel theorem, which states that "a subset of the reals is compact if and only if it is closed and bounded." Heine also worked on Legendre polynomials, Bessel functions, problems of continuity, and Lamé functions.

Charles Hermite
1822-1901

French mathematician who dealt chiefly with Abelian and elliptic functions as well as the theory of numbers. During his lifetime, he held positions at the finest schools of Paris—Ecole Polytechnique, Collège de France, Ecole Normale Supérieure, and the Sorbonne University. Of all his students, the best known is Henri Poincaré (1854-1912), a brilliant mathematician-physicist fond of the philosophy of science. Hermite is often recognized as one of the greatest mathematicians of the nineteenth century.

Ludwig Otto Hesse
1811-1874

German mathematician who introduced the Hessian determinant while investigating the properties of cubic and quadratic curves. Hesse taught physics and chemistry prior to his career in mathematics, then was appointed to positions at Heidelberg and Munich as a professor of mathematics. Hesse's main body of work was in developing theories of invariants and algebraic functions. His theory of homogeneous forms was published with that of Arthur Cayley's.

David Hilbert
1862-1943

German mathematician whose work in geometry was more influential than any mathematician since Euclid. His work helped to put geometry on a formal footing, while his work in other areas proved equally fundamental. Hilbert also made significant contributions to mathematical physics, including work that paralleled Albert Einstein's during the development of general relativity theory. Although his contributions are too numerous to do more than simply mention here, he is remembered as one of the greatest mathematicians in the history of the field.

George William Hill
1838-1914

American mathematician and physicist who performed landmark work in solving problems in orbital mechanics. Hill worked extensively on problems of orbits, including those of the Moon and the moons of Jupiter. His contributions to mathematical astronomy and celestial mechanics, including the very difficult three- and four-body problems, were considered major and fundamental by many leading mathematicians of the day. Hill was elected a Fellow of the Royal Society and awarded the Copley medal. He was also elected President of the American Mathematical Society.

(Marie Ennemond) Camille Jordan
1838-1922

French mathematician who worked primarily in topology (the study of surfaces), introducing many important concepts in the field. Jordan taught at the École Polytechnique and the Collège de France, and his work inspired students Sophus Lie and Felix Klein in their future work as important mathematicians. Jordan was also interested in the theory of finite groups; some of his work in this area was motivated by his interest in crystal structures.

Thomas Penyngton Kirkman
1806-1895

English mathematician whose most important work involved various aspects of polyhedra (shapes with multiple sides). His other work involved group theory and, in his late 70s, the theory of knots. However, despite his many serious contributions to mathematics, Kirkman is perhaps best known for the "Fifteen Schoolgirls Problem," in which he states, "If fifteen young ladies of a school walk out three abreast for seven days in succession: it is required to arrange them so that no two shall walk abreast more than once."

Sofia Vasilyevna Kovalevskaya
1850-1891

Russian mathematician who, despite the limitations imposed upon women scientists of her time, made major contributions to the studies of partial differential equations, celestial mechanics, and the studies of rigid bodies. Married to paleontologist Vladimir Kovalevsky in order to travel abroad to study, she impressed her professors with her extraordinary talents in mathematics. Unfortunately, she died of influenza and pneumonia at the age of 41, at the height of her career.

Leopold Kronecker
1823-1891

German mathematician whose primary contributions dealt with elliptic functions and algebraic theory. Independently wealthy, Kronecker did not have to work and pursued mathematics for the love of the field. He became the editor of Crelle's journal after Crelle's death in 1855. Although he made significant contributions to mathematical research, he fell out of favor with his colleagues when he refused to accept infinite numbers, transcendental numbers, irrational numbers, and other similar concepts.

Ernst Eduard Kummer
1810-1893

German mathematician who worked on the theory of functions and developed the concept of "ideal numbers," which played a major role in most subsequent attempts to solve Fermat's Last Theorem. This work also proved important to other areas of mathematics and, for his contribution, Kummer was awarded the Grand Prize by the Paris Academy of Sciences in 1857. Renowned as a fine teacher of mathematics, he was also elected a Fellow of the Royal Society and a member of the Berlin Academy of Sciences.

Gabriel Lamé
1795-1870

French mathematician who made important contributions in the areas of differential geometry and number theory. Lamé also worked on Fermat's Last Theorem, engineering mathematics,

and elasticity, as well as studying the properties of diffusion in crystalline solids. His work in elasticity is honored by two constants named after him, and in other branches of mathematics by the Lamé curves that bear his name.

Adrien-Marie Legendre
1752-1833

French mathematician who proved that there are an infinity of prime numbers in any arithmetic progression whose first term and increment are relatively prime and who obtained many results in elliptic function theory and celestial mechanics. In 1805 Legendre published the method of least squares in a study on the orbit of comets. What we know as Euclidean geometry owes much of its form to the textbook Legendre wrote for the children of the French Revolution.

Emile Michel Hyacinthe Lemoine
1840-1912

French mathematician who did important work in geometry. Lemoine's contributions were underappreciated during his lifetime, in part because his methods—though shorter, more sophisticated, and more elegant than others that existed at that time—were difficult to understand and to follow. In addition to his mathematical skills, Lemoine was an accomplished musician, joining a respected chamber orchestra. He also worked as a civil engineer and edited a mathematical journal for several years before his death.

(Marius) Sophus Lie
1842-1899

Norwegian mathematician who developed the mathematical formalism known as Lie algebra and made significant contributions in the area of differential and partial differential equations. Lie was a long-time friend and scientific collaborator with Felix Klein; their early collaboration on problems of transformation groups influenced much of Klein's later work. Lie served as the chair of mathematics at Leipzig for nearly a decade, returning to his native Kristiania (later renamed Oslo) shortly before his death.

Carl Louis Ferdinand von Lindemann
1852-1939

German mathematician who was the first to prove π is a transcendental number (a number that is not the root of any algebraic equation with rational coefficients). Lindemann's work was primarily in geometry and analysis, in which, among other accomplishments, he proved that it is impossible to "square the circle"—that is, to construct a square with an area identical to a given circle using only a ruler and compass.

Rudolf Otto Sigismund Lipschitz
1832-1903

German mathematician whose most important work was in the areas of quadratic differential forms and mechanics in physics. His work on Hamilton-Jacobi methods of integrating equations of motion were readily applied to celestial mechanics to great effect. He also described the "Lipschitz condition"—an inequality guaranteeing a unique solution to certain differential equations. Lipschitz was also the first to apply a branch of mathematics called "Clifford algebras" to problems involving rotations of objects in Euclidean spaces.

Johann Benedict Listing
1808-1882

German mathematician who coined the term "topology" for the study of surfaces. During his inquiries into topology, Listing independently discovered the properties of the Möbius strip, established by August Möbius in 1865. Listing also produced major works on the mathematical descriptions of optics and other works of applied (as opposed to "pure") mathematics.

(Victor Mayer) Amédée Mannheim
1831-1906

French mathematician who developed the style of slide rule that would become the standard for over 100 years. Although Mannheim did not invent the slide rule (versions had existed for some years), he was responsible for standardizing it in the form that became commonplace until the popularization of pocket calculators in the 1970s. Mannheim, an officer with the French Artillery, also studied surfaces and Chasles's theory of transformations.

Andrei Andreevich Markov
1856-1922

Russian mathematician who was the first to give a rigorous proof of the central limit theorem and who discovered that this theorem and the law of large numbers also hold for certain dependent trials. A probabilistic process in which the probability of an outcome depends only on the outcome of the previous trial is now called the Markov process. This is used to model a wide range of physical phenomena, from gas diffusion to traffic problems.

Hermann Minkowski
1864-1909

German mathematician whose four-dimensional geometry of space and time (known since as "Minkowski space") influenced the development of Albert Einstein's general theory of relativity. Minkowski's contributions also included the use of geometric methods in number theory. He was born of German parents in what was then part of the Russian Empire (now Lithuania) and educated in Germany. He taught mathematics at universities in Bonn, Königsberg, Zurich, and Göttingen. Minkowski's major work was *Raum und Zeit* (*Space and Time*), published 1907.

Magnus Gösta Mittag-Leffler
1846-1927

Swedish mathematician who founded and edited the journal *Acta Mathematica* for 45 years, and who made important contributions in the area of mathematical analysis. His contributions included work on the general theory of functions and investigations into the relationship between dependent and independent variables. Mittag-Leffler was not only a gifted mathematician, but possessed the finest mathematical library in the world and was widely respected by his peers both professionally and personally. During his life, he was honored by virtually every mathematical society in the world.

August Ferdinand Möbius
1790-1868

German mathematician and theoretical astronomer best remembered for his developments in topology, including the one-sided figure since known as the "Möbius strip." Möbius taught astronomy at the University of Leipzig, and supervised the establishment of its observatory. His astronomical works include *Die Elemente der Mechanik des Himmels* (*The Elements of Celestial Mechanics*, 1848). In mathematics, his *Der Barycentrische Calcul* (*The Calculus of Centers of Gravity*, 1827) played an important part in the subsequent development of projective geometry.

Gaspard Monge
1746-1818

French mathematician who contributed to the study of projective geometry. Monge not only invented descriptive geometry, but pioneered the development of analytical geometry. His name appears often in histories of the French Revolution, during which time he helped establish the metric system and the renowned Ecole

Polytechnique in Paris. He was granted the title Comte de Péluse by Napoleon in 1808.

Eliakim Hastings Moore
1862-1932

American mathematician who, in addition to his important work in algebra and group theory, helped to found an American school of mathematics on equal terms with the superior scholarship of foreign universities. Under Moore's leadership, the University of Chicago became the leading American university for the study of mathematics and one of the foremost schools in the world. Moore was elected president of the American Mathematical Society in 1901 and edited the *Transactions of the American Mathematical Society* from 1899 until 1907.

Max Noether
1844-1921

German mathematician who was one of the leaders in the study of algebraic geometry in the nineteenth century. Following the lead of Luigi Cremona, Noether studied the invariant properties of algebraic functions during certain transformations. Noether's daughter, Emmy, followed in Noether's footsteps, conducting researches into many of the same areas of mathematics and developing general formulations of some of his more important results. Noether was physically handicapped his entire life due to an attack of polio as a youth.

Moritz Pasch
1843-1930

German mathematician who discovered many details of Euclid's work that had gone unnoticed for over 2000 years. Pasch also argued that geometers (those who study geometry) placed excessive reliance on intuition rather than formal mathematical proofs, claiming that the principle of duality contradicted physical intuition in some areas of geometry, such as the distinction between points and lines. These ideas had a significant influence on David Hilbert during his later work.

George Peacock
1791-1858

English mathematician who contributed to the areas of calculus and algebra theory. Educated at home until shortly before entering Cambridge, Peacock became a tutor and lecturer at Trinity College. Peacock worked to introduce advanced methods of calculus to English mathematicians, winning a reputation as a reformer that would

stay with him his entire life. In addition to his mathematical reforms, he worked to reform statutes at Cambridge University and was a member of governmental reform committees as well.

Benjamin Peirce
1809-1880

Benjamin Peirce is generally regarded as the first American research mathematician. Peirce was a professor of mathematics and astronomy at Harvard from 1833 to 1880. He also served in the important position of superintendent of the Coast Survey from 1867 to 1874. Peirce's most important work in mathematics was *Linear Associative Algebra,* in which he investigated various possible systems of algebra. Written and circulated privately in 1870, *Linear Associative Algebra* was published posthumously in 1881.

Charles Sanders Peirce
1839-1914

American mathematician who contributed to the study of associative algebras, mathematical logic, and set theory, extending the work of his father, Benjamin Peirce. Charles Peirce was born in Cambridge, Massachusetts, and attended Harvard University. He worked for many years on the Coast and Geodetic Survey. Between 1879 and 1884 he gave occasional courses on logic at Johns Hopkins University, though never held an academic post during his career.

Siméon Denis Poisson
1781-1840

French mathematician important for advances in the Fourier series and for his work on definite integral theory. Poisson began by studying medicine, changing to mathematics in 1798. A year later, he wrote his first paper, which was on finite integrals. Studying under two of the day's premier mathematicians, Pierre Laplace and Joseph Lagrange, Poisson quickly became recognized for his work in both mathematics and physics. Poisson published significant research papers in the areas of mathematics, electricity and magnetism, and astronomy during his career.

Gregorio Ricci-Curbastro
1853-1925

Italian mathematician who invented the field of absolute differential calculus. This was a more general extension of work initiated by Carl Gauss and, as the foundation of tensor analysis, was used extensively by Albert Einstein in developing the theory of relativity. Ricci-Curbastro collaborated extensively in his later years with his student Tullio Levi-Civita. One of his most important papers (and the only one in which he used only the name "Ricci") was requested by Felix Klein and acknowledged as a necessary piece of work.

Paolo Ruffini
1765-1822

Italian mathematician and physician who was the first to prove that quintic equations (equations in which one term is raised to the fifth power) cannot be solved using only radicals (square roots). This important proof, which flew in the face of accepted mathematical thinking, was not well received by mathematicians of the day; with the exception of Augustin Cauchy, on whom it had a tremendous impact. In developing his proof, Ruffini also laid the foundations for modern group theory, which did not then exist.

Carle David Tolmé Runge
1856-1927

German mathematician and physicist who made several discoveries toward finding solutions for some types of algebraic equations. Starting his professional life as a mathematician, Runge published several papers in Gösta Mittag-Leffler's journal, *Acta Mathematica*. Later, after accepting a position at Hanover, Runge began studying the properties of spectral lines from various elements, eventually showing that helium has spectral lines distinct from those of hydrogen. This helped convince chemists that hydrogen and helium were separate elements.

George Salmon
1819-1904

Irish mathematician who is best known for his series of textbooks written between 1848 and 1862. Salmon also worked with Arthur Cayley on the properties of cubic surfaces, and made other discoveries involving surfaces, including ruled surfaces (surfaces covered with a regular gridwork of lines).

Friedrich Wilhelm Karl Ernst Schröder
1841-1902

German mathematician best known for his work on ordinal numbers and ordered sets. Schröder was also one of the first mathematicians to use the term mathematical logic, and his studies in this area were heavily influenced by George Boole's logic system. Schröder's detailed work on algebraic logic was used in developing modern algebraic theory as well as in providing the foundation for modern lattice theory.

Nermann Cäsar Hannibal Schubert
1848-1911

German mathematician whose discoveries in enumerative geometry were long overlooked. Enumerative geometry is the branch of mathematics that looks at the portions of algebraic geometry having a finite number of solutions. Schubert established a system to solve enumerative geometry problems using methods developed by Michel Chasles. The relative lack of mathematical rigor led to a long period in which his results were overlooked, but many have been examined more thoroughly today and found to be correct.

Hermann Amandus Schwarz
1843-1921

German mathematician who worked on determining surfaces of least area and the conformal mapping of polyhedral surfaces onto spherical surfaces. His work on least-area (or minimal) surfaces was a problem of the calculus of variations. Such surfaces are used today as, among other things, the tops of large tented structures and are also called "soap bubble" surfaces. Schwartz was also a member of the local Volunteer Fire Brigade and assisted the local railway stationmaster.

William Shanks
1812-1882

English mathematician who calculated extraordinarily precise values of π, e, and Euler's constant, g. Shanks's calculations of π, carried out manually to 707 decimal places, were unfortunately found to be correct for only the first 527 decimal places. Similarly, a minor mistake in his calculations of e resulted in verification of only the first 528 places. Shanks also published a table of primes up to the number 60,000 and calculated the natural logarithms of several numbers to more than 130 decimal places.

Mary Somerville
1780-1872

Scottish mathematician and scientific writer who published papers in physics and became well known for her 1831 translation of Pierre Laplace's *Mécanique Céleste*. One of the few female mathematicians of her era, Somerville became interested in mathematics while studying Euclid with the help of her younger brother's tutor. A strong advocate of women's education, she published her own treatise *Finite Differences* and several other books on popular science.

Oxford honored her by naming one of its colleges after her.

Karl Georg Christian von Staudt
1798-1867

German mathematician and astronomer who did important work in the branch of mathematics known as projective geometry. His *Geometrie der Lage,* published in 1847, helped to free projective geometry from any metrical basis, and further work presented geometric solutions to quadratic equations. Staudt also determined the orbit of a comet while working on his doctorate. He was a professor at the Polytechnic School at Nuremberg and, later, at the University of Erlangen.

Jakob Steiner
1796-1863

Swiss mathematician who was one of the greatest contributors in the field of projective geometry. Steiner made several important contributions to geometry, including the Steiner surface (which has an infinite number of conic sections) and the Poncelet-Steiner theorem, which states that only one given circle and one straight edge are required for any Euclidean construction. Contrary to feelings expressed by Moritz Pasch, Steiner felt that excessive calculation replaced thinking, but the practice of geometry stimulated it.

Thomas Jan Stieltjes
1856-1894

Dutch mathematician who is often called the father of the analytic theory of continued fractions. Stieltjes had an undistinguished academic career, relying upon his father to help him find a job after failing to complete his university studies. However, as a mathematician, he excelled, making fundamentally important contributions in differential equation theory, interpolation, elliptic functions, divergent series, and the gamma function. His most important work, published after his untimely death, was awarded a prize by the Académie des Sciences.

Jacques Charles Francois Sturm
1803-1855

Swiss mathematician and physicist who is best known for his part of the Sturm-Liouville problem in differential equations. Sturm worked in a number of mathematical specialties, including duplicating some of Augustin Cauchy's results, albeit in a mathematically superior manner. Sturm also worked extensively with problems in physics, making the first accurate measurement

of the velocity of sound in water. Sturm's work in heat diffusion was never published, but led to other mathematical results that later proved important.

Peter Ludwig Mejdell Sylow
1832-1918

Norwegian mathematician whose work in finite group theory was nothing short of profound. Sylow's reputation is based almost entirely on a single ten-page paper that appeared in 1872, in which he proves three theorems in finite group theory. These theorems form the foundation of the field and are used in virtually all work in this area of mathematics. Based on his reputation, Sophus Lie arranged to have a special chair created for Sylow at Christiania University, where Sylow taught from 1898 until his death.

James Joseph Sylvester
1814-1897

English mathematician who laid the basis for the fields of combinatorics and abstract algebra. He was the first Jew to achieve eminence in the English mathematical community, although he spent a number of years in the United States and created the first research mathematics department there at Johns Hopkins University. He was influential in the coining of mathematical terminology in the new fields that he had helped to create.

Peter Guthrie Tait
1831-1901

Scottish mathematician and natural philosopher who worked with William Thomson (later Lord Kelvin) on the theory of knots. Tait published tables of the properties of knots in support of Thomson's theory that held molecules were simply vortices or knots in the ether thought to pervade space. Tait also studied thermal conductivity and thermoelectricity, and wrote a paper on calculating the trajectory of golf balls. In 1860 he was awarded a position at Edinburgh, beating out James Clerk Maxwell, another candidate for the chair.

Paul Tannery
1843-1904

French mathematician and scholar who he became interested in the works of Greek mathematics and their corresponding philosophies. His use of Greek numerals in his historical research became so natural to him that he came to believe they had worthy advantages over the system in use at that time. Tannery became well known for his contributions to the 12 volume *Oeuvres de Descartes* (1897-1913).

John Venn
1834-1923

English mathematician and minister best known for developing Venn diagrams to graphically represent sets and their intersections. In addition to set theory, Venn studied logic and probability theory, and lectured in Moral Science at Cambridge University. Economist John Maynard Keynes described his "Symbolic Logic" as "probably his most enduring work on logic," and it is still considered an important achievement today. Venn co-authored a book on the history of Cambridge University and also invented a machine for bowling cricket balls.

Giuseppe Veronese
1854-1917

Italian mathematician who made important contributions to the study of projective geometry. At the early age of 26, Veronese described an n-dimensional hypothesis which showed that simplifications could be obtained in passing to high dimensions. He also noted that when a simple surface in high dimension was projected onto 3-D space, difficulties would arise. Subsequent mathematicians, such as Italians C. F. Manara and C. Segre, would continue to discuss and publish papers on Veronese's work well into the twentieth century.

Vito Volterra
1860-1940

Italian mathematician whose most important work was in the area of integral equations. Volterra showed how some of the work of William Rowan Hamilton and Karl Jacobi could be extended from mechanics to other problems in mathematical physics. He also published papers in which he introduced equations used in the biological study of predator-prey relationships. Refusing to take an oath of allegiance to the Fascist Italian government in 1931, Volterra was forced to leave the University of Rome and lived abroad for most of the rest of his life.

Léon Walras
1834-1910

Dutch mathematician who was instrumental in laying the foundation for the mathematical study of economics. His *Elements of Pure Economics* and other works attempted to establish economics as a branch of mathematics, while also advocating state nationalization of land and the abolition of taxes. His work was well received in America and Italy, but was spurned in Britain and France, though later received posthumous appreciation.

Pierre Laurent Wantzel
1814-1848

French mathematician who earned fame for his work on solving equations using radicals. He also published proofs of several famous problems in geometry, verifying that certain problems could not be solved using only a ruler and compass. Wantzel excelled in other areas as well, including physics, history, philosophy, and writing. He died at the early age of 34, apparently due to overwork as his body was unable to keep pace with the rigors of his mind.

William Whewell
1794-1866

British scientist and theologist who in 1840 wrote, "We need very much a name to describe a cultivator of science in general. I should incline to call him a scientist." This statement would later be recognized as the first written use of the term "scientist." Moreover, with a three-volume book entitled *History of the Inductive Sciences,* Whewell is best remembered as one of the first scholars to make a systematic treatise on the history and philosophy of science.

Grace Emily Chisholm Young
1868-1944

English mathematician who worked on spherical trigonometry and the foundations of calculus. Young was dissuaded from studying medicine by her family, taking up mathematics instead. With her husband, mathematician William Young, she published over 200 papers and several books that were highly respected; most of these, however, were published only under his name because women were not yet accepted in the sciences. She was awarded the Gamble Prize by Girton College for her contributions to mathematics.

Hieronymous Georg Zeuthen
1839-1920

Danish mathematician best known for his studies of medieval and ancient Greek mathematics. In addition to his historical studies, Zeuthen made important discoveries in algebraic geometry, developing the enumerative calculus first proposed by Michel Chasles. However, the movement of geometry towards mathematical rigor caused this work to be overlooked for many years until some of the more interesting results were later verified.

Bibliography of Primary Sources

Books

Boole, George. *An Investigation into the Laws of Thought.* 1854. Boole here offered one of the first sustained treatments of symbolic logic.

De Morgan, Augustus. *Formal Logic.* 1847. In this work De Morgan indicated how logical combinations, written in symbolic form, could be transformed into equivalent forms.

Fourier, Jean-Baptiste Joseph. *Théorie analytique de la chaleur* (Analytical Theory of Heat). 1822. An important work in which Fourier showed how the conduction of heat in solid bodies could be analyzed in terms of series with sines and cosines as terms. The implications of this method were much wider-ranging than heat conduction, affecting data analysis techniques in many scientific and engineering fields. It also had a great influence on the theory of functions of a real variable.

Frege, Gottlob. *Foundations of Arithmetic.* 1884. In this work Frege attempted to prove that all arithmetic reasoning could be established on the basis of logic alone. Frege defined the equality of numbers in terms of the establishment of a one-to-one correspondence between elements of two sets. He also proposed that each of the individual integers might be interpreted as the class or infinite set of sets containing all sets with the same number of members.

Gauss, Carl Friedrich. *Disquisitiones Arithmeticae.* 1801. Published when Gauss was only 24, this far-ranging treatise on number theory had an enormous impact on mathematics in the nineteenth and twentieth centuries.

Laplace, Pierre Simon. *Théorie Analytique des Probabilités.* (Analytical Theory of Probability). 1812. An influential work from the French mathematician and astronomer.

Lobachevsky, Nikolai. *Geometrical Researches on the Theory of Parallels.* 1840. Lobachevsky, who was an important pioneer in non-Euclidean geometry, was praised by Carl Friedrich Gauss for this work.

Peano, Guiseppe. *The Principles of Arithmetic.* 1889. Peano in this book introduced symbol manipulation rules equivalent to those of Gottlob Frege but in far more tractable notation, and in particular presented a set of axioms for the natural numbers that conclude with a statement of the principle of induction.

Peirce, Benjamin. *Linear Associative Algebra.* 1881. In this posthumously published work Peirce investigated various possible systems of algebra.

Staudt, Karl Georg Christian von. *Geometrie der Lage* (Geometry of Projection). 1847. Von Staudt here established an axiomatic system for projective geometry. This axiomatic system did not make use of distance; instead, von Staudt was able to create within the axiomatic system a notion of coordinates intrinsic to geometry, as opposed to the extrinsic notion of coordinates based on distance.

Other

De Morgan, Augustus. "Induction (Mathematics)." 1838. This article is noteworthy as the work in which De Morgan named the principle of induction.

Frege, Gottlob. *Begriffsschrift* (Concept Language). 1879. In this 88-page pamphlet, Frege presented an entirely symbolic language that could express the logical possibilities in any situation. Frege made a number of important innovations. In particular, Frege's formalism allowed for statements that include quantifiers, such as "for all" or "for some." Frege also provided for statements that included variables, that is, symbols that could stand for other concepts. Unfortunately, Frege's notation, which made use of vertical and horizontal lines, proved quite unwieldy and not well suited for publication.

Green, George. "Essay on the Application of Mathematical Analysis to the Theories of Electricity and Magnetism." 1828. An important work in the burgeoning nineteenth-century field of electromagnetism.

Klein, Felix. "Erlanger Programm." 1872. In this lecture upon his appointment to the University of Erlangen, Klein offered a unifying theory for the various geometrical studies of the time.

Poncelet, Jean-Victor. "Traité des propriétés projectives des figures." ("Treatise on the Projective Properties of Figures.") 1822. This work set the stage for a reformulation and unification of geometry that was to culminate in the work of Felix Klein in 1872. The most important study that Poncelet undertook in "Traité" was the study of central projections.

Quetelet, Lambert Adolphe Jacques. "A Treatise on Man and the Development of his Faculties: An Essay on Social Physics." 1835. In this famous 1835 treatise, noted statistician Quetelet expanded upon his concept of the "average man," a fictitious being for whom everything happens according to the average results obtained for the whole society.

Medicine

Chronology

1800 Humphry Davy, an English chemist, introduces the first chemical anesthetic, nitrous oxide or laughing gas.

1805 Morphine, the active ingredient in opium, is isolated by German chemist Friedrich Wilhelm Adam Serturner, who thus lays the groundwork for alkaloid chemistry.

1819 René Laënnec, a French physician, invents the stethoscope.

1825 French physician Pierre-Fidèle Bretonneau performs the world's first successful tracheotomy for croup.

1846 American surgeon John Collins Warren becomes the first physician to use ether as an anesthetic.

1854 While caring for British patients in the Crimean War, Florence Nightingale begins to establish new standards for nursing and patient care.

1858 First edition of *Gray's Anatomy* published.

1865 Laying the groundwork for antiseptic surgery, English surgeon Joseph Lister uses carbolic acid (phenol) to prevent infection during an operation on a compound fracture.

1878 Patrick Manson, a Scottish physician, shows that the tropical disease elephantiasis is transmitted by mosquitoes.

1882 German bacteriologist Robert Koch announces discovery of the tuberculosis bacillus, proving the existence of pathological agents.

1895 Wilhelm Konrad Röntgen, a German physicist, discovers x rays.

1899 The German pharmaceutical firm of Farbenfabriken Bayer introduces aspirin.

Overview:
Medicine 1800-1899

Despite the exciting advances that took place in science and medicine in the seventeenth and eighteenth centuries, it was only in the nineteenth century that medicine itself became scientific. This was largely the result of the integration of the natural sciences into medical theory. During the eighteenth century the foundations of scientific medicine were first established. The ideas of the Enlightenment had inspired the search for rational systems of medicine, as well as practical means of preventing disease and improving human welfare. Social and medical reformers argued that scientific investigations of the abominable conditions of cities, navies, armies, and prisons, as well as the human body and pathological signs and symptoms, could improve the health and prosperity of society as a whole. Advocates of public health and preventive medicine, like Johann Peter Frank (1745-1821), sometimes urged states to adopt authoritarian methods to accomplish their goals and ideals. By studying the lives of peasants and workers, reformers hoped to make physicians and philosophers see how diseases were generated by a social system that kept whole classes of people in conditions of permanent misery.

Giovanni Battista Morgagni (1682-1771), pioneer of pathological anatomy and author of *On the Seat and Cause of Disease* (1761) established the existence of correlations between clinical symptoms and postmortem findings. Morgagni's research helped establish a new epoch in medical science and a new attitude toward specific diagnostic and surgical interventions. His work encouraged scientists to find ways of detecting hidden anatomical lesions in living patients. This goal was realized by the chest percussion studies of Leopold Auenbrugger (1722-1809), the invention of the stethoscope by René Laënnec (1781-1826), the introduction of increasingly sophisticated medical instruments, the establishment of "hospital medicine" and "pathological anatomy" at the Paris Hospital, the "tissue theory" of Marie François Xavier Bichat (1771-1802), the "numerical method" (clinical statistics) of Pierre Charles Alexandre Louis (1787-1872), and so forth. Although hospital reform was difficult and expensive, the hospital was transformed into the new center of medical treatment, teaching, and research. Large urban hospitals offered unprecedented opportunities for clinical experimentation, autopsies, and statistical studies. As hospitals assumed a more important role in the care of the patient, especially in growing urban areas, nursing emerged as a respectable profession for women. Gaining admission to the medical profession itself was very difficult for women, but Elizabeth Blackwell (1821-1910) and others demonstrated that women could practice medicine, establish clinics, hospitals, and medical colleges.

Although nutrition is generally regarded as a twentieth century science, the belief that health and long life depend on the regulation of food and drink is one of the most ancient and universal principles of medical philosophy. The chemical revolution of the eighteenth century challenged traditional ways of classifying foods. By the end of the nineteenth century these chemical categories were giving way to a new physiological concept of the role of food substances in the "animal economy." The modern science of nutrition grew out of efforts to understand and isolate the dietary factors that prevented deficiency diseases, but this required considerable progress in chemistry. Nevertheless, the scurvy experiments of James Lind (1716-1794) proved it was possible to prevent diseases by specific changes in diet. Lind tested possible antiscorbutics in a controlled dietary experiment and demonstrated that oranges and lemons cured scurvy. Nevertheless, lemons did not become part of standard rations in the American Navy until 1812. During the nineteenth century, the threat of infectious diseases diverted attention from dietary and degenerative diseases. But in the twentieth century, the chronic diseases, especially those that seem to be related to diet and obesity, have overshadowed the threat of infectious disease.

Perhaps the greatest medical achievement of the Age of Enlightenment was the discovery that inoculation and vaccination could prevent epidemic smallpox. Smallpox was such a dangerous and widespread threat that it was called "the most terrible of all the ministers of death." In many parts of Asia, India, Turkey, and Africa, folk healers attempted to protect people from virulent smallpox by "inoculation," that is, deliberately giving them a mild case of the disease with the aim of stimulating the body's resistance against subsequent exposures. European doctors dismissed these practices as barbaric superstitions, but during the eighteenth century increasing interest in natural "curiosities" led to closer observation of folk medicine. Smallpox inoculation gave medical practitioners and public health officials unprecedented responsibility for the control of epidemic disease. Weighing the risks and benefits of inoculation became an awesome responsibility for parents. Inoculation also paved the way for the rapid acceptance of vaccination. Edward Jenner (1749-1823) tested the folk belief that cowpox, a mild disease, provided protection against smallpox. In 1798 Jenner published an account of his experiments. Despite the medical profession's tendency to resist new ideas and methods, Jennerian vaccination spread throughout the world within a decade. Although debates about the safety and efficacy of preventive vaccines have raged ever since the first experiments on smallpox inoculation and vaccination, early in the nineteenth century some physicians predicted that vaccination would soon eradicate smallpox. In 1958 the World Health Organization (WHO) adopted a Smallpox Eradication Program that led to the end of smallpox in 1977. Public health authorities hoped that the lessons learned in the smallpox campaign would lead to global immunization programs for controlling diphtheria, whooping cough, tetanus, measles, poliomyelitis, and tuberculosis.

Historians have called malaria the most devastating disease in history. Even at the end of the twentieth century malaria is still a major public health threat in many parts of the world. Seventeenth-century scientists discovered that quinine was a specific remedy for malaria, and quinine became one of the "tools of empire" that made European exploitation and colonization of Africa and much of Asia possible. The isolation of quinine in 1820 was one of the great achievements of nineteenth-century chemistry. Some nineteenth-century scientists predicted the imminent conquest of malaria, but indifference to "tropical medicine" has proved to be as pernicious a disorder as malaria, yellow fever, and other diseases of the tropics. As the construction of the Panama Canal demonstrated, scientific knowledge could be used to bring "tropical fevers" under control, if the economic and political incentives were sufficient. At the end of the twentieth century, the central problem in tropical medicine remains the same: the impoverished nations that need medicines and vaccines cannot afford them, and the wealthy nations which can afford to develop remedies for so-called tropical diseases have little motivation to do so.

Despite considerable difficulty, nineteenth-century clinical medicine was ultimately transformed by the integration of the great discoveries of the basic sciences with the traditional foundations of medical science, that is, clinical observation and autopsy. With the development of new instruments and ways of looking at the human body and pathological signs and symptoms, specialization became a fundamental aspect of the medical profession. Throughout most of history, specialists had been regarded as little more than quacks and empiricists. Fundamental changes in ways of looking at the body followed the establishment of the cell theory by Matthias Schleiden (1804-1881) and Theodor Schwann (1810-1882), cellular pathology, by Rudolf Virchow (1821-1902), and the experimental physiology of François Magendie (1783-1855) and Claude Bernard (1813-1878). Anesthesia and antisepsis transformed the ancient art of surgery, but surgeons could not yet control shock and blood loss. Advances in therapeutics, however, would be comparatively modest until the twentieth century when antimicrobial drugs were discovered. Scientists could, however, point to innovative diagnostic tests, vaccines, antitoxins, synthetic antipyretic drugs, serum therapy, and the search for the "magic bullets" that would mimic the body's own immune responses as major achievements.

Certainly, one of the great advances of nineteenth century was the establishment of the germ theory of disease by Louis Pasteur (1822-1895), Robert Koch (1843-1910), and others. Isolating the microbial agents that caused infectious diseases and elucidating their means of transmission promised great advances in the control, prevention, and treatment of epidemic diseases. Based on the fundamentals of germ theory, Paul Ehrlich (1854-1915) established the basis of chemotherapy, and along with Ilya Metchnikoff (1845-1916), the science of immunology. The work of Ehrlich, Emil von Behring (1854-1917), and Shibasaburo Kitasato (1852-1931) led to the production of antitoxins for diphtheria, tetanus, and other diseases. Koch's discovery of the microbe that causes tuberculosis and Pasteur's development of a vaccine to treat rabies were among the most dramatic achievements by the pioneers of germ theory. One unfortunate consequence of the great achievements of modern medicine is the widespread misconception that the infectious diseases have been conquered and that the chronic, degenerative diseases are our only remaining medical problems. Understanding the complex relationships that link health, disease, demography, geography, ecology, and economics, and the differences in patterns of disease found in wealthy nations and developing nations, remains a major challenge. The emergence of new diseases, such as AIDS, in the late twentieth century demonstrates the need for a global and historic perspective in medicine and in the biomedical sciences.

LOIS N. MAGNER

René Laënnec Revolutionizes the Diagnosis of Chest Diseases with His Invention of the Stethoscope

Overview

In 1816 French physician René Laënnec began using a device of his own invention to listen to the sounds of the living heart and lungs. This simple advance revolutionized the diagnosis of chest diseases and later contributed to understanding their pathology and therapeutics—that is, what they are and how to treat them.

Background

Some of the best observers in the history of medicine lived in the early nineteenth century. The state of medical science then was such that physicians could cure almost nothing but could recognize and describe a great deal that would escape physicians today. Using only their five senses, their skill in observation and description was generally more acute than that of today's doctors, who have at their service arsenals of diagnostic instruments and therefore do not have to depend so much upon direct physical examination.

Even the best physical examination, however, can provide only a superficial, incomplete account of the health of the patient. Doctors have always sought better methods to examine patients and diagnose disease. Sometimes these better methods involve new instruments, sometimes new techniques, sometimes both.

In the 1750s Viennese physician Leopold Auenbrugger (1722-1809) discovered that the healthy chest and the diseased chest sound different when struck. The healthy chest, dry and full of air, sounds like a cloth-covered drum. The diseased chest, which may contain various thick fluids, sounds more muffled. Auenbrugger developed a technique of striking the patient's chest gently but firmly with his fingers, and he learned to recognize the many meanings of the different sounds in different parts of the chest. He called this technique "percussion" and published his results in Latin in 1761.

The medical community did not think highly of percussion until Napoleon Bonaparte's personal physician, Baron Jean Nicolas Corvisart des Marets (1755-1821), published his French translation of Auenbrugger's treatise in 1808. After that percussion was universally accepted, almost overnight. One of Corvisart's students was René-Théophile-Hyacinthe Laënnec (1781-1826).

As diagnosticians, both Corvisart and Laënnec understood the importance of listening carefully to the patient's chest. They also realized that percussion alone was not sufficient. They practiced a technique called "auscultation," or listening directly to chest sounds, which in those days meant that the physician had to put his ear on the patient's body.

Auscultation has been known since antiquity. The founder of the Western medical tradition, Hippocrates (460-375 B.C.), advocated it. The problem with auscultation was not to convince doctors of its value. Rather, the problem was mechanical: how to get the ear close enough. Hearing the chest clearly and accurately was difficult with fat patients. A patient might have lice or skin conditions that would spread to the doctor if he got too close. A woman might feel offended with the doctor's ear on or near her breasts. In those days the social demands of modesty and delicacy dissuaded male physicians from close physical contact with their female patients.

One day in 1816 Laënnec was frustrated trying to examine an unusually fat woman with symptoms of heart disease. Her gender, her obesity, and the size of her breasts prevented him from

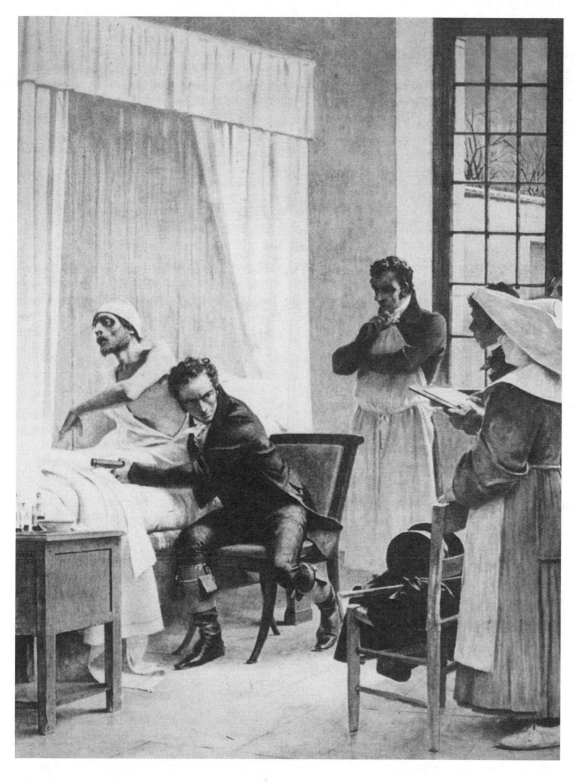

René Laënnec at the Necker Hospital in Paris, where he introduced the stethoscope. *(J.L.-Charmet; Academie Nationale de Medicine. Reproduced by permission.)*

listening to her heart. Almost instinctively he rolled up a few sheets of paper into a tight tube, put one end of this cylinder to his ear and the other on her chest. What he heard amazed him. The chest sounds were louder and clearer than he had ever heard them before, and the background noise was mostly eliminated when he blocked his other ear. This was the first stethoscope.

Laënnec named the instrument from two Greek words, *stêthos* ("chest") and *skopos* ("one who watches"), because he and Corvisart thought that it almost enabled them to "see" inside the chest. Laënnec called the technique of using the stethoscope "mediate auscultation" and distinguished it from "immediate auscultation," or putting the ear directly on the patient.

Impact

Laënnec instantly recognized the great advance he had created in diagnostic and clinical medicine, but he did not publish his results right away. Instead, he spent the next few years gathering data on the many different chest sounds and what they meant. He studied the chest sounds of dying patients and then verified his conjectures at autopsy. Thus, he could correlate sounds heard through the stethoscope with the conditions of diseased tissue in cadavers.

When he was ready, he introduced his discovery in 1819 in a massive two-volume work: *De l'auscultation médiate, ou traité du diagnostic des maladies des poumons et du coeur* ("On Mediate Auscultation, or: A Treatise on the Diagnostics of the Diseases of the Lungs and Heart"). This was soon translated into English by Sir John Forbes as *A Treatise on the Diseases of the Chest, in Which They Are Described According to Their Anatomical Characters, and their Diagnosis, Established on a New Principle by Means of Acoustick Instruments.* The second edition of Laënnec's book (1826) was even more thorough in depicting pathological, anatomical, and therapeutic findings and quickly became a classic in the English-speaking world when Forbes published his translation in 1827.

The principle of the stethoscope, namely, that sound travels better through solids than through air was known long before Laënnec. But he was the first to use this principle effectively in the practice of medicine. His challenge from 1816 to 1819 was to demonstrate to other physicians that the sounds heard with a stethoscope could provide more accurate diagnosis than those heard with the unaided ear. In other words, he had to prove that mediate auscultation was better medically than immediate auscultation. His arguments soon won many prominent converts who refined his research.

Laënnec's countryman Victor Collin (dates unknown) wrote *Des diverses méthodes d'exploration de la poitrine et de leur application au diagnostic de ses maladies* ("On the Various Methods of Exploring the Chest and Their Application to the Diagnostics of Chest Diseases"), which was translated into English by W. N. Ryland as *Manual for the Use of the Stethoscope: A Short Treatise on the Different Methods of Investigating the Diseases of the Chest.* Pierre Adolphe Piorry (1794-1879) combined the techniques of percussion and mediate auscultation and reported his results in *De la percussion médiate* ("On Mediate Percussion"). Jean-Baptiste Bouillaud (1796-1881) used the stethoscope to gather data for his classic description of endocarditis in *Traité clinique des maladies du coeur* ("Clinical Treatise on Heart Diseases."

Among the first of Laënnec's followers outside France was Irish physician William Stokes (1804-1878), whose books include *An Introduction to the Use of the Stethoscope* (1825) and *A Treatise on the Diagnosis and Treatment of Diseases of the Chest* (1837). Bohemian diagnostician Josef Skoda (1805-1881) made further advances in the classification and interpretation of chest sounds. His 1839 work *Abhandlung über Perkussion und Auskultation* ("Treatise on Percussion and Auscultation") has been translated into many languages and is useful for clinicians even today. The intricate descriptions of bronchitis and several other ailments by Philadelphia physician William Wood Gerhard (1809-1872) in *Lectures on the Diagnosis, Pathology, and Treatment of the Diseases of the Chest* (1842) remain unsurpassed for their accuracy and detail. Henry Ingersoll Bowditch (1808-1892) established the stethoscope as a diagnostic tool in America with *The Young Stethoscopist; or, The Student's Aid to Auscultation* (1846).

The published work of Laënnec, Forbes, Collin, Piorry, Bouillaud, Stokes, Skoda, Gerhard, Bowditch, and others was widely available to the medical community and clearly demonstrated the great medical value of the stethoscope. But many physicians worldwide were still reluctant to accept its use in their own practices. This reluctance was because of the shape and inconvenience of the instrument.

Laënnec had promptly replaced his earliest paper stethoscopes with wooden ones, about the same size and shape as a modern flashlight. Some of his successors used hollow wooden tubes for lung sounds and solid wooden tubes for heart sounds. Some used wooden tubes with ivory or brass attachments. But these were all "monaural" or "one-ear" stethoscopes. The physician could listen with only one ear and could not look directly at the patient while using this kind of instrument.

To try to overcome the significant disadvantages of the monaural stethoscope, British physician Nicholas Comins (dates unknown) proposed his design for a "binaural" or "two-ear" stethoscope in 1829, but it did not transmit sound as well as the monaural wooden tube and was soon abandoned. Many other clumsy attempts at a binaural stethoscope appeared in the next two decades.

The modern stethoscope is binaural and has a chestpiece consisting of two sides: a hollow "bell" for isolating low-frequency sounds and a flat "diaphragm" for isolating high-frequency sounds. About 30-40 cm. of flexible tubing runs between the bell/diaphragm assembly and the two earpieces. This now familiar shape did not begin to emerge until the 1850s. Through the independent efforts of British physician Arthur Leared (1822-1879), New York physician George P. Cammann (1804-1863), and others, practical binaural stethoscopes became commonly available after about 1856.

After the dawn of the binaural stethoscope, Austin Flint (1812-1886) made so much progress using mediate auscultation as a diagnostic tool that he was praised as "the American Laënnec." Among his important books in this field are *Physical Exploration and Diagnosis of Diseases Affecting the Respiratory Organs* (1856); *A Practical Treatise on the Diagnosis, Pathology, and Treatment of the Diseases of the Heart* (1859); and *A Manual of Auscultation and Percussion, Embrac-* *ing the Physical Diagnosis of Diseases of the Lungs and Heart, and of Thoracic Aneurism* (1876).

ERIC V.D. LUFT

Further Reading

Books

Allison, Linda. *The Stethoscope Book.* Reading, MA: Addison-Wesley, 1991.

Duffin, Jacalyn Mary. *To See with a Better Eye: A Life of Laënnec.* Princeton, NJ: Princeton University Press, 1998.

Kervran, Roger. *Laënnec: His Life and Times.* New York: Pergamon, 1960.

Marks, Geoffrey. *The Amazing Stethoscope.* New York: Messner, 1971.

Marks, Geoffrey. *The Story of the Stethoscope.* Folkestone, England: Bailey and Swinfen, 1972.

Reiser, Stanley Joel. "The Stethoscope and the Detection of Pathology by Sound." In *Medicine and the Reign of Technology.* Cambridge: Cambridge University Press, 1978.

Rogers, Spencer Lee. *The Monaural Stethoscope* San Diego, CA: Museum of Man, 1972.

Other

Smith, Hugh Macgregor. *The Transmission of Medical Knowledge: The Introduction and Acceptance of Mediate Auscultation in Great Britain, 1816-1843.* Thesis (Ph.D.). Minneapolis: University of Minnesota Press, 1998.

State University of New York at Stony Brook. *Stethoscope.* Videocassette. Philadelphia: Lippincott, 1980.

Human Digestion Studied by William Beaumont, Theodor Schwann, Claude Bernard, and William Prout

Overview

At the beginning of the 1800s the process of digestion was a mystery wrapped in conjecture and debate. During the nineteenth century, however, four men contributed important pieces toward solving the puzzle of digestion. American William Beaumont (1785-1853) first observed the workings of a living person's stomach in a patient with a gunshot wound that did not heal. Englishman William Prout (1785-1850) showed that hydrochloric acid was in digestive juice. German physiologist Theodor Schwann (1810-1882) discovered pepsin, the enzyme responsible for digestion in the stomach. French investigator Claude Bernard (1813-1878) uncovered the roles of the pancreas and liver in digestion and showed that the major organ of digestion was not the stomach but the small intestine. By the end of the century the work of these four had established that digestion does not happen in the stomach alone but is a complex process beginning with saliva in the mouth and involving the entire digestive tract.

Background

The nineteenth century began with controversy about the nature of body physiology and the meaning and nature of life and disease. Many physiologists believed in vitalism, a doctrine based on the view that a "vital" or spiritual force causes life, prompting them to argue that a process like digestion could not be described in

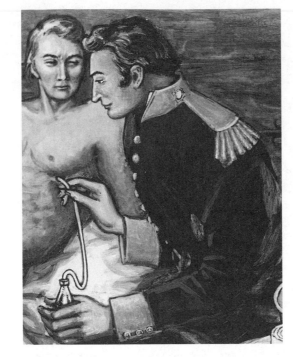

William Beaumont's treatment of Alexis St. Martin helped him make pioneering advances in the understanding of human digestion. *(UPI/Corbis-Bettmann. Reproduced by permission.)*

chemical or mechanical terms. Vitalism was in vogue during this time and would remain popular for several decades.

Interest in digestion had roots in classical Greece. Explanations of digestion included stomach heat, putrefaction, grinding, and fermentation. The seventeenth and eighteenth centuries spawned a great debate between those who believed it was a chemical process and those who insisted it was a grinding, mechanical process.

Seventeenth-century researcher Jan Baptiste van Helmont (1579-1644) had proposed that chemical action digested food by fermentation. Frenchman Rene de Reaumur (1683-1757) and Italian Lazzaro Spallanzani (1720-1799) experimented not only on animals and birds but also on themselves and argued that digestion was chemical. But the vitalists ridiculed their ideas, contending that in no way could human processes be described in such unspiritual terms.

At the turn of the nineteenth century European researchers were very interested in digestion and bitter controversy raged, especially in France. A well-known textbook on physiology by French professor Francois Magendie (1783-1855) argued that digestive juice was not a solvent and that any presence of acid in the stomach was caused by the breaking down of food or by

saliva. The argument was so intense during the 1820s that the French Academy of Science sponsored a contest on the process of digestion in animals. At this time American medicine lagged behind and, consequently, no major American researchers contributed to this debate in Europe.

Such was the background in the early years of the nineteenth century. The vitalists were in command. Controversy over chemical versus mechanical explanations was burning in Europe. Monistic theories of disease—the idea that all diseases have one cause and should be treated by bleeding and purging—were the major influences on the practice of medicine. Doctors had little formal training and became licensed by serving an apprenticeship with another doctor. Into this atmosphere came the four men whose research challenged these beliefs and set the stage for our understanding of the digestive system today.

Impact

On June 6, 1822, William Beaumont, a frontier army doctor, was called to a fur company store in Mackinac Island, Michigan Territory, to treat a Canadian trapper who had been shot at close range. Little did Beaumont realize that the call would give him the opportunity to change the course of knowledge about digestion, as well as lead to the development of experimental medicine.

The shot had created a wound the size of a man's hand in the abdominal area and, in spite of great effort to close it, left a hole in the stomach, called a gastric fistula. Beaumont thought surely his patient, Alexis St. Martin, would die, but within the year he was recovering and in good health— with the open hole still in his stomach. Beaumont could look directly into his stomach and observe its motion. He could pour food and drink in and siphon out the contents. Beaumont hired St. Martin to work for him so that he could continue his experiments.

In England William Prout, a physician turned chemist, investigated the gastric juices of animals with brilliant experiments. In 1824 he extracted stomach juices and demonstrated it contained hydrochloric acid. When he published his work, his contemporaries could hardly believe that such a strong acid could exist in the stomach of organisms and not cause harm. However, such was Prout's credibility that in 1827 they accepted his research into the digestion of food nutrients.

Prout divided food into categories: water, carbohydrates, fats, and proteins. While many of his ideas were based on speculation, he did publish an analysis of the saccharinous or carbohydrate class. Nutritionists today still use his classifications.

Beaumont, in America, had no knowledge of the controversy and interest in digestion in Europe. However, he did have access to a living human stomach and could go beyond what they could do. Although he had no research experience or training, Beaumont was a careful observer, wrote in his journals, and proceeded in an orderly manner.

On August 1, 1825, he began his controlled experiments by suspending cooked beef, salted beef, salt pork, raw beef, corned beef, stale bread, and cabbage in the stomach on silk strings and closing the hole with a bandage. St. Martin continued his work around the house. At one, two, and three o'clock Beaumont observed the way each item was digesting and carefully recorded his observations. When St. Martin complained of sickness, Beaumont observed a number of white spots in the stomach. He later realized he was looking face to face at indigestion. As might be expected, St. Martin tired of his role as a human guinea pig and needed some persuasion to continue.

In 1833 Beaumont published his findings in the book *The Experiments and Observations of the Gastric Juice and Physiology of Digestion.* After performing some 200 experiments over a period of 10 years, Beaumont listed 51 conclusions regarding the chemical nature of digestion. Primary among these was that the stomach secreted gastric juice from folds in the lining and that gastric juice was the agent of chemical breakdown. He described the inner coat of the stomach as pale pink covered with a mucous coat that changes appearance when diseased. The stomach moves sideways and up and down to churn its contents. He also described how alcohol causes gastritis or inflammation to the stomach lining. He found that vegetables were less digestible than other foods and that milk coagulates early in digestion.

Beaumont had revealed more about the stomach than had been known before. He earned recognition both in Europe and the United States. He also noted an important factor that he did not quite understand. An unknown substance was present that researchers in the United States could not identify because of the lack of organic chemistry analysis. That discovery would be the next piece of the puzzle.

German physiologist Theodor Schwann, working with the famous physiologist Johannes Mueller (1801-1858) in Berlin, became very interested in investigating digestive processes. He isolated a substance from the stomach that was separate from hydrochloric acid and called that substance pepsin. This turned out to be Beaumont's unknown factor. Pepsin, the first enzyme to be prepared from animal tissue, works with the hydrochloric acid to break down protein.

Schwann later discovered the muscular nature of the esophagus, noting that it contained striated muscle and acted as a pipe to move food from the mouth to the stomach. He also was the first to use the term metabolism to describe chemical changes in living tissue and applied the idea of cell theory to animals.

Claude Bernard, a French physiologist, developed an early interest in digestion while working as an assistant to François Magendie. Fascinated by Beaumont's research, Bernard replicated the gastric fistulas in animals. His wife and daughter, along with other antivivisectionists, strongly opposed his experiments on live animals .

One day Bernard noted that laboratory rabbits were passing clear urine like meat-eating animals. He assumed the animals had not been fed and were digesting their own tissues. He fed meat to the animals and studied the pancreas in autopsy. He found that pancreatic juice breaks down fat molecules into fatty acids and glycerol. While most of the research previously assumed all digestion took place the stomach, he showed the small intestine to be the major organ of digestion. Later, he found the nerves that control the digestive process.

His work on the pancreas led to a second great discovery—the role of the liver in digestion. He isolated glycogen, a white starchy substance, and determined the complex substance was made by the liver, stored as a reserve of carbohydrates, then released to keep a constant blood sugar level. In 1865 he wrote a textbook called *An Introduction of the Study of Experimental Medicine,* in which he argued that the precepts of vitalism do not explain life and urged that all medicine be based on methodical and experimental processes.

The work of these four men laid the foundation for the understanding of digestion and the treatment of its many complex diseases. By the end of the century, vitalism was waning, living systems were explained by physical and chemical processes, and experimental medicine using the scientific method was advancing.

EVELYN B. KELLY

Further Reading

Brock, W. H. *From Protyle to Proton: William Prout and the Nature of Matter.* Bristol: A. Hilger, 1985.

Horsman, Reginald. *Frontier Doctor William Beaumont America's First Great Medical Scientist.* Columbia and London: University of Missouri Press, 1996.

Tarshis, Jerome. *Claude Bernard: Father of Experimental Medicine.* New York: Dial Press, 1968.

Virtanen, Reino. *Claude Bernard and His Place in the History of Ideas.* Lincoln: University of Nebraska Press, 1960.

The Establishment of Schools for the Disabled

Overview

During the nineteenth century the number of educational institutions in the United States grew rapidly. In addition to elementary schools, high schools, and colleges, schools devoted to the training and welfare of disabled children were established. Special schools were created to assist in the training and education of children who were blind, deaf, and mentally handicapped. The attempt to educate the handicapped was related to major nineteenth-century reform movements that fought against slavery and other forms of institutionalized injustice, mistreatment, and neglect.

Background

Education for the Deaf

Children who are severely deaf are unable to learn from the spoken word and may not develop spoken language. Such children are considered mute through deafness. The lack of normal communication may lead to isolation and great difficulty in education. Efforts to teach deaf students to read, write, speak, read lips, and use sign language were reported in the seventeenth century. Thomas Braidwood (1715-1806) established the first schools for the deaf in Edinburgh and London. Other schools soon followed. Schools for the deaf adopted a mixture of approaches to instruction; some specialized in silent methods, but most used the oral method, which called for lip reading and the use of speech.

In the United States the teaching of deaf-mutes has been traced back to the work of Philip Nelson in late-seventeenth-century Massachusetts. However, the establishment of schools for the deaf did not take place until the early nineteenth century. Francis Green of Boston was an important leader in efforts to educate deaf children. His deaf son had been sent to Braidwood's school in Edinburgh. Green conducted a survey to estimate the number of deaf children in Massachusetts and concluded that a special school for the deaf was needed. In 1812 John Braidwood, a grandson of Thomas Braidwood, began teaching deaf children in Virginia. This led to the establishment of a school for the deaf.

Following a census of the deaf in Connecticut in 1812, a group of Hartford citizens organized a society for the instruction of the deaf and sent the Reverend Thomas Hopkins Gallaudet (1787-1851) to study the methods used in Paris and Great Britain. Gallaudet had studied theology at Andover but decided to devote his life to the education of deaf-mutes. In 1817 Gallaudet and Laurent Clerc, a deaf instructor from Paris, established the American Asylum for the Deaf and Dumb in Hartford, an institution that is regarded as a landmark in the education of the handicapped. An appropriation from the state of Connecticut represented the first time any of the states had provided such financial support to a school for the handicapped. A major contribution from the federal government, in the form of a grant of public land, followed in 1819. Gallaudet's decision to adopt the sign language method used in Paris had a profound impact on the direction of deaf education in the United States. The method of instruction at the school was American Sign Language, with a manual alphabet, and writing. Ill health forced Gallaudet to retire in 1830.

In 1810 the Reverend John Stafford discovered a number of deaf children in New York City almshouses and attempted to teach them. This led to the founding of the New York Institution for the Deaf in 1818 with 62 students. In Philadelphia, David Seixas began teaching deaf children in 1820 and established a school one year later. The Hartford school sent Laurent Clerc to assist him. Other states soon established schools for the deaf. By the 1860s there were more than 20 schools for the deaf in the United States. Until 1867 all of the American schools for the deaf used the manual system of instruction. Oral instruction was adopted in 1867 by the Clarke school in Northampton, Massachusetts, and the Institution for the Improved Instruction for the Deaf (later the Lexington School for the Deaf) in New York City. Thus, the division between supporters of the manual (sign language) system and those who favored speech and lip reading was established in the nineteenth century. Many educators were influenced by Thomas Hopkins Gallaudet's son Edward Miner Gallaudet (1837-1917), who supported teaching speech to deaf children. Gallaudet had observed this approach during a tour of European schools in 1867.

A compromise in 1886 urged instructors of the deaf to teach all students to speak and read lips, in addition to the use of sign language. This was called the "Combined System." The establishment of day schools for the deaf, however, favored the use of oralism. The first of such schools was the Horace Mann school in Boston, which was founded in 1868. Sarah Fuller, who gave Helen Keller her first speech lessons, served as principle for 41 years. By 1894 there were at least 15 day schools for the deaf and many more were soon established. Students who attended such schools were able to enroll in regular high schools and colleges. In practice, schools for the deaf followed three methods of education: oral, manual, and combined, but methods of instruction had to be adapted to the needs of individual students.

Education for the Blind

The first efforts to educate and train the blind are attributed to Valentin Haüy (1745-1822), who has been called the "father and apostle of the blind." In 1784 Haüy established the National Institution for Blind Children in Paris. By the early nineteenth century news of Haüy's success in teaching the blind to read led educators in other countries to establish similar schools. Schools were opened in Liverpool (1791), London (1799), Vienna (1805), Berlin (1806), Amsterdam (1808), Stockholm (1808), and Zurich (1809), among others.

Three schools for the blind were established almost simultaneously in the United States. American educator and physician Samuel Gridley Howe (1801-1876) directed the New England Asylum for the Blind (1832), later called the Perkins School for the Blind. In 1831 Howe was asked to establish an asylum for the blind. To prepare himself Howe studied the methods of instruction used in European schools. He began teaching blind children at his father's house in Boston in 1832. This small venture became the Perkins school.

Howe and his wife, Julia Ward Howe (author of "The Battle Hymn of the Republic" and first president of the New England Woman's Suffrage Association) were leaders of the anti-slavery movement. Along with the well-known educational reformer Horace Mann (1796-1859), Howe worked to establish public education in Massachusetts as well as special educational institutions for the mentally ill, the retarded, and the blind. Mann and Howe also fought to improve conditions in insane asylums and prisons. In 1837 Massachusetts created the nation's first State Board of Education and selected Mann to be the board's first secretary.

The remarkable story of Laura Dewey Bridgman (1829-1889), a deaf-mute, brought considerable attention to the Perkins school. An attack of scarlet fever at the age of two destroyed Bridgman's sight and hearing. She entered the Perkins school in 1837, and Howe began to teach her to recognize the words for common objects and then the individual letters. Famed English writer Charles Dickens visited the school in 1842 and wrote enthusiastically about Howe's success with Bridgman. Bridgman lived at the Perkins school until her death. Her story was regarded as a landmark in the education of the deaf and blind and was told in several biographies. Although both Howe and Mann used Bridgman as proof that all children were born with a natural potential for learning, they tended to exaggerate her accomplishments and minimized the difficulties experienced in her education.

A school for the blind was opened in New York (1832) by John D. Russ, and another in Philadelphia (1833) by Julius R. Friedlander. These schools were supported by private contributions. In 1837 a state-supported school for the blind was opened in Ohio. Because educators believed that blind children required special training before reaching school age, little children were originally placed in residential nursery schools. Educators later decided that it was better to keep blind children at home if the parents could provide the proper environment. Some schools specialized in the care and education of deaf-blind children.

Systems of using tangible letters, such as wooden blocks or cast metal letters, for the blind appear as early as the sixteenth century. However, Haüy's method of printing in relief is regarded as his most important contribution to the blind. James Gall of Edinburgh produced the first book for the blind in Great Britain in 1827. The first American book for the blind was printed in Philadelphia in 1833, using a system invented by Friedlander. Many books were printed at the Perkins School using a system developed by Howe. Despite the significance of these early systems, they were generally quite difficult to use and were eventually abandoned. Louis Braille (1809-52), a blind teacher at the National Institution for Blind Children (Paris), introduced a dot system that the blind could write as well as read. William B. Wait at the New York Institute developed the "New York point" system for the Education of the Blind in the 1860s. Joel W. Smith, a blind teacher living in Boston, invented American Braille in the 1870s. These systems were used in most American schools for the blind until 1916, when a modified version of the Braille alphabet was officially adopted.

Education for the Mentally Retarded

Howe conducted important investigations concerning the condition and treatment of mentally impaired children and lobbied strenuously for legislation providing for aid and education for the blind, the deaf, and the mentally retarded. As a result of reformist calls for humane and productive care of the feeble-minded, the governor of Massachusetts established a "Commission of Inquiry into the Conditions of the Idiots of the Commonwealth." Howe wrote an influential supplement to the 1848 report, entitled "On the Causes of Idiocy." This classic work was important in framing the debate about mental retardation as a social problem, raising concerns about the influence of heredity on disease and behavior and establishing the statistical dimensions of the perceived problem.

Scientists had become interested in the problem of educating mentally retarded children at the turn of the century, when Jean Marc Gaspart Itard, a French doctor, published his classic book, *The Wild Boy of Aveyron* (1801). Over a five-year period Itard had attempted to train and educate a young boy found naked in the woods of Aveyron. Itard's disciple, Édouard Séguin, introduced a system of physical and sensory activities to stimulate the development of mental processes. His system influenced the work of Maria Montessori, an Italian pediatrician who introduced her own methods of training mentally retarded children in the 1890s. The "Montessori system" was later adapted to the education of normal children.

The merits of institutionalization and deinstitutionalization of the handicapped and mentally retarded was already subject to debate by the mid-nineteenth century. Hervey B. Wilbur founded a private residential school for the mentally retarded, known as the Barre School, in 1848. Two years later Howe established the Massachusetts School for Idiotic and Feeble Minded Youth, the first residential public school for the mentally retarded in the United States. Similar residential schools for the training of the mentally retarded were soon established in other states. In 1896 the first public school special class for the education of the mentally retarded was established.

Impact

The establishment of specialized schools, institutions, asylums, and hospitals was, at least in part, the outcome of the nineteenth-century tendency to use institutional solutions to solve social problems. This resulted in the transfer of functions that had previously been considered part of the private family sphere into public institutions administered by a newly emerging class of "experts" and professionals. Indeed, before 1800 confinement or residential instruction of the mentally ill and the disabled was very rare. The establishment of specialized schools for the disabled in the first half of the nineteenth century reflected a remarkably optimistic view of the advancement of knowledge and the possibility of ameliorating or even curing infirmities of mind and body. Although the early attempts to assist the disabled were often disappointing and failed to establish a path from dependency to independence, such schools called attention to the complex problems involved and provided a foundation for more sophisticated approaches.

LOIS N. MAGNER

Further Reading

Books

Bender, Ruth E. *The Conquest of Deafness: A History of the Long Struggle to Make Possible Normal Living to Those*

Handicapped by Lack of Normal Hearing. Danville, IL: Interstate Printers & Publishers, 1981.

Richards, Laura Elizabeth Howe. *Samuel Gridley Howe.* New York: D. Appleton-Century Company, 1935.

Schwartz, Harold L. *Samuel Gridley Howe, Social Reformer, 1801-1876.* Cambridge, MA: Harvard University Press, 1956.

Trent, James W., Jr. *Inventing the Feeble Mind: A History of Mental Retardation in the United States.* Berkeley: University of California Press, 1994.

Tyor, Peter L. and Leland W. Bell. *Caring for the Retarded in America: A History.* Westport, CT: Greenwood, 1984.

Periodicals

Bledsoe, C.W. "Dr. Samuel Gridley Howe and the Family Tree of Residential Schools." *Journal of Visual Impairment & Blindness* 87 (1993): 174-6. (A Special Issue On Residential Schools: Past, Present, Future.)

Gardner, J. F. "The Era of Optimism, 1850-1870: A Preliminary Reappraisal."*Mental Retardation* 31 (1993): 89-95.

Knoll, James. "Samuel Gridley Howe and Burton Blatt on True Common Sense." *Mental Retardation* 34 (1996). 257-9.

Valentine, Phyllis Klein. "A Nineteenth-Century Experiment in Education of the Handicapped: The American Asylum for the Deaf and Dumb." *New England Quarterly* 64 (1991): 355-75.

Medical Education for Women during the Nineteenth Century

Overview

Elizabeth Blackwell, who earned a medical degree in 1849, was a pioneer in the struggle to open the medical profession to women. Because almost all nineteenth century medical schools admitted only male students, women physicians established several separate medical schools, including the Women's Medical College of Pennsylvania, which began preparatory classes in 1848. Although most of the women's medical colleges were short-lived, their establishment was instrumental in helping women gain access to medical education and to subsequent careers in the medical profession.

Background

In 1849 Elizabeth Blackwell (1821-1910) earned a place in history when she became the first woman in the United States to earn a medical degree. Blackwell was a pioneer in the very difficult battle to open the medical profession to women. During this time period, the United States was going through great social changes. Social reformers agitated for the abolition of slavery, the rights of women, and for new approaches to health and healing. Although women had served as healers and midwives in America since the Colonial era, they had been excluded from formal medical training. The women's rights movement encouraged women to educate themselves about health and hygiene, to enter the professions, and to dispel the widespread belief that women were naturally sickly and frail.

Opposition to the dominant medical theories and practices of the nineteenth century led to the growth of alternative medical sects, such as homeopathy, hyropathy, botanical medicine, eclecticism, and osteopathy. The irregular medical sects were particularly attractive to women patients, and the irregular medical schools were more willing to accept women students.

In 1847, when Blackwell began to apply to medical schools, all the regular schools admitted only men, but a few years later the first female medical colleges were being established and a few schools had become coeducational. As early as 1842 a group of physicians in Philadelphia were considering establishing a medical school for women. Three members of the first faculty began teaching preparatory classes to women in 1848 so that they would be ready to enroll when the Female Medical College of Pennsylvania was formally chartered in 1850. Forty women applied for admission: eight planned to earn a medical diploma, while the rest expected to audit the lectures. The established medical community of Pennsylvania attempted to destroy the college by passing resolutions against the professors and the graduates of the school. The State Medical Society and several county medical societies decreed that women were unfit to study or practice medicine and that some of the professors were irregular practitioners. Despite continuing struggles for recognition, the school was forced to close during the Civil War. After the war the school reopened as the Woman's Medical College of Pennsylvania. Dr. Ann Preston, a graduate of the school and a woman's rights advocate, served as dean of the college and led the fight to gain recognition from the medical societies of Pennsylvania. The county medical societies did not rescind the old resolutions until 1874.

A questionnaire sent to 244 graduates of the school in 1881 by Dean Rachel Bodley provided valuable information about their work, financial, social, and marital status. Of the 189 women who responded, 166 were practicing medicine, usually with an emphasis on gynecology and

ALICE HAMILTON AND HARVARD

❧

In 1893 Alice Hamilton (1869-1970) earned her M.D. from the coeducational medical school of the University of Michigan. Influenced by her experiences at Hull House in Chicago, she applied her medical skills to helping workers in the dangerous trades through her pioneering work in industrial hygiene and industrial toxicology. She was invited to join the faculty of Harvard University as assistant professor of industrial medicine in 1919, because she was the only qualified candidate. To protect Harvard's tradition of male privilege, however, Hamilton was informed that she must never attempt to use the Harvard Club, which had no ladies' entrance at the time, or demand her quota of football tickets, nor was she to embarrass the faculty by marching in the commencement procession and sitting on the platform. Each year she received a printed invitation to the commencement, but her invitation always included the warning that "under no circumstances may a woman sit on the platform." Hamilton was the author of *Industrial Poison in the United States* (1925), *Industrial Toxicology* (1934), and the autobiography *Exploring the Dangerous Trades* (1943).

obstetrics. Surprisingly, 150 women reported receiving "cordial social recognition," while only seven said that their reception had been negative. About one-third were members of a state, county, or local medical society. Most of the practitioners were quite successful, as measured by the average income for the group, which was about $3,000. Several women reported an annual income between $15,000 and $20,000. Many graduates had successfully combined marriage and motherhood with the practice of medicine. The success of these graduates might reflect the

special nature of students who had overcome great obstacles to enter the medical profession.

Even after Blackwell had broken one barrier to opening the medical profession, women who graduated from medical schools found themselves facing additional obstacles, including exclusion from hospital training and practice. In response, Elizabeth Blackwell, along with her sister Dr. Emily Blackwell (1826-1910) and Dr. Marie Zakrzewska, who had earned medical degrees from the Cleveland Medical College, Ohio, founded the Woman's Medical College of the New York Infirmary (1868), a college that was dedicated to the highest standards. While many medical schools required only five months of instruction for two years, the Woman's Medical College established a three-and-a half-year program with a graded curriculum. Students had to pass both entrance and graduation examinations.

In 1848 a group of Boston physicians opened the New England Female Medical College. The goal of the new college was to teach medicine and midwifery (obstetrics) to women. After years of struggling to survive, the College merged with the Medical Department of Boston University, which was a homeopathic institution. Not surprisingly, Harvard had refused to absorb the New England Female Medical College. After teaching at the New England Female Medical College, Dr. Marie Zakrzewska, a strict advocate of orthodox medicine, left to establish the New England Hospital for Women and Children. The hospital served as a major training institution for women doctors and a school for nurses. Dr. Zakrzewska was especially committed to the special female character and social mission of the New England Hospital.

The Woman's Medical College of the New York Infirmary and the Woman's Medical College of Pennsylvania sent many graduates to the New England Hospital for further training. By 1895 these women were joined by a few who had studied at Ontario Medical College for Women in Toronto or the Woman's Medical College in Chicago. A very small number came from the coeducational University of Michigan. The Woman's Medical College of Pennsylvania, the Woman's Medical College in Baltimore, and the New England Hospital continued to grow in the twentieth century and provided separate medical education and training for women. The Woman's Medical College of Pennsylvania was the last surviving school of its kind, but in 1969 the school became coeducational.

Impact

By the 1860s women could earn medical degrees from at least three schools for women and from several schools that became coeducational. The number of women doctors increased substantially between 1870 and 1900, from about 500 to about 7,000. Even Harvard's medical school was affected by social reform movements and, in 1850, accepted a female and three black students. Unfortunately, a student riot in protest led the four candidates to withdraw. After this incident Harvard remained closed to women medical students until 1945.

After the Civil War medical schools proliferated, including female medical schools. About twenty such schools opened between 1850 and 1895, but most were short-lived. Nevertheless, in the 1880s even the *Journal of the American Medical Association* (*JAMA*) acknowledged that some of the women's medical colleges had prospered while maintaining high standards. A report in 1885 called attention to the high standards of the Woman's Medical College of Philadelphia, which had pioneered in offering a three-year graded curriculum and an eight-month school year. Moreover, graduates of the women's colleges did not do worse on medical examinations than graduates of men's colleges. Despite evidence that favored women's medical colleges and their graduates, the editors of *JAMA* continued to oppose higher education for women on the grounds that the energy women expended in pursuing professional goals would interfere with their proper reproductive functions.

By the late nineteenth century coeducation was becoming more common in higher education. This was especially true of the newer schools and state universities, especially those schools west of the Appalachians. For example, women were admitted to the medical departments of Syracuse University (1870), the University of Michigan (1871), and the University of California (1874). In the District of Columbia, both white and black women were able to earn medical degrees at Howard University. Financial problems in the 1880s forced Georgetown and George Washington universities to admit women. Once the economic crisis had been resolved, women were no longer accepted. Women graduates still faced obstacles to professional advancement: internships, residencies, clinical positions, and membership in medical societies were closed to them. In response, women doctors had to open their own dispensaries, clinics, and hospitals.

Johns Hopkins was the first elite eastern school to accept women. In 1892 the university was committed to establishing a model medical school, but it did not have sufficient funds. A group of prominent women raised $100,000 and offered the money to Johns Hopkins if women were admitted on the same basis as men. The trustees offered to open the school to women if the grant were increased to $500,000. When the women succeeded in raising the money, they insisted on raising admission standards to include a bachelor's degree, knowledge of French and German, and premedical studies. The faculty and trustees considered these criteria too demanding but reluctantly agreed. This decision had a profound impact on the future of medical education in the United States: the medical college of Johns Hopkins would serve as an ideal and a model for the nation. Generally, women welcomed the evolution of specific, even rigorous, criteria for admission to medical school. In 1896 33% of the students at Johns Hopkins Medical School were women. Women apparently thought that if formal rules were applied uniformly they would gain access to schools that had previously excluded them through informal and unwritten codes.

As more medical schools became coeducational, women greeted these changes as signs of inevitable progress and assumed that separate schools for women were no longer needed. In the 1890s women doctors were optimistic that their cause had prevailed and that women would be integrated into all medical schools. Optimism was one of the factors that led to the virtual disappearance of separate medical colleges for women. But the "Golden Age" for women doctors did not last long. As the medical marketplace became more competitive and professional societies strengthened their position, a reaction against women practitioners grew. Soon women found themselves subjected to new barriers and obstacles. Worse yet, when women's schools closed or merged with nominally coeducational schools, women faculty found themselves unemployed or given only informal positions and subjected to degrading restrictions. Many of these displaced women doctors dedicated themselves to progressive reform movements and social medicine and public health campaigns.

Educational reforms, stricter state licensing, and financial difficulties also contributed to the closing of many medical colleges at the turn of the century. Institutional backlash can be considered one of the causes of the decline in the percentage of women physicians after about 1900. Many previously coeducational schools returned to the policy of excluding women. Four percent of all medical graduates in 1905 were women, but women constituted only 2.6% of medical graduates in 1915. In 1955 less than 5% of medical graduates were women. At Johns Hopkins, the percentage of women students dropped from 33% in 1896 to 10% in 1916. Women who did get through medical school then found it virtually impossible to get into hospital internship and residency programs.

By the 1970s agitation for changing the education climate for women resulted in a heightened awareness about sex discrimination and some decrease in openly discriminatory policies. Between 1970 and 1975 the number of women in medical schools tripled. By the 1990s about 40% of medical students and 18% of practicing physicians were women.

LOIS N. MAGNER

Further Reading

Abraham, Ruth J., ed. "Send Us a Lady Physician": Women Doctors in America 1835-1920. New York: W. W. Norton & Company, 1985.

Alsop, Gulielma Fell. History of the Woman's Medical College, Philadelphia, Pennsylvania, 1850-1950. Philadelphia: Lippincott, 1950.

Blackwell, Elizabeth. Pioneer Work in Opening the Medical Profession to Women: Autobiographical Sketches by Dr. Elizabeth Blackwell. Introduction by Dr. Mary Roth Walsh. New York: Schocken, 1914. Reprint, 1977.

Blustein, Bonnie Ellen. Educating for Health and Prevention: A History of the Department of Community and Preventive Medicine of the (Woman's) Medical College of Pennsylvania. Canton, MA: Watson Publishing International, Science History Publications, 1993.

Bonner, Thomas Neville. To the Ends of the Earth: Women's Search for Education in Medicine. Cambridge, MA: Harvard University Press, 1992.

Corea, Gena. The Hidden Malpractice: How American Medicine Treats Women as Patients and Professionals. New York: William Morrow, 1977.

Drachman, Virginia G. Hospital with a Heart: Women Doctors and the Paradox of Separatism at the New England Hospital, 1862-1969. Ithaca, NY: Cornell University Press, 1984.

Furst, Lillian R., ed. Women Healers and Physicians: Climbing a Long Hill. Lexington, KY: University Press of Kentucky, 1997.

Hurd-Mead, Kate Campbell. Medical Women of America: A Short History of the Pioneer Medical Women of America and of a Few of Their Colleagues in England. New York: Froben Press, 1933.

Morantz-Sanchez, Regina Markell. Sympathy and Science: Women Physicians in American Medicine. New York: Oxford University Press, 1986.

More, Ellen Singer and Milligan, Maureen A., eds. The Empathic Practitioner: Empathy, Gender, and Medicine. New Brunswick, NJ: Rutgers University Press, 1995.

Walsh, Mary Roth. "Doctors Wanted: No Women Need Apply": Sexual Barriers in the Medical Profession, 1835-1975. New Haven, CT: Yale University Press, 1977.

Cholera Epidemics:
Five Pandemics in the Nineteenth Century

Overview

During the nineteenth century, five cholera pandemics swept through India, Asia, Europe, and North America, infecting huge segments of the population and killing millions of people; a *pandemic* is defined as an epidemic encompassing a wide geographical area. Millions more would undoubtedly have perished were it not for the work of Doctors John Snow (1813-1858) and Robert Koch (1843-1910), whose research helped isolate the cause of the disease. Their discovery of the link between cholera and poor sanitation practices would forever change public health policies in Europe and North America.

Background

For centuries the deadly cholera disease festered on the Indian subcontinent. As early as 2,500 years ago, Sanskrit writings tell of the spread of an illness similar in symptoms to cholera. In 1503 the explorer Correia reported that 20,000 men in the army of the Sovereign of Calicut came down with a "disease, sudden-like, which struck with pain in the belly, so that a man did not last out eight hours at a time." Over the next 300 years, epidemics would periodically emerge and spread throughout India and the East, but it was not until 1817 that the world would come to know the devastating effects of cholera.

In that year, outbreaks erupted in several Indian provinces. Among the first victims were soldiers in the English army barracks of Fort William in Calcutta. Once the men were infected, the onset of the sickness was rapid—and painful. Victims would experience vomiting and severe diarrhea, accompanied by terrible cramps. Their bodies would soon dehydrate, and death would often occur in the space of hours. Within weeks, 5,000 British soldiers had succumbed to cholera.

The sickness quickly spread throughout India. Over the next several months, 25,000 of Calcutta's residents were treated for cholera—4,000 died. By 1818 it had moved across India, striking an estimated 7.5% of the exposed population. Now, for the first time, the disease escaped the continent. With the advent of new roads, railroad tracks, and faster ships, it spread like wildfire. Over the next three years cholera was carried by merchant ships to China, Japan, and Southeast Asia. Pilgrims traveling to and from Mecca spread the disease throughout the Arab world. By 1821 it had invaded Java and Persia. In 1823 Egypt and Syria fell victim. Thus began the first cholera pandemic.

In 1824 another epidemic outbreak of cholera developed in India's Ganges Delta. By 1829 it had moved through Persia and to the shores of the Caspian Sea. In that same year, it expanded north and west into Russia, where the Russian army carried it into Poland. In 1831-1832 it reached England and France. Within 18 days of its arrival, the French reported at least 7,000 deaths. Next, cholera crossed the Atlantic by way of immigrants, infecting thousands in Canada and the United States.

By the end of the nineteenth century, three more major pandemics would make their way around the world, spreading the disease across Europe, North America, India, and Asia. By the end of the century, millions would lose their lives.

Impact

In 1831, during the second pandemic, cholera reached the shores of England. In the port town of Sunderland, old man Sproat, aged 60, and the young, healthy William Sproat, Jr., were among the first to die. Doctors from around the country descended on Sunderland, trying to ascertain a cause. Their diagnosis—the victims were either old and feeble, or predisposed to sickness. Although local coal merchants and traders warned that cholera was being carried to England on ships from India, the doctors disagreed. They firmly believed that it was not contagious.

At that time, a young doctor named John Snow was beginning his career as an apprentice in Newcastle-on-Tyne. He watched as patients continued to come down with the symptoms of the disease, and was horrified by what he saw. When cholera recurred in 1848, Snow began to chronicle its spread. He carefully followed its origination in India and the path it took westward, towards London and Paris. In 1849 he published a pamphlet entitled *On the Mode of Communication of Cholera,* which outlined his belief that poor sanitation and unclean water were directly linked to the epidemic.

During the nineteenth century, sanitary conditions in the world's big cities were abysmal. The Industrial Revolution had led to the growth of urban tenements, where fresh water was hard to come by and sewers were simply elongated cesspools. Refuse was shoveled by hand, and barrels of excrement were often thrown outside, even flung from open windows. London got much of its water from the polluted Thames River or from filthy wells. A neighborhood of 20 to 30 families would draw their water from a single pump two or three times a week. If the pump was not working, they would often reuse unclean water.

In the mid 1850s Snow carried out a detailed investigation of the cholera epidemic. He was able to demonstrate that over 500 cases in central London could be traced to a single source—a public well known as the Broad Street Pump. Everyone who drank the water was coming down with the disease. He asked local authorities to remove the pump handle, and as soon as they complied, cholera quickly disappeared from the area. Snow soon traced other outbreaks to the Thames and to other polluted wells. He reached the conclusion that the disease was being transmitted from person to person through the water supply. He found that the water was being contaminated by untreated sewage tainted with the cholera virus, which permeated the ground and polluted wells. Snow's discovery of the link between contaminated water and cholera created a model for the understanding of disease transmission, and was the first step leading to a public understanding for the need for proper sanitation. But the actual cause of the disease was not discovered until many years later by the German bacteriologist Robert Koch.

In 1884, while examining Calcutta's water supply, Koch observed a unique, comma-shaped bacteria. When he examined the fecal matter of 70 people who had died of cholera, he found the same bacteria in all of the victims. For the first time, he was able to isolate the exact cause of death as the *Vibrio cholerae* bacteria, or cholera.

Koch confirmed Snow's water-cholera relationship hypothesis, and discovered that the disease was ingested by victims, either from fecal-infected water, or through eating shellfish, vegetables or fruits onto which flies had dropped infected human fecal matter. From his research, doctors learned of the necessity to isolate cholera victims in order to stem the spread of the disease.

More importantly, once the link between sanitation and cholera was established, the way was paved for public health policies, which had not previously been in existence. In the 1840s the General Board of Health was established in England, calling for better methods of refuse collection, new sewers, and the destruction of slums. The Public Health Bill of 1848 set up a central authority to ensure that homes had proper drainage and that local water supplies were clean. When cholera struck that country for the third time in 1868, improvements in the water system led to far fewer infections.

England's success was recognized by governments around the world. In New York City, where a large immigrant population had brought the disease in, a health board was created and port quarantines were enforced. The U.S. federal government soon created a Public Health Service, as did governments throughout North America and Europe. Water and waste systems continued to improve throughout the century, and in the 1890s when cholera struck the world for the fifth time, those continents remained virtually unscathed.

The discovery of a treatment for cholera, however, took much longer. In 1830 the German chemist R. Hermann first discovered that a change in the blood's fluid balance had an effect on cholera excretions. But it wasn't until the early 1900s that the replenishment of water and salt was used to cure the dehydrating effects of the disease.

The cholera pandemics of the nineteenth century had a devastating effect on much of the world. Britain lost an estimated 130,000 people over the course of five epidemics. In India, cholera claimed more than 25 million lives from the 1800s to the early part of the twentieth century. Although modern sanitation and advancements in the treatment and prevention of cholera have reigned in the disease, the risk is far from over. The World Health Organization (WHO) estimates that 78% of the population in underdeveloped countries is without clean water and adequate waste disposal. As recently as 1991 and 1993, epidemics raged through Central America and Southeast Asia. War, famine, flood, and the lack of clean water in developing nations keeps the risk of another outbreak ever present.

STEPHANIE WATSON

Further Reading

Dury, Michael. *The Return of the Plague.* Gill & MacMillan, 1979.

Karlen, Arno. *Man and Microbes.* New York: Quantum Research Associates, Inc., 1995.

Koch, Robert, and K. Codell Carter. *Essays of Robert Koch.* Greenwood Publishing Group, 1987.

Kohn, George C., ed. *Encyclopedia of Plague and Pestilence.* New York: Facts on File, 1997.

Watts, Sheldon. *Epidemics and History: Disease, Power and Imperialism.* New Haven and London: Yale University Press, 1997.

Modern Anesthesia Is Developed

Overview

The discovery and development of anesthesia is one of the most important medical discoveries of the nineteenth century. It not only relieved pain, it also allowed doctors to perform life-saving surgical procedures, thereby increasing a person's lifespan and quality of life.

Background

Although pain was always assumed to be inevitable during surgery, many substances and concoctions were used to relieve pain over the centuries. Most frequently various plant extracts such as opium or preparations containing alcohol were used as anesthetics. Sometimes these

This rendering by an unknown artist depicts James Young Simpson's accidental discovery of chloroform. (*Wellcome Institute. Reproduced by permission.*)

substances, along with the speed of a skillful surgeon, helped the patient survive painful procedures. Yet such methods of pain relief were inconsistent at best.

Events leading to modern surgical pain relief began in England in the 1770s. Joseph Priestley (1733-1804), a minister, author, and scientist, isolated several gases, including oxygen and nitrous oxide, or "laughing gas." In one of his publications Priestley expressed his hope that there might be medical uses for these gases. By the early 1780s several physicians, including Thomas Beddoes (1760-1808) in Bristol, were researching such uses. In 1799 Beddoes opened the Pneumatic Medical Institute to study the gases, and hired a young man named Humphry Davy (1778-1829) as Research Director. Davy and others did extensive research on nitrous oxide and in a book about their research (published in 1800) Davy suggested that inhalation of the gas might relieve pain during certain types of surgery. Beddoes and Davy also noticed that breathing nitrous oxide relieved the pain of a toothache.

In 1823 English physician Henry Hill Hickman began a deliberate search for an anesthetic. He used high doses of carbon dioxide during surgeries in mice and dogs. In his published account Hickman noted the lack of response to

an incision in these animals. Hickman had indeed discovered inhalation anesthesia, yet he was ignored by other physicians and scientists of his day.

In the United States in the small town of Jefferson, Georgia, physician Crawford Long (1815-1878) performed several surgeries beginning in March 1842 using ether vapor as an anesthetic. Ether had been known since the sixteenth century, and its vapor was often inhaled at "ether frolics" or parties to produce a brief feeling of euphoria. Long had received excellent medical training for his day, graduating from the prestigious University of Pennsylvania Medical School at a time when most physicians in America were trained by apprenticeship to other physicians. Yet Long failed to report or publicize his monumental achievement until years later. Apparently, since surgery was so rare in his practice, Long did not immediately realize the importance of what he had discovered.

Impact

Finally, in 1845 and 1846, two American dentists brought anesthesia to mainstream medical practice. Horace Wells had a successful practice in Hartford, Connecticut, when he attended a nitrous oxide demonstration on the night of December 10, 1844. At that time various showmen toured England and the United States giving public exhibitions of the exhilarating or intoxicating effects of breathing this gas. On that night Gardner Qunciy Colton invited audience members to join him onstage and breath nitrous oxide. One young man who did so, Samuel Cooley, injured himself in the excitement but apparently felt no pain. Wells immediately realized the potential of nitrous oxide in medical practice. The next day Wells invited Colton to give him the gas while John Riggs, Wells's former pupil, pulled a wisdom tooth. Wells felt no pain, and began to use the gas in his practice.

In January 1845, Wells traveled to Boston to demonstrate this new method of pain relief at the Massachusetts General Hospital. Wells appeared before the surgery class of John Collins Warren, one of America's best-known surgeons, and attempted a tooth extraction on a patient who had breathed nitrous oxide. Unfortunately, the patient cried out in apparent pain, and Wells's demonstration was considered a failure.

William Thomas Green Morton (1819-1868), a Boston dentist and former pupil and business associate of Wells's, correctly decided that Wells had the right idea but the wrong gas. With the help of a Harvard Medical School professor named Charles T. Jackson, Morton began experiments with another gas, sulfuric ether. After successful attempts on various animals and dental patients, Morton was ready for a surgical patient. Despite the earlier failure of Wells, Warren allowed Morton to demonstrate his method at the hospital in front of staff physicians and medical students. On October 16, 1846, a young man named Edward Gilbert Abbott had a tumor removed from his neck. Morton let Abbott breath the gas from a small apparatus; after four or five minutes, he told Warren the patient was ready. Although Abbott cried out and moved during the surgery, afterward he denied feeling any pain. The next day a tumor removal on another patient under ether was successful. In the following months word of this procedure spread around the world. By the end of 1847 books and pamphlets about ether anesthesia appeared in the United States and many countries in Europe.

Although anesthesia made a great contribution to medicine, medical practice during the rest of the nineteenth century changed slowly. Surgical mortality, or deaths, remained high until late in the century, due primarily to the slow acceptance of the germ theory and the need for cleanliness in operating rooms and hospitals. Some physicians were also reluctant to use anesthesia due to fears of possible side effects.

During the second half of the century, anesthetic practice slowly improved and expanded. A Scottish obstetrician named James Young Simpson (1811-1870) tried using ether to relieve his patient's pain during childbirth. Unhappy with ether's smell and its lengthy period of action, Simpson experimented with a number of other substances and finally discovered the anesthetic properties of chloroform. He quickly introduced this gas into his practice, and soon chloroform was being used in surgery as well. Ether and chloroform remained in common use around the world until the 1950s, when use of halothane, a synthetic gas, became widespread.

Englishman John Snow (1813-1858) is often called the first anesthesiologist—a physician whose practice consists of giving anesthesia in the operating room and monitoring the patient's vital signs. Snow never actually limited his practice to anesthesia; few physicians in either England or the United States could support themselves that way until well into the twentieth century. Yet Snow was considered a superb physician whose interest in anesthesia included a desire to understand how the gases affected the body. Snow did extensive research and left many published accounts of his work. After Snow's death in 1858, Joseph T. Clover (1825-1882) became one of the best-known figures in England associated with anesthesia. Clover designed a portable machine to administer gases that remained in use decades after his death. Another Englishman, Frederick Hewitt (1857-1916), designed an early anesthesia machine to administer variable portions of nitrous oxide and oxygen, a combination used widely for dental procedures and short surgeries from the late nineteenth century until today.

Regional and local anesthesia, forms of pain relief and prevention that do not require the patient to be unconscious, as with general anesthesia, were also discovered during the nineteenth century. In the 1880s Carl Koller (1857-1944), an associate of Sigmund Freud's (1856-1939) in Vienna, discovered in their work with cocaine that the drug made his tongue numb. Koller quickly attempted experiments on animals and realized that use of cocaine prevented pain. After a friend reported Koller's results at a medical conference in Germany, word quickly spread around Europe and the United States.

Over the next few years numerous physicians used injections of cocaine in the eye, mouth, and other areas of the body to block nerve impulses. By the end of the century, cocaine was used for spinal anesthesia by August Bier in Germany. Since 1900 many other drugs have been developed to replace cocaine. Today regional and local anesthesia are frequently used during surgeries and after surgeries to manage pain.

Another nineteenth-century innovation related to anesthesia is intubation. This procedure, in which a flexible breathing tube is inserted through the mouth into the windpipe, assists the anesthesiologist by creating a dependable airway during an operation. Scottish surgeon William Macewan first used this procedure in July 1878. However, interest by Macewan and others in this procedure was sporadic until an American physician, Joseph O'Dwyer, began using it on children with diphtheria to assist their breathing. O'Dwyer

designed several tubes that were soon adopted by New Orleans surgeon Rudolph Matas. However, several decades would pass before intubation would become a standard technique.

The discovery and development of anesthesia is one of several important medical developments of the nineteenth century. Pain, a constant for most of human existence, can now be controlled and prevented in many situations.

A.J. WRIGHT

Further Reading

Duncum, Barbara M. *The Development of Inhalation Anesthesia.* London: Oxford University Press, 1947.

Keys, Thomas E. *The History of Surgical Anesthesia.* Park Ridge, IL: Wood Library-Museum of Anesthesiology, 1996.

Rushman, G.B., N.J.H. Davies, and R.S. Atkinson. *A Short History of Anesthesia: The First 150 Years.* Boston: Butterworth Heinemann, 1996.

Antiseptic and Aseptic Techniques Are Developed

Overview

The development of antiseptic and aseptic techniques had a dramatic impact on the health and life of those living in the late-nineteenth century. Essentially, these techniques combat the growth and transmission of harmful organisms. Antisepsis, meaning the topical destruction of bacteria, was developed as an offshoot of French bacteriologist Louis Pasteur's germ theory. Asepsis, meaning the absence of harmful organisms, was a later refinement of antisepsis and led to the development of modern surgery. Both techniques vastly reduced infection rates and, therefore, increased survivability from trauma or disease. Antisepsis and asepsis influenced industry and accompanied cultural changes during the latter Victorian period and into the Industrial Revolution.

Background

English surgeon Joseph Lister (1827-1912) pioneered modern antisepsis. Based on Pasteur's earlier studies of fermentation and putrefaction, Lister reasoned that tissue breakdown from infection was caused by minute organisms. Lister developed an antibacterial solution contain-

ing carbolic acid, first spraying it in the air and then on his surgical instruments. When Lister first applied his antibacterial solution directly to compound fracture wounds in 1865, he observed that its use resulted in dramatically lower rates of infection. Lister subsequently championed cleanliness in the surgical operative area and embraced the burgeoning antiseptic techniques contributed by his contemporaries. Lister's reputation allowed him to serve as surgeon to Queen Victoria, who granted him the title of Baron in 1897.

Though successful, early antiseptic techniques were not immediately adopted by practicing physicians. Pasteur's germ theory initially created confusion among the medical community, some of whom doubted its clinical significance. As early as 1846 Austrian physician Ignaz Semmelweis (1818-1865) suggested that physician handwashing between attending patients lowered infection rates. While working in obstetrics at the Vienna General Hospital, Semmelweis observed high mortality rates among mothers with post-partum fever, or puerperal sepsis, in one of the units where physicians and medical students attended births, performed

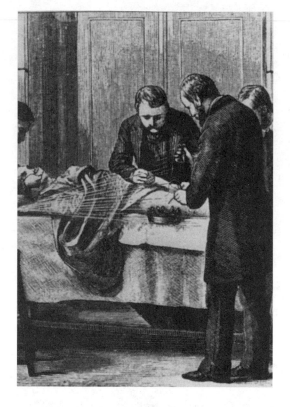

This 1882 engraving shows Joseph Lister using a carbolic spray during a surgical operation. Lister's advances in antiseptic techniques helped revolutionize surgery during the nineteenth century.
(*Fotomas/Barnaby's. Reproduced by permission.*)

surgeries, and conducted autopsies. Semmelweis also observed that in another unit where midwives attended births, mortality was significantly lower. When a colleague of Semmelweis died of sepsis after puncturing his finger during an autopsy, Semmelweis connected the two events. Semmelweis concluded that contamination with the infectious material itself might spread a disease from person to person. To break this chain of infection he insisted that physicians wash their hands after autopsies with a lime chloride solution and wash with soap and water between attending births. Although mortality rates from puerperal sepsis plummeted, Semmelweis' techniques were initially discounted. Hospital practice was reluctant to change, and the development of antiseptic technique suffered a setback. Semmelweis died impoverished in a mental institution in 1865.

By 1880, based on the work of Lister, infections and particularly post-surgical infections continued to decline. Within a generation antisepsis was refined to asepsis, meaning the absence of harmful organisms. Asepsis is achieved mainly through sterilization. German physician Ernst von Bergmann (1836-1907) made a major breakthrough in asepsis when he introduced steam sterilization of surgical instruments in 1885. Born in Latvia, Bergmann was professor of surgery at the universities of Berlin and Würzburg. He is also credited with introducing the sterilization of wound dressings and other medical equipment used during surgical operations. Bergmann's method used steam under pressure and is the basis for modern sterilization procedures. American surgeon William Stuart Halstead (1852-1922) introduced sterile rubber gloves to surgery in 1898. Spurred by his fiancé's complaint that continued antiseptic use and handwashing were irritating to the hands, Halstead's innovation further minimized the opportunity for cross contamination between surgeon and patient. Halstead was the first professor of surgery at Johns Hopkins University, where he revolutionized surgical technique, including the meticulous handling of tissues during surgery.

Impact

The nineteenth-century development of antisepsis and aseptic technique laid the foundation for the ascent of modern surgical technique. Prior to Lister's discovery of antisepsis, almost 80% of surgical patients contacted gangrene from operations performed in rooms with poor ventilation and crowded by observers. Surgeons wore street clothes, sometimes with aprons, seldom changing either between patients. At best, instrument preparation consisted only of washing with soap and water. Sawdust from mill floors was used both as wound dressing and as an absorbent material for the surgical floor. Surgery was the last resort in the physician's armamentarium (methods of treatment) and was most often performed in conjunction with a traumatic or wartime injury. Unreliable anesthesia and battlefield conditions created urgency that did little to foster antisepsis. Amputated limbs and tissues were stored haphazardly, often collected in a central area and disposed of only at the end of the day. Autopsies were performed in the same area as surgeries. Amid these conditions disease flourished and Lister and his contemporaries campaigned for change.

With Bergmann's refinement of antisepsis to aseptic techniques, surgery saw an unprecedented boom in innovation, beginning in 1890 and continuing into the early twentieth century. Low mortality rates and advances in anesthesia made surgery an option less dire. As more surgeries were performed, knowledge of anatomy and especially physiology increased. This new knowledge led to the effective performance of surgery for treatment of internal diseases. The body's reaction to injury became better understood, which aided in often-performed trauma surgery. Surgeries that were previously assumed impossible became practical realities. Bergmann, a neurosurgeon, was among these pioneers, introducing stringent standards of surgical asepsis while performing procedures on the brain and spinal cord. Some physicians began to advocate use of presumed preventive surgeries such as circumcision of the newborn and removal of the adenoids during childhood. Hospitals dedicated rooms solely for the purpose of surgery, where aseptic techniques were maintained.

Nursing also contributed to the lower mortality rates of the time. Nurses incorporated post-surgical antiseptic and aseptic techniques while caring for their patients, and they advocated improvements in hospital environmental cleanliness, while delivering and teaching enhanced personal hygiene for those in their care. Hospitals began to be perceived as sanitary institutions in which to benefit from the latest medical care rather than the crowded, dreadful institutions of the past. The number of hospitals greatly increased from 1870 until after the turn of the century to accommodate those patients benefiting from new medical advances made possible by the discovery of aseptic techniques. Medical professions also experienced a boom as universities and hospitals trained physicians and nurses. Medical and surgical research entered into an exuberant phase of innovation.

By 1880 principles of antisepsis and the germ theory influenced social culture in the Western world. Cleanliness was perceived as essential to both personal and social responsibility. As a result of these new sensitivities, nineteenth-century society rediscovered and returned to ancient traditions (often grounded in religious traditions) that called for frequent bathing. Dirt was perceived to be hazardous. The middle class embraced cleanliness as a method of displaying prosperity. The later Victorian period was an age concerned with class stratification, and standards of cleanliness became a measure of not only social standing but moral character as well. Refinement and spirituality were associated with cleanliness, while the unclean were described as base and animalistic.

Some normal human bodily functions became associated with dirt, thereby carrying moral significance as well. Healthy genital organs were considered contaminated due to their function.

Physicians often adopted this prevailing attitude, sometimes labeling normal body secretions as infectious material. This provided the context for the endorsement of universal circumcision as preventive medicine. Maternal mortality in childbirth also continued, with efforts to further reduce it sometimes hampered by moralists. Many physicians and theologians still considered puerperal fever as a manifestation of God's intent to punish woman during childbirth. Although Listerian antisepsis had a dramatic initial effect on lowering maternal mortality, some hospitals still had 7% mortality rates as late as 1895. Women who had their babies at home attended by physicians or midwives greatly outnumbered women who gave birth in hospitals.

The emphasis on antisepsis also influenced industry at the dawn of the industrial age. The pharmaceutical industry developed new medicines and manufactured them under antiseptic conditions. Pasteurization, a process of semi-sterilization that eliminates bacteria responsible for spoilage, saved the wine industry in Europe. The availability and distribution of milk greatly increased, as did its consumption, after the pasteurization process was applied. Improved canning methods led to increased distribution and longer shelf life of foods. With improved food preservation came greater population mobility. Many who previously farmed for their sustenance moved to the cities for opportunities with burgeoning industry. Growing cities assumed the challenge of maintaining a water supply free from harmful organisms and developing an organized waste disposal system. Agencies in urban centers arose to recognize, track, and research trends in infections and matters of public health. Many of these, such as the Pasteur Institute in Paris and the Lister Institute of Preventive Medicine in London, still exist today.

BRENDA WILMOTH LERNER

Further Reading

Haeger, Knut. *The Illustrated History of Surgery.* New York: Bell Publishing, 1988.

Tomes, Nancy. *The Gospel of Germs: Men, Women, and the Microbe in American Life.* Cambridge, MA: Harvard University Press, 1998.

Birth of the Nursing Profession

Overview

Led by the pioneering efforts of Florence Nightingale, the nursing profession arose during the middle and latter part of the 1800s. The establishment of a professional nursing corps in Europe and the United States dramatically impacted the medical profession and society in general, improving standards of care and providing an avenue for women to enter the work force.

Background

Although largely undocumented, some form of nursing has been practiced for thousands of years. Women's traditional roles of caregivers at home led naturally to an interest in medicine. Methods of nursing rose from two circles: one scientific, the other religious and social. Egyptians hired midwives to assist with childbirth. Emperors' wives tended to the ill in ancient Rome.

During the Middle Ages, crusaders left thousands of sick and injured behind in their quest to take back the Holy Land from the Moslems. Monks and knights took care of the sick when the fighting subsided. These men were known as "knight hospitallers." Many found that they preferred the role of nurse over solider, and nursing became their trade.

At the turn of the seventeenth century, however, a dark veil fell upon this early form of nursing. Monasteries to care for the ill were abolished by the Protestants and replaced by workhouses and almshouses for the poor. Places to nurse the sick were few and far between; those that did exist bore abysmal conditions. The period from 1600 to 1850 became known as the "Dark Age of Nursing."

Prisoners and elderly prostitutes provided much of the care in the remaining institutions. Overwhelming poverty, filth, and disease marred this early age of nursing. Those dumped in hospitals suffered immeasurable pain and suffering and, eventually, death. These horrific conditions lasted for several hundred years.

Nursing, as we know it today, began during the mid-1800s with a single woman's crusade to reform treatment of the ill during the Crimean War. In 1854 Sir Sidney Herbert, British Secretary of War, wrote to hospital reform advocate Florence Nightingale (1830-1910) asking her to lead a group of nurses to Scutari, Turkey, to tend to wounded soldiers. The Crimean War and Nightingale's involvement in it would forever change the course of nursing education.

Nightingale and her handpicked team of 38 women arrived at Scutari on November 4, 1854. Although they had heard horrifying reports of sickness from war correspondents, Nightingale and her team were unprepared for the misery and neglect that awaited them. Wounded soldiers lay scattered on filthy floors; cholera and typhus had spread through the unventilated and unsanitary wards like wildfire. The lack of sewers and laundering facilities, combined with a disorganized medical service, resulted in the death of one out of every two solders. Some 10,000 men required desperate medical care.

Nightingale forbid her team to care for the soldiers until they were specifically asked to do so by the doctors. In the week that followed she and the nurses set up a kitchen using their own supplies and provided meals to the ill. But the situation continued to worsen and doctors had to ask for help. Finally, Nightingale's crew took on the task for which they were trained. She enlisted the help of all the able bodied to dig latrines, clean barracks, wash laundry, and feed the soldiers. Nurses, at last, were able to care for the sick.

Nightingale tackled the task with almost fanatical devotion. She and her nurses brought remarkable improvement. The death rate of wounded soldiers soon dropped to an amazing 2.2 percent. By the end of the Crimean War the death rate was just one percent.

While ensuring soldiers were taken care of, Nightingale worked furiously to raise funds so the war hospital could function effectively. Yet Nightingale's most significant contribution at Scutari is said to be her profound sympathy for the suffering. Guided by a single lamp, she visited the soldiers late at night, attending not only to their physical needs, but looking out for their social welfare as well. "The Lady with the Lamp" made sure that, for the first time, wounded soldiers received sick pay.

At the close of the war in 1860, Nightingale founded the Nightingale School and Home for Nurses at Saint Thomas's Hospital in London. The school's establishment marked the beginning of a formalized and professional system of nursing instruction.

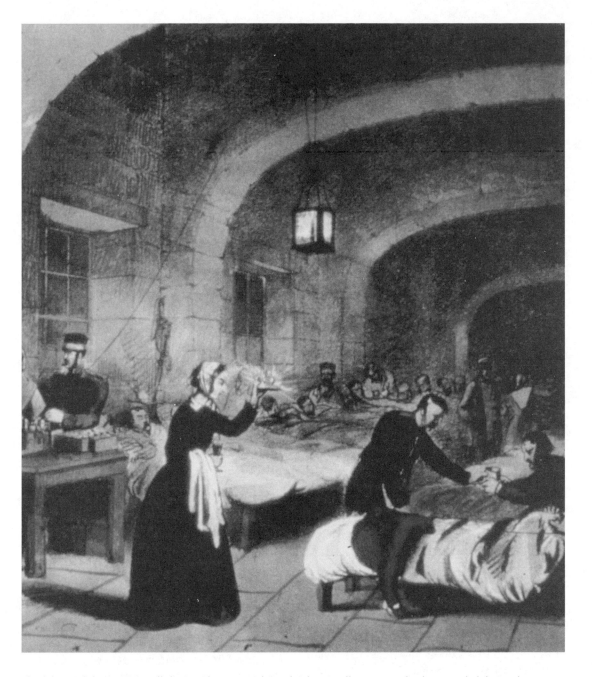

This lithograph by J.A. Benwell depicts Florence Nightingale's heroic efforts to care for the wounded during the Crimean War, an effort that helped establish the nursing profession as a legitimate part of human medicine. (*Wellcome Institute. Reproduced by permission.*)

After Nightingale finished her rounds in Europe, another young woman began similar work on the battlefields of the United States. At the outbreak of the Civil War, Clara Barton resigned her job at the U.S. Patent Office and volunteered to distribute supplies to wounded soldiers. In 1862 Barton was granted unprecedented permission to deliver supplies directly to the front lines, which she did without fail. Two years after the war, Barton was named superintendent of Union nurses.

Years after the war, Barton learned about the Treaty of Geneva, which provided relief for sick and wounded soldiers. Twelve nations had signed the treaty, but the United States had refused. Barton vowed to investigate why. It was during this time that Barton learned about the Red Cross.

Barton's crusade for the Treaty of Geneva and the Red Cross began in 1873—the same year the first nursing school opened in the United States. She moved to Washington, D.C., where she successfully lobbied for her causes. The American Red Cross formed in 1881, and the U.S. signed the Geneva agreement the following year. Barton remained Red Cross president until 1904.

Impact

The impact of the efforts of Nightingale, Barton, and others to establish nursing as a legitimate part of the medical profession was enormous. In addition to improved standards of care, not only on battlefields but in hospitals as well, the rise of nursing provided an important venue whereby women could enter the workforce. Before this could happen, however, the nursing profession had to institute training and education standards.

During the time of Nightingale and Barton, nurses were largely untrained personnel. Female nurses fought an uphill battle for recognition of their profession. Many considered nursing a lowly job and assumed its practitioners were dubious at best. But Nightingale believed nursing to be a suitable career for women, and she worked toward ensuring that the relationship between nurses and doctors was a professional one. In 1873 in a letter offering advice to nursing students, Nightingale wrote "nursing is most truly said to be a high calling, an honourable calling." By the end of the nineteenth century, the idea that nurses needed to be educated and trained had spread to much of the Western world.

Inspired by Nightingale's founding of the first nursing school, other schools were established in the United States in New York, Massachusetts, and Connecticut. The first nursing school in the United States, New York's Bellevue Hospital, opened its doors on May 1, 1873, with little or no fanfare. The Connecticut Training School in New Haven and the Boston Training School at Massachusetts General Hospital started that same year. Five years later, in 1878, the New England Training School for Women and Children was founded in Boston.

Linda Richards became the first American woman to graduate from nursing school. She went on to become superintendent of the new Boston Training School and later founded several other training schools, including one in Japan.

Six years later, Mary Eliza Mahoney became the first African American to receive a nursing degree. The Massachusetts-born hospital worker received her diploma from the New England Hospital for Women and Children. Medical, Sur-

gical, and Maternity Nursing were required courses during the first year, while the last four months were dedicated to home care, or Private Duty Nursing. Only 4 of 18 women who started the rigorous course with Mahoney graduated. Her high level of performance thwarted racial bias and paved the way for other African American women to enter the profession.

The nursing profession still lacked prestige, however. Male physicians often treated nurses with disrespect, others half-heartedly accepted them under the condition that they "stayed in their places." But despite the dominating male attitude, men, too, entered the nursing profession. The Mills School for Nursing and St. Vincent's Hospital School for Men were started in New York in 1888.

At the very end of the nineteenth century, historical and social events occurred that supported the rise of modern nursing. As medical science advanced and new hospital facilities opened, professional nurses slowly gained acceptance. During the early twentieth century, the army and navy nurse corps were established, increasing the need for professional training. Florence Aby Blanchfield, the first commissioned female in the U.S. Army, served as superintendent of the Army Nurse Corps (ANC) during World War II. Blanchfield commanded more than fifty thousand army nurses and began posting them near the front war lines to provide surgical nursing care.

In 1894 the Superintendents of Female Nursing Schools gathered in New York for their first annual meeting. Early nursing leaders in the United States were intelligent women who, like Florence Nightingale, possessed a great talent for organization. All shared a vision of what they thought nursing should be and they organized groups of dedicated women to implement the actions that would make the vision a reality.

In 1896 Isabel Hampton Robb, a graduate of Bellevue Training School in New York, organized the American Nurses Association and the National League for Nurses. Three years later Lavinia Dock formed the International Council of Nurses (ICN.) ICN promoted a four-fold function of nursing—promote health, prevent illness, restore health, and alleviate suffering.

Soon, nurses began performing tasks once reserved solely for physicians. Many followed doctors into specialties, such as pediatric nursing.

Today in the United States, nursing candidates must graduate from a state-approved nursing school. Options available are two-year associate degree program, a three-year diploma program (usually hospital-based), or a four-year university program. Throughout the world, registered nurses form the largest group of healthcare workers. In 1999, 1.8 million people worked as RNs in the United States. Historically, more women than men entered the nursing profession, but between 1980 and 1992 the American Nurses Association reported a 97% increase in the number of men entering the profession.

KELLI A. MILLER

Further Reading

Books

Brown, Pam. *Florence Nightingale*. Gareth Stevens, Inc., 1989.

Kalisch, Philip A., and Beatrice J. Kalisch *The Advance of American Nursing*. Philadelphia: Lippincott Williams & Wilkins Publishers, 1995.

Yost, Edna. *American Women of Nursing*. Philadelphia: J.B. Lippincott, Co., 1965.

Other

American Association for the History of Nursing. http://www.aahn.org/index.html

American Nurses Association. http://www.ana.org/

The Internet's Nursing Resource. http://www.wwnurse. com/

The Florence Nightingale Museum. http://www.florence-nightingale.co.uk/

Koch's Postulates: Robert Koch Demonstrates That a Particular Organism Causes a Particular Disease

Overview

Robert Koch was the first scientist to firmly establish the link between germs and disease. A doctor with a small rural practice in Germany, Koch's interest in microscopic studies led him eventually to identify the bacteria that causes the disease anthrax, which was then a common killer of sheep and cows and occasionally farmers. From there, Koch worked to develop the means by which he could prove without a doubt that these organisms were indeed to blame, the basic steps of which are now known as Koch's postulates. According to these postulates, or rules, a microbe may be proved as the cause of a disease if and only if:

1) It is found in all cases of the disease.

2) It may be isolated from the diseased body and grown artificially in a laboratory setting, creating what is called a pure culture.

3) The cultured bacteria can then be injected into a healthy animal and cause the same disease to appear in it.

4) The same bacteria can then be again found in and isolated from the newly infected lab animal.

Koch's work on anthrax was published in 1876 and touched off a revolution in medical knowledge. By the turn of the century, scientists working primarily under the tutelage of Koch and Louis Pasteur had identified most major bacilli, including those responsible for anthrax, gonorrhea, pneumonia, typhoid fever, septicemia, tuberculosis, cholera, plague, tetanus, diphtheria, and meningitis.

Background

As a young man Robert Koch (1843-1910) dreamed of working as a ship's doctor and traveling to distant and exotic places. Upon finishing his medical degree, however, he instead married and settled down to private practice in the small German town of Wollstein. Koch's work in studying bacteria began as a part time hobby after he received a microscope as a gift.

Koch began his work at a time when microbes were beginning to be studied but were still largely mysterious. The microscopic world had only been discovered to exist by Anton van Leeuwenhoek (1632-1723) a century or so earlier. Interest in this tiny world was renewed in the early nineteenth century by Louis Pasteur (1822-1895), a French chemist. In 1854 Pasteur began to study fermentation. Asked for help by local alcohol distillers who were having problems producing consistent quantities and qualities of alcohol from sugar beets, he found fermentation to be a result of the actions of active yeasts. In addition, he found that when these yeasts were crowded out by other microbes, the alcohol would not ferment properly. Along the way, he discovered that heating the alcohol to a certain temperature would kill off the offending microbes, allowing the yeasts to do their work and ensuring a consistent quality. This process of heat sterilization is now known as *pasteurization*.

Pasteur went on to settle an old and persistent dispute over spontaneous generation. For centuries, people had believed that many kinds of small creatures (for example, maggots found in rotting meat) developed spontaneously. The

PETTENKOFFER DISPUTES GERM THEORY AND LIVES TO TELL ABOUT IT

~

Robert Koch's work on the germ theory was not universally accepted by all scientists. When Koch announced he had isolated and identified the cholera vibrio, his findings were in particular disputed by a fellow German, Max von Pettenkoffer (1818-1901). Pettenkoffer firmly believed in the miasmatic theory of cholera, which held that putrefaction of the air caused the disease. He invited Koch to send him some of these supposed causative agents, which Koch obligingly did. Koch received the following message from Pettenkoffer:

Herr Doctor Pettenkoffer presents his compliments to Herr Doctor Professor Koch and thanks him for the flask containing the so-called cholera vibrios, which he was kind enough to send. Herr Doctor Pettenkoffer has now drunk the entire contents and is happy to be able to inform Herr Doctor Professor Koch that he remains in his usual good health. [quoted in Roy Porter, *The Greatest Benefit to Mankind,* W.W. Norton, 1996]

It remains a great mystery why Pettenkoffer did not become ill after swallowing the samples. Fortunately, this episode did not provide a setback to the advancement of the germ theory.

discovery of the microbial world had renewed belief in this process, for such creatures seemed far too tiny, chaotic, and ubiquitous to be generated in any other manner. Pasteur devised a test in which sterilized broth was exposed only to air filtered through cotton. Broth left standing in open air quickly teemed with microbes, but that in filtered air remained sterile. In this way, he proved that microbes did not simply appear spontaneously but were airborne.

The idea that these microbes could be responsible for diseases had been suggested but in the mid-nineteenth century was still hotly contested. Opponents believed that most diseases were caused by a corruption of the air or genetic dispositions, not by small microbes that were passed from one person to another. Those who accepted that microbes might be to blame for disease were further split between those who thought microbes were capable of transforming themselves to cause more than one disease, and those who insisted that for each disease there was a specific cause. Pasteur, a believer in the germ theory himself, gave this idea another push forward when he was called upon to help save France's silkworm industry. In 1865 an unknown disease was killing off silkworms in enormous numbers, threatening the economy of many towns. Pasteur, having helped improve the beer and wine industry through his research, was asked to apply himself to the problem of the silkworm. After several years of study he was finally able to establish that two separate parasitic organisms were to blame for the silkworm deaths. Thus Pasteur showed the importance of microbes in affecting the world and established a link between microbes and disease.

The final step was taken by Robert Koch, who emerged on the world stage in 1876 with systematic proofs that small rod-shaped bacilli were the cause of anthrax. Koch had received a microscope as a birthday gift from his wife, and he began to spend his free time between patients examining samples of all sorts under it. Eventually, he studied a drop of blood that had come from a sheep recently killed by anthrax. In addition to being deadly, anthrax was a mysterious disease that seemed to arise from nowhere. Certain fields were avoided and believed to be cursed because upon taking a healthy herd to graze in them, the farmers would find any number dead the next day. Koch found in this drop of blood some unusual rod-shaped microbes, which he correctly deduced to be the cause of anthrax. With remarkable caution, dedication, and intelligence, he set about figuring out how to prove beyond a doubt that these were indeed to blame for the disease. The process by which he proved this has become known as Koch's postulates. First, he found the tiny rods in all cases of animals he examined that had died of anthrax. Next, he reasoned that if these rods were indeed to blame for the disease, then putting them into a healthy animal should cause that animal to become ill. Using available resources, Koch carefully took a bit of fluid containing these rods and inserted it into the tail of a mouse. When the mouse began showing symptoms of anthrax and shortly died, he knew he had succeeded in passing the disease. He then dissected the mouse to see if he could still find the same little rods in the mouse's blood, which he did. Koch then continued his tests by taking a bit of fluid from the newly infected mouse and using it to infect another healthy one. In this manner, Koch continued until he had successfully infected 20 generations of mice. This multi-step process became the means by which specific bacteria were proven to cause certain specific diseases.

Impact

Koch's work brought him instant acclaim, but it also had a dramatic impact on medical science. His postulates set the method for all future identifications of specific microbes and helped to demonstrate that each disease was caused uniquely by a single microbe. Given a laboratory in Berlin and several assistants to work with in 1880, Koch began to devote himself full time to microscopic studies. His Berlin lab became something of a rival to that of Louis Pasteur in Paris, who was also continuing his microscopic studies. Pasteur had turned his attention from silkworms to human diseases, working on rabies. Unable to identify the causative agent in rabies, he nonetheless succeeded in developing a vaccine by 1885.

Koch's discovery of anthrax set off a race to discover the bacteria responsible for other diseases. In 1882 he succeeded in isolating the bacteria that causes tuberculosis (*Mycobacterium tuberculosis*). The next year, a severe cholera epidemic broke out in Egypt and threatened to spread across the Mediterranean Ocean to Europe. Both Germany and France sent teams to Alexandria to try and discover the responsible germ. Sadly, Louis Thuillier, one of Pasteur's team members, would contract the disease and die, a stark reminder of the hazards of working in continuous close proximity to deadly microbes. Koch headed up his own team and succeeded in isolating the cholera vibrio (*Vibrio cholerae*). He followed up this discovery by traveling to India, where cholera was endemic, confirming his earlier identification and discovering infected water supplies to be the cause of its transmission.

The two labs were not entirely competitive, however, and did use one another's work to build up new discoveries. The bacillus responsible for diphtheria (*Corynebacterium diphtheriae*) had been identified in 1883 by Theodor Albrecht Edwin Klebs (1834-1913), a scientist who studied with neither Koch nor Pasteur. Shortly thereafter, Friedrich Löffler (1852-1915), working in Koch's lab, was able to culture it. Working in Paris, Émile Roux (1853-1933) and Alexandre Yersin (1863-1943) found that the bacteria put out a toxin, which was actually responsible for causing the disease. Back in Berlin, Emile von Behring (1854-1917) picked up the work on diphtheria in an attempt to find a means of preventing it from infecting children. Assisted by the Japanese scientist Shibasaburo Kitasato (1852-1931), Behring found that blood serum from infected animals had the power to neutralize those toxins, and he therefore named it antitoxin. Antitoxin proved to be of little use in preventing diphtheria infections, as it protected for only a few weeks. It did prove useful as cure, however, as was proved in 1891. Behring first tested his antitoxin on a patient, albeit somewhat reluctantly, on Christmas night. A child who was severely ill and sure to die within days was injected with this antitoxin and made a miraculous recovery. Behring would go on to receive the Nobel Prize in medicine in 1901 for his work on serum therapy.

The idea of blood serum as an antitoxin proved useful in another disease, tetanus. At the same time that he was working with Behring on diphtheria, Kitasato also succeeded in culturing the tetanus bacillus (*Clostridium tetani*) and discovered that it too used a toxin. Kitasato's work on developing the tetanus antitoxin proved highly useful during World War I. Tetanus, like anthrax, resides in soils and can enter the body through open cuts or wounds. During the war, antitoxin was routinely given to all wounded men, and the death rate dropped dramatically.

Kitasato was involved in one last race to identify a microbe at the very end of the century. An outbreak of bubonic plague in Hong Kong in 1894 prompted both Kitasato (Robert Koch's student) and Alexandre Yersin (Louis Pasteur's student) to set up temporary labs and attempt to identify the bacillus. The discovery of it came simultaneously for both, and while the name of the bacillus is now *Yersinia pestis*, Kitasato is also acknowledged for his contemporaneous discovery.

Koch's careful studies of microbes proved their connections to disease and helped shape modern germ theory. Along with Louis Pasteur in France, Koch became a leader in the science of bacteriology. While the two men were never friends, they inspired each other to further achievements and each trained numerous students who carried on their work.

KRISTY WILSON BOWERS

Further Reading

Books

Brock, Thomas D. *Robert Koch: A Life in Medicine and Bacteriology*. New York: Science Tech Publishers, 1988.

de Kruif, Paul. *Microbe Hunters*. New York: Harcourt Brace & Company, 1926.

Lechevalier, Hubert A. and Morris Solotorovsky. *Three Centuries of Microbiology*. New York: Dover Publications, 1974.

Rosen, George. *A History of Public Health*. Baltimore: Johns Hopkins University Press, 1993.

Other

"Robert Koch." Nobel Prize Web Site. http://www.nobel.se/laureates/medicine-1905-1-bio.html

The Battle against Tuberculosis: Robert Koch, the Development of TB Sanitariums, and the Enactment of Public Health Measures

Overview

During the late nineteenth century, tuberculosis, the "white plague," was the leading cause of death throughout the United States and Western Europe. Then, in 1882, a German physician named Robert Koch (1843-1910) discovered the bacterium that causes tuberculosis. With proof that the disease is caused by an infectious agent, doctors and public health officials could turn their attention toward preventing and curing it. An enormous industry sprang up around tuberculosis sanitariums—facilities where infected people would go for months or years, in the hopes that healthy living would cure them. At the same time, cities took the first steps toward implementing public health measures designed to control the spread of disease. Many of these new public health policies were controversial and unpopular, but the growth of the public health movement is credited with slowing the spread of tuberculosis throughout Europe and North America.

Background

Tuberculosis, or TB, is a bacterial disease that is spread by the respiratory route from person to person. People become infected when they breathe in air exhaled by someone with an active case of the disease. In most cases, prolonged exposure is necessary for a person to become infected. The tuberculosis bacterium can lie dormant for years. Once it becomes active, however, the bacterium infects and kills lung tissue; if the infection becomes severe enough, the patient dies.

Tuberculosis has been around at least since the Ice Age, and it exists in every region of the world; evidence of the disease has even been found in the spinal tissue of some Egyptian mummies. But there was never an epidemic like the one of the late nineteenth century. By the mid-1800s industrialization had sent throngs of people to cities across Europe and North America in search of work. Conditions in cities became crowded and the inhabitants suffered from poor nutrition, lack of hygiene, and overwork—the perfect scenario for the spread of disease. Tuberculosis spread rapidly through the packed cities and soon became epidemic. In the late 1800s about one in four people was infected with "consumption," as TB was often called. Almost every family had lost at least one member to the disease.

At the same time, there was a revolution going on in the study of diseases, especially those caused by microorganisms. In 1863 a French parasitologist named Casimir-Joseph Davaine (1812-1882) was the first to observe microscopic organisms in the blood of people suffering from certain diseases. He later observed similar organisms in the blood of sheep suffering from anthrax, a deadly disease of livestock. Davaine theorized that those organisms were in fact the cause of disease in animals and humans, but he had no proof that this was the case.

In 1876 Koch devised a method for growing the anthrax organisms in the laboratory. He then inoculated mice with these anthrax spores, and the mice became sick. Koch had proven that the microorganisms did in fact cause disease. For several years after this discovery Koch worked to perfect his methods of culturing bacteria outside of the human body; he also pioneered techniques for staining bacteria so that they could be seen under a microscope and photographed.

In 1882, convinced that tuberculosis was also caused by a microorganism, Koch turned his attention to finding it. He took samples of diseased lung tissue from tubercular patients, cultured them in a dish, and waited. Eventually the experiments worked—Koch was able to isolate and identify the bacterium that causes tuberculosis—*Mycobacterium tuberculosis*. In later years Koch attempted to develop a tuberculosis vaccine out of dead bacteria, which he called tuberculin, but it did not work. However, it was later discovered that an injection of tuberculin could be used to diagnose TB; when the dead bacteria are injected into an infected person, a rash develops near the site of the injection. This method of diagnosis is still used today.

Impact

Up until the time of Koch's discoveries it was widely thought that people who were infected with tuberculosis were weak or ignorant or had done something to bring the disease upon themselves; many considered it "God's will" that certain people became infected. The idea that some other organism—a germ—could make people sick had never been considered. By finding the organism that caused tuberculosis, Koch made it possible to prevent the spread of the disease and to start the search for a cure. TB was no longer God's will, it was something that could be—if not cured—at least controlled by man.

Once science and medicine accepted Koch's "germ theory," the war on germs began. Major cities, including New York, Chicago, and Philadelphia, formed the first Boards of Health. The role of the boards was to try to stop disease by halting the spread of germs—including the tuberculosis bacterium. Since there were not yet any drugs that could cure tuberculosis, officials focused on enforcing standards of hygiene and controlling the behavior of the public.

One of the most aggressive soldiers in this new public health movement was the General Medical Officer of New York City, a man named Hermann M. Biggs. Biggs believed that education and preventative health measures were the only way to fight disease. He made a map of New York

City, pinpointing every single case of tuberculosis. He encouraged the disinfection of rooms where tubercular patients had stayed, started a program for registering all TB patients with the city, and he tried to require hospitals to isolate people with TB away from other patients. Biggs was praised for his thoroughness, but many of his policies were

TUBERCULOSIS AND THE ARTS

Considering how widespread the tuberculosis epidemic was throughout Europe and North America in the nineteenth century, it is no surprise that many famous people were stricken with the disease—especially artists and writers. The list of notables who died of TB during this time includes Anton Chekhov (1904), Frederick Chopin (1849), John Keats (1821), Robert Louis Stevenson (1894), and Anne, Charlotte, and Emily Brontë (who died between 1821 and 1855).

Some philosophers and physicians of the time developed a theory that "consumption" actually sought out the artistic, and for a time TB infection became almost fashionable. It was thought of as a disease that only refined, spiritual people contracted. As the infection progressed, patients became thinner, more delicate, and more fragile, which was seen as further proof of their spirituality. Lord Byron wrote of tuberculosis: "I should like to die of consumption. The ladies would all say, 'Look at that poor Lord Byron, how interesting he looks in dying!'"

Because tuberculosis was such a part of life, it also became rich fodder for many nineteenth-century artists. In Giacomo Puccini's opera *La Bohème* about a group of young artists in Paris, the heroine, Mimi, collapses and dies of consumption. The title character in French author Alexandre Dumas's *Camille* coughs up blood and dies in the arms of her beloved, Armand. Camille is based on a real-life friend of Dumas's who died at the age of 24. The Norwegian artist Edvard Munch was greatly influenced by the deaths of his mother and older sister from tuberculosis. His paintings "The Dead Mother" and "Death in the Sickroom" are directly based on his experiences.

controversial. Because the poor were more susceptible to tuberculosis due to their living conditions, many viewed public health measures as a conspiracy against them. Other officials worried that strict government enforcement of health measures would turn cities into police states where sick people were considered criminals.

In 1893 Biggs established Blackwell's Island Hospital, a hospital for tuberculosis patients only, in accordance with his belief that those afflicted with TB should be kept apart from the rest of society until they recovered or died. It was not the first tuberculosis hospital—disease-specific hospitals had sprung up all over the United States and Europe starting around 1880. The treatments available to doctors were limited and low-tech, but they did what they could to help patients get stronger and stay as healthy as possible. Treatment at a tuberculosis hospital mostly involved rest, relaxation, a special diet, and a very long stay. Even if the disease was in its early stages, a minimum hospital stay of at least two years was prescribed; it often took much longer than that to get a tubercular patient healthy enough to leave. Even so, years in a hospital carried no guarantee of a cure or even long-term improvement. With no antibiotics to truly cure the disease, the most a person could hope for was some degree of healing.

While some patients were forced into urban tuberculosis hospitals, countless others were willingly fleeing to sanatoriums in the country. The first TB sanatorium was started by Edward Livingston Trudeau, a wealthy New York doctor who himself was diagnosed with tuberculosis in 1873. As his health deteriorated, Trudeau accepted that he was going to die and decided to spend his last days in the Adirondack Mountains. To his surprise, he got better. He became convinced that the climate there was the cause of his "cure," and started a sanatorium for other tuberculosis sufferers outside Saranac Lake in 1884. He started off treating just two patients, but soon the Lake Saranac Sanatorium grew to the size of a small town, with 36 buildings, including laboratories, patients' quarters, stables, and a post office.

Before long sanatoriums became a huge industry, particularly in the United States. They were built in places that seemed "healthy": on beaches, in the mountains, near lakes, in the desert. In fact, parts of the United States that were barely inhabited before suddenly boomed with the sanatorium craze; it is estimated that 60 percent of Colorado was first settled by sanatorium patients and their relatives.

No matter where they were located, all sanatoriums promised the same things—a place where tuberculosis patients could get fresh air, good food, rest, and exercise, all under the supervision of a medical staff. Like the urban TB hospitals, sanatoriums prescribed long stays—anywhere from two to 20 years. But the infected went willingly; many thought it was their last chance to recover, and often it was. But despite the resort-like setting, a stay at a sanatorium was not an extended vacation. The rules were strict and everything was regulated—when patients woke and went to bed, when and what they ate, who they spoke to, when they were allowed to go to the bathroom, even what they read. Different sanatoria had different theories on what kind of treatment would cure TB. Some swore by bizarre diets of raw eggs and cream, others believed that patients should be kept cold all the time—even making them sleep outside in winter.

Although none of those methods could be proven to restore health to tuberculosis patients, many did get well, and sanatoriums and tuberculosis hospitals proved critical to public health efforts—they took people who were sick with a contagious disease and put them in places where they couldn't spread that disease. The quarantining of patients was to be instrumental in the eventual decline of infectious tuberculosis. But while there is little doubt that the public health movement was good for society, questions still remain about how the movement affected the rights of the individual.

GERI CLARK

Further Reading

Brock, Thomas D. *Robert Koch: A Life in Medicine and Bacteriology.* New York: ASM Press, 1988.

Daniel, Thomas M. *Captain of Death: The Story of Tuberculosis.* Rochester, NY: University of Rochester Press, 1997.

Ellsion, David L. *Healing Tuberculosis in the Woods.* Westport, CT: Greenwood Publishing Group, 1994.

Feldberg, Georgina D. *Disease and Class: Tuberculosis and the Shaping of Modern North American Society.* Livingston, NJ: Rutgers University Press, 1995.

Ryan, Frank. *The Forgotten Plague: How the Battle Against Tuberculosis was Won—and Lost.* Boston: Little, Brown, 1992.

Deviancy to Mental Illness: Nineteenth-Century Developments in the Care of the Mentally Ill

Overview

At the opening of the nineteenth century the view of insanity was just beginning to shift from unacceptable deviancy to a form of treatable illness. During earlier periods those with mental illness were held in suspicion of demon possession and other unearthly states. Public mistrust and fear of the insane resulted in their confinement, perpetually shackled in miserable cells. In the spirit of the Enlightenment an objective assessment of insanity permitted social and medical efforts to examine the causes and possible cures of the afflicted. A leading goal in the care and treatment of mental illness was to shepherd sufferers back into to the fold of normal society in the interest of preserving social order. Though the causes of mental illness were not identified and few cures emerged, the new empirical approach to the care and treatment of the mentally ill punctuated the beginning of an era that witnessed vast improvements on their behalf. By the end of the nineteenth century medical science had devoted prodigious efforts towards the study of mental illness, and fervent social reformers sought improvements for the care of the mentally ill.

Background

French physician Philippe Pinel (1745-1826) revolutionized the care of mentally ill. Pinel, while working in Paris at the turn of the nineteenth century, championed the notions that the environment, not intangible forces, was a causal factor for insanity and that mental illness was curable. Pinel's work set forth a novel trend in the treatment of the mentally ill, since he helped to disentangle the causes of mental illness from abstract deviancy. Since the environment was significant to mental health in Pinel's framework, he changed the asylum environment by releasing the mentally ill from their shackles. Pinel also introduced *traitement moral* (moral treatment), which established a social interaction-oriented approach to the treatment of mental illness for the newly released patients. With *traitement moral* Pinel practiced empirical observation of and interaction with patients, in a form of trial and error, to uncover patterns of lucidity and triggers for normal and abnormal states in his patients. Although Pinel denied the benefits of restraints and corporal punishment for the mentally ill, he did condone verbal chastisement and threats to maintain control over patients. Though most of Pinel's assertions were not original, his approach to and execution of treatment for the mentally ill opened the door for changes and improvements in the care of the mentally ill.

Along with Philippe Pinel, Italian physician Vincenzio Chiarugi (1759-1820) and American physician Benjamin Rush (1745-1813) worked in their respective countries to reform the care of the insane by treating the condition as an illness, rather than deviancy, during the early nineteenth century. The efforts of Quaker merchant William Tuke (1732-1822), however, proved more influential than the work of either Chiarugi or Rush. At England's York Retreat Tuke continued the Quaker tradition of moral reform for dependent groups by implementing new practices in the care of the mentally ill. Like Pinel, Tuke shirked the use of restraints and corporal punishment as a means to control inmates. Instead, the mentally ill were likened to children, and as such, the use of threats and intimidation produced the desired beneficial results. Tuke's model left open the possibility that patients would recover given the opportunity. The expenditure of patience and time-consuming observation paid off with patients released as cured. Asylums for the mentally ill garnered a fresh air of legitimacy as a result of Pinel and Tuke's work. With large asylums held in higher esteem, local authorities were offered a new respectable place to relocate the mentally ill.

Throughout the nineteenth century the proponents of various schools of thought investigated diagnostic models for mental illness. Swiss theologian Johann Kasper Lavater (1741-1801) advocated the classification of facial features to evaluate character and personality in the field of physiognomy. In the same vein Austrian anatomist Johann Spurzheim (1776-1832) advocated phrenology, the study of head shape, to diagnosis mental problems. Although physiognomy and phrenology were debunked by the close of the nineteenth century, the tangible aspects of both methods offered stiff competition to the emerging field of psychiatry, which did

This painting by Charles Muller illustrates Philippe Pinel's pioneering efforts to improve care for the mentally ill. *(J.-L. Charmet; Academie Nationale de Medicine. Reproduced by permission.)*

not link physical characteristics with mental, personal, or emotional states. Instead, early psychiatrists, who initially were physicians working in asylums, followed empirical observations of patient behavior and generally linked patients' conditions with physical (accident, brain lesion) or moral (excessive behaviors like greed or masturbation) phenomena. Unlike twentieth-century psychiatry, which generated a plethora of diagnostic terms for neuroses, nineteenth-century psychiatry was more conservative and held to only five categories of mental illness: mania, melancholia, monomania, dementia, and idiocy.

Impact

Despite the advances made by Pinel and his contemporaries, improvements for the majority of the mentally ill progressed slowly throughout the nineteenth century. Physical restraint of the mentally ill with muffs, shackles, collars, cribs, and strong chairs declined during the nineteenth century but never disappeared entirely. Physicians in the United States and England participated in lengthy debates on the value of restraint. Where American physicians believed temporary restraint was useful to calm excited patients, English physicians insisted all forms of restraint were unnecessary and associated the American use of restraints to its legacy of slavery. Other antiquated

ideas persisted into but eventually perished during the nineteenth century. For example, the long-held myth that the insane did not suffer from the elements led to horrid conditions with inmates exposed to the extremes of weather conditions without the benefit of clothing or heat. Fortunately, that myth was put to rest by the midpoint of the century, due to the acceptance of mental illness as a treatable condition. While physical restraint declined and humane living conditions came into vogue, chemical restraint (ether, chloroform, etc.) emerged as a non-physical means to restrain excited patients. Chemical restraint, however, lost popularity almost as quickly as it came into fashion.

In the United States and Europe Pinel's *traitement moral* was adopted and implemented in the construction of asylums and the care of the mentally ill. Outside of France *traitement moral* was translated to "moral treatment," which included "morality" along with Pinel's empirical methods. By and large the incorporation of morality affected the design and atmosphere of nineteenth-century asylums. The basic structure of nineteenth-century asylums was modeled after the family unit. This system raised inmates to the status of children and the asylum superintendents to the role of caring parents. In this scheme the goal was for the parental superintendents to raise the childlike inmates to the plateau of the sound mind. In the family model labor therapy provided a handy diversion for inmates. Nearly ubiquitous in Europe and the United States by the mid-nineteenth century, labor therapy quickly gained popularity under the rubric of moral treatment since it set inmates to work, thereby modeling the productive activities of the outside world. Also in the effort re-create society, the cottage system in asylum design reflected small communities with clusters of cottages housing mentally ill patients under treatment.

In the United States Unitarian social reformer Dorothea L. Dix (1802-1887) proved to be the greatest advocate for improved conditions for the mentally ill. Dix was influenced by William Tuke's methods during her visits at the York Retreat in England. Upon her return to America Dix spearheaded a campaign to create federal asylums for the care of the mentally ill. To provide evidence of the need for federal responsibility, Dix systematically visited county poorhouses in several American states and reported the lamentable conditions to the U.S. Congress. Dix claimed the mentally ill suffered a variety of horrors and immoral conditions in local poorhouses throughout the country. And, while some houses were better than others, Dix insisted the range in quality was unacceptable. Dix argued the need for one standard of care that was defined by the national government. That standard, Dix hoped, would insure equal care to all the mentally ill in asylums. The plan for a federal system, however, was thwarted as too ambitious. Dix's next attempt was to lobby individual state legislatures with the intent of creating state level statutes. Dix was more successful on that front. In New York, for example, the legislature passed the Willard Act of 1865, which mandated state level care of insane paupers, which was later strengthened with the State Care Act of 1890.

A notable nineteenth-century advancement in the care of the mentally ill was the recognition of the possibility of recovery. By the mid-point of the nineteenth century, many asylums were built around the expectation that at least some patients would recover and others may not. The assignment of "acute" or "chronic" insanity to an inmate often dictated their life course. Acute insanity was generally an arbitrary one-to-two-year period in which a variety of treatments were attempted. If the patient recovered during that period, they were released from the asylum. If an inmate's mental illness extended beyond the arbitrary period, their condition was re-classed as chronic insanity, at which time chronic patients were transferred to an asylum for the chronically insane. Asylums for the chronically insane served a custodial role and offered limited forms of treatment. And, fulfilling caretaker's expectations, recovery rates at asylums for the chronically insane were very low. Institutions for the chronically insane drew negative attention since their inmate population was rather permanent, in comparison to acute asylums where inmates were in and out within two years whether they were cured or not. Thus, by the end of the nineteenth century, chronic institutions—with overcrowded wards filled with long-tenured inmates seemingly lost and forgotten in the fray—represented a dismal counterpoint to the otherwise positive developments in the treatment of the mentally ill during the century.

SHAWN M. PHILLIPS

Further Reading

Dwyer, Ellen. *Homes for the Mad: Life Inside Two Nineteenth-Century Asylums.* New Brunswick, NJ: Rutgers University Press, 1987.

Foucault, Michel. *Madness and Civilization: A History of Insanity in the Age of Reason.* New York: Vintage Books, 1973.

Grob, Gerald. *The Mad Among Us: A History of the Care of America's Mentally Ill.* Cambridge, MA: Harvard University Press, 1994.

The Development of New Systems of Alternative Medicine: Homeopathy, Osteopathy, Chiropractic Medicine, and Hydrotherapy

Overview

Alternative medicine enjoyed considerable popularity in the 1800s. Both medical practitioners and patients began to understand the possibility of healing the body without invading it, and all the common methods of alternative medicine approached healing in this noninvasive way. The most popular and prevalent alternative medicine of the nineteenth century was homeopathy. Osteopathy also grew to be an accepted alternative to traditional, or allopathic, medicine. What is now considered chiropractic medicine was created in 1895. The benefits of hydrotherapy were also enjoyed in the nineteenth century.

Background

Homeopathy

A German physician named Samuel Hahnemann (1755-1843) founded the practice of homeopathy near the beginning of the nineteenth century. He joined the Greek terms *homoios* and *pathos,* meaning, respectively, "similar" and "suffering." This juxtaposition is the basis of homeopathy, which treats disease by introducing remedies that create symptoms that are similar to those caused by the disease itself.

Hahnemann made this discovery while he was translating a book written by Scottish physician William Cullen (1710-1790). In the book, Cullen explained that a certain herb, Peruvian bark, was useful in treating malaria because it was bitter. This didn't seem logical to Hahnemann, since many other herbs that were bitter did not aid in the curing of malaria. He experimented on his own body by taking doses of Peruvian bark. After repeating the dosage, he began to suffer from fever, chills, and other symptoms usually associated with malaria. From this he deduced that Peruvian bark helped treat malaria because it and malaria had similar effects on humans. This is where the "similar" or "homeo" part of homeopathy originated.

Almost since its creation, homeopathy has been controversial. Traditional physicians, whom Hahnemann named "allopaths," disliked the practice because it contradicted their own, which was then partly composed of a practice known as "bleeding" the patient with leeches. Allopathic physicians felt threatened by the encroaching homeopathic physicians, and some traditional doctors admitted that they feared economic repercussions due to the competition. Pharmacists, then called apothecaries, also resented the evolution of homeopathy. Allopathic medicine demanded that more intense and frequent drugs be used to treat patients; this meant more profit for the apothecary. Homeopathic doctors prescribed simpler remedies, never in combination.

One apothecary so disliked Hahnemann and his work that he began deliberately filling Hahnemann's prescriptions incorrectly. When Hahnemann discovered this, he began dispensing the remedies himself, which was at that time illegal in Germany. Hahnemann was discovered, tried, and found guilty. In order to continue practicing homeopathy, he moved to Kothen, where the Grand Duke, himself a supporter of homeopathy, allowed him special permission to practice and dispense.

A Dutch homeopath named Hans Gram emigrated to the United States in 1825, bringing with him his medical beliefs. The practice of homeopathy flourished in the United States, and in 1844 the nation's first medical society, the American Institute of Homeopathy, was created. Two years later the American Medical Association was founded. These two organizations helped fuel the increasing animosity between homeopaths and allopaths. Homeopaths were spurned from professional medical societies, and physicians even created an ethical code that prohibited professional contact with homeopaths. This antagonism spread to Europe, and the future of homeopathy was threatened.

Perhaps one of the reasons why homeopathy was controversial was its inclusion and acceptance of women practitioners decades before woman's suffrage. Boston Female Medical College, founded in 1848, was dedicated to the study of homeopathy. The college merged with another homeopathic institution, Boston College, in 1873. The American Institute of Homeopathy admitted female homeopaths as early as 1871; the American Medical Association did not follow suit until 1915.

This photograph of a homeopathic pharmacy in Philadephia around 1890 shows the popularity of herbal cures in America in the latter part of the nineteenth century. (*Hahnemann Archives. Reproduced by permission.*)

Homeopathy survived the approbation of medical doctors in large part due to its effectiveness. Death rates for epidemics such as cholera and yellow fever were 30% lower in homeopathic hospitals. The knowledge that homeopathy was effective was indicated in the policy of several insurance companies to offer discounts to their patients who chose homeopathic medicine.

In 1899 there were in the United States over 20 homeopathic medical schools, 100 hospitals, and 1,000 pharmacies. Homeopathy was praised by famous nineteenth-century American writers such as Mark Twain, Henry James, Louisa May Alcott, Henry Wadsworth Longfellow, and Nathaniel Hawthorne.

Osteopathy

The osteopathic philosophy states that if the body is functioning in harmony, no disease or discomfort will occur. Should disease occur, regaining the body's harmony should cure the ailment. Osteopathic doctors liken the human body to a machine. The founder of osteopathy,

Dr. Andrew Still (1828-1917), wrote: "As long as the human machine is in order, it will perform the function that it should. When every part of the machine is properly adjusted and in perfect harmony, health will hold dominion over the human organism by laws as natural and as immutable as the laws of gravity." "Nature knows perfectly your powers, plans, and purposes," he asserted.

Still founded osteopathy in 1874, when he opened his practice in Kansas. He opened the first school of osteopathy in Kirksville, Missouri, in 1892. Four years later, Vermont became the first state to declare that osteopathy was legally its own school of medicine. Still was motivated to find an alternative to conventional medicine when, as a trained doctor, he watched three of his children die of viral meningitis. His religious and spiritual beliefs led him to develop osteopathy, which he named after the Greek word *osteo*, meaning bone. Since his theories revolved around the structure of the body and the relationship of the body's structure and function, "osteopathy" was a logical choice.

Osteopathy differs from conventional medicine because it asserts that, rather than expose the body to outside or unnatural remedies in order to cure it, the osteopathic doctor should seek to regulate the body's functions by natural means. Rather than focusing on the sickness, an osteopath focuses on strengthening the healthy parts of the body by increasing nutrients, blood flow, and nerve impulses.

Osteopathy maintains that certain elements called *life essentials* are needed to maintain harmony in the body. These life essentials are food, water, light, air, heat, exercise, protection, and rest. The study of osteopathy integrates a thorough knowledge of pathology, anatomy, and physiology. Osteopaths believe that the body functions as a unit, and that its structure and its function are inexorably intertwined. Osteopaths also assert that, given the life essentials to maintain balance, the body is capable of regulating and, ideally, curing itself.

After being founded by Still in the United States, osteopathy was carried to Australia and Great Britain in the last years of the nineteenth century.

Chiropractic Medicine

Unlike homeopathy and osteopathy, the history of chiropractic medicine dates back to the early Egyptians, who routinely practiced spinal adjustments.

In 1895, however, an important change in the philosophy of chiropractic medicine occurred. Daniel David Palmer (1845-1913), a physiologist and anatomist, founded what is considered modern chiropractic medicine. He believed that all living creatures are endowed with an "innate intelligence" that regulates the body's vital functions. Rather than treat symptoms, Palmer sought to remove interference within the nervous system, thus allowing a being's "innate intelligence" to function operatively. This belief system became popular with those who wished for less invasive procedures than were offered by conventional medicine. Palmer stated that chiropractic medicine revolves around "the science of life, the knowledge of how organisms act in health and disease, and also the art of adjusting the neuroskeleton."

Hydrotherapy

Like chiropractic medicine, the use of hydrotherapy dates back to early civilizations. Romans built spas for therapeutic baths and utilized natural mineral water to aid in healing.

In Europe in the nineteenth century, intricate spas were built amid beautiful settings with lakes, mountains, and forests. Mineral baths became common cures for a myriad of ailments. People believed that the minerals magnesium, potassium, sulfur, and calcium were particularly beneficial. The mineral baths in Bath, England, became incredibly popular. The rejuvenating possibilities of hydrotherapy spread to the United States as well. Like other avenues of alternative medicine, hydrotherapy offered a noninvasive method of healing.

Impact

Alternative medicine affected thousands of people in the nineteenth century, people who sought to heal themselves without the more drastic approaches of conventional medicine. Conventional medicine did continue to be the most common wisdom sought by the ill, however, and conventional doctors, under the umbrella of the American Medical Association, increasingly gained respect and power as the century progressed. The animosity between conventional and alternative medicine caused the latter to slip into obscurity as the next century began. Still, the teachings of Hahnemann, Still, and Palmer affected large numbers of people throughout the world, and set the stage for a resurgence of alternative medicine in the future.

CAROLYN CRANE LOVE

Further Reading

Cook, Trevor M. *Samuel Hahnemann: The Founder of Homeopathic Medicine.* Wellingborough, UK: Thorsons, 1981.

Kaufman, Martin. *Homeopathy in America.* Baltimore: Johns Hopkins Press, 1971.

Palmer, Daniel David. *The Science of Chiropractic: Its Principles and Adjustments.* Davenport, IA: The Palmer School of Chiropractic, 1906.

Sutherland, William Garner. *Teachings in the Science of Osteopathy.* Edited by Anne L. Wales. Cambridge, MA: Rudra Press, 1990.

Tropical Disease in the Nineteenth Century

Overview

The nineteenth century was a period of expansion of western imperialist power into Africa, Asia, the Middle East, Central and South America, and other regions of the world. This involved "population mixing" where one group of people met and lived side by side with another. Inevitably each group met a new "germ pool"—diseases specific to the other group which were now shared. Because the principles of sanitation—provision of clean water and disposal of sewage, etc.—were not fully understood yet by either the newcomers or those who were indigenous to the newly penetrated territories, neither group protected itself properly against infection and cross infection from communicable diseases.

Background

In most cases, the ways in which germs traveled from person to person and caused illnesses were not yet understood in the nineteenth century. The interest of colonists in developing the resources of new territories to meet the needs of the new industries of the West also meant that they introduced different forms of cultivation, such as farming cash crops and the intensified mining of natural resources. In many cases these new activities upset the balance between the people living in a particular environment and the cycle of nature. Sometimes this population mixing, combined with new forms of economic activity, caused epidemics of diseases. Often the new arrivals actually brought diseases with them that the local populations did not have a strong resistance to.

In Somaliland, for instance, smallpox, cholera, influenza, venereal diseases, and tuberculosis epidemics occurred soon after the arrival of colonists from Britain. In Mauritius the arrival of French soldiers who traveled via India with workers brought from India introduced new diseases to the local people for which they had not developed a resistance. The transmission of tropical diseases was a two way process. During the American Civil War (1861-65) one half of the white troops and four out of five of the black troops in the Union Army suffered from malaria. Cholera, typhoid, and other diseases were present in both Europe and North America throughout the nineteenth century and it was not until the science of epidemiology was developed early in the twentieth century that the processes of epidemics were understood and could be combated.

Impact

The building of new roads, railways, public buildings, and plantations required a different type of work force than in traditional societies, and sometimes the colonists resorted to forced labor and cheap labor to undertake these new projects. It took many years to realize that workers camps could themselves become hotbeds of infection, especially if people were poorly fed and housed. The creation of new centers of commerce and agriculture caused "internal migrations," where people from one part of a territory migrated toward the work opportunities. This process also involved population mixing. Sometimes the new cash crops actually caused people to give up their traditional agriculture, leaving the population poorly fed while the new crops were exported to the colonizing countries.

Even the introduction of money as the prevailing form of exchange upset the traditional balance of many societies that had relied on bartering goods to meet people's needs. Sometimes the new money was used for alcohol and sugar-based foods rather than the traditional balanced

Ronald Ross with his wife outside the laboratory of a Calcutta hospital, where he completed his theory on the mosquito transmission of malaria. *(Wellcome Institute. Reproduced by permission.)*

support of their families took to alcohol which only increased the problems. In 1863 Richard Burton (1821-1890), the explorer, described the Lagos Government house in Nigeria as "a corrugated iron coffin or a plank lined morgue containing a dead Governor once a year."

Many missionaries and other church-based medical personnel did try to improve the welfare of the people in these territories. They offered a different type of medicine than that which the traditional healers practiced and often there were clashes between the two cultures. But there were some areas of agreement. The Yoruba people of Nigeria, for instance, believed that many illnesses were caused by worms or "insects" that were so small that they resemble what came to be known to the tropical medical researchers as bacteria. Many of the people living in the tropical cultures made good use of herbal remedies from the plants available to them, and some of these have since been recognized in Western medicine. Dr. David Livingstone (1813-73), one of the early Western missionaries and explorers in Africa, was always interested to learn about these medicines. These health problems led to the development of "tropical medicine" as a field of research and health as doctors tried to understand the dynamics of the epidemics and communicable diseases.

One of the most important researchers in theses early years of tropical medicine was Patrick Manson (1844-1922), who graduated from Aberdeen University Medical School in Scotland then went to China where he stayed for nearly a quarter of a century, studying the local diseases. In 1878 he observed that filariae, the worms that cause elephantiasis in man, pass part of their life cycle in the *Culex* mosquito. This made him realize that the parasites that cause many of the tropical diseases are transmitted by other carriers, or vectors. In 1894 he realized that the parasite of malaria also passes part of its life cycle in the mosquito and this was confirmed by another great researcher in the tropical diseases, Ronald Ross (1857-1932), at the end of the nineteenth century. These observations led the governments concerned to undertake mosquito control programs that are still the basis of prevention of tropical diseases. The building of the Panama Canal, for instance, could only be undertaken once it was understood that the laborers had to be protected from mosquitoes that were spreading yellow fever. Yellow fever is caused by a virus transmitted by the bite of the female *Aedes aegypti* mosquito

diets. In countries of the South Pacific, for instance, these changes resulted in widespread diabetes and obesity. In the Songea District of Tanzania so many men left to work in the new plantations and industries elsewhere that the health of the women and children of the families left behind were endangered by malnutrition. In Mauritius, the colonial economy resulted in the country being very dependent on sugar as its one cash crop for export and the population relying on imported food rather than the traditional products of its diet.

By the middle of the nineteenth century there were Western medical doctors and missionaries present in these territories, but their first preoccupation was usually the health of the colonists rather than the colonized people. The armies of the occupying powers had to be kept healthy if they were to be effective. The civil servants and merchants who worked in the territories were likely to fall ill. Usually they had to send their children home after a year or two to be raised by relatives because life was so hazardous in the tropical countries. Often men who were left to work in these conditions without the

which breeds in stagnant water near human habitations. Another form of this disease is spread by another species of mosquitoes that live in the trees of tropical jungles.

Ronald Ross, an Englishman born in Almora, India, studied malaria in that country as a member of the Indian Medical Service during the last 20 years of the nineteenth century. In 1897 he observed the malarial parasite in the stomach of the *Anopheles* mosquito. In West Africa he identified the mosquito that transmits African fever. He eventually received the Nobel Prize in 1902 for his work on malaria, and became a professor at the newly organized Ross Institute and Hospital for Tropical Diseases in London. He was sometimes known as "The Mosquito Man."

Another Scot who was to make a major contribution to the understanding of tropical diseases was William Boog Leishman (1865-1926), who graduated in medicine from Glasgow University then went to India as part of the Army Medical Service. He undertook studies with his microscope of enteric fever and kala azar. When he returned to England he worked on inoculation against typhoid fever, and developed a process of staining blood samples for malaria diagnosis.

In 1873 Dr. Armauer Hansen (1841-1912), a Norwegian, discovered the bacillus that causes leprosy, and nine years later Robert Koch (1843-1910) proved that a related microbe caused tuberculosis. Koch and other German researchers were convinced that infectious diseases were caused by living, parasitic organisms that were passing from animal species to humans. His work on anthrax bacilli was published in 1876 and caused a revolution in the study of tropical diseases. In 1883 he went to Egypt as leader of the German Cholera Commission to study the transmission of that disease, and formulated a series of rules for the control of such epidemics which form the basis of strategies still used today. In 1891 he became director of the new Institute for Infectious Diseases at the Medical Faculty of Berlin. In 1896 he went to South Africa and pioneered the inoculation of healthy animals with material taken from animals already suffering from diseases. He also worked in India and Africa on malaria, blackwater fever, and plague at the end of the nineteenth century. He was awarded the Nobel Prize for Physiology or Medicine in 1905.

However, it was not until well into the twentieth century that effective preventive measures and cures became available, and the field of medical microbiology became a major contributor to the understanding of tropical and other diseases.

SUE RABBITT ROFF

Further Reading

Arnold, David. *Warm Climates and Western Medicine: The Emergence of Tropical Medicine 1500-1900.* Clio Medica/The Wellcome Institute Series in the History of Medicine, no. 35. Rodopi-USA/Canada, 1996.

Curtin, Philip D. *Death by Migration: Europe's Encounter with the Tropical World in the Nineteenth Century.* Cambridge: Cambridge University Press, 1990.

Nineteenth-Century Biological Theories on Race

Overview

In 1848 the whole of Europe was plagued by revolution. Fueled by nationalist and ethnic individual interests, these revolutions were a symptom of change in how men viewed the concept of race. Drawing upon biological theories of the day, the European revolutionaries of the mid-nineteenth century formulated a social theory of race that served their nationalist interests. In a mesh of scientific inquiry and political dogma, the terms race and ethnicity were used interchangeably. Although the conflicts brought upon by the transfer, and in most cases misappropriation, of scientific ideas to the social concepts of race were not limited to the nineteenth century, this era spawned many of the modern conceptions and misconceptions concerning race.

Background

As European contact with distant places and different peoples became commonplace, the desire to explain both physical and cultural differences increased. European conquistadors and explorers documented the physical characteristics of native peoples they encountered in the Americas. Centuries later, American slave owners and European colonists in Africa both mentioned in their writings what they perceived as characteristic traits of the native African peoples. Moreover, the era of colonization sparked a general philosophical and theological interest in describing distant groups of people.

It was not, however, until the late-eighteenth century that scientific, or methodological, processes for studying diverse populations emerged. With the introduction of the Linnean binomial system of classification in the latter-half of the 1700s, the question of species variation and race became a highly controversial topic within the fields of biology and medicine.

In the early half of the nineteenth century, theories about the origins of human lineage were dependent upon Linnean models. These groupings defined types of men by outward physical characteristics, with a special emphasis placed on skin pigmentation. The Linnean model of variation, however, was based on two erroneous assumptions. The first of these inaccurate assumptions denied the extinction of species (i.e., there had never been a destruction of species). The second faulty assumption asserted that species were inherently stable and immutable (i.e., not prone to change).

Reconciling the variations among different groups of peoples thus raised the question of whether each race constituted a unique species, or sub-species, of *Homo sapiens* (i.e., mankind).

Impact

In 1800 French anatomist Georges Cuvier (1769-1832) introduced a new concept into the accepted model of human typology. Cuvier used the term "hidden agent" in an attempt to mesh popular theological and medical theories. In essence Cuvier and other scientists began to attribute differences in physical appearance among the races to environmental factors. Cuvier noted that "dark men of Africa" were more suited to tropical climates while European colonists in a similar climate were "prone to fevers" and "not of long life." This reliance on climate, geography, and social structure led to the increasing intertwining of the concepts of variety and race.

Environmental influences on development were essentially based upon the work of French anatomist Jean-Baptiste Lamarck (1744-1829),

who attempted to explain such things as why giraffe had long necks. Lamarck reasoned that a giraffe, by stretching its neck to get leaves, actually made its neck lengthen and that this longer neck was then somehow passed on to offspring. Accordingly, the long-necked giraffe resulted from generation after generation of giraffe stretching their necks to reach higher into trees for food.

It is now well established that Lamarck's theory of evolution by acquired characteristics was simply wrong. Individual traits are, for the most part, determined by an inherited code contained in the DNA of each cell and are not influenced in any meaningful way by use or disuse. Further, Darwinian natural selection—a fundamental part of the theory of evolution—more accurately explains the long necks of the giraffe as a physical adaptation that allowed exploitation of a readily available food supply that, in turn, resulted in enhanced reproductive success for "long necks."

By 1820 the static models of species and race were no longer the accepted scientific norm. Phylogeny, the evolutionary process that produces a certain species, became the focus of biological explanations of variety among humans. While the advancement of physiological theories opened the possibility that biological diversity was the result of an evolutionary process, the theory in its infancy tried to establish a hierarchy of species through which every animal evolved during its gestation period. According to phylogenic theory, ontogeny—the life history of a given individual—repeats phylogeny. In essence this theory proposed that animal embryos had to evolve in the womb through "lower" phylogenic stages—including that of the adult form, of fish, reptiles, birds, and mammals respectively—before taking its final form. More subtle differentiation in species was made once the embryo was "characteristically human." With disastrous social consequences, the erroneous notion of a hierarchy of species was applied to human society as well. Stratification of humans based upon race was made easier by use of justifications drawn from faulty science (e.g., a Caucasoid fetus was thought to have evolved through all of the races of men and was, therefore, the most "evolved").

In his 1828 publication *History of the Development of Animals,* Estonian biologist Karl Ernst von Baer (1792-1876) demolished the notion that "ontogeny recapitulates (repeats) phylogeny" with his detailed writings and research in embryology. Von Baer noted that at various stages of development the embryos of any animal species, including man, resembled a primitive form of the adult of the species. The important philosophical and scientific contribution of von Baer's work was to establish that man did not go through primitive evolutionary stages and that, in fact, there was no progression through any species hierarchy of human development. In essence von Baer's work established that there were no "less evolved races" of humans.

Concurrent with emerging phylogical studies, several physicians set out to carefully define variation among peoples in terms of their anatomical differences and commonalties. Nineteenth-century scientific and medical literature became filled with detailed observations of the external appearance of physical differences (e.g., facial features and skull sizes) of the races. Dubious craniological studies dwelled upon not only identifying common traits among members of race groups, but also measured supposed skull capacity and brain size. Equating larger skull size with greater brain capacity, later nineteenth-century adherents of such hypotheses compounded the scientific error by expanding the concept into an explanation of sociological differences (e.g., linking brain size with cultural advancement). Countless anatomical studies turned the uncontrolled and disputed measurements that indicated Caucasoid skulls were larger than those of Africans, Asians, or indigenous Americans into a justification for social dominance based on the mistaken assertion that so-called "species related" factors were the key biological determinant of the cultural worth and potential of various races.

During the nineteenth century men were classified by their race or type and were said to "be of the blood of their race." "Blood," distinct from its more common biological meaning, was the name given to the mysterious medium that carried hereditary materials, or that which defines particular individual attributes. Accordingly "blood theory" held that hereditary materials were infinitely divisible and children were the result of an equal mixing of the genes of the parents. The offspring of two racially similar parents was "whole-blooded," while the child of two racially distinct parents was referred to as a "half-blood." "Whole-bloodedness" was a central concern to race theorists who believed that constant interbreeding between the races could result in the disappearance of certain hereditary materials or even whole races. Despite the formation of more complex models of heredity in the mid-nineteenth century and the advent of the study of genetics, the principle of divisible hereditary materials remained a part of medical race theory until the end of the century. As consequence, blood theory models had a profound influence on legal and social definitions of race—definitions that persist in some forms to the present day.

The latter decades of the nineteenth century cradled infant concepts of evolution, speculation, genetics, and modern twentieth century racism. British naturalist Charles Darwin's publication of *The Origin of the Species* in 1859 changed both the historical and biological concept of man. While previous models were open to change and evolution within the different species, Darwin's theory traced the biological development of different species from common ancestors through an ongoing evolutionary process hinged on the principle of natural selection, often referred to as "survival of the fittest." Darwinian theory moved away from the blood theory idea of degeneration—the parceling out of hereditary materials until their influence is minute—and formulated an evolutionary scenario in which certain traits were selected by the ability of their carrier to thrive and reproduce. Incorporating the natural, biological, medical, and social sciences, Darwin's work was and remains a seminal scientific, biological, and anthropological text.

Although gaining nearly universal scientific acceptance, Darwinian (i.e., evolutionary) theories of speciation subsequently came into conflict with the Communist social philosophies of Karl Marx, who promoted the idea that man was largely a product of his own will. Incredibly, the outdated theories of race and speciation based on erroneous nineteenth-century Lamarckian concepts actually flourished once again during Stalin's totalitarian Soviet Union of the mid-twentieth century. Ten years after the October 1917 revolution in Russia, an uneducated plant-breeder in the struggling Soviet Union named Trofim Denisovich Lysenko (1898-1976) came to control almost all of Soviet science. Lysenko's suppression of Soviet science was based on discarded Lamarckian concepts of evolution by acquired characteristics (i.e., that organisms evolved through the acquisition of traits that they needed and used). As a puppet of Stalin, Lysenko and his errant scientific ideas found political favor with such force that they nearly destroyed Soviet agriculture.

Twentieth-century evolutionary theory—especially as it applied to race differences—gained acceptance (prior to a detailed understanding of the genetic mechanisms) because it provided an explanation of variation within and among populations. Differences among human populations are small when measured against the wide variation of physical differences within populations. Modern biologists explain these differences as having wide and divergent relationships to genes located on chromosomes. Importantly, this argues that the differences between races in one characteristic (e.g., skin pigmentation) is not likely to be useful for predicting differences in other characteristics (e.g., intelligence) and therefore provides potent scientific argument against racism (i.e., stereotyping based on race).

ADRIENNE WILMOTH LERNER

Further Reading

Montagu, Ashley. *Man's Most Dangerous Myth: The Fallacy of Race*. New York: Columbia University Press, 1942.

Western Missionaries Spread Western Medicine Around the World

Overview

Missionary activities expanded dramatically during the nineteenth century, introducing Western medicine to the "uttermost ends of the earth." These movements stemmed from the period at the end of the eighteenth century known as the "Great Awakening," when powerful evangelists converted large masses to Christianity. To propagate the Christian faith, groups organized to send people to other countries. The term *missionary* derives from the Latin word *mittere,* meaning *to send,* and refers to sending one on an errand.

During the nineteenth century, the evangelical mission of the church converged with the precepts of Western culture and experimental medicine. Missionaries, such as David Livingstone (1813-1873) in Africa and Adoniram Judson (1788-1850) in Burma, were not only preachers of the gospel but explorers who disseminated medical knowledge to the new cultures they encountered. Inland missions, such as those in the Asia, established schools that trained native doctors. Missionaries contributed directly toward the expansion of Western medicine and thinking, and laid the foundation for health care systems of the twentieth century throughout the world.

Background

The nineteenth-century merger of missionary zeal and medicine had roots in two separate traditions—the Evangelical church with Christ as healer and the development of scientific medicine. In Christian history, plagues and illness have been a flagship for missionary causes. The incurable diseases that struck the Roman empire during the first three centuries A.D. provided major thrusts toward establishing Christianity as a world force. Jewish communities were recognized for their morality and their care for the sick and poor. When Christianity emerged from the Jewish community, the same concerns were carried on. In disasters such as the plagues of Orosius in 125 A.D. and Antoninus 164-180 A.D., Christians cared for the sick and dying and gained respect of the community. Although Christians had compassion, the Church itself actively repressed scientific thought and discovery for over 1,000 years. Medical advances were stifled until the end of the fifteenth century.

The precepts of Western medicine can be traced to the age of scientific thought that flourished between 1550 to 1700. Some very specific ideas about the body and disease developed that distinguished Western medicine from long-standing traditional medicine and Eastern philosophy. With the transformation of thought about the nature of the physical world during this time, traditions of Greek and Islamic Arabic medicine were replaced by the findings of experimental science. The British scientist William Harvey (1578-1657), who discovered the circulation of the blood, along with French philosopher Rene Descartes (1596-1650), usually are credited with this beginning. The development of Western medicine was characterized by the following: 1) the idea that the human body is a machine, a reflection of the Western conception of a division or duality of mind and body (Eastern medicine, on the other hand, asserted the unity of mind and body); 2) a close association with the use of tools and devices; 3) the development of germ theory and efforts to identify specific conditions and their corresponding diseases; 4) concentration on diagnosis and treatment rather than on promoting health or preventing illness; and 5) emphasis on direct observation of patients. In the decades after the Great Awakening, missionary activity carried these ideas and medical approaches throughout the world.

Impact

The prophetic world vision developed when ministers began preaching that God had commanded Christians to evangelize the world. The father of modern missions was Englishman William Carey (1761-1834), a village cobbler who became a scholar and linguist. He founded the first foreign missionary organization, the Baptist Missionary Society, in 1792. When he set sail for India in June 1793, there was much excitement in both England and America. On February 6, 1812, five American missionaries were ordained. Four months later Adoniram Judson arrived in India. The movement built up steam, then sent forth missionaries from English-speaking countries throughout the world, bringing with them Western culture. Carey worked for 30 years to organize a growing network of missionaries. As a student of botany and agriculture, he also spread scientific thinking in his mission schools.

Throughout history, missionaries from all religions have transplanted both their religious and secular cultures. Christian missions of the nineteenth century especially sought to impose their own culture and beliefs upon foreign converts. Missionaries taught Western education, medicine, architecture, music, work habits, and dress. Evangelical missionaries were generally highly trained and might be called to work in various areas. They started schools, promoted agricultural improvements, taught hygiene, and an advanced standard of living. In this way, Western missionaries promoted what was known about Western medicine.

Even those not trained as physicians became disseminators of medicine. The evangelicals began preaching to call sinners to God, which then evolved into social concern. A wide variety of churches were involved as these evangelists went to foreign fields. In 1797 the Netherlands

Missionary Society was formed, and in 1810 the American Board of Commissioners. Roman Catholics also began a new thrust. For example, Joseph Damien de Veuster (1840-1888) was a Belgian Catholic missionary to a leper colony in Hawaii. Known as Father Damien, he cared for the lepers' needs and gave them medical aid until he himself died of leprosy.

In the 1820s and 1830s overseas missions became a regular part of English and Scottish church life. The relationship between colonial expansion and missions is complex. Missionaries were simply following the flag of their countries and were not working to expand new territories. However, the missionary movement and imperial or colonial expansion ended up as fellow travelers.

Many countries were interested in China as a mission field. The Chinese Evangelization Society sent Dr. James Hudson Taylor to found the China Inland Mission in 1865. His goal was to take the gospel to every part of the Chinese empire, which had just opened up to the West. Late in the nineteenth century Canadian missionaries organized medical training in Chengdu, Sichuan, a province in southwest China. Five Western mission boards joined to begin a medical and dental school. Missionaries were agents of change in China. They have even been referred to as "Evangelists of Science." These colleges provided an institution for the interaction of the two cultures and transmission of Western medical knowledge. The Chinese medical elite provided the backbone through the upheavals in China in the twentieth century.

Jesuit missionaries and Dutch physicians took Western medicine to Japan in the sixteenth and seventeenth centuries. European books on anatomy and medicine written in the eighteenth century set the stage for a Japanese text on physiology in 1836. Like China, the medical schools founded by missionaries developed an elite group of medical personnel dedicated to Western medicine. James Curtis Hepburn (1815-1911) was a Presbyterian missionary doctor to Singapore when he developed malaria. He became one of the first missionaries to Japan. During the last part of the eighteenth century the Japanese government had encouraged the westernization of Japanese medicine. Japanese scientists contributed great medical breakthroughs, such as the discovery of the plague bacillus in 1894 and a dysentery bacillus in 1897. Alexander Duff (1806-1876) established the Vellore Medical College in 1860 for education of Indian nurses and women physicians. A majority of India's nurses have been Christians.

The scramble for influence and the colonization of Africa brought in missionaries. David Livingstone was from a poor Scottish family and struggled to gain medical qualifications to become a medical missionary. He was sent by the London Missionary Society in 1841 and became most famous as an explorer. Traveling widely through Central Africa on waterways such as the Zambezi River, he mapped more uncharted territory than any white man and was honored as the hero who found Victoria Falls. His travels spread Western medicine to major areas of the continent. Missionaries like Livingstone and Robert Moffat (1795-1883) vigorously opposed the slave trade and were influential in anti-slavery movements.

Nineteenth century missions had their first notable successes in Polynesia, Madagascar, and the East Indies. Although India and China became only one to two percent Christian, the missionaries still impacted medicine. In Burma, Korea, Ceylon, and Indonesia, significant churches developed. Africa proved difficult because so many missionaries died of tropical diseases, which Western medicine was not developed enough to combat. Wilfred Thomas Grenfel (1865-1940) was a British medical missionary who established a chain of hospitals and nursing centers in Labrador.

It is hard to realize that in the nineteenth century missionaries were at the forefront of public awareness. In the twentieth century they have been stereotyped and satirized as long-frocked, plain people with safari hats. One typical secular joke showed a missionary in a pot about to be cooked by savages, with a caption reading, "Would you care to say grace?" Also, some of the missionaries enthralled readers back home with biographies of exotic and amusing encounters with "savages." Missionaries of the century garnered large crowds and much attention back home, very similar to astronauts of today. When they spoke, auditoriums were packed. By the end of the nineteenth century, missionaries had spread into almost every conceivable part of the world. With them was carried Western influence and Western medicine.

EVELYN B. KELLY

Further Reading

Choa, C. H. *Heal the Sick Was Their Motto: The Protestant Medical Movement in China*. London: Coronet, 1990.

Latourette, Kenneth Scott. *History of the Expansion of Christianity*. Grand Rapids, MI: Zondervan, 1970.

Minden, Karen. *Bamboo Stone: The Evolution of a Chinese Medical Elite*. Toronto: University of Toronto Press, 1994.

Nichols, C. S. *David Livingstone*. International Publications, 1996.

Wellman, Sam. *David Livingstone: Missionary and Explorer*. New York: Chelsea House, 1998.

Nineteenth-century Views of the Female Body and Their Impact on Women in Society

Overview

The nineteenth-century "woman question" was the theme of endless books by male physicians, scientists, and philosophers. American physicians argued that woman, by her very nature, was condemned to weakness and sickness because female physiology was inherently pathological. In the 1870s, doctors increasingly focused on the threat of higher education for women, especially in co-educational institutions. Doctors claimed that the strain of brainwork during puberty interfered with the proper development of the female reproductive system.

Background

During the nineteenth century, the United States experienced major economic, demographic, political, and social changes. The challenge to traditional social roles raised by some women met with considerable resistance from conservative factions that feared such changes. As nineteenth-century women began to participate in reform movements and moral crusades, they found that to be effective participants they needed education, financial independence, and control over their reproductive functions.

Periods of interest in the fundamental "nature" of women and the factors that affect their health generally correspond to times of social stress, dislocation, and activism. Thus, in the late twentieth century the "woman question" in nineteenth-century medicine became a subject of great interest to social historians. Some scholars saw evidence of a male conspiracy against women as patients and practitioners, with women as victims and men as oppressors. Others emphasized the complexity of historical change and cultural development, and established more sophisticated approaches to analyzing how the subordinate status of women has influenced their lives.

The importance of science in the nineteenth century led to the use of "scientific" arguments to rationalize and legitimate Victorian culture when traditional social and economic patterns were under attack. Women's attempts to challenge tradition led traditionalists to use medical and biological arguments to rationalize traditional sex roles and to block women's access to education, birth control, the professions, and so forth. Doctors portrayed themselves as scientists, experts, advocates of reason and morality, natural law, and evolution, but they accepted traditional views about the nature of woman. Because woman was seen as the product of her peculiar biology, her natural state was to be nurturing, passive, emotional, spiritual, frail, and even sickly. In comparison to man, woman was said to have a smaller head and brain, weaker muscles and nervous system, and an intuitive, but primitive moral sense that limited her role in society to the domestic sphere.

Woman's whole being, especially her central nervous system, was said to be controlled by her uterus and ovaries. Female biological and social life were naturally "internal," defined by the uterus and the home. All female disease, mental and physical, could be ascribed either directly to the uterus and ovaries, or indirectly to the "reflex irritation" model, that is, disorders of the reproductive system caused pathological changes in other parts of the female body.

Impact

When physicians entered the nineteenth-century debate about the nature of woman and the woman's rights movement, they claimed that their special knowledge of female physiology gave their ideas and prejudices the status of objective and irrefutable scientific truth. However, despite their assertion of scientific expertise, their arguments were not based on scientific findings. For example, physicians did not understand the menstrual cycle, but assumed that ovulation, conception, and menstruation occurred at the same time. Many eminent physicians argued that female physiology, including the menstrual cycle, was in itself pathological and that women were naturally weak and sick. The best known proponent of this rationale was Edward H. Clarke, author of an influential book entitled *Sex in Education: or, a Fair Chance for the Girls* (1874).

In addition to his large private practice in Boston, Clarke was a Harvard professor, and a leader in the battle to prevent the admission of female students to Harvard. He was especially interested in diseases of the nerves, and blamed the woman's rights movement for many nervous disorders. Clarke subscribed to the prevailing idea that the human body was a closed system with a limited "energy bank" or "nervous force." This concept rationalized the idea of conflict between brain and uterus. The body was a battlefield where all organs fought for a share of limited resources, but the struggle between the brain and the female sex organs was particularly dangerous.

It followed that women would graduate from college, if they survived the ordeal at all, as sterile sickly invalids. Only total rest during the menstrual period could allow proper development of the female reproductive system. According to Clarke, the "intellectual force" expended by girls studying Latin or mathematics destroyed significant numbers of brain cells, in addition to decreasing fertility. If they were not totally infertile, educated women would face dangerous pregnancies and deliveries because they had smaller pelvises and their babies had bigger brains. Eventually the rigors of co-education would sterilize women of the upper classes and educated women would disappear because they produced no offspring. Only women who had been protected from the dangers of brainwork during puberty would remain. As Clarke concluded, separate schools were needed for girls that would establish a mandatory rest period of four days per month to accommodate the menstrual cycle.

As evidence, Clarke presented the sad case history of Miss D., who entered Vassar at fourteen, where she started to faint in the gymnasium due to exercising during her periods. She graduated before age nineteen, but suffered from dysmenorrhea, constipation, hysteria, nervousness, headaches, invalidism and a flat chest. Another unfortunate student died soon after graduation; the post-mortem revealed a worn-out brain.

Even though statistical studies from the late nineteenth century found that college women were as healthy as other women were, many doctors ignored the evidence and continued to follow the teachings of Dr. Clarke. Tests of motor and mental skills of men and women found no special effects associated with the female reproductive cycle. Critics of the teachings of Dr. Clarke argued that doctors who shared his beliefs were simply prejudiced and influenced by the fact that the women they saw as patients were indeed sickly. They simply did not see healthy, educated women. Female researchers argued that it was possible that because of bad diet, lack of fresh air and exercise, tight corsets, and restrictive clothing, nineteenth-century girls were often sickly. Indeed, as the female waistline was allowed to get larger, while clothing became less restrictive and cumbersome, and girls were allowed to get fresh air and exercise, their health improved.

Some skeptics argued that the great oversupply of doctors in the late nineteenth century led to the diagnosis of more sickness, such as chronic, but non-fatal "female complaints," especially among upper-class women. Indeed, doctors typically argued that rich women were the most delicate and sickly. Servants, factory workers, and other poor women did not seem to need a week of rest during their menses. Mary Putnam Jacobi (1842-1906), an eminent physician, argued that women were diagnosed as perpetual invalids because doctors saw them as lucrative patients. In 1876 Jacobi won the prestigious Boylston Prize from Harvard for her response to the Clarke hypothesis. Her book, *The Question of Rest for Women During Menstruation,* demonstrated that education and professional work did not damage women's health. Other critics argued that Clarke's work was not a scientific treatise, but a polemic against admitting women into education on an equal basis with men. Certainly, many women were not as healthy as they could be, but the true remedy for them was more education, not less, especially

education about human physiology. Representatives of many colleges studied Clarke's claims and argued that their women students were very healthy. Moreover, the resident physician at Vassar College found no evidence for the existence of Dr. Clarke's unfortunate Miss D.

Alternative healers and health reformer advocates, dismissed by orthodox medical practitioners as quacks and cultists, attracted many female followers with their argument that disease was preventable. Although many varieties of health reformers flourished in nineteenth-century America, most shared a basic belief that better diet, fresh air, exercise, and education in the "laws" of human health and physiology would lead to health for both men and women.

LOIS N. MAGNER

Further Reading

Apple, Rima D., ed. *Women, Health, and Medicine in America: A Historical Handbook.* New Brunswick, NJ: Rutgers University Press, 1992.

Barker-Benfield, Graham J. *Horrors of the Half-Known Life: Male Attitudes Toward Women and Sexuality in Nineteenth-century America.* New York: Harper & Row, 1976.

Ehrenreich, Barbara and English, Deirdre. *For Her Own Good: 150 Years of the Experts' Advice to Women.* New York: Anchor/Doubleday, 1978.

Leavitt, Judith Walzer, ed. *Women and Health in America: Historical Readings.* Madison, WI: University of Wisconsin Press, 1984.

Shorter, Edward. *A History of Women's Bodies.* New York: Basic Books, 1982.

Thompson, Lana. *The Wandering Womb: A Cultural History of Outrageous Beliefs About Women.* Amherst, NY: Prometheus Books, 1999.

Verbrugge, Martha H. *Able-Bodied Womanhood: Personal Health and Social Change in Nineteenth-Century Boston.* New York: Oxford University Press, 1988.

Vertinsky, Patricia A. *The Eternally Wounded Woman: Women, Doctors, and Exercise in the Late Nineteenth Century.* Urbana, IL: University of Illinois Press, 1995.

Wijngaard, Marianne van den. *Reinventing the Sexes: The Biomedical Construction of Femininity and Masculinity.* Bloomington, IN: Indiana University Press, 1997.

Phrenology in Nineteenth-Century Britain and America

Overview

Phrenology was an attempt in the early nineteenth century to make judgments about a person's characteristics by measuring the surface of his or her skull. It was enormously popular in Britain during this time, symbolizing the progressive nature of science during the Industrial Revolution. It was popular to a lesser extent in America. Although the theory was flawed, it is regarded as an important step towards modern theories of localization of brain function.

Background

Phrenology is the theory that judgments about a person's character and mental capacities can be made by studying the surface of his or her skull. Franz Joesph Gall (1758-1828), the inventor the theory, believed that the brain was divided into specific areas and that each area was responsible for a human characteristic such as pride or wit. The size of each area was linked to the "power of manifestation" of that trait. Put simply, Gall believed those people with larger "pride areas" of the brain, for example, were more proud. He reasoned that it was possible to judge the size of an area, and hence to make judgments about a person's character, by examining the surface of the skull directly above that area.

Phrenology thus provided a map of the human skull that related different areas to different human characteristics. The map was often illustrated by literally drawing the divisions onto a model of a head and naming the different areas. Known as phrenological heads, these can still be seen in junk shops today.

The medical theory of the eighteenth century seems, at first glance, to be very different from Gall's ideas at the beginning of the nineteenth. Medical thought in the eighteenth century still owed much to the second-century physician and philosopher Galen of Pergamon (A.D. 138-201) and the even earlier writings of Hippocrates (c. 460-377 B.C.). In the Galenic theory the liquid components of the body—the humors—were of primary importance. The solid organs merely processed the humors.

However, changes in the way the body was being understood preceded the popularization of phrenology. Throughout the eighteenth century there was a shift in emphasis away from the humors towards the solid organs of the body. In neurophysiology this change was linked to theories of the seventeenth-century French philosopher and mathematician René Descartes (1596-1650). Descartes identified the pineal gland in the brain as being the point at which the immaterial soul acted upon the human body. Galen's theory included a liquid—the "vital spirit"—that differentiated humans from animals, a notion that Descartes rejected.

Eighteenth-century medicine and philosophy were also concerned with the search for the "sensorium commune," the area within the brain "where all forms of sensation were correlated." Although not everyone perceived the sensorium commune as interacting with the soul, it had a similar status to the Cartesian vision of the pineal gland. Together, these and other theories were moving medical beliefs about human cognition and emotion away from the humors and towards a brain-centered theory.

These theories did not necessarily conflict with the idea of "soul" or "will." Such ideas permeated medical thought at the time and allowed medicine to investigate the human body without challenging the status of religion. To advocate a materialist view of the human mind, where all thought and actions have their causes in the physical working of the brain, would have been a direct attack on the church.

At the same time that physiologists were asserting the importance of the brain, the first attempts were being made to understand human behavior in what would now be described as psychological terms. The eighteenth century was a period of intense interest in the relationship between brain and behavior. Gall's theory was unique in terms of its vision and its subsequent success, but Gall was not alone in attempting to explain human behavior in terms of the human brain.

Likewise, although phrenology was the first theory to link areas of the brain to specific human traits, the idea that judgments about character could be made by studying the human body—known as physiognomy—was part of the folk culture of the time. Many biographers of Gall have noted that while still at school he observed that the more able of his classmates tended to have large and protruding eyes. Although this is clearly in keeping with the theories he went on to develop, it would also have made sense within the "common sense" thinking of the eighteenth century.

Phrenology differed from other psyognomic theories in many ways, but to understand why it was so much more successful that its rivals, it is necessary to understand the social climate of early nineteenth-century Europe.

Throughout Europe at this time the traditional centers of power, the aristocracy and the church, were coming under attack. The Enlightenment, an intellectual movement of the previous century, had promoted the use of the scientific method and reason and encouraged the questioning of traditional beliefs about religion.

The violent events of the French Revolution (1789-1799) were still fresh in the minds of the European aristocracy. In Britain, the only European country in which phrenology was widely popularized, a revolution of a different kind was occurring. The introduction of industrial methods of production was bringing about changes in the way society was structured. The Industrial Revolution was increasing the numbers of the skilled professional middle class, such as medical professionals and factory managers, and the unskilled urban workers.

Living conditions for the urban working classes were bad, with low wages, long hours of often hard labor and poor housing and health. The concerns of the new middle classes were considerably different. They were anxious to establish their place in the new social hierarchy, to distance themselves from the workers in the new factories without challenging the status of the aristocracy. It was within this climate that Johann Gaspar Spurzheim (1776-1832), the major populariser of phrenology within Britain, first came to London in 1814.

Impact

The influence of phrenology on all levels of British society during the nineteenth century was significant. It was by far the most well-known scientific theory of the century among the middle and working classes. The most popular account of phrenology sold almost twice as many copies as Charles Darwin's seminal book about natural selection, *On the Origin of Species*.

Spurzheim took phrenology to America, where it also enjoyed widespread success. In both countries it helped shape ideas about brain and behavior, education and health.

It also played an important role in the development of modern theories of localization of brain function.

One of the reasons phrenology was successful in Britain was that it meant different things to different groups. Spurzheim had changed the names of the different phrenological areas that Gall had defined, increasing their number and making the characteristics sound more positive. Although Gall and Spurzheim claimed that the phrenological areas were the result of empirical studies, the fact that areas could be redefined shows the flexibility of the theory.

Promoted via the literary and philosophical societies of middle-class Britain, phrenology boomed between 1820 and 1830. For the new middle classes, phrenology embodied the progressive nature of science in general. Phrenological journals, modeled on established scientific journals, were published, and the array of skulls, models, and charts that accompanied phrenological lectures helped to add to its aura of objectivity and rationality.

Science was seen as symbolic of a new society based on reason rather than religion and privilege. That is not to say the middle classes wished for a revolution. Members of the literary and philosophical societies of the time benefited from an association with science that raised them above the working classes, while the ban on discussion of politics and religion at these societies ensured that they did not threaten the existing social order.

For the laboring classes phrenology offered the chance of self-improvement. Phrenology declared human differences to be innate, but it also offered the chance to change one's character through education. For the working classes this offered the potential of social advancement, while for the middle classes it gave a "natural" explanation for the difference between the two classes.

Phrenology and phrenologists played an important role in promoting an education for the working classes that embodied this idea of natural differences. The Mechanics Institutes, first established during the 1820s, were a response to middle-class fear of growing unrest among the urban workers. Designed as places of education for the working classes, these emphasized practical skills suitable for the worker's role in society. Phrenology was also taught at these institutes.

The use of phrenology to justify the existing social order was understandably not accepted by the more radical members of the working classes. Some viewed phrenology as materialistic—and hence denying the existence of God—and used it to attack the conventional religious views of the middle and upper classes. But it was never used as effectively by these radicals as it was by the middle classes. This was in part because its message of individual self-improvement contradicted the ideals of the socialist movement, the major response of the working classes to the political economy of the time.

The status of phrenology amongst the middle classes declined from 1830 onwards. The values that they had used phrenology to promote were now becoming dominant. At the same time some middle-class phrenologists were extolling a more radical materialistic interpretation of the theory. This, combined with the proliferation of street vendors offering phrenological examinations, hastened its demise among the middle classes. Phrenology was attacked on grounds of scientific inaccuracy before and after its heyday, but it cannot be said that any new science contributed to its demise.

Among the working classes phrenology had become enormously popular, gaining a status similar to that enjoyed by horoscopes today. It influence was to remain strong until the beginning of the twentieth century.

Phrenology never had such a wide influence in America as it did in Britain. Following a lecture tour by Spurzheim's in 1832, phrenological societies were set up in Boston and Philadelphia. Spurzheim died during the Boston leg of his tour and his subsequent martyrdom probably helped promote these and other phrenological societies. Unlike in Britain, the success of phrenology in America was not due to these societies (many of which quickly ceased to exist) but to traveling "salesmen" who offered phrenological examinations and lectures for a small fee. Most famous of these were the Fowler brothers. After a successful tour they set up business in New York, a business that lasted until the beginning of the twentieth century. The younger of the brothers visited England in 1860 and briefly revived the then-flagging popularization of phrenology in that country.

The social influence of phrenology differs markedly between Britain and America, but the impact of phrenology was felt beyond these societies. In terms of the way we view human character and the human mind, phrenology had an enormous impact. Although the British practitioners of phrenology were keen to stress that it was not a materialist theory, it was often perceived as such. Gall's lectures in Vienna were banned by the Austrian Emperor because of this. Phrenology's limited success in other European countries was in part due to its materialist connotations.

Phrenology also influenced scientific views of the human mind. Gall was not the first person to link behavior with physiology, but he was the first to localize function to specific areas of the human brain and to back this up with empirical evidence (albeit collected in an unscientific manner). He also differed from other phsyiognomists in believing that many different aspects of what makes us human—mental abilities, emotions, and personality—could be understood by studying the brain.

Gall's theory that these functions could be identified with different areas of the brain—known as localization of function—was not accepted during his lifetime. While phrenology enjoyed enormous popular support in Britain, the scientific ideas behind the theory were rejected by the academic community in Paris, where Gall lived for the time. But despite this rejection, phrenology influenced physiologists such as Jean Baptiste Bouillard (1796-1881) and Paul Broca (1824-1880), who went on to establish the theory of localization of function in the second half of the nineteenth century. Modern neuroscience, using new brain imaging techniques, has subsequently shown that different areas of the brain are indeed specialized for specific tasks such as vision and motor control.

JIM GILES

Further Reading

Books
Cooter, Roger. *The Cultural Meaning of Popular Science.* Cambridge: Cambridge University Press, 1984.

Periodicals
Barker, Fred G. "Phineas among the Phrenologists: The American Crowbar Case and the Nineteenth-century Theories of Cerebral Localization." *Journal of Neurosurgery* 82 (April): 672-682.

Greenblatt, Samuel H. "Phrenology in the Science and Culture of the Nineteenth Century." *Neurosurgery* 37 (October 1995): 790-805.

Hodson, Derek and Bob Prophet. "A Bumpy Start to Science Education." *New Scientist* 111 (14 August 1986): 25-28.

Other
The History of Phrenology on the Web: http://www.jm vanwyhe.freeserve.co.uk.

The Birth of a Profession:
Dentistry in the Nineteenth Century

Overview

The nineteenth century saw the rise of dentistry as a distinct profession, with its own practitioners, techniques, and standards. The emphasis of dental care shifted from simply removing painful teeth to trying to avoid extractions by filling cavities. By the end of the century, preventive dentistry sought ways to keep the cavities from developing in the first place. Nineteenth-century dentists were the first professionals to use anesthetic drugs, a development that made modern surgery possible.

Background

The writings of ancient Greek and medieval Moorish physicians described treatments for dental problems such as teething and diseases of the mouth. But the solution to most dental pain was to extract the offending tooth. Early physicians sometimes performed this procedure, but so did barbers and blacksmiths. False teeth were luxury items fashioned by jewelers and other skilled craftsmen. Folk healers, with their poultices and potions, were another option if care by a physician was unavailable or unaffordable. Some of their simple remedies, such as gargling with salt water, are still used today.

Like so many other fields, dentistry was put on a more scientific basis in the eighteenth century. The invention of the microscope in the 1700s had provided the means for understanding the structure of the tooth. French surgeon Pierre Fauchard developed a number of new techniques for repairing, replacing, and straightening teeth. He also coined the term "surgeon-dentiste," from which we get the word "dentist." Important texts on dentistry were published in both France and England.

With the new energy and optimism brought by American independence, the leadership in dentistry began passing to the New World. Many skilled European dentists crossed the Atlantic, to take advantage of increased economic opportunities or in flight from the bloody French Revolution. They trained American dentists in the most advanced techniques of the time. The famous patriot Paul Revere was a silversmith, and also made false teeth. He initiated the science of forensic dentistry by recognizing his own work and thereby identifying the skull of Dr. Joseph Warren, killed at the Battle of Bunker Hill. The first native-born American dentist, Josiah Flagg, practiced a wide variety of dental techniques including gold fillings, orthodontics, root canal, and even oral surgery to correct harelip. He also invented the first dental chair, by fitting an ordinary wooden chair with an adjustable headrest and an extended arm to hold his instruments.

Dentists were generally trained by the apprentice system, but there was nothing stopping anyone from taking out newspaper ads claiming to be an "operator on the tooth." Itinerant practitioners roamed the countryside, announced by posters and handbills and stopping a few days at a time. While many were skilled and reputable dentists dedicated to serving a far-flung population, others were quacks, glad to leave the results of their ineptitude behind as they moved to the next town.

Organized dentistry arose in the late 1830s as a movement by trained dentists to separate themselves from the charlatans. After several unsuccessful attempts, the first nationwide organization of dentists, the American Society of Dental Surgeons (ASDS), was established in 1840. It soon took over the publication of the first authoritative dental periodical, the *American Journal of Dental Science,* which had begun publishing in 1839. The first dental college in the world opened in Baltimore in 1840, offering the new degree of "Doctor of Dental Surgery," or D.D.S. The faculty of the Baltimore College of Dental Surgery consisted of two physicians and two dentists, and the course of study lasted two years, just as did medical training at that time.

Impact

With the organization of the dental profession came wider dissemination of new techniques. At the same time, pressure increased for standardization, as a forum now existed for debate about which techniques were the best and which should be rejected. One of the early battles concerned the best way to fill teeth.

While the emphasis in dentistry was now moving from extracting to saving decaying teeth, restorative techniques were in their infancy.

Before a tooth can be filled, all the decay must be removed. Dentists of the early nineteenth century were limited to manual drills, which were twirled between the thumb and forefinger. It wasn't until 1871 that James Beall Morrison patented a much faster foot-operated treadle drill, modeled after the workings of early sewing machines. The next year, an electrical drill was patented by George F. Green. But since most dental offices were not electrified at that time, the innovation was adopted only gradually. High-speed drills, which reduced pain from heat and pressure, were not introduced until the 1950s.

Many types of materials were tried for filling teeth. In mid-century, a compound based on tree resin was marketed as "Hill's Stopping." But resin-based materials were not strong enough to provide a permanent chewing surface. Some dentists experimented with molten metal, but its heat damaged the tooth. Eventually, gold foil became the standard among reputable dentists.

In 1833 a pair of French brothers named Crawcour came to the United States touting their "Royal Mineral Succedaneum" for filling teeth. The material was an amalgam, or pasty mixture of silver and mercury. Consideration of the amalgam on its merits was confounded by the terrible reputation of the Crawcour brothers. They left decay in teeth they filled, and indulged in outrageous advertising. Finally, they were run out of the country. Some dentists wanted to experiment with amalgam, because gold foil was expensive and difficult to work with. Still, tainted by association with its promoters, amalgam was rejected by the ASDS, which for a time required its members to sign a pledge forswearing the new material. The conflict split the profession. Many dentists refused to sign the pledge, because certain difficult jobs could be done better with amalgam, and some of their patients could not afford gold.

By 1850 the pledge was rescinded, but the damage was done, and the ASDS disbanded in 1856. The American Dental Association (ADA) was formed in 1859 by a merger of several smaller groups. As a national organization, its growth was interrupted by the Civil War, but after several reorganizations and a few name changes, it became today's ADA in 1922.

In the 1870s the dental community finally came to the conclusion that different filling materials could be preferred for different situations. Gold foil was used for some restorative procedures, and amalgam for others. The amalgam formula devised in 1895 by G. V. Black to provide the best combination of hardness, molding ability, and thermal properties is very similar to that used today.

In addition to electricity, plumbing was another important innovation in nineteenth-century dental offices. Without plumbing, patients used brass spittoons that had to be cleaned by hand. With running water came the Whitcomb Fountain Spittoon in 1867. Water flowed continuously into the spittoon bowl to keep it clean, and drinking water was supplied from a pipe. In keeping with the Victorian fashion of dressing up utilitarian objects, the pipe was shaped like a swan. Similarly, dental chairs of the day were often made of carved mahogany with plush upholstery.

Even today, some patients' teeth are beyond repair, and of course this was true in the nineteenth century as well. Early dentures were made from ivory or animal bone, which tended to discolor and absorb odors. Another alternative was to use teeth from animals or from human cadavers, but many people found this option unappealing. In the eighteenth century, French porcelain became the material of choice, but the dentures were molded as one piece, and tended to distort in the firing process.

In 1808 Giuseppangelo Fonzi invented "terro-metallic incorruptibles," the forerunners of modern dentures. Fonzi molded individual porcelain teeth, each with an embedded platinum pin for soldering into a gold or silver denture base. In 1851 John Allen of Cincinnati patented more natural-looking "continuous-gum" dentures, with pink porcelain gums into which were fastened a few teeth in a row; these gum sections were then attached to the denture base. Around the same time, rubber products began to be used to produce a denture base that was much more comfortable and about one third the price of those made from precious metals.

Dentists pioneered one development with such extraordinary impact that it became known as "dentistry's gift to medicine." That innovation was anesthesia, without which modern surgery would never have developed. Before anesthesia, surgeons performed operations such as amputation of limbs as quickly as they could, while the patient experienced indescribable pain. Some chose death by gangrene rather than endure it. Others died of the shock. Abdominal surgery was almost unheard of.

In the early nineteenth century, the euphoria-inducing properties of substances such as nitrous oxide, or "laughing gas" and ether, were known, but used only for entertainment at "ether frolics" and sideshows. During an exhibition at a fair in 1844, dentist Horace Wells (1815-1848) saw an acquaintance under the influence of laughing gas. Stumbling across the stage, the man injured his shin without appearing to notice it. Wells subsequently tried using laughing gas while having a tooth extracted, and felt no pain. Two years later, another dentist, William Morton (1819-1868), gave a public demonstration of the use of ether during a surgical operation performed by a colleague. Ether, being a more powerful agent with which deep unconsciousness could be induced, became the surgical anesthetic of choice for many years, while nitrous oxide became standard in dentistry.

At the very end of the century, the invention of radiography allowed dentists to see inside patients' teeth and diagnose cavities before they caused major damage. New knowledge in microbiology also provided understanding of the bacteria that cause tooth decay. These bacteria digest the sugars and starches on the teeth, producing acids and causing the enamel to deteriorate. Once this happens, other bacteria can penetrate deeper into the tooth. Armed with this information, dentists began seeking to prevent the decay process by encouraging better dental hygiene practices and developing new tooth powders and pastes.

Women and African-Americans were practicing dentistry in the nineteenth century, although the organized profession was slow to recognize this. While neither group had adequate access to formal education, a few women and hundreds of blacks were trained under the apprentice system. One such woman, Emeline Roberts, was trained by her husband and continued their practice when he died in 1864, running it on her own for 60 years. She was not admitted to the Connecticut State Dental Society until 1893. The first woman in the world to receive a degree in dentistry was Lucy Beaman Hobbs. She graduated from the Ohio College of Dental Surgery in 1866.

The first black man to receive a D.D.S. degree was Robert Tanner Freeman, who graduated from Harvard University's School of Dental Medicine in 1869 with its first class of six students. But until 1954, when the United States Supreme Court ruled educational segregation unconstitutional, almost all African-American dentists were trained at two all-black schools, at Howard University in Washington, D.C., and Meharry Medical College in Nashville, Tennessee.

SHERRI CHASIN CALVO

Further Reading

Glenner, Richard A., Audrey B. Davis, and Stanley B. Burns. *American Dentist: A Pictorial History with a Presentation of Early Dental Photography in America.* Osceola, WI: Pictorial Histories Publishing Company, 1991.

Ring, Malvin E. *Dentistry: An Illustrated History.* New York: Harry N. Abrams, Inc., 1992.

Wynbrandt, James. *The Excruciating History of Dentistry: Toothsome Tales and Oral Oddities from Babylon to Braces.* New York: St. Martins Press, 1998.

Medicine in Warfare
in the Nineteenth Century

Overview

It is often said that war is the best medical school. The battlefields supplied young, often inexperienced surgeons with what seemed to be an unlimited number of cases. Military physicians were called on to deal with endless cases of camp fevers and dysenteries. Although some of the conditions encountered by military surgeons and physicians are peculiar to warfare, many of the lessons they learned and the training they obtained in military service were later applied to civilian life. Military medicine underwent many changes and reforms of techniques, organization, and personnel during the nineteenth century in response to experiences gained in the numerous wars fought in Europe and America.

Background

Napoleon's favorite surgeon, Dominique-Jean Larrey (1766-1842), provides a good example of the advances achieved in military medicine and surgery. Baron Larrey, a French military surgeon, is primarily remembered for introducing the *ambulances volantes* in 1793. The moving ambulance made possible the rapid removal of the injured from the battlefield, while there was still hope of saving their lives. More importantly, however, Larrey was the founder a new era in military surgery. A strong, skillful, and daring surgeon, he was the first surgeon to amputate through the hip joint. He was also willing to cut into the pericardium (the membrane surrounding the heart) to remove fluid and pus from the pericardial cavity. Although the process known as débridement (the complete removal of dead or dying tissue and contaminants from a wound) had been recognized as early as the sixteenth century, it was generally ignored until Larrey's teacher Pierre Joseph Desault (1738-1795) realized its importance. During the Napoleonic Wars, Larrey insisted on rigorous attention to this procedure. Another method that Larrey used for cleaning wounds is known as maggot therapy. This technique was as effective as it was repulsive. Impressed by the cleanliness of wounds infested with maggots, Larrey realized that maggots improved wound healing by removing dead tissues. The intrepid Larrey traveled with Napoleon's army through Germany, Austria, Spain, and Russia. As a token of his respect and affection, Napoleon left his favorite surgeon 100,000 francs in his will.

Several simple, practical additions to the articles available to doctors were introduced by Johannes Freidrich August von Esmarch (1823-1908), a German military surgeon. In 1869 Esmarch devised the first "field dressing." Esmarch insisted that all soldiers should carry a first-aid kit and that each man should be taught how to use it. He also devised a rubber bandage (1873), now known as Esmarch's bandage, which is used to render a limb bloodless before amputation. The bandage is wrapped tightly around a limb to force the blood out so that operations can be carried out with relatively little loss of blood. During the Franco-Prussian War in 1870, Esmarch served as surgeon-general of the Prussian forces.

Another innovation that arose from nineteenth-century warfare was the recognition of the value of competent, trained, and well-organized female nurses. Florence Nightingale (1820-1910), who is mainly associated with the professionalization of nursing, was also deeply involved in the reform of military medicine. Nightingale demonstrated the value of good nursing during the Crimean War, fought in Turkey after England and France declared war on Russia in 1854 for its attack on Turkey. Med-

ical preparations for the war were totally inadequate and the medical situation became a major scandal and disgrace in England. The conditions for sick and wounded soldiers were appalling because of the incompetence and indifference of the military and political authorities. The sick were left to suffer and die in filthy, inadequate hospitals. Epidemics of cholera, dysentery, and fever overwhelmed the doctors and orderlies who were supposed to care for the sick and wounded. Drugs, soap, anesthetics, drinking water, and decent food were almost nonexistent. The standard diet for the sick consisted of salt-pork or beef, hard biscuits, and bad coffee. The sick and wounded were evacuated to the notoriously filthy base hospital at Scutari. Amputations, without anesthetics, were performed in the middle of the wards that housed the sick and the dying. Many more men had died from disease than from battlefield injuries by the time Nightingale came to the Crimea with 38 nurses and important supplies for the patients and the hospital. These dedicated and hardworking women were not initially welcomed by the military authorities, but eventually Nightingale proved the value of good nursing. Her first priority was to clean up the hospital and bring order to hospital administration, including a routine of round-the-clock nursing care for the patients. She used her own money and funds collected in a public appeal, launched by the *London Times*, to establish kitchens and a laundry. When Nightingale arrived, the sanitary situation of the Scutari hospital building was abominable. Sewers were inadequate and filth flooded the wards and contaminated the water supply. Under Nightingale's direction, the mortality rate decreased from 42 percent to about 2 percent. Although nursing was an old profession, it was not until the Crimean War, through the work of Nightingale, that the military authorities actually realized the value of good nursing services.

Impact

The American Civil War was also marked by the inadequacy of medical services. Surgeons and drugs were in short supply and antisepsis virtually unknown. Despite the American invention of anesthesia in the 1840s, surgeons generally amputated without anesthetic. Many thought that the heart could best stand the stress of surgery if it was performed while the patient was still in a state of "battle tension." Interestingly, a manual of military surgery written by J. J. Chisholm, a surgeon in the Confederate Army, recommended the use of curare for tetanus. The Crimean War had provided some lessons for improvements in hygiene and sanitation, such as the importance of adequate rations and nursing services. Nevertheless, for a war fought before the acceptance of the germ theory of disease, basic field hygiene and sanitation and post-surgical care remained major problems. As was usually the case, diseases such as dysentery, tetanus (lockjaw), and typhoid killed more soldiers than battle injuries.

The Union Army did, however, achieve significant organizational reforms and advances in the management of mass casualties before the end of the war. Surgeon-General William A Hammond, who was appointed in 1862, reformed the supply system, worked with the U.S. Sanitary Commission, and built large general hospitals. Jonathan Letterman, who served as medical director of the Army of the Potomac, built on the work of Baron Larrey and instituted his own innovative procedures and policies. Letterman advanced procedures for the use of ambulances to evacuate the wounded, required progressive surgical triage and treatment, centralized field medical logistics, instituted preventive medical inspections, established large mobile field hospitals in tents, simplified methods for the collection of medical data, and so forth. Indeed, the "Letterman system" became the foundation of modern medical systems in all armies.

In 1862 Hammond established the Army Medical Museum in Washington, D.C., to collect and study pathological specimens from military hospitals. Eventually the museum became the Armed Forces Institute of Pathology. Hammond ordered the writing of a comprehensive medical and surgical history of the War of the Rebellion. The completion of this monumental six-volume history took 18 years. At the time of its publication it was the most complete report of the medical data gathered during any war in history. Another contribution to medical history from this period was the creation of the Library of the Office of the Surgeon General (later the National Library of Medicine) by John Shaw Billings (1838-1913). Under Billings's direction this library became the largest collection of medical literature in the world. Billings also established

the invaluable *Index Catalogue of the Library of the Office of the Surgeon.*

Another lasting innovation of the nineteenth century evolved from Henri Dunant's experience at the battle at Solferino in 1859, where the forces of Italy and France defeated Austria. Dunant (1828-1910), a wealthy Swiss banker, was shocked by the plight of the wounded, including at least 40,000 injured men who were left lying on the battlefield without aid or medical attention. Dunant's account of his experiences, *A Memory of Solferino* (1862), was widely read and very influential. Dunant called for the establishment of a voluntary aid organization to provide care for wounded soldiers. At the Geneva Convention of 1864 representatives from 16 countries drew up the system of rules for the care of the wounded and the protection of hospitals and medical personnel that eventually resulted in the formation of the Red Cross. In 1901 Dunant was awarded the Nobel Peace Prize.

LOIS N. MAGNER

Further Reading

Ashburn, Percy Moreau. *A History of the Medical Department of the United States Army.* Boston: Houghton Mifflin, 1929.

Bengston, Bradley P., and Julian E. Kuz. *Photographic Atlas of Civil War Injuries: Photographs of Surgical Cases and Specimens of Orthopaedic Injuries and Treatments During the Civil War.* Otis Historical Archives. Grand Rapids, MI: Medical Staff Press, 1996.

Chapman, Charles B. *Order Out of Chaos: John Shaw Billings and America's Coming of Age.* Boston: Boston Medical Library, 1994.

Dunant, Henry. *A Memory of Solferino.* Geneva: International Committee of the Red Cross, 1986.

Freemon, Frank R. *Microbes and Minnie Balls: An Annotated Bibliography of Civil War Medicine.* Rutherford, NJ: Fairleigh Dickinson University Press, 1993.

Hutchinson, John F. *Champions of Charity: War and the Rise of the Red Cross.* Boulder, CO: Westview Press, 1997.

Miles, Wyndam. *A History of the National Library of Medicine, the Nation's Treasury of Medical Knowledge.* Washington, DC: Government Printing Office, 1982.

Robertson, James, Jr., ed. *The Medical and Surgical History of the Civil War.* 15 vols. 1883. Reprint of *Medical and Surgical History of the War of the Rebellion.* Wilmington, NC: Broadfoot Publishing Co., 1992.

Modern Surgery Developed

Overview

Surgery is an ancient branch of medicine, but it was not until the nineteenth century that doctors learned to apply practical and effective measures for controlling pain and preventing surgical infection. New techniques for anesthesia and antisepsis were key elements of the transformation of the ancient art of surgery into one of the most respected and powerful areas of medical specialization. With dependable methods for the control of pain, infection, and bleeding, surgeons were able to go beyond the treatment of wounds, fractures, dislocations and carry out new and daring operations on the interior of the body.

Background

The transformation of surgery seems to have occurred with remarkable speed in the nineteenth century when surgeons were given the tools to overcome two of the great obstacles to major operative procedures: pain during surgery and the life-threatening infections that often followed surgery. These obstacles were largely overcome in the nineteenth century with the introduction of anesthesia (methods of controlling pain) in the 1840s and antisepsis (methods of fighting infection) in the 1860s.

A closer examination of the evolution of surgery, however, suggests a more complex explanation for the remarkable progress that seemed so revolutionary. Traumatic injuries, wounds, ulcers, skin diseases, fractures, dislocations, bladder stones, urinary disorders, and venereal diseases had been treated by surgeons, with some success, for hundreds of years. Nevertheless, the status of the surgeon was traditionally lower than that of the physician. Medicine was regarded as a learned profession, while surgery was a mere technique that required training and experience rather than learning. During the eighteenth century, however, progress in anatomical investigation, and the acceptance of a new approach to pathology, provided an intellectual framework for advances in surgery. The transformation of surgery in the nineteenth century closed the gap between medicine and surgery and established the basis for the modernization of a powerful and unified medical profession.

Although various narcotics have been used in religious and healing rituals for thousands of years, Dr. Oliver Wendell Holmes (1809-1894) reflected conventional medical wisdom when he said that nature offered only three natural anesthetics: sleep, fainting, and death. The preparations used to induce ceremonial intoxication would not satisfy the criteria for anesthetic agents that were established in the nineteenth century: relief of pain must be inevitable, complete, and safe. Impure mixtures of drugs were acceptable for ceremonial purposes, but they could cause unpredictable and dangerous effects in a person undergoing surgery. Ancient methods were too unpredictable to fit the criteria for modern surgical anesthesia, but the world of drug lore provided examples of potentially useful anesthetic agents. Ancient soporific and narcotic potions contained opium, mandrake, henbane, wine, marijuana, hellebore, belladonna, henbane, jimsonweed, hemlock, and other dangerous drugs. Curare, an arrow poison used by South American Indians, does not relieve pain, but it is useful in modern surgery because it causes muscle relaxation.

In the eighteenth century, the chemical revolution provided a new series of agents that could be used as painkillers. Joseph Priestly (1733-1804), the discoverer of oxygen, also discovered nitrous oxide, or "laughing gas." Humphry Davy (1778-1829) inhaled nitrous oxide while suffering from toothache in 1795 and noted that the pain caused almost disappeared. He suggested that nitrous oxide might be useful during surgical operations. Davy's associate Michael Faraday (1791-1867) discovered the soporific effect of ether vapor during experiments on various gases.

Impact

Many individuals were involved in the discovery of inhalation anesthesia in the nineteenth century and several of them became embroiled in bitter priority battles. Horace Wells (1815-1848) and William Thomas Green Morton (1819-1868) were dentists who shared a successful partnership. Wells recognized the anesthetic properties of nitrous oxide, but his public attempt to demonstrate the effectiveness of his method was unsuccessful. Morton decided to search for another anesthetic agent and in 1846 successfully demonstrated the value of ethyl ether for inhalation anesthesia. Charles T. Jackson (1805-1880), chemist and physician, later claimed that he had discovered ether anesthesia and had instructed Morton in its use. While this priority battle raged in New England, Georgia physician Crawford Williamson Long (1815-1878) announced that he had operated under ether anesthesia before Morton.

After confirming the benefits of ether anesthesia, James Young Simpson (1811-1870), one of Scotland's leading surgeons and obstetricians, discovered that inhaling chloroform produced a sense of euphoria as well as loss of consciousness. Chloroform was easier to administer than ether, but was also more dangerous. Within two years of Morton's first public demonstration, ether, nitrous oxide, chloroform, and other anesthetics were widely used in dentistry, obstetrics, and surgery. Most doctors cautiously accepted anesthesia as a "mixed blessing" which had to be used cautiously and even selectively.

The safety of anesthesia was not the only point of contention. The attack on the use of anesthetics in childbirth was particularly virulent. Clergymen denounced the use of obstetrical anesthesia and advised women to endure the pains of childbirth with patience and fortitude, because the Bible said that Eve was condemned to bring forth children in sorrow. Some obstetricians argued that labor contractions were identical to labor pains; therefore, without pain normal births would be impossible. Simpson was able to address his critics with both theological and scientific rebuttals. The curse in Genesis, he argued, had been revoked in Deuteronomy: "The Lord will bless the fruit of the womb and the land." Moreover, the word translated as "sorrow" in the case of Eve's punishment was really the word for "labor," either in farming or childbirth. Furthermore, he continued, God had established the principle of anesthesia when he caused a deep sleep to fall upon Adam before operating on his rib. When John Snow (1813-1858) successfully administered chloroform to Queen Victoria in 1853 during the birth of her eighth child, the issue of whether a proper lady would accept anesthesia was quickly settled.

With proper management, inhalation anesthesia was generally safe, complete, and inevitable. However, general anesthesia is not suitable for all operations. Fortunately, chemical agents suitable for use as local anesthetics and instruments for their delivery were available by the time the concept of surgical anesthesia had been accepted. Friedrich Wilhelm Sertürner (1783-1841) had turned crude opium into crystals of morphine and Charles Gabriel Pravaz and Alexander Wood (1725-1884) had invented the modern hypodermic syringe. Cocaine had been used by the ancient Incas of Peru for surgical operations, including trepanation. Coca leaves were also used to fight pain, hunger, nausea, and fatigue. Although Europeans quickly took up the Native American custom of smoking tobacco, they ignored coca until the nineteenth century. By the time Carl Koller (1857-1944) and Sigmund Freud (1856-1939) began their experiments on cocaine, chemists had isolated various alkaloids from coca leaves. Koller used cocaine for the relief of eye diseases such as trachoma and iritis. After learning about Koller's experiments, William Stewart Halsted (1852-1922), one of New York's leading surgeons, developed sophisticated techniques for achieving local anesthesia by injecting cocaine solutions into the appropriate nerves.

The specific effect of anesthesia on the frequency of operations has been a matter of debate, but the evidence suggests that anesthesia did expand the amount of surgery performed. In part, the rise in surgical cases was an outgrowth of urbanization, industrialization, and concomitant changes in the role of the hospital. The increase in gynecological surgery, especially ovariotomy, was especially problematic; gynecological surgeons claimed that these operations could cure insanity and various nonspecific "female problems." Some critics were convinced that surgeons who operated in teaching hospitals performed operations not because they expected to save patients, but because impoverished hospital patients were viewed as "teaching material."

As surgical operations became more sophisticated, post-surgical infections assumed epidemic proportions throughout the hospitals of Europe. Some scholars believe that the notorious rise in post-surgical infections was more closely associated with industrialization and urban poverty than anesthesia. Indeed, veterinary surgery was relatively free of the problem of wound infection, although such operations were generally carried out under primitive conditions. Although the cause of wound infection was not clearly understood until the developments of scientific germ theory, "uncleanliness" had been a major suspect since the time of Hippocrates (460?-377? B.C.). However, wound infection was such a common occurrence that many surgeons considered it essentially a normal part of wound healing.

The evolution of the hospital into a center for medical education and research may have been a major factor in the appalling mortality rates of nineteenth-century hospitals. Descriptions of major hospitals invariably refer to the overcrowding, stench, and filth of the wards. The introduction of the "antiseptic system" by Joseph Lister (1827-1912) can be regarded as a major factor in breaking the link between hospital surgery and post-surgical infections. Lister was an experimental scientist, as well as a talented surgeon, who appreciated the insights of Louis Pasteur (1822-1895) concerning the relationship between germs and disease and applied this theory to the problem of wound infection. Lister chose compound fractures for his experiments, because life-threatening infections were frequent complications of open or compound fractures. The prognosis for compound fractures was so poor that amputation was considered a reasonable course of treatment. Lister tested the effects of carbolic acid, which had been used as a disinfectant for sewers and garbage dumps. In 1865 he successfully treated an eleven-year-old boy with a compound fracture of the leg. Further refinements of the antiseptic system led to effective treatments for a variety of life-threatening conditions. Although few Americans today are familiar with the work of Lister, some vague memory of "Lister the germ-killer" survives in advertisements for "Listerine." In 1879 this "germ-fighter" was sold to doctors and dentists as a general antiseptic. By 1879, Pasteur and Charles Chamberland (1851-1908) had demonstrated that heat sterilization was superior to chemical disinfection of surgical instruments. Chamberland's autoclave, a device for sterilization by moist heat under pressure, was in general use in bacteriology laboratories by 1883.

As surgeons mastered the use of anesthesia and antisepsis, operations that had once been impossible became routine. Christian Albert Theodor Billroth (1829-1894) is considered the founder of modern abdominal surgery, as well as

the founder of the Vienna School of Surgery. Among his pioneering operations were the resection of the esophagus in 1872 and the complete removal of a cancerous larynx in 1873. He spent ten years developing methods for removing parts of the intestine and sewing the cut ends together (intestinal suture). In 1881 he performed a very dangerous operation to remove a cancerous pylorus (part of the stomach). The patient survived the operation, but died of cancer four months later because the cancer had spread to the liver.

After the basic concept of antisepsis had been accepted, additional safety measures, such as surgical gloves, gowns, caps, and masks were added to the full "aseptic ritual" and spectators were banished from the operating room.

LOIS N. MAGNER

Further Reading

Keys, Thomas E. *The History of Surgical Anesthesia.* New York: Dover, 1963.

Ludovici, L. J. *The Discovery of Anaesthesia.* New York: Thomas Y. Crowell, 1961.

Magner, Lois N. *A History of Medicine.* New York: Dekker, 1992.

Pernick, Martin S. *A Calculus of Suffering: Pain, Professionalism, and Anesthesia in 19th-Century America.* New York: Columbia University Press, 1985.

Ravitch, Mark M. *A Century of Surgery, 1880-1980.* 2 vols. Philadelphia: J.B. Lippincott, 1982.

Rutkow, Ira M., and Stanley B. Burns. *American Surgery: An Illustrated History.* Philadelphia: Lippincott-Raven, 1998.

Rutkow, Ira M. *The History of Surgery in the United States, 1775-1900.* San Francisco: Norman, 1988.

Sykes, William Stanley. *Essays on the First Hundred Years of Anaesthesia.* 3 vols. Chicago: American Society of Anesthesiologists, 1982.

Wangensteen, Owen H., and Sarah D. Wangensteen. *The Rise of Surgery From Empiric Craft to Scientific Discipline.* Minneapolis: University Minnesota Press, 1978.

Nineteenth-Century Developments
Related to Sight and the Eye

Overview

While the human eye has always inspired poets and writers, this small organ was studied by ancient scholars and physicians long before the advent of modern science. Many of the words

related to the study of the eye derive from the Greek (*ophthalmos*) and Latin (*opticus*) words for eye. Thus, the study of the eye is called ophthalmology, and the measure of the eye, optometry. An ophthalmologist is a doctor of medicine who diagnoses and treats diseases of the eye. An optometrist has been specially trained to give eye health and vision examinations.

While much of the structure of the eye was known at the beginning of the nineteenth century, the science developed so rapidly that its history is essentially a succession of biographies of scientists who added pieces to the puzzle. Several lesser-known physicians also developed advances in diagnoses and equipment. By the end of the nineteenth century, the knowledge of eye structure and function set the stage for the unprecedented treatments, such as laser surgery, and other advances during the twentieth century.

Background

Ophthalmology is an old science. Ancient Egyptians recognized and treated eye disorders, and surgeons in Greece operated on the eye. Aulus Celsus, a first-century A.D. Roman, summarized knowledge about the eye in his *De medicina*. According to legend, the Chinese invented glasses perhaps as early as 500 B.C. When Marco Polo (1254-1324) visited China in 1275, he saw many Chinese with glasses. During the golden age of Arabic medicine around 1000 A.D., famed physicians studied the eye and systematically treated eye diseases. A Jewish physician Isaac ben Soloman Israeli (832?-932?) was a noted oculist who maintained a practice outside Cairo. He wrote eight medical works in Arabic, including a treatise on ophthalmology. Paramedic oculists in the Middle Ages traveled around to different communities treating the eye and peddling crude glasses. Many of these were fitted so badly that they did more harm than good. Although many of these itinerants used magic and quackery, they are sometimes credited with founding the practice of eye care. A sixteenth-century German oculist, Geor Barisch, also wrote on eye diseases.

However, little basic knowledge was added about the eye until the eighteenth century, when knowledge emerged to set the stage for nineteenth-century advances. With new procedures for grinding lenses, glasses became more accurate. Benjamin Franklin (1706-1790) is credited with developing bifocals. In 1738 a procedure was developed for the correction of strabismus, or crossed eyes. Night blindness was discussed in 1767. In 1750 glaucoma was fully described, and in 1794 colorblindness was recognized.

Impact

At the beginning of the nineteenth century, interest in the eye and its diseases was increasing. The first course in ophthalmology was offered in 1803 at the University of Göttingen. The London Eye Infirmary was begun in 1805 and operated as a teaching hospital, establishing the first modern medical specialization. Sir William Lawrence (1783-1867) was a British eye surgeon who practiced at this infirmary. A pioneering surgeon and investigator, Lawrence wrote *Treatise on the Venereal Diseases of the Eye* in 1830 and *Treatise on the Eye* in 1833. These works were outstanding for their time.

One of the first to affect nineteenth-century investigation of visual phenomena was a Czech physiologist Jan Evangelista Purkinje 1787-1869). Receiving his medical degree in 1818, he was very interested in subjective visual happenings. He based his theories of natural science on observation and experiments. Lacking the facilities in Prague for investigation, he began self-observation of errors in perception and sensations that seemed to have no physical cause. He determined that these occurrences are not by chance, but are related to the structure and function of the eye and its nerve connections to the brain. Purkinje became interested in shadows of one's own retinal vessels as night approaches. In 1855 a contemporary, Heinrich Mueller, confirmed that rods and cones were related to the phenomena Purkinje described. Purkinje also linked visual errors or illusions to connections in the brain. Studying the importance of reflex images in ophthalmology, he recognized that the interior of the eye could be examined by a light reflected into a concave lens. Later, Hermann Helmholtz (1821-1894) would use this principle to develop the ophthalmoscope. Purkinje also studied dizziness and vertigo, and the effects of drugs such as digitalis and belladonna on the eye.

In 1838 German physiologist Ernst Wilhelm von Brucke (1819-1892) chose for his doctoral dissertation to disprove vitalism, the idea that a vital force was the basis of all life. He theorized that osmosis in living things was not related to a vital force but was measurable and explained by chemical and physical forces. Choosing the eye as his central subject, he compared the reflection of light from the background of eyes of different vertebrates and also studied stereoscopic vision, afterimages, and optical media. He showed diffuse reflected light could make the human eye glow, like the eyes of many animals. Helmholtz used this principle in his ophthalmoscope. Brucke's book *Anatomical Description of the Human Eye* (1847) became a standard work for physicians. A ciliary muscle in the eye that Brucke first described bears his name.

Johannes Peter Müller (1801-1858), a German physician, was a founder of experimental physiology. He taught and trained several great German scientists of the nineteenth century. He had many interests and wrote a famous *Handbook of Human Physiology* in 1841, which was a standard text for several generations of students. Sight was one of his many interests, and he determined that a sensation of light is caused by electrical stimulation of the optic nerve.

Helmholtz contributed greatly to knowledge about the sense of sight. While he was being trained as a military surgeon, he was fortunate to have Müller as his teacher. Müller was a vitalist, and Helmholtz sought to prove him wrong by investigating the principle that he called conservation of energy. He is credited with striking at the heart of vitalism when he showed that matter is neither created not destroyed. Helmholtz was also proficient in mathematics and physics and was one of the first to apply physics to life processes. He was the first to measure the speed of impulses and for ten years, from 1856-66, wrote his multivolume *Physiological Optics*.

In 1851, building upon knowledge of several predecessors, Helmholtz invented the ophthalmoscope. This devise for inspecting the interior of the eye consisted of a small mirror to direct a strong light into the eye. When the light reflects off the retina and back through a small hole in the ophthalmoscope, the physician can see a magnified image of the structures at the back of the eye. The ophthalmoscope revealed such structures as the optic disk, retina, blood vessels, macula, and choroid.

Albrecht von Graefe (1828-1870) is recognized as the father of modern ophthalmology. The son of the famous German pioneer of plastic surgery, Karl von Graefe (1787-1840), he dedicated himself to diseases of the eye. In 1850 he founded one of Europe's first eye clinics. Recognizing the importance of Helmholtz's ophthalmoscope, he promoted its use in his clinic. Working with the physicians at the University of Berlin, Graefe developed surgical procedures for many eye defects. In 1857 he removed a part of the iris, a procedure known as an iridectomy, to alleviate the pressure of glaucoma. Glaucoma results when pressure builds up from within the fluid of the eye, affecting the optic nerve. In 1860 he connected several types of blindness to an inflammation of the optic nerve, or optic neuritis. Concerned about cataracts, he developed a procedure for removing the lens in 1867.

Graefe's name was given to a procedure he developed relating to Graves disease, a condition of hyperthyroidism where the eyes seem to bulge out. "Graefe's sign" is the failure of the upper lid to follow a downward movement of the eyeball when a person changes from looking up to looking down. Graefe wrote a seven-volume *Manual of Comprehensive Ophthalmology* from 1874-80.

Several physicians in different fields contributed to procedures for preventing disorders. Karl Credé (1819-1922), a German gynecologist, became concerned that mothers who had the sexually transmitted disease gonorrhea would infect their babies, who then became blind. Credé, looking at some of the work of Joseph Lister (1827-1912) and others using chemicals to kill contaminants, found that one percent silver nitrate dropped into the eyes would prevent gonorrheal ophthalmia. The procedure is still used routinely in hospitals today.

By the end of the century, ophthalmology was developing into a well-respected specialty. Allvar Gullstrand (1862-1930), a Swedish physician, was head of an ophthalmology clinic in Stockholm. In 1894 he became a professor of ophthalmology at the University of Uppsala. His greatest achievements were in the field of optics and his studies of the eye as an optical system. His work in astigmatism and how the eye accommodates though layers of crystalline lens led to an accurate model of the eye. He re-edited and updated Helmholtz's *Handbook of Physiological Optics*. Combining a microscope with a slit lamp, he was able to determine the exact location of a foreign particle in the eye. Gullstrand wrote many papers that pushed the knowledge of ophthalmology into the next century. He was awarded a Nobel Prize in medicine in 1911 for his investigations of the dioptics of the eye.

The nineteenth century saw an explosion of knowledge about the eye because of advances in clinical and experimental medicine. These developments transformed the oldest discipline in medicine into an exciting specialty.

EVELYN B. KELLY

Further Reading

Classel, Gary. *The Eye Book: The Complete Guide to Eye Disorders and Health.* Baltimore: Johns Hopkins University Press, 1998.

Collins, James. *Your Eyes: An Owner's Guide.* Englewood Cliffs, NJ: Prentice Hall, 1995.

Vander, James. *Ophthalmology Secrets.* St. Louis: Hanley and Belfus, 1998.

The Study of Human Heredity and Eugenics during the Nineteenth Century, Focusing on the Work of Francis Galton

Overview

Francis Galton (1822-1910) first coined the term eugenics in 1883. It stems from the Greek word *eugenes,* meaning good in birth. Though Galton defined the term eugenics rather broadly, he essentially intended the term to mean the science of improving human stock. In other words, he intended to the give groups of people, or races, he viewed as most suitable a chance to prevail over those he viewed as less suitable. Thus eugenics was to become a study dedicated to improving human beings through selective breeding, that is by encouraging the best or most fit members of society to breed more while inhibiting or preventing those that were deemed undesirable or less suitable from having children.

Background

Though Galton was the first person to use the actual term eugenics, one of the first people to write about and promote racial superiority in modern times was Arthur de Gobineau (1816-1882), in his book *Inequality of the Races* (1856). Gobineau insisted that the inferiority and superiority of races is both biological and hereditary. Gobineau was also among those who promoted the notion of the superiority of the Aryan race.

In mid-nineteenth century Germany, Ernst Haeckel (1834-1919) promoted the work of Charles Darwin (1809-1882). Haeckel was a believer in German and Aryan greatness and he built on an already growing German interest in spirituality, mysticism, and myth, that placed man on the top of the evolutionary scale. Aryan man, according to Haeckel, was at the pinnacle of the scale. He promoted the idea that there was a very real need to breed more from the Nordic races and less from everyone else for the sake of a better world. He referred to this notion as Monism, the concept of being at one with the universe.

An early English eugenicist, Herbert Spencer (1820-1903), applied the Darwinian notion of evolution to the social sciences. Spencer was also heavily influenced by the writings of Thomas Malthus (1766-1834) who promoted the idea that there would soon be too many people alive and that this number of people could not be supported by the resources that the world had to offer. Spencer combined these ideas into a theory in which he purported that "survival of the fittest" (Spencer's phrase, not Darwin's) depended then on the people who were able to evolve upward in society to prosper.

Galton, a cousin of Darwin's, was one of the first to recognize the implications of Darwin's theory of evolution for humankind. In this new post-Darwinian world, Galton hoped that with the advanced knowledge of science, scientific principles would be used to improve both individuals and the whole of society. Galton approached this idea in a somewhat biased manner as he firmly believed that the upper classes were superior in every way. Galton had noted that many of the more important and influential people in English society were related. Even in his own family he realized there were a significant number of people whom he deemed to be functioning at genius level. Galton published his study in 1869 in a book entitled *Hereditary Genius,* which provided statistical background for his theory.

Galton's aim was to create a population of intellectually superior men and women. The problem with the world, according to Galton, was that there were too many poor people and too many undesirables who were wiping out the slower breeding desirables. It should be noted that Galton and his wife were not able to have children and this fact may have played a part in his thinking. Thus Galton's work sought a way to quicken evolution. He wanted humankind to have all the immediate advantages of something that, if left to nature, could take hundreds of years. In short, what Galton believed was that if the animal world could be bred for strength and agility, people could also be bred to improve the physical and mental traits of future generations. Galton hoped for a eugenic society based on the laws of heredity, which would be accepted as morally and scientifically correct. These laws would then become a moral imperative for people who wished to breed.

Early in his career Galton had gone so far as to suggest governmental intervention in order to speed up the process of selective breeding, though soon gave up on this idea. Instead, Galton was to conclude that if people were properly educated in the advantages of eugenics they would choose this way as the only way for society. People would choose to mate in a way that would be advantageous for the future of society and for their children. Towards the end of his life Galton was to outline his idea of an ideal eugenic society in his novel *Kantsaywhere*. In this fictional work Galton imagined an island where individuals are paired with mates by the intellectual elite, according to their physical and intellectual ability. Everyone on the island accepts this, knowing that this action is for the good of the island community as a whole.

Galton's biggest problem in his study was that he did not know what governed heredity. After his statistical study of the English upper classes he decided that heredity was governed by what he called relationships, but he never understood the science behind these, as the rediscovery of Mendel's laws of heredity did not occur until 1900.

Other early eugenicists who were influenced by Galton were Walter Weldon and Karl Pearson (1857-1936). Weldon worked in the area of biometrics—the application of statistics to biology, while Pearson used mathematical tools to refine the work of both Galton and Weldon. Thus Weldon provided the eugenics movement with the necessary statistical basis for social action while Pearson wished to encourage research so that laws of natural selection could be better understood. Unlike Galton, Pearson believed that no amount of education could be of help to the desirable. Instead, he forwarded a more aggressive policy that proposed that more intelligent people would be bred.

Impact

Prior to 1900, eugenics was considered mostly a "positive" or "soft" theory, an intellectual exercise and nothing more. However, after 1900 eugenics developed and was considered "negative" or "hard," as it became an active and sometimes politically and socially powerful movement. Scientific knowledge replaced Lamarck's notion of inheritance of acquired characteristics and the importance of the environment, and replaced it with the Weismann/Mendel germ-plasm model, in which the environment had no effect.

The largest impact of eugenics can be seen in the establishment of the various eugenic societies in Europe and the United States, and in the influence these societies had on legislation that was passed in various countries. In Ger-

many, social Darwinism had become popular through the writings of Haeckel. Later, through the writings of Wilhelm Schallmayer and Alfred Ploetz, the notion of state interference became more widespread. Schallmayer and Ploetz proposed that in the newly created Germany the good of the state should take precedence over the individual. Thus they suggested a form of voluntary eugenics that was to be encouraged by the state. This linking of the importance of the good of the state with the limited value of the individual is important in German history. After the defeat of Germany in the First World War, the good of the state was to take precedence as Germany tried to rebuild.

Unlike Germany, there was no monolithic movement in the United States. The notion of eugenics gained support from a broad group of people ranging from conservatives to New Dealers to Socialists. In 1905 Charles Davenport (1866-1944) founded Cold Springs Harbor, the first large-scale eugenics research center in the United States. Davenport studied over 400,000 family pedigrees and he helped support eugenic family planning as well as intelligence testing.

Another effect of eugenics in the United States was the proposed emigration restriction placed on people coming from so-called inferior places, especially Italy and a number of eastern European countries. The movement also was instrumental in passing sterilization laws in about thirty states. These laws, which were designed to ensure that people with mental retardation, mental disease, and some physical diseases, such as epilepsy, would not procreate. Though these laws were to remain on the statute books until the 1970s, for the most part the notions of the eugenics movement were discredited by the 1930s. First, it became obvious that even mass sterilization would not reduce the passing on of so-called bad traits from one generation to the next. Also, social workers and anthropologists were now working closely with the lower classes and their research showed that these classes were not necessarily intellectually or socially inferior; instead, they simply lacked the means and methods to improve themselves. Secondly, support of eugenics was seen as being too close to German philosophy and, during World War II, this was viewed as un-American. Finally, when the reality of the horrors of the Holocaust became known, eugenics was no longer seen as a theory that had any advantages for a progressive caring society.

BRID C. NICHOLSON

Further Reading

Burleigh, Michael. *Death and Deliverance.* Cambridge: Cambridge University Press, 1994.

Cowen, Ruth Schwartz. *Sir Francis Galton and the Study of Heredity in the Nineteenth Century.* Ann Arbor, MI, 1969.

Degler, Carl. *In Search of Human Nature: The Decline and Revival of Darwinianism in American Social Thought.* New York: Oxford University Press, 1991.

Forrest, Derek W. *Francis Galton: The Life and Work of a Victorian Genius.* New York, 1974.

Himmelfarb, Gertrude. *Darwin and the Darwinian Revolution.* New York: Alfred Knopf, 1985.

Kelves, Daniel. *In the Name of Eugenics.* Cambridge, MA: Harvard University Press, 1985.

Paul, Diane. *Controlling Human Heredity: 1865 to the Present.* New York: Humanities University Press, 1995.

The Field of Public Health Emerges in Response to Epidemic Diseases

Overview

Public health broadly combines efforts towards ensuring physical health through medical research, city planning, regulations in the workplace, and sanitation. The field of public health emerged in the nineteenth century as a response to a surge of epidemic diseases, which were largely caused by the Industrial Revolution in the Western world and the new living conditions that industrial culture created. The concentration of the poor in urban centers provoked epidemic diseases such as cholera, typhoid, and yellow fever, and social reformers and the scientific community invested in medical and environmental safeguards to protect their populations from these widespread diseases. Over the course of the century, discoveries of the causes of disease primarily in France, Great Britain, Germany, and the United States transformed the field of public health into a socio-medical operation necessary to the health of modern communities.

Background

The philosophical roots of the public health movement are the liberal politics of the Age of Enlightenment, an eighteenth-century period notable for its optimistic belief in social progress and the perfectibility of humankind. Political revolutions in France and America inspired confidence in man's ability to transform his culture into a liberated community free from the tradition of a ruling aristocracy. This increased sense of power among the middle-classes encouraged sympathy for the underclass—the sick, the poor, children, and the elderly—which in turn motivated social reformers to work towards improving the lives of these vulnerable members of the population.

No official tradition of public health existed in the eighteenth century, largely because pre-industrial societies were relatively isolated from each other, which reduced the spread of diseases that devastated whole populations in the nineteenth century. But the rapid rise of population and the development of scientific fields in the second half of the eighteenth century set the stage for the public health movement. Reform of social services such as hospitals, prisons, and orphanages in the late 1700s anticipated the total overhaul of public institutions that most Western countries would effect in the century to come.

Optimism in science, too, encouraged members of the scientific and medical communities to apply their knowledge towards lofty goals, such as the total eradication of disease. Perhaps the greatest achievements of the late eighteenth century were the new understanding of hygiene in the spread of disease, and English physician Edward Jenner's (1749-1823) development of a vaccine for smallpox in 1796. Jenner's vaccine became a well-established method of preventative medicine by the century's end, but it principally was used by the middle and upper classes. Attention to the health of the poor would be a primary project of the public health movement in the nineteenth century.

The social and economic changes of the early nineteenth century in the Western world explain the development of public health. The shift from an agricultural economy to an industrial one set into place the conditions for the deadly rise of epidemic diseases. Industrialism initiated the immigration of numerous people from non-industrialized regions to new urban centers, and refugee populations were a major source of the passage of disease. Commercial

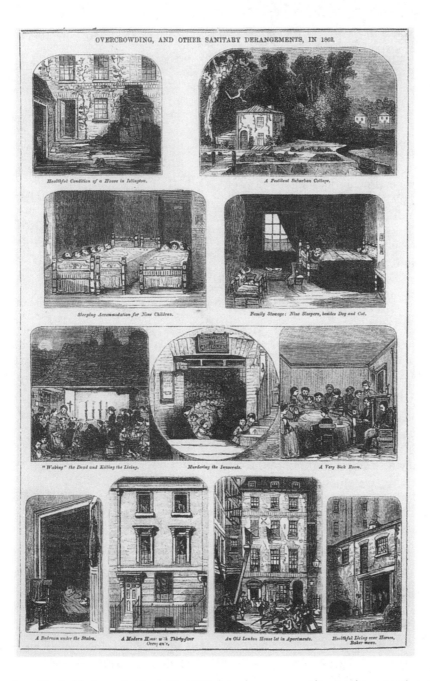

This engraving from *The Builder* magazine (June 14, 1862) shows how overcrowding in urban areas during the Industrial Revolution created major health problems. *(Wellcome Institute. Reproduced by permission.)*

ventures and the development of railways and shipping lines also facilitated the spread of diseases. Most importantly, industrialism concentrated great numbers of people in hastily constructed cities, and the crowded, often filthy living conditions were ideal breeding grounds for infectious disease and conditions such as malnutrition, dehydration, and infant mortality.

Impact

Most historians place the beginning of public health in France and Great Britain in the early nineteenth century. At this time, important social and scientific connections were made between poverty and disease. England's Sanitary Movement in the 1830s and 1840s best represents early efforts to unite health improvements with government policy. Led by English bureaucrat Edwin Chadwick (1800-1890), this movement focused upon making sanitary improvements to urban slums in England and Wales. England's Public Health Act of 1848 appointed medical health officers in each district of the country. Political quarrels existed, though, over the degree to which the British state should control individual health, and many thought that laws such as compulsory vaccination were infringements on personal liberty.

In France former army doctor Louis-René Villermé (1782-1863) led the public health movement in the first half of the century. In 1826 he published a statistical study of mortality in Parisian neighborhoods, concluding that disease is not a natural condition but an effect of poverty—and as such, it could be controlled. Villermé's work drew as much upon economic theory as it did upon medicine. Founded in political economy, which takes a qualitative approach to human happiness and wealth, he determined that the poor could be "civilized" away from their wretched way of life and made both richer and healthier. This process would also contribute to France's wealth, as less money would be required to fight sickness and build hospitals. Villermé's studies helped to pass laws limiting child labor (1841) and to establish a public health advisory board (1848).

The United States developed a health bureaucracy based upon the English model. In 1866 New York was the first state to develop a health department, complete with local health boards set up in each town. In 1872 the American Public Health Association was organized. The second half of the century in America witnessed a huge increase in immigration and population, and public health efforts multiplied as the country struggled to restrict communicable disease in urban slums. The American public health movement embodied medical science and evangelical piety, and the idea that "cleanliness was next to godliness" was used as a tool to educate the poor.

The middle of the century documents the beginnings of an international health organization. In 1851 the first international sanitary conference was held in Paris, and it was attended by delegates from several Western European and Mediterranean countries. In 1864 the Red Cross was established at the Geneva Convention. Although originally conceived as an international organization devoted to providing relief to wounded soldiers, the Red Cross later turned to the alleviation of human suffering in general.

One of the century's great contributors to public health was Max von Pettenkofer (1818-1901), a German physician and professor of medical chemistry. Von Pettenkofer underestimated the germ theory of disease and argued that disease was spread through environmental factors such as air-born miasma, soil conditions, clothing, and climate changes. Even though he was later proved wrong, the changes von Pettenkofer proposed to Munich's sewage and water supply virtually wiped out typhus in that city. He was so assured of his theory of disease that in order to disprove the germ theory, he famously swallowed a liquid containing what was said to be the bacteria of cholera. While he did experience some stomach discomfort, von Pettenkofer denied that these symptoms were cholera. Nevertheless, his contribution to municipal sanitation established Munich as one of the world centers of the study of sanitation, and he is credited with raising sanitation to a legitimate science.

The incidence of cholera in the nineteenth century provides a case study of the need for comprehensive public health across national and class borders. Epidemic outbreaks in 1831, 1848, 1854, 1866, 1871, and 1892 killed hundreds of thousands. The disease, which causes nausea, vomiting, and then dehydration, struck and killed fast—in as little as three hours. As English epidemiologist John Simon (1813-1858) proved, cholera is carried in drinking water and is often caused by the presence of sewage—a predicament all too common in slums where as many as 100 people shared one toilet. Military troops sometimes caught cholera, too, as their camps matched the crowded environments of the slums. Non-medical responses, such as the national day for "fasting and humiliation" declared in England's Parliament to combat cholera (1832), showed that modern medical theories of causation still operated alongside traditional religious concepts of disease as sin.

Tropical medicine became a specialty field after numerous white colonists sickened and died during their imperial ventures in Asia, India, and Latin America. The tropics became known as the "white man's grave" because Europeans were vulnerable to strains of disease and the changes of climate that did not exist in their home countries. Western medicine in these provinces, though, primarily served the colonial population and not the native peoples.

The growth of hospitals during the nineteenth century illustrates the influence of public health efforts. Major breakthroughs in the understanding of germs and hygiene radically changed the sanitary environment of hospitals. Before these discoveries, a patient might be prepared for surgery simply by having her clothes removed. English nurse Florence Nightingale (1820-1910) helped prove that a clean environment drastically curbed the spread of infection. When Hungarian pathologist Ignaz Philipp Semmelweis (1818-1848) demanded that surgeons wash their hands with chlorinated lime, the death rate in his ward dropped to just 3%.

Women's struggle for liberation from the tradition of homemaking was one social change that impacted the rise of public health. In England Nightingale's elevation of nurses from underpaid hospital maids to skilled professionals opened up one of the first respectable professions for women outside the home. Women's organizations throughout Western Europe and America sent trained female volunteers into slums to spread knowledge about disease, hygiene, and nutrition.

By the end of the century in Europe and America, most major cities had developed modern sewage and drainage systems, garbage collection, and hospitals. Due to a century of work in public health, Western populations were better educated about sanitation, pregnancy and natal care, disease transmission, and nutrition. Larger cities often boasted soup kitchens and free health clinics for the poor. Still, the severe health risks caused by continuing urban growth meant that public health measures would need to keep adapting to the challenges of modern industrial culture in the twentieth century.

TABITHA SPARKS

Further Reading

Duffy, John. *The Sanitarians: A History of American Public Health.* Chicago: University of Illinois Press, 1990.

Fraser, Derek. *The Evolution of the British Welfare State.* London: MacMillan, 1984.

Porter, Dorothy. *Health, Civilization and the State.* New York: Routledge, 1999.

Porter, Roy, ed. *The Cambridge Illustrated History of Medicine.* New York: Cambridge University Press, 1996.

Rosen, George. *A History of Public Health.* Baltimore: Johns Hopkins University Press, 1993.

Biographical Sketches

Clara (Clarissa Harlowe) Barton
1821-1912
American Founder of the American Red Cross, Teacher and Social Reformer

Clara Barton served as a nurse during the American Civil War, but her primary contributions to the war effort were to the monumental task of obtaining supplies, organizing relief efforts for wounded soldiers, and identifying the dead and wounded. After the war, she created an organization to search for missing men. However, Clara Barton is best known for her role in founding the American Red Cross.

Barton was born in Oxford, Massachusetts. She was the youngest child of Stephen Barton, a prosperous farmer and leader in public affairs. Clara's early education was principally at home under the direction of her brothers and sisters. When Barton was sixteen she became a schoolteacher. She later completed her own education at Clinton, New York. She opened the first free school in Bordentown with only six pupils, but at the end of the year the school was serving six hundred students.

Clara Barton. *(The Library of Congress. Reproduced by permission.)*

In 1854 she moved to Washington, D.C., where Charles Mason, the Commissioner of Patents, appointed her to the first independent government clerkship held by a woman. She resigned from the Patent Office at the outbreak of the Civil War and devoted herself to the relief of sick and wounded soldiers, often working with the field surgeons not far from the battlefield. Her admirers said that, despite her love of the Union cause, when she worked in the battlefield, she never discriminated between soldiers from the North or the South. Although she did not yet know about the Red Cross of Geneva, she was already an advocate of total neutrality in war relief efforts.

After the war Barton went to Andersonville, the notorious Confederate prison camp in Georgia, to identify and mark the graves of the Union soldiers who had died there. She was also involved in the establishment of the first national cemetery and a bureau of records in Washington, D.C., that was dedicated to the search for missing men. In 1869 a full report was completed and accepted by Congress.

In 1869, in failing health, Barton went to Switzerland to rest and recover. While in Europe she became involved in relief efforts for victims of the Franco-Prussian War. Her work with victims of the war led to an association with the International Red Cross. Gustave Moynier, the President of the International Committee of the Red Cross, asked Barton to serve as the official bearer of an invitation to the President of the United States to endorse the articles of the Geneva Convention. Barton returned to the United States in 1873 and worked towards that goal, but President Rutherford Hayes ignored the invitation. His successor, President James A. Garfield, offered to support the initiative. Unfortunately, Garfield was assassinated before any action could be taken. Chester A. Arthur, the next President, urged Barton to form a Red Cross Organization in preparation for American accession to the Geneva Treaty.

The American National Red Cross Society was established in 1881, with Barton as its president. Later that year she presented the invitation to endorse the Treaty of Geneva to President Arthur, who urged Congress to do so. In 1882 President Arthur signed the document that made the United States a member of the International Red Cross. Barton served as president of the American Red Cross until 1904.

The Red Cross eventually evolved from a society that distributed only war relief to one that also responded to peacetime disasters, such as floods, fires, earthquakes, famines, storms, and epidemic diseases. This addition to the Geneva Convention has often been called the "American amendment," but Barton's influence on this transition appears to have been exaggerated by her American supporters. Indeed, the Red Cross organizations of some other nations had been involved in peacetime disaster relief long before the Geneva Convention was officially amended. Barton's writings include an official *History of the Red Cross* (1882), *The Red Cross in Peace and War* (1898), *A Story of the Red Cross* (1904), and the autobiographical *Story of My childhood* (1907).

LOIS N. MAGNER

William Beaumont
1785-1853
American Physician and Physiologist

William Beaumont, physician and physiologist, achieved international fame for the research on human digestion he performed on Alexis St. Martin, who became known as "the man with a hole in his stomach." Beaumont's work is important not only in terms of his scientific observations, but as a landmark in the history of human experimentation and biomedical ethics.

Beaumont was born in Lebanon, Connecticut, into a poor farming family. Although Beaumont had little formal education, he was able to leave the family farm at age twenty-one and become schoolmaster in the village of Champlain, New York. His teaching position allowed him to save enough money to become a medical apprentice to Dr. Benjamin Chandler. The year that Beaumont spent with Chandler was his only formal medical training. Nevertheless, Beaumont joined the army as surgeon's mate during the War of 1812. Attempts to establish a private practice after the war were unsuccessful and Beaumont reenlisted in the Medical Department of the army.

Beaumont was sent to Fort Mackinac, which was then a remote army post on the Western frontier. Mackinac Island, in the straits of the Great Lakes, was an outpost of the American Fur Company. In addition to his work as post surgeon, Beaumont was allowed to establish a private medical practice in order to earn enough money to marry Deborah Green Platt and support a family.

On June 6, 1822, Alexis St. Martin, a young French Canadian, was accidentally shot in the abdomen at very close range. Beaumont thought the wound would be fatal, but he cared for St. Martin to the best of his ability with poultices of flour, charcoal, yeast, and hot water. He changed the dressings frequently, cleaned the wound, removed debris, and bled the patient to fight against fever. Surprisingly, St. Martin survived, but all attempts to close the wound were unsuccessful. Rather than allow the young man to make the difficult journey back to Canada, Beaumont hired St. Martin as a household servant. Although Beaumont was largely self-taught in physiology, as well as medicine, he soon realized that St. Martin's permanent gastrostomy (new opening into the stomach) provided a unique opportunity to study digestion in a healthy human being. He was able to conduct experiments that tested many contemporary theories of human digestion. Beaumont was able to insert and remove various kinds of foods from St. Martin's stomach and determine how long it took to digest them. Beaumont paid a small wage to St. Martin and planned to conduct lecture tours to demonstrate his experiments, but St. Martin frequently ran away. In 1832 Beaumont and St. Martin signed a contract that gave Beaumont the exclusive right to perform experiments on St. Martin. This document was the first such contract in the history of human scientific experimentation.

In 1833 Beaumont published his landmark work on human digestion, *Experiments and Observations on the Gastric Juice and the Physiology of Digestion.* In his introduction, Beaumont assured the reader that the experimental procedures had caused no harm to St. Martin. He also emphasized his lack of formal training and claimed that this allowed him to make observations and conclusions without the distortion caused by allegiance to previous theories. His book provides a detailed case history of experimental results, a review of the scientific literature related to his researches, and various attacks on scientists who had reached conclusions different from his own. Beaumont is remembered as the first investigator to provide a detailed, experimentally based description of normal human digestion. Despite his lack of formal training, he became a pioneer of physiology and an outstanding figure in American medical and scientific history.

LOIS N. MAGNER

Emil Adolphe von Behring
1854-1917
Prussian Surgeon and Bacteriologist

Trained as a military surgeon, Emil von Behring later became interested in microbi-

ology and received a position in the Berlin laboratory of Robert Koch (1843-1910). His work with Shibasaburo Kitasato (1852-1931) on diphtheria showed the efficacy of blood serum as a treatment. He is perhaps best known for his dramatic Christmas night rescue of a child dying of diphtheria by giving him serum antitoxin, thus proving its clinical value.

Behring was born in Hansdorf, Deutsch-Eylau, Prussia (today part of Poland) in 1854, one of thirteen children. Financially unable to attend the university there, Behring enrolled instead in the Army Medical College in Berlin, where he received his degree in 1878. He was then obliged

WILLIAM BEAUMONT
AND MEDICAL MALPRACTICE

～

The last decades of William Beaumont's life were embittered by his involvement in two medical malpractice battles. Beaumont had unsuccessfully attempted to save the life of a man who had been attacked by Darnes Davis, a carpenter. Davis had struck his victim on the head with an iron cane. Beaumont attempted to relieve cranial pressure by performing a trephination (removal of a circular piece of bone). When the case came to trial in 1840, Davis's lawyers argued that Beaumont had caused the death by drilling a hole into the victim's skull in order to perform experiments on the brain, just as he had left a hole in St. Martin's stomach in order to do experiments on the digestion. Only four years later, Beaumont was involved in a medical malpractice lawsuit filed by Mary Dugan. Although Beaumont was acquitted, the case created a great deal of hostility in the medical community of St. Louis.

to serve in the military and was sent to Poland to work. While there, he became interested in septic diseases. He began several years of research on iodoform, a crystalline compound of iodine. He published a study in 1882, asserting that while iodoform did not kill infectious microbes, it did have possibilities for neutralizing their poisons or toxins. As the German government became aware of Behring's studies and interests, they supported his further training in bacterial research, sending him first to work with C. Binz in Bonn, then in 1888 to the Institute of Hygiene, where he worked with the renowned Dr. Robert Koch.

Emil von Behring. *(The Library of Congress. Reproduced by permission.)*

After several years there Behring moved with Koch to the Institute for Infectious Diseases. In 1894 he became Professor of Hygiene at the University of Halle, and the next year he was appointed Director of the Hygenic Institute at Marburg. In 1896 Behring married Else Spinola, with whom he had seven children. He died in Marburg on March 31, 1917.

Behring's work in Berlin under Robert Koch centered on serum antitoxins for diphtheria, tetanus, and tuberculosis. Koch, along with French chemist Louis Pasteur (1822-1895), had come to international attention for his meticulous development of the methods and means for studying, isolating and growing the microbes that caused infectious diseases. In 1877 Koch had found the cause of anthrax, a deadly disease of cattle and sheep. This was followed by his identification of the germs causing tuberculosis (1882) and cholera (1883). Behring began his work on diphtheria, picking up on the discovery already made by Émile Roux (1853-1933) and Alexandre Yersin (1863-1943) of the Pasteur Institute that the diphtheria bacillus caused illness and death not in and of itself, but rather by means of a toxin that it emitted. Working with Japanese scientist Shibasaburo Kitasato, Behring began experimenting on guinea pigs in an attempt to find a chemical that might be injected into infected animals that would kill off the dis-

ease-killing bacilli without harming the animal. In the course of these experiments, Behring noticed several animals that recovered from a diphtheria infection. Working with these animals, he found them immune to the disease—injections of large doses of the bacillus did not harm them. This led to the discovery that their blood serum, the clearish liquid that will separate out from blood when it is left to stand, did not kill the bacilli themselves but did neutralize their toxin. Behring named this substance antitoxin. Behring and Kitasato published their results in 1890.

Behring continued his studies of antitoxin, moving from guinea pigs to larger animals including goats and sheep. Eventually, he found that horses could be made immune to the disease, thereby producing large quantities of antitoxin. Experiments showed that this antitoxin protected animals from diphtheria, but for only a few weeks, making it impractical to use. Behring then turned his thoughts from using the antitoxin as a preventive measure to using it as a curative one. The first clinical trial was performed, very dramatically, on Christmas night in 1891. A dose of antitoxin was given to a child in a Berlin clinic as a last-ditch effort to save him. It was a success. After this, Behring devoted his time to working out a reliable means of measuring and producing exact strengths of antitoxin, a feat that was finally accomplished with the assistance of Paul Ehrlich (1854-1915).

Behring devoted the rest of his life to refining the use of antitoxin. He spent several years studying tuberculosis as well as diphtheria. In 1914 he founded a laboratory for the production of sera and vaccines, a capital venture that provided for his financial well being. He received numerous awards and honors in his lifetime, including honorary membership in numerous National Societies. He was awarded the title of Professor in 1893, given a noble title in 1901, and awarded the Nobel Prize in medicine in 1901.

KRISTY WILSON BOWERS

Claude Bernard
1813-1878
French Physiologist and Physician

Claude Bernard's research transformed many areas of physiology and demonstrated that many vital functions could be understood in term of chemistry rather than as aspects of animated anatomy. His most significant discoveries included the glycogenic function of the liver, the role of the pancreatic juices in digestion, the functions of the vasomoter nerves, and the nature of the action of curare, carbon monoxide, and other poisons. Bernard believed that his demonstration of the glycogenic function of the liver was his most important piece of work, but the implications of his researches eventually revolutionized ideas about metabolism. Perhaps his most important contribution was his theoretical framework. He based this research career on his concept of determinism, that is, faith in the experimental method and its applicability to physiology, the science of life. Bernard insisted that instead of continuing ancient disputes about "vitalism" and "mechanism," scientists should analyze and compare relationships among phenomena in living beings and the inanimate world.

Born into a poor peasant family, Bernard was fortunate to have received instruction in classical subjects from the parish priest. After more advanced studies, Bernard taught language and mathematics at a Jesuit school while tutoring private pupils. Financial difficulties forced him to take a position as assistant to an apothecary at Lyons, but he found the work boring and dreamed of writing great plays for the Parisian theater. Soon after he arrived in Paris, he was advised to find another profession if he wanted to make a living. Bernard chose medicine and attended the Parisian School of Medicine. Finding himself unsuited to the private practice of medicine, he became assistant to the distinguished physiologist François Magendie (1783-1855). Eventually, Bernard replaced Magendie as professor.

Bernard's studies of sugar in the blood of carnivores proved that, contrary to prevailing theory, animal blood contains sugar even when it was not supplied by foodstuffs. In tests of the theory that sugar absorbed from food was destroyed when it passed through the liver, or lungs, or some other tissue, Bernard put dogs on a carbohydrate diet for several days and then killed the animals immediately after feeding. Large amounts of sugar appeared in the hepatic veins. To his surprise, animals in the control group, which had been fed only meat, had large amounts of sugar in their hepatic veins, but not in the intestines. Bernard had discovered gluconeogenesis, that is, the conversion of other substances into glucose in the liver. Further work led to the discovery of glycogen (the carbohydrate storage polymer of animals), as well as the synthesis and breakdown of glycogen. The

Claude Bernard. *(The Library of Congress. Reproduced by permission.)*

investigation of glucose metabolism led to the concept of the "internal secretions" which were products transmitted directly into the blood instead of being poured out to the exterior of the gland or organ secreting them.

Bernard realized that the metabolic theory of the cell put forth by Theodor Schwann (1810-1882) was applicable to the fundamental problem of physiology, the relationship between the cells and their immediate environment. Bernard believed that he was the first scientist to insist that complex animals had two environments: an external environment in which the organism lived and an internal environment in which the cells functioned. Ultimately, vital phenomena occurred within the fluid internal environment bathing all the anatomical elements of the tissues. This was the basis of Bernard's well-known dictum: "The constancy of the internal *milieu* is the condition for free and independent life."

When Bernard was 47 years of age, exhaustion and illness forced him into a period of rest and reflection during which he wrote about the broader implications of his work. His *Introduction to the Study of Experimental Medicine* is a remarkably lucid and widely read text. A close examination of his research notebooks, however, indicates that the path to each of his discoveries was much more confused and tortuous than the published accounts admit. His work was extended by other scientists, most notably by Lawrence J. Henderson (1878-1942) and Walter Bradford Cannon (1871-1945) who emphasized the concept of the constancy of the internal environment and coined the word "homeostasis" to describe the conditions that maintained the constancy of the interior environment.

LOIS N. MAGNER

Elizabeth Blackwell
1821-1910
English American Physician and Medical Educator

Elizabeth Blackwell was a pioneer in opening the medical profession to women and served as an inspiration to generations of American girls. Although Blackwell is often described as the first woman doctor in America, this is not strictly true. Other women had practiced medicine and obstetrics, but Blackwell was the first woman in American history to earn a degree from an orthodox medical college.

Elizabeth Blackwell was born in Bristol, England, into a large, well-to-do family. When a fire destroyed her father's business in 1832, the family immigrated to America. Eventually, the Blackwells settled in Cincinnati, Ohio, where Samuel Blackwell died in 1838, leaving his widow with nine children and little money. The oldest girls supported the family by running a boarding school until their brothers established themselves in business. The Blackwells were interested in many social reform movements, including the abolitionist crusade, the women's rights movement, education for women, and the New England Transcendental Movement.

In her autobiography Elizabeth said that she decided to become a doctor after a woman friend, who was dying of a painful "female disease," told her that her suffering would have been mitigated if she could have seen a "lady doctor." Although many medical practitioners called themselves "doctor" after private studies or apprenticeships, Elizabeth was convinced that she must attend a legitimate medical college.

In 1847, when Blackwell began to apply to medical schools, none of the established medical schools admitted women. Many schools rejected her even though admission standards for nineteenth century medical schools were notoriously low. Some physicians suggested that she disguise herself as a man and enter medical school, but

Elizabeth Blackwell. *(Museum of Finnish Architecture. Reproduced by permission.)*

Blackwell saw her struggle as a moral crusade that had to win public approval. Finally, Geneva Medical College, New York, a small country school, accepted her. Being the only female at Geneva Medical College presented many difficulties, but Blackwell was able to persuade the professors that she should be allowed to attend all lectures and demonstrations. Blackwell's success did not convince the College to accept other female students. Indeed, Geneva Medical College rejected Elizabeth's sister Emily (1826-1910). (In 1854 Emily graduated from Cleveland Medical College.)

In 1849 Blackwell was awarded her diploma of Doctor of Medicine. Like many other American medical graduates, she went to Europe for further training and clinical experience. While she was treating a baby with an eye infection, some infectious fluid spurted into her eyes. This resulted in a serious infection and permanent damage to one eye. In England Blackwell met Florence Nightingale (1820-1910), who became a major influence on her ideas about sanitation and proper hospital administration.

Convinced that opportunities for women doctors in America were improving, Blackwell decided to settle in New York. In 1853 Blackwell began her battle to establish a dispensary and hospital where women physicians could obtain clinical experience while serving the poor. Her sister Emily and Dr. Maria Zakrzewska soon joined her. The New York Infirmary for Women and Children was a pioneering effort and the first time a hospital was conducted entirely by women. The hospital was needed because female medical graduates were denied essential hospital experience and instruction. In 1868 the Infirmary established a medical school for women. The Infirmary's school established strict entrance examinations and emphasized clinical training.

Elizabeth Blackwell served as one of the few role models available to American women interested in medical careers, but she refused to allow the Women's Rights Movement of her time to divert her from her crusade for women in the medical profession. She returned to England to continue the battle and, in 1859, became the first woman listed in the Medical Register of the United Kingdom. Her successes stimulated the work of Dr. Elizabeth Garrett Anderson and Dr. Sophia Jex-Blake (1840-1912), who established the London School of Medicine.

When Elizabeth Blackwell assessed the progress of women in medicine in 1869, she was entirely optimistic about the future. In 1899, convinced that special women's colleges were unnecessary, Emily Blackwell closed the Woman's Medical College of the New York Infirmary. Unfortunately, the 1890s proved to be a very brief "Golden Age" for women physicians.

LOIS N. MAGNER

Jean Martin Charcot
1825-1893
French Physician, Neurologist and Teacher

Jean-Martin Charcot was a pioneer of modern neurology and psychotherapy. Psychologists remember him for his investigations of hypnosis and hysteria, and his influence on Sigmund Freud (1856-1939). Among his contemporaries Charcot was highly respected as a teacher at La Salpêtrière and the creator of a unique neurological clinic. A prolific author and indefatigable researcher, he published many important clinical descriptions of nervous disorders and histological observations of lesions associated with various disorders.

Charcot was born in Paris, where he studied medicine and established an international reputation as a clinician, neurologist, and teacher. Although Charcot failed the competitive examinations in 1847, he became an Interne at the

Salpêtrière, a major Parisian hospital and mental asylum, the next year. In 1853 he was awarded his M.D. for a thesis based on studies that differentiated rheumatoid arthritis from gout and other diseases of the joints. His mentor was French physician G. B. A. Duchenne de Boulongne (1806-1875). In 1859 Charcot published a study of intermittent claudication in humans. He was among the first to report on this phenomenon. Although Charcot is mainly remembered for his work in neurology, he remained interested in the degenerative diseases of the joints. He also described various aspects of liver and thyroid disease. In 1863 Charcot and André Victor Cornil published important observations of the renal lesions in gout. Five years later Charcot published a detailed description of a form of arthritis associated with the form of neurosyphilis known as tabes dorsalis. The condition is now known by his name, and tabetic joints are known as "Charcot's joints."

When Charcot joined the staff of the Salpêtrière he began a systematic clinical study of the unclassified, unknown chronic conditions that afflicted many of the hospital's long-term patients. He also performed painstaking postmortem examinations in order to correlate lesions found at death with the clinical symptoms observed in patients. This research program resulted in classical descriptions of diseases such as cerebral hemorrhage, diseases of the aged, chronic diseases, Menière's syndrome (a form of vertigo), multiple sclerosis, amyotrophic lateral sclerosis, Charcot-Marie-Tooth disease, tabes dorsalis (neurosyphilis), infantile paralysis, and hysteria. Despite his dedication to pathological research, Charcot was generally opposed to animal experimentation. Nevertheless, he was not a rigid antivivisectionist and he defended Louis Pasteur's experiments with rabies vaccinations. In 1872 Charcot was appointed Professor of Pathologic Anatomy at the Sorbonne. Ten years later he became the first Professor of Disease of the Nervous System at Paris. His landmark work *Lectures on the Diseases of the Nervous System* was published in 1877. His published lectures provide an invaluable overview of his work, thought, and approach to teaching.

Like his contemporaries J. Hughlings Jackson (1835-1911) and Paul Broca (1824-1880), Charcot was intrigued by the problem of cerebral localization and aphasia. Charcot and Jean Albert Pitres published a series of papers that provided rigorous proof of the existence of the cortical motor center in human beings. Charcot differentiated between Aran-Duchenne type muscular atrophy and amyotrophic lateral sclerosis, now known as Lou Gehrigs's Disease, but previous called Charcot's disease. In 1886 Pierre Marie (1853-1940) and Charcot published the first description of the form of muscular atrophy now known as Charcot-Marie-Tooth disease.

Although his use of hypnotism and his dramatic demonstrations of hysteria were greeted with some skepticism, including the possibility that Charcot had suggested or provoked the "crises" and symptoms in their patients, his basic ideas about hysteria provided fruitful insights into the psychopathology of traumatic experiences. Charcot thought that hysteria was a progressive, irreversible mental disease that might be caused by a weak neurological system and triggered by traumatic events. According to Charcot, only hysterics could be hypnotized, because the condition was itself similar to an attack of hysteria. Ultimately, even though many of his most distinguished students concluded that hypnosis was a psychological phenomenon rather than a neurological condition, his methods and ideas had a profound impact on the development of psychoanalysis.

Charcot was a talented artist and caricaturist as well as an outstanding scientist. Many of his drawings were preserved at the Salpêtrière. Charcot collaborated with Paul Marie Louis Pierre Richer, an artist at the Salpêtrière, in the development of books on disease and deformity as portrayed by artists. These books had a significant influence on subsequent studies of the relationship between medicine and art. Charcot's complete works were published in nine volumes between 1888 and 1894.

LOIS N. MAGNER

Paul Ehrlich
1854-1915
German Bacteriologist

Paul Ehrlich is recognized as the founder of hematology (the study of the blood) and chemotherapy (the treatment of disease with chemicals.) His many accomplishments include the development of the side-chain theory of immunity, the use of dyes to treat tropical diseases, and the discovery of a treatment for the dreaded disease syphilis. Ehrlich was awarded the 1908 Nobel Prize in medicine for advancing the field of immunology and contributing to the production of a serum for diphtheria.

Paul Ehrlich. *(The Library of Congress. Reproduced by permission.)*

Ehrlich was born on March 14, 1854, into a respected middle-class Jewish family in Strehlen, Silesia, Germany (now part of Poland.) Although he was an excellent pupil, Ehrlich found school a dutiful bore until his cousin, a bacteriologist, introduced him to dyes and the world of the microscope. He developed such an all-consuming interest in staining that fellow students teased him about it. He continued his dye investigations through his university work at Breslau, Straubourg, and Freiberg. When he graduated with a degree in medicine from Leipzig in 1878, he was already a respected authority on dyes and received an appointment to the prestigious Berlin Charite' Hospital.

At Charite' he became so deeply engrossed with the investigation of blood through the medium of staining that he seldom ventured out of the hospital. He developed dyes like methylene blue to stain bacteria and used aniline dyes to stain blood cells. His staining techniques advanced the diagnosis of the tuberculosis bacillus discovered by Robert Koch. By 1883 Ehrlich had published 37 scientific studies.

In 1884, at 30 years of age, he became a professor at Berlin University where he joined a group of distinguished scientists who were making Berlin the world center of research in bacteriology. In 1897 Ehrlich was appointed to a separate institute for serum research, which he described as "small but my own." Here, he perfected the famous side-chain theory of immunity, which states that each cell has a series of side chains or receptors which not only absorb nutrients but react to toxins. During exposure to the toxin, the cell is "trained" to produce excess side chains that are released into the bloodstream as antibodies. If the cell survives, at the next exposure, antibodies link up with the toxin very much like a lock and key fit together.

Ehrlich was barely settled into his small institute when he was invited to join Frankfurt's Royal Institute for Experimental Therapy in 1899. He initiated a study of cancer, or malignant tumors, and the possible use of chemicals to treat cancer and pathogenic organisms. He tested hundreds of chemicals looking for the "magic bullet" that would target a specific organism.

When he became interested in one-celled parasites, such as the cause of African sleeping sickness, he returned to his old love—dyes. While testing mice infected with the tropical disease he found that cured mice responded to compounds containing nitrogen and the element arsenic.

The spirochete *Treponema pallidum* had just been identified as the cause of the sexually transmitted disease syphilis. For three years he synthesized compounds. After 605 failed attempts, Preparation 606—later called Salvarsan—proved to be the "magic bullet" for syphilis. Following extensive experiments with mice and rats, he cautiously began testing human subjects. Salvarsan was finally released for use in 1910. Because the compound contained arsenic, questions about side effects emerged, and Ehrlich was attacked—even accused of fraud. However, Salvarsan was finally accepted in 1914 as a great medical breakthrough.

The strain took its toll on Ehrlich, and he suffered a series of strokes. He was buried in the Jewish cemetery in Frankfurt on August 23, 1915. Ehrlich was known for his remarkable energy, his work ethic, as well as his kindness and modesty. His wife, Helwig, whom he married at age 28, was concerned about his health habits of eating little and smoking twenty-five strong cigars a day. The Frankfurt street on which the institute was located was renamed Paul Ehrlich Strasse for the famous scientist. When the National Socialists (Nazis) came to power in 1938, they changed the name to eliminate anything Jewish. However, after World War II, Poland renamed Strehlen, his hometown, to Ehrlichstadt in honor of its famous son.

EVELYN B. KELLY

Carlos Juan Finlay
1833-1915
Cuban Physician, Epidemiologist and Public Health Reformer

Although the American physician Walter Reed (1851-1902) is generally associated with the discovery of the means of transmission of yellow fever, it was the Cuban physician Carlos Juan Finlay who first provided evidence that the *Aedes egypti* mosquito served as the vector of the disease. Yellow fever was one of the most feared epidemic diseases of the nineteenth century. An understanding of the means of transmission of the disease led to a fairly high level of control over the disease and was instrumental in the successful completion of the Panama Canal.

Finlay was born in Camagüey, Cuba. His father was a Scottish physician and his mother was French. Finlay attended schools in France and Germany before enrolling at Jefferson Medical College in Philadelphia. While a medical student, Finlay became interested in Professor John Kearsly Mitchell's suggestion that an unseen botanical agent was the cause of tropical fevers. After obtaining his M.D. degree in 1855, Finlay continued his training in Paris. He returned to Cuba in 1863, where he established a private practice and pursued his interest in public health medicine.

Yellow fever was one of the most feared epidemic diseases of the nineteenth century. The disease was endemic in the Caribbean islands, but outbreaks also occurred in North America. Almost nothing was known about the cause of the disease or its means of transmission, and preventive measures seemed futile. In 1879 an official U.S. Yellow Fever Commission was sent to Cuba for an investigation. The Spanish government appointed Finlay to serve as the local liaison officer. After his participation in the work of the commission, Finlay developed new ideas about the disease and began his preliminary investigations.

At the International Sanitary Conference held in Washington, D.C., in 1881 Finlay pointed out the futility of the sanitary measures being used to fight the spread of yellow fever. He believed that an intermediate agent transmitted the disease from the sick to the healthy and that effective control measures would require the destruction of this vector. Finlay suspected that a mosquito served as the vector of yellow fever. After considering the characteristics of various species, he decided to investigate the *Culex* mosquito (*Aedes aegypti*), a nocturnal insect with a short flight span. Finlay collected eggs, hatched them, and allowed selected mosquitoes to bite patients with yellow fever. To confirm his hypothesis he needed non-immune volunteers who would allow themselves to be bitten by the infected insects.

Finlay obtained permission from the military authorities to inoculate soldiers who had recently arrived from Spain. Finlay eventually inoculated over one hundred volunteers, but the results were not always consistent. In 1894 Finlay presented his results and his recommendations for the control of yellow fever at the Eighth World Congress of Hygiene and Demography. He suggested that patients should be isolated, houses should be fumigated to keep away mosquitoes, and potential mosquito breeding sites should be eliminated.

In 1897 Giuseppe Sanarelli claimed to have discovered a bacterial agent that caused yellow fever. The Surgeon General of the U.S. Army appointed a Board to study the infectious diseases of Cuba. The Board included Major Walter Reed, James Carroll (1854-1907), Jesse Lazear (1866-1900), and Aristides Agramonte (1869-1931). The Board quickly discovered that Sanarelli's alleged causal agent was simply a bacillus frequently found in cadavers.

On August 1, 1900, the members of the Board visited Finlay, who gave them the eggs of the mosquito that seemed to be the vector of yellow fever. Lazear hatched the eggs and allowed the mosquitoes to bite yellow fever victims. Lazear then allowed the mosquitoes to bite several volunteers, including Carroll and himself. Carroll recovered from a severe case of yellow fever, but Lazear died. In October 1900, at a meeting of the American Public Health Association, Reed announced that a mosquito was the intermediate agent of yellow fever.

When he returned to Havana, Reed directed a program of controlled experiments that provided proof for Finlay's hypothesis. Leonard Wood (1860-1927), a physician who was serving as Military Governor of Cuba, called the confirmation of Finlay's doctrine the most important step forward in medicine since the discovery of the smallpox vaccination by Edward Jenner (1749-1823). Major William Gorgas (1854-1920) also praised Finlay and used the methods Finlay had suggested in the battle against yellow fever in Cuba and Panama.

LOIS N. MAGNER

Sigmund Freud
1856-1939
Austrian Psychiatrist

Sigmund Freud forever altered the ways humankind looked at its own thought processes. His groundbreaking work in psychology aroused intense feelings and controversy. Though some of his theories have been disputed, no one questions his influence on civilization. Many of his concepts—such as Oedipus complex, sibling rivalry, libido, transference, death wish, and telling speech blunders (dubbed Freudian slips)—have become part of popular culture and psychological understanding.

Freud was born in Morovia (now Czechoslovakia). His father was a wool merchant. His family moved to Vienna when he was four years old, and he considered himself an Austrian. His family was Jewish.

When he was nine years old, Freud entered high school, which was usual in Austria. He graduated at the top of his class, and soon entered medical school in Vienna in 1881. Ironically, he was not interested in medicine. There were, however, a limited number of fields open to Jews. He chose medicine so that he could "gain knowledge about human nature." Due to his photographic memory, medical school was rather easy for him. He passed his exams with top scores.

Shortly thereafter Freud met Martha Bernays. The two wanted to marry but he was too poor. In 1885 he traveled to Paris to study with the famous neurologist Jean Martin Charcot (1825-1893), who was working with patients suffering from a mental illness known as *hysteria*. Freud returned to Austria in 1886, married Martha Bernays, and began his own work with hysterical patients. He began to use the word *pyschoanalysis* to describe his treatment of patients. At first his work met with hostility from other doctors.

By 1910 Freud's theories about the human mind were gaining recognition. Two of his students—Alfred Adler (1870-1937) and Carl Gustav Jung (1875-1961)—abandoned Freud's philosophies to develop their own theories of psychology. Freud's ideas were constantly evolving; he was continually revising his theories and writings, the most important of which include *The Interpretation of Dreams* (1900), *Three Essays on the Theory of Sexuality* (1905), and *The Ego and The Id* (1923).

Sigmund Freud.*(The Library of Congress. Reproduced by permission.)*

Freud's theories of behavior stemmed from a connection between the conscious and the unconscious. He believed that ideas in the unconscious mind controlled behavior, and that dream interpretation was a key to unlocking the unconscious mind. He was controversial because of his beliefs that sexual feelings affected behavior from an early age. Some of his philosophies seem to stem from personal experience. He experienced intense sibling rivalry when his younger brother was born, and later extreme guilt when the infant died. He later realized that his unconscious feelings of being neglected by his mother led to his actions of contempt for his brother. Freud also experienced sexual feelings for his mother as a child, and recurring dreams of his own inadequacy. These personal feelings are mirrored in his psychological philosophies.

The same year *The Ego and The Id* was published, Freud noticed a growth in his jaw. He was diagnosed with cancer of the mouth. Although his condition was extremely painful, he continued his work. He underwent over 30 operations for his cancer. When the Nazis took control of Austria in 1938, Freud, his wife, and six children fled to safety in England. A year later, in London on September 23, with the help of a compassionate physician who provided "adequate sedation," Freud died.

CAROLYN CRANE LOVE

Elizabeth Fry was one of the most important prison reformers of the nineteenth century. She also helped to reform the British hospital system and the treatment of the mentally ill.

Fry was born Elizabeth Gurney, the daughter of a wealthy banker and merchant. The Gurneys were members of the Society of Friends, a religious group also known as Quakers. Upholding the belief that all humans are equal in the sight of God, the Quakers were the first religious group to denounce slavery. They were also concerned about the welfare of prisoners. Early Quakers had been imprisoned for their beliefs and experienced the horrible conditions of incarceration firsthand. They also believed that there is something of God in everyone, even in criminals, therefore the goal of prison should be reformation, not simply punishment. Fry's religious background had much to do with her enthusiasm for reform.

In 1798 Fry met with an American Quaker named William Savery, who inspired her to devote her energies to helping those in need. She began by setting up a Sunday school in her home, where she taught local children to read. Also in 1798, Elizabeth met her husband, Joseph Fry, who was from another wealthy Quaker family. They married in 1800 and their first child was born the next year. Between 1801 and 1812, Elizabeth had eight children.

In 1813 Stephen Grellet, a friend of the Fry family, visited Newgate, London's chief prison, where prisoners were held before execution or transportation. Grellet was shocked by conditions in the women's section, where he found prisoners sleeping on bare stone floors and newborn babies without clothing. He informed Fry, and she visited the prison the next day, taking warm clothing for the babies and clean straw for sick prisoners to lie on.

Fry's visit to Newgate was the beginning of her life's work, but for family reasons, including the birth of two more children, she did not visit Newgate again until 1816. During this visit, she suggested that the prisoners set up a school for the children. Although the women were eager to follow this suggestion, Fry could not obtain backing for the school until she established the Association for the Improvement of the Female Prisoners in Newgate, a committee of twelve women. As well as founding the school, the association appointed a woman matron to supervise the woman prisoners (formerly guarded by men) and provided materials and instruction for the women to sew quilts and other items they could sell to buy food, clothing, and fresh straw for bedding. Members of the association also took turns visiting the prison every day to read from the Bible.

Fry's brother-in-law, Thomas Fowell Buxton, joined the association in 1817, and published a book based on his investigations at Newgate. When he was elected to Parliament in 1818, he promoted Fry's work. She gave evidence to a parliamentary committee investigating London prisons, becoming the first woman asked to do so. Although impressed with her work, Parliament did not respond to Fry's concerns until 1823, when legislation was introduced to provide for regular visits from prison chaplains, women warders for women prisoners, and payment for jailers (who had depended on fees from prisoners).

Beginning in 1818, Elizabeth and the association also visited convict ships. Those convicted of minor crimes were often transported, or shipped to British colonies, especially Australia. The prisoners were held aboard ships in the Thames River for six weeks before sailing. During that time, members of the association visited the ship every day, set up a school, and supplied each prisoner with materials for making patchwork quilts during the voyage. Over the next 20 years, Fry visited and organized 106 convict ships, every ship that carried women prisoners.

Fry's reforms were not limited to prisons. She set up the Brighton District Visiting Society to provide help and comfort for the poor, and soon similar societies were established throughout Britain and in Europe. She campaigned for improvements in the treatment of mental patients. Fry also founded a training school for nurses, and her views on their training influenced Florence Nightingale.

Fry was unusually influential for a woman of her day. Even the Queen of England met with her and donated to her projects. And, unlike those of many other reformers, her suggestions were acted upon throughout most of Europe during her lifetime.

JANET BALE

Francis Galton
1822-1911
English Biologist and Statistician

Francis Galton has been called the last of the gentleman scientists—men who dabbled in

Francis Galton. *(The Library of Congress. Reproduced by permission.)*

science as a hobby rather than a profession. Galton was brilliant and, with hundreds of publications to his name, prolific, but he was also a dilettante—his inquiries ranged from the effectiveness of prayer to the body weights of British nobles. Many of his ventures were successful: he pioneered the use of fingerprints for identification, tested Darwin's theories, and discovered the anticyclone. But his most notable contribution to science is a source of more infamy than fame: eugenics, the study of improving a human population by selective breeding, a cause that would later be championed by bigots and genocides.

It is understandable that Galton would have latched on to the idea that intelligence and talent are hereditary. He himself came from an eminent family—Charles Darwin (1809-1882) was his cousin, Erasmus Darwin (1731-1802) his grandfather. Francis showed intelligence, even genius, from an early age; by four he could multiply, tell time, and read English, Latin, and some French. He entered medical school at age 16 but used it partly as a chance to systematically sample the pharmaceuticals cabinet, in alphabetical order (he stopped at C, when a dose of Croton oil made him vomit). Even the Croton oil was not as harmful as the grueling schedule, though, and when his father's death left him with a substantial inheritance, Galton was quick to leave school.

After school Galton made a name for himself as an explorer, earning a gold medal from the Royal Geographical Society (of which he would later become a fellow) for his excursions into uncharted southern Africa. In 1853, having satisfied at least some of his wanderlust, he married Louisa Butler and gradually began dabbling in science. He was the first to use statistics to determine correlation in biology—the phenomenon where if one variable, like arm length, varies, another one, like leg length, will vary in the same way (a person with long arms will, on average, have longer legs than a person with short arms). This concept was in many ways revolutionary, especially in Galton's own work, which ofte involved correlating people's intelligence with that of their ancestors or offspring.

The experiments with hereditary intelligence represented the part of his work Galton was most passionate about. He was generally interested in the budding science of genetics; one of Galton's most notable experiments involved disproving his cousin Charles Darwin's theory that hereditary traits were passed down by tiny particles called "gemmules" that were found in every part of the body. Galton believed that talent and intelligence—what he termed "eminence"—was one of these heritable traits. He studied England's most prominent families and found that any individual's eminence was directly related to that of his parents and grandparents. This led him to propose a program of eugenics, assuring that notable families would continue to produce notable children. Statistically, he reasoned, it would take many times more pairings of unremarkable people to produce as many eminent children as one pair of geniuses, so in order for a society to grow in eminence certain people must be encouraged to reproduce while others are discouraged.

Galton countered his own plan of selective breeding by dying childless in 1911. He did leave behind a chair in eugenics at University College, London, endowed in his will; a number of works, like his study of fingerprints, that would represent his lasting impact on the scientific world; and a brainchild, eugenics, that would be come to represent the depths of ignorance.

JESSICA BRYN HENIG

Samuel Hahnemann
1755-1843
German Physician

Samuel Hahnemann is regarded as the founder of homeopathy, a controversial type of medicine that revolves around the principal that "like cures like," or that a disease can be cured by medications that produce the symptoms of that disease in healthy people.

Hahemann was born in Saxony (now part of Germany) in 1755. This was the time of the Enlightenment, a political and intellectual movement that swept through Europe encouraging freedom of thought, religion, and education. Hahnemann's family was poor, but stressed learning. His father taught him never to learn passively, but to question everything he was told. The young Hahnemann read everything he could get his hands on, and by the time he was 24, could read and write at least seven languages, and had read almost every medical text written in Europe.

Hahnemann became a doctor in 1791, at the age of 36, and practiced conventional medicine for nine years. Even so, he was always known as an avid experimenter, chemist, and rebel who was unafraid to speak his mind or challenge conventional wisdom. He was also unafraid to experiment on himself. In 1790 he was testing a theory that Peruvian bark (also called quinine) was useful in treating malaria. Hahnemann gave himself repeated small doses of quinine and noticed that he started to suffer fever, chills, and other symptoms of malaria. He concluded that the reason that quinine was useful against the disease was because it caused symptoms similar to the disease itself.

Hahnemann called his new theory "homeopathy," (from the Greek *homoios*, which means "similar," and *pathos*, which means "suffering") and from then on practiced medicine on the premise that like treats like. He began doing experiments with many conventional medicines of the time to prove his theory. In 1810 Hahnemann published the first edition of *The Organon*, which defined his homeopathic philosophy of medicine. The same year, Napoleon attacked Hahnemann's hometown of Leipzig, killing 80,000 people and leaving behind a typhus epidemic. Hahnemann had great success treating the survivors with homeopathic remedies, and his reputation as a healer spread.

Nevertheless, Hahnemann was ridiculed by the medical and scientific establishments for his ideas. He was particularly disliked by the apothecaries because his treatments called for using only one medication at a time, and in small doses, causing the apothecaries to make very little money from his patients. In addition, the apothecaries did not always make preparations as precisely as Hahnemann's treatments required, and he even accused some of giving patients the wrong prescriptions. Hahnemann soon began to dispense his own medicines, which was illegal in Germany at the time. The apothecaries brought charges against Hahnemann and had him arrested.

But Hahnemann continued his work. By 1821 he had proven 66 homeopathic remedies and had published a reference work called *Materia Medica Pura*. In 1831 a cholera epidemic swept through central Europe, killing thousands of people. Hahnemann instituted the first widespread use of homeopathy to fight the epidemic, and achieved a 96% cure rate (as opposed to the 41% cure rate achieved by allopathic doctors of the time.)

In 1834 Hahnemann met and married Marie Melanie d'Hervilly, a socialite from Paris who was less than half his age. They worked together in his practice in Paris until July 2, 1843, when Hahnemann died at the age of 88. By the time of his death, homeopathy was slowly spreading throughout Europe and North America, despite the mainstream medical profession's opposition. Homeopathic medicine was especially popular with royalty, artists, and other celebrities. Hahnemann's homeopathic methods persist today, and many of his original remedies are still prescribed by homeopaths throughout the world.

GERI CLARK

Shibasaburo Kitasato
1852-1931
Japanese Physician and Bacteriologist

Shibasaburo Kitasato was a Japanese physician who became interested in studying microbes and their link to diseases. Under government sponsorship, he spent six years in Berlin working with Robert Koch (1843-1910). Kitasato is best remembered for his work on tetanus and diphtheria, successfully growing the first pure culture of the tetanus bacillus in 1889. Working with Emil von Behring (1854-1917), Kitasato demonstrated the power of blood serum as an antitoxin for treating both these diseases. In 1894 Kitasato discovered, simultaneously with Alexandre Yersin (1863-1943), the bacillus that causes bubonic plague.

Born in 1852 on Japan's southern island of Kyushu, Kitasato attended medical school first in Kumamoto and later in Tokyo. After graduat-

ing he went to work for the Central Sanitary Bureau. There he worked in the field of public health, studying cholera epidemics and working to prevent outbreaks of infectious diseases. In 1886 Kitasato was sponsored by his government to study and work in Berlin at the laboratory of the world famous Robert Koch. Koch had come to public attention in recent years for his groundbreaking work in developing the field of bacteriology. Koch had spent years perfecting methods for isolating and culturing particular germs, tracking down the microbes responsible for causing certain diseases. He had already established the sources of anthrax (1877), tuberculosis (1882), and cholera (1883) when Kitasato arrived in his lab.

Kitasato learned from Koch the careful and painstaking methods for culturing, studying, and experimenting with various microbes. His first major achievement came in 1889, when he succeeded in culturing the bacteria that causes lockjaw, the tetanus bacillus (*Clostridium tetani*). This was remarkable because Kitasato was the first to discover a means of growing anaerobic bacteria, germs that grow without contact with air. In addition, working with fellow researcher Emil von Behring, Kitasato discovered that the symptoms of tetanus were produced by a toxin put out by this microbe. The next year, after many experiments, Kitasato and Behring found that blood serum, the clearish liquid that separates out from blood when it is allowed to stand, from animals who had been infected and recovered worked as an antitoxin. By injecting this antitoxin into infected animals, their symptoms were cleared and recovery assured. The impact of this discovery was proven during World War I. Tetanus infections are picked up when open wounds come into contact with infected soil, and prior to this time soldiers died in large numbers as a result. Given as a precaution to virtually all wounded soldiers, the death from tetanus dropped dramatically during the war.

Upon returning to Japan in 1892, Kitasato established his own laboratory in Tokyo for continuing his studies of microbes and infectious diseases. This became the Institute for Infectious Diseases, and Kitasato received government funding to carry out his research and train other scientists.

In 1894 Kitasato was sent to Hong Kong, where an outbreak of bubonic plague was causing a high death toll. Kitasato set up a laboratory there and set to work tracking down the bacillus responsible. He succeeded in isolating the bacillus at virtually the same time as another scientist, Swiss researcher Alexandre Yersin. Yersin, trained in the French laboratory of Louis Pasteur, named the bacillus after his mentor, calling it *Pasteurella pestis*. This name remained official until 1971, when it was changed to recognize Yersin. It is now officially known as *Yersinia pestis*, though the old name is still frequently cited. Kitasato is given credit for the simultaneous discovery.

In 1914 the Japanese government made Kitasato's Institute for Infectious Diseases part of the Ministry of Education, a move that Kitasato disagreed with. He felt that his work had practical applications for public health and therefore should have been a part of the hygiene department under the Ministry of the Interior. As a result, Kitasato and his staff resigned their positions. He then went on to found a private laboratory known as the Kitasato Institute. This Institute is a non-profit organization still conducting important research today. At the same time as he was organizing his new institute, Kitasato accepted the job of creating and organizing a new medical faculty for the University of Keio.

In recognition of his contributions to the control of infectious diseases, Kitasato received numerous honors and awards not only in his own country but from many others as well. Upon leaving Germany to return to Japan, he was awarded the title of "Professor," the first non-German to receive this honor. He turned down several offers from abroad, opting to return to work in his native Japan. In 1917 he was appointed to the House of Peers by the Emperor Taisho and raised to the title of Baron in 1924. He died at home in 1931.

KRISTY WILSON BOWERS

Robert Heinrich Hermann Koch
1843-1910
German Physician

Robert Koch was a key player in the field of bacteriology and, ultimately, hygiene and public health. His brilliant research uncovered the causes of anthrax, tuberculosis, and cholera, and led to the development of Koch's Postulates, a series of guidelines for the study of infectious diseases that are still used today.

Koch was born on December 11, 1843, in Clausthal, Germany, a mining town in the Harz Mountains. As a boy he loved natural science

Robert Koch. *(The Library of Congress. Reproduced by permission.)*

and collected rocks and plants and dreamed of being a great explorer. He studied at the University of Göttingen and, after interning in Hamburg, married and settled down to become a country doctor. Upon receiving a microscope for his 28th birthday, he began to study many things, including anthrax, a dreaded disease of warm-blooded animals, especially sheep. Isolating the organism from the blood, he grew the microbes in various media and discovered a stage where the anthrax bacilli have translucent coverings called spores. These spores could remain in the ground for years, then, when conditions were right, emerge into the rod-shaped organisms that cause the disease. Koch perfected the idea that these organisms could be cultured in pure form. In 1877 he published a work describing how thin smears of bacteria could be fixed on slides by gentle heat. His work was illustrated with excellent photomicrographs. He also invented the hanging drop, in which bacteria are suspended in a nutrient solution on a glass slide. Studying various wounds, Koch infected healthy animals with six different bacteria, demonstrating that a particular bacteria causes a specific disease.

Koch's scientific investigations earned him a position in Berlin in the German Health Office. He devised a new method of growing organisms on gelatin and then transferring pure colonies to test tubes of nutrient broth. Modifying the staining techniques developed by Paul Ehrlich (1854-1915), Koch worked to find the cause of tuberculosis. The bacillus that causes tuberculosis is very difficult to grow in pure culture, but Koch devised a technique of successive media and finally succeeded in growing them. On March 24, 1882, he announced to the Physiological Society of Berlin that he had isolated the bacillus that causes tuberculosis.

Koch devised a series of four basic principles, known as Koch's Postulates, by which a particular bacterium may be decisively linked to a specific disease. First, the bacterium in question must be present in every case of the infectious disease. Second, it must be possible to cultivate this bacterium in a pure culture. Third, a laboratory animal inoculated with the pure culture of the bacterium must develop the disease. Fourth, the bacterium must be recovered from the inoculated and infected test animal and reproduced again in a pure culture. If all of these criteria are met, it may be deduced that the suspect bacterium is indeed the cause of the disease.

When a serious cholera epidemic struck in 1883, Koch participated in a medical expedition to Egypt as a member of a German government commission. He soon isolated a small comma-shaped bacillus that he believed to be the cause. When the outbreak subsided he went to India, where cholera was endemic, and found the disease was transmitted by drinking water, food, and clothing.

Winning much acclaim and a large monetary prize, Koch returned to Germany to continue research as the director of the Institute of Hygiene. He concentrated again on tuberculosis and, in 1890, announced the discovery of tuberculin, which he hoped would be a cure. Tuberculin proved a failure as a cure for the disease, but is still used today as a diagnostic test for tuberculosis.

After his disappointment and embarrassment with tuberculin, Koch expanded his research into a variety of diseases, including leprosy, hinderpest, bubonic plague, surra, Texas fever, and malaria. He led expeditions in East and West Africa to study tropical diseases. In 1905 he received the Nobel Prize in medicine for his investigations and discoveries in relation to tuberculosis. He died on May 10, 1910, in Baden-Baden, Germany, of a heart attack.

EVELYN B. KELLY

Richard Krafft-Ebing
1840-1902
German Neuropsychologist

Richard Krafft-Ebing was a pioneer in the field of sexual psychopathology. He wrote the first-ever series of case studies of deviant sexual behavior, *Psychopathia Sexualis,* and was also a well-respected forensic psychologist.

Krafft-Ebing was born in Mannheim and educated at the University of Heidleberg in Germany. He was appointed professor of psychiatry at the University of Strasbourg at the relatively young age of 32. A year later he was appointed the director of the Feldhof Asylum near Graz, Austria, and from 1892 until his death he was the head of the psychiatry department at the University of Vienna.

Early in his career, Krafft-Ebing focused his work on the study of sexual behavior, a new field at that time. In 1886 he published *Psychopathia Sexualis,* a collection of 238 case histories of what he called "sexually abnormal people." It was in this book that the terms "sadism" and "masochism" were first used, the former to refer to behavior where sexual pleasure is associated with physical cruelty to the partner; and the latter in respect to the association of sexual pleasure with humiliation or abuse. "Sadism" was coined in honor of the Marquis de Sade, and "masochism" comes from the name of Leopold von Sacher-Masoch, a well-known novelist of the time who was said to enjoy being humiliated by women.

Psychopathia Sexualis also contains the first-ever classification of sexual behaviors, and was the first work to discuss subjects such as fetishism, exhibitionism, homosexuality, pedophilia, and autoerotism. Krafft-Ebing was also one of the first psychologists to advance the belief that sexual orientation is biologically based, as opposed to a matter of choice. The book was a huge success in the academic community and Krafft-Ebing added to and edited it 12 times over the years. The final version was completed in 1906, four years after his death. His ideas also unleashed a flood of interest in the examination of sexual behavior, and soon many studies appeared on the subject throughout Europe, and in Germany in particular. Krafft-Ebing's work was also widely read by the public, many of whom adopted it as a work of popular pornography. *Psychopathia Sexualis* is still in print and is still used today, though some of Krafft-Ebing's ideas are now disputed.

Though Krafft-Ebing is best known for beginning the study of sexual behavior, his work in psychiatry, criminology, and forensic psychopathology also helped advance psychology as a clinical science. He was also a forensic psychologist who investigated the legal and genetic aspects of criminal behavior and was often consulted by the courts as an expert witness. In addition, he made early discoveries about such neurological conditions as epilepsy, and found a relationship between syphilis and general paralysis. He investigated paranoia and was an early proponent of the use of hypnotism as a form of psychotherapy and a way of treating the mentally ill.

Krafft-Ebing continued to research and teach at the University of Austria until his death. He died in 1902 in Mariagrun, Austria.

GERI CLARK

René-Théophile-Hyacinthe Laënnec
1781-1826
French Physician and Inventor

In 1816 René Laënnec invented the stethoscope. His 1819 book about his use of this instrument inaugurated the modern era of diagnosing chest, heart, and lung diseases accurately and scientifically.

Laënnec was born in Quimper, a small town on the seacoast of West Brittany, France. He was baptized "Théophile-René-Marie-Hyacinthe Laënnec," but was known professionally as either "René-Théophile-Hyacinthe Laënnec," which appears on his tombstone, or simply "R.-T.-H. Laënnec."

Laënnec's mother died of tuberculosis when he was five. His father had little interest in his children and in 1788 sent him to live with his uncle Guillaume, a prominent physician in Nantes. Laënnec studied hard and aimed to become an engineer, but the French Revolution violently interrupted his education. Especially in 1793-94 Nantes was a bloodbath, as Jean-Baptiste Carrier suppressed the Vendée revolts. Several times Guillaume narrowly escaped execution.

By 1795 the worst was over. Laënnec, influenced by both his recent war experience and the pervasiveness of tuberculosis in his own family, now decided on a career in medicine. At the age of fourteen he was apprenticed to his uncle as a surgeon in several hospitals in Nantes. He wandered around northwestern France for a few

years and served briefly in Napoleon's army before moving to Paris to begin serious medical study in 1801.

He learned dissection under Baron Guillaume Dupuytren (1777-1835) at the École Pratique but soon became the star pupil of Baron Jean Nicolas Corvisart des Marets (1755-1821) at the Hôpital de la Charité. He received his M.D. in 1803 with the top prize in medicine and the only prize in surgery. He soon became known as one of the best clinicians and medical scientists in France. Through the influence of Corvisart, he secured prestigious appointments in Paris, such as visiting physician at the Hôpital Necker. For the rest of his life he divided his time between working in Paris and vacationing for his health in Brittany. He had always been weak and sickly, suffering from asthma, tuberculosis, and migraine headaches.

Besides inventing the stethoscope, Laënnec's greatest contribution to medicine was helping to redefine the concept of disease itself. Before his time, scientists did not know clearly what a disease was. Some, such as William Cullen (1710-1790), thought that diseases were separate "things" that could be classified like animals or plants. Others, such as Benjamin Rush (1745-1813), thought that there was only one disease—fever—and that the various so-called diseases were just aspects of it.

But in France Laënnec and many of his teachers, colleagues, and students shifted the focus of medical research from the whole disease to the particular organs or parts of the body that were affected by it. They recognized that the respective anatomies of diseased and healthy persons differ in specific, measurable ways. Thus, they centered their attention on the "lesion," that is, on any change from a normal anatomical structure. They discovered that such changes are usually harmful. They performed autopsies to correlate their observations of diseases in living patients with their knowledge of anatomy.

Using autopsies and dissection to study disease was a relatively new idea in Laënnec's time. This branch of medical research became known as "pathological anatomy," and its understanding of individual lesions is now fundamental to the understanding of disease.

Several lesions are named after him, such as "Laënnec's thrombus," a blood clot in the heart, and "Laënnec's cirrhosis," a certain kind of progressive liver destruction.

Laënnec's pathological research contributed most to our knowledge of respiratory tract diseases, especially tuberculosis. Ironically, it was this disease that killed him in 1826.

ERIC V.D. LUFT

Baron Dominique-Jean Larrey
1766-1842
French Surgeon

Dominique-Jean Larrey was a French military surgeon who served during the French Revolution and during the reign of Napoleon. He is credited with introducing many of the methods of modern military field medicine, including field hospitals, ambulances, and triage on the battlefield.

Larrey was born at Baudean, Hautes-Pyrenees, France, in 1766. His parents were very poor, so the local priest took pity on Dominique-Jean and sent him to school. When he was 13 his father died, and Larrey was sent to Toulouse to live with his uncle, Dr. Oscar Larrey, a noted surgeon. At 21, Larrey moved to Paris and joined the French Navy. Five years later he joined the army and served in northern France as a field surgeon during the French Revolution.

Larrey saw almost immediately that there was a need for rapid evacuation of wounded soldiers from the battlefield, and he designed two types of what he called "flying ambulances." One kind was a small, two-wheeled carriage drawn by two horses that could carry two casualties. It had a folding door and a removable floor covered with a mattress. On the sides of the carriage were pockets that held medical instruments and supplies. The second kind of ambulance was a four-wheeled carriage drawn by four horses that could carry four casualties. It had springs in the body to absorb some of the shock and make the ride more comfortable for the wounded.

Larrey also established new surgical guidelines for treating wounded in the field. At that time, soldiers who required an amputation were left alone until the battle was over and they could be attended to. Larrey noticed, however, that delaying amputation was worse for the wounded, as they had more pain, bled more, and died more often from their wounds. His new guidelines dictated immediate amputation in cases of shattered limbs or joints, broken small bones or nerves, or major muscle or artery loss. He also changed the way that wounds were cleaned on the battlefield—only water was to be used, and the wounds were dressed with adhesive bandages that let them drain. These measures reduced the number of serious infections and deaths among wounded soldiers. Larrey also improved his own surgical skill during this time. It was said that he could amputate a leg in one minute and an arm in 17 seconds.

In 1795 Larrey was assigned to the forces of Napoleon Bonaparte (1796-1821). Napoleon was a great supporter of Larrey's surgical methods, and encouraged him to refine his ideas even further. During this time, Larrey perfected the organization of the flying ambulance corps. He placed them as close to the front line of battle as possible, and their first job was to transport the wounded from the battlefield to the hospital, without regard for rank. The idea that the most seriously injured man would be helped first, no matter what his rank or distinction, was a novel concept. Larrey also established the first field hospitals staffed by surgeons who could perform more delicate operations.

In 1797 Napoleon was sent to Egypt to open a land route to India, and he took Larrey with him. In Egypt, Larrey adapted the flying ambulance for desert warfare by replacing the horse-drawn carriage with camels, which carried medical supplies and litters for transporting wounded.

When Larrey returned to France, Napoleon had established a new government, with himself as leader. He appointed Larrey Surgeon General to the Imperial Guard of France. Larrey remained with Napoleon throughout his reign as Emperor, and was himself wounded twice—at Austerlitz and Waterloo. After Napoleon's defeat at Waterloo, Larrey was captured by Prussian troops, but his life was spared by the Prussian commander. Larrey, who was known for treating enemy wounded as well, had saved the life of the commander's son during an earlier battle.

Larrey returned to France and settled down to teach in the military hospital and to write. He died in 1842 at the age of 76, shortly after returning from a tour of French military hospitals in northern Africa.

GERI CLARK

Charles Louis Alphonse Laveran
1845-1922
French Physician, Military Surgeon and Parasitologist

Alphonse Laveran was a French surgeon who was awarded the Nobel Prize for medicine in 1907 for his discovery, and subsequent

In 1880, through research done on his own, Laveran noticed that the black pigment in a malaria patient's blood contained an abundance of single-cell organisms with flagella, or whip-like tails that help them to move around. At the time, the medical community had made discoveries indicating that bacteria was the cause of disease and were beginning to see malaria as bacteria-related. However, this protozoan was not a bacterium, and it seemed to act as a parasite on the red blood cells of the patients. Laveran recorded that the protozoan would enter the blood cells, grow to almost the size of the cell itself and then divide into spores, reproductive cells produced by protozoa to make more cells. The spores, after destroying the blood cell, would invade unaffected cells, beginning the process again. The dark quality of the patient's blood turned out to be the waste produced by the protozoan.

Laveran's discovery that malaria was caused by these single-cell animals did not lead immediately to fame and fortune. Indeed, his military superiors did not acknowledge this achievement and overlooked him for promotion. In 1884 he served as Professor of Military Hygiene at Val-de-Grâce; he became chief surgeon in 1891and director of the Eleventh Corp's medical service in 1894. Laveran left the military in 1896 to join with the Pasteur Institute, where his inquisitive mind and his desire to research were welcomed. While at the Institute, he carried out more research on parasitic blood diseases and finally in 1907 was awarded the Nobel Prize for his work. The prize came with a substantial amount of money; Laveran used these funds to open a tropical medicine laboratory at the Pasteur Institute. The following year he organized the Societe de Pathologie Exotique and was its president until 1920. Laveran died in Paris, the place of his birth, in 1922. His work with malaria led a later scientist, Sir Ronald Ross (1857-1932), to discover that the malaria parasite is transmitted by mosquito from human to human. This discovery provided the opportunity to control the disease.

MICHAEL T. YANCEY

Alphonse Laveran. *(The Library of Congress. Reproduced by permission.)*

research, that disease could be spread by single-cell protozoa in the blood system. His continuing research following this breakthrough included diseases caused by other single-cell animals in the blood system.

Laveran was born in Paris in 1845. The son of an army surgeon, he moved with his family to Algeria in 1850; the family returned to Paris in 1855. Laveran was a bright student who attended two well-known medical schools in the city of Strasbourg. In 1867 he graduated with an M.D.; his doctoral dissertation concerned the regeneration of damaged nerves. Soon after, with the outbreak of the Franco-Prussian war, a war between Germany and France, Laveran followed his father's footsteps and went into the army as a surgeon.

Following the war in 1874, Laveran sat for a very competitive examination and was subsequently appointed Professor of Military Medicine at the Ecole du Val-de-Grâce. Between 1878 and 1883 he was stationed in Algeria, where he began to treat and observe patients suffering from the disease malaria. Malaria, also known as "marsh fever" at the time, had afflicted mankind for thousands of years, and the hypotheses regarding its causes were many. One of these was that bad air, especially air over swamps or marshes, caused the disease. This is where the term malaria originated (*mal aria* is the Latin term for bad air).

Sir Joseph Lister
1827-1912
British Surgeon

Joseph Lister is recognized as the father of antiseptic surgery. His insistence that a surgeon must protect the wound against outside organisms is one of the fundamental guiding principles of modern surgery.

Lister was born April 5, 1827, in Upton, Essex, England. His parents, Joseph Jackson and Isabella Harris Lister, took a great interest in their son's education. They instructed him and sent him to Quaker schools that emphasized natural history and science. At age 16 he decided medicine would be his career.

Lister graduated from King's College, London, and became a house surgeon at University Hospital in 1852. Edinburgh, Scotland, was recognized as an ancient medical center, and Lister was appointed as assistant to James Symes, the best surgeon of the day. He later married Symes's daughter. Lister received an appointment to the Edinburgh Infirmary in 1856, and later to the new Glasgow Royal Infirmary in 1861. As a surgeon Lister was concerned that over half of the patients who underwent amputations died. He noted that simple bone fractures, which had unbroken skin, almost always healed without any problems; however, compound fractures, in which the broken bone punctured through the skin and was exposed, often resulted in hospital gangrene or other infections that caused the patient to die.

In the 1860s little was known about microbes and the causes of infection. Explanations ranged from miasma, or bad air, to the explosion of tissue when exposed to air. Lister read Pasteur's studies of airborne microbes and was convinced that it was not the air itself, but organisms from the air. Louis Pasteur (1822-1895) used heat to kill microbes, but since this would not be practical in an operating room, he began to search for a chemical to use. He read about how carbolic acid (phenol) had been used to purify sewage at Carlisle, England, and hit upon the idea of using this compound in the operating room.

On August 12, 1865, Lister performed a historic operation that ushered in a new era of surgery. An eleven-year-old boy named James Greenlees was run over by the wheel of a wagon and had a compound fracture. Lister applied carbolic acid to the wound, dressed, and splinted it carefully. Six weeks after surgery James walked out into the world and into surgical history. Not only did Lister put the phenol solution on wounds, he soaked instruments and anything coming into contract with the wound. He even developed a carbolic-acid atomizer to spray the room, an idea he later discarded. Another contribution to surgery was the use of strong, anti-

Joseph Lister. *(The Library of Congress. Reproduced by permission.)*

septic catgut ligatures that were stored in carbolic acid.

In 1867 Lister published a series of cases showing how surgical mortality fell dramatically in his Male Accident Ward. In 1869 Lister became head of clinical surgery at Edinburgh and the happiest years of his life followed. The Germans had experimented with antisepsis during the Franco-Prussian War and his clinics were packed with visitors and students. He was invited to lecture at prominent centers in Germany.

Lister's work was not appreciated in the United States and England where opposition to the germ theory remained steadfast. However, in 1877 Lister was offered a position at King's College, London, and again the opportunity to make surgical history. At that time the standard procedure for treating a fractured patella involved forcing the simple fracture into a compound fracture, often resulting in death from infection. Lister performed a spectacularly successful and widely publicized procedure whereby he wired the patella using antisepsis. This operation marked a turning point in the acceptance of germ theory and antisepsis among physicians.

Lister enjoyed a privilege denied many scientific innovators; he saw his principles accepted during his lifetime and was honored with the title of baronet in 1883. He was also appointed as one of the twelve original members of the Order of Merit in 1902. The Lister Institute of Preventive Medicine was founded in 1891.

Lister was a humble, religious, and unassuming man, uninterested in financial gain or fame. After the death of his wife in 1893, he retired from surgery and, at his death in 1912, was almost completely blind and deaf.

EVELYN B. KELLY

Pierre Charles Alexandre Louis
1787-1872
French Physician and Statistician

Pierre Charles Alexandre Louis is best known for an approach to medicine known as the "numerical system." Louis's admirers credited him with establishing medicine as an exact science by demonstrating the value of the statistical method in diagnostics and therapeutics. Louis's meticulous work on tuberculosis and typhoid fever provided statistical studies of the major symptoms of the diseases and the postmortem lesions associated with them. By systematically testing the effectiveness of therapeutic phlebotomy, Louis demonstrated that the timing and quantity of bloodletting had no impact on the course or mortality of the diseases for which it was commonly employed.

The son of a vineyard keeper, Louis grew up in the Champagne region of France. After a few years of law school, he became interested in medicine and studied at Rheims and Paris. He was awarded the M.D. from Paris in 1813 and joined an official mission to Russia where he worked as an itinerant doctor before establishing a medical practice in Odessa. He returned to Paris for further training after finding himself helpless during a diphtheria epidemic and began his systematic analysis of clinical findings in the wards of the Charité, one of the major hospitals of Paris.

Most of Louis's professional life was spent in Paris at La Pitié Hospital. He was greatly respected and influential as a clinician and a teacher, but he never became a member of the Paris Faculty. He was especially popular with American students who regarded him with reverence bordering on idolatry, according to Oliver Wendell Holmes (1809-1894).

Like other members of the Paris clinical school of thought, Louis was opposed to all systematic ideas and medical practices based only on theories. His disdain of theory was so great

Pierre Louis. *(Wellcome Institute. Reproduced by permission.)*

that he has been called a "radical empiricist." Louis sought truth through observation, experiment, pathological anatomy, empiricism, and statistical methods.

In *Anatomical-Pathological Researches on Phthisis* (1825) Louis summarized his observations on phthisis (pulmonary tuberculosis). In addition to his analysis of 358 dissections and almost 2000 clinical cases, Louis discussed epidemiological and public health issues. His observations led him to formulate "Louis's Law," which asserts that tuberculosis usually originates in the left lung and that primary lesions in the lungs precede tuberculous lesions found elsewhere in the body.

Like most nineteenth-century families, Louis's was directly affected by tuberculosis. His only son died of the disease in 1828. Once again Louis left Paris, this time on an expedition to Gibraltar to investigate yellow fever. In 1839 he published his *Anatomical, Pathological and Therapeutic Researches on the Yellow Fever of Gibraltar of 1828*. In 1829 Louis returned to Paris and resumed his statistical analysis of his clinical observations.

Louis's landmark work *Researches on the Effect of Bleeding in Various Inflammatory Diseases, and on the Action of Emetics and Cupping in Pneumonia* was published in France in 1835, and in an English translation in 1836. Louis was a member of a generation of French doctors who had become skeptical about the efficacy of traditional drugs and therapies. Most of the doctors of the Paris Clinical School turned their attention to diagnosis, arrived at by a combination of physical examination and autopsy. Louis encouraged the advocates of "therapeutic skepticism" with his demonstration that bloodletting had little or no beneficial effect on various conditions.

Typhoid fever, one of the most important disease problems and diagnostic puzzles of the early nineteenth century, was characterized in Louis's landmark treatise on the anatomy, pathology, and therapeutics of the disease. *Researches Anatomical, Pathological and Therapeutic on the Disease Known under the Names Gastro-Enteric, Putrid Fever, Adynamic, Ataxic or Typhoid Fever* (1829) presented the classic pathological description of the disease. The book established typhoid fever as a specific disease identified at autopsy by characteristic lesions of the spleen and mesenteric glands, and Peyer's patches. Louis was the first to describe the characteristic lenticular rose spots. The question of whether typhoid fever was different from typhus was later answered by one of his many devoted American students, W. W. Gerhard.

LOIS N. MAGNER

Elie Metchnikoff
1845-1916
Russian Biologist, Bacteriologist and Pathologist

Noble laureate Elie Metchnikoff made significant contributions to biology and medicine. He won the Nobel Prize in medicine (with Paul Ehrlich) in 1908 for his theory of immunity. Metchnikoff was the first to discover that immunity stems from the action of white corpuscles in the blood that devour foreign bodies such as bacteria, and that inflammation in infected parts of the body is the tissue's defensive reaction to irritation and germs. Metchnikoff also demonstrated key similarities in the embryonic development of invertebrate and vertebrate animals and is considered one of the founders of comparative pathology and evolutionary embryology.

Metchnikoff was born on May 15, 1845, in the western area of the Russian Empire known as Ukraine. The son of a noble landowner, he demonstrated from an early age a passionate interest in science. He used his first microscope at age fifteen, and thereby began a lifetime study

Elie Metchnikoff. *(The Library of Congress. Reproduced by permission.)*

of microorganisms. By the time he was nineteen, he was a published author and a university graduate. Upon completion of Kharkov University in 1864, he went abroad to work with leading scientists in Germany and in Italy, where he helped develop a new field—comparative evolutionary embryology.

After receiving both a master's and a doctoral degree from St. Petersburg University, Metchnikoff taught zoology from 1870 to 1882 at the University of Novorossiia in Odessa. In 1886 Metchnikoff was appointed director of the first Russian bacteriological station, an institution established to study and prepare inoculations for infectious diseases such as rabies, tuberculosis, and cholera. Two years later Metchnikoff joined the Pasteur Institute, which was founded in Paris by the famous bacteriologist Louis Pasteur (1822-1895), and here he spent the rest of his life teaching and conducting research.

Metchnikoff devoted much of his time after 1881 to developing and then defending his groundbreaking phagocytic, or cellular, theory of immunity. Contrary to established theories, which linked immunity with chemical properties of blood serum, Metchnikoff demonstrated that it was phagocytosis—the absorption and digestion of microbes by amoeba-like specialized cells (phagocytes) in animals—along with inflammation that produce recovery and immunity from disease. Despite early opposition, Metchnikoff's theory is now a basic principle of immunology; moreover, his findings deepened understanding of how and why natural and artificial vaccinations work.

Metchnikoff believed deeply in public service and public health. He gave lectures and wrote popular articles on hygiene and medicine. He developed broad research interests, including a study of the aging process and senility, and contributed to the development of gerontology in Russia. Metchnikoff proposed that old age and death occur prematurely in humans due to the intoxication of the body by intestinal bacteria, and he recommended that people sterilize food, limit meat consumption, and use fermented milk products. His studies on aging led him to investigate arteriosclerosis and diseases caused by intestinal microbes, including typhoid fever, infantile cholera, and syphilis. In 1903 he and a fellow researcher at the Pasteur Institute, M. Roux, were the first to produce syphilis experimentally in monkeys, which enhanced the ability of researchers to diagnose the disease and test potential treatments and vaccines. The study of syphilis, which until this time had remained purely clinical, could now become an experimental science.

While abroad Metchnikoff stayed in close contact with Russian scientists, and many leading Russian microbiologists and epidemiologists came to the Pasteur Institute for training. He is credited with building the first Russian school of microbiologists, immunologists, and pathologists. In addition to the Nobel Prize, Metchnikoff received numerous awards, including an honorary doctorate from Cambridge University, and was elected to various international academies of science, scientific societies, and institutes.

On July 15, 1916, Metchnikoff died in Paris at the age of seventy-one.

ELAINE M. MACKINNON

Florence Nightingale
1820-1910
English Nurse

Driven by a message from God to nurse the sick, Florence Nightingale became known as the mother of the modern nursing profession. She single-handedly changed the way English hospitals functioned and the way nurses worked within them. Her reforms spread throughout the world.

Nightingale was born to aristocratic parents in May 1820, in Florence, Italy. Her parents chose to name her after the city of her birth. Once she was one year old, Nightingale's parents decided to return to their native England.

Typical of wealthy English families, the Nightingales had more than one estate. They summered at Lea Hurst, a relatively small house with 15 bedrooms, and lived the rest of the year at Embley Park, their larger home. They also spent each spring and fall visiting London. The Nightingales were extremely social, and Florence and her older sister Parthenope grew up within a flurry of balls, banquets, and social gatherings.

When Nightingale was 24, she claimed to receive a clear message that she was to nurse the sick in service to God. In 1844, however, nursing was not an honorable profession. Hospitals were scenes of incredible filth and associated with despicable people. Her parents forbid her to follow her calling and restricted her actions. She lived the next several years in misery. Late at night she would read articles she had secretly received. From these she became an expert on hospital conditions and reform. She took advantage of rare opportunities to nurse ailing relatives and neighbors. These were her only fleeting moments of happiness and fulfillment.

Relatives and friends finally helped Nightingale convince her family that she should be allowed to become supervisor of the Institution for the Care of Sick Gentlewomen in Distressed Circumstances. Over strenuous objections, the Nightingale's permitted her to take the position. At the age of 32, Nightingale was finally allowed to answer God's call.

Within a few months details of Nightingale's reputation had spread to the British government. After England and France declared war on Russia, beginning the Crimean War, Nightingale was asked to travel to Turkey to lead a group of nurses in the care of battle-injured soldiers. The conditions were horrible. Men screamed in agony and were left to suffer, disease raged in every corridor, and soldiers lived in their own filth. Although she encountered such severe conditions, within a few months she had largely sanitized the hospitals, found clean bedding and clothing for the men, and greatly increased the comfort of all involved. Those around her called her gift "Nightingale power." Soldiers called her "the lady with the lamp" because late at night she wandered the halls while carrying a lamp, making sure the men were quiet and able to rest.

Nightingale returned to England a hero, having revolutionized nursing. In the midst of this triumph, Nightingale's health collapsed. She proceeded to write letters and reports from her bed and was an invalid much of the rest of her life. It was later thought that she suffered from Chronic Fatigue Syndrome, though at the time it was considered to be exhaustion from the war, and perhaps overexposure to disease. In spite of her illness, she continued to work for more extensive sanitation and reform in hospitals.

For the last 14 years of her life Nightingale did not leave her bedroom. When she was 87 she received the Order of Merit from King Edward VII. It was the first time the award had ever been given to a woman. Three years later, at age 90, she died, completely blind and no longer lucid. She willed her body to science and requested that only a simple cross mark her grave.

CAROLYN CRANE LOVE

Sir William Osler
1849-1919

Canadian Physician and Professor of Medicine

Canadian William Osler influenced the establishment of medical education in Canada, the United States, and Great Britain. An outstanding clinician, he sought to lead physicians away from textbooks and to the bedsides of patients. He also foresaw the rise of medical specialists, a trend that did not take root until the end of the nineteenth century when improved travel allowed patients to more conveniently reach a specialist.

Born July 12, 1849, in Bond Head, Canada, Osler was the youngest of nine children of an Anglican missionary. He considered becoming a priest but decided upon medicine, receiving his doctor of medicine at McGill University, Montreal, Quebec, in 1872. He was recognized early in his career for his investigations of blood corpuscles and platelets, the cells responsible for blood clotting. He continued his education in London, Berlin, and Vienna, and returned to Canada as a general practitioner in Dondas and lecturer at McGill. His great interest was pathology, and he spent much time in the postmortem laboratory. However, he always emphasized that physicians should get to know the patients themselves. He constantly repeated that it is more important to know what sort of patient has the disease than to know what sort of disease the patient has. At that time Western physicians were moving toward depersonalization, placing greater focus on the afflictions—lesions and diseases—of the body rather then the person.

In 1884 Osler accepted a position at the University of Pennsylvania, Philadelphia. There he was given carte blanche to develop a curriculum in clinical medicine. In Philadelphia he helped organize the Association of American Physicians. In 1888 Osler was invited and accepted an invitation to become the first professor of medicine at the new Johns Hopkins University in Baltimore. There he was instrumental in developing the curriculum and clinical program that made that institution one of the most famous medical schools in the world. Students were encouraged to study their patients in the wards and laboratory, and then to present their findings to a head doctor and to other students. This pattern of help and consultation spread to other medical schools. During the first few years, while there were still no students, Osler wrote his famous text, *The Principles and Practice of Medicine* (1892), which encouraged medical students to go into clinical medicine.

In 1897 Frederick T. Gates (1853-1929), a newly appointed director of a philanthropic foundation, read the textbook and used the principles as a basis for the establishment of the Rockefeller Institute of Medical Research in New York.

While visiting Oxford in 1904, Osler was appointed Regius Professor of Medicine, an honor reserved for British citizens; Osler qualified through his Canadian citizenship. There Osler used his time to teach, lecture, and write prolifically. He was honored by receiving the baronet title, becoming Sir William Osler.

Osler was an energetic man with a dynamic and magnetic personality. His goal was to develop the human and caring side of medicine. He married Grace Goss, widow of a colleague in Philadelphia, and great-granddaughter of Paul Revere. When their one son, Revere, was killed in World War I, Osler became very depressed and lost his spirit. Osler died December 29, 1919, in Oxford, of pneumonia. His outstanding library as well as his personal books and papers were given to McGill University.

Medical terminology has immortalized Osler in the following terms: Osler's nodes, red and tender swellings on the hands related to heart problems; Osler-Vaquez disease, a build up of red blood cells or polycythemia; Osler's maneuver, a way of compressing the radial artery; Osler-Rendu-Weber disease, a hereditary disorder marked by nosebleeds.

EVELYN B. KELLY

Philippe Pinel
1745-1826
French Physician and Psychiatrist

Philippe Pinel revolutionized the treatment of the insane by insisting upon "moral management" instead of violence and incarceration. He was among the first psychiatric doctors who believed that mental illness was curable and that mental patients would respond positively to sympathetic care. Today, Pinel is credited with inventing the mental hospital, and with it, the idea that mental illness can be treated.

The son of a doctor of modest income and the eldest of seven children, Pinel was born in a small village in southwest France. He studied mathematics at the university at Toulouse before taking a medical degree at Montpelier. Pinel's early career as a physician was spent in a private family clinic, and in 1793 he was appointed physician at Bicêtre, a public mental asylum that housed 800 men. Despite the great impact he was to have upon the mentally ill and the field of psychiatry in general, Pinel was described as shy and retiring.

With his appointment at Bicêtre, Pinel earned the title of "the liberator of the insane." He freed the patients—really inmates—from their chains and dungeons, stopped clinicians from punitive measures such as whipping and bloodletting, and ordered that no further violence be used against the patients. Before Pinel, the insane had been considered a threat to society because the nature of their illnesses was not yet understood. Insanity was considered to be a biological disease, and as such, not treatable with the medical technologies and theories of the day. But Pinel argued that insanity was often a curable disorder. Some of Pinel's contemporary psychiatrists, including William Tuke in England and Chiarugi in Italy, were also exploring humane methods of treating the mentally ill.

The empathy that Pinel extended to sufferers of mental illness directly reflects the ideals of the French Revolution (1787-1799), which effected the near-total transformation of French politics and culture at this time. Revolutionary ideals included the concept that all men were created equal and should alike be allowed to strive for liberty. Pinel extended these concepts to the mentally ill, so that his liberation of the insane served as a powerful symbol of the "New France." Further, the zeal for social reform and sympathy for the disadvantaged spurred by revolutionary politics also inspired Pinel's decision to improve the living conditions of the mentally ill.

Pinel's method of curing mental illness followed a simple philosophy that at the time was a radical departure from his culture's approach to insanity. At Bicêtre, and later at the insane asylum for women at Salpêtrière, he demonstrated that deranged patients often responded sanely if they were treated as capable of "normal" behavior. His method of therapy worked towards the total rehabilitation of a patient, which was marked by his or her re-entrance to society. In the hospitals he oversaw, Pinel confirmed that a respectful approach to a patient and his or her problems earned the patient's trust and therefore enabled the doctor to learn more about the nature of the illness. He insisted that a doctor to the insane must live among them to be able to sufficiently observe and learn about their problems.

Pinel's achievements include the publication in 1798 of *Philosophical Classification of Diseases,* which described many psychotic symptoms, and the 1801 publication of *Treatise on Moral Management,* which contains the basic philosophy of his approach to insanity. During his career, besides serving at the hospitals at Bicêtre and Salpêtrière, Pinel was elected to a professorship of hygiene and pathology at the Ecole de Medicine and named as a consulting physician to Napoleon. His most famous student, Esquirol, continued the tradition of "moral management" and furthered the work that Pinel started by observing and classifying different mental illnesses.

TABITHA SPARKS

Wilhelm Conrad Röntgen
1845-1923
German Physicist

Wilhelm Röntgen wasn't allowed to graduate from high school because he got into trouble with a friend over a caricature of one of his teachers. But he won the first Nobel Prize for Physics in 1901 and had a very distinguished career in German universities. With a German father and Dutch mother, Röntgen grew up in Holland. Later, he went to a polytechnical school in Zurich, Switzerland, where he got a diploma in mechanical engineering. He married Anna Bertha Ludwig and was head of his department at the University of Würzburg when he noticed something important about the cathode rays with which he was working.

Sir William Crookes (1832-1919) had developed the Crookes tube in 1876 in which the pressure in a vacuum was reduced to the point that cathode rays shot straight across the tube and hit the wall opposite, causing it to glow with a greenish fluorescence. He noticed that wrapped and unexposed photographic plates left near his tubes became fogged, but he didn't understand why this happened. In 1895 Röntgen was experimenting with the Crookes tube in Würzburg, Germany, and observed the fluorescence of a barium platinocyanide screen that happened to be in its path. Putting things like books and cards between the tube and the wall seemed not to vary the effect. This suggested to Röntgen that it wasn't light or ultraviolet rays causing the fluorescence; it was found that x rays arise wherever cathode rays encounter solids and that the effect varies with the atomic weight of the target. The speed of the cathode particles also affects the penetrating effect of the x rays. X rays are a form of invisible, highly penetrating electromagnetic radiation with much shorter wavelengths or higher frequency than visible light. Their wavelength range is from less than a billionth of an inch to less than a trillionth of an inch. Scientists eventually realized that x rays are produced when high-energy electrons from a heated filament cathode strike the surface of a target.

Röntgen was greatly aided in his work by the newly emerging science of photography, which by the end of the nineteenth century was well advanced. Louis Daguerre and his colleagues had "fixed" an image from a camera obscura by adding mercury to the silver compounds being used, and the English inventor Fox Talbot (1800-1877) had also fixed his images on sensitized paper. These developments led to the production of "dry plates" in the 1870s and the emergence of the Kodak Company, which made cameras affordable to the general population. The first moving ("motion") picture was presented about the same time that Röntgen was making his observations about the effects of an unknown form of "radiation." Röntgen was an experimentalist: he experimented with different ideas and combinations of effects in order to understand what was happening, rather than forming a hypothesis from what was already known and then trying to confirm it in his laboratory. So he used photography to capture pictures of the effects of different experiments he undertook with the cathode rays of the Crooke's tube on different substances. This was possible because photography is the art of capturing different amounts of light as they impact on surfaces and substances, and Röntgen was interested in understanding this new form of creating fluorescent light on objects that were several feet away from the source of the energy. Röntgen explained his scientific process by saying "I didn't think, I investigated," but of course he was "playing around" with different techniques and technologies that had been developed in the later years of the nineteenth century. Using a form of Crooke's tube that had been adapted by his colleague at the University of Würzburg, Philipp Lenard (1862-1947; winner of the 1905 Nobel Prize for Physics for his work on electrons and atoms), Röntgen took a photograph of the skeleton of his wife's hand, with her wedding ring clearly visible but none of the flesh or veins. This photograph started the era of radioactivity.

SUE RABBITT ROFF

Ronald Ross
1857-1932
British Physician, Parasitologist and Epidemiologist

Ronald Ross was awarded the Nobel Prize for medicine in 1902 for elucidating the role of the *Anopheles* mosquito in the transmission of malaria. By explaining the complex life history of the malarial parasite, he made it possible to understand aspects of the problems of malarial fevers that had confounded physicians and scientists for hundreds of years. Millions of people throughout the world suffered from malaria during Ross's time and many areas were virtually uninhabitable because of malarial fevers. Moreover, his work made it possible for scientists to discover the role of insect vectors in the transmission of many other diseases.

Ross, the son of a military officer, was born in Nepal. In 1865 he was sent to England to attend school. Although Ross was primarily interested in poetry, art, and music, his father insisted that he study medicine. In 1874 he was enrolled in St. Bartholomew's Hospital, London. Although he passed the examination for membership in the Royal College of Surgeons, he failed the examination for the Society of Apothecaries. Rather than appeal to his father for further support, he took a position as a ship's surgeon for a year and a half. After this experience he passed the examination for licentiate of the Society and Apothecaries in 1881 and was able to join the Indian Medical Service. The position gave him ample time to devote to poetry and mathematics.

He was allowed a furlough in England in 1888-89 and used the time to earn a Diploma of Public Health. During this period he married Rosa Bloxam, with whom he had four children. When Ross returned to India he became interested in the malarial parasite, which had been discovered in 1880 by Alphonse Laveran (1845-1922). Ross's initial attempts to find the parasite in blood smears taken from malaria patients were unsuccessful, but during another furlough in London in 1894, Patrick Manson (1844-1922) taught him how to identify the *Plasmodium*. Manson also discussed his belief that mosquitoes transmitted malaria. Ross was convinced that he would be able to find evidence in India that would confirm Manson's hypothesis. After two years of fruitless work, in 1897 he found the malarial parasite in the stomach wall of a brown dapple-winged mosquito that is now known as *Anopheles*. With help from Manson, Ross was able to obtain a research leave from the Indian Medical Service in order to carry out further research in Calcutta. Using bird malarial as a model system, Ross was able to demonstrate the developmental steps that the malarial parasite undergoes within mosquitoes and birds.

In 1899 Ross retired from the Indian Medical Service, which had repeatedly obstructed his malaria research, and was appointed to a lectureship at the new Liverpool School of Tropical Medicine. Ross developed mathematical models to help explain the epidemiology of malaria. He served as a consulting advisor on malaria prevention throughout the world, traveling to Sierra Leone, Mauritius, Greece, and other areas where malaria was a major public health threat. While in Sierra Leone, he was able to demonstrate the life cycle of human malaria. Although the problem of malaria was complex, Ross optimistically believed that mosquito control was the key to prevention and that it could be accomplished simply and inexpensively. These ideas were most fully explained in his book *Prevention of Malaria* (1910), but his *Memoirs* (1923) provide a more detailed account of his career and his bitter priority battle with the Italian parasitologist Batttista Grassi (1854-1925), who claimed to have discovered the malaria transmission cycle before Ross.

In 1912 Ross settled in London and was able to devote more of his time to literature and poetry, the fields he had originally hoped to pursue. His novels and poems received little encouragement, but the establishment of the Ross Institute, which later became part of the London School of Hygiene and Tropical Medicine, memorialized his scientific work.

LOIS N. MAGNER

Ignaz Phillip Semmelweis
1818-1865
Hungarian Physician

Ignaz Semmelweis was the first to recognize that the hands of physicians carried the dreaded puerperal, or childbirth, fever. Although his observations were scoffed at and largely ignored during his lifetime, Semmelweis introduced aseptic techniques into medical practice that later proved fundamental to the profession.

Born July 1, 1818, in Buda (now Budapest), Austria-Hungary, Semmelweis attended the University of Pest and graduated a doctor of medicine from the University of Vienna in 1844. He also received a master's degree in midwifery and surgical training, as well as instruction in diagnostic and statistical methods. While at the First Obstetrical Clinic at the university teaching center at Vienna General Hospital, where he received his first appointment, Semmelweis's duties included the instruction of medical students, surgical procedures, and clinical examination.

Semmelweis was puzzled by the problem of puerperal fever, a condition related to childbirth where the mother, after several days, developed a high fever then died. While most women delivered babies at homes, those who came to the maternity hospitals faced mortality rates as high as 25 to 30 percent. The chief of the hospitals accepted these deaths as unavoidable, attributing it to overcrowding, onset of lactation, poor ventilation, or just bad air, called miasma.

Semmelweis observed that the death rate among women who delivered in a second clinic, designed for the training of midwifes, was two to three times lower than those in his clinic, where they were examined by medical students. In the medical school all patients who died were placed in a room for post-mortem examination. A friend, Jakob Kolletschka, died after accidentally puncturing his hand during a post-mortem examination. Autopsy revealed the same pathology as those who were dying with puerperal fever. Semmelweis made the connection that something on the hands of the doctors was carrying disease to healthy patients.

In May 1847, he devised a system where doctors and students would wash their hands in a solution of chlorinated lime after autopsy and before examination of patients. The staff protested, but Semmelweis was able to enforce the policy. In only one month puerperal fever had declined to less than two percent. He expanded the treatment to all instruments coming in contact with patients in labor.

Unfortunately, Semmelweis lost his position at the hospital after participating in a liberal revolution that swept Europe in 1848. He returned to Pest and private practice in 1850, married, and eventually had five children.

However, when an outbreak of puerperal fever was devastating the hospital in Pest, Semmelweis requested to be put in charge of the department. Implementing his techniques, he saw a reduction to less than one percent, while hospitals in Prague and Vienna were still experiencing up to 15 percent mortality.

In 1855 Semmelweis became a professor of obstetrics at the University of Pest and his techniques were accepted throughout Hungary. In 1861 he published a major paper, "Etiology, Understanding, and Preventing of Childbed Fever," that displayed his meticulous research. However, other countries, especially Germany, remained disdainful and critical of hand-washing and dismissed Semmelweis's writings and papers. Increasingly bitter and frustrated, in 1863 Semmelweis began to suffer from mental illness and was committed to an asylum. He died two weeks later. Ironically, his death was the result of a blood poisoning condition resembling puerperal fever, caused by a surgical accident.

Semmelweis's basic insights about the transmission of contagious diseases, underappreciated during his life, later influenced the work of Joseph Lister (1827-1912) and contributed to the germ theory of Louis Pasteur (1822-1895).

EVELYN B. KELLY

James Young Simpson
1811-1870
British Physician

James Young Simpson was one of the most prominent British physicians of his time. His contributions included many published papers and pamphlets, new surgical procedures, and obstetric forceps still in use today. Simpson campaigned for improvements in medical practice, education, and the design of hospitals. His practice was so well known that patients came from continental Europe to see him. Simpson also discovered and promoted chloroform's anesthetic properties and its use in obstetrics.

Simpson was born on June 7, 1811, in Bathgate, a village between the large Scottish cities of Glasgow and Edinburgh. He was the eighth child of David Simpson, a baker, and Mary Jarvis Simpson; the latter died when he was nine years old. Simpson attended a local school and showed enough promise by age 14 that his family scraped together enough money to send him 18 miles away to the University of Edinburgh. At the university Simpson studied the standard curriculum of rhetoric, mathematics, literature, Greek, and Latin.

Before the end of his first year, a friend persuaded Simpson to attend lectures on anatomy offered by Robert Knox (1791-1862). There Simpson found his calling and began medical studies by 1828. In two years Simpson had finished his courses and begun work on his doctoral thesis when a former pathology professor suggested he specialize in obstetrics. After more coursework, Simpson embarked on the standard tour of European medical centers such as London, Oxford, Paris, and Brussels.

After these travels, Simpson returned to Edinburgh, started an obstetric practice, and began teaching privately. His reputation in both areas grew so quickly that in 1835 he was elected president of the Royal Medical Society and his first speech in office attracted both local and international attention. Yet, despite these achievements and his busy practice and teaching duties, Simpson made little money and had no chance for advancement in his situation. So in 1839, when James Hamilton resigned from the chair of midwifery (obstetrics) at the university, Simpson began an intense campaign for the post despite his youth and relative inexperience. Since his bachelor status might be held against him, Simpson quickly married Jessie Grindlay, a distant relative he had known for several years and whose father had helped defray the cost of his university education. Simpson put his new bride to work cataloging more than 700 medical books and pieces of equipment he had purchased to impress the city officials who would choose Hamilton's successor. Despite serious opposition, Simpson's efforts worked. On February 14, 1840, he became "Professor of Medicine and Midwifery and of the Diseases of Women and Children." Although deep in debt from campaigning, Simpson had a prestigious position that would assure his future fame and fortune.

By the time of his death in London on May 6, 1870, Simpson had become a giant figure in the British medical establishment. According to contemporary accounts, more than 30,000 people lined the streets of Edinburgh as his funeral procession passed. Simpson certainly deserved such adulation. He was acclaimed by patients and students as an excellent clinician and teacher. Simpson designed the long obstetric forceps (that still bear his name) to aid in delivery. He introduced iron wire sutures into surgery, as well as acupressure, a method to stop hemorrhage, or bleeding. Simpson was also known as a harsh critic of hospital conditions of his day, and for his professional publications on fetal pathology and medical history. In recognition of his achievements, Simpson was appointed one of the Queen's physicians for Scotland in 1847; and in 1866 Queen Victoria made him a baronet. Yet Simpson's greatest medical achievement was the discovery and promotion of the anesthetic properties of chloroform.

Inhalation of gases as a therapy for various diseases had been tried since the 1780s, but not until the early 1840s was the concept successfully used for the relief of surgical pain. In 1842 physician Crawford Long (1815-1878) used ether on a few patients in Georgia, and four years later dentist William Morton (1819-1868) administered ether for surgical patients in Boston. Morton's demonstrations, held at the prestigious Massachusetts General Hospital, introduced inhalation anesthesia to the world; within months, anesthesia with ether was being used in countries all over Europe.

On January 19, 1847—just three months after Morton's efforts in Boston and about a month after ether was first used for surgery in Great Britain—Simpson used ether in an obstetric case. Long, in Georgia, began a similar use at about the same time. By the fall of 1847, Simpson began searching for a better agent. Ether had a disagreeable smell and tended to irritate the lungs of patients. Simpson and several friends experimented with various gases before hitting upon chloroform in early November. Simpson quickly began to support use of this gas instead of ether in obstetrics. His reputation and promotional talents eventually helped overcome opposition to anesthesia generally and to its use in obstetrics.

Chloroform and ether have not been used as human anesthetics since the 1950s; in the past few decades synthetic gases with fewer side effects have replaced the older agents. Yet Simpson's work a century and a half ago legitimized the use of medical interventions to relieve the pain of labor. Millions of women around the world whose labor pains have been eased by various types of anesthesia have benefited from Simpson's groundbreaking efforts.

A. J. WRIGHT

John Snow
1813-1858
English Physician, Epidemiologist and Anesthetist

John Snow correctly concluded that cholera was transmitted through the ingestion of contaminated water. Snow came to this conclusion by conducting several epidemiological studies in London, England. He also is known as the father of the science of anesthesiology.

Snow rose to prominence from rather humble beginnings as a farmer's son. He was born in York, England, in 1813, attended local schools, and served as an apprentice to a surgeon. This informal training led him to work as a surgeon's assistant until 1836 when he moved to London in order to pursue the formal study of medicine at the Windmill Street School of Medicine and the Westminster Hospital. Snow was a member of the Westminster Medical Society and gave speeches on typhus, alcoholism, and respiration at the club's meetings. He also was interested in the study of toxicology, or the science of poisons and their effects on living organisms. Throughout his life, Snow advocated vegetarianism and temperance (abstinence from alcohol), two movements that had adherents in both the United States and England in the first half of the nineteenth century.

Snow's fame rests on his study of cholera, a communicable and highly fatal disease that wreaks havoc on the gastrointestinal tract. Its symptoms include dehydration, paleness, and diarrhea. The disease is transmitted by ingesting food or water contaminated with the cholera bacillus. Although Snow was unaware of the existence of this microorganism (known by scientists as *Vibrio cholerae* and only discovered in 1884), he surmised that men and women contracted the disease by drinking infected water. His colleagues incorrectly believed that the disease was a fever brought on by the rotting of vegetation. Snow published his findings in 1849 in a work entitled *On the Mode of Communication of Cholera*. He examined the incidence, distribution, and communication of cholera in a number of London neighborhoods during the cholera epidemics that took place in England in the mid-nineteenth century. The main victims of the disease were the working and lower classes. Snow showed that infected water supplies spread the disease to the neighborhoods that used the water.

Snow's reputation as a scientist not only depended on his study of cholera but also on his investigation into the effects of chloroform and ether. He developed a fine reputation as an anesthetist as a result of his scientific experiments and his medical practice. Because of his knowledge of the subject and his experience, he was chosen to anesthetize Queen Victoria during the births of three of her children, Princes Arthur and Leopold and Princess Beatrice. The application of chloroform during childbirth became popular after Queen Victoria agreed to undergo the procedure. Consequently, the pains of labor and delivery were diminished for many women.

KAROL K. WEAVER

Andrew Taylor Still
1828-1917
American Physician

Andrew Still, a trained medical doctor, created a new branch of alternative medicine, osteopathy, which grew to be one of the more accepted alternatives to conventional medicine.

Still was born in Virginia on August 6, 1828. His father was also a medical doctor. Still received his formal medical training and degree at the College of Physicians and Surgeons in Kansas City, Missouri.

After he was married and had several children, a tragedy occurred that altered the course of his life. His family contracted viral meningitis, and three of his children died of the disease. Still felt helpless as he, a physician, watched them suffer and finally die. He began to envision a method of practicing medicine that would be more instrumental in saving lives and maintaining health.

As he began his search, he combined his medical knowledge with his religious and spiritual beliefs, which told him that nature held the ultimate power and knowledge of the universe.

His spiritual beliefs are aptly capsulated by his words: "Nature knows perfectly your powers, plans, and purposes." Because he believed this, he thought that if allowed and in balance, the body would function perfectly. He believed that the structure of the body and the body's function were keenly related. Thus, he named this type of alternative medicine osteopathy. "Osteo" is the Greek prefix for bone; Still related it to the musculoskeletal system.

Still founded the American Academy of Osteopathy in his hometown of Kirkland, Missouri, in 1874. As he conveyed his vision of medicine to others, he attracted followers. Among them were his surviving sons. Once he became famous, medical professionals attempted to lure him to large cities to open academies and give lectures, but he preferred his quiet life in the Midwest. He was known there for his genius in osteopathy and his eccentric habits of dress and personality.

In 1896 Vermont became the first state to officially consider osteopathy a legally distinct school of medicine. Within 100 years, osteopathic physicians would practice in every state and throughout the world and would be considered, by license and practice, equal to medical doctors.

CAROLYN CRANE LOVE

Louis-René Villermé
1782-1863
French Physician and Social Reformer

Louis-René Villermé was a physician who devoted his life to working for the relief of the suffering among impoverished workers. Early in his career, he decided to abandon private practice and devote himself to social and scientific research to benefit the working class. He was especially appalled by the horrors of child labor and worked to establish laws to prohibit the exploitation of children in factories and mines.

After attending a day school in Lardy, Villermé moved on to a college in Paris and decided to study medicine at the new School of Health. In 1803, during the war between France and Great Britain, many medical students were called to serve as military doctors. Villermé was appointed as surgeon and soon rose to the rank of surgeon-lieutenant. He served with distinction on battlefields in Germany, Poland, and Spain.

At the end of the war, Villermé returned to Paris to work on his doctoral thesis. While practicing medicine in Paris, he learned that a colleague had claimed credit for an operation that Villermé had invented. Although the Academy of Sciences later validated Villermé's priority claim, he was so distressed by the experience that he decided to abandon private practice and devote himself to social and scientific research.

Villermé's widely read book *The Prisons, What They Are and What They Have to Be* described the horrors of war and the dreadful treatment of prisoners. He also became concerned with the suffering of workers. His research involved the analysis of working conditions, mortality rates, and the relationship between diseases and particular occupations. These studies represented a new stage in the exploration of the relationship between human health and industrial conditions. Descriptive studies gave way to statistical tables of mortality rates in hospitals for various categories, such as sex and occupation. In general the results demonstrated that mortality rates decreased when the wages of workers increased.

Villermé's papers analyzed the relationships among mortality, health, and social classes, birth distributions in urban and rural areas, and the influence of income, occupation, and living conditions on health. His research examined the mortality rates of prisoners, the height and weight of French men, the health of agricultural workers, vaccination, moral hygiene, almshouses, hospital architecture, and the distribution of poverty and wealth in civilized nations. In 1832 he devoted his attention to the battle against the cholera epidemics that were terrifying Europe.

When the Royal Academy of Moral and Political Sciences decided to provide a substantial grant for a study of the condition of the working classes, Villermé and Benoiston de Chateauneuf were asked to carry out the inquiry. Villermé studied industries and workers, interviewed judges, doctors, manufacturers, workers, and went into factories, workshops, and the homes of workers. He noted working conditions, temperature, lighting, vibrations, dust, dangerous postures, and the monotony of repetitious movements. He demonstrated differences between the conditions endured by industrial and service workers. His descriptions of the textile factories, which were then a major French industry, were especially detailed and precise. The harsh working conditions for women in many factories and the exploitation of very young children particularly disturbed him.

For each district, Villermé prepared detailed tables which summarized the wages for different kinds of work and the prices of necessities. These reports appeared as *Statistical Studies of the Physical and Moral State of Workers in the Production of Cotton, Wool, and Silk*. Villermé urged the government to impose "a law of humanity" to restrict child labor in factories and mines. The first French law to regulate child labor was quite inadequate; it permitted the employment of children as young as at eight years of age. In 1850 the legal age was raised to ten; in 1874 it was raised to twelve.

A paper entitled "Accidents produced in industrial work-shops by mechanical engines" demonstrates Villermé's skill as a safety engineer and mechanic. He wrote about the construction of "clean machines" that would not produce dusts, inquiries concerning accidents and injuries, protective fittings, grates, railings, cages, straps, safe ways of moving and using machines, the construction of well-lit work places, and the need for safety inspectors. Villermé's research was an important forerunner of modern epidemiology and occupational medicine.

LOIS N. MAGNER

Rudolf Carl Virchow
1821-1902
German Physician, Pathologist and Anthropologist

Rudolf Virchow was one of the most prominent German physicians of the nineteenth century, and his success reflected the rising influence and organization of the German medical community after 1840. Working as an activist for public health, Virchow merged the social and political reform movements with the developing German medical community, and he believed that medicine was the ultimate science of man.

In 1839, when Virchow began studying medicine at Berlin's Friedrich-Wilhelms Institut, German medicine was mostly theoretical and did not deal extensively with the clinical problems or experimental techniques. Virchow, however, was trained by Johannes Müller and Johann L. Schönlein, two of the first German teachers to promote experimental laboratory methods, physical diagnostic methods, and epidemiological analyses. After he completed his medical degree at the University of Berlin in 1843, Virchow repaid his military commitment by serving as the surgeon at the Charité Hospital in Berlin. His doctoral dissertation was on the corneal manifestations of rheumatic disease, and during his work at the Charité Hospital he performed microscop-

Rudolf Virchow. *(The Library of Congress. Reproduced by permission.)*

ic studies on vascular inflammation and the problems of thrombosis and embolism.

Beginning around 1845 Virchow articulated a new vision for the German medical community that he believed would shift it away from being a largely theoretical activity. In speeches at the Friedrich-Wilhelms Institut he asserted that doctors could advance their science by making clinical observations, performing animal experimentation, and studying microscopic pathological anatomy.

His work during the typhus epidemic in 1848, when he saw economically and socially disadvantaged Poles suffer disproportionately from the disease, encouraged Virchow to integrate his liberal social and political views with his work as a physician. He called for the enactment of new political, educational, and economic reforms, which he justified on the basis of their potential to improve public health, and he asserted that every individual had a constitutional right to good health. The defeat of these proposals, combined with his attacks on his older colleagues' promotion of humoralism (the idea that the body's health depended on various "humors" within), led to Virchow's ouster from the Charité Hospital in 1849.

Virchow worked at the University of Würzburg as a pathological anatomist through-

out the first half of the 1850s. These years marked the height of his scientific work and teaching activities, and his students included Edwin Klebs (1833-1913), Ernst Haeckel (1834-1919), and Adolf Kussmaul (1822-1902). He continued this work when, in 1856, he accepted an invitation to become the director of the newly established Pathological Institute. As director, Virchow resumed his work linking the advancement of the German medical community with political reformism. He highlighted the value of clinical observation and experimentation, and he condemned the "speculative" methods of earlier medical doctors. Newly improved microscopes and biochemical techniques allowed Virchow to reduce processes to the cellular level and effectively modernize medicine by attacking the existing humoral and neural physiopatholgical interpretations of disease.

The last decades of Virchow's life were largely devoted to his promotion of anthropology. He encouraged work in physical anthropology by studying the physical characteristics of Germans and by performing a racial survey of German schoolchildren. His work led him to argue that there was not one German race, but rather a mixture of different types organized under one social and political system. Throughout the 1870s and 1880s he participated in archeological projects in Pomerania, Hissarlik, and Egypt. His interests in anthropology and his substantial political and professional influence led to his role as cofounder of the German Anthropological Society in 1869. He was also instrumental in the building of the Berlin Ethnological Museum and the Museum of German Folklore.

MARK A. LARGENT

Biographical Mentions

Thomas Addison
1793-1860

British physician for whom Addison's disease, a metabolic dysfunction of the adrenal glands, and Addison's (or pernicious) anemia are named. In 1855 Addison was the first to identify a set of symptoms that correlated to a disease of an endocrine gland.

Elizabeth Garrett Anderson
1836-1917

English physician whose efforts to obtain professional medical education and a license to practice paved the way for British women to receive accredited training and to work as doctors. When refused admission to medical schools in England, Anderson studied privately and became a doctor by passing the Apothecaries exam in 1865. In 1866 she set up a clinic for women in London which became a small hospital—the first in England to have women doctors. She also founded the London Medical School for women in 1883, and was elected mayor of Aldeburgh in 1908, becoming the first woman mayor in England.

Gabriel Andral
1797-1876

French physician who pioneered the field of blood pathology and wrote the first textbook on internal medicine. Andral studied the blood of both animals and humans, and was the first to describe the proportions of the constituents of normal blood. He noted that disease affects the blood and that bodily functions were interdependent. This insight fueled his vigorous opposition to bloodletting, a practice performed by most physicians during his day. Andral earned his medical degree from the University of Paris in 1821 and was appointed professor of medicine in 1828, eventually becoming head of the pathology department.

Alexander Thomas Augusta
1825-1890

American physician who was the first black surgeon in the U.S. Army. Born a freedman in Norfolk, Virginia, Augusta studied under private tutors and, in 1856, earned a medical degree from Trinity Medical College in Toronto. During the American Civil War, Augusta was appointed surgeon of colored volunteers with the rank of major. He later became the first black person to head a major hospital and formed the nucleus of the new medical school at Howard University. Augusta remained on the faculty until 1877, when he returned to private practice in Washington, DC. He was buried in Arlington National Cemetery.

Alexander Bain
1818-1903

Scottish philosopher and psychologist who first used the term "trial and error." A professor of logic at the University of Aberdeen, Scotland,

from 1860 until 1880, Bain advanced the study of psychology as a science by asserting the link between mental association and physiological processes. Bain authored the first psychology textbooks written in English, *The Senses and the Intellect* (1859) and *The Emotions and the Will* (1859), and founded the first psychological journal, *Mind* (1876).

Samuel Siegfried Carl von Basch
1837-1905

Czech-Austrian physician who developed the first device for measuring blood pressure without cutting a blood vessel. Working in Leipzig and Berlin with other gifted physiologists, Basch became interested in the workings of the heart and blood. In 1876 he designed a small bulb with a balloon-like diaphragm stretched across the bottom that connected to a manometer, an instrument for measuring the pressure of gases and liquids. The diaphragm pressed on the artery until the pulse stopped and at this point indicated pressure. Though not accurate compared to modern instruments, the idea of the sphygmomanometer for measuring blood pressure without an invasive procedure was a medical breakthrough.

Marie François Xavier Bichat
1771-1802

French physician and pathologist who is regarded as the founder of tissue theory and animal histology. Bichat performed over 600 autopsies in an attempt to understand the fine structure of the human body. Bichat argued that the organs of the body should be studied in terms of their fundamental structural and vital elements, which he called tissues. The basic actions of tissues were explained in terms of their ability to react to stimuli (irritability) and the ability to perceive stimuli (sensibility).

John Shaw Billings
1838-1913

American surgeon and librarian who is credited with developing the first system for organizing medical literature in the United States. Between 1864 and 1895 Billings developed the Army Medical Library, which eventually became the National Library of Medicine, the world's largest medical reference library. In 1879 he began the *Index Medicus*, a monthly guide to medical literature, which is still in use in the United States today. Billings spent the last 17 years of his life as the first director of the New York Public Library.

Christian Albert Theodor Billroth
1829-1894

Austrian surgeon regarded as the founder of modern abdominal surgery. Billroth received his degree from Berlin in 1852 and held teaching positions at Zurich and Vienna, where he made major contributions to surgical practice. Modern surgery was just beginning, and Billroth was especially interested in wound fever, which he believed was caused by organisms. He insisted on regular temperature taking and was one of the first to introduce antisepsis in continental Europe. In 1881 he made surgical history by removing part of a cancerous stomach.

Henry Ingersoll Bowditch
1808-1892

American physician whose fame is primarily because of a single article, "On Pleuristic Effusions, and the Necessity of Paracentesis for their Removal," published in 1852. In this article Bowditch promoted the use of a suction pump in chest surgery. He also wrote an influential book about the stethoscope; actively promoted hygiene, sanitation, public health, and preventive medicine; and contributed to the understanding of pulmonary tuberculosis, then known as "consumption" or "phthisis."

Sir William Bowman
1816-1892

English surgeon and histologist known for his studies of the structure and function of the kidney, eye, and striated muscles. Bowman was appointed to King's College Hospital, London. The development of the microscope enabled him to investigate tissue structure. He found that urine was a by-product of blood filtration in the kidney and that a capsule surrounding the kidney cells, or nephrons, is of prime importance in the process. This structure is now called Bowman's capsule. Later, Bowman became an outstanding expert in eye surgery and diseases. He received the honor of baronet in 1884.

Hermann Brehmer
1826-1889

German physician who established sanitariums to cure tuberculosis. Contracting tuberculosis as a young botany student, Brehmer was instructed by his physician to seek a healthier climate. He journeyed to the Himalayas to study plants and to rid himself of the disease. When he returned cured, he decided to study medicine and wrote a doctoral dissertation, "Tuberculosis is a Curable Disease," that outlined plans for a treatment facility featuring proper nutrition, fresh air, exercise, rest, and good care. In 1859 he established a sanitarium among the fir trees of Gorbersdorf, Silesia, in central Europe. The success of his facility inspired the formation of similar sanitariums in hundreds of mountain and seaside resorts worldwide.

Pierre-Fidele Bretonneau
1778-1862

French physician who performed the first successful tracheotomy, a surgical procedure involving the creation of an incision in the windpipe to make an artificial breathing tube. He also recognized and named the disease diphtheria. Bretonneau received a medical degree in Paris in 1815 and became head of the hospital at Tours in 1816. He determined that specific organisms caused specific diseases, foreshadowing the germ theory of Louis Pasteur (1822-1895), and recognized that typhoid fever and typhus were separate diseases. Bretonneau was also convinced that diphtheria was contagious and tried in vain to infect animals, a feat later accomplished by other scientists.

Richard Bright
1789-1858

English physician and anatomist who described the form of chronic nephritis now known as Bright's disease. Bright made major contributions to the study of morbid anatomy of the living patient. His studies of the retention of urea in the body fluids of patients with kidney failure helped to demonstrate the value of biochemical studies of disease. His studies of various brain and abdominal disorders were also of great value in clinical medicine.

Pierre Paul Broca
1824-1880

French physician and neurologist who is regarded as the founder of physical anthropology. While serving as a surgeon in Paris, he was primarily responsible for the establishment of the Institute of Anthropology. An autopsy on a patient who had been unable to speak for many years led Broca to suggest that a specific area of the brain was the "motor center" for speech. This area was later designated as Broca's convolution, or Broca's center, and the speech disorder was called "aphasia."

David Bruce
1855-1931

British physician and pathologist who is best known for his landmark studies of parasitology, especially sleeping-sickness. In 1887, Bruce reported that a bacterial agent causes Malta fever. His studies of nagana, a disease of domestic animals in South Africa, showed that the disease was transmitted by the bite of an infected tsetse fly. This insight led to his discovery that sleeping sickness (a disease of humans that is caused by a microbial agent known as a trypanosome) was also transmitted by the tsetse fly.

Ernst Wilhelm von Brücke
1819-1892

German physiologist who helped introduce new experimental methods into medical research. He was an early advocate of animal experimentation and the integration of chemistry and physics into biological research. Brücke's research included studies of skeletal muscle structure, vision, and the mechanism of speech. He was also very interested in art and wrote of the relationship of the physiology of vision to painting.

Hans Buchner
1850-1902

German bacteriologist and immunologist who first discovered substances in the blood that killed bacteria, now known as gamma globulins. Buchner devised a method for culturing anaerobic bacteria, a type of bacteria that grow in the absence of air. Born in Munich into a family that encouraged science, he was the brother of Nobel laureate chemist Eduard Buchner. Hans Buchner attended the Universities of Munich and Leipzig, receiving his medical degree in 1874. He worked as an army doctor during his early life, and later became surgeon general of the German army. He was also a professor at the University of Munich, where he taught until his death.

Joseph-Bienaime Caventou
1795-1877

French pharmacist known for his research on phyto (plant) chemicals and for discoveries in alkaloid chemistry. Caventou advanced clinical medicine from crude plant extracts to the use of natural and synthetic compounds. The son of a pharmacist, he completed his internship in hospital pharmacy and received an appointment to Saint-Antoine Hospital. From 1817-20 he and fellow scientist Pierre-Joseph Pelletier (1788-1842) discovered many natural products, including chlorophyll, the green pigment in plants. They also recognized the nature of morphine, strychnine, and caffeine. Their most dramatic discovery was quinine, a drug derived from the cinchona tree that found use as a treatment for malaria. By age 26 Caventou was established as a gifted investigator and, in 1830, became professor of chemistry, a post he held until his retirement in 1859.

Karl Crede
1819-1892

German physician known for his work in gynecology. He developed a method of expelling the placenta by pressing the thumb downward in pressure on the uterus through the abdominal wall. This maneuver, applied in the direction of the birth canal, is called Crede's method. Also, he developed a method for removing urine through a flaccid bladder, one without muscle tone. However, his best known contribution is the application of 1% silver nitrate into the eyes of newborns to prevent blindness in the baby if the mother had gonorrhea.

Rebecca Lee Crumpler
1833-?

American physician who was the first African American woman to earn a medical degree. She was born in Richmond, Virginia, and raised in Pennsylvania by an aunt who cared for the sick in their community. Inspired by her aunt, Crumpler became a nurse in Massachusetts in 1852. She entered New England Female Medical College in Boston and received her Doctress of Medicine degree on March 1, 1864. Dr. Crumpler practiced medicine in Boston until the end of the Civil War, then moved back to Richmond where she worked with newly freed slaves. Crumpler later returned to Boston and in 1833 wrote a book on the medical care of women and children.

Dorothea Lynde Dix
1802-1877

American who led the crusade to build state hospitals for the mentally ill. In 1841 Dix visited a correctional facility in Massachusetts and was stunned by the treatment of the mentally ill. She lobbied the United States legislature to improve prison conditions and insane asylums and requested funds for an institution specially designed to treat the mentally ill in Massachusetts. She did the same in state after state, traveling thousands of miles alone throughout the U.S. and Europe until she was 80. Dix also served as superintendent of the U.S. Army Nurses during the Civil War.

Daniel Drake
1785-1852

American physician and medical geographer who founded the Ohio Medical College. In the early 1800s Drake traveled through the interior of the United States, gathering data on the customs, diet, and diseases of people living on the frontier. His research culminated in the most important work on malaria published to that time, *A Systematic Treatise . . . on the Principal Diseases of the Interior Valley of North America*. In 1819 Drake founded the Ohio Medical College (now the University of Cincinnati College of Medicine) based on his belief in the importance of providing medical students with a hospital-based education.

Guillaume Benjamin Amand Duchenne
1806-1875

French physician and neurologist noted for his research on muscular diseases, and for pioneering the use of electricity in the diagnosis and treatment of disease. Duchenne was credited with discovering several neuromuscular disorders, including Duchenne Muscular Dystrophy (DMD), named after him; DMD is a degenerative disorder of the spinal cord which can lead to paralysis in its victims. Duchenne also pioneered the use of electronic stimuli in the treatment of muscle dysfunction.

Jean Henri Dunant
1828-1910

Swiss banker and philanthropist whose plea for better care for the war wounded led to the foundation of the International Red Cross. After witnessing Napoleon's bloody battle at Solferino, Italy, Dunant published *Un Souvenir de Solferino* (*A Memory of Solferino*) in 1862, asking for the creation of an international organization to care for those wounded in war. The following year, two international conferences were held and the International Red Cross was established. Dunant gave the Red Cross its insignia (a red cross on a white background) and helped to fund the organization. In 1901 he was awarded the first Nobel Peace Prize, which he shared with Frédéric Passy.

Mary Anna (alt. Elson, Mariana) Elson
1833-1884

German-born American physician who was among the first female graduates of the Women's College of Philadelphia, and the first Jewish woman to complete her studies at that school. In 1859, a time when few women were entering the medical profession, Elson earned a medical degree and went on to establish a successful career as a physician in Philadelphia, and later in Indiana.

Theodor Escherich
1857-1911

German bacteriologist who first isolated the *E. coli* bacteria. Escherich first discovered the bacteria *Bacterium coli* (later renamed *Escherichia coli* in his honor) in the intestinal tract of humans and animals. While the benign form of the bacterium is used in the digestive process, harmful strains of *E. coli* can contaminate beef, fruit juice, and other foods, causing severe food poisoning symptoms. Escherich was also a pioneering pediatrician who worked to improve child care, especially infant hygiene and nutrition.

Johann Friedrich August von Esmarch
1823-1908

German surgeon and inventor of Esmarch's apparatus, a rubber bandage that was used as a tourniquet to perform bloodless surgery. Von Esmarch's experience as an army surgeon inspired his development of his surgical techniques, which included his design of a triangular cloth bandage, carried by soldiers, and his promotion of the aspectic handling of wounds. His textbook, *Surgical Techniques* (1901), became a classic in the field of general surgery.

Niels Ryberg Finsen
1860-1904

Danish physician best known for developing the Finsen treatment, which used artificial light to treat skin conditions. The Finsen lamp (developed in 1893) directed a carbon arc emitting ultraviolet rays on the skin, primarily treating lupus. The light penetrates a small area of skin through quartz telescopes, while a flow of cold water keeps the skin cool. Finsen founded the Finsen Institute in Copenhagen, for the study of phototherapy and skin conditions.

Reginald Huber Fitz
1843-1913

American pathologist who studied in Berlin under Rudolf Virchow and was instrumental in advancing the understanding of the cell as a component of disease. Fitz's systematic use of clinical and bedside observation led him to his greatest contributions to medicine, the characterization of acute appendicitis and acute pancreatitis. A Professor at Harvard Medical School, Fitz was known for his extensive clinical studies and for his microscopic examination of tissues.

Franz Joseph Gall
1758-1828

German physiologist and anatomist who was the originator of phrenology, the attempt to determine intelligence and personality from the shape of the skull. Gall was convinced that certain mental functions were centered in specific parts of the brain, and that the shape of a person's skull showed the degree of development of the various parts of his or her brain. His ideas were condemned by the Austrian government as antireligious and he was forced out of the country in 1805. Though phrenology has been discredited, Gall's ideas about localized brain functions were later proven.

William Crawford Gorgas
1854-1920

American physician who made major contributions to public health reform, urban and military sanitation, and the control of yellow fever. As chief of sanitation in Havana, Cuba, Gorgas worked with Dr. Carlos Juan Finlay, Dr. Walter Reed, and others to prove that yellow fever was transmitted by the bite of infected mosquitoes. Utilizing this insight to control yellow fever in Panama, Gorgas was largely responsible for overcoming one of the most serious obstacles to the construction of the Panama Canal.

Albrecht von Graefe
1828-1870

German surgeon and ophthalmologist who founded the university surgical clinic in Berlin, Germany. A competitor of the state Charite hospital, Graefe's free clinic was open to any patient who sought treatment for diseases of the eye. A notable surgeon, Graefe developed an operation to treat glaucoma. He was interested in ways to more unintrusively inspect the eye of the living patient, and he pioneered the clinical use of the opthalmosocope.

Henry Gray
1827-1861

British anatomist who wrote an influential book about anatomy that continues to be used today. Gray was a teacher of anatomy and curator of the medical museum at St. George's Hospital in London. Gray, with the aid of colleague and illustrator H. Vandyke Carter, authored and published *Gray's Anatomy* in 1858. The work

was compiled from Gray's own dissection research and remains. Organized by terms rather than sections of the body, Gray's innovative work has, with several revisions, remained the definitive text on the subject of anatomy.

William Gull
1816-1890

British physician and lecturer in natural philosophy who rose to national prominence when he was called upon to treat the Prince of Wales for typhus. Before becoming the Physician-in-Ordinary to Queen Victoria in 1872, Gull served as a the Fullerian Professor of Physiology at Guy's Hospital in London.

Jakob von Heine
1800-1879

German scientist and physician who was the first to clearly describe poliomyelitis (commonly known as polio), an infectious viral disease of the central nervous system. In a book published in Germany in 1840, Heine correctly asserted that polio was a contagious disease (contracted by mouth) and articulated a treatment regimen used well into the twentieth century.

Hermann von Helmholtz
1821-1894

German theoretical physicist and experimental physiologist whose studies of animal heat led to the formulation of the fundamental principle of the conservation of energy. He was a remarkably versatile scientist, as demonstrated by his researches on physiological optics, acoustics, color blindness, electricity, and magnetism. His research on physiological optics led to his invention of the ophthalmoscope, an instrument used to examine the inside of the eye.

Friedrich Gustav Jakob Henle
1809-1885

German anatomist and histologist who authored one of the outstanding anatomical systems of the nineteenth century. He was the first to describe the epithelia of the intestines and the skin. Many anatomical structures are named for Henle, including parts of the kidney, uterus, and the root sheath of hair. His 1840 essay on the possible relationship between microbes and disease is regarded as an early version of the germ theory of disease.

Thomas Hodgkin
1798-1866

English physician and scholar who first described the lymphatic disease that bears his name. In January 1832 Hodgkin cited several cases of a disease that affected the lymph system in his paper "On the Morbid Appearances of the Absorbent Glands and Spleen." The disease, a malignant growth of cells in the lymph system, was thereafter referred to as Hodgkin's disease. After his paper was published, it was determined that some of the cases Hodgkin described were not Hodgkin's disease but instead other lymphatic disorders with similar characteristics. These disorders are now called non-Hodgkin's lymphoma.

Abraham Jacobi
1830-1919

German-American physician who was a founder of pediatrics as a field of medical specialization in the United States. In 1860 he was appointed to the chair of diseases of children at New York Medical College. Jacobi founded the first pediatric clinic in New York in 1862 and wrote extensively on the diseases of infancy and childhood. He was one of the first American physicians to use the diphtheria antitoxin.

Sophia Louisa Jex-Blake
1840-1912

English physician who was responsible for 1876 legislation permitting women in Britain to accept medical degrees and to receive licenses to practice medicine. Jex-Blake studied medicine at Edinburgh University in Scotland and passed her exams, but the university did not allow women to accept medical degrees; she eventually qualified as a doctor in 1877. Jex-Blake also helped establish a medical school for women in London in 1874, and another in Edinburgh in 1886.

Emeline Jones
?-?

The first female dentist in the United States. Her interest in dentistry began when she married a dentist at age 18, though, as her enthusiasm was discouraged, she studied in secret. Once her skill was obvious, however, her husband made her his business partner; she continued to practice alone after his death. Jones worked as a dentist for over 50 years, often traveling to see patients. In 1893 she became the first woman elected to the Connecticut State Dental Society.

Robert Knox
1859-?

Surgeon and anatomist who became prominent in the United States for his work with human anatomy. He received his degree in medicine from the University of Virginia in 1882, and did postgraduate work in medicine in New York. In 1902 he became the chief surgeon of the Southern Pacific Railroad. He presided over the Texas State Medical Association as its president.

Carl Koller
1857-1944

Austrian physician and ophthalmic surgeon who refined the use of cocaine as a local anesthetic, after Sigmund Freud suggested that possible use for the substance. Koller used cocaine in a drop form, applying drops to the eye prior to surgery. He received his medical degree from the University of Austria in 1882. Later, he served as an ophthalmic surgeon at Mt. Sinai hospital.

Rudolph Albert von Kölliker
1817-1905

Swiss physiologist, anatomist, biologist, and zoologist who made landmark achievements through his use of the microscope. Kölliker is famous for his knowledge of histology, a branch of anatomy involving study of the minute structure of plant and animal tissues. His memoir on cephalopods (marine mollusks) became a classic in that field. He investigated and made clear what tissues composed arteries, muscles, skin, bone, and teeth. The fields of embryology and microbiology were furthered due to his work. He received England's highest honor, the Copley Medal, in 1897.

Ernst Krakowizer
1822-1875

Austrian physician and author who emigrated to the United States in 1848, after his involvement in an insurrection in Padua, Italy. He practiced medicine in Brooklyn until retiring in New York. He was involved in restructuring Bellevue Medical College in 1874, and in creating the German dispensary. He practiced medicine at Mount Sinai and other hospitals, and published articles in several medical periodicals.

Adolf Kussmaul
1822-1902

German linguist and physician whose interest in the causes of human communication disorders motivated him to author *Die Storungen der Sprache* ("The Impediments of Speech") in 1877. Kussmaul's work offered some of the first clinical descriptions concerning the physiology of language and served as a foundation for the development of modern logopedics (studies of language pathology).

Emanuel Libman
1872-1946

German physician who was one of the foremost clinicians and pathologists of his time. With his research of blood cultures and subacute bacterial endocarditas (later known in its non-bacterial form as Libman-Sacks disease), Libman had profound influence on the study of internal medicine, particularly bacteriology.

Justus von Liebig
1803-1873

German chemist who was one of the leading researchers and chemical educators of the nineteenth century. He improved analytical methods for organic compounds, but is best known for his pioneering researches in agricultural chemistry, nutrition, and other aspects of the chemistry of living things. His theory of chemical fermentation led to a controversy with Louis Pasteur (1822-1895) about the role of microorganisms in fermentation and putrefaction.

Crawford W. Long
1815-1878

American physician who was among the first American doctors to use ethyl ether as a general anesthetic agent for surgical operations. Although Long used ether anesthesia in several operations as early as 1842 and 1843, he did not publish or present papers on his discovery until 1848. He attempted to assert his priority in 1854, but William Morton's demonstration of ether anesthesia in 1846 at the Massachusetts General Hospital was already well known by that time.

Karl Friedrich Wilhelm Ludwig
1816-1895

German physiologist who made landmark studies of the circulation, invented the kymograph and a blood-pump for sampling gases in blood, and pioneered the graphic method in physiology. Ludwig contributed to our understanding of the secretion of the urine, introduced revolutionary methods for the study of the respiration, blood circulation, and the nerves. He was also a popular and generous teacher who allowed his students to publish collaborative research under their own names.

François Magendie
1783-1855

French physician and physiologist who is considered the father of experimental pharmacology. Magendie conducted experiments on animals and on himself, including studies of arrow poisons from Java and Borneo. He studied the mechanism of action of various emetics and discovered emetine. His studies of the motor and sensory roots in spinal nerves became the subject of a priority battle with Sir Charles Bell (1774-1842).

Sir Patrick Manson
1844-1922

Scottish parasitologist who founded the field of tropical medicine. While practicing medicine in China, Manson studied elephantiasis, a lymphatic disease that causes the tissues of the limbs and genitals to become extremely swollen. In 1877 he showed that the cause of this disease, the parasitic *Filaria bancrofti* worm, is spread to humans by mosquitoes. Manson's discovery—that an insect can carry a parasite which it transmits to humans—helped Ronald Ross (1857-1932) to identify the parasite that causes malaria, also a mosquito-borne disease.

Etienne-Jules Marey
1830-1904

French physician and physiologist who invented the modern cine camera and developed the new field of scientific cinematography. His interest in analyzing the mechanics of physiological movements led to several valuable recording instruments, including the Marey *tambour,* various specialized cameras for recording the movement of humans and other animals, and high-speed and time-lapse photography techniques.

Ephraim McDowell
1771-1830

American surgeon who in 1809 removed a twenty-two-and-a-half-pound ovarian tumor from a 47-year-old woman in the backwoods of Kentucky without anesthesia. This was not the first such operation, but the first to succeed by skill rather than luck. The patient survived 29 years after her surgery. McDowell's report of his results in "Three Cases of Extirpation of Diseased Ovaria," published in 1817. The article established ovariectomy as a standard surgical procedure.

Samuel Morton
1799-1851

American physician who was a pioneering physical anthropologist. Morton earned medical degrees from the Universities of Pennsylvania and Edinburgh. He wrote anatomy and pathology texts and was active in the Philadelphia Association for Medical Instruction and Academy of Natural Sciences. His most significant studies focused on craniology. He measured one thousand human skulls from around the world and published books in which he asserted the differentiation of five races. Slavery advocates embraced Morton's ideas as scientific support for racism.

William Thomas Green Morton
1819-1868

American dentist who made pioneering contributions to surgical anesthesia. Morton's partner in a successful dental practice, Horace Wells (1815-1848), experimented on the anesthetic properties of nitrous oxide. Charles T. Jackson (1805-1880), Morton's teacher at Harvard Medical School, suggested the use of ether as "toothache drops." After conducting experiments on himself and his dog, Morton publicly demonstrated ether anesthesia at Massachusetts General Hospital on October 16, 1846. He and Jackson became involved in a battle for priority.

Valentine Mott
1785-1865

American physician who pioneered vascular surgery. Mott earned a medical degree from New York's Columbia College, studied surgical techniques abroad in London and Edinburgh, then returned to the United States to teach at Columbia and establish a private practice. He tied arteries to treat patients suffering aneurysms and also developed novel methods to excise cancerous tumors, reconstruct facial and spinal deformities, and remove bladder stones. Named chairman of surgery at New York University, Mott also advised Civil War medical administrators.

Johannes Peter Müller
1801-1858

German physician, comparative anatomist, and physiologist whose name is immortalized in several of the anatomical entities he described, including the "Müllerian duct." Müller was a pioneer in applying the microscope to pathological research. His research on the effect of stimuli on the sense organs led to his Law of Specific Nerve Energies. His experiments on the direction of nerve impulses in spinal nerves confirmed the so-called Bell-Magendie Law and advanced understanding of reflex action. Many of Müller's pupils became outstanding scientists.

Albert Niemann
1834-1921

German physician who was the first person to isolate cocaine. While a graduate student at the University of Göttingen in 1860, he was able to isolate a white, odorless, crystalline substance from the Peruvian coca leaf that he subsequently named "cocaine," short for cocaine hydrochloride. He explained the extraction of cocaine in his doctoral thesis; this was printed in the *American Journal of Pharmacy* in 1861. In 1914 Niemann described a form of xanthronatosis later called Niemann-Pick disease, a fatal infant disease characterized by mental and physical deterioration.

Elisha North
1771-1843

American physician who wrote *A Treatise on a Malignant Epidemic, Commonly Called Spotted Fever* (1811), the first important book about cerebrospinal meningitis. North recommended taking the patient's temperature as part of routine diagnosis sixty years before Carl Reinhold August Wunderlich brought the clinical thermometer into general use.

Louis Pasteur
1822-1895

French chemist and biologist who founded the science of microbiology. He developed the germ theory of disease that illustrated how diseases are spread by bacteria. Pasteur also proved that microbes could be weakened in a laboratory and then placed in an animal's body to create resistance, or immunity, to the microbe. Pasteur used this process of vaccination to protect sheep against the deadly anthrax disease before going on to develop vaccines for human diseases, including rabies. In the early 1960s, Pasteur invented pasteurization, the process that prevents milk from souring by heating it to a high temperature and pressure before bottling.

Ivan Petrovich Pavlov
1849-1936

Russian psychologist and physiologist who is best known for his theory of conditioned and unconditioned reflexes, which he developed on the basis of experiments with dogs and chimpanzees. Elected in 1907 to the Russian Academy of Sciences, Pavlov also made contributions to medical science and chemistry, and in 1904 he received the Nobel Prize for his work on digestive enzymes. Pavlov's research greatly advanced physiology and behavioral psychology and deepened understanding of the learning process.

Jules Emile Péan
1830-1898

French surgeon who is regarded as one of the founders of modern gynecological surgery. He developed techniques of vaginal hysterectomy, ovariotomy, and gasterectomy. He invented the hemostatic forceps, a clamp device to stop bleeding during surgery. Péan performed the first operation to correct diverticulitis of the bladder, and he also practiced aseptic surgery, or surgery free of disease-causing microorganisms, before microorganisms were understood.

William Pepper
1843-1898

American physician who pioneered the formation of university hospitals. A graduate of the University of Pennsylvania, Pepper served as a physician at Philadelphia hospitals, lectured, wrote medical texts, and was the curator of local medical collections. He helped plan and raise funds for the University of Pennsylvania hospital, then the prototype for clinical teaching in American. As that university's provost, Pepper supported curriculum reforms and set educational standards adopted nationally.

Max Josef von Pettenkofer
1818-1901

German physician who was a public health reformer and a founder of modern hygiene. Pettenkofer served as the first Professor of Hygiene at Munich. Although he was particularly interested in the origin of infectious diseases, he did not accept the germ theory of disease. Pettenkofer believed that poisonous miasmata (bad airs) in combination with polluted ground water spread infectious diseases. Therefore, he promoted the implementation of modern water and sewage systems as effective means of fighting epidemic diseases.

Wilhelm Petters
1824-1889

Czech physician who first showed that the urine of diabetics contains large amounts of acetone. The level of acetone is important in determining the severity, and subsequent treatment, of the disease. In the diabetic patient, the lack of cellular glucose initiates a process in which fats and proteins are broken down and the byproducts burned for energy; this results in the production of acetone and a "sweetness of breath" that is often mistaken for alcohol.

Lydia Estes Pinkham
1819-1883

American feminist who manufactured a home remedy for female reproductive pain that earned millions of dollars. In 1875 Pinkham began marketing her Vegetable Compound, consisting of unicorn root, pleurisy root, and alcohol. The medicine grew popular with women who were reluctant to consult male physicians about "female problems." Sales of the medicine boomed, even after Pinkham's family reduced the compound's alcohol content to save money. Medical science found no therapeutic value in the ingredients.

Charles Gabriel Pravaz
1791-1853

French physician who invented the modern hypodermic needle and the modern galvanocautery. In 1853 Pravaz published a description of the first hypodermic syringe, with its hollow metal needle. Alexander Wood (1725-1884) independently invented a similar device during the same year. With the hypodermic syringe, physicians had a new means of administering drugs, especially the highly purified drugs that had become available through the work of nineteenth-century chemists. Injections of morphine were subsequently used to provide local anesthesia.

William Prout
1785-1850

English physician and chemist who formulated Prout's Hypothesis, which states that all atomic weights are whole numbers. His contributions to medicine were the result of his experiments and lectures on animal chemistry, which was then a largely unexplored field. Prout analyzed various natural products, such as urea and uric acid in urine, and suggested that excretory products were the result of the breakdown of tissues.

Isaac Ray
1807-1881

American physician and pioneer of psychiatry, among the leading authorities of his day on the relation between mental illness and the law. Ray was co-founder (1844) of the Association of Medical Superintendents of American Institutions for the Insane, later the American Psychiatric Association; superintendent (1845-1866) of Butler Hospital in Providence, Rhode Island; and author of *A Treatise on the Medical Jurisprudence of Insanity* (1838), *Mental Hygiene* (1863), and *Contributions to Mental Pathology* (1873).

Pierre François Olive Rayer
1793-1867

French physician who wrote a major contribution to dermatology, *Traité théorique et pratique des maladies de la peau* (1826-1827), and a landmark study of kidney disease, *Traité des maladies des reins, et des altérations de la sécrétion urinaire* (1839-1841). Both of these works include gigantic pathological atlases. His meticulous research furthered the understanding of albuminuria, the pituitary gland, tuberculosis, obesity, diabetes, human glanders, human farcy, anthrax, bacterial infections, and parasitic diseases such as schistosomiasis.

Walter Reed
1851-1902

Medical officer in the United States Army who helped demonstrate how to control yellow and typhoid fevers. Between 1898 and 1901 Reed led two commissions to study the origin and spread of infectious epidemics in army camps. His experiments proved that flies were the predominant carriers of typhoid fever and that unsanitary conditions helped spread it. Reed's experiments focusing on yellow fever established that the bite of certain mosquitoes transmitted the disease. His team conducted a series of daring experiments in which physicians and soldiers volunteered to be infected by yellow fever germs, so that they could determine the course of the disease and how it might be controlled.

Pierre Paul Émile Roux
1853-1933

French physician and bacteriologist who contributed to the development of the modern germ theory of disease while working with Louis Pasteur (1822-1895). Roux worked on vaccines for anthrax and rabies. Experiments conducted by Roux and Alexandre Yersin (1863-1943) demonstrated that the diphtheria bacillus produced a toxin that could by itself cause the disease. Roux was also involved in the development of an antitoxic serum for the prevention and treatment of tetanus.

Louis Albert Sayre
1820-1900

American physician who advanced orthopedic surgery in the United States. Sayre completed a medical degree at New York City's College of Physicians and Surgeons and became the first chair of orthopedic surgery in the country, teaching at Bellevue Hospital's medical college. He devised a plaster cast for spinal curvatures and a shoe for club feet. Sayre wrote several books and traveled to Europe, sharing his innovative methods. He was president of the American Medical Association in 1880.

Theodor Schwann
1868-1938

German biologist who made landmark contributions to the development of histology and cell theory. Schwann's studies of yeast fermentation provided evidence against the doctrine of spontaneous generation and led to progress in understanding cellular metabolism. His studies of the notochord led him to propose that the nucleated entities he had observed were the animal counterparts of the plant cells that had been studied by M. J. Schleiden (1804-1881). Schwann concluded that plants and animals were composed of cells and cell products.

James Marion Sims
1813-1883

American gynecologist and surgeon who introduced silver sutures in the 1850s to reduce the risk of infection, improved or originated several complicated gynecological operations, pioneered treatments for infertility, modified the vaginal speculum, and reformed the techniques of gynecological examination. A southerner and a Confederate sympathizer, he voluntarily spent the American Civil War in England and France, where he wrote his most influential work, *Clinical Notes on Uterine Surgery* (1866).

John Stearns
1770-1848

American physician who professionalized medicine in New York. A Yale graduate, Stearns attended medical lectures at the University of Pennsylvania. He established a practice in Waterford, New York, in 1793 and founded the Saratoga County Medical Society. Stearns proposed forming a state medical society, received an honorary medical degree, and was chosen as the New York Academy of Medicine's first president. He also promoted the pharmaceutical qualities of fungus-produced ergot.

Jean Antoine Villemin
1827-1892

French physician who made landmark contributions to the study of the infectious nature of tuberculosis. Villemin demonstrated that tuberculosis could be transmitted to previously healthy animals by experimental inoculation. In his early experiments he inoculated rabbits with tuberculous material taken from a lethal human case. Initial attempts to repeat Villemin's results were unsuccessful, but later researches vindicated his conclusions.

Thomas Watson
1792-?

English physician who in 1843 was the first to propose the use of rubber gloves in surgery. This advancement in medicine reduced dramatically the risk of infection and the spread of microorganisms during surgical procedures. He received his medical degree from Cambridge in 1825. Following graduation, he practiced medicine in London until 1870. He was also a professor and lecturer of medicine at Royal College and Kings College in London.

Carl Friedrich Otto Westphal
1833-1890

German psychiatrist known for his research on various psychoses; these are mental disorders in which the patient has lost contact with reality. He is credited with giving the first description of agoraphobia, or the fear of open places. Westphal was also the first to introduce the term paranoia, a mental condition used to describe delusions of suspicion and jealousy.

Daniel Hale Williams
1858-1931

African-American physician who performed one of the earliest open-heart operations. In 1893 a young stabbing victim was brought to Provident Hospital in Chicago. Williams opened the chest, exposed the heart, and sutured the wound in the pericardium, the membrane that covers the heart. Although at the time operating on the heart was considered impossible, the patient survived. Williams contributed to improvements in Freedmen's Hospital in Washington, DC, and other aspects of medical care and medical education for African-Americans.

Erastus Bradley Wolcott
1804-1880

American surgeon who was the first physician to excise a human kidney. Wolcott received a surgical degree from the College of Physicians and Surgeons of Western New York. Moving to Fort Mackinaw, Wisconsin, to serve as an Army surgeon, he established a practice in Milwaukee where he performed the kidney operation. Wolcott was the state militia's surgeon general and, after the Civil War, Congress named him director of the National Home for Disabled Volunteer Soldiers.

Carl Reinhold August Wunderlich
1815-1877

German physician who was the first to describe fever as a symptom of disease and not a disease itself. He made more than one million measurements of individual body temperatures to determine that the "normal" healthy temperature is 98.6 degrees Fahrenheit. Wunderlich urged other doctors to carefully monitor fevers in order to chart the course of a patients' disease. But until the invention of the portable clinical thermometer in 1877, most physicians had no way of carrying out Wunderlich's recommendations.

Wilhelm Wundt
1832-1920

German physiologist and psychologist who is known as the founder of experimental psychology. In 1862 he offered the first course ever taught in scientific psychology, and in 1879 set up the first psychology laboratory—located at the University of Leipzig, where he carried out experiments to determine the dimensions of feeling and perception. Two years later he started the first psychology journal, *Philosophische Studien* ("Philosophical Studies"). Until Wundt's time, psychology had been considered a branch of philosophy, not a science.

Alexandre Yersin
1863-1943

Swiss bacteriologist who discovered the bacillus that causes bubonic plague. During an outbreak in Hong Kong in 1894 Yersin isolated the plague bacillus from plague buboes. One year later he inoculated animals with anti-plague vaccine. Yersin also helped elucidate the role played by healthy human carriers in the transmission of epidemic diseases. Yersin and Pierre Roux (1853-1933) proved that the diphtheria bacillus produces a toxin that causes the signs of the disease.

Thomas Young
1773-1829

English physicist and physician who is primarily remembered for his contributions to the wave theory of light. A remarkably versatile scientist, Young is also honored as the founder of modern physiological optics. His biomedical research included studies of astigmatism, the problem of refraction, color perception, and the mechanism by which the eye accommodates and focuses light rays on the retina of the eye. In addition, Young contributed to the decipherment of Egyptian hieroglyphics, particularly the Rosetta Stone.

Marie Elizabeth Zakrzewska
1829-1902

German-born American physician who joined Elizabeth Blackwell (1821-1910) in the battle to reform medical education and to open the medical profession to women. Zakrzewska studied and taught midwifery at the Berlin School for Midwives before coming to America and earning a medical degree from the Cleveland Medical College. After gaining experience as general manager and resident physician at Blackwell's New York Infirmary and College for Women, she moved to Boston and established the New England Hospital for Women and Children.

Bibliography of Primary Sources

~

Books

Beaumont, William. *The Experiments and Observations of the Gastric Juice and Physiology of Digestion.* 1833. Beaumont's 1833 work discusses his pioneering efforts to understand the workings of the stomach.

Bernard, Claude. *An Introduction of the Study of Experimental Medicine.* 1865. Bernard, considered the founder of experimental medicine and a pioneer in understanding digestion, summed up his life's work in this book.

Blackwell, Elizabeth. *Pioneer Work in Opening the Medical Profession to Women: Autobiographical Sketches by Dr. Elizabeth Blackwell.* 1914. A memoir by one of the most important women in the medical field during the nineteenth and early twentieth centuries.

Esmarch, Johann Friedrich August von. *Surgical Techniques.* 1901. This textbook, which grew out of von Esmarch's experience as an army surgeon, is a classic in the field of general surgery.

Gray, Henry. *Gray's Anatomy.* Illustrated by H. Vandyke Carter. 1858. This work was compiled from Gray's own dissection research and remains. Organized by terms rather than sections of the body, Gray's innovative work has, with several revisions, remained the definitive text on the subject of anatomy.

Hahnemann, Samuel. *The Organon.* 1810. This book introduced and defined Hahnemann's homeopathic philosophy of medicine.

Krafft-Ebing, Richard. *Psychopathia Sexualis.* 1886. This was the first collection of case studies dealing with sexually abnormal behavior, including definitions of sadism and masochism.

Kussmaul, Adolf. *Die Storungen der Sprache.* ("The Impediments of Speech.") 1877. Kussmaul's work offered some of the first clinical descriptions concerning the physiology of language and served as a foundation for the development of modern logopedics (studies of language pathology).

Laënnec, René. *De l'auscultation médiate, ou traité du diagnostic des maladies des poumons et du coeur.* (On Mediate Auscultation, or: A Treatise on the Diagnostics of the Diseases of the Lungs and Heart.) 1819. This massive two-volume work introduced the world to Laënnec's landmark invention of the stethoscope. The second edition of Laënnec's book (1826) was even more thorough in depicting pathological, anatomical, and therapeutic findings, and quickly became a classic in the English-speaking world when a translation was published in 1827.

Larrey, Dominique Jean, Baron. *Mémoires de chirurgie militaire, et campagnes.* 4 vols. 1812-17. Larrey, principal surgeon of the French army under Napolean Bonaparte, discussed his battlefield experiences in this memoir.

Louis, Pierre Charles Alexandre. *Researches Anatomical, Pathological and Therapeutic on the Disease Known under the Names Gastro-Enteric, Putrid Fever, Adynamic, Ataxic or Typhoid Fever.* 1829. This work established the classic pathological description of typhoid fever, including its distinction as a specific disease identified by characteristic lesions of the spleen and mesenteric glands, and Peyer's patches.

Nightingale, Florence. *Notes on Matters Affecting the Health, Efficiency and Hospital Administration of the British Army.* 1858. This book grew out of extensive testimony given by Nightingale about her experiences during the Crimean War, during which her efforts as a nurse dramatically improved the recovery rates of British soldiers.

North, Elisha. *A Treatise on a Malignant Epidemic, Commonly Called Spotted Fever.* 1811. This was the first important book about cerebrospinal meningitis.

Osler, William. *The Principles and Practice of Medicine.* 1892. Osler's 1892 work was a leading textbook in medical education.

Pinel, Philippe. *Philosophical Classification of Diseases.* 1798. In this work Pinel first described many psychotic symptoms.

Pinel, Philippe. *Treatise on Moral Management.* 1801. This 1801 work contains the basic philosophy of Pinel's approach to insanity.

Rayer, Pierre François Olive. *Traité des maladies des reins, et des altérations de la sécrétion urinaire.* 1839-41. This work was a landmark study of kidney disease. The book included a gigantic pathological atlas.

Rayer, Pierre François Olive. *Traité théorique et pratique des maladies de la peau.* 1826-27. Rayer's 1826-27 work was a major contribution to dermatology. The book included a large pathological atlas.

Snow, John. *On the Mode of Communication of Cholera.* 1849. Snow in this work examined the incidence, distribution, and communication of cholera in a number of London neighborhoods during the cholera epidemics that took place in England in the mid-nineteenth century. The main victims of the disease were the working and lower classes. Snow showed that infected water supplies spread the disease to the neighborhoods that used the water.

Woodward, J. J., C. Smart, G. A. Otis, and D. L. Huntington. *The Medical and Surgical History of the War of the*

Rebellion, 1861-65. 6 vols. 1870-88. A medical history of the American Civil War.

Periodicals

Bowditch, Henry Ingersoll. "On Pleuristic Effusions, and the Necessity of Paracentesis for their Removal." 1852. Bowditch's fame rests primarily on this article, which promoted the use of a suction pump in chest surgery.

Hodgkin, Thomas. "On the Morbid Appearances of the Absorbent Glands and Spleen." 1832. In this paper Hodgkin cited several cases of a disease that affected the lymph system.

McDowell, Ephraim. "Three Cases of Extirpation of Diseased Ovaria." 1817. McDowell's article established ovariectomy as a standard surgical procedure. Included in the piece is his account of an 1809 operation in which he removed a 22.5-pound ovarian tumor from a 47-year-old woman in the backwoods of Kentucky without anesthesia. This was not the first such operation, but the first to succeed by skill rather than luck. The patient survived 29 years after her surgery.

Physical Sciences

Chronology

1800 Electrolysis, and with it electrochemistry, pioneered by English scientists William Nicholson and Anthony Carlisle, who show that the hydrogen and oxygen in water can be separated by an electrical current.

1801 Italian astronomer Giuseppi Piazzi is the first to discover an asteroid; he names it Ceres.

1824 In his *Reflexions sur la Puissance Motrice du Feu,* French physicist Nicolas Leonard Sadi Carnot becomes the first to analyze the interrelation between heat and work, thus providing a framework for the science of thermodynamics.

1827 German physicist Georg Simon Ohm develops Ohm's law, which states that the amount of current passing through a wire is directly proportional to the thickness of the wire.

1830 Charles Lyell publishes the first volume of *The Principles of Geology,* a seminal work that would influence Charles Darwin, among others.

1840 Swiss-American geologist Jean Louis Rodolphe Agassiz puts forth his theory concerning a past Ice Age and the movement of glaciers, ideas that would not be accepted for some 25 years.

1842 Austrian physicist Christian Johann Doppler demonstrates that the frequency of sound waves varies as the source moves closer to or further from the listener (the Doppler Effect), and predicts that light waves will behave in the same manner.

1846 French astronomer Urbain-Jean-Joseph Leverrier and English astronomer John Couch Adams discover the planet Neptune.

1848 Scottish physicist William Thomson, better known as Lord Kelvin, puts forth the idea of an absolute zero temperature at which the motion of gas molecules stops; this becomes the basis for the Kelvin scale.

1869 First periodic table, which arranges the elements in order of atomic weight and predicts the existence of undiscovered elements, created by Russian chemist Dmitri Ivanovich Mendeleyev.

1873 James Clerk Maxwell publishes *Treatise on Electricity and Magnetism,* a landmark work that brings together the three principal fields of physics: electricity, magnetism, and light.

1894 British scientists John William Strutt (Baron Rayleigh) and William Ramsay isolate argon, the first in a series of rare or inert gases.

Overview:
Physical Sciences 1800-1899

The end of the eighteenth century found two basic Newtonian ideas triumphant. Scientists carefully experimented with natural phenomena and measured important features of these happenings, trying all the while to establish quantitative relations among the changing elements, The paradigm for this is found in Isaac Newton's (1642-1727) landmark 1704 work, *Opticks*. Another Newtonian paradigm was his theory of gravitation. Here one takes a mechanical model of small particles with forces of attraction or repulsion between any two of them, and then proceeds to deduce their behavior following from some basic force laws. Both the Newtonian experimental method and the Newtonian mechanical model of the universe were widely accepted by those investigating the physical world.

Physics

An amazing discovery occurred at the very beginning of the nineteenth century: electricity could be generated and made to flow in a current. Sir Humphry Davy (1778-1829) immediately used this current of electricity to analyze the chemical composition of a number of substances. A little later, Hans Christian Oersted (1777-1851), a Danish scientist, noticed that flowing electricity created magnetic effects, which was followed by the discovery by British scientist Michael Faraday (1791-1867)that moving magnets could conversely create electric currents. All of this experimental work called for a theoretical explanation, which scientists required to be mathematically rigorous and precise. The grand theory of James Clerk Maxwell (1831-1879) did just that in 1862.

Scientists had known many of the properties of light for a long time. Yet early in the nineteenth century several novel properties were discovered. The most important of these were that: 1) beams of light can "interfere" with each other; and 2) reflected light is polarized (which is why some sunglasses can reduce the glare of reflected sunlight). These observations were instrumental in leading to a new wave theory of light, which ultimately replaced the earlier Newtonian particle theory. Both Thomas Young (1773-1829) and Augustin Fresnel (1788-1827) are credited

with the development of this wave theory. It was Maxwell later in the century who declared that light waves in a medium called the ether could be represented as electromagnetic vibrations in this ether.

Maxwell's theory also implied that oscillating electric charges would produce invisible electromagnetic waves traveling through the surrounding space. Heinrich Hertz (1857-1894) devised experiments to produce and detect these waves, which are the basis for radio and television today.

No new properties of heat were discovered at the beginning of the century. Instead, scientists investigated the way that heat engines were capable of doing work, either pumping water from flooded mines or turning a huge flywheel, which in turn could power smaller machines. French scientist Sadi Carnot (1796-1832) published his theories about ideal, perfectly efficient engines early in the century. The focus on the conversion of heat to mechanical work led English scientist James Joule (1818-1889) to investigate the reverse process—namely, the conversion of mechanical work to heat. These researches (and many others) led to the concept of "energy," which can take many forms, heat being one of them. The notion of energy was introduced with the Law of Conservation of Energy, which states that energy is neither created nor destroyed; it only changes from one form to another.

It is common knowledge that heat flows from hotter to cooler bodies. In the course of investigating, for instance, how the temperature changes in time along a metal rod with one end in a furnace, Joseph Fourier (1768-1830) used a mathematical approach that had far-reaching consequences for the use of mathematical techniques in the physical sciences. This natural tendency of heat to flow from hotter to cooler bodies was described by another important law propounded by Rudolf Clausius (1822-1888) towards the end of the century. The assertion that in a closed system the total amount of heat available to do work gradually gets smaller is known as The Second Law of Thermodynamics.

At the beginning of the century most scientists considered heat to be a weightless fluid. Electricity too was thought to be a fluid. Yet as time went on, more and more scientists attempted to give a theoretical explanation of all observable effects in terms of small unobservable particles of matter and their interactions according to mechanical laws. Thus, the pressure of an enclosed gas was attributed to the impact of minute particles on the wall of the container. Temperature likewise was the observable effect of the energy of the constituent particles making up the gas. But in order to give a complete mathematical treatment of the behavior of these collections of particles, scientists had to appeal to statistical averages and probabilities. This was a novel approach to explaining the observable world, because the earlier Newtonian model relied on strictly deterministic connections between causes and effects. American scientist Josiah Willard Gibbs (1839-1903) played a major role in this development, as did German scientist Ludwig Boltzmann (1844-1906).

Chemistry

The Newtonian model, which saw the world as composed of small particles of matter attracting and repelling one another according to strict mathematical laws, was extremely fruitful. At the beginning of the century John Dalton (1766-1844) was inspired to develop the atomic theory of chemical substances. In this theory chemical elements were each composed of qualitatively similar atoms, and atoms combined in a fixed ratio to make chemical compounds. The atoms of different elements differed by weight, so the atoms of any one element all had the same weight and the molecules of any distinct chemical compound also had the same weight, because the proportions of the atoms in the compound were fixed. Yet no one was able to arrive at consistent results using this atomic theory until Amedeo Avogadro (1776-1856). He proposed that equal volumes of gases at the same temperature and pressure have the same number of molecules. Once that principle was understood late in the century, the relative weights of molecules and thus atoms could be fixed. This led quickly to Dmitri Mendeleyev's (1834-1907) table arranging the chemical elements by weight in a pattern of recurring properties.

Even before these results, it was discovered that substances with the same chemical composition sometimes had different properties. This was especially evident in the so-called organic chemicals—those that contain carbon. This led to the recognition that the structure of molecules—how the atoms are arranged—is as important as the

chemical composition. All these developments led to the astonishing development of chemical science, with its delicate instrumentation and fine measuring devices, and chemical engineering, with its abundance of synthetic substances.

Earth Sciences

An overview of the nineteenth century would not be complete without a reference to the important developments in geology that turned this discipline from an apology for the Biblical account of creation into a science of discovery. The evidence for glaciation was recognized. The layers of different sorts of rocks were traced and plotted from country to neighboring country. The identification of these strata was made possible by the correlation of the fossils found within them. And the fossil record helped to establish a time order for geologic processes. The length of time for geological changes to occur was a discovery that surprised many people: the earth was much older than anyone had thought.

Astronomy

Astronomy, which began the century carefully mapping the positions of heavenly bodies by means of larger and better calibrated telescopes, discovered in the middle of the century that starlight could be analyzed by a spectroscope, and thus the chemical composition of the stars could be determined. Techniques for measuring the distance to stars were also developed. The newly developed technology of photography helped enormously in these tasks. As geology expanded the time scale of our world, so did astronomy expand the spatial dimensions of our universe.

Conclusion

As the century drew to a close, many physical scientists felt that nearly everything had been discovered about the physical world. The only thing left to do was to apply what we had already learned about the world. Except, as the scientist William Thomson (better known as Lord Kelvin; 1824-1907) pointed out, there were two tiny clouds darkening the prospects for all the wonderful mechanical explanations of the world: the difficulty of accounting for the motion of Earth through the ether, and the inability to account for the energy distribution of certain sorts of radiation. We shall see that these two "clouds" were in reality the two doors to the new physical sciences of the twentieth century.

MORTON L. SCHAGRIN

Revival of the Wave Theory of Light in the Early Nineteenth Century

Overview

The nature of light is a very old issue in the history of science, dating back at least to Greek times. The prevalent belief among eighteenth-century natural philosophers was that light was made up of particles, not waves. The revival of the wave theory of light in the early nineteenth century is largely a tale of two cities, with Thomas Young (1773-1829) discovering the law of interference in London followed by the development of the mathematical wave theory by Augustin Fresnel (1788-1827) in Paris. The rise of the wave theory during the first three decades of the century is often regarded as a revolution in science. It exemplified a new style of scientific reasoning, with abstract mathematical models taking precedence over intuitive mechanical analogies for light. The ensuing debate on the validity of the new theory led to a closer examination of the standards and goals of scientific research. The domain of optics was redefined during this time, with the study of the physical aspects of light coming to be recognized as important for its own sake, apart from its relevance for theology or vision.

Background

Light and vision were intimately connected to the ancient Greeks and Arabs. The tactile theory, which held that our vision was initiated by our eyes reaching out to "touch" or feel something at a distance, gradually lost ground to the emission theory, which postulated that vision resulted from illuminated objects emitting energy that was sensed by our eyes. The nature of the emitted light occupied Renaissance thinkers in Europe, with early views envisioning light as a stream of particles, perhaps supported by the ether, an invisible medium thought to permeate empty space and all transparent materials.

Using the principle that light rays take the path that minimize their travel time, Pierre de Fermat (1601-1665) accounted for the phenomenon of refraction, the bending of light at the boundary between two transparent media such as air and water. An analogous principle for light waves was introduced by Christiaan Huygens (1629-1695), who considered spherical pulses of light propagating through an elastic ether.

Huygens's principle would form the basis of the wave theory developed by Fresnel a century and a half later.

The void in the wave theory during the interim period is usually attributed to the influential legacy of Sir Isaac Newton (1642-1727), who preferred to think of light as made up of corpuscles, or particles, that were governed by the laws of motion that carry his name. Newton observed concentric fringe patterns in the reflection of light from a spherical glass surface—known as Newton's rings—but he failed to recognize the signature of wave interference in this phenomenon. His bias against the wave theory was grounded in the belief that light traveled in straight lines, forming geometrical shadows of sharp objects. Observations to the contrary, in which light diffracted around an object to form complex patterns near the edge of the shadow, were known to Newton but failed to convince him of the wave hypothesis.

Most eighteenth-century natural philosophers concerned with the nature of light thought, as Newton did, that light was composed of individual particles subject to mechanical forces and inertia. A less rigorous view held it to be more like a fluid of particles, subject to collective motion in the ether, with analogies drawn to heat and fire, sometimes with Biblical connotations. The wave theory of light was largely ignored during this century, with some exceptions. The mathematician Leonhard Euler (1707-1783), for instance, advocated a vibration theory based on a comparison of light and sound. Just as sound travels by vibrations in the air that are longitudinal, or parallel to its motion, Euler conceived of light as longitudinal vibrations of an ethereal medium.

Young himself believed in the vibration theory at the turn of the nineteenth century. Trained as a physician, his early research on human vision and acoustics led him to consider the physical nature of light and sound. He devised several experiments to test his views on light, the most famous of which is the double-slit experiment that carries his name. He considered light from an aperture incident on two evenly placed slits on an otherwise opaque screen. The light emerging from the two slits formed fringes of alternating bright and dark bands on an observation screen. Young identified this periodic pattern with wave interference, the light waves from the two slits superposing to annul or enhance each other, much like two overlapping ripples in a water tank. He measured the fringe spacing for different colors, affirming Euler's conjecture that the color of light is connected with the frequency of the ethereal vibrations. Young also recognized the role of interference in the formation of Newton's rings.

Despite Young's successes with the vibration theory, he was unable to tame the phenomenon of double refraction, long known to be an embarrassment to both particle and wave views of light. This was the tendency of a beam of light to refract into two distinct beams upon entering certain crystals, such as Iceland spar (calcite), with the relative intensities of the two beams depending on the angle of entry into the crystal. Neither particle nor vibration theory could explain how light could be "sided" like this, preferring one angle to another, as neither particles nor sound waves shared this property. This property of the polarization, or sided-ness, of light was shown by Étienne Malus (1775-1812) in 1810 to be associated with reflection as well, as differently polarized beams reflected by different amounts from a mirror based on their angle of incidence, a phenomenon known as partial reflection. Polarization provided the bridge between the vibration theory of Young and the true wave theory developed by Fresnel in the second decade of the century.

Fresnel began investigating diffraction phenomena in 1814, leaning toward a wave theory of light and advancing the notion that the high frequency of the wave oscillations perhaps accounted for the near-straight-line motion of light. Being mathematically inclined, Fresnel sought to construct a theory of diffraction based on Huygens's principle, allowing for each point on the wave front to be a source of spherical waves that interfered with one another. This marked the true beginnings of a mathematical theory of wave propagation. Fresnel used the principle in its generality, with a continuum of points on each wave front generating secondary wavelets that interfered over a full range of phases, not just two. Using analytical calculus, Fresnel was able to derive formulae for several diffracting geometries, including diffraction through a narrow slit and around an opaque disc, winning him a prize from the Paris Académie in 1819.

As early as 1817 Young suggested to Fresnel that perhaps the polarization of light could also be explained if one considered transverse waves, where the ether oscillated perpendicular to the direction of travel. Transversality would give light a two-sidedness, since there are two independent directions along which a wave could oscillate perpendicular to its motion. This was suggested by an experiment that Fresnel did in 1819, along with Dominique-François Arago (1786-1853), in which it was found that differently polarized beams of light did not interfere with one another. By 1822 Fresnel was able to incorporate transverse waves into his theory and

THE INVENTION OF THE KALEIDOSCOPE

~

A kaleidoscope is an optical instrument containing mirrors placed at special angles to form multiple, symmetrical reflections of light. Colored glass or plastic, or liquid mixtures of oil and water are sometimes used to create changing, colorful patterns. Meaning "a beautiful form to see" in Greek, the kaleidoscope has offered inspiration to generations of artists, designers, and musicians. It was invented in 1816 by Scottish physicist David Brewster, who wrote: "If it be true that there are harmonic colors which inspire more pleasure by their combination than others; that dull and gloomy masses, moving slowly before the eye, excite feelings of sadness and distress; and that the aerial tracery of light and evanescent forms, enriched with lively colors, are capable of inspiring us with cheerfulness and gaiety; then it is unquestionable, that, by a skillful combination of these passing visions, the mind may derive a degree of pleasure far superior to that which arises from the immediate impression which they make upon the organ of vision." Nearly 200,000 kaleidoscopes were sold in Paris and London within three months, but Sir Brewster was unable to profit from the sales, as he was unsuccessful in enforcing his patent on the instrument.

produce convincing explanations for double refraction and partial reflection, with the two beams of light corresponding to the two transverse polarizations of the wave. The wave theory quickly gained in reputation after this period.

Impact

When Young first spoke in support of an analogy between light and sound waves before the Royal Society of London in 1800, his implicit rejection of Newton's views on light did not go

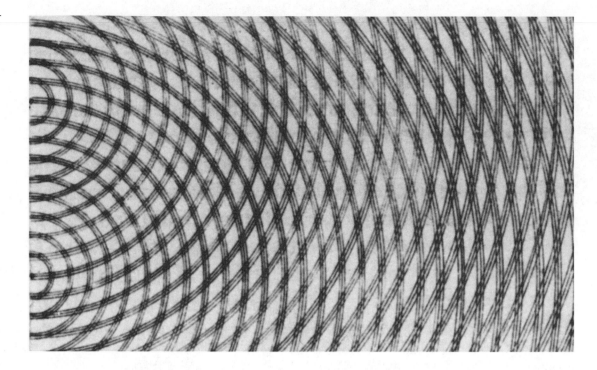

Thomas Young's 1807 representation of waves in a ripple tank.

over well with his English audience. His later expositions on interference in the double-slit experiment met with disbelief. The idea that a screen uniformly illuminated by a single aperture could develop dark fringes with the introduction of a second aperture—that the addition of *more* light could result in *less* illumination—was hard to accept, especially for those not used to thinking about light as a wave. A similar difficulty arose with Fresnel's theory of diffraction, with one of the judges on his 1819 prize committee, Siméon-Denis Poisson (1781-1840), highlighting the seemingly absurd fact that his theory implied a bright spot at the center of the shadow of an illuminated opaque disc, something that Arago immediately verified.

The situation changed dramatically in the 1820s with an increasing number of scientists adopting the wave theory of light. Fresnel's wave theory won support more readily than the vibration theories of Euler and Young, for several reasons. The replacement of longitudinal waves with transverse waves allowed polarization to be incorporated into a wave description. The theory gave concrete numerical predictions that could be tested readily, including phenomena like diffraction and double refraction that were hard to reconcile with the particle view. Also significant was the fact that the wave theory was an axiomatic theory founded essentially on Huygens's principle, rather than a set of ad hoc hypotheses characteristic of

the particle theories, and this found increasing resonance among the scientists of the 1830s, especially the younger generation.

Analogies between light and other phenomena played a less pivotal role in the new wave theory. Particle theories compared light to material bodies, subject to mechanical forces, or else envisioned light as a fluid akin to heat and electricity, also modeled as fluids at the time. The vibration theory was based on the analogy with sound, invoking material properties in the ether like density and elasticity to explain the vibrations. By contrast, Fresnel's wave theory emphasized methodology, working out the mathematical consequences of an analytic principle. While its proponents continued to invoke the ether to interpret the theory and facilitate its assumptions, the formalism of the theory did not stand or fall with a particular interpretation. Indeed, the formalism was not restricted to light waves but could be used to explain all wave phenomena.

Although the wave theory dominated optical science after 1830, there remained a few critics who could not embrace its premises and who continued to seek explanations in terms of particles and rays. The dispute was on the nature of scientific laws and their relation to empirical facts, centering on the wave theory. Physical optics in the early nineteenth century was essentially an inductive science, like thermodynamics or chemistry, consisting of a collection of disparate observations in need of a unifying theoretical description. Although the wave theory provided a coherent description that could be tested, it relied on abstract hypotheses like Huygens's principle and transversality that were themselves not immediately testable, much less the properties of the ether that sustained them. This bothered those who thought that scientific hypotheses or laws should be both necessary and sufficient in explaining all relevant phenomena. Scottish physicist David Brewster (1781-1868), for instance, saw the abstract premises of the wave theory as superfluous and unwarranted and preferred a simpler induction from known facts. The wave theorists, on the other hand, put more stock in the unifying power of their theory, allowing it to gradually gain confidence by new experiments, implying a subtle shift in methodology. The transversality of the light wave, for instance, would eventually gain more physical significance in the work of James Clerk Maxwell (1831-1879), with associations to electricity and magnetism.

Optics had a different connotation for both scientists and their lay audience by the end of the wave revolution. Whereas eighteenth-century treatises and lectures in optics might include sections on theology and vision in their discussion of light, the nineteenth-century textbooks tended to treat the physical aspects of light exclusively, with increasing use of mathematics for the wave theory. Gradually, the empirical and theoretical aspects of light started to take center stage, with less concern for its sociological or physiological ramifications. The upshot was a specialization of the field, with less participation by the lay audience in scientific discourses and a more mathematically trained scientific elite.

It is often remarked that the modern discipline of theoretical physics became distinct from natural philosophy in the early nineteenth century, with increasing emphasis on the use of advanced mathematics to describe physical theories. A larger revolution was indeed happening in physical science at the time, with fields becoming increasingly specialized and autonomous, new scientific methods being expounded, and changes in the training and career patterns of the scientists. Many of these changes were integral to the rise of the wave theory of light.

ASHOK MUTHUKRISHNAN

Further Reading

Books

Buchwald, Jed Z. *The Rise of the Wave Theory of Light: Optical Theory and Experiment in the Early Nineteenth Century.* Chicago: University of Chicago Press, 1989.

Cantor, Geoffrey N. *Optics after Newton: Theories of Light in Britain and Ireland, 1704-1840.* Manchester, UK: Manchester University Press, 1983.

Hecht, Eugene. *Optics.* 3rd ed. Reading, MA: Addison-Wesley, 1998.

Zajonc, Arthur. *Catching the Light: The Entwined History of Light and Mind.* New York: Bantam Books, 1993.

Periodicals

Fresnel, Augustin J. "Mémoire sur la diffraction de la lumière." *Annales de Chimie et de Physique* 11 (1819): 246-96 and 337-78.

Young, Thomas. "On the Theory of Light and Colours." *Philosophical Transactions* 92 (1802): 12-48.

Nineteenth-Century Development of the Concept of Energy

Overview

The concept of energy is fundamental to the understanding of all physical motion, whether in nature or derived from humanity's technologies. Nature's examples of energy are familiar enough to anyone: pounding ocean surf, volcanic eruptions, wind and electrical storms, and even the beating down of the Sun's rays. These and many other displays have intrigued and influenced humanity in its development and applications of the concept of energy in nature and in the laboratory. The theory of energy and its conservation was long in coming but has led to numerous practical technological applications.

Background

The word energy comes from Greek, meaning work, and the concept was explored progressively by medieval thinkers (momentum), Galileo (1564-1642; force acting on a body), Isaac Newton (1642-1727; gravitational force and the laws of basic mechanics), and other seventeenth- and then eighteenth-century scientific thinkers. However, the modern conceptualization of energy, the delineation of its several forms, and the finalization of most physical laws governing it are products of the nineteenth century.

There are six types of basic energy: mechanical, heat and light (both part of radiant), chemical, electrical, and nuclear. All were identified and explored in the nineteenth century, with the last of them, nuclear, just being on the threshold of definition when the century ended. Incipient concepts of conservation of energy touched on previously would find realization through nineteenth-century scientific experimentation of the various energy forms, which proved their transformation principles of one into the other, further proving that energy can neither be created nor destroyed—that is, energy is conserved.

The term "energy" came into use in 1807, coined by an English physician and physicist named Thomas Young (1773-1829) as a definition of the ability or capacity to do work. This is still fundamental to understanding the concept of energy. Young's definition was in regard to perhaps the simplest example of energy, mechanical energy, an example being an object pushed, pulled, raised, or lowered. Work is performed on the object—or the object has realized energy—and the object goes from a state of rest to one of motion.

The cumulative knowledge in the understanding of energy began its progression with seventeenth- and eighteenth-century investigators. The term for energy at that time was "vis viva," or "life force," reflecting the idea of physical motion involved. Dutch scientist Christiaan Huygens (1629-1665), studying colliding objects, concluded that the force of the objects after the collision changed but was not lost—this was an early concept of the conservation of energy. Other steps in the theory were studying the processes of heating, cooling, and chemical change (such as combustion). These were recognized as part of physical motion, but even by the early nineteenth century the understanding was at a formative stage. It was only after the middle of the previous century that Scottish chemist Joseph Black (1728-1799) had finally distinguished between heat and temperature.

Attempts to explain energy progressed with more sophisticated observations and experiments of physical phenomena. That there was a unity to all forces began with the idea of a single energy-producing substance, first being the phlogiston theory (1697), where "phlogiston" was defined as a weightless element present in any combustible matter. This was used to describe all physical and chemical phenomena. By late in the next century phlogiston was forgotten due to Black's caloric theory, which defined heat as an "expansive fluid" in matter called "caloric," a type of weightless particle that was attracted to matter and raised its temperature—that is, imparted energy. The idea enabled Black to explain how matter could change state with his concept of "latent heat"—water containing more caloric could change to steam (heated) or ice (cooled).

The nineteenth century was a period of prevalent experimentation with supposed energy forms. The heating of gases, explaining expansion and contraction—as well as combustion (using different chemical gases)—held prime clues to the nature of energy. Electrical investigations were also much in vogue in the nineteenth century. Electrical current was thought of as a fluid. In fact, the fluid idea of force and energy—as caloric—persisted for a long time in regard to electricity and magnetism. Radiant energy from light—the Sun being the oldest recognized—was one area that the caloric theory could not interpret.

In 1798 American scientist Benjamin Thompson (also known as Count Rumford, 1753-1814) experimented with heating metal and then water by mechanical work without the use of fire. He concluded that heating was accomplished by a mechanical motion that heated the particles of the object itself. In 1824 the son of a famous French Napoleonic general, Nicolas Sardi Carnot (1796-1832), wrote a paper called "Reflections on the Motive Power of Heat" in which he discovered by simple heat engine experiments that heating had a definite direction: the imparting of energy was done from a higher temperature to a lower one. That is, heat always moved from a warmer source to a colder one—-an expression of what would be called the Second Law of Thermodynamics. But, like Rumford's experiments (forgotten for nearly 50 years), Carnot's conclusion was not appreciated for some 25 years.

During that time two other thinkers independently continued with the same research Rumford had done on the relationship of heat and work. English scientist James P. Joule (1818-1888) did extensive experiments from 1839 that proved the equivalence of mechanical work and heat. Joule was evidently the first investigator who realized that heat was a form of energy (other mechanical experiments would also reveal that frictional loss was heat and thus energy). His original experiments were in terms of electrical work (1840) producing heat. Here was yet another clue to the transformation of energy—mechanical to electrical to heat. About the same time German physician Julius R. Mayer (1814-1878) was reaching the same conclusion that there was a quantitatively fixed relation between mechanical work and heat without benefit of experiments but by deductions from physiological observations. He then provided an estimate of the equivalence by turning to data on the specific heats of gases (1842). In an 1845 paper he related his findings in relation to what was again the conclusion of the conservation of energy by saying that energy in all its forms was preserved. Mayer also provided many examples of how this could be applied. His ideas were doubted, yet he and Joule are credited with the same theory.

Both Joule and Mayer provided a mechanical, or more generally, a dynamic theory of heat, in which heat was not a special substance but the energy of motion—what would be called kinetic energy. Mayer's investigations with gas pointed again to this important medium of experiment for resolving the nature of energy by studying the relationship between various dynamic means of heating.

Two of Mayer's contemporaries were prime proponents in this quest. Rudolf Clausius (1822-1888) was a German academic theoretical physicist, and William Thomson (also known as Lord Kelvin, 1824-1907) was an academic mathematician and experimental physicist. Clausius and Thomson resolved Carnot's interpretation of the Second Law of Thermodynamics with the prevalent mechanical theory. In 1851 Thomson added a paper ("On the Dynamical Theory of Heat") that established the many applications available via the laws of thermodynamics. The next year he wrote on the concepts of available heat and the dissipation of heat in mechanical work, topics that Clausius did not pursue until after he had rendered the Second Law of Thermodynamics mathematically (1854) and reasoned through a series of papers in which he called the dissipation of heat *entropy* (from the Greek for "change"). He went on to further define concepts of the kinetic theory of gases already inaugurated by Joule in 1848.

During the mid-1800s, one of the most distinguished scientific minds of the century was also studying these questions of energy in relation to physical motion at the molecular level. Hermann von Helmholtz (1821-1894) was a German physiologist and physicist and the most well rounded in his research of any nineteenth-century scientist. It was Helmholtz who provided the most comprehensive treatment of the conservation of energy (he still used the term "force" from the old "living force" idea) in an 1847 paper. He rendered the theory mathematically and introduced the fundamental concept that the conservation of energy was based on only two basic physical abstractions: force and matter.

Another outstanding nineteenth-century scientist, James Clerk Maxwell (1831-1879) pursued the interrelationship of electromagnetic energies. His field theory provided the link of electromagnetic forces in space to the nature of light energy and paved the way for the research of the late century that delved into the forces and energy of the atom itself. Here, once again, the theory of the conservation of energy provided a template for future research and technology application.

Impact

The concept of energy proved to be an invaluable and adaptable foundation to explore a progressively complex physical world manifested in the phenomenon of motion. The most important guiding principle in that pursuit of energy delineation has been the theory of the conservation of energy.

The practical applications of the understanding of energy and energy conservation were already at work in the modification and improvement of steam engines and the instruments measuring the parameters of those engines. Subsequently, the need to improve the efficiency in the energy cycle with steam led to the greater efficiency of the steam turbine. Later, the chambered engine medium for providing energy advanced to the internal combustion engine in a similar progression of sought-for efficiency. This led to advances in heating, ventilating, and refrigerating systems. And power plant applications of energy to provide greater yield led to natural gas, electric, and hydroelectric systems. Finally, as the nineteenth century gave way to the twentieth, the advance of energy technology from scientific knowledge applied to modern warfare on the eve of World War I raised inevitable moral questions of the right and wrong of such applications—questions that have continued to the present day.

WILLIAM J. MCPEAK

Further Reading

Feynman, Richard P., Robert E. Leighton, and Matthew Sands. *The Feynman Lectures on Physics.* 3 vols. Reading, MA: Addison-Wesley, 1963-65.

Resnick, Robert, and David Halliday. *Physics for Students of Science and Engineering.* 2 vols. New York: Wiley, 1960.

The Michelson-Morley Experiment, the Luminiferous Ether, and Precision Measurement

Overview

In 1887 Albert A. Michelson (1852-1931) and Edward W. Morley (1838-1923) performed what has become one of the most famous physics experiments in history. Using an extremely sensitive optical instrument—the interferometer—they attempted to measure Earth's velocity with respect to the luminiferous ether, a hypothetical substance that most nineteenth-century physicists believed necessary for the propagation of light. Against all expectations, their experiment yielded a negative result, indicating no motion of Earth relative to the ether. Ether theories were modified to account for this null-result, but no fully satisfactory solution presented itself until the introduction of Albert Einstein's special theory of relativity in 1905.

Background

The optical experiments of Thomas Young (1773-1829) and Augustin de Fresnel (1788-1827) at the beginning of the nineteenth century helped revived the wave theory of light. As with other wave phenomena—like sound waves in air and ocean waves in water—light waves were thought to require a medium of transmission. This medium was called the luminiferous (light bearing) ether.

An important nineteenth-century scientific question was the relationship between the ether and material bodies moving through it. Young believed that matter passed freely through the ether without in anyway disturbing it. This seemed necessary for ether-wave theories to explain stellar aberration. Discovered in the eighteenth century by James Bradley (1693-1762), aberration is the apparent displacement of a star

Albert Michelson. *(The Library of Congress. Reproduced by permission.)*

from its actual position due to the combined velocity of Earth and starlight. It was thought that if Earth's motion disturbed the ether, then starlight would be deflected in a manner inconsistent with this well-known effect.

Dominique François Arago's 1810 failure to measure Earth's velocity relative to the ether challenged Young's conclusions. Arago (1786-1853) accepted that the velocity of light, c, was constant in the ether and could only be measured to be c if one were at rest relative to the ether. He reasoned that motion through the ether with velocity *v* in the same or opposite direction as a beam of light, such as Earth's motion away from certain stars and toward others in its solar orbit, should yield light velocity measurements smaller, *c-v*, or greater, *c+v*, respectively. Knowing that light beams with different velocities refract differently, Arago designed an experiment to observe this difference. However, no such effect revealed itself. This null-result suggested the existence of a stagnant layer of ether near Earth's surface. If true, this would have undermined Young's explanation of aberration.

In 1818 Fresnel introduced his hypothesis of partial ether drag, which reconciled Young's position with Arago's result. Fresnel argued that transparent bodies dragged a permanent amount of ether within them in proportion to the square of their refractive indices. This altered the velocity of light by just enough so that optical experiments like Arago's could not detect Earth's motion relative to the ether. Fresnel's hypothesis further entailed that the bulk of ether remained undisturbed by material bodies, which agreed with Young's aberration explanation.

According to Fresnel's theory, Earth's motion through the ether was in principle detectable. However, James Clerk Maxwell (1831-1879) noted in his 1878 *Encyclopaedia Britannica* article "Ether" that the expected effect was too small to observe with existing optical instruments. When Michelson learned of Maxwell's views in 1879, he took-up the challenge of designing a sufficiently precise instrument for which he was later to coin the name "interferometer."

As the name suggests, Michelson's instrument exploits optical interference. A coherent beam of light is produced and then split in two. Each beam is directed along one of two mutually perpendicular interferometer arms, then reflected by a mirror back along its path to recombine with the other beam. The recombined beam is then directed to an observing telescope, where a pattern of alternating light and dark regions—known as interference fringes—is produced. The mirrors are adjusted so a fringe center falls on the telescope's fiducial mark. Differences in the velocity of light between the interferometer arms means light beams arrive at the telescope slightly out of phase, causing the interference fringes to shift with respect to the fiducial mark. If Earth moves relative to the ether, such fringe-shifts should be observed while the interferometer rotates.

The experiment was performed in 1881 at the Royal Astrophysical Observatory in Potsdam,

Germany. No fringe-shifts of the expected magnitude were observed. Michelson suggested that this null-result refuted Fresnel's ether theory. However, this conclusion proved unwarranted. Alfred Potier, and later H. A. Lorentz (1853-1928), noticed an error in Michelson's theoretical calculations. Correct calculations indicated fringe-shift magnitudes smaller than originally predicted, in fact, smaller than could be observed with the 1881 interferometer.

In 1887 Michelson and Edward Morley repeated the Potsdam experiment with better experimental controls and a more sensitive instrument. The two most important changes were an increase in the optical paths lengths—making the interferometer ten times more sensitive—and the instrument being mounting on a massive stone floating in mercury, which insulted it from vibrations and allowed more accurate fringe-shift reading while the instrument rotated. Although Michelson and Morley expected a fringe-shift, they once again obtained a null-result.

Impact

Contrary to popular belief, the Michelson-Morley null-result was not considered a serious threat to ether theories, nor was it taken as proof that the velocity of light was absolutely constant. Though the null-result puzzled ether theorists, there was no sense of crisis in the physics community. Indeed, consensus had it that the result would eventually be explained within the ether framework.

One failed attempt to explain the null-result was a modified Stokes' ether-drag theory. George Stokes' (1819-1903) theory, requiring complete ether entertainment at Earth's surface, accounted for the null-result. However, in 1892 Lorentz pointed out inconsistencies in Stokes' explanation of aberration. Furthermore, Oliver Lodge's (1851-1940) 1892 experiments failed to detect any ether-drag near rapidly moving disks.

The most promising explanation within the ether paradigm was the Fitzgerald-Lorentz contraction hypothesis. Originally published in 1889 by G. F. Fitzgerald (1851-1901), the contraction hypothesis states that as an interferometer moves through the ether its arms shrink in the direction of motion by just the amount necessary to cause a null-result. Fitzgerald's proposal was motivated by Oliver Heaviside's (1850-1925) 1888 discovery that the electromagnetic field of a moving charge shrinks by exactly this amount. Although he had no electron theory of matter, Fitzgerald thought it reasonable to assume the intermolecular forces of the interferometer arms to be electromagnetic in nature. Thus, they could be expected to vary in accordance with Heaviside's result. Fitzgerald felt sure such a variation would cause the interferometer arms to shrink the required length.

The contraction hypothesis implied that the observed velocity of light would be constant. However, this constancy was not absolute. For Fitzgerald, Lorentz, and other ether theorists, the null-result did not mean light was always and everywhere traveling with the same velocity. It only appeared so because the interferometer arms had shrunk. In principle, an instrument either not susceptible to shrinking or vastly more sensitive would be able to detect Earth's motion. Attempts to measure an ether-drift continued into the late 1920s and beyond, always with null-results.

Lorentz independently proposed the contraction hypothesis in 1892 and attempted to justify it in terms of his electron theory. Its initial viability faded as Lorentz was forced to make more complex and implausible assumptions. This changed in 1905 with publication of Albert Einstein's (1879-1955) special theory of relativity. As Einstein himself noted, his theory provided "an amazingly simple summary and generalization of hypotheses which had previously been independent from one another."

The Michelson-Morley experiment had only a minor and indirect role in the genesis of Einstein's special theory of relativity. It did, however, play a significant role in convincing many physicists of the theory's validity, supporting as it did Einstein's postulate of the absolute constancy of the velocity of light.

Though the interferometer failed in the purpose for which it was created, it nevertheless remains one of the most sensitive and versatile instruments ever created. Michelson used the instrument to measure the gravitational constant, indices of refraction, soap film thickness, coefficients of expansion, and to test screw pitch uniformity, analysis of spectral lines, and stellar aberration, all with a precision never before achieved. Two of these measurement standout: his progressively better measurements of the velocity of light and his measurement of the world meter standard in terms of light waves. The latter was considered by his contemporaries vastly more significant than his ether null-result.

STEPHEN D. NORTON

Further Reading

Books

Buchwald, Jed Z. "The Michelson Experiment in the Light of Electromagnetic Theory before 1900." In *The Michelson Era in American Science 1870-1930,* edited by S. Goldberg and R. Stuewer. New York: AIP, 1988.

Holton, Gerald. *Thematic Origins of Scientific Thought, Kepler to Einstein.* Rev. ed. Cambridge, MA: Harvard University Press, 1988.

Swenson, Loyd S., Jr. *The Ethereal Aether: A History of the Michelson-Morley-Miller Aether-Drift Experiments, 1880-1930.* Austin, TX: University of Texas Press, 1972.

Whittaker, Sir Edmund. *A History of the Theories of the Aether and Electricity.* 2 vols. New York: Dover Publications, 1953.

Periodicals

Holton, Gerald. "Einstein, Michelson, and the 'Crucial' Experiment." *Isis* 60 (1969): 133-197.

Hunt, Bruce J. "The Origins of the Fitzgerald Contraction." *British Journal for the History of Science* 21 (1988): 67-76.

Michelson, Albert Abraham. "The Relative Motion of the Earth and the Luminiferous Ether." *American Journal of Science* 22 (1881): 120-129.

Michelson, A. A. and Edward Williams Morley. "On the Relative Motion of the Earth and the Luminiferous Ether." *American Journal of Science* 34 (1887): 333-345.

Wilson, David B. "George Gabriel Stokes on Stellar Aberration and the Luminiferous Ether." *British Journal for the History of Science* 6 (1972): 57-72.

Heinrich Hertz Produces and Detects Radio Waves in 1888

Overview

In 1888 German physicist Heinrich Hertz (1857-1894) produced and detected electromagnetic waves in his laboratory. His goal was to verify some of the predictions about these waves that had been made by Scottish physicist James Clerk Maxwell (1831-1879). Of course, simply producing electromagnetic waves was not sufficient unless they could be detected, too. What Hertz did not realize at the time is that his discovery not only verified and validated Maxwell's work, but it also made possible the later invention of radio, television, radar, and other devices that depend on the production and detection of electromagnetic radiation.

Background

Early theories in physics assumed that all actions required some sort of direct contact or influence to make things happen. A hand pushing a ball or a wall stopping a ball are examples of this. However, gravity and magnetism seemed to violate this concept by seemingly allowing action at a distance without direct physical contact between objects. This led to speculation that various "ethers" existed that filled space so that, for example, magnetic forces would act on the ether that would, in turn, act on a piece of iron to pull it towards the magnet. By the mid-1800s, physicists had developed a theory involving separate ethers for the transmission of heat, gravity, static electricity, magnetism, and other phenomena that seemed to embody action at a distance.

In the 1840s Michael Faraday (1791-1867) developed the concept of a physical field in which each point had specific properties relating to forces acting on bodies within that field. A gravitational field, for example, is defined by the strength of the gravitational force and its direction at every point in the field. At the surface of the earth, the gravitational field points directly at the center of the earth with an acceleration of 9.8 meters per second per second (a force of one gravity). Electrical and magnetic fields are similar in nature.

The next step was made in the 1860s by Maxwell, who showed that electricity and magnetism are related and that interactions between these two forces will produce what is called electromagnetic radiation. First, Maxwell showed that light is a form of electromagnetic radiation. By explaining heat as a form of electromagnetic radiation similar to light, Maxwell was able to combine many of the ethers into one—the one he thought was needed to transmit electromagnetic waves. (We now know this ether is not present, as shown in 1887 by the Michelson-Morley experiment.) At the same time, Maxwell's equations suggested that electromagnetic radiation could have either longer or shorter wavelengths than light.

Maxwell developed one of several competing theories involving fields to explain electrical and magnetic action at a distance. Others were developed by Helmholtz, Faraday, and Wilhelm Weber (1804-1891). Studying each of these theories in turn, Hertz saw some similarities that apparently escaped other physicists of the day, leading him to speculate about the nature of electromagnetism. While working with equipment designed to test some properties of electromagnetic fields, Hertz accidentally developed a crude oscillating circuit that transmitted electromagnetic energy to a similar circuit used as a monitor. Eventually Hertz realized that these devices could be used to reliably produce and detect oscillating electromagnetic waves—radio waves—that traveled through space.

Impact

The lasting importance of Hertz's discovery cannot be overstated. Consider the use to which radio and other electromagnetic waves are put today: radio, television, radar, food preparation, welding, heat sealing, magnetic resonance imaging, radio astronomy, and navigation are only a few of the applications.

It should be noted, however, that radar waves are generated in a different manner than are radio waves. Specifically, radio waves are generated by inducing electromagnetic oscillations in an antenna that are then broadcast to distant receiving stations. By contrast, radar waves are generated in a device called a cavity magnetron that is very different from an antenna. However, radar (an acronym for radio detection and ranging) is possible without this device, and, in any event, without having first discovered how to produce and detect radio waves, radar would not have been possible at all. Similarly, the widespread use of radio waves for communication across long distances depended on the invention of the vacuum tube in 1907 by Lee De Forest (1873-1961). But the discovery of ways to generate and receive radio waves was still a necessary prerequisite for radio communications.

The impact of Hertz's discovery is easily recognized in the following categories of use: communications, science, industry, and military.

The most obvious impact of generating and receiving radio waves is in communications. Although not originally envisioned by Hertz, it took only six years for Italian engineer Guglielmo Marconi (1874-1937) to construct a simple device that used radio waves to ring a bell. In 1901 Marconi successfully received a radio transmission sent from England in Newfoundland.

Transmission of voices and music by AM radio followed in 1906, less than 20 years after Hertz's initial success. Other inventions followed, including television, communications satellites, and so forth, each simplifying a formerly difficult task—staying in touch over long distances. Prior to radio, communication beyond one's town was difficult and, for most people, rare. Hertz, while not directly involved in changing this, certainly took the first steps by showing it was possible to generate and receive waves that could travel so far so quickly. It may be a cliché to say that radio and television have made the world a smaller place, but it is a cliché because it is true.

The arena of science was profoundly affected by Hertz's discoveries as well, and in a number of ways. At the time he announced his results, they had the effect of stimulating research into many aspects of electrodynamics and electromagnetism. This, in turn, contributed to the revolution in physics that was already looming on the horizon. Hertz died before the landmark discoveries that initiated this revolution, but his work was important. Radio astronomy, which has taught us much about the nature of the universe, is entirely dependent on receiving and interpreting radio waves from outer space. Our current theory of the formation of the universe, the Big Bang theory, was strengthened immeasurably by the discovery of the cosmic microwave background radiation field, discovered as a result of investigations into improving radio communications. Much medical research and treatment utilizes magnetic resonance imaging (MRI) that uses radio waves as part of the imaging process. Radar waves, a form of radio frequency radiation, have been bounced off the moon, Venus, Mercury, and a number of asteroids to learn their distances and to map their surfaces. Radar is also used extensively in weather research, helping to predict and analyze incipient storms. And, of course, deep-space probes convey their information and receive instructions via radio signals.

In industry, radio and other electromagnetic waves are used frequently, too. Microwave ovens use radio-frequency radiation to cook food, while other microwave devices are used to weld plastics, and seal bags. The use of radar for air traffic is well known, of course, as is its use for police speed traps. Radio frequency radiation is also used for joining metals in some industries.

Finally, there are innumerable military uses of electromagnetic radiation. One unintended outcome of developing radio communications was to take a great deal of authority and autonomy away from ships' captains. Previously, a captain at sea was almost a minor deity, alone in command of his ship and able to use his full discretion in carrying out his orders. The implementation of radio contact enabled his superiors to remain in contact at virtually all times, following his progress, second-guessing his decisions, giving additional instructions, and so forth. Losing this degree of autonomy upset many captains, but the strategic and tactical advantages overshadowed this, and the ability to coordinate the actions of many ships over thousands of square miles of ocean were immense. The development of smaller radio sets brought these same advantages to the battlefield, changing the nature of warfare and leading to today's emphasis on battlefield information. Add to that the uses to which the military has put radar (detection of enemy units, proximity fuses, electronic countermeasures, to name but a few) and it is apparent that electromagnetic radiation is fundamental to today's military forces worldwide.

In summary, the societal impacts of Hertz's research came quickly, were far-reaching, and may be considered ongoing if taking into account the still expanding fields in which radio, radar, and other high-frequency electromagnetic radiation are used. While not as powerful as the development of electronics, it can be argued that radio has transformed more lives than electronics because of the relative ubiquity of radios compared to computers in the world. Virtually every person in the world has reasonable access to radios (excepting, of course, the small percentage of people who live in very remote and primitive areas); the same can hardly be said of computers. Indeed, the most important indication of the importance of radio to modern society lies in the degree to which we take it for granted. Few question the ability to turn on the radio to hear music or news. Picking up a cordless telephone, a cellular phone, or a walkie-talkie are routine events for most, and we accept as routine that we can see news or sporting events occurring anywhere in the world in real time. Any of these technological commonplaces would have been considered minor miracles prior to Hertz's discoveries.

P. ANDREW KARAM

Further Reading

Buchwald, Jed. *The Creation of Scientific Effects: Heinrich Hertz and Electric Waves*. Chicago: University of Chicago Press, 1994.

Hellemans, Alexander, and Bryan Bunch. *The Timetables of Science*. New York: Simon and Schuster, 1988.

The Discovery of Radioactivity: Gateway to Twentieth-Century Physics

Overview

Radioactivity was one of several discoveries made at the turn of the twentieth century that led to revolutionary changes in physics. Unlike some discoveries, it was completely unexpected. The discoverer was looking for something else when he found it, the scientific world initially ignored it, and most of its ramifications were not apparent until much later. As radioactivity gradually transmuted into nuclear physics, its impact reverberated far beyond the confines of physics, forever changing society in its wake. The discovery of radioactivity changed our ideas about matter and energy and of causality's place in the universe. It led to further discoveries and to advances in instrumentation, medicine, and energy production. It increased opportunities for women in science. Radioactivity introduced new health hazards, and its military applications permanently changed world politics. Applications of radioactivity created ethical problems which have yet to be resolved.

Antoine Becquerel. *(The Library of Congress. Reproduced by permission.)*

Background

None of this was foreshadowed at the start of 1896, when the scientific world was agog at reports from Germany of a new invisible radiation which penetrated opaque bodies. The first x-ray photo of the bones in a human hand mesmerized professors and the public alike. These rays seemed to come from the phosphorescent screens used to detect cathode rays (later identified as electron beams), a popular and controversial topic in the late nineteenth century.

It was natural to wonder whether other phosphorescent substances gave off invisible penetrating rays. One of the scientists who was impressed by the x rays, Antoine Henri Becquerel (1852-1908), had inherited a collection of phosphorescent minerals assembled by his father, a leading expert on optical luminescence. Becquerel was the third generation of a family of famous physicists who were professors at the Natural History Museum in Paris, and he had established his reputation with researches on optical phenomena. He returned to the Museum and began testing the minerals.

First Becquerel would wrap a photographic plate in black paper in order to block visible light. After placing a mineral on the paper, he would expose it to sunlight in order to make it phosphoresce. Then he would wait to see whether an image would form on the plate. Most of the minerals had no effect, but a uranium compound made a strong image on the plate. One day he set out a sample containing uranium, but the sun appeared only intermittently. When the weather did not improve, he finally developed the plate, and to his great surprise saw a sharp image on it!

This did not make sense, because phosphorescent materials needed light in order to glow. Perhaps there had been enough light on the cloudy days after all. But when he kept the sample in a light proof box, it still marked the plate. Apparently these invisible rays did not require light. What they did seem to require was uranium, since everything Becquerel tested that contained uranium worked, while the other minerals did not. (Exceptions were later attributed to errors.) Uranium metal worked even better than uranium compounds—and metals did not phosphoresce. Still, for some time Becquerel believed the rays he had found were a kind of invisible light.

Becquerel published his findings in 1896 and 1897, but most scientists were not very interested, since their journals were being flooded with reports on various kinds of invisible rays. Satisfied that he had established his discovery, Becquerel investigated a different topic for the next year and a half. An engineer in London, Silvanius P. Thompson (1851-1916), had also found in 1896 that uranium gave off invisible rays, but after he learned that Becquerel had already published this result, Thompson likewise dropped this topic.

The uranium rays nevertheless caught the attention of a young Polish student in Paris. Maria Sklodowska Curie (1867-1934), who had recently married the French physicist Pierre Curie (1859-1906), was looking for a subject for her doctoral thesis. She decided to search for other elements that might give off invisible rays, naming this property *radioactivity*. Becquerel had shown that uranium rays had electrical effects, and Curie used this process (later called ionization) to test mineral samples. First she found that thorium gave off rays, but G. C. Schmidt in Germany had already published this finding. Then she noticed that pitchblende, a uranium ore, emitted more radiation than uranium itself. Several new elements had been discovered during the late nineteenth century, and Curie wondered if one might be hidden in the mineral pitchblende. That prospect was so tantalizing that her husband decided to join her in the search. After backbreaking labor, and with the help of the chemist Gustav Bémont, the Curies announced the discovery of two new elements, which they named polonium and radium (1898).

Impact

This finding startled the scientific world, and soon more researchers were investigating the new elements and the radiations they emitted. Having more powerful sources made it easier to do experiments, and the electrical method allowed more sensitive and precise measurements than the cruder photographic method. Industrialists developed factories to process uranium ores and the market for uranium ores burgeoned.

Some scientists doubted the Curies' findings, and even questioned the existence of radioactivity. Marie Curie worked for years to obtain sufficient radium and polonium to determine their atomic weights. This feat convinced the skeptics, and eventually the electrical methods pioneered by radioactivity researches were accepted by the rest of the scientific community.

New methods led to advances in instrumentation, which in turn led to further discoveries about atomic structure, subatomic particles, and cosmic rays.

While some were not sure of radioactivity's existence, others wanted to elevate it to a universal property of matter. As reports poured in from across Europe and beyond of radioactivity detected in springs, soils, snow, air, in fact almost everywhere, the hypothesis of universal radioactivity seemed plausible. Eventually experimenters found that radioactivity in the environment came from traces of radioactive elements, rather than from universal radioactivity. These studies contributed to the later discovery of cosmic rays.

At first most scientists believed the new rays were x rays. In 1898 a young physicist working in Canada, Ernest Rutherford (1871-1937), found that two types of rays (he believed two types of x rays) were emitted by uranium. In 1899 researchers in Germany, Austria, and France showed independently that some of the rays were actually charged particles, later known as electrons. Later three types of rays were identified. Two (alpha and beta rays) were shown to be particulate; the gamma rays were similar to x rays.

The idea that atoms could spontaneously lose part of their material substance stoked speculations about atomic transmutation, which mingled with various forms of spiritualism which were circulating in the popular press at the turn of the century. More reputable scientific opinion predicted conversions between matter and energy; could all the material world be nothing more than energy forms? Nineteenth-century electrical theory, and later Albert Einstein's special theory of relativity, gave formulas for computing the conversion of mass to energy, but instruments were not sensitive enough to test these predictions with radioactive substances.

From the beginning scientists had been puzzled by radioactivity's persistence, and by their inability to affect it. Phosphorescence eventually disappears if the phosphor is not reexposed to light, yet uranium's activity seemed to continue unremittingly, year after year. With the discovery of radium, which gave off much more energy per gram than uranium, the question became critical. What was radioactivity's energy source? Researchers took samples deep inside a mine, enclosed them in lead, sealed them from light, heated them, cooled them, altered them physically and chemically, yet the rays always continued unabated, carrying huge amounts of energy, many times more than any known chemical reaction. Radioactivity seemed to violate the principle of conservation of energy.

By 1899 scientists realized that some radioactive bodies gradually lose their activity. Rutherford determined that this loss followed an exponential law. All bodies gradually lose their radioactivity, but for some elements the loss was not detected because the process can take up to billions of years. The transmutation theory of radioactivity was published by Rutherford and Frederick Soddy (1877-1956) in 1903. This theory states that radioactivity's energy comes from the radioactive atoms themselves as they change themselves into new elements. Thus the law of energy conservation was preserved. Much later atomic energy was used for nuclear reactors and atomic bombs.

The transmutation theory caused scientists to change their ideas about atoms and about the energy available in matter. Atoms were not unchangeable, they contained huge stores of energy, and they were built out of smaller particles. The exponential law of decay meant that probability theory could be used to describe radioactivity. The realization that atomic disintegration was a random process changed physicist's ideas about causality in nature, affecting in turn areas as distant as modern art and literature. Radioactivity altered ideas about the earth's age, and later provided methods for measuring it.

Radioactivity's biological effects were scarcely recognized when the nineteenth century came to a close. Experimenters had noticed that radium caused burns, but many decades elapsed and many lives were lost before radiation induced illnesses were identified and adequate safety precautions were adopted. Scientists learned that radiation caused mutations in plants and animals. During the twentieth century radioactivity was used to treat cancer, detect illnesses, study physiological processes, and sterilize foods.

Radioactivity opened new career paths for both students and established scientists. Because of the progressive attitudes of leaders in the field, and because of Marie Curie's example, an unusual number of women did research in radioactivity. During the 1930s the field of radioactivity gradually turned into nuclear physics, which later produced the subfield of particle physics.

Advances in nuclear physics brought new ethical, social, and political concerns. Was it right to use, or even to create, nuclear weapons? Would the human race eventually destroy itself in a nuclear war? Could we deter such a war by stockpiling nuclear weapons?

Privileged by hindsight, it seems easy to look back and trace many modern developments to Becquerel's experiments with uranium. Yet no one in the nineteenth century, certainly not Becquerel himself, could have foreseen the consequences of those experiments. Nineteenth-century researchers worked with theories, instruments, laboratories, and expectations quite different from those of today. By 1900, radioactivity was only a minor subspecialty of physics, and nearly everything for which we recognize it today was yet to materialize.

MARJORIE C. MALLEY

Further Reading

Books

Badash, Lawrence. *Radioactivity in America.* Baltimore and London: Johns Hopkins, 1979.

Mladjenović, Milorad. *The History of Early Nuclear Physics (1896-1931).* Singapore: World Scientific, 1992.

Pais, Abraham. *Inward Bound.* New York: Oxford, 1986.

Phillips, Melba Newell, ed. *Physics History from AAPT Journals.* College Park, MD: AAPT, 1985.

Quinn, Susan. *Marie Curie.* New York: Simon and Schuster, 1995.

Rayner-Canham, Marelene F., and Geoffrey W. Rayner-Canham. *A Devotion to their Science: Pioneer Women of Radioactivity.* Philadelphia: Chemical Heritage Foundation, 1997.

Romer, Alfred. *The Discovery of Radioactivity and Transmutation.* New York: Dover, 1964.

Romer, Alfred. *Radiochemistry and the Discovery of Isotopes.* New York: Dover, 1970.

Wilson, David. *Rutherford: Simple Genius.* Cambridge, MA: MIT Press, 1983.

Periodicals

Boudia, Soraya. "Marie Curie: Scientific Entrepreneur." *Physics World* 11 (December 1998): 35-9.

Jauncey, G. E. M. "The Early Years of Radioactivity." *American Journal of Physics* 14 (1946): 226-41.

J. J. Thomson, the Discovery of the Electron, and the Study of Atomic Structure

Overview

Late in the nineteenth century physicists were working hard to understand the properties of electricity and the nature of matter. Both subjects were transformed by the experiments of J. J. Thomson, who in 1897 showed the existence of the charged particles that came to be known as electrons. Along with the nearly contemporaneous discoveries of radioactivity and x rays, the discovery of the electron focused the attention of scientists on the problem of atomic structure, as well as on ways to put these invisible phenomena to use with inventions such as radio and television.

Background

Joseph John Thomson (1856-1940) spent his professional life at England's Cambridge University, where he passed in four years from prize-winning student (he was ranked second "wrangler" in the prestigious "mathematical tripos" examination in 1880) to head professor at the Cavendish Laboratory—a position previously held by James Clerk Maxwell (1831-1879) and Lord Rayleigh (1842-1919). Maxwell, who first put forth the theory of an electromagnetic field, set up the Cavendish Laboratory in 1874 as a place to pursue investigations in experimental physics and to provide electrical standards for industry. Although he died in 1879, his influence continued to be felt there among the physicists. Mathematical physics had long been established at Cambridge, and while Maxwell, Thomson, and most other Cambridge physicists continued to work successfully in this tradition, the Cavendish Laboratory helped make Cambridge an important center for experimental investigations as well. It remained the preeminent center for the study of subatomic physics into the early decades of the twentieth century.

Although Maxwell himself did not set an explicit agenda for his successors, in the period following his death many of Cambridge's physicists devoted themselves to developing, testing, and expanding Maxwell's theory of electricity and magnetism. An important part of this program was the effort to establish a microscopic basis for electromagnetic phenomena. This would enable physicists not only to describe the behavior of electricity, but also to understand how its phenomena were produced. In his own work, Maxwell had made use of two important research strategies: he derived mathematical relationships between measurable quantities, and he also created analogies or models between electrical effects and well-understood mechanical devices. Neither of these methods required an understanding of the nature of the phenomena themselves. Thomson and his colleagues began from Maxwell's framework, but the goals of their research came to include the understanding of fundamental atomic structure and processes.

Soon after taking over his position as head of the Cavendish Laboratory, Thomson began to study the discharge of electricity in gases. He sought to test various ideas about the dissociation of molecules in an electric field. The earliest experiments consisted of passing a discharge between two large parallel plates that functioned as electrodes, in a container filled with gas and connected to a vacuum pump that could vary the pressure within the container. These experiments, performed with many different gases, provided numerous observations but ultimately led to more puzzles around the relationships among electric current, gaseous discharge, and the chemical combination of atoms.

Thomson refined these investigations by attempting to study the same phenomenon, the discharge of electricity through a gas, with a different experimental arrangement. He filled a bulb with gas at a low pressure, and surrounded it with an electrical coil. The coil produced an electrical field within the bulb that was simpler to study than the one produced by an electrode, and Thomson was optimistic about its results. But by 1892, Thomson was frustrated with difficulties in attaining quantitative measurements in his experiments, and was once again looking for a new approach. He turned for awhile to the study of the electrolysis of steam, analyzing the appearance of hydrogen and oxygen under various experimental conditions. This gave him useful insights into the variable charge on atoms, and encouraged him that his experimental program was indeed elucidating the relationship between the processes of chemical combination and electromagnetism.

During 1895 and 1896, in a scientific atmosphere enriched by reports of the discovery of x rays and "Becquerel rays" or radioactivity, Thomson brought the study of cathode rays into his research on electricity and gases. He had studied cathode rays previously, trying to provide a theoretical framework to describe the rays. Now he entered the ongoing debate about whether these rays were themselves streams of charged particles, or some kind of disturbance in the underlying electrical "ether." Thomson sought to show that cathode rays could be deflected by a magnetic field—an observation that would confirm that the rays were charged particles. He did this by directing a stream of cathode rays through narrow slits into the field between two charged plates, and then measuring the stream's deflection. These experiments not only showed that the cathode rays were indeed charged particles, but also allowed Thomson to determine the ratio between the charge carried by the particles and their mass. The ratio of charge to mass turned out to be much higher than expected, and even more surprising, to be independent of the nature of the gas in the tube. Thus, these particles, or corpuscles as Thomson initially referred to them, must have a very small mass, and he also argued that since they seemed to be the same in all gases, they must actually be part of atoms themselves. Thomson had identified one of the primary building blocks of matter—what came to be called the electron.

Although Thomson's role in the discovery of the electron is indisputably central, it would be a mistake to believe it to be exclusively his accomplishment. In addition to the other researchers at the Cavendish Laboratory, important work on the discharge of electricity through gases and the nature of cathode rays was done by Arthur Schuster (1851-1934), Philipp Lenard (1862-1947), and Heinrich Hertz (1857-1894); Hendrik Lorentz (1853-1928) pursued Thomson's discovery, improved upon his measurements and endowed the corpuscles with the name "electron."

Impact

J. J. Thomson's identification of the electron in 1897 focused new attention on questions of atomic structure. Thomson conjectured that the electron was a fundamental building block of matter or atoms, and along with his colleagues at Cambridge attempted to build upon his discovery in order to model atomic structure with theoretical speculations and extensive experimental investigations, particularly scattering experiments. They struggled to explain many observations, such as the nature of positive charge, the relation between number of electrons and atomic weight, and the mechanical stability and chemical properties of atoms. While the Cambridge scientists and others working within the framework they had established came up with models of the atom that successfully accounted for many of these phenomena, the behavior of atoms came to be explained much more effectively as physicists adopted the ideas of quantum science beginning about 1912.

Other investigations also built upon Thomson's discovery. Further research by Thomson, as well as work by Henri Becquerel (1852-1908), Lenard, Ernst Rutherford (1871-1937), and others, helped to show that the electron identified by Thomson was the same as the negatively charged particles observed in phenomena such as radioactivity and the photoelectric effect. American scientist Robert Millikan (1868-1953) improved upon Thomson's measurement of the charge on the electron by observing the motion of charged oil drops. By the 1920s, scientists were studying electrons within the framework of quantum physics, and began to explore the theory that electrons behaved not only as particles but also as waves. Several Nobel Prizes were given for early research related to the discovery and study of the electron, including one to Thomson in 1906 and to Millikan in 1923. As testimony to Thomson's influence as a teacher, seven of his research assistants also went on to win Nobel Prizes for physical research.

The impact of the discovery of the electron extended far beyond science. Throughout the nineteenth century, research into electrical phenomena had been intertwined with efforts to advance practical uses of electricity such as the telegraph and electrical power. The investigations of Thomson's era helped bring about the rapid invention and development of "wireless telegraphy," or radio, and led to the invention of television and later the development of microwave technologies such as radar. Radio arose in part from investigations into the nature of the electromagnetic "ether" or atmosphere, a subject that Thomson also addressed in his research. The invention of television is more directly indebted to the discovery of the electron, as electronic television is based on cathode ray tubes in which a beam of electrons is aimed at a screen. While Thomson's experiments and theories did not result directly in any of these inventions, his contributions advanced understanding of the nature and behavior of electrical processes and atomic structure, making such technological developments easier and faster.

LOREN BUTLER FEFFER

Further Reading

Books

Dahl, Per F. *Flash of the Cathode Rays: A History of J. J. Thomson's Electron.* Bristol: Institute of Physics Publishing, 1997.

Davis, E. A., and Isobel Falconer. *J. J. Thomson and the Discovery of the Electron.* London: Taylor & Francis, 1997.

Springford, Michael, ed. *Electron: A Centenary Volume.* Cambridge: Cambridge University Press, 1997.

Periodicals

Chayut, Michael. "J. J. Thomson: The Discovery of the Electron and the Chemists." *Annals of Science* 48 (1991): 527-544.

Falconer, Isobel. "Corpuscles, Electrons and Cathode Rays: J. J. Thomson and the 'Discovery of the Electron.'" *British Journal for the History of Science* 20 (1987): 241-276.

Feffer, Stuart M. "Arthur Schuster, J. J. Thomson, and the Discovery of the Electron." *Historical Studies in the Physical Sciences* 20 (1989) 33-61.

Heilbron, John L. "J. J. Thomson and the Bohr Atom." *Physics Today* 30 (1977): 23-30.

Kragh, Helge. "J. J. Thomson, the Electron, and Atomic Architecture." *Physics Teacher* 35 (1997): 328-332.

Unification: Nineteenth-Century Advances in Electromagnetism

Overview

Advances in nineteenth-century concepts of electromagnetism moved rapidly from experimental novelties to prominent and practical applications. At the start of the century gas and oil lamps burned in homes, but by the end of the century electric light bulbs illuminated an increasing number of electrified homes. By mid-century (1865) a telegraph cable connected the United States and England. Yet, within a few decades, even this magnificent technological achievement was eclipsed by advancements in electromagnetic theory that spurred the discovery and development of the radio waves that sparked a twentieth-century communications revolution. So rapid were the advances in electromagnetism that by the end of the nineteenth century high energy electromagnetic radiation in the form of x rays was used to diagnose injury. The mathematical unification of nineteenth century experimental work in electromagnetism profoundly shaped the relativity and quantum theories of twentieth-century physics.

Background

In the late-eighteenth and nineteenth centuries philosophical and religious ideas led many scientists to accept the argument that seemingly separate forces of nature (e.g., electricity, magnetism, light, etc.) shared a common and fundamental source. In addition, profound philosophical and scientific questions posed by Isaac

Michael Faraday's magneto-electric induction device was the first mechanical electrical generator. *(Corbis-Bettmann. Reproduced by permission.)*

Newton's *Opticks* (published in 1704) regarding the nature of light still dominated the nineteenth century intellectual landscape. Accordingly, in addition to a search for a common source of all natural phenomena, an elusive "ether" through which light could pass was thought necessary to explain the wave-like behavior of light.

The discovery of the relationship between electricity and magnetism at the end of the eighteenth century and the beginning of the nineteenth century was hampered by a rift in the descriptions and models of nature used by mathematicians and experimentalists. To a significant extent, advances in electromagnetic theory during the nineteenth century mirrored unification of these approaches. The culmination of this merger was Scottish physicist James Clerk Maxwell's (1831-1879) development of a set of equations that accurately described electromagnetic phenomena better than any previous non-mathematical model.

The development of Maxwell's equations embodied the mathematical genius of the German mathematician Carl Friedrich Gauss (1777-1855), the reasonings and laboratory work of French scientist André Marie Ampère (1775-1836), the observations of Danish scientist Hans Christian Oersted (1777-1851), and a wealth of experimental evidence provided by English physicist and chemist Michael Faraday (1791-1867).

Although improved mathematical reasoning allowed for deeper understandings of electrical and magnetic experimentation, the emergence of technological society drove the translation and articulation of electromagnetism into the practical world of technological innovation. Simple observations made by scientists—including Benjamin Franklin—during the mid-eighteenth century continued to intrigue scientists and inventors who sought to carry on Franklin's practical descriptions of electrical phenomena. In addition, as the nineteenth century progressed electricity itself came to play an important role as increasingly technological societies attempted to develop the machines and tools needed to meet the needs of rapidly expanding populations and burgeoning urban societies.

Impact

In 1820 Oersted demonstrated the relationship of magnetism to electricity by placing a wire near a magnetic compass. When an electric current was applied to the wire the compass needle showed a deflection characteristic of a changing magnetic field. Inspired by Oersted's demonstrations, a year later Faraday—the devout son of a blacksmith—proved his genius in the practical world of laboratory experimentation by developing a "rotator" now credited as the first electric motor. Faraday's initial apparatus consisted of an electrical current carrying wire rotating about a

magnet. Subsequently, Faraday also clearly demonstrated the converse induction of current by rotating magnets about the wire. Although it was another half century before their widespread production, the first practical electric motors were all designed according to principals documented by Faraday. Faraday's method to produce electric current with magnets—known as electromagnetic induction—is a method still used by modern power generators.

Faraday's subsequent publication of his work with electromagnetic induction in 1831 formed the basis of a collection of papers eventually published as *Experimental Researches in Electricity*. Faradays' work became the standard authoritative reference for nineteenth century scientists and is credited as inspiring and guiding inventors such as Thomas Edison (1847-1931).

During the last decades of the nineteenth century, electric motors drove an increasing number of time- and labor-saving machines that ranged from powerful industrial hoists to personal sewing machines. Electric motors proved safer to manage and more productive than steam or fuel burning engines. In turn, the need for the production of electrical power spawned the construction of dynamos, central power stations, and elaborate electrical distribution systems.

Ampère, a professor of mechanics at the Ecole Polytechnique in Paris, was another influential nineteenth-century scientist influenced by Oersted's observations. Ampère's subsequent influence on the theoretical development of electromagnetic theory is thought by many historians of science to be similar to the influence of Newton's contributions to a functional understanding of gravity. Ampère deepened and tightened the relationship between electrical and magnetic phenomena through a series of brilliantly devised experiments that demonstrated the fundamental principles of electrodynamics (the effects generated by electrical current). Although Ampère made a number of experimental revelations of his own, it was his mathematical brilliance that laid the foundation for subsequent development of electromagnetic theory by quantifying and translating physical electromagnetic phenomena observed by Faraday and other experimentalists into the language of mathematical formulations.

The culminating fusion of nineteenth-century experimentation and mathematical abstraction of electromagnetism came with the development of Maxwell's equations of the electromagnetic field. These four famous equations united concepts regarding electricity, magnetism, and light. Maxwell's equations were, however, more than mere mathematical interpretations of experimental results. By developing precise formulas with enormous predictive power, Maxwell set the stage for the formation of quantum and relativity theory. Twentieth-century giants such as Max Planck (1858-1947), Albert Einstein (1879-1955), and Niels Bohr (1885-1962) all credited Maxwell with laying the foundations for modern physics.

Maxwell collected and first published his electromagnetic field equations in 1864. By 1873 Maxwell's publication, *Electricity and Magnetism,* fully articulated the known laws of electromagnetism. Perhaps most importantly, Maxwell propositions regarding the propagation of electricity and magnetism resulted in the theory of the electromagnetic wave, and thereby allowed the unification of known electrical and magnetic phenomena into the electromagnetic spectrum.

Although empirical proof of the existence of the electrons did not come until the end of the nineteenth century, Maxwell's equations established that the electric charge is a source of an electric field and that electric lines of force begin and end on electric charges (though this is not necessarily true in a changing magnetic field). In addition, prior to Maxwell's equations it was thought that all waves required a medium of propagation. Maxwell's equations established for scientists that, no matter how counter-intuitive, electromagnetic waves do not require such a medium. That an "ether" or transmission medium was unnecessary for the propagation of electromagnetic radiation (e.g., light) was subsequently demonstrated by the ingenious experiments of Albert Michelson (1852-1931)and Edward Morley (1838-1923).

With his electromagnetic field equations, Maxwell was able to calculate the speed of electromagnetic propagation. When Maxwell's calculated speed of electromagnetic propagation fit well with experimental determinations of the speed of light, Maxwell and other scientists realized that visible light was simply a part of an electromagnetic spectrum. Subsequently, not only did the emerging concept of an electromagnetic spectrum explain much of the phenomena associated with visible light, it also predicted that visible light was only a small part of the spectrum. Based on this insight German physi-

cist Heinrich Rudolf Hertz (1857-1894) in 1888 demonstrated the existence of radio waves.

Hertz regarded Maxwell's equations as a path to a "kingdom" or "great domain" of electricity in which all electromagnetic radiation is understood to be but a slightly differing manifestation of the same electromagnetic phenomena—differing only in terms of wavelength and frequency. Exploration of the electromagnetic spectrum resulted in both theoretical and practical advances. Near the end of the nineteenth century Wilhelm Roentgen's discovery of high energy electromagnetic radiation in the form of x rays, for example, found its first practical medical use.

Maxwell's mathematical unification of experimental work in electromagnetism laid the foundation for the development of relativity and quantum theory. The equations remain a powerful tool to understand electromagnetic fields and waves. Indeed, the equations still have many practical applications, including the design of electrical transmission lines and electromagnetic (e.g., radio, television, microwave, etc.) antenna.

Although the development of radio was largely accomplished in the early years of the twentieth century, its genesis was in the advancement of the nineteenth-century understanding of electromagnetism. Not until the development of the Internet nearly a century later would another technology besides radio so completely tear down the walls of geographic distance in human society. For the first time humans could communicate in their language over long distances with immediacy and spontaneity. Divergent and diverse societies became, for the first time, united in an increasingly global civilization.

The study of the interaction of the fundamental forces of nature—including electromagnetism—dominates many modern research programs.

K. LEE LERNER

Further Reading

Goldman, Martin. *The Demon in the Aether: The Story of James Clerk Maxwell.* Edinburgh: Paul Harris Publishing, 1983.

Tolstoy, Ivan. *James Clerk Maxwell: A Biography.* Chicago: University of Chicago Press, 1981.

Whittaker, Edmund. *A History of the Theories of Aether and Electricity.* New York: Harper & Brothers, 1951-53.

The Replacement of Caloric Theory by a Mechanical Theory of Heat

Overview

The nineteenth century witnessed a definitive resolution of questions regarding the nature of heat. The flow of heat was recognized as one way in which systems could exchange energy with their environment, and the equivalent of heat in mechanical work was determined. While heat flow into a system increased its store of energy, only a fraction of the energy could be recovered as useful work. An additional quantity, entropy, that determined the direction of heat flow was introduced and ultimately interpreted as an increase in the randomness of molecular energy. The mechanical theory of heat firmly established the atomic hypothesis—that all aspects of the behavior of matter can be explained by the interaction of component atoms and their parts—as a scientific fact. In so doing, it increased confidence that all natural phenomena could ultimately be explained on an atomistic basis, an advance that modern historians consider one of the most important in modern science, on a par with another great nineteenth-century achievement—Charles Darwin's groundbreaking theory of evolution.

Background

At the beginning of the nineteenth century, the majority of scientific thinkers accepted some form of the caloric theory as an explanation for the phenomena associated with heat. Caloric was considered to be a massless fluid that could flow between bodies, a fluid not unlike the electric fluid proposed by Benjamin Franklin or a comparable magnetic fluid, also in vogue. In some versions of the theory, the fluid was composed of discrete particles. The great French chemist Antoine Laurent Lavoisier (1743-1794) devoted the first few sections of his definitive text, *The Elements of Chemistry,* to the interaction of caloric with matter. The caloric theory was in accord with a great many observations, including the existence of heat capacities that governed the exchange of heat between bodies, the latent heat absorbed by substances in phase changes, and the release of heat in chemical reactions. Although some phenomena, such as the production of heat by friction, lacked a satisfactory explanation within the theory, adherents of the theory assumed that one would eventually be found.

The caloric theory provided the basis for French engineer Sadi Carnot's essay "Reflections on the Motive Power of Heat." In this work Carnot (1796-1832) analyzed the behavior of an idealized steam engine in which, in each cycle, heat was withdrawn from a reservoir of caloric at an elevated temperature, did work by expanding inside a piston, and was discharged to a second reservoir, the exhaust, at a fixed lower temperature. Carnot showed that the amount of mechanical work that can be obtained from a given amount of heat energy by such a device is independent of the material that does the physical work. Instead, it depends only on the temperatures of the two reservoirs. The connection between heat and motion was elaborated by German physicist Julius Robert Mayer (1814-1878) in a paper published in 1842, generally considered to be the first clear statement of the principle of conservation of energy. Mayer's interest in the subject dates from his observation, made in 1840 while working as a ship's surgeon in the tropics, that the venous blood of men in the tropics was a brighter color than that in northern latitudes, suggesting that less energy was needed from the oxidation of food to maintain body temperature in warmer climates. In his 1842 paper Mayer described the interconvertibility of kinetic potential and heat energy and provided an estimate of the mechanical equivalent of heat, the amount of mechanical energy needed to produce a fixed rise in temperature in a given mass of water. This was refined in a series of experiments by the highly respected English amateur scientist James Joule (1818-1889) beginning in the following year.

With a quantitative relation between heat and mechanical energy established, the caloric theory became far less credible. German physicist Rudolf Clausius (1822-1888) noted that Carnot's analysis, which assumed that the amount of caloric extracted from the heat source equaled that delivered to the exhaust in each cycle, was incorrect, in that part of the heat energy had been converted into work. In his paper of 1850 he corrected Carnot's error by assuming that the amount of energy which could not be converted to work in normal operation of the engine would be the same as the amount that would be extracted from the exhaust in each cycle if the engine were run in reverse. He found that the minimum amount of energy per cycle discharged to the exhaust was in the same ratio to the absolute temperature of the exhaust as the amount of energy taken in during each cycle was to the absolute temperature of the hotter reservoir. In 1865 Clausius introduced the concept of entropy change, to denote the ratio of heat energy exchanged to absolute temperature, the importance of this ratio having been noticed by Lord Kelvin (1824-1907) at about the same time. Clausius noted that energetically allowed processes would occur spontaneously in nature only if the entropy of the universe as a whole were to increase. He and Kelvin were thus among the first to suggest that the final state of the universe would be one of maximum entropy.

With heat energy understood to be essentially mechanical in nature, it remained for physicists to provide an interpretation of the energy and entropy in terms of molecular properties. Joule, Scottish physicist J. C. Maxwell (1831-1879), and Austrian physicist Ludwig Boltzmann (1844-1906), who together established the distribution of molecular speeds in a gas and found that the average kinetic energy of gas molecules is directly proportional to the temperature, quickly resolved the question as to energy. Boltzmann also provided a molecular interpretation for entropy, proposing that it was proportional to the number of different ways in which the random kinetic energy could be distributed among the molecules in a system. While controversial for a time, this introduction of statistical concepts into physical theory was elaborated, especially by American physicist J. Willard Gibbs (1839-1903), into a new discipline. The statistical mechanics of Gibbs and Boltzmann, brought into its current form after the development of quantum mechanics in the twentieth century, is now universally accepted as providing the needed connection between thermal phenomena and the underlying dynamics of matter on the atomic level.

Impact

The replacement of the caloric theory by a mechanical theory of heat illustrates a number of issues in the history and philosophy of science. One important point is that theory is never fully determined by observation and experiment. Any given set of observations will be explainable by more than one theory. The caloric theory worked well enough for many aspects of the

behavior of gases. Theories may be modified or discarded, however, not only when discordant observations are made but also when it becomes possible to unify the theories of phenomena previously considered unrelated. Such unifications include that of celestial and terrestrial mechanics accomplished by Isaac Newton (1642-1727) in the seventeenth century and that of optics, electricity, and magnetism accomplished by Maxwell in the nineteenth century, as well as that of heat and mechanics as discussed above.

In 1917 American biochemist and philosopher L. J. Henderson wrote that "Science owes more to the steam engine than the steam engine owes to science." The industrial revolution and the accompanying concentration of production and wealth was in large part driven by the steam engine, a device that made it possible to convert heat energy into mechanical energy, to power the looms and lathes of industrial nations. The new industries required not only mechanical energy, but also new materials and an improved understanding of the thermal properties of matter and the process of combustion. Thus, the technology of heat energy raced ahead of the basic scientific understanding of heat, but many important advances in physical theory occurred nonetheless as the result of the very practical need to be able to build bigger and more efficient heat engines.

From a philosophical and religious standpoint, the success of the mechanical theory of heat represented a redefinition of materialism. The atomic hypothesis, originally enunciated by the philosophers Democritus and Leucippus in the fifth century BC and integrated into the materialist philosophy of Epicurus, had long been considered suspect by Christian theologians who feared that it allowed no place for spirit. As the various imponderable fluids of the eighteenth century—caloric, electricity, and magnetism—were given atomistic interpretations, the Cartesian dualism of body and soul was largely abandoned within science in favor of a purely mechanistic view. At the same time, a number of modern empirically minded philosophers, including the Austrian Ernst Mach (1838-1916), were critical of the atomic theory for another reason: they opposed treating as real entities that were in principle not observable. Boltzmann, in particular, was severely criticized, a factor that may have led to his suicide in 1906.

From a purely scientific standpoint, American physicist Richard Feynman (1918-1988) has described the fact that all matter is composed of atoms as the single most important scientific discovery of our civilization. Its consequences are very far reaching, and most practicing scientists now accept it as a universally accepted truth. Thus, it is generally believed among scientists that all aspects of the behavior of matter, from the flow of heat between bodies to thought and emotion in humans, can be explained—if not predicted—by the interaction of the component atoms and their parts. The discovery that energy, although conserved, would become increasingly unavailable in time due to the increasing entropy of the universe led to some unpleasant conclusion as to the long-term fate of the universe. According to Clausius and Kelvin, the universe will inevitably evolve into a uniform thin gas of atoms at a constant temperature, and nothing within the scope of nature can be done to prevent it. While the Christian churches maintained faith in an eventual "new heaven and new earth" to be brought about by direct divine intervention, for the majority of non-believers

the inevitable "heat death" of the universe added to a sense of the pointlessness of existence. More recent developments in mechanics and cosmology allow some room for alternatives to the heat death scenario, however.

DONALD R. FRANCESCHETTI

Further Reading

Bronowski, Jacob. *The Ascent of Man.* Boston: Little Brown, 1995.

Burke, James. *Connections.* Boston: Little Brown, 1995.

Feynman, Richard P., R. B. Leighton, and Matthew Sands. *The Feynman Lectures in Physics.* 3 vols. Reading, MA: Addison Wesley, 1963-65.

Fox, Robert. *The Caloric Theory of Gases from Lavoisier to Regnault.* London: Oxford University Press, 1971.

Ihde, Aaron J. *The Development of Modern Chemistry.* New York: Harper and Row, 1964.

Magie, William F., ed. *A Sourcebook in Physics.* Cambridge, MA: Harvard University Press, 1963.

Nineteenth-Century Advances in the Mathematical Theory and Understanding of Sound

Overview

The nineteenth century saw the development of the mathematical techniques and experimental methods needed to understand the vibrations of objects and the motion of sound waves in air and other media. These developments were summarized and integrated by John Strutt (1842-1919), better known as Lord Rayleigh, in his *The Theory of Sound,* first published in 1877. Raleigh's text remains one of the fundamental reference works for acoustical scientists and engineers.

Background

The history of the science of acoustics is somewhat unusual in that, unlike the case of heat or projectile motion, the essential nature of sound has been correctly understood since the time of the ancient Greeks. They identified the origin of sound with the vibration of bodies and understood it to be transmitted through the air in some fashion. The fact that vibrating strings under tension produced harmonious sounds if their lengths were in simple numerical ratios was known to the disciples of Greek mathematician Pythagoras (580?-500? B.C.).

The development of musical instruments, begun in antiquity, advanced rapidly in the Renaissance and subsequent baroque and classical periods. Scientific understanding of sound lagged behind, however, for lack of a means of recording sound and the mathematics needed to describe its origin and movement. Galileo (1564-1642) included a discussion of vibrating bodies in his 1638 *Discourses Concerning Two New Sciences,* relating vibration to the behavior of the pendulum. Robert Hooke (1635-1703), who discovered the fundamental law of elasticity, tried to study the vibration of sound sources. Isaac Newton (1642-1727) included a somewhat incorrect discussion of the motion of sound waves in his *Principia Mathematica,* published in 1687.

The description of the vibrating string drew the attention of some of the best mathematicians of the seventeenth and eighteenth centuries, including Jakob Bernoulli (1654-1705), Jean d'Alembert (1717-1783), Leonhard Euler (1707-1783), and Joseph Louis Lagrange (1736-1813). The description of vibration required generalizing the calculus, which had been invented by

Newton, to describe the rates of change of quantities in time, to a calculus of more than one variable, which could describe quantities that were functions of both time and position. The necessary methodology was provided by French engineer Jean Baptiste Fourier (1768-1830) in his *Analytical Theory of Heat,* published in 1822. Although Fourier concerned himself primarily with the flow of heat through matter, the mathematics of heat flow presented the same problems as that of sound, and he was able to show how any solution of the corresponding equations defined over a region of space could be expressed as a sum of sine- and cosine-like functions, each having its own characteristic variation with time.

Hermann Helmholtz (1821-1894) was a military surgeon who became professor of physiology at the University of Königsberg and later professor of physics at the University of Berlin. His early career was devoted to a study of the physical senses and the nervous system. He was the first to measure the speed of a nerve impulse. His interest turned to acoustics in 1853, apparently after spotting some mathematical errors in a paper on the subject. He identified the cochlea as the organ of hearing and proposed a plausible mechanism for hearing whereby sounds are detected by the resonance of parts of this structure with the incoming sound wave. He then explained how the quality of musical tones resulted from there being a combination of different frequencies. His book summarizing this work, *On the Sensations of Tone,* was first published in 1863.

Helmholtz's book stimulated interest in acoustics on the part of numerous physicists, among them Strutt, who would soon inherit the title and estate of his father, becoming the third Baron Rayleigh. Apparently, Strutt read Helmholtz's work upon the suggestion of Professor Donkin, an astronomer at Oxford University who recommended that he should acquire a reading knowledge of German. Throughout his career Lord Rayleigh worked in several areas of physics, publishing over 400 scientific papers, 128 of them dealing with acoustical phenomena. He began writing his two-volume treatise *The Theory of Sound* while recuperating from an illness in 1872, and saw its publication in 1877, with a revised and enlarged edition incorporating many new developments in 1894.

Rayleigh's treatise is largely a work of synthesis, integrating his own discoveries with those of others. It deals both with the origin of sound and its transmission through the air and other media. As Professor R. B. Lindsay wrote in the preface to the first American edition, "Lord Rayleigh appeared on the acoustical scene when the time was precisely ripe for a synthesis of experimental phenomena, much of which was, however too idealized for practical application." Much of Rayleigh's book deals with approximation methods for dealing with vibrations and wave motion in situations in which the equations of motion can not be solved in any simple mathematical form. Rayleigh's careful organization, clarity of expression, and balance between mathematical elegance and the pragmatic realities of laboratory investigation assured the book's value to generations of future physicists.

Impact

Modern acoustics is the study of the generation, transmission, and detection of sound waves travelling through gaseous, liquid, or solid media, including the reflection and refraction of sound waves at the boundaries between different media. Rayleigh's treatise provides a compact and thorough summary of the first two areas, generation and transmission, and in its second edition makes a connection with the emerging technology of the telephone.

The need for understanding the fundamentals of sound production, transmission, and detection was to grow rapidly following the appearance of Rayleigh's treatise in 1877, the same year that Thomas Edison (1847-1931) demonstrated the first phonograph. The previous year, a United States Patent had been issued to Alexander Graham Bell (1847-1922) for a "device to transmit speech sounds over electric wires." Fourteen years later, Carnegie Hall in New York City opened for its first concert, a building considered nearly perfect for musical performances without electronic amplification due to its special acoustic design. With the invention of the vacuum tube triode by the American Lee De Forest (1873-1961) in 1907 it became possible to amplify sounds to levels audible throughout a large area, the acoustics of loudspeakers and ever-larger auditoriums became matters of immediate practical concern. The first motion picture with sound was demonstrated in 1926.

The use of ultrasound, or sound waves traveling above the range of human hearing, has come to be very important in many areas of technology. Military applications of ultrasound generation and detection were developed during the First World War. SONAR (sound navigation and ranging) was needed to track submarines under water. Depth sounders, which bounce sound waves from the water bottom, are an important byproduct for civilian use. Ultrasound is routinely used in industry to nondestructively test for defects in materials. It has also proven its value in medical diagnosis, as a portable means of imaging soft tissues without the radiation hazard inherent in x-rays. Expectant mothers are now routinely examined by ultrasound to check the health of the developing fetus. Frequently they receive a videotape or computer-generated image to take with them, and sometimes these are carried and displayed like baby photos by the proud parents to be.

With the invention of the transistor by John Bardeen (1908-1991), Walter Brattain (1902-1987), and William Shockley (1910-1989) in 1947, it became possible to build very small sound amplification systems, systems small enough to be carried about by the hearing impaired. Since the 1950s portable hearing aids have gotten smaller and smaller with many modern aids fitting entirely within the ear canal.

Acoustical techniques are also employed in seismology, the study of vibrations of the earth, and particularly of earthquakes. Seismic waves are typically subsonic, or lower in frequency than the range of human hearing. An earthquake gives rise to several different types of acoustic waves that arrive at points in the Earth's surface at different times. At first, the acoustic data from identified earthquakes was used to refine the model of the Earth's inner structure. Currently seismic data is used to locate and profile new earthquakes. Seismic monitoring is also used to identify belowground explosions of nuclear weapons.

The mathematical analysis of sound waves can be used in modern electronic technology to synthesize both speech and musical sounds. Thus the approach begun by Helmholtz and Rayleigh has, in a sense, come full circle to produce computers that can speak and listen, as well as the keyboards and synthesizers that are ubiquitous in contemporary music.

DONALD R. FRANCESCHETTI

Further Reading

Helmholtz, Hermann L. F. *On the Sensations of Tone.* New York: Dover Publications, 1954.

Hunt, Frederick V. *Origins in Acoustics: The Science of Sound From Antiquity to Newton.* New Haven: Yale University Press, 1978.

Levinson, Thomas. *Measure for Measure: A Musical History of Science.* New York: Simon & Schuster, 1994.

Strutt, John William. *The Theory of Sound.* 2 vols. New York: Dover Publications, 1945.

Leverrier, Adams, and the Mathematical Discovery of Neptune

Overview

The planet Neptune was discovered in 1846 following laborious calculations by Englishman John Couch Adams (1819-1892) and Frenchman Urbain Leverrier (1811-1877). These astronomers, attempting to explain deviations noted in the orbit of Uranus, independently and nearly simultaneously predicted the location of Neptune, which was then located with little trouble by German astronomer Johann Gottfried Galle (1812-1910). This discovery was viewed as a triumph of mathematics and physics, winning acclaim for both Adams and Leverrier.

Background

Through most of human history mankind lived in a solar system with seven "planets"—Mercury, Venus, the Moon, Mars, Jupiter, Saturn, and the Sun—that revolved around the Earth. The work of Galileo (1564-1642) and Nicolaus Copernicus (1473-1543) showed this notion to be false, but the Copernican solar system still ended at the orbit of Saturn until William Herschel (1738-1822) discovered an additional planet, Uranus, in 1781. Strictly speaking, Herschel was not the first person to see Uranus; at least 20 "pre-discovery" observations were uncovered following its official discovery, but it was not recognized as a planet. As a sixth magnitude object, Uranus is barely visible to the naked eye, so it must have been seen intermittently by ancient stargazers. Herschel was simply the first to recognize it for what it was.

A photograph of Neptune. The planet's discovery was a key event in nineteenth-century astronomy. *(AP/Wide World. Reproduced by permission.)*

Occurring less than 100 years after Isaac Newton (1642-1727) published his theories of gravitation and motion, astronomers immediately used Newton's laws and the celestial mechanics of Johannes Kepler (1571-1630) to check the observed motion of Uranus against predictions. Expected to confirm Newton, these observations instead showed discrepancies that worsened with time. The dilemma posed was that either Newton's laws were incorrect or some force was pulling Uranus from its predicted path. Even when accounting for the gravitational attraction of other major planets the discrepancy was not fully resolved. The only known force that could account for the noted discrepancies was another planet even further from the sun than Uranus. Finally, in 1846, two mathematicians, Leverrier of France and Adams of England, nearly simultaneously determined that another planet must be causing the noted discrepancies. Their independent calculations agreed closely with respect to the new planet's distance from the sun, orbital period, mass, and location in the sky.

Both Leverrier and Adams had difficulty persuading their colleagues to search for the new planet. Adams ran into opposition from Sir George Airy (1801-1892), the Astronomer Royal of Great Britain, who questioned whether it was possible to discover a planet simply by performing calculations. At the same time, Leverrier's unpopularity with French astronomers led to repeated refusals to search the sky at the calculated coordinates. Finally, Leverrier persuaded Galle, a German astronomer, to search for the new planet while, in England, Airy agreed that Professor James Challis (1803-1882) should try to locate it. Aided by a set of new and unpublished star charts of the ecliptic (the plane in which the planets lie), Galle and his assistant Heinrich d'Arrest (1822-1875) found the new planet in less than an hour on their first night of observations. Further observations confirmed the discovery.

It is interesting, in retrospect, to note that both Leverrier's and Adams's solutions for Neptune's orbit were incorrect. They both assumed Neptune to lie further from the sun than it actually does, leading, in turn, to erroneous calculations of Neptune's actual orbit. In fact, while the calculated position was correct, had the search taken place even a year earlier or later, Neptune would not have been discovered so readily and both Leverrier and Adams might well be unknown today except as historical footnotes. These inaccuracies are best summarized by a comment made by a *Scientific American* editor:

Leverrier's planet in the end matched neither the orbit, size, location or any other significant characteristic of the planet Neptune, but he still garners most of the credit for discovering it.

It is also worth noting that, after Neptune's mass and orbit were calculated, they turned out to be insufficient to account for all of the discrepancies in Uranus's motion and, in turn, Neptune appeared to have discrepancies in its orbit. This spurred the searches culminating in Pluto's discovery in 1930. However, since Pluto is not large enough to cause Neptune and Uranus to diverge from their orbits, some astronomers speculated the existence of still more planets beyond Pluto. Hence, Pluto's discovery, too, seems to be more remarkable coincidence than testimony to mathematical prowess. More recent work suggests that these orbital discrepancies do not actually exist and are due instead to plotting the planets' positions on the inexact star charts that existed until recently.

Impact

Neptune's discovery resonated on several levels in the world of the mid-1800s. 1) It expanded the size of the solar system by a factor of 50%; 2) It served to validate Newton's and Kepler's laws of physics and celestial mechanics; 3) It heated up the Anglo-French scientific rivalry; and 4) It made the universe seem more understandable and predictable.

Of Leverrier's part in this discovery, the reigning French astronomer Francois Arago (1786-1853) said:

> In the eyes of all impartial men, this discovery will remain one of the most magnificent triumphs of theoretical astronomy, one of the glories of the Académie and one of the most beautiful distinctions of our country.

> Another of Leverrier's colleagues proclaimed ". . . he discovered a star with his pen, without any instruments other than the strength of his calculations alone."

The comments are indicative of the scientific impact of this discovery; Neptune was found because it was calculated that it *had* to be there, and it was. This was taken to indicate that scientists were finally starting to understand the laws that made the cosmos work. Scientists took this as validation of Newtonian physics, Keplerian celestial mechanics, and the sheer power of science properly applied. It is also noteworthy that Neptune is the only planet discovered in this manner. Uranus, the first telescopically discovered planet was found as the result of a search, but one that was not mathematically directed. A position was calculated for Pluto, but Pluto was found only after a laborious search and was not nearly as close to the calculated position as was Neptune.

The discovery of Neptune raised additional questions about the place of God in the universe. The discoveries of "deep time," fossils, and Uniformitarianism by geologists reduced the authority once held by the Book of Genesis as an accurate accounting of the early history of the Earth. Similarly, the discovery of Copernican astronomy, the mapping of the heavens, and the more recent discoveries of the sheer size and age of the universe seem to leave less necessity for an almighty Creator to form, maintain, and rule the universe. Instead, the universe seemed to become increasingly mechanistic as the powers of scientific prediction and explanation increased.

In scientific circles, this successful prediction led to continuing efforts to mathematically "discover" still more planets. Noting discrepancies in Mercury's orbit, Leverrier calculated the orbital parameters of an inner planet he named Vulcan. In spite of repeated sightings of candidates, it was eventually determined that Vulcan does not exist. In a similar vein, many astronomers have determined orbits for planets beyond Neptune, suggesting from one to four trans-Neptunian planets might exist. In fact, the paths of the Pioneer and Voyager space probes were tracked as they left the solar system to determine if any of the probes were gravitationally influenced by large unknown planets in the outer solar system. To date, no studies have shown the existence of a single large planet beyond Neptune, although a number of trans-Plutonian asteroids have been located.

The events of 1846 significantly increased the status of science, rather than religion, as a model of explanation. The Enlightenment of the eighteenth century had started such a process long before Neptune's existence was even postulated, let alone confirmed. However, the manner in which Neptune's position was predicted and confirmed served notice that science was beginning to assume increasing importance in its ability to predict and explain the workings of our planet and universe.

The impact to the average person was less significant, but still noticeable. One indication of this is the headline of the London *Times* on October 1, 1846: "Leverrier's planet found." Both England and France found reason to exhibit national pride in the achievements of their scientists and, in fact, relations between these two nations were strained much more than relations between Adams and Leverrier until it was decided to credit both men with Neptune's discovery. It is also noteworthy that, even today, publications of the Royal Greenwich Observatory bemoan the fact that Neptune's discovery must be shared with someone from another country and examine the events that kept Britain from having sole credit for the discovery.

P. ANDREW KARAM

Further Reading

Kippenhahn, Rudolf. *Bound to the Sun.* New York: W. H. Freeman and Company, 1989.

Miner, Ellis D. *Uranus, the Planet, Rings, and Satellites.* New York: Ellis Horwood Publishers, 1990.

Heavenly Rocks: Asteroids Discovered and Meteorites Explained

Overview

Giuseppe Piazzi discovered the first asteroid, Ceres, on New Year's Day, 1801, then lost it when it traveled behind the Sun. Luckily, a new and valuable mathematical method, least squares, allowed Ceres to be rediscovered. This success established the reputation of Carl Friedrich Gauss, one of history's greatest mathematicians. Just two years later in 1803, the view of the Solar System was further expanded by a study that forced science to accept the idea that space debris came to Earth. French physicist Jean-Baptiste Biot investigated a rock that fell from the heavens, and his report convinced skeptics that meteorites came from space. By the century's end, meteorites were accepted as extraterrestrial in origin, and the first steps were taken toward proving that craters were on Earth as well as on the Moon.

Background

Giuseppe Piazzi (1746-1826) was an eminent astronomer who directed two prominent Italian observatories in Naples and Palermo. He was not looking for an asteroid (more properly, minor planet) in 1801. Rather, he was doing a systematic study of stars, work that ultimately resulted in his cataloging of the positions of 7,646 stars. His knowledge of an upcoming search for a new planet led him to correctly understand what it was he had accidentally discovered during his search, Ceres (named for the patron goddess of his native Sicily), the largest asteroid. The search, and his conclusion, were both prompted by a numerical coincidence that is still unexplained.

In 1766 J. D. Titius discovered a curious regularity to the placement of the order of the planets. He constructed his series of numbers in this way: Take 0 and add 3, then double the number for each succeeding number in the series so that you get 0, 3, 6, 12, 24, etc. Finally, take each of these numbers, add 4, and then divide by 10. The third number in the series is one. The third planet from the Sun, Earth, by definition, is one astronomical unit (AU) away. The first number in the series, 0.40, is a good match for Mercury (0.39 AU). In fact, the rest of six known planets — Venus, Mars, Jupiter and

Saturn — all match up fairly well if you ignore the curious gap at 2.8 AU. Johan Bode (1747-1826), a German astronomer, popularized this formula (now known as Bode's law) in 1772. It might have faded into obscurity, but William Herschel (1738-1822) made a discovery that changed the scientific world.

In 1781 the scientific world was complacent. Isaac Newton's physics was king, and, for many scientists, it explained the universe. In the view of much of the scientific community, everything that was important had already been discovered. Their jobs involved merely working out the details. Then William Herschel looked at Bode's law and wondered: Was there another planet to discover? The numbers told him where to point his telescope. He didn't find anything at 2.8 AU, but he did find something by looking further out: Neptune, a new planet, and, at 19.2 AU, a better match to its spot in the series of Bode's law than Saturn was.

Herschel's discovery excited the whole scientific community; it quite literally opened up new worlds. Astronomy was no less energized. In 1800 German astronomers under Baron von Zach were organizing to conduct a thorough search for a planet near 2.8 AU. Before they could commence their work, however, Piazzi had announced his discovery.

While looking for the planet, Piazzi spotted what he thought might be an uncharted star because it was a simple point of light. (Asteroids, meaning "star-like," got their name from Herschel because of this appearance.) To Piazzi's surprise, he found that it had moved with his next observation. It wasn't a star, and it wasn't a comet, either. Comets typically appear fuzzy because of their escaping gases. This image was sharp. At the same time, it didn't have the disc shape of a planet. Whatever it was, Piazzi knew it was right where the Germans intended to look, at 2.8 AU. Piazzi reported his findings to the scientific community knowing they would find the object's position to be significant. Unfortunately, he was only able to make a few observations before the astronomical body was lost behind the Sun.

In 1801 no means existed for predicting where Ceres would reemerge from behind the

Sun, and it seemed likely that Piazzi would never see it again. Fortunately, the problem caught the interest of one of the most brilliant mathematicians ever. Carl Friedrich Gauss (1777-1855) was still in his early twenties when, from just three recorded observations, he made a prediction of where astronomers would find Ceres. Late in 1801 Baron von Zach and Heinrich Olbers (1758-1840) relocated the minor planet right where Gauss said it would be.

Just two years later in 1803, what was probably another asteroid accounted for another challenge to astronomy. The question in this case wasn't where the rock—a meteorite—was, but where it came from. Reports of rocks falling out of the sky extend throughout recorded history, but the scientific community did not believe these fantastic tales. Even when respected physicist Ernst Chladni (1756-1827), who collected meteorites, suggested they were the debris from an exploded planet, the notion was still considered nonsense. The kindest explanation was that the rocks fell from a short distance, having been thrown up from the Earth by a great force. In 1807 in America eminent Yale chemist Benjamin Silliman (1779-1864) and a colleague reported that they had seen a meteorite fall and found it. Even this strong a testimony met resistance. United States President Thomas Jefferson, an amateur scientist and a committed rationalist, declared that it was easier to believe that two Yankee professors were lying than that rocks fell from the sky.

By then, however, the point had already been won in Europe. In 1803 there were reports of a recent meteorite fall in France. The French intellectual community, fiercely rational, decided to settle the question of the rock's origin once and for all. They sent Jean-Baptiste Biot (1774-1862), one of their best young scientists and a protégé of the powerful and influential astronomer Pierre-Simon Laplace (1749-1827), to listen to the reports and to examine the rock. Biot, going against the flow of the scientific community, declared his finding that the meteorite had indeed fallen from the heavens.

Impact

The discoveries of Piazzi and Biot led to renewed wonder and curiosity about the heavens. Asteroids were different from anything else in the known heavens, and within a few years three more had been discovered, two of them by Heinrich Olbers. Today we know there are thousands of asteroids, mostly in the asteroid belt at

about 2.8 AU. This belt became one of the prime points of evidence that caused the scientific community to accept the theory for which Laplace is perhaps most famous, the nebular hypothesis for the origin of the Solar System. By this hypothesis, a rotating cloud of matter cooled and contracted to form the Sun and its planets. The belt represents matter from this earlier stage, and now it is generally believed that the asteroids did not come together because of the disruptive gravitational forces of its huge neighbor, Jupiter.

Asteroid hunting captured the public imagination during the nineteenth century—300 were found before the advent of photographic astronomy in 1891. One of the oddest "finds" was by Frederic Petit. In 1846 he claimed to have discovered Earth's second moon, a captured asteroid. The details of the report were somewhat absurd, with a regular close approach to Earth by the asteroid of just 7 mi (11.4 km). Petit was tenacious. Fifteen years later he was using his second moon to explain anomalies in Lunar motion. Most scientists ignored him, but Jules Verne did not. In Verne's 1865 novel, *From the Earth to the Moon*, Petit's asteroid makes a frighteningly close approach to the heroes and becomes a reference point that allows them to determine their position.

Mathematical prediction seemed to take a giant step forward with the second success of Bode's law, but this triumph was short lived. Astronomers hunted near 38.8 AU for the next planet in the series in vain. In 1846, after precisely calculating the orbit of Uranus and looking at how the actual orbit deviated, Urbain Leverrier (1811-1877) found Neptune at 30.06 AU. This planet was too close to the Sun and too large to explain away, and it broke the series. To date there is no explanation for the success (or failure) of Bode's law, and it is largely considered to be merely a numerical coincidence.

Perhaps the largest impact of the fresh look at the heavens at the beginning of the nineteenth century was the world's discovery of a genius, Carl Friedrich Gauss. His creation of the method of linear least square alone has had a profound effect on all observational science. Not only has it provided a convenient and reliable way of reducing error in natural sciences, such as chemistry and physics, but it has become one of the standards for assessing the legitimacy of observations in the social sciences, such as psychology and anthropology.

The work on Ceres also helped Gauss to continue his career after his sponsor died in 1807. His contribution to astronomy had enhanced his reputation and helped win him a position as professor of astronomy and director of the observatory at the University of Göttingen even though his nation was at war. He did not disappoint his sponsors. He went on to show how probability could be represented as a bell-shaped curve and to make contributions to geometry, the understanding of fluids, gravitation, and magnetism.

Biot's report on meteorites meant that it was possible, without traveling to other planets or collecting moon rocks, for scientists to chemically analyze extraterrestrial material and get a better understanding of the Solar System. Meteorites also brought into question the origin of craters. Craters on the Moon (and on Earth) had always been assumed to be of volcanic origin. In 1873 Richard Proctor was the first to suggest that the craters on the Moon came from bombardment of rocks, an idea that quickly gained broad acceptance. The first study of an Earth crater, Barringer Meteor Crater in Arizona, began just before the end of the nineteenth century. Oddly enough, the original work indicated that the crater did not originate with a meteorite. However, early in the twentieth century this position was reversed.

Biot's report changed more than the scientific community's view of meteorites. It helped to open scientists up to the reports of non-scientists, which, among other things, has led to the discovery of new medicines and new animal species.

PETER J. ANDREWS

Further Reading

Asimov, Isaac. *Isaac Asimov's Biographical Encyclopedia of Science and Technology.* New York: Doubleday and Co., 1976.

Friedlander, Michael W. *Astronomy: From Stonehenge to Quasars.* Englewood Cliffs, NJ: Prentice-Hall, 1985.

Harrington, Philip S. *Touring the Universe Through Binoculars : A Complete Astronomer's Guidebook.* New York: John Wiley and Sons, 1990.

Miller, Ron, and William K. Hartmann. *The Grand Tour: A Traveler's Guide to the Solar System.* New York: Workman Publishing Co., 1993.

Nineteenth-Century Developments in Measuring the Locations and Distances of Celestial Bodies

Overview

At the beginning of the nineteenth century, astronomers knew how far away some objects in the solar system are, but not how far away any of the stars are. During the nineteenth century they measured accurate positions for thousands of stars and gathered their data in catalogs. Their measurements were precise enough to reveal tiny shifts in stellar positions. These shifts resulted not from the motions of the stars but from the motions of Earth in its orbit, and this knowledge allowed several astronomers in the 1830s to measure the distances to several of the closest stars. These distance measurements were the beginning of the work that is at the heart of modern cosmology: the discovery of the size and shape of the universe.

Background

During the nineteenth century, universities and observatories became important resources that allowed scientists to carry out large projects like star catalogs. The first three people to measure the distance to a star—Friedrich Bessel, Thomas Henderson, and F. G. W. Struve—all worked for organizations that supported their work and supplied the fine instrumentation they used. Astronomers were able to use the resources of their schools and observatories to work together to produce accurate and complete catalogs of the stars. All of this astronomical activity led to several important discoveries of how the positions of the stars in the sky change as a result of the motion of Earth around the Sun.

One of the effects of the way that Earth orbits the Sun is that over the course of a year nearby stars show a tiny annual back-and-forth motion called parallax. Parallax provides the most direct way to find the distance to a star. You see something like parallax when you walk from one place to another and notice that a nearby object like a tree seems to be in a differ-

Friedrich Bessel. *(The Library of Congress. Reproduced by permission.)*

ent position against more distant background objects, say to the right of a building rather than lined up with the building. The farther you walk and the closer the tree is to you, the more it will seem to shift against the background. For many years astronomers knew that stars should show the same kind of change in position, and if they could measure a star's parallax they could calculate its distance. The distance we move (from one end of Earth's orbit to the other) is the same for all stars, and the closer a star is to us, the more parallax we should see compared to more distant stars. We see parallax not because the stars themselves are moving, but because our view of them is. The apparent changes in a star's position are far too small to be visible to the naked eye, or even to be easily measurable. Astronomers from Galileo onward tried to measure parallax, but it wasn't until the 1830s that astronomers had the tools that could do the job.

Before the apparent motion due to parallax could be measured, astronomers had to identify and sort out other effects of Earth's motion. James Bradley (1693-1762) discovered small effects called aberration and nutation that other astronomers then took into account in making their parallax measurements. Without taking these tiny effects precisely into account, astronomers would have made inaccurate measurements. Furthermore, many star catalogs

recorded the motion of a star across the sky, called its proper motion. Proper motion is due to the motion of the star relative to the Sun. Astronomers measuring the distance to stars chose to observe stars with a large proper motion, guessing that these stars might be closer and show a larger (and easier to measure) parallax. Without the catalogs to point them toward likely targets, their work would have been much more time consuming and difficult.

Astronomers in the nineteenth century also investigated the Sun. Joseph von Fraunhofer (1787-1826) studied the Sun and developed the heliometer, an instrument that measures the width of the Sun on the sky very precisely. Friedrich Bessel (1784-1846) used this instrument to measure a star's parallax and calculate its distance.

Bessel used the heliometer to measure the position of a star called 61 Cygni relative to a dimmer star nearby. His own catalog showed 61 Cygni to have a large proper motion. He guessed (correctly, as it turned out) that the dimmer star was much further away and would show no measurable parallax, and so could be considered an unmoving reference point for measuring 61 Cygni's parallax. He compared the positions of the two stars at different times of the year, six months apart (when Earth was at opposite ends of its orbit). He was able to measure 61 Cygni's parallax and calculate its distance, and he found that 61 Cygni was about ten light years away (a figure only about 6% higher than the currently accepted distance). At around the same time, two other astronomers announced that they had used parallax to calculate stellar distances as well. Thomas Henderson measured the distance to Alpha Centauri (the closest star to the Sun), and F. G. W. Struve (1793-1864) did the same for Vega. By the end of the century astronomers had used parallax to measure the distance to approximately thirty stars.

Impact

Bessel's work, along with that of Struve and Henderson, proved that astronomers could measure the distances to stars using parallax. Other parallax measurements followed, first a trickle and later a flood. Struve, like Bessel, used a relatively sophisticated instrument that was not widely available at other observatories, but Henderson used a fairly common piece of equipment, so other observatories were able to begin making their own parallax measurements. In the late 1880s astronomers in England began to use photography to measure parallax and found that it took less work and was more accurate. In the twentieth century a satellite measured the parallax of thousands to stars.

As scientists measured parallaxes and calculated distances, they realized that the brightest stars were not, as you might expect, always the closest. This was a clue to them that not all stars are the same. If a star has no measurable parallax, it must be relatively far away. If it still appears very bright to us despite the distance, it must be a true powerhouse of a star. Later astronomers used other tools to classify stars based on their brightness and other properties. Their work relied upon this discovery about differences between stars.

Distance measurements, along with the positional information contained in the great catalogs of the nineteenth century, enabled astronomers by the end of the century to begin to try to map out the parts of the universe that we could see. Although astronomers were not yet sure that the Milky Way was a stellar system separate from other systems (which we call galaxies today), they were able, in the 1870s, to see that the Milky Way rotated.

Once astronomers began to measure the distances to stars, they could start to use a technique known as the standard candle. Sometimes all the stars (or galaxies) of a particular type are of roughly the same brightness, and that type of star or galaxy can be used as a standard candle. Any time you see a star of that type, you know how bright it really is. This is different from how bright it appears from Earth. The brightness that we see in the night sky is a combination of the star's true brightness and its distance from us. The true brightness, the apparent brightness in the sky, and the distance are like three pieces of a puzzle; if you have any two of them, you know what the third piece is. So if you know what type a star is, and that type has a constant true brightness, you can measure the apparent brightness that we see and calculate its distance. This method is a powerful one for determining

cosmic distances, and it is still being refined today. But the first step in getting it to work is to find the distance of at least a few "standard candle" stars so that you can use the distance and the apparent brightness to find the true brightness. Thus, distances found through parallax are the first step in figuring out distances to much more faraway stars and eventually galaxies.

In the late twentieth century a European satellite named Hipparcos mapped the skies with unprecedented accuracy and measured the parallaxes of thousands of stars. Since Hipparcos was above the distorting effects of Earth's atmosphere, it could measure very tiny changes in position. The data from Hipparcos are being used to answer some important questions in cosmology by recalculating the distance to some of the known "standard candle" stars, which are crucial links to determining the age and size of the universe.

Using various standard candles, astronomers in the twentieth century extended their distance scale until now we measure distances to galaxies that are billions of light years away. The rate at which these galaxies are moving away from us is relatively easy to measure. The distances are harder to measure but are very valuable to know because they help astronomers determine how fast the universe is expanding and how old it is, both necessary to explaining the origin and life history of the universe.

MARY HROVAT

Further Reading

Books

Ferguson, Kitty. *Measuring the Universe: Our Historic Quest to Chart the Horizons of Space and Time.* New York: Walker & Co., 1999.

Periodicals

Lovi, George. "Rambling through the Skies." *Sky & Telescope* 84 (September 1988): 275-276.

Roth, Joshua and Roger W. Sinnott. "Our Nearest Celestial Neighbors." *Sky & Telescope* 92 (October 1996): 32-34.

Trefil, James. "Puzzling Out Parallax." *Astronomy* 26 (September 1998): 46-51.

A New View of the Universe: Photography and Spectroscopy in Nineteenth-Century Astronomy

Overview

The development of photography and spectroscopy in the nineteenth century allowed astronomers to record and analyze the light coming from stars and other celestial objects. This transformed astronomy from a purely descriptive science to a systematic study of the behavior of these objects, laying the foundation of the discipline we now call astrophysics. The realization that the stars are made of elements also found on Earth, and that the Sun is actually a rather ordinary star, changed the way we look at ourselves and the Universe.

Background

Leonardo da Vinci (1452-1519) and other Renaissance scientists experimented with early cameras—optical devices for projecting an image onto a surface. However, at that time there was no way to preserve the image. In 1727 Johann Schulze (1687-1744), a German physicist, discovered that silver salts are sensitive to light and created images using them. A century later, the French physicist Joseph Nicephore Niepce (1765-1833) found a way to "fix" these images onto a metal plate so that they could be kept indefinitely. The earliest surviving photograph, of the view out Niepce's window, was made in 1826. Subsequent improvements by chemists improved the process of making photographs, as well as enhanced their quality and durability.

Astronomers were quick to recognize the potential for photography to aid and publicize their work. At the 1862 World Exhibition in London, stereoscopic "three-dimensional" photographs of lunar craters and sunspots were a great success. The needs of astronomers also helped to drive photographic technology. Images took so long to develop with the earliest photographic chemicals that fast-moving shutters were not necessary for ordinary photography. However, they were invented by the Englishman Warren de la Rue (1815-1889) to enable photographs of the Sun to be taken without burning the film.

Photographic images were found to have a number of advantages over direct observation, in addition to the obvious benefit as records of permanence and portability. Photographic film can detect finer gradations of light than can the eye, allowing greater detail to be observed. Film is also sensitive to parts of the electromagnetic spectrum outside the human visual range, such as ultraviolet, and can be adjusted to emphasize particular wavelengths or colors. With a long exposure, film can detect objects far too faint for the human eye to see. Eventually photography became so essential to astronomical observation that professional astronomers rarely observed directly with the eye. Most of their work was done by studying photographic plates. This was the case until fairly recently as electronic imaging and computer analysis became the new standard for observation.

The other major new astronomical technique used during the nineteenth century was spectroscopy. During the 1660s Isaac Newton (1642-1727) had shown that the light from the Sun could be broken up into a continuous spectrum like a rainbow by using a prism. Later, it was discovered that the same effect could be achieved and more precisely measured using a diffraction grating with thousands of tiny grooves or slits. In 1802 William Hyde Wollaston (1766-1828) observed a few dark lines breaking up the solar spectrum; he assumed that these were the boundaries between colors. But beginning in 1814 the Munich optician Joseph von Fraunhofer (1787-1826) mapped hundreds of these tiny gaps, which came to be called Fraunhofer lines.

Meanwhile, in the laboratory scientists turned their spectroscopes to as many sources of light as they could find. They soon realized that passing light through a gas produced dark lines at various points on the spectrum, while burning a substance produced a spectrum with bright lines superimposed upon it. In 1859 Gustav Robert Kirchoff (1824-1887) of Heidelberg recognized that the dark and bright lines at a given wavelength were produced by the same materials. If the material was burned, bright lines were seen. If light was passed through it, there would instead be a dark line in the same place on the spectrum. The key was the temperature. A hot gas gives off light at particular wavelengths, producing bright emission lines. If the gas is cooler than the light coming from behind it, it instead absorbs light at its characteristic frequencies, and dark absorption lines are seen. The wavelengths of the lines are actually the result of changes in atomic and molecular energy levels, concepts unknown in the nineteenth century. But well before the real reason for absorption and emission lines was understood, astronomers were using them to gain a great deal of information about the composition and temperature of objects in the sky.

Impact

By the middle of the nineteenth century astronomers were taking photographs of the Sun, Moon, and stars. During the solar eclipse of 1860, de la Rue and the Italian astronomer and Jesuit priest Pietro Angelo Secchi (1818-1878) both obtained good results, which together revealed the outer corona of the Sun and the solar prominences, or great eruptions extending from the surface. While these had been observed before, some astronomers had argued that they were optical illusions; this theory was now disproved.

The permanent and detailed records of star fields that accumulated on photographic plates, including several systematic sweeps of the heavens, allowed stellar photometry, the measurement of light from the stars, to be put on a scientific basis for the first time. Previously, stellar magnitudes, or brightnesses, were estimated visually by comparing one star to another in a 2000-year-old system originated by the great astronomer Hipparchus (c.190-c.120 B.C.). Now, with special filters and plates, more accurate catalogs of stars could be produced.

Locations of objects could also be pinpointed more precisely and compared in successive photographs of the same region. Many variable stars were discovered, the brightness of which changed over time. Searching large numbers of photographic plates for small changes in detailed star fields proved a labor-intensive task, and many astronomers delegated this exacting work to female assistants. This provided an entry path for a number of women into a previously all-male field.

Since the turn of the nineteenth century it had been known that there existed a number of minor planets, or asteroids, in the large gap between the orbits of Mars and Jupiter. A few had actually been observed through telescopes and their orbits calculated. In 1891 Max Wolf (1863-1932) of Heidelberg applied photography to the search by using long exposures. An asteroid, so much closer to Earth than the stars in the background, moves fast enough across the field to leave a short but noticeable track on the plate. With this method asteroids were catalogued by the hundreds over the next few years.

Spectroscopic techniques were quickly recognized as being important in astronomy because of the ability to study hot gaseous objects like the Sun and stars. The absorption lines observed by Fraunhofer indicated that sunlight comes from hot gases beneath the surface of the Sun, and is absorbed at some wavelengths by cooler gases higher up. Thus spectroscopy provided clues to the layered structure of the gases in the Sun, which would be better understood in the twentieth century in terms of the nuclear reactions that take place in stellar cores.

Known substances were studied in the laboratory so that the positions of their lines could be compared to those observed in space. By the end of the century, dozens of elements had been identified in the Sun, including hydrogen, sodium, calcium, magnesium, carbon, and iron. The eclipse of 1868 had allowed astronomers to take advantage of the blocked solar disc to take the spectrum of the outer regions of the Sun. The English astronomer Sir Joseph Lockyer (1836-1920) found that, as well as hydrogen, there was another previously unknown element in the Sun, which he called helium after *helios,* the Greek name for the Sun. Helium was not discovered on the Earth until 1895, when it was found in the mineral clevite.

Soon astronomers learned to adjust their spectroscopes so that they could be focused on a specific region of the Sun without waiting for an eclipse. More frequent observations made clear the dynamic nature of the Sun, as sunspots waxed and waned, and disturbances in the outer atmosphere were associated with changes in the lines that were seen.

With better focusing techniques, spectra could be obtained from more distant stars. As early as 1823, Fraunhofer had written that other stars had lines in their spectra similar to that of the Sun. In 1864 Sir William Huggins (1824-1910) had identified these lines with elements on Earth. The fact that the Sun and stars are made of the same materials we find around us, and that there exist a great many stars similar to the Sun, contributed to a shift away from the generally accepted worldview that our Solar System is unique in character. Astronomers began to classify the stars in terms of the details of their spectra, and made inferences as to their stage of development, the beginning of the concept that the large-scale Universe has changed over time rather than having originated as we see it today.

Of course, there are other objects in the sky besides stars. In 1864 Giovanni Battista Donati (1826-1873) first applied spectroscopy to the observation of a comet, detecting lines but making no identification. Four years later Huggins saw similar lines and realized that they indicated the presence of hydrocarbons. The discovery of simple organic compounds in these "visitors" from outer space would lead some to speculate that life may have come to Earth as the result of an impact. However, this theory just removed the riddle of life's origins on Earth to an unknown outer space location.

Other objects that elicited great curiosity were the nebulae, or clouds. In 1781 the French observer Charles Messier (1730-1817) recorded the positions of 103 of these fuzzy patches in the sky so that he wouldn't confuse them with the comets he was seeking. The eighteenth-century philosopher Immanuel Kant (1724-1804) and a few other advanced thinkers considered the idea that these nebulae might be "island universes" outside the Milky Way. This first hypothesis of the existence of other galaxies was later rejected by prominent astronomers like William Herschel (1738-1822). Herschel's three great catalogues of nebulae included some that seemed to be halos of what he called "shining fluid" surrounding stars, and he assumed that this was the material of which all the nebulae were made.

Later generations of astronomers were to discover that, without a reliable distance scale, the early catalogues of nebulae lumped together separate entities such as the hot gases from exploding stars relatively nearby, clusters of stars within the Milky Way, and fuzzy glimpses of distant galaxies. The better resolution and understanding provided by nineteenth-century photographic and spectroscopic observations helped to sort out these different types of objects.

Another major contribution of spectroscopy to the science of astronomy was the information it provided about the motion of objects. Most people are familiar with the way the pitch of an ambulance siren falls as it recedes into the distance. This is an example in sound waves of an effect that Christian Doppler (1803-1853) first detailed for light in 1842. He showed that as a luminous body approaches, the light it gives off decreases in wavelength, shifting toward the blue end of the spectrum. Conversely, as it recedes, the light shifts toward the red. Changes in the expected positions of lines in a spectrum could therefore be interpreted as indicating motion in the object producing those lines.

Applying Doppler's principle to observations of the Sun, astronomers clocked violent disturbances in its outer atmosphere, which reached speeds of 300 miles per second. The rotation of the Sun was also confirmed by observing a red shift at one edge and a blue shift at the other. Many stars were discovered to be binaries, showing pairs of lines that shifted as the two stars orbited each other, receding from and approaching us in turn.

In 1868 Huggins observed a shift in a hydrogen absorption line in the spectrum of Sirius, and interpreted it as indicating that the star was moving away from the Solar System at a considerable speed. Other such observations followed, with everything seeming to be flying away from us. It was left to the twentieth-century astronomer Edwin Hubble (1889-1953) to realize that this was evidence of an expanding Universe, with all its components getting farther from each other like dots on the surface of a balloon being blown up. Projecting this scenario backward in time, scientists began to postulate the theory that the Universe resulted from a massive explosion, or "Big Bang," billions of years ago.

SHERRI CHASIN CALVO

Further Reading

Clerke, Agnes M. *A Popular History of Astronomy During the 19th Century.* 1908. Reprint, St. Claire Shores, MI: Scholarly Press, 1977.

Cohen, Bernard. *Aspects of Astronomy in America in the Nineteenth Century: An Original Anthology.* North Stratford, NH: Ayer Company, 1980.

Crowe, Michael J. *Modern Theories of the Universe: From Herschel to Hubble.* Mineola, NY: Dover, 1994.

Schaaf, Larry J. *Out of the Shadows: Herschel, Talbot, and the Invention of Photography.* New Haven: Yale University Press, 1992.

Taton, Rene, et al. *Planetary Astronomy from the Renaissance to the Rise of Astrophysics: Part B: The Eighteenth and Nineteenth Centuries.* Cambridge: Cambridge University Press, 1995.

Nineteenth-Century Efforts to Catalog Stars

Overview

Among the many achievements of nineteenth-century astronomy was the production of a large number of comprehensive and accurate star catalogs. Building upon the less precise observations of previous centuries, European astronomers in the nineteenth century took advantage of better telescopes and new technologies to produce their improved catalogs. Yet their projects to map, chart, and catalog stars affected more than just the science of astronomy. Their efforts would help advance navigation and exploration and promote international co-operation among astronomers of different nationalities during a period of tension between the nations of Europe.

Background

The production of star catalogs is an ancient practice. Many pre-historic societies around the world had some method of charting or recording the movements of the stars, planets, the Moon, and the Sun. These societies even learned that it was possible to navigate by using the stars. The early Greeks produced star catalogs from naked eye observations that were used for centuries in Europe, Asia, and Africa. Centuries later Nicholas Copernicus (1473-1543) argued that Earth goes around the Sun instead of the Sun going around Earth. One implication of this new theory was that the stars must be much farther away from Earth than was previously thought. After Copernicus European astronomers began to study the stars more carefully, cataloging them with newfound interest. Even before the use of telescopes, astronomers such as Tycho Brahe (1546-1601) of Denmark and his assistant Johann Kepler (1571-1630) of Germany gathered tremendous amounts of data on star positions. Galileo Galilei (1564-1642) of Italy, perhaps the first European to turn a telescope on the stars, immediately recognized the advantage of using telescopes to explore and chart the heavens.

If the centuries following Copernicus saw an increase in the exploration of the heavens, it is interesting to note that there was a corresponding increase in the exploration of Earth. After Christopher Columbus's rediscovery of the New World in 1492, European nations began to explore and conquer the rest of the world. The ability of their ships to navigate the oceans was crucial to the task of building their overseas empires. Having accurate star catalogs was an important part of navigation for powerful countries such as Spain and England, who developed have the largest empires.

England provides us with a good example of the connection between navigation and astronomy. In the seventeenth and eighteenth centuries English astronomers John Flamsteed (1646-1719), Edmond Halley (1656-1742), and James Bradley (1692-1762) all helped to compile star catalogs that were used to aid English ships as they explored the globe. Each of these three astronomers worked as England's Astronomer Royal at some point in his career, a position that in part required the astronomer to provide the English navy with star charts for navigational purposes. Bradley in particular made some of the most accurate and extensive star observations of his day between the years 1750 and 1762. In this way the work of these English astronomers was closely tied to their country's tasks of exploration, navigation, and empire-building.

Also during the seventeenth and eighteenth centuries, newer, larger telescopes enabled astronomers such as William Herschel (1738-1822) to see farther than ever before. Some objects that previously looked like individual stars turned out to be double stars, clusters of many stars, or even nebulae (interstellar gas or dust). Both William Herschel and Charles Messier (1730-1817) produced catalogs of star clusters and nebulae, adding to the numerous star catalogs already available. Before the nineteenth century, then, there had been many star catalogs produced in Europe. Nineteenth-century astronomers would take these catalogs, build upon them, refine their measurements for accuracy, and vastly increase the number of stars and other objects cataloged.

Impact

The centuries before 1800 was a time when the cataloging of stars was a somewhat disorganized, if still productive, activity. The nineteenth century was a time of increased organization and production. With improved skills and better technologies, nineteenth century astronomers began to look at the catalogs of previous centuries and redo them with more accuracy. Two important astronomers did this: Friedrich Bessel (1784-1846) and John Herschel (1792-1871).

Friedrich Bessel started out as a merchant's clerk in Germany. But he wanted to become a sea-trader so he could have a better life, so he taught himself navigation and astronomy, the two skills needed by a sea-trader. Soon he became an expert in astronomical theory and went to work for a number of different German observatories. At one of these observatories he obtained the catalog made by James Bradley (mentioned above). Bradley's eighteenth-century observations of stars were the best available. With new instruments and mathematical techniques Bessel set about reworking Bradley's observations to make them more accurate. When he was done he had produced a catalog of more than 3,000 stars. This catalog, which he called the *Fundamenta astronomiae,* would be one of the most important star catalogs produced in the nineteenth century.

John Herschel also took the catalogs of a previous astronomer and reworked them. This astronomer was his father, William Herschel. William had cataloged thousands of double stars, star clusters, and nebulae, but only those visible in the Northern Hemisphere. John Herschel, who grew up surrounded by his father's telescopes, decided to complete his father's work by cataloging the night sky of the Southern Hemisphere. First he resurveyed the northern heavens, finding many objects that his father had overlooked. In 1833 he published a new catalog of the northern heavens that contained 2306 objects.

John then decided to go to the Southern Hemisphere to catalog the double stars, star clusters, and nebulae there. For more than four years he lived in Cape Town, an English colony at the southern tip of Africa. Although Cape Town had it's own observatory, built to help the English navigate the oceans, John had brought his own telescopes to Africa. In addition to cataloging stars and other objects in the southern sky, he spent time exploring the country, hunting for plants and animals, and watching the English and other Europeans interact with the native Africans. He explored both the heavens and the earth at Cape Town, and he considered himself to be a hunter of stars and nebulae as well as plants and animals.

But when John Herschel saw how poorly the Europeans treated the native Africans he changed this view. After he had observed how some cruel Europeans hunted the African peo-

ples, and how the Africans in turn would attack the Europeans, he no longer wanted to think of himself as a hunter. He began to see his search for stars and nebulae to be more like that of a farmer who harvests crops rather than a hunter who kills his prey. Here, then, we have an example of a nineteenth-century astronomer whose interaction with a different culture caused him to change the way he looked at astronomy. When he was done his observations became the most important catalog ever produced of the objects in the southern heavens. Astronomers around the world would use it for years to come.

John Herschel and Friedrich Bessel are just two examples of astronomers who produced important catalogs in the nineteenth century. The second half of the nineteenth century saw the further production of still more massive star catalogs. Each of these contained thousands of entries for the positions of stars, star clusters, and nebulae. The nebulae were of special interest to astronomers, and in 1864 John Herschel published his *General Catalog,* which contained positions for 5,000 of these objects. In 1888 John Dreyer (1852-1926) published the *New General Catalog,* a larger catalog of more than 13,000 nebulae. Not everyone was concerned with nebulae, however. Two German astronomers managed to produce a gigantic catalog containing nearly 325,000 individual stars. This catalog, done by Friedrich Argelander (1799-1875) and Eduard Schönfeld (1828-1891), was called the *Bonner Durchmusterung* and is still used today.

We have seen that nineteenth-century catalogs included many objects besides individual stars. The observation and cataloging of double stars, star clusters, and nebulae was an important part of nineteenth-century astronomy. It is also important to remember that all of the catalogs produced represent the work of many astronomers from different centuries and different countries. Because most catalogs were built upon the work of previous astronomers, no single astronomer could ever hope to be as productive as an entire community. The efforts made to catalog stars in the nineteenth century were co-operative; they were not always the work of one or a few astronomers. There are many examples of this co-operation.

In 1868 Friedrich Argelander persuaded the German Astronomical Society to lead an international program, the goal of which was to determine star positions to a high degree of precision. This project would involve about twenty observatories and resulted in considerable international cooperation among the world-wide astronomical community. And in Paris in 1887 an international congress of astronomers agreed to begin a survey of the entire night sky using the relatively new technology of photography. Eighteen observatories around the world participated, and many stars never before seen were cataloged. Both of these international projects took place at a time when the European political situation was going from bad to worse.

Near the end of the nineteenth century the nations of Europe were arguing among themselves over the ownership of other parts of the world. But in these international astronomical projects of 1868 and 1887 we can see that an international community of European astronomers was able to work together. They united the many distant observatories, those controlled by different countries, in the common project of observing the stars and improving star catalogs. The production of star catalogs in the nineteenth century was, in the end, the result of an astronomical community that cut across time and across national borders.

STEVE RUSKIN

Further Reading

Books

Dewhirst, David and Michael Hoskin. "The Message of Starlight: The Rise of Astrophysics." In *The Cambridge Concise History of Astronomy,* edited by Michael Hoskin. New York: Cambridge University Press, 1999.

Dick, Steven J. "Astronomical Catalogues." In *History of Astronomy: An Encyclopedia,* edited by John Lankford. New York: Garland Publishing, 1997.

Hoskin, Michael. "The Astronomy of the Universe of Stars." In *The Cambridge Concise History of Astronomy,* edited by Michael Hoskin. New York: Cambridge University Press, 1999.

Periodicals

Green-Musselman, Elizabeth. "Swords into Ploughshares: John Herschel's Progressive View of Astronomical and Imperial Governance." *British Journal for the History of Science* 31 (1988): 419-35.

John Dalton Proposes His Atomic Theory and Lays the Foundation of Modern Chemistry

Overview

As the nineteenth century dawned a significant problem that remained in the chemical sciences was the ultimate nature of matter. Was matter continuous and therefore had no finer structure or was it discontinuous and thus made of tiny particles? The chemical revolution due to the work of Antoine Lavoisier (1743-1794) and his circle that had occurred in the last two decades of the eighteenth century had clarified the concept of what elements are, developed a comprehensive and consistent vocabulary of chemistry, and led to the introduction of quantitative methods in chemical investigations. However, to fully understand the nature of chemical reactions one needed to have a way to visualize how the elements combined together. The atomic theory of matter as proposed by John Dalton in his New System of Chemical Philosophy (Part I,1808; Part II,1810) was the first successful attempt to solve this problem.

Background

The concept that matter may ultimately be composed of particles originated in Greek natural philosophy. In the fifth century B.C. Democritus (c. 460-370 B.C.) proposed that matter was composed of individual indestructible particles (called "atoms" in Greek for "uncuttable") and that the size and shape of these particles were responsible for the properties of matter. The atomic theory of the Greek philosophers lacked any evidence based upon observation, measurements, and testing by experiment. These ideas, though interesting, could not be considered a scientific theory.

The atomic concept was rejected by most Greek philosophers, particularly Aristotle (384-322 B.C.), because of the paradox that these atoms had no sensible properties—yet they had to be responsible for all the properties of matter that one could sense, such as an object being hot. The concept of atoms would also mean that there were possibly an infinite number of primary substances in nature. This was in direct conflict with the idea of the four elements—earth, air, fire, and water—being the primary building blocks of everything on Earth. A further problem was that if matter was particulate, then there would be spaces or voids between matter, which would make motion impossible. Finally, if matter was made up of atoms, then a purely mechanical explanation of human actions and behavior would be possible. By the Middle Ages such an explanation was rejected because it introduced the possibility that human actions were not set in motion by a divine being. A revival of atomism would have to wait until the rise of experimental science in the seventeenth century.

Robert Boyle (1627-1691) in his *Skeptical Chemist* (1661) proposed that all matter is composed of solid particles that can be rearranged to form new substances. What differentiated these different types matter was their size, shape, and structural pattern. Isaac Newton (1642-1727) in his *Opticks* (1704) also proposed a particulate view of matter, and he further proposed that there were strong short range forces that existed between these particles that could be of an attractive or repulsive nature. This was used by him to explain why some chemical reactions occurred and others did not. As was the case with the Greeks, Boyle, Newton, and others had no evidence to back up their claims. However, their view that matter is particulate signaled an increasing consensus among scientists of the era—a consensus that would make the theory proposed by Dalton much more readily acceptable.

John Dalton (1766-1844) was a most unlikely person to develop the atomic theory. Born into a devout Quaker family in a rural area of northwest England, he was drawn early in life to an interest in the natural sciences. His formal education was spotty and he was basically self-taught. He was for a time the equivalent of a high school teacher in Manchester, England. He quit classroom teaching in 1800 to provide private instruction in the sciences and mathematics in Manchester, which he did for the balance of his life.

The impetus for the development of the atomic theory was Dalton's life-long interest in meteorology and the study of gases. This interest had developed from his association in his youth with a fellow Quaker, John Gough, who provided Dalton with most of his formal education in the sciences and mathematics.

Dalton developed his atomic theory as a way of trying to answer certain questions about the atmosphere. In the eighteenth century it was shown that the atmosphere was a mixture of gases rather then a single substance. The identities of many of these gases had only had been recently established. Dalton wondered if the atmosphere was a simple mixture of gases such as oxygen, nitrogen, carbon dioxide, and water vapor, or perhaps if there was some type of chemical reaction that occurred between these gases. Since the atmosphere appeared to be a homogenous mixture of gases, the consensus view at the end of the eighteenth century was that the various components were chemically combined and dissolved in the water vapor. Further evidence for this view was that if the atmosphere was a simple (i.e. physical) mixture of components, then one would expect that the various gases would settle out according to their weights, with the heaviest closest to the surface and the lightest on top. Since this is not what was found, it seemed logical that air was a chemical compound.

Dalton believed that the atmosphere was a physical mixture based upon his belief that water vapor could not be combined chemically with the gases in the air. Dalton viewed matter as composed of spherical particles and believed that these particles or atoms contained a shield of heat around them. This was essential for Dalton to explain why unlike particles tended to repel each other and thus produce a physical mixture of gases in the atmosphere. The idea of the shell of heat, or "caloric" as it was called, was incorporated from Lavoisier's model of the gaseous state and the belief that heat was a material element. Dalton used the term *atoms* for these particles in order to show that the original concept had originated in Greek natural philosophy.

Each atom in nature had its own size, which was a function of the volume and the radius of its shell of heat. Dalton formulated these ideas between 1801 and 1803, but evidence was lacking. This was supplied by a close friend of Dalton in Manchester, William Henry (1774-1836). Henry found that if you kept the temperature of a liquid such as water constant, the amount of a non-reactive gas that could be dissolved in it would increase as you increased the pressure place upon it. This led Dalton in 1803 to suspect that it was the weight of the particles of the gases that was the key determining factor. He measured the relative weights of various gases from their composition and presented his first table of relative atomic weights for a variety of gases and other substances.

Up to this point Dalton had shown little interest in chemistry, but by 1804 he realized that if atoms were considered to be the ultimate particle in nature and that each atom had its own particular weight, this could explain observations that had been made concerning the composition of compounds. As methods for the analysis of compounds had been refined, it was found that a compound always had the same composition— no matter if it was obtained from natural sources or made synthetically. Thus, if one analyzed rain water against water made by combining hydrogen and oxygen together in a laboratory, one would find that there would be 11.2% hydrogen and 88.8% oxygen in each case. This became known as the law of definite proportions, and it was the atomic theory that showed why this was so. If hydrogen and oxygen combine in a 1:1 ratio, then an atom of oxygen must be eight times heavier than an atom of hydrogen in order to get the constant composition of water.

Impact

The ability of Dalton's atomic theory to explain the law of definite proportions was only the beginning of its impact on the field of chemistry. Another chemical problem that Dalton was able to solve using the atomic theory was the observation that a particular element such as nitrogen, for example, could combine with oxygen and form a series of unique compounds containing nitrogen and oxygen. Analysis of these compounds showed that there was a regular relationship between the amount of nitrogen that combined with oxygen. Thus, if you have a fixed amount of nitrogen, the amount of oxygen combined would stand in a series of whole numbers—i.e. 1:1,1:2,1:3, and so on. This came to be known as the law of multiple proportions and was a puzzle until Dalton. He was able to explain multiple proportions by assuming that an atom of nitrogen could react with one, two, or more atoms of oxygen to form a series of compounds.

Dalton's explanation of multiple proportions was used by William Hyde Wollaston (1766-1828) and Thomas Thomson in 1807 to explain the relationship of potassium bicarbonate and potassium carbonate. Thomson, who had discussed the atomic theory with Dalton in 1804 when he visited Manchester, was so impressed by how it solved his problem that he arranged for Dalton to lecture at the universities of Edinburgh and Glasgow on his views of chemical reactions using the atomic theory. These lectures given in 1807 were well received and ultimately formed the basis for the *New System of Chemical Philosophy,* the first part published in 1808. The atomic theory itself only occupied five out of the 916 pages of this first part of the *New System.*

In general, the impact of the atomic theory can be summarized as follows:

1) The definition of an element as being made of atoms, and the idea that each atom has its own unique properties, led at last to a clear understanding of what an element is.

2) Since there were no limits on the number of different atoms possible in nature, then it seemed perfectly reasonable that there were elements that had not yet been discovered. This led to the search for new elements, a search that occupied many chemists during the nineteenth century and that led to the discovery of numerous elements.

3) Since elements are made of atoms and many different elements seem to have similar chemical properties, this raised the question of why certain groups of elements were similar in nature while others differed greatly. This contributed to the development of schemes to classify elements, an effort that culminated in Dmitri Mendeleyev's (1834-1907) first periodic table of the elements in 1869.

4) Since atoms combine together to form molecules, the atomic theory stimulated investigations as to the reasons that certain atoms combine together and others do no. This led to the development of theories of chemical bonding in the nineteenth century.

Dalton's concept that matter was made up of these incompressible particles surrounded by an atmosphere of heat was difficult for nineteenth-century chemists to accept. However, if one used the concept of atomic weight as a tool, then the synthesis of compounds was made much easier. The atomic model in which atoms always retained their identity showed why the law of conservation of mass existed in nature and why transmutation of the elements could not occur: an atom of lead could never be transformed into an atom of gold, since atoms always maintained their identity no matter what you did to them. The atomic theory maintains that there are as many atoms as there are elements in nature and that within an element all the atoms are the same.

There was much skepticism concerning the atomic theory for several reasons. The most obvious was the inability of Dalton to physically demonstrate the presence of atoms in matter.

Even if you could not show their physical presence, however, the relative weights of the atoms were useful as a means of chemical synthesis. It was in this light that the atomic theory had its greatest usefulness in the first half of the nineteenth century. When Dalton was awarded the Royal Society medal in 1826, it was for his "development of the theory of definite proportions, usually called the atomic theory of chemistry."

MARTIN D. SALTZMAN

Further Reading

Brock, William H. *The Norton History of Chemistry.* New York: Norton, 1993.

Knight, D.M. *Atoms and Elements.* London: Hutchinson, 1967.

Rocke, A.J. *Chemical Atomism in the Nineteenth Century.* Columbus, OH: Ohio State University Press, 1984.

Development of Physical Chemistry during the Nineteenth Century

Overview

Before the Scientific Revolution of the sixteenth through the eighteenth centuries, all the natural sciences were grouped together under the heading of "natural philosophy." But by 1800 many had become separate disciplines, with distinct subject matters and investigative methods. During the later nineteenth century, however, it became increasingly clear that discoveries in one science often had major consequences for another, and many scientists sought to recover the previous unity between the sciences. One result was the rise of "physical chemistry," which united ideas and techniques from both physics and chemistry.

Background

By the early nineteenth century, physics and chemistry had segregated into two distinct fields. Physics studied the general motions of bodies in space and time according to mathematical laws due to the action of various "forces." Chemistry focused on the discovery, analysis, synthesis, and qualitative description of particular species of matter, chiefly elements and compounds. Heat, light, electricity, and magnetism, once believed to be "imponderable" (weightless) fluids and hence chemical substances, were now thought by many to be only apparent effects of the motions of microscopic particles and invisible forces, and thus a part of physics instead. Attempts to develop a quantitative "force" chemistry, which would explain reactions by mathematical laws of atomic motions due to gravitational or electrical forces of attraction and repulsion, failed.

This situation changed beginning in the 1840s, when the new concept of "energy" began to replace the older one of "force." Like force, energy could be measured quantitatively and its

effects expressed in mathematical equations; unlike force, it could be applied far more generally to subjects not involving any apparent mechanical motions or forces, and it more clearly distinguished between measurement of capacity and intensity factors (e.g., heat vs. temperature). By measuring the energy content of a substance or system, as well as changes in that content, physicists could better explain both the generation or absorption of heat and the production and alteration of electrical currents. This made it possible once again for chemists to study the effects of these currents on chemical reactions.

A key step forward was the formulation between 1842 and 1865 of the first two laws of thermodynamics—thermodynamics being the study of motion of heat and its capacity to do "work," such as producing light, electricity, or magnetic current. The first law of thermodynamics states that the energy content of a closed system (one isolated from outside energy sources) is constant, so that forms of energy within it can only be transformed into one another. The second law of thermodynamics states that no process of energy transformation for work is perfectly efficient; some energy is always randomly dissipated as heat, so that the "entropy" or degree of disorder in a closed system always increases. Together these laws mean that energy cannot be created from nothing, and that no transformation process can occur indefinitely without the input of additional outside energy.

Impact

The physical laws of thermodynamics initially promoted development of a field called "thermochemistry." In the 1840s Germain Hess (1802-1850) announced two new chemical laws, the law of constant summation of heat and law of thermoneutrality. These laws were quickly extended by Julius Thomsen (1826-1909) and Marcellin Berthelot (1827-1907). Thomsen determined the heats of neutralization for numerous acid-base reactions producing neutral salts, and he used galvanic cells (electrochemical batteries) to calculate the work necessary to decompose various chemical compounds. He also proposed that the evolution of heat accompanying a chemical reaction is a quantitative measure of the chemical affinity (degree of reactivity) of the reactants.

Berthelot introduced the terms "exothermic" and "endothermic" to describe chemical reactions that evolve and absorb heat. He also measured heats of combustion for many reactions and framed three laws governing all chemical processes:

1) The heat change in a chemical reaction equals the amount of work done internally in a chemical system.

2) Assuming no external work is done, the heat evolved or absorbed by a reaction depends only on the initial and final states of the reactants and products.

3) Every chemical change in a closed system yields the products accompanied by the greatest evolution of heat (the "principle of maximum work").

While Thomsen and Berthelot's laws proved to be true only for certain exothermic reactions, they provided chemists with the means for predicting and measuring the courses and products for many reactions.

Electrochemistry also made key contributions to the development of physical chemistry. Humphrey Davy (1778-1829) and Jacob Berzelius (1779-1848) used galvanic cells to decompose compounds into simpler components, by passing electrical currents through water-based solutions and collecting the reaction products at the oppositely charged wire poles. Between 1832 and 1834, Michael Faraday (1791-1867) discovered the two basic laws of electrolysis: 1) the amount of a reaction product collected (as solid or liquid) or evolved (as gas) at an electrode is a product of the strength of the electrical current and the time; and 2) it also depends upon the equivalent weights (ratios of combination) of the reaction products. Faraday also introduced the terms "electrolysis" to describe this process; "electrode" for the poles; "cathode" and "anode" for the positive and negative poles; "ion" (Greek for "wanderer") for the

dissociated components that migrate to the opposite poles; and "cation" and "anion" for the ions attracted to the cathode and anode.

Further important advances in electrochemistry occurred in Germany. Between 1853 and 1859 Johann Wilhelm Hittorf (1824-1914) discovered that the concentration at the electrodes of salts in solutions, and the rates of cation and anion migration, differ between the cathode and anode. From 1867 to 1876 Friedrich Kohlrausch (1840-1910) used alternating rather than direct current to perform solution electrolysis, thereby minimizing the decomposition of salts in solutions and obtaining more accurate measurements of effective currents.

A third source of inspiration for physical chemistry was the kinetic theory of motion by gas particles. Alexander Williamson (1824-1904) proposed in 1850 a dynamical "kinetic" theory of chemical equilibrium, resulting from a balance between ongoing dissociations and re-associations of molecules and elements, instead of the older static theories of a balance between opposing gravitational or electrical forces. Rudolf Clausius (1822-1888), Leopold Pfaundler (1839-1920), August Horstmann (1842-1929), and Peter Waage (1833-1900) and Otto Guldberg (1836-1902) made further advances in this area. The latter pair, working between 1864 and 1879, collaborated to formulate the law of mass action, stating that a reaction in solution reaches an equilibrium point expressed by a mathematical constant.

The formal establishment of physical chemistry as a distinct scientific field was the joint effort of Jacobus van't Hoff (1852-1911), Svante Arrhenius (1859-1927) and Wilhelm Ostwald (1853-1932). In 1884 van't Hoff published his groundbreaking research on the behavior of solute particles in solutions. Drawing upon the work of Williamson, Clausius, Pfaundler, Horstmann, and Guldberg and Waage, he proved theoretically that the properties of solutions could be explained by the kinetic dissociation and association of some of the solute particles into molecules, whose activities obey the law of mass action and whose degrees of dissociation increase with the solution temperature.

That same year Svante Arrhenius completed a doctoral dissertation in which he argued that electrolysis could be explained by assuming that individual ions carry specific units of electrical charge. The amount of current passing through an electrolytic solution thus varied with the degree of dissociation of the solute, as a function of mass action rather than of electrical charge itself. When Arrhenius's skeptical professors gave him only a barely passing grade, he sent copies of his dissertation to several chemists, including van't Hoff and Ostwald, hoping for a more favorable response.

Between 1876 and 1884 Ostwald had shown that physical properties of solutions, particularly heats of reaction and electrical conductivity, could be used to calculate the speed and degree of completion of chemical reactions. Upon reading Arrhenius's dissertation and van't Hoff's book at almost the same time, Ostwald realized that their theories explained his own experimental results, and he established contact with both of them. Arrhenius joined Ostwald's laboratory, and in 1887 Ostwald and van't Hoff founded a new scientific periodical, the *Zeitschrift für physikalische Chemie* ("Journal of Physical Chemistry"), to propagate their research and views, with Ostwald as chief editor.

Between 1884 and 1887, drawing upon advances by Arrhenius and Friedrich Pfeffer (1845-1920), van't Hoff derived a chemical solution law ($PV = iRT$) that was analogous to the well-know gas law in physics ($PV = nRT$), where P = pressure, V = volume, n = number of moles of gas particles, R = the gas constant, and T = temperature. van't Hoff replaced the n with i, where i = the ionic dissociation constant, which differs for each solution. Further important points were van't Hoff's chemical use of both laws of thermodynamics, to explain the conversion of electrical energy into heat and the evolution of heat as signifying a shift in the point of reaction equilibrium, and his refutation of Berthelot's principle of maximum work by demonstrating that such reactions are reversible. In late 1887 and early 1888, Arrhenius and Ostwald made further key advances in this area.

The final area of early research in physical chemistry concerned reaction kinetics and catalysis. Between 1849 and 1854 Ludwig Wilhelmy (1812-1864) used polarized light to study the rate of inversion of the sucrose (table sugar) molecule between two forms, and found it to be related directly to temperature and inversely to concentration in solution. In 1884 van't Hoff established these relations in mathematical form and related them to changes in the internal energy contents of solutions. During the 1890s Ostwald defined a catalyst as "a substance that changes the velocity of a reaction without itself being changed in the process." He demonstrated that catalysts work by lowering the energy barriers necessary for a reaction to occur, but without changing the energy relations between the initial reactants and final products.

Calculation of energy factors thus proved important for studying the activity of chemical systems. Hermann von Helmholtz (1821-1894) and Josiah Willard Gibbs (1839-1903) both derived important formulas related to energy factors in chemical systems. Gibbs also introduced the concept of a chemical potential, or threshold of activity for chemical energy, and formulated the "phase rule" for studying physical changes of state between solid, liquid, and gaseous phases due to chemical reactions. Ostwald furthered the application of energy relations to chemical systems by translating Gibbs's little-known articles into German and republishing them in 1892.

The rise of physical chemistry was not simply a string of unbroken triumphs, however. Detailed explanations of the underlying mechanisms of ionic dissociation, electrolysis, reaction kinetics, and catalysis could not be offered until the advent of modern atomic theory, which made it possible to explain these in terms of molecular structure and the gain and loss of electrons by atoms. Ironically, instead of being a "general science" that would reunite physics and chemistry, as Ostwald hoped, physical chemistry instead developed into an independent discipline bridging the persisting gap between those fields.

Nevertheless, the successes were extraordinary. Arrhenius, van't Hoff, and Ostwald won three of the first nine Nobel prizes in chemistry. Today, physical chemistry ranks beside organic chemistry as one of the two leading areas of chemical research. Its basic concepts of ionic dissociation, energy relations, reaction kinetics, and catalysis are fundamental to all modern chemical research and industrial applications. Its success in melding ideas and discoveries from physics and chemistry inspired the creation of numerous other scientific cross-disciplines, such as biochemistry, biophysics, geochemistry, and geophysics.

JAMES A. ALTENA

Further Reading

Books

Barkan, Diana K. "A Usable Past: Creating Disciplinary Space for Physical Chemistry." In *The Invention of Physical Science: Intersections of Mathematics, Theology and Natural Philosophy Since the Seventeenth Century,* edited by Mary Jo Nye, et al. Boston: D. Reidel, 1992.

Clark, Peter. "Atomism vs. Thermodynamics." In *Method and Appraisal in the Physical Sciences,* edited by C. Howson. Cambridge: Cambridge University Press, 1976.

Dolby, Richard G. A. "The Case of Physical Chemistry." In *Perspectives on the Emergence of Scientific Disciplines,* edited by Gerard Lemaine. Chicago: Aldine, 1976.

Hiebert, Erwin N. "Developments in Physical Chemistry at the Turn of the Century." In *Science, Technology, and Society in the Time of Alfred Nobel,* edited by Carl G. Bernard, et al. New York: Pergamon Press, 1982.

Laidler, Keith. *The World of Physical Chemistry.* Oxford: Oxford University Press, 1993.

Nye, Mary Jo. *Before Big Science: The Pursuit of Modern Chemistry and Physics, 1800-1940.* New York: Twayne Publishers, 1996.

Nye, Mary Jo. *From Chemical Philosophy to Theoretical Chemistry: Dynamics of Matter and Dynamics of Disci-* *plines, 1800-1950.* Berkeley: University of California Press, 1993.

Servos, John. *Physical Chemistry from Ostwald to Pauling: The Making of a Science in America.* Princeton, NJ: Princeton University Press, 1990.

Periodicals

Dolby, Richard G. A. "Debates Over the Theory of Solution: A Study of Dissent in Physical Chemistry in the English-Speaking World in the Late Nineteenth and Early Twentieth Centuries." *Historical Studies in the Physical Sciences* 7 (1976): 297-404.

Dolby, Richard G. A. "Thermochemistry versus Thermodynamics: The Nineteenth Century Controversy." *History of Science* 22 (1984): 375-400.

Other

Root-Bernstein, Robert. "The Ionists: Founding Physical Chemistry, 1872-1890." Ph. D. dissertation, Princeton University, 1980. [Available from University Microfilms International in Ann Arbor, MI.]

Finding Order Among the Elements

Overview

The periodic table is one of the most fundamental tools of chemistry. It summarizes information about each element and reveals how the elements are chemically related to one another. The first widely accepted periodic table was published in 1869 by a chemistry professor named Dmitri Mendeleyev. He began work on his table hoping to help his students to learn about the elements. He ended up by creating a classification system that helped chemists predict new elements and that led to the discovery of the particles that make up atoms.

Background

One of the key developments that led to the periodic table was the determination of the atomic weights of the elements. In 1805 English chemist John Dalton (1766-1844) stated that every atom of an element has the same weight. This idea implied that it would be possible to measure the atomic weights of the elements. (It is now known that Dalton's hypothesis is incorrect; not all atoms of an element have the same weight. Atomic weights can be measured, but they usually represent an average.)

In 1809 Joseph Louis Gay-Lussac (1778-1850) proposed that when gases undergo a chemical reaction, they do so in simple whole number ratios of their volumes. For example, he

Mendeleyev's groundbreaking first periodic table. *(The Library of Congress. Reproduced by permission.)*

showed that dinitrogen oxide (N_2O) was formed from two volumes of nitrogen to one volume of oxygen. Jöns Jacob Berzelius (1779-1848), a Swedish scientist, used the ideas of Dalton and Gay-Lussac to determine the atomic weights of the 69 elements known at that time. He did so by measuring the relative volume of oxygen with which various elements could combine. He could then infer the atomic weight of the elements from these measurements of volume. He published quite accurate tables of atomic weights in 1818 and 1826.

Berzelius was also largely responsible for the fact that the ideas of Amedeo Avogadro (1776-1856) went unnoticed for almost half a century. In 1811 Avogadro had hypothesized that equal volumes of gases contain equal numbers of molecules. If this were the case, the relative atomic weights of gases would be fairly simple to determine. For instance, if one liter of oxygen weighed approximately 16 times the weight of one liter of hydrogen, then it could be concluded that one atom of oxygen weighs about 16 times as much as one atom of hydrogen. Avogadro also proposed that two atoms of an element may combine to form one molecule of gas, as in oxygen (O_2) and nitrogen (N_2). It was on this point that Berzelius strongly disagreed with Avogadro. As a result Berzelius and many other scientists of the time wrote incorrect formulas for many compounds. For instance, Berzelius wrote the formula of water as HO rather than H_2O. These incorrect formulas sometimes resulted in incorrect measurements of atomic weight.

In 1858 Italian chemist Stanislao Cannizzaro (1826-1910) showed that the atomic weights of the elements in a molecule could indeed be calculated by applying Avogadro's hypothesis. In addition, he helped to define the difference between atomic weight (that of an atom, such as H) and molecular weight (that of a molecule, such as H_2). Cannizzaro brought forward Avogadro's hypothesis at the first international meeting of chemists. This meeting, called the Karlsruhe Conference, was held in Heidelburg, Germany, in 1860. After Cannizzaro's presentation, Avogadro's hypotheses became widely accepted within a few years.

One of the scientists in attendance at the Karlsruhe Conference was Dmitri Mendeleyev (1834-1907). Mendeleyev was a Russian chemist who happened to be studying in Europe at the time. Mendeleyev made the acquaintance of Cannizzaro, from whom he obtained measurements of atomic weights and a familiarity with the ideas of Avogadro. After returning to Russia Mendeleyev began searching for a logical way to organize the elements. Eventually, he noticed that when he arranged the elements in order of increasing atomic weight, similar elements appeared at regular, or periodic, intervals. Mendeleyev used his observations to make a table that reflected this pattern.

In March 1869 Mendeleyev presented his table to the Russian Chemical Society. A revised version followed two years later. The atomic weight of the elements in his table increased from left to right across each row, or period. Each column, or group, of the table contained elements with similar properties. The heading of each column indicated the valency of the elements in that group. An element's valency is responsible for the way in which it combines with other atoms. Thus, Mendeleyev's table

AN UNFORTUNATE METAPHOR

〜

In 1864, five years prior to the publication of Mendeleyev's table, an English chemist named John Newlands suggested that when the elements were arranged according to atomic weight, similar chemical properties would be found with every eighth element. He named his hypothesis the "law of octaves," comparing the pattern of eight elements to that of the eight notes of the musical scale. The basic idea behind his hypothesis—a connection between atomic weight and chemical properties—was valid, as Mendeleyev was soon to show. However, he lacked sufficient evidence to support his claim, and his comparison of elements to musical notes was ridiculed as being "mystical" and unscientific. One chemist even asked him sarcastically if he found similar patterns by arranging the elements alphabetically. However, Newlands' idea was not nearly as farfetched as that of another chemist, who claimed to have found a connection between atomic weight and the distances of the planets from the sun.

showed that an element's chemical properties, as seen through its valency, were a function of its atomic weight. Mendeleyev called this relationship the periodic law, and his table became known as the periodic table.

Impact

As he worked on his table, Mendeleyev began to suspect that there were elements that had yet to be discovered. (In fact, at the time his original table was published, scientists had described only 69 of the 112 chemical elements known today.) He left blanks in his table to accommodate these undiscovered elements and even made predictions of their properties based on those of neighboring elements. In all, Mendeleyev made predictions for ten new elements, seven of which eventually proved to be correct.

Three of these elements, which Mendeleyev called eka-boron, eka-aluminum, and eka-silicon, were discovered within 20 years of the publication of his first table. In 1875 Paul Émile Lecoq de Boisbaudran (1838-1912), a French chemist, discovered the element gallium, which turned out to be Mendeleyev's eka-aluminum. Mendeleyev had predicted an atomic weight of 68 and a specific gravity of 5.9 for this element. (The specific gravity of an element is its density divided by the density of water.) When measured experimentally, gallium's atomic weight is 69.72 and its specific gravity is 5.94—a close match to Mendeleyev's predictions. Lars Fredrick Nilson discovered scandium in 1879, which Per Teodor Cleve (1840-1905) identified as Mendeleyev's eka-boron, and Clemens Winkler (1838-1904) discovered germanium, Mendeleyev's eka-silicon, in 1886. Germanium's specific gravity was measured at 5.469, which agreed nicely with Mendeleyev's prediction of 5.5. However, some of the chemical reactions Mendeleyev predicted for these three elements turned out not to be as accurate. So although these discoveries confirmed Mendeleyev's system and showed that it could be used as a practical research tool, they also demonstrated the need for obtaining experimental data.

Despite his successful predictions, Mendeleyev's periodic law did not always hold true. For instance, according to his law it would be impossible for two elements to have the same atomic weight. However, the elements nickel and cobalt as well as the elements ruthenium and rhodium were assigned identical atomic weights in his table. In addition, the difference in atomic weight between consecutive elements in the table was inconsistent. The smallest difference appeared to be about 0.25 unit and the greatest seemed to be about 4 to 5 units. If an element's properties were truly dependent on its atomic weight, it seemed as if atomic weight should change by the same amount between consecutive elements. Another problem was that Mendeleyev had to switch some elements in order to make them fit his pattern. For example, he positioned tellurium to the left of iodine even though iodine was assigned an atomic weight of 127 and tellurium was assigned an atomic weight of 128.

Mendeleyev's table also raised the question of what is added to an element when its atomic weight increases. The pattern Mendeleyev had discovered in the atomic weights of the elements gave support to the idea that the elements had the same origin. According to this idea, each of the elements is made from the same basic particles—the same "stuff"—but in a unique arrangement or quantity. For instance, atoms of gold and silver would be made of the same types of particles, but gold atoms would have a different arrangement or a different number of particles than those of silver.

In the late 1800s several scientists began to speculate about the types of subatomic particles that might exist. However, many other scientists had problems with the idea of so-called "ultimate particles." They argued that atoms of elements with a high atomic weight would have to consist of hundreds of similar particles, which was thought to be a highly unstable situation. In addition, it was argued that if there were particles smaller than atoms, one element might be able to change into another, which was generally believed to be impossible. This debate was not settled until the twentieth century, when the existence of subatomic particles (protons, neutrons, and electrons) was finally confirmed. (Scientists now know that strong forces within an atom's nucleus allow similarly charged protons to be packed closely together, and in some cases, radioactive decay does result in one element changing into another.)

Once subatomic particles were discovered, it was realized that the periodic law holds true for atomic *number* rather than atomic *weight*. The atomic number of an element is equal to the number of protons or the number of electrons in a neutral atom. It is the number of electrons that governs an atom's chemical and physical properties, and the modern periodic table is based on increasing atomic number rather than increasing atomic weight.

When the elements are arranged by atomic number, consecutive elements differ by one proton and one electron. The difference in neutrons between consecutive elements, however, does not follow a set pattern. It is for this reason that Mendeleyev's periodic law does not always hold true. Atomic weight is based in part on the number of neutrons in an atom, so a law based on atomic weight will sometimes be inconsistent.

The discovery of the noble gases also posed a temporary threat to the periodic table. William Ramsay (1852-1916) and Lord Rayleigh (1842-1919) discovered the noble gas argon in 1894. Mendeleyev doubted that the men had actually discovered a new element, mainly because there was no space for it in his periodic table. Ramsay suggested that argon should be placed in its own group after chlorine and before potassium. However, argon's atomic weight was less than that of potassium, so this idea was not seen as acceptable at the time. Within four years, however, Ramsay had isolated the noble gases krypton, neon, and xenon. (Helium had been discovered earlier by examining the light of stars.) These elements helped to fill up the rightmost column of the table, showing that Ramsay's original suggestion was correct and that the essence of the periodic law was not violated. Today, the periodic table contains spaces for 118 elements, and it is used in classrooms and laboratories throughout the world.

STACEY R. MURRAY

Further Reading

Books

Cobb, Cathy, and Harold Goldwhite. *Creations of Fire: Chemistry's Lively History from Alchemy to the Atomic Age.* New York: Plenum Press, 1995.

Newton, David E. *The Chemical Elements.* New York: Franklin Watts, Inc., 1994.

Other

Dmitry Mendeleev Online. www.chem.msu.su/misc/mendeleev/welcome.html

WebElements periodic table of the elements. www.webelements.com

Nineteenth-Century Advances in Understanding Gases, Culminating in William Ramsey's Discovery of Inert Gases in the 1890s

Overview

The nineteenth century saw many advances in our understanding of gases, their behavior, and their uses. Charles, Gay-Lussac, Faraday, Avogadro, and others contributed to this knowledge, formulating some of the basic laws that are known to govern the behavior of gases under a variety of physical and chemical conditions. This knowledge, the ability to liquefy and to compress gases, and other advances have led, in turn, to a great many inventions and discoveries that have profoundly impacted the life of virtually everyone on Earth.

Background

It was not until the early seventeenth century that scientists began to realize that gases exist as a separate form of matter. In the mid seventeenth century (1660-1662) several scientists independently determined that, as the pressure on a gas changes, so does its volume. This fact, now called Boyle's law (or, in continental Europe, Mariotte's law after Edme Mariotte, who discovered it independently) states that $PV = K$, a constant, for any gas. That is, the absolute pressure P multiplied by the volume V is constant. This means that, if a gas is compressed (decreasing the volume) the pressure must increase by the same relative percentage so that this equation remains constant. Put another way, reducing the volume by one half must cause the pressure to double.

During the remainder of the seventeenth and eighteenth centuries the primary work done in these areas was in the area of separating and identifying various gases from air. During this time oxygen, nitrogen, and hydrogen were among the gases isolated and identified as separate chemical elements.

The early part of the nineteenth century saw a renewed interest in the physical relationships that govern the behavior of gases as well as ways to manipulate them. Charles' law (also known as Gay-Lussac's law) states the relationship between gas temperature and pressure, noting that pressure increases as temperature is raised. Avogadro combined Boyle and Charles' laws into a single equation, the Ideal Gas Law, which is stated $PV = nRT$. In this law, n refers to the number of moles (or molecules) of gas present, R is the universal gas constant, and T is the temperature of the gas on the absolute scale, starting at absolute zero. This law, while it does not work at extremes of temperature or pressure, is nonetheless useful and accurate under most conditions and is still widely used by physicists and chemists today.

In addition to understanding the physics and chemistry of gases, scientists were learning to manipulate gases. In 1823 Faraday liquefied chlorine and, soon afterwards, liquefied carbon dioxide and hydrogen chloride. Others learned to compress gases, distill off specific gases from liquid air, and more.

As these manipulations were taking place, others were identifying new gases, notable for their apparent refusal to participate in chemical reactions. Sir William Ramsay first isolated argon from liquefied air in 1894. In quick succession, Ramsay also identified helium (detected in 1868 through spectroscopic observation of the Sun), neon, krypton, and xenon and, with the identification of radon in 1900, all the known noble gases. Ramsay later won the Nobel Prize in Chemistry (1904) for these discoveries.

Impact

The impact of these discoveries falls into two general areas: 1. Scientific study of gases led to many advances, particularly in the studies of thermodynamics, the periodic table, atomic structure, and 2. Industrial and commercial inventions and advances, including compressed and liquefied gases for industrial use, liquefied gases for rocket propulsion, development of refrigeration and air conditioning, and lights using ionized noble gases.

Each of these has had far-reaching consequences that are described in more detail in the following paragraphs.

The work of Avogadro, Charles, and Gay-Lussac all noted the relationship between the pressure and temperature of a gas. All of these scientists realized that this relationship depended on some absolute temperature, that gas temperature and pressure depended in some way on the average speed of gas particles (either atoms or molecules), and that there must exist a temperature at which these motions must stop. William Thomson (later Lord Kelvin) determined the temperature at which all molecular motions must stop, the temperature now known as absolute zero because it is thought to be impossible for colder temperatures to exist. Recent investigations at such low temperatures have led to the creation of the Bose-Einstein Condensate, a long-sought condition in which large numbers of atoms act as a single large atom. This, in turn, has shed new light on the properties of matter under extremely cold conditions and has resulted in the creation of "atomic lasers." It may also lead to new, highly accurate

clocks that might replace current atomic clocks by being both smaller and more accurate.

The properties of the noble gases were not predicted by the periodic table, but their occurrence certainly supported the version published by Mendeleev in 1869 because they appeared at precise intervals. Their extreme unwillingness to undergo chemical reactions eventually led to theories of electron shells, suggesting that atoms react to form chemical compounds, sharing electrons so that the outermost shells contain eight electrons. This realization is one of the cornerstones of chemistry and helped form, in turn, theories of atomic structure that were so important during the revolution in physics that took place in the early twentieth century. In addition, understanding that outer electron shells contain up to eight electrons suggested that other shells may also contain specific numbers of electrons. Understanding this provided a physical explanation for the periodic table—the various rows and groups reflect the filling of electron shells in each successive element. The innermost shell, containing only two electrons, is completely filled with helium, a noble gas. The next shell, with eight electrons, is filled sequentially by the next group of elements, from lithium to neon, another noble gas. This theory has proved extremely successful in explaining the behavior of various elements, why chemical reactions take place, and why the periodic table looks as it does.

As important as these scientific advances were, the applications of research into gases are probably even more important to most people on earth. Compressed gases are used in scientific research, industry, and more. Inert gases are equally important. Argon is used in welding to flood the work area during TIG (tungsten-inert gas) and MIG (metal-inert gas) welding to reduce oxidation of the metals being welded, making a stronger weld. Argon and helium are used as inert atmospheres to keep reactive or sensitive items from chemical harm. Neon, krypton, and xenon are used for lighting, and argon is used to fill Geiger-Müller tubes that detect radiation. Argon is used to keep reactive metals such as titanium, uranium, plutonium, and zirconium from oxidizing or catching fire during processing.

Another area in which knowledge of gases has proved a boon to humanity is in the development of air conditioning and refrigeration. These technologies depend on the behavior of gases under carefully controlled variations of pressure and were not developed until 1876, when changes in gas temperature at different pressures could be predicted and quantified.

In both cases, a cold refrigerant gas passes through a heat exchanger inside the refrigerator (or freezer) unit where the cold refrigerant absorbs heat. This warms it up and passes on to the compressor. Here, the warm gas is compressed, achieving a high pressure and high temperature. In this state it is passed through a heat exchanger (which, in a household refrigerator can be seen on the back of the unit). In the heat exchanger, the refrigerant is hotter than the air around it, so it gives off heat to the atmosphere, cooling it down and liquefying it. The next step is to pass the warm liquid through a small thermal expansion valve (sometimes called a TXV). This valve is almost completely shut, so the pressure on one side is much lower than on the compressor side. Because of this, the liquid that passes through the valve turns into a gas immediately and, in accordance with the Ideal Gas Law, it cools by the same proportion. So, if pressure goes down by a factor of 10, the temperature drops by the same amount, usually to below freezing for both refrigerators and freezers.

The impact of refrigeration and air conditioning on human society has been great. Air conditioning has made it possible to inhabit many parts of the earth that were formerly inhospitable or that were very difficult to live in. Such areas include many of the tropical regions, much of Australia, and the desert areas. Human life was possible in these areas, of course, but was never comfortable. It has also made possible parts of the industrial and technological revolutions since human operators of high-temperature equipment require cool temperatures and because many pieces of electronic equipment require precisely controlled environments, regardless of exterior weather.

In addition, these technologies have been used to preserve foods more effectively. In the early twentieth century it was not uncommon for households to shop daily or, at most, every few days for fresh produce, meats, and dairy products because there was no way to keep food fresh. A great deal of food grew moldy or rotted in storage, and food could only be transported a short distance for similar reasons. The development of relatively cheap refrigeration technology has made it possible to shop less frequently, to store foods almost indefinitely in the frozen state, to ship foods around the world, and to minimize waste during shipping and storage. Indeed, some foods such as ice cream only exist in the frozen state and owe their current popularity to the existence of cheap freezers. Like advances in air conditioning, these, too, are due in large part to the greater understanding of gases that took place during the nineteenth century.

P. ANDREW KARAM

Further Reading

Atkins, P.W. *The Periodic Kingdom.* New York: Basic Books, 1995.

Elaboration of the Elements: Nineteenth-Century Advances in Chemistry, Electrochemistry, and Spectroscopy

Overview

By the end of the nineteenth century advances in electrochemical and spectroscopic methods, together with the discovery of elements exhibiting unique radioactive properties, worked to transform mankind's view of the cosmos on both the celestial and subatomic scale. By the end of the century the elements and matter comprising all things could no longer be viewed as immutable. Moreover, the rapid incorporation of the powerful properties of radioactive elements into medical practice established a course followed with increasing regularity and rapidity throughout the twentieth century. Although the composition and nature of radioactive elements was, at best, superficially understood, the practical benefits to be derived by society from their use forced their incorporation into technology far ahead of the pace of scientific understanding.

Background

The dramatic rise of scientific methodology and experimentation during the later half of the eighteenth century set the stage for the fundamental advances in chemistry and physics made during the nineteenth century. In less than a century, European society moved from an understanding of the chemical elements grounded in mysticism to an understanding of the relationships between elements found in a modern periodic table. During the eighteenth century laws against witchcraft were repealed, and near the close of the century an industrial revolution began to sweep across Europe. At the same time scientific disagreements became less grounded in religious disputes and more focused on differing interpretations of fundamental laws as they applied to practical inventions.

During the eighteenth century there was a steady march of discovery with regard to the chemical elements. Isolations of hydrogen and oxygen allowed for the formation of water from its elemental components. Nineteenth-century scientists built experiments on new-found familiarity with elements such as nitrogen, beryllium, chromium, and titanium.

Although a number of important discoveries took place with regard to elements during the eighteenth century, the nature of the interactions between elements continued to vex chemists. English scientist John Dalton's (1766-1844) explanation of chemical atomic theory at the beginning of the nineteenth century opened the door to a new understanding of chemistry based on the interactions of the elements. Subsequent nineteenth-century advances in electrochemistry provided chemists with powerful new tools of analysis and simultaneously dispelled the notion that simple electrostatic forces hold all elements together in compounds.

Advances in spectroscopy awaited further refinement of renowned English physicist Sir Isaac Newton's (1642-1727) postulates published in his 1704 work, *Opticks,* wherein Newton had described a rainbow of colors by passing sunlight through a prism (the principal component in the development of the spectroscope). At the dawn of the nineteenth century the "corpuscular" theory of light (i.e., that light was composed of particles) traveling through an "ether" still dominated scientific circles. English scientist William Wollaston (1766-1828) and Bavarian scientist Joseph Fraunhofer (1787-1826) laid additional foundations upon which subsequent nineteenth-century advances in spectroscopy arose by using prisms to investigate the colors emitted by various elements. Swiss mathematician Leonard Euler (1707-1783) set forth formulations for calculating the refraction

of light using waves—each wavelength represented a particular color—that are still accepted as correct. In 1801 the publication of an experiment demonstrating the wave nature of light set the stage for a century-long battle over the duality of light (that it has both a wave and particle nature).

Impact

By the mid-nineteenth century, chemistry was in need of organization. New elements were being discovered at an increasing pace. Accordingly, the challenge for nineteenth-century chemists and physicists was to find a key to understand the increasing volume of experimental evidence regarding the properties of the elements. In 1869 the independent development of the periodic law and tables by the Russian chemist Dmitri Mendeleyev (1834-1907) and German chemist Julius Meyer (1830-1895) brought long-sought order and understanding to the elements.

Mendeleyev and Meyer did not work in a vacuum. English chemist J. A. R. Newlands (1837-1898) had already published several works that suggested the existance of relationships among families of elements, including his "law of octaves" hypothesis. Mendeleyev's periodic chart of elements, however, spurred important discoveries and isolation of chemical elements. Most importantly, Mendeleyev's table provided for the successful prediction of the existence of new elements, and these predictions proved true with the discovery of gallium (1875), scandium (1879), and germanium (1885).

By the end of the nineteenth century the organization of the elements was so complete that British physicists Lord Rayleigh (born John William Strutt, 1842-1919) and William Ramsay (1852-1916) were able to expand the periodic table and to predict the existence and properties of the noble gases argon and neon.

Nineteenth-century advances, however, were not limited to mere identification and isolation of the elements. By 1845 German chemist Adolph Kolbe (1818-1884) synthesized an organic compound, and in 1861 another German chemist, Friedrich Kekulé (1829-1896), related the properties of molecules to their geometric shape. These advances led to the development of wholly new materials (e.g., plastics and celluloids) that had a dramatic impact on a society in the midst of industrial revolution.

The most revolutionary development with regard to the elucidation of the elements during the nineteenth century came in the waning years of the century. In 1895 Wilhelm Röntgen (1845-1923) published a paper titled "On a New Kind of Rays." Röntgen's work offered the first description of x rays and offered compelling photographs of a human hand. The scientific world quickly grasped the importance of Röntgen's discovery. At a meeting of the French Academy of Science, Henri Becquerel (1852-1908) observed the pictures taken by Röntgen of bones in the hand. Within months Becquerel presented two important reports concerning "uranium rays" back to the Academy. Becquerel—who was initially working with phosphorescence— described the phenomena that later came to understood as radioactivity. Less than two years later two other French scientists, Pierre (1859-1906) and Marie Curie (1867-1934), announced the discovery of the radioactive elements polonium and radium. Marie Curie then set out on a systematic search for radioactive elements and was able, eventually, to document the discovery of radioactivity in uranium and thorium minerals.

As the nineteenth century drew to a close, Ernest Rutherford (1871-1937), using an electrometer, identified two types of radioactivity, which he labeled alpha radiation and beta radiation. Rutherford actually thought he had discovered a new type of x ray. Subsequently, alpha and beta radiation were understood to be particles. Alpha radiation is composed of alpha particles (the nucleus of helium). Because alpha radiation is easily stopped, alpha radiation- emitting elements are usually not dangerous to biological organisms (e.g., humans) unless the emitting element actually enters the organism. Beta radiation is composed of a stream of electrons (electrons were discovered by J. J. Thomson in 1897) or positively charged particles called positrons.

The impact of the discovery of radioactive elements produced immediate and dramatic changes in society. Within a few years, before the close of the century, high-energy electromagnetic radiation in the form of x rays, made possible by the discovery of radioactive elements, was used by physicians to diagnose injury. More importantly, the rapid incorporation of x rays into technology established a precedent increasingly followed throughout the twentieth century. Although the composition and nature of radioactive elements was not fully understood, the practical benefits to be derived by society outweighed scientific prudence.

Italian scientist Alessandro Volta's (1745-1827) discovery in 1800 of a battery using discs of silver and zinc gave rise to the voltaic pile or the first true batteries. Building on Volta's concepts, English chemist Humphry Davy (1778-1829) first produced sodium from the electrolysis of molten sodium hydroxide in 1807. Subsequently, Davy isolated potassium, another alkali metal, from potassium hydroxide in the same year. Lithium was discovered in 1817.

Throughout the nineteenth century the development and contribution of electrochemistry paralleled growing understanding of electrical phenomena. In 1834 English experimental chemist Michael Faraday (1791-1867) published "On Electrical Decomposition," in which he presented terms such as anion, cation, electrodes, anodes, and cathodes that persist in modern usage. A half-century later electrochemistry proved that it had progressed from novelty to primary research tool with the publication of Svante Arrhenius's (1859-1927) work on dissociation and electrolyte solutions. Not only did this work contribute to the development of acid-base chemistry, it also opened new avenues of biological and medical research.

Studies of the spectra of elements and compounds led to further discoveries. German chemist Robert Bunsen's (1811-1899) improvements to the famous laboratory burner that bears his name allowed for the development of new methods for the analysis of the elemental structure of compounds. Working with Russian-born scientist Gustav Kirchhoff (1824-1887), Bunsen made possible flame analysis (a technique now commonly known as atomic emission spectroscopy or AES) and established the fundamental principles and techniques of spectroscopy. Bunsen examined the spectra (i.e., component colors) emitted when a substance was subjected to intense flame. Bunsen's keen observation that flamed elements emit light only at specific wavelengths—and that each element produces a characteristic spectra—along with Kirchhoff's work on black body radiation set the stage for the subsequent twentieth-century development of quantum theory. Using his own spectroscopic techniques, Bunsen also discovered the elements cesium and rubidium.

Using the spectroscopic techniques pioneered by Bunsen, other nineteenth-century scientists began to deduce the chemical composition of stars. These discoveries were of profound philosophical importance to society because they proved that Earth did not lie in a privileged or unique portion of the universe. Indeed, the elements found on Earth—particularly those associated with life—were found to be commonplace in the cosmos.

In 1868 French astronomer P. J. C. Janssen (1824-1907) and English astronomer Norman Lockyer (1836-1920) used spectroscopic analysis to identify helium on the Sun. For the first time an element was first discovered outside the confines of Earth.

Spectroscopy gained wide use. In 1885 Swiss scientist Johann Balmer (1825-1898) set forth formulae that described the observed spectrum of hydrogen that Niels Bohr (1885-1962) subsequently used in the twentieth century to develop his model of the atom. In 1890 Dutch physicist Hendrick Lorentz (1853-1928) proposed that atoms produced visible light by oscillations. In 1896 another Dutch physicist, Pieter Zeeman (1865-1943), discovered a splitting of spectral lines in a gas subjected to a magnetic field (now termed the Zeeman effect).

Appropriate to the century that saw the development of evolutionary theory, nineteenth-century advancements in chemistry and the discovery of new elements—especially radioactive elements—profoundly changed the philosophical notions held since the ancient Greeks that the elements that comprised the cosmos, and all living things therein, were unchanging.

K. LEE LERNER

Further Reading

Bronowski, Jacob. *The Ascent of Man*. Boston: Little, Brown, 1973.

Seaborg, G. T. *Elements of the Universe*. New York: Dutton, 1958.

Stwertka, Albert. *A Guide to the Elements*. Rev. ed. New York: Oxford University Press, 1996.

French Mineralogist René Just Haüy Founds the Science of Crystallography with the Publication of *Treatise of Mineralogy*

Overview

In 1801 René Haüy, a French mineralogist, described one of the first coherent theories of crystal structure, published as the *Treatise of Mineralogy*. From this start grew the science of crystallography, the study of crystals and their structure, growth, and form. Since that time, the science of crystallography has matured and developed new tools, including x-ray diffraction, to study the crystals from quartz to DNA. Haüy's description included an explanation for many phenomena that had been commented on, but never explained, and allowed crystallography to take its place as an undisputed scientific field rather than a simple description of shapes and forms.

Background

Although depictions of minerals appear in paintings and other artwork dating back as far as several thousand years, the first work on mineralogy was not written until the third century BC. For two thousand years, mineralogy remained more descriptive than scientific as it remained unable to predict any properties or to do more than describe the properties of a mineral. In 1669 Nicolaus Steno (1638-1686) noted that the angles between adjacent faces in quartz crystals were always the same, regardless of the crystal's size. Then, in 1783, Rome de l'Isle measured interfacial angles on a number of crystals, confirming and expanding Steno's work. Soon after, Haüy developed a mathematical theory of crystal structure that turned out to be remarkably accurate and gave crystallography a legitimate place among the sciences.

According to the most popular story, Haüy dropped a calcite crystal and, when it broke, noticed that each small piece had faces similar to those of the larger crystal. Further attempts to cleave the smaller crystals along the same lines as the larger one proved that the interfacial angles (the angles between each set of faces) remained the same no matter how small the crystals were made. Haüy deduced that this was because a "unit cell" (which Haüy called an integral molecule) existed that was the same shape as any of the visible crystals and that the crystals were all made up of these invisible unit cells that stacked together to form the larger crystals. He further suggested that the unit cells were made of small cubes that, by stacking in specific arrangements, could be made to form any of the crystal shapes seen in nature. By changing the manner in which these unit cells were stacked, all the different faces and interfacial angles of the crystals then known (and all known today) could be explained.

While, like the story of Isaac Newton (1642-1727) and the apple, this incident may or may not have actually happened, there is no doubt that Haüy's theory of crystallography gave the field the ability to explain many observations and to make predictions that were later shown to be correct. Haüy's later work dealt with further explanation of his findings and developing a mathematical framework for his theories. He also helped develop a set of conventions for describing crystal structures that allowed scientists from anywhere in the world to understand the work performed by others. This helped to put crystallography on a more formal footing and facilitated scientific collaborations.

Although the existence of atoms was still not fully accepted at this time, Haüy's theory demanded some sort of internal structure to crystals that made possible the outward forms we see. We now know that atoms are present in a precise pattern that corresponds to the outward shape of the crystal. Even on the atomic level, then, Haüy's theory remains accurate.

Work by other mineralogists followed quickly. William Wollaston (1766-1828) developed a reflecting goniometer in 1809 that made possible very rapid and precise measurements of a crystal's interfacial angles, giving accurate and much needed information on crystal structure. Measurements made with Wollaston's device and the chemical theory developed by Jöns Jakob Berzelius (1779-1848), a Swedish chemist, were the last ingredients needed for crystallography to become a full-fledged field of scientific inquiry.

Impact

Perhaps the simplest way to state the impact of Haüy's discoveries is to point out that no fewer than 20 Nobel prizes have been awarded for work based in whole or in part on principles stated by him or made possible by his work. These include the discovery of x-ray diffraction, the first description of the structure of DNA, discovery of the fullerene form of carbon, understanding of the structure and function of many biologically important compounds (including enzymes, vitamins, and many proteins), and much more.

It is important to note that, at first, Haüy's work was simply a mathematical description of what the insides of crystals might look like. The theory gave results that seemed to match what sci-

entists saw in nature, but it was not necessarily correct. It was not until the advent of x-ray diffraction and full acceptance of the atomic theory of matter in the early twentieth century that scientists agreed that Haüy's theory of crystal structure actually provided a very good match to reality.

Another advancement was Haüy's description of the different crystal systems, or basic shapes that crystals can take. These systems are monoclinic, triclinic, orthorhombic, isometric (of which cubic symmetry is one possibility), hexagonal, and tetragonal. For example, the hexagonal system of crystals, to which corundum and quartz belong, is characterized by six faces surrounding a central axis, while the isometric system consists of cubical unit cells. Halite (salt) is an example of the isometric system and each salt crystal is either a cube or is comprised of several joined cubes.

At the time Haüy announced his discoveries they had very little immediate practical impact, and virtually no impact for anyone other than crystallographers. Scientists were delighted to have a new explanatory theory to test and use, but there was little impact on everyday life. As time went on, the interfacial angles of most crystals were measured and described, the mathematics of crystallography was refined, and the field languished for a time. Other discoveries of crystal properties were used in other branches of geology, but the very small-scale study of crystals awaited the discovery of x-ray diffraction. Once this was discovered in 1912 by Max von Laue (1879-1960), crystallography took off, becoming ever more powerful as a tool to describe the forms taken by natural substances and to predict the properties of materials based on the orderly arrangement of atoms within a substance. Even substances that are normally liquid at room temperature can be studied by freezing them or otherwise inducing them to form crystals.

In fact, x-ray diffraction is still an important tool in studying crystals, to the point that linear accelerators have been constructed for the sole purpose of generating intense x-rays for more detailed studies of crystals and crystalline solids. These studies are expected to yield valuable scientific and commercial knowledge in a number of areas.

Of the discoveries dependent on the science of crystallography, perhaps the most important was the discovery of the structure of DNA. DNA was first synthesized and crystallized in the lab by Arthur Kornberg (1918-) in the 1950s. A few years later, the study of crystallized DNA through x-ray diffraction helped James Watson (1928-) and Francis Crick (1916-) deduce the structure and many of the functions of DNA, winning them the Nobel Prize and making possible all of the recent advances in genetic science that have since taken place. Similar studies have helped scientists to determine the structure of many biologically important substances, including vitamin B_{12}, certain enzymes and amino acids, some proteins, and similar compounds. In addition, crystallizing and studying some viruses and viral proteins has given researchers more information about how viruses can make people ill. This information, in turn, helps to develop more effective vaccines or treatments. Likewise, the study of some crystallized toxic agents can also lead to similar benefits.

Outside of biology, the study of crystals has proven fruitful as well. Liquid crystals are important in the displays of digital watches, calculators, and many computer monitors. Most minerals and synthetic compounds are subject to x-ray diffraction to determine their crystal structure, which can suggest possible new uses for materials as well as identifying possible weaknesses. X-ray diffraction is routinely used to identify minerals that would otherwise be impossible to identify, especially microscopic minerals such as those in soil. In fact, soil mineralogy, a field of increasing importance, depends very heavily on the study of crystals so small they can only be seen with an electron microscope. However, even at this size, x-ray diffraction reveals the interfacial angles, the spacing of planes of atoms within the crystal, and other important information. This information, in turn, can be used to determine the soil's suitability for farming, supporting buildings, or other uses.

In addition, the crystal structures of some minerals are being used to make them more useful to people. Some clays, for example, can be treated so that they will absorb oils and other organic substances, helping to decontaminate parts of the environment. Other crystalline substances have been investigated to try to better understand their electrical or structural properties, while x-ray diffraction studies have been instrumental in designing new generations of drugs and understanding the functioning of many biological compounds. In short, the field of crystallography continues to provide valuable insights into the functioning of our world and should do so for many years to come.

P. ANDREW KARAM

Further Reading

Books

Klein, Cornelius, and Cornelius Hurlbutt. *Manual of Mineralogy.* New York: John Wiley & Sons, 1998.

Sands, Donald E. *Introduction to Crystallography.* New York: Dover, 1994.

Other

International Union of Crystallography. http://www. iucr.ac.uk/welcome.html

The American Crystallographic Association. http://nexus. hwi.buffalo.edu/ACA/

William Smith Uses Fossils to Determine the Order of the Strata in England and Helps Develop the Science of Stratigraphy

Overview

Stratigraphy, the study of rock strata, emerged at the beginning of the nineteenth century for both scientific and economic reasons. Although William Smith was the first to use fossils to trace a long series of strata over a large area, his practical rather than scientific approach meant that his work was not influential outside of England. Georges Cuvier and Alexandre Brongniart, however, were well established in the scientific community, and their slightly later research became the basis for most future work on fossils and geological history. Nevertheless, William Smith is usually remembered as the father of English geology and as a pioneer of stratigraphy.

Background

In the late eighteenth century scientists began to ask a new kind of question about the structure of Earth's crust. Before, they had only been interested in the physical and chemical origins of rocks and the valuable minerals that they contained. Now, knowing the historical order in which strata (layers of rock) had been deposited became an important part of geology, too. The addition of this new emphasis was primarily due to a German mining teacher named Abraham Gottlob Werner (1749-1817).

The order of Earth's strata had been studied from time to time on a local basis in previous centuries, but Werner was the first to endorse a general system that divided all rocks into four basic "formations," according to how long ago they had been deposited. (Since new rocks were almost always formed on top of old ones, the age of any layer was determined by its position relative to others.) As he was a teacher, his 1787 "short classification" was very influential and was used all over Europe. However, because Werner's categories were so broad, geologists still had to find a way to sort out the sequence of layers within each formation and to match up strata occurring in different geographical locations. The age of a rock could not be known from its appearance and chemistry alone, because the same types (such as limestone, clay, and sandstone) occurred over and over again throughout the strata. Some geologists suspected that fossils, if examined more closely, might offer a way to get around this problem.

Before 1800 fossils had generally not been studied by people who were interested in determining the order of strata. Geologists felt that any fossil could come from a rock of any age, because Christian belief held that the same animals had existed unchanged for all time. Early in the nineteenth century, though, some biologists in Germany and France began to claim that many fossils came from species that no longer existed. If extinction was real, then the animals that had existed in the past must have differed from period to period. Fossils could thus be used to date rocks and to determine whether strata in widely separated locations were part of the same formation.

Ironically, the first person to apply this principle to give a detailed geological description of a vast area knew little if anything of the scientific theories involved. William Smith (1769-1839) was an English surveyor and engineer who acquired an extensive firsthand knowledge of the geology of England while engaged on numerous contracts from the 1790s onwards. By 1799 he realized that fossils were the key to determining the order of the "Secondary" (now called Mesozoic) rocks around Bath, in southwestern England. He was then able to show that the same

William Smith. *(Archive Photos, Inc. Reproduced by permission.)*

strata could be found across the country, following a diagonal line to the northeast, thus demonstrating England's regular and predictable geological structure. Lacking any formal scientific training, however, Smith was not interested in the fossils themselves or in understanding the historical development of Earth. His interest in geology came from economic motives instead.

In Smith's time the Industrial Revolution was well underway in England. In this period of economic and technological growth and change, there was a large demand for mineral resources such as coal, metals, and building materials. Large landowners hired men like Smith to help them find valuable deposits on their property, and so it was important for him to be able to predict what sort of rocks would be found in a given location, even at a great depth underground. This task required knowledge of the order in which the strata occurred as well as the ability to tell where in this sequence a previously unexamined mass of rock belonged. Not surprisingly, these objectives were the same as Werner's, since practical concerns dominated in both cases. An important difference, though, was that on the European continent subjects such as mining geology were taught in state-run technical schools, while in Britain engineers and surveyors like Smith received only a practical, on-the-job education. Smith's social and intellectual background played a major role in determining the impact and influence of his work.

Impact

William Smith exhibited his maps and fossil collections to practical and gentlemanly audiences alike on numerous occasions in the first decade of the nineteenth century. He was particularly aided by the patronage of Sir Joseph Banks (1743-1820), president of the Royal Society of London. For various reasons, however, it was not until 1815 that Smith published his fully colored, large-sized map of the strata of England, with written commentary in following years. Although Smith's map was generally acknowledged to be a great accomplishment, it was based on work that he had done almost twenty years earlier, without reference to academic geology. In the meantime, two distinguished French scientists had also discovered the value of fossils to stratigraphy and had already communicated their work to a scientific audience. The impact of Smith's contribution cannot be understood without reference to their work.

Georges Cuvier (1769-1832) was an important and influential biologist who was already famous for his work on fossils and extinction; Alexandre Brongniart (1770-1847) was a geology professor and the director of the state porcelain factory, which gave him an interest in finding deposits of material useful for making ceramics. Working together in the region around Paris from roughly 1804 to 1808, they combined their skills to establish a detailed stratigraphic sequence of Tertiary rocks, based on the different fossils found in each layer. They reported their results promptly in scientific journals, including an English translation in 1810 and a geological map in 1811.

Cuvier and Brongniart planned, executed, and communicated their research in a framework that made it easy for other scientists to understand and apply their conclusions. In addition, they used stratigraphy not just to locate mineral resources but to understand the geological and biological history of an area. For the first time it was possible to see what organic life had been like in the distant past and how it had changed in response to changes in the physical environment.

By the time William Smith finally published his map, four years after Cuvier and Brongniart's, the next generation of English geologists was already beginning to take an interest in the work

of the French school. These scientists, exemplified by the Geological Society of London (GSL), which had been founded in 1807, were concerned with historical rather than practical questions. Thus, Smith's accomplishment, although it was a first, was not directly influential on geological thinking in either Britain or continental Europe. However, it should not be thus inferred that his impact was entirely negligible.

Smith did make an effort to communicate his work, if not through scientific journals. He spread his ideas through personal contact with other geologists, including his nephew John Phillips (1800-1874), who eventually became a professor at Oxford. Smith's methods were widely appreciated by practical men like himself, who used them in their regular work. Smith's methods also sometimes provided crucial information about local stratigraphy to the gentlemen scientists who were responsible for the major debates and developments in early nineteenth-century British geology.

Furthermore, in seeking to determine the order of the more complex strata that underlay the rocks studied by Smith, high-profile members of the GSL such as Adam Sedgwick (1785-1873) and R. I. Murchison (1792-1871) were continuing a project that he had begun, even if their terms and ideas were completely different. Although the gentlemanly GSL could not accept someone who labored for a living as a member, Sedgwick and Murchison acknowledged their own appreciation of Smith's work, and his impact on them, by awarding him in 1831 the GSL's first Wollaston medal and by calling him the "father of English geology."

Smith's 1815 map also had some impact. It covered a much longer sequence of strata and a much larger geographical area than Cuvier and Brongniart's. It also used an innovative color-shading scheme to convey the three-dimensional structure of the strata, a method that was technically superior but impractically expensive. Less usefully, Smith named the strata according to his own personal terminology rather than attempting to connect them with already described formations elsewhere in Europe. This was one way in which his work had very limited impact, since geologists wanted to establish historical interpretations and mapping standards that would be applicable everywhere. Finally, if nothing else, Smith's map demonstrated the basic value of fossils as a powerful tool for solving stratigraphical problems. For example, it is likely that Brongniart himself saw a preliminary version of the map while visiting Joseph Banks on a trip to London in 1802 and was encouraged by the possibility of such a project.

The development of stratigraphy was arguably the most important event in the history of early nineteenth-century geology. With this tool, geologists were able to construct a complete, synthetic history of Earth, realizing for the first time just how long a period of time the fossil record represented. This theme was taken up by Charles Lyell (1797-1875), and it later found a place in the theory of natural selection, implying that the ultimate intellectual and social impact of the use of fossils in stratigraphy have been truly immense. More immediately, though, there were real economic benefits of knowing where to look, and where not to look, for coal and other resources. The desire to find and exploit mineral wealth motivated the establishment of numerous state geological surveys around the world from the 1830s onwards, and the ability to make accurate and useful maps was an important part of this movement.

The use of fossils in stratigraphy was thus valuable to different people for different reasons. William Smith's contribution is best measured by his own practical standards, according to which it was a notable success. Cuvier and

Brongniart's work was equally successful by their newly developed historical standards, which were also quickly becoming those of the geological community at large.

BRIAN C. SHIPLEY

Further Reading

Books

Gohau, G. *A History of Geology.* Rev. and trans. by A. V. and M. Carozzi. New Brunswick, NJ: Rutgers University Press, 1990.

Laudan, Rachel. *From Mineralogy to Geology: The Foundations of a Science, 1650-1830.* Chicago: University of Chicago Press, 1987.

Oldroyd, David. *Thinking About the Earth: A History of Ideas in Geology.* London: Athlone, 1996.

Porter, Roy. *The Making of Geology: Earth Science in Britain 1660-1815.* Cambridge: Cambridge University Press, 1977.

Rudwick, Martin J. S. *The Great Devonian Controversy.* Chicago: University of Chicago Press, 1985.

Torrens, Hugh S. "Patronage and Problems: Banks and the Earth Sciences." In *Sir Joseph Banks: A Global Perspective,* edited by R. Banks, et al. Kew, UK: Royal Botanic Gardens, 1994.

Periodicals

Rudwick, Martin J. S. "Cuvier and Brongniart, William Smith, and the Reconstruction of Geohistory." *Earth Sciences History* 15 (1996): 25-36.

Charles Lyell Publishes *The Principles of Geology* (1830-33), in Which He Proposes the Actual Age of Earth to be Several Hundred Million Years.

Overview

Until relatively recently most people assumed that Earth was relatively young. Charles Lyell, in his work *The Principles of Geology,* was the first to be taken seriously when proposing an ancient Earth. This, in turn, opened the way for future scientific work, including the entire structure of modern geology, the theory of evolution, and concepts of "deep time" in many branches of science.

Background

Throughout most of human history, people assumed that Earth and all its inhabitants were recently created. In the Western world, the Bible was thought to represent the literal truth about

Earth's creation, leading most to conclude that Earth was around 6,000 years old. The concept of an older Earth was not unknown in scientific circles at the beginning of the nineteenth century; Scottish geologist James Hutton had earlier proposed an ancient Earth, but one that was fundamentally different from what we recognize today. Hutton's vision of Earth was of an endless cycle of sediments from the land filling the oceans, turning to rock, and uplifting to form new continents, which then eroded to start the cycle again. Most scientists believed in an old Earth too but felt that most features formed suddenly and catastrophically. The majority of the public, however, believed in a young Earth. These two beliefs—in a young Earth and in Cat-

Charles Lyell. *(The Library of Congress. Reproduced by permission.)*

astrophism—were the generally accepted truths as the nineteenth century opened.

These beliefs led inevitably to others. A young Earth required direct creation of all plants and animals present today, because there was insufficient time for evolution to work. This gave too little time for intelligence to develop, suggesting that humans were created intelligent, having the right to conquer the world. In geology and astronomy, the world and the universe must have achieved their current forms as the result of direct creation or by catastrophic events such as the biblical flood. This view of the universe made humans special beings, created in the image of God and placed squarely at the center of the universe.

Some discoveries, however, started scientists on the path to believing that there might be more to the history of Earth. Galileo, Copernicus, and others removed Earth from the center of the universe, while Hutton's ancient Earth further shook conventional thinking regarding direct creation. Another breakthrough was the understanding that fossils were not only the remains of animals, but that some represented animals that no longer existed. At first these were dismissed as animals that failed to survive the biblical flood, but further discoveries that seemed to show species changing, one into the other, and the discovery of fossils on even the highest mountains slowly began to convince

people that poor swimming ability was not the only reason for these creatures' extinction.

In the late 1700s and early 1800s an English canal digger, William "Strata" Smith, noticed a consistent relationship between certain rock layers and the fossils found in them. This led him to suggest that fossils could be used to reliably date rock layers relative to each other, even in areas in which beds of rock may be inclined or overturned.

These scientific discoveries, along with their philosophical and theological implications, occupied the thoughts of religious leaders, theologians, and scientists for some time, setting the stage for the final dethronement of humans as special beings, looked upon with special favor by a divine creator.

Impact

In 1830 Charles Lyell began publication of his three-volume work, *Principles of Geology.* Drawing upon the Enlightenment tenet that one must follow scientific evidence no matter where it leads, Lyell looked carefully at the evidence and determined that the history of Earth differed radically from the biblical version. This led him to three extremely important conclusions:

1. Earth is much more ancient than can be explained by a literal reading of the Bible.

2. The processes that have formed Earth and the Universe are the same as processes taking place today. This principle—that physical processes are uniform in nature across time—is known as Uniformitarianism.

3. Most of the processes that have shaped Earth and, by extension, the universe, take place over a very long period of time by means of gradual, almost imperceptible, change.

Lyell's conclusions had a strong and immediate impact in scientific and religious circles, an impact that continues in some measure to the present.

From a scientific standpoint Lyell's conclusions were astounding. Science now had vast amounts of time to play with, to fill with events, and to populate with ideas. Under the precepts of Uniformitarianism, scientists could make assumptions about how quickly sediments accumulated or how slowly rock eroded. This, in turn, led to the idea of geologic ages and, coupled with Nicolaus Steno's principle of superposition (the assumption that rocks on top were probably deposited last), allowed geologists to start constructing a geologic history of various regions and, eventually, of Earth itself. Theoretically, scientists could work out the history of Earth from scientific principles rather than by relying on the Bible and other similar documents, thus encouraging natural rather than supernatural explanations for the phenomena that surround us and that shaped Earth.

But Lyell's conclusions reached farther than science. They affected the way that people viewed the Bible and the trust that could be placed in it. Concluding that Earth is ancient (Lyell's estimate was several hundred million years old, as compared to the currently accepted age of 4.6 billion years) directly contradicted the book of Genesis. If accepted, one could no longer believe in the Bible as the literal word of God or as an accurate history of Earth and the universe. Instead, one had to believe that the Bible either contained factual mistakes or that it was allegorical. Either of these alternatives carried profound implications for Western religions.

Finally, Lyell's conclusion that Earth and the universe that surround us are the results of long-term, imperceptible change contradicts both biblical and Catastrophic beliefs. This conclusion led other scientists—notably Charles Darwin and Alfred Wallace—to explore the idea that humans might also be the result of millions of years of imperceptible changes in some other organism. Without the time and the concept of constant, imperceptible change that Lyell's vision gives us, evolution is impossible.

Lyell made some mistakes in constructing his theories. The most serious came from applying his concept of Uniformitarianism too generally. Not only did Lyell state that geological and physical processes were identical at all times, but he also stated that life processes were uniform over time. This led him to the conclusion that, at some time in the future, extinct animals would again walk (or swim or fly) Earth as geologic history repeated itself in a great cycle. Lyell eventually understood this to be erroneous, after Darwin and Wallace published their works on evolution.

It is also worth pointing out that Uniformitarianism has its limits. Recent research indicates that catastrophe does play a part in the history of Earth. Meteor impacts are perhaps the best-known examples, but large volcanic eruptions, the filling of the Mediterranean Sea, the emptying of Lake Missoula to form the Scab Lands of Eastern Washington State, and other sudden events are either global or local catastrophes. An important distinction, however, is that the physical processes leading to these catastrophic events remain constant through time. Or, put another way, the clock may strike suddenly on the hour, but it ticks continually.

Because of his work, Lyell is often referred to as the father of modern geology. He made a methodical effort to lay out an almost heretical theory in meticulous detail, providing overwhelming evidence of its veracity. His conclusions directly influenced Darwin's thinking (Darwin actually read Lyell's books while sailing to South America on the *Beagle*), setting the stage for the latter scientist's theory of evolution, perhaps the key scientific advance of the nineteenth century. In addition, Lyell's painstaking work encouraged Darwin to write with the same level of detail, which, in turn, gave his *Origin of Species* added authority. Lyell's scientific style continues to influence the nature of research and scientific writing to this day.

Lyell's work did not have an immediate impact on the average person living in the early nineteenth century. Most people even in his native England had little idea who Lyell was or that anyone was challenging the biblical account of the creation of Earth. However, these matters concerned clergymen, scientists, and the aristocracy, those having the time to think about such matters and the education to appreciate the questions raised. In time, especially after the publication of Darwin's work, an amalgamation of Darwin's theory and Lyell's chronology became better known among the general public.

The subsequent impact of Lyell's work was enormous. Many of the most successful scientific theories of the last 160 years have depended in part on the framework constructed by Lyell. Evolution theory, continental drift, the formation of the universe, and other concepts we take for granted depend on almost imperceptible change over eons. In astronomy, geology, biology, and other sciences, Lyell's concept of uniformity through time is assumed as a given, with some constraints. And these subsequent theories, in turn, have given us the framework of modern scientific knowledge.

P. ANDREW KARAM

Further Reading

Cloud, Preston. *Oasis in Space*. New York: W.W. Norton & Company, 1988.

Emiliani, Cesare. *Earth*. New York: Cambridge University Press, 1992.

Gould, Stephen Jay. *Time's Arrow, Time's Circle*. Cambridge, MA: Harvard University Press, 1988.

Schopf, James. *Cradle of Life*. Princeton, NJ: Princeton University Press, 1999.

The Discovery of Global Ice Ages
by Louis Agassiz

Overview

Today, the concept of thick ice sheets covering large portions of the globe is a familiar one. We now know that ice sheets advance and retreat, altering landscape and climate as they do so. This knowledge, however, is relatively recent, the result of a great deal of geological deduction on the part of the Swiss scientists Johann von Charpentier (1786-1855) and Jean Louis Agassiz (1807-1873). Charpentier first advanced a reasonable scientific explanation of a recent ice age to explain many of the phenomena found in the Alps, while Agassiz fought to have the idea win acceptance in both the scientific and popular arenas. Acceptance of this theory has changed the way we view Earth and its climate, including recent debates on global warming.

Background

For centuries it had been noticed by scientists and local residents that many areas of northern Europe and North America possessed unusual jumbles of sand, gravel, mud, and silt that showed no consistent layering. Many of these jumbles contained large rocks that were obviously brought from some other location and left, presumably deposited by some outside agent.

Initial explanations ranged widely and included some ideas that now seem silly. The most widespread explanations referred to the biblical flood and assumed that these deposits were simply left behind when the flood waters receded from the face of the Earth. A variant on this theory held that many glacial phenomena resulted from icebergs afloat in the post-flood waters, gouging rocks as they floated in shallow waters and depositing rocks and other sediments frozen into their undersides. In fact, the term "glacial drift" as a synonym for what is now known as "till" is a carry-over from such theories. Other explanations included suggestions that glacial erratics (the large rocks apparently brought from some distance) were launched by underground pressure, as though shot from a cannon; that water had suddenly issued forth from now-lost caverns; or that large amounts of water had recently condensed out of the atmosphere and left these deposits. Finally, in the 1820s, several Swiss men began to formulate some other ideas.

The first of these was a chamois hunter named Jean-Pierre Perraudin, who convinced engineer Ignace Venetz (1788-1859) that glacial marks in Alpine valleys were, indeed, left by previous larger glaciers. Venetz expanded on this idea over the next several years, albeit without much success, but was able to convince Charpentier of its essential accuracy. In 1836, Charpentier took Agassiz on a trip through the Alps, which proved utterly convincing. Agassiz not only accepted Venetz's glacial theories, but expanded them further to suggest that, in addition to the expansion of Alpine glaciers, other glaciers descended from the north to cover virtually all of Europe and North America. Unlike Charpentier, Agassiz was an active and ultimately convincing advocate of the glacial theory, although it took many years for the scientific community to fully accept the idea of nearly two miles (3.2 km) of ice sitting atop most of Europe and North America. Since that time, geologists have also come to realize that Earth has undergone many ice ages, going back more than two billion years.

Impact

The theory of ice ages has impacted both the science community and the public in a number of ways. For scientists, understanding ice ages has provided useful tools for better understanding geology, climate, and our current world. The public has shown a great interest in the concept of ice ages and has gained a better appreciation for the variations in climate that the Earth undergoes, even without human intervention. This, in turn, has led to an increased awareness of the possible impacts of humanity on the environment.

Understanding that Earth had undergone one relatively recent ice age helped to explain many problematic geologic deposits found in the Alps, northern Europe, and North America. It did not explain many other, similar deposits found elsewhere. Accepting a single ice age, however, made it easier to accept multiple ice ages, and this has become the standard explanation for the great many glacier-related deposits found on Earth. In fact, it is now assumed that at least four major glaciations have occurred in the northern hemisphere over the past million years or so. In addition, ancient glacial deposits and other evidence suggest strongly that ice ages are not limited to the current era but have, instead, occurred periodically throughout the history of the Earth. This, in turn, causes scientists to wonder why.

Questioning the origins of ice ages has been a fruitful endeavor. The periodicity of northern hemisphere ice advances led to the suggestion by the Serbian geophysicist Milutin Milankovich (1879-1958) that regular variations in the inclination of the Earth's axis coupled with regular orbital variations and other factors periodically coincide to lower global temperature long enough to start a glacial advance. Geologists—once the theory of plate tectonics was accepted—suggested that large continents periodically congregate near one of the poles, causing the land to cool and spurring glacial advances. Others feel that long-term variations in weather may follow changes in solar activity, the passage of the solar system through interstellar dust clouds, or other extraterrestrial events. All of these suggestions have resulted in research that, even if inconclusive with respect to ice ages, has led to a better understanding of our climate and factors that may influence it. Perhaps the single most important outcome of all these research efforts, however, is the realization that Earth's climate does change dramatically over time.

Research efforts at the present time are aimed at determining reasons for sudden climatic shifts, because of fears of global warming caused by human activities. The reasoning is that, if we can understand what caused temperatures to change, we can better understand whether we can cause global warming. However, with all of these studies, it must first be acknowledged that the typical temperature of Earth is much warmer than present global temperatures. In fact, we are currently in what is known as an interglacial period, meaning that the glaciers have temporarily retreated, but there is no reason to assume that they will not again advance in the future. On the other hand, it is also entirely possible that the most recent ice age has, in fact, ended, in which case we would expect global temperatures to begin rising, ice caps to begin melting, and glaciers to be retreating. While scientists have not agreed on the time, magnitude, or direction of these changes (i.e., whether the glaciers will advance again or continue their recent retreat), they unanimously agree that there will again be a major climatic change. There is a growing realization that the climate changes and that human actions may be responsible. This, in turn, is likely to encourage further policies, such as low-emissions vehicles, substitute fuels, and so forth, that are designed to reduce the likelihood and magnitude of future global warming.

Whether the glaciers will melt or advance a fifth time is still not known with any degree of certainty. Similarly, whether human activities will hasten an advance, prevent it, or have no impact remains unknown. What is known is that ice once covered large parts of the Earth and is now gone, leaving irrefutable proof of its existence and power.

P. ANDREW KARAM

Further Reading

Benn, Douglas. *Glaciers and Glaciation.* Edward Arnold Press, 1998.

Hallam, A. *Great Geological Controversies.* Oxford: Oxford Science Publications, 1989.

Hambrey, Michael. *Glaciers.* Cambridge: Cambridge University Press, 1992.

Women Scientists in the Nineteenth-Century Physical Sciences

Overview

At the beginning of the nineteenth century, it was uncommon for women to be active in the sciences. While most science was still done by men as the century drew to a close, women had found opportunities for work in the physical sciences. New educational opportunities helped women find jobs in astronomy, chemistry, and physics. In addition, women educated themselves or worked with family members to do scientific research and prepare publications in the sciences, either as amateurs—like many of the scientists at the time—or as professional scientists.

Background

During the nineteenth century, women's opportunities to go to school expanded. New colleges gave women the background and skills they

Maria Mitchell. *(The Library of Congress. Reproduced by permission.)*

needed to perform scientific research and also to find jobs in the sciences (for example, performing calculations at observatories). These colleges also gave women faculty the chance to do scientific research. Schools that did not formally admit women would sometimes allow women to attend lectures and gain an education (although not always a degree) that way.

Women also educated themselves for scientific careers, though this was a more difficult route. In addition, some women were educated by their families and some worked with family members (fathers, brothers, or husbands) to do scientific research. University-educated women also sometimes found it possible to collaborate with a spouse.

Some fields provided more entry points for women than others. Astronomy provided many ways for women to find work, although often in a supporting role rather than as original researchers. The rise of photography toward the end of the century gave women the chance to work with photographs while avoiding the long nights of observing that were thought to be too demanding for women. As observatories began huge catalogs of data about stars, women were able to make the calculations and classifications that allowed the catalogs to be completed.

Geology required fieldwork, which could be difficult for women to pursue, but it also allowed women to work as scientific illustrators and as amateur fossil collectors who could contribute to the growing body of knowledge on geological features. Chemistry and physics often required laboratories and equipment that women frequently did not have access to, but for some women a college education provided a way to break into laboratory research in these sciences.

Impact

Several schools for women were founded in the United States during the nineteenth century. Women faculty at these and other institutions prepared other women for careers in science. Sarah Whiting is notable for her work at Wellesley, where she created the first undergraduate physics laboratory for women and later outfitted an observatory, the result of a gift from Mrs. J. C. Whitin. At Vassar, Maria Mitchell (1818-1889) also educated future scientists and ran the observatory. In her early adulthood, Mitchell made astronomical observations for the U.S. Coast Survey with her father and made a comet discovery that won her a gold medal from the King of Denmark. Margaret Maltby, a physicist who did research in Germany, taught physics at several colleges and taught the first course in the physics of music.

American and European universities produced some well-known women scientists. Marie Curie (1867-1934) is possibly the most famous. She was for many years the only person to win two Nobel prizes in separate fields (one for physics and one for chemistry). She worked with her husband, Pierre Curie (1859-1906), to discover the radioactive elements polonium and radium and to study radioactivity, at the time a new property of matter about which scientists knew very little. Another European woman, Yulua Lermontova was one of the few nineteenth-century women to get a doctorate from a German university. She did research on several topics, including the chemical properties of petroleum.

Even women with university degrees often found it hard to break into the more prominent fields of organic and inorganic chemistry, but did valuable work in other areas. Ellen Swallow Richards (1842-1911) began studying chemistry at the Massachusetts Institute of Technology as a special student before MIT formally admitted women. She was their first woman graduate in 1873. Her work in chemistry focused on practical applications, including municipal sanitation and home economics. She worked in fields as diverse as nutrition and oceanography, and was important in the early study of ecology. Rachel Lloyd, probably the first American woman to receive a Ph.D. in chemistry, studied agricultural chemistry, in particular the properties and uses of sugar beets.

Dorothea Klumpke Roberts was a high-profile astronomer of her day. Educated at the University of Paris, she went on to work at the Paris Observatory as their first woman staff member. There she supervised the Observatory's part of the sky-mapping project called *Le Carte du Ciel*. She became the first woman to make airborne astronomical observations when she observed the Leonid meteor shower on a hot-air balloon ride in 1891. She also lectured and published on astronomical subjects and, with her husband, took many astronomical photographs in the early part of the twentieth century.

There were not many formally trained female geologists in the nineteenth century. One of the few was Florence Bascom, who received a Ph.D. in geology late in the century from the University of Wisconsin (by special dispensation; again this was before women were officially admitted). She was the first woman to work as a geologist for the U.S. Geological Survey. She mapped rock formations in Maryland, Pennsylvania, and New Jersey for the Survey and also taught at Ohio State University and later at Bryn Mawr.

Self-education was a more difficult process, but several self-educated women succeeded in doing notable scientific research. Helen Abbott Michael studied the chemistry and evolution of plants without the benefit of a university education. She had some medical training but did not complete an M.D. until after she began her plant research. She established the first scheme for the biochemical classification of plants. Agnes Pockels, also self-educated in her field with some help from her brother, did important work in the chemistry of surface layers (for example, contaminants on the surface of water).

New scientific projects offered jobs for women with college degrees as well as those who were self-educated. Observatories working on large star catalogs hired women to work as "computers" to prepare the data for the catalogs, and also to study photographic plates and classify stars. For example, Edward Pickering (1846-1919) at the Harvard College Observatory hired many women to work on the Henry Draper catalog. Working for him, Williamina Fleming (1857-1911) established a system of classifying stars that was refined later by Annie Cannon

(1863-1941) and is the system still in use today. Antonia Maury (1866-1952) also worked for Pickering on classifying stars and on studies of the variable star Beta Lyrae. Most of her classification system is not used anymore, but she did establish a lettering system for classifying parts of a star's spectrum that can help to identify giant and supergiant stars.

Some women were able to work with family members who were scientists. Caroline Herschel (1750-1848), the discoverer of eight comets and the first woman to discover a comet, assisted her brother William Herschel (1738-1822) in his astronomical observations. Born in Germany, she visited William in England to pursue musical interests and ended up staying for the rest of her life, assisting in his scientific work. She recorded observations, made calculations, and helped him make the telescopes they used for their observations. She continued to work on her own for 26 years after his death.

Margaret Huggins (1848-1915) was a skilled astronomer, performing visual and photographic observations. She and her husband were both active in the emerging study of the spectra of stars, and worked together on an atlas of stellar spectra. Mary Somerville (1780-1872) was also supported by her husband in her scientific work, although he was not himself a scientist. She wrote on the solar spectrum and the motions of celestial bodies, as well as a book on the connections between the physical sciences.

Family connections helped several women in the field of geology. Mary Anning (1799-1847) worked with her father in his fossil-collecting business, which she carried on after his death. She had no formal training, but made several important discoveries of dinosaur skeletons. Orra White Hitchcock also did geological work as an illustrator. In the 1830s and 1840s she provided geological drawings for her husband, Edward Hitchcock (1793-1864), the first state geologist of Massachusetts. She was among the first of several women who used a skill that was considered appropriate for women—drawing and painting—to do scientific work.

These and other women of the nineteenth century did important scientific work, sometimes under difficult conditions, and paved the way for the women of the twentieth century to enter the sciences in greater numbers.

MARY HROVAT

Further Reading

Books

Creese, Mary R. S. *Ladies in the Laboratory?: American and British Women in Science, 1800-1900: A Survey of Their Contributions to Research.* Lanham, MD: Scarecrow Press, 1998.

Ogilvie, Marilyn Bailey. *Women in Science: Antiquity Through the Nineteenth Century: A Biographical Dictionary with Annotated Bibliography.* Cambridge, MA: MIT Press, 1986.

Shearer, Benjamin F., and Barbara S. Shearer, eds. *Notable Women in the Physical Sciences: A Biographical Dictionary.* Westport, CT: Greenwood Press, 1997.

Veglahn, Nancy. *Women Scientists.* New York: Facts on File, 1991.

Periodicals

Dobson, Andrea K., and Katherine Bracher. "Urania's Heritage: A Historical Introduction to Women in Astronomy." *Mercury* 21 (January-February 1992): 4-15.

Kehoe. "Brilliant Science, Bitter Scandal: The Life of Marie Curie." *Biography* 3 (July 1999): 88-94.

Opalko, J. "Maria Mitchell's Haunting Legacy." *Sky & Telescope* 83 (May 1992): 505-507.

Spradley, J. L. "The Industrious Mrs. Fleming." *Astronomy* 18 (July 1990): 48-51.

Tickell, Crispin. "Princess of Paleontology." *Nature* 400, no. 6742 (22 July 1999): 321.

The Transformation of the Physical Sciences into Professions During the Nineteenth Century

Overview

For today's students, pursuing a career in science is as normal as undertaking the study of law or medicine. Yet, before 1800 it would have seemed almost inconceivable that one day people could earn a living practicing science. In fact, the word "scientist" was not even invented until 1833. During the nineteenth century, however, the proliferation of scientific societies, journals, and opportunities for advanced research and education lead to the professionalization of the physical sciences and the formalized structure of specialization, publication, and academic training that characterizes modern scientific scholarship.

Background

The primary characteristics of all professions are authority and autonomy. Only members of a profession are granted the privilege of determining what is true or valid within that field. For instance, only astronomers can determine whether a new discovery represents a pulsar or a quasar, just as only physicians can determine whether a growth is cancerous or benign. Once a determination has been made, it can only be legitimately challenged by another member of the same profession. The means by which a field gains this authority and autonomy generally revolve around coming to an agreement about what constitutes expertise in the field and creating positions in which a person can earn a living by practicing a profession.

During the nineteenth century four key elements came together to allow science to become a profession: 1) creation of salaried positions for work within a scientific field; 2) academic instruction at advanced levels; 3) formation of specialized societies and meetings; and 4) publication of new research in peer-reviewed journals. These elements arose at different times within various scientific fields, and across national borders. Between 1800 and 1900, however, almost all the physical sciences achieved these necessary elements and gained the status of professions.

In order to understand the developments in science in the nineteenth century, it is necessary to place the events in context. In the seventeenth and eighteenth centuries science was practiced by individuals who collected specimens, performed experiments, and communicated their discoveries with each other through informal correspondence. A few important organizations existed which served to give cohesion to this scattered group of natural philosophers. The Royal Society of London and the Paris Academy of Sciences held meetings at which members discussed scientific topics. The publications and correspondence of these organizations were very important for keeping members in touch with current discoveries and theories. As important as these societies were, they were not professional societies, because membership was not based on expertise in the field and many of the topics discussed bore little resemblance to science. In fact, many members of the Royal Society were wealthy aristocrats who considered it more of a gentlemen's club than a scientific society.

Impact

This informal organization of science might have continued indefinitely except for the changing role of science in society. Chemistry and physics were vital to the new technology and new industries arising during the Industrial Revolution in the late eighteenth and early nineteenth centuries. Advances in astronomy and new discoveries in magnetism were crucial for nautical navigation during this age of exploration and colonialism. As science became more important for the economy and military, the need for trained scientists encouraged the growth of educational and salary opportunities for scientists, along with the creation of methods for evaluating expertise. The creation of educational, salary, and evaluation opportunities came about through the development of scientific societies, journals and conferences, and university-supported science studies.

Although the Royal Society of London and the Paris Academy of Sciences had provided a foundation for gatherings of scientifically inclined individuals, by the nineteenth century there were numerous factors that promoted the growth of new societies. Primary among these were specialization and locality. As interest in science grew and more and more people outside of the big cities became involved in science, local societies were born throughout Europe and America. The Academy of Natural Sciences of Philadelphia and the Manchester Philosophical Society are examples of the many local societies formed to meet the needs of individuals who either did not have the social prestige necessary to be admitted to the Royal Society or who simply wished to be able to gather with other scientifically-inclined people near to home.

In addition to these many general science local societies, the nineteenth century saw the birth of specialized societies. Whereas the Royal Society and the Paris Academy of Sciences discussed all scientific topics, these specialized societies were dedicated to individual branches of science. The Astronomical Society of London was founded in 1820 to meet the needs of British astronomers, and the Astronomische Gesellschaft was formed in Germany in 1863. The first national chemical society was formed in England in 1841, and others were soon founded in other countries. In time, even more specialized societies arose as scientific fields split into narrower categories such as physical chemistry, organic chemistry, inorganic chemistry, and so on. Societies allowed science to evolve from individuals pursuing science independently, to groups organized around a locality or a specific field. In time, these loosely organized groups gained national and even international cohesion.

Journals and conferences cemented the organization of science through widespread dissemination of new research and bringing together scientists at annual meetings. Most journals began as the published records of society meetings and articles submitted by members. The prestige and quality of the society, its members, and the editor of the journal combined to determine the prestige of the journal. While many society journals were only of local interest and ceased publication with the end of the society, others became widely read and are primary research publications in their respective fields to this day. For example, four of the major chemistry research journals today—the *Journal* of the American Chemical Society, the *Berichte* of the Deutsche chemische Gesellschaft, the *Journal* of the Chemical Society of London, and the *Bulletin* of the Société chimique de Paris—began as society publications in the nineteenth century. Journals played an especially important role in the professionalization of science by presenting a space in which new research could be presented, reviewed, and spread throughout the scientific community.

At the beginning of the nineteenth century new scientific discoveries and research were published in many different kinds of periodicals. Articles about important scientific advances were likely to be found in general intellectual magazines next to literary stories and philosophical essays. As society journals became more prominent and gained prestige, it became more common for scientists to publish in these specialized journals. With more and more contributions, editors of the most prestigious journals had to select which articles to publish. Articles with the most interesting, accurate, and important information were the most likely to see print. Once published, articles were subject to intense scrutiny by the rest of the scientific community and reviews were often published in succeeding issues which either praised or condemned earlier articles. By the end of the nineteenth century, journals had replaced this informal system with a formal system of subjecting every submitted article to review by established and respected scientists before publication. This was the beginning of the peer-review system which is in place today. In this way journals became not only the main conduit for spreading scientific knowledge, but one of the primary tools for evaluating the credentials and merit of scientists and their work.

Like journals, conferences became another tool for disseminating new information and establishing scientific credentials. In Germany, scientists had begun to gather annually at a meeting called the Association for the Advancement of Science. A group of British scholars, recognizing the benefit of creating a society which was not as exclusive as the Royal Society and which catered to the outlying communities, formed the British Association for the Advancement of Science in 1831. The American Association for the Advancement of Science followed suit in 1848. These societies held annual meetings in cities and towns throughout the countries. During these meetings, scientists were given the opportunity to give talks and demonstrations about their research. Other societies began holding conferences and these soon became an important means for a scientist to make a name for himself and introduce his ideas or discoveries into the scientific community.

Although societies, journals, and conferences were necessary elements in the professionalization of science, the single most important development in the nineteenth century was the creation of university-supported science studies. The first university teaching and research lab was the lab of Justus Von Liebig (1803-1873), chemistry professor at Giessen in Germany. Before Liebig's lab, most advanced science education had taken place between individual masters and students. Undergraduate studies in science were available at most universities, but Liebig introduced the idea of teaching advanced science and supporting organized research at the university in 1827. In his lab, students took classes in advanced chemistry and undertook systematic research under the direction of established scientists. For the first time, students worked individually on distinct research projects which all addressed some aspect of a larger question posed by the professor in charge of the lab. This method benefited both the professor and the students. The professor was able to explore many different aspects of research within his field by delegating projects to his students. The students were able to learn advanced theory and experimental techniques under the guidance of one of the top scientists in the field, while working on individual research projects. Other universities soon followed the example of Giessen. In terms of aiding professionalization, this new system of labs created many new jobs for scientists and produced qualified individuals to fill the new positions. At last, students graduating with degrees in science could count on being able to find a salaried position that would allow them to earn money through teaching and doing research in their fields. Industry also hired many of these new graduates, knowing they had been appropriately trained and evaluated by their advisors.

Together, the growth of university-supported science studies, peer-reviewed journals, scientific conferences, and societies provided the elements necessary for the physical sciences to become professions by the end of the nineteenth century. By 1900 astronomers, chemists, and physicists were trained in university labs, gained employment in academia or industry teaching and doing research, and presented their new discoveries in journals and at conferences. Through education and publication, scientists evaluated each other and determined who was to be considered "legitimate." These elements of professional science are still in place today. The organization that the scientific community created in the nineteenth century gave scientists the autonomy and authority over their fields which enables them to be considered professionals.

REBECCA B. KINRAIDE

Further Reading

Books

Oleson, Alexandra, and Sanborn Brown, eds. *The Pursuit of Knowledge in the Early American Republic.* Baltimore: Johns Hopkins University Press, 1976.

Periodicals

Ben-David, Joseph. "The Profession of Science and its Powers." *Minerva* 10 (1972): 362-383.

Shils, Edward. "The Profession of Science." *The Advancement of Science* 24 (1968): 469-480.

John Couch Adams
1819-1892
English Astronomer

John Couch Adams was born to a farming family in rural Landeast, Cornwall. His mathematical abilities were evident at an early age, impressing teachers and earning him a scholarship at St. John's College, Cambridge. Graduating in 1843 with high marks, he had already become interested in the problem of Uranus's orbit, which varied from what was predicted using the laws developed by Johannes Kepler (1571-1630) and Isaac Newton (1642-1727). Studying these discrepancies, Adams determined that there must be an additional, then undiscovered, planet beyond Uranus. This meant that when the unconfirmed planet was ahead of Uranus in its orbit, it would pull on Uranus, speeding it up in its orbit. Similarly, when Uranus pulled ahead of this planet, it would be slowed in its orbit. These were the discrepancies Adams and others had noted. The alternative, that Kepler's and Newton's laws were incorrect, was not conceivable since they worked so well for other planets.

Adams's study of the problem involved writing and solving sets of equations with up to 27 unknown terms, a laborious and difficult task. In order to solve this problem, Adams assumed the new planet was at twice the distance from the sun as Uranus (we now know this to be an incorrect assumption as Neptune lies at only about half again the distance from the sun as Uranus). He solved many of the equations and developed a detailed conceptual understanding of the problem in his head, later writing everything down.

Completing his calculations in October, 1845, Adams tried several times to meet with Sir George Airy (1801-1892), one of reigning English astronomers at that time. After several unsuccessful attempts to meet directly with Airy (he neglected to make an appointment, then returned during their meal), Adams left his work for review. Airy, not convinced of the accuracy of Adams's assumption regarding the new planet's distance, wrote Adams to question this but did not receive a response. Meanwhile, papers published by French astronomer Urbain Jean Joseph Leverrier (1811-1877) convinced both English and German astronomers to search for Neptune,

which was discovered by Johann Gottfried Galle (1812-1910) in September, 1846.

There was some discussion regarding allocation of credit for Neptune's discovery. Adams performed the calculations first, but they were not published and eventually led to nothing. Leverrier, second to calculate Neptune's position, did publish his calculations, but Galle was the first to actually see Neptune. Ultimately all three men shared credit for this discovery.

Following the discovery of Neptune, Adams was recognized in Britain as a brilliant astronomer and was awarded a Fellowship at St. John's. Offered a knighthood at one point by Queen Victoria, he refused, fearing he did not have the financial resources needed for the life style demanded of a knight. He did accept an appointment as Lowndean Professor of Astronomy and Geometry at Cambridge and, in 1861, was appointed director of the Cambridge Observatory. Although elected president of the Royal Astronomical Society twice, he turned down the post of Astronomer Royal when Airy retired. He also found time to marry; in 1863 he wed Eliza Bruce.

Throughout his life Adams remained sincere, modest, and self-effacing, in spite of his mathematical and scientific genius. After his death in 1892, he was memorialized with a tablet at Westminster Abbey, near that of Sir Isaac Newton. Adams is currently regarded as the greatest English astronomer and mathematician since Newton.

P. ANDREW KARAM

Jean Louis Rodolphe Agassiz
1807-1873
Swiss-American Naturalist,
Paleontologist and Glaciologist

Jean Louis Agassiz was one of the foremost natural scientists of his time. He was particularly well-known for his work with fossil fishes and for his advocacy of the idea of global ice ages. Although he accepted the idea that species could become extinct, he resisted the theory of evolution, believing instead in the divine creation of species.

Louis Agassiz, the son of a Swiss minister, was born in French Switzerland in 1807. Like many of his contemporaries, he attended univer-

Jean Louis Agassiz. *(The Library of Congress. Reproduced by permission.)*

met with a great degree of skepticism at first, it was later found to be accurate.

Agassiz moved to the United States in 1846, becoming a professor at Harvard University and starting the Museum of Comparative Zoology. In the United States, he was a great proponent of advancement of the sciences, urging the formation of what is now the National Academy of Sciences (of which he was a founding member). He also devoted a great deal of energy to raising funds for scientific research in the United States, attempting to raise the profile of the sciences in his adopted country.

In spite of his research in the evolution of fossil fishes, Agassiz remained a staunch opponent of evolution until his death. He firmly believed in the divine creation of life, although he admitted that species did become extinct from time to time. Towards the end of his life it was a disappointment to him that even his son, Alexander, was a proponent of evolution, but his own views remained unwavering. Given this, it is ironic that Agassiz's own work ended up supporting evolutionary theory. In fact, he is well-known for his work showing that, to some extent, the development of an animal fetus within the womb shows some of the evolutionary steps that the animal's ancestors passed through during their evolution. For example, developing human fetuses have, at various times in their development, tails, gills, and webbed fingers. Agassiz himself felt that this observation was his single greatest contribution to science, and Charles Darwin (1809-1882) and others used this information to help buttress the (then) new theory of evolutionary change. Agassiz died in 1873 at the age of 66.

P. ANDREW KARAM

sity in both Switzerland and Germany, graduating with a degree in medicine in 1830. After graduation he traveled to Paris to study comparative anatomy under Georges Cuvier (1769-1832), a versatile French scientist. Cuvier was impressed with Agassiz's work on fossil fish, and this approval helped to set the course of much of Agassiz's future work. Although Couvier died only six months after their first meeting, Agassiz always considered himself Couvier's intellectual heir and took it upon himself to defend Couvier's work for the rest of his life. Agassiz published his masterwork on fossil fishes by basing it, in part, on notes given to him by Couvier.

As a junior professor at the Lyceum of Neuchâtel, Agassiz became interested in signs that glaciers had once moved far below any existing glaciers. Traveling with his friend, Johann von Charpentier (1786-1855), Agassiz became convinced of the accuracy of Charpentier's hypothesis that deep ice had transported many materials from a great distance. Expanding on this theory, Agassiz realized that many of the surface features in England, Scotland, and much of Europe were consistent with the existence of continental glaciation, leading to a theory of ice ages that he propounded with great enthusiasm. Later, on a trip to North America, he expanded this theory to the New World, suggesting the existence of a global ice age in the recent geologic past. Although this theory

Dominique François Jean Arago
1786-1853
French Astronomer and Physicist

Dominique Arago was a French astronomer and physicist who, in his scientific career, made important discoveries in the areas of optics, electromagnetic radiation, and served as director of the Paris Observatory. He was also politically active, serving as Minister of War and Marine for the provisional government following the 1848 revolution.

Arago was born in Estagel, France, and studied at the École Polytechnique in Paris, where he became a professor of analytical geometry at the

Arago was also active in other areas of physics. In particular, he discovered the phenomenon of magnetic induction by showing that a rotating copper disk will deflect a magnetic compass needle suspended above it. This principle is widely used in many electrical generators and motors today. He also worked with the French physicist Jean Baptiste Biot (1774-1862) to accurately measure arc lengths on the earth, proposing this as the basis for standardizing units of length in the metric system.

In astronomy, Arago's field of primary interest, he was no less active. His best-known discovery is of the sun's chromosphere, the lower atmosphere just above the visible surface of the sun. He showed it to be composed chiefly of hydrogen gas, now known to be the primary constituent of the sun and all other stars. Later in his career, he inspired his student Urbain Leverrier (1811-1877) to investigate some irregularities in the orbit of Uranus. These investigations eventually led to the discovery of the planet Neptune, the first planet to be discovered by mathematical calculation and prediction.

Later in his career, Arago became increasingly involved in politics, first as director of the Paris Observatory and secretary for the Académie des Sciences, and later as an active participant in the Revolution of 1830. Named the Minister of War and Marine by the provisional government after the 1848 revolution, Arago was responsible for eliminating slavery from the French colonies and participated in extending universal suffrage to adult French men.

During his career Arago was elected a Fellow of the Royal Society and was awarded the Copley Medal by that body in 1825. He is remembered by Crater Arago on the moon and by a commemorative plaque on the Eiffel Tower. Arago died in 1853 in Paris, shortly after seeing experimental confirmation of his predictions about light's velocity changes in different materials.

P. ANDREW KARAM

Dominique Arago. *(The Library of Congress. Reproduced by permission.)*

age of 23. Arago's first major contributions to science came in 1811 when, working with Augustin Fresnel (1788-1827), he discovered that two beams of light that are polarized in perpendicular directions do not interfere with each other. This curious result, difficult to explain if light was composed of particles, suggested strongly that light consisted, instead, of waves and had a large impact on the study of light for many years. Another of Arago's triumphs involved verification that a bright spot of light does indeed appear in the center of the shadow cast by a circular obstruction (such as a disk). This was predicted by Fresnel in a paper on the diffraction of light, assuming that light consisted of waves and not the particles conjectured by Isaac Newton (1642-1727). Roundly criticized by Siméon Poisson (1781-1840) for this prediction, Arago showed the effect to be real, further helping to confirm the wavelike nature of light.

Later, in 1838, Arago suggested that the velocity of light might be slower as it passes into denser media, such as water or glass. He was unable to perform this experiment due to failing eyesight, but Armand Fizeau (1819-1896) and Jean Foucault (1819-1868) carried out the experiment in 1850 and proved Arago correct. This principle helps to explain the refraction of light in various media such as lenses and diamonds.

Amedeo Avogadro Conte di Quaregna
1776-1856
Italian Chemist

Amedeo Avogadro was an Italian lawyer, chemist, and physicist. He is best known for determining what is known as Avogadro's number, a physical and chemical constant used extensively in chemistry and physics calculations, including those involving gases.

The son of a lawyer and senator, Avogadro was born in 1776 in Turin, Italy. Avogadro began his career by earning a doctorate in law in 1796 and working as a lawyer for three years. In 1800 he began studying mathematics and physics with a private tutor, deciding to make his career in natural science instead of politics as was expected of him. He was appointed a professor of natural philosophy at the College of Vercelli in 1809, later earning an appointment as professor of mathematical physics.

Inspired by the work of Alessandro Volta (1745-1827), Avogadro began working on some problems in physics with his brother, Felice. Their work in this area earned Avogadro a nomination to the Royal Academy of Science of Turin, a great honor. This recognition gained him a position as demonstrator at the Royal College of the Provinces and convinced him that his future was in science rather than law.

Avogadro's most important accomplishment, however, was the research leading to his hypothesis that equal volumes of gas at a given temperature contain the same number of molecules. Avogadro identified this number as 6.023 x 10^{23} molecules in one gram molecule (or mole) of any given substance. Although taken for granted today, this was revolutionary and controversial at the time and was not accepted until 50 years after Avogadro's death.

Avogadro was attempting to explain an observation made by Joseph Gay-Lussac (1778-1850) that combining equal volumes of gas under some circumstances did not lead to any measurable increase in the total gas volume. Avogadro explained this by positing that the chemical reactions formed another gas and that equal volumes of any gas at the same temperature and pressure, regardless of chemical composition, contain the same number of molecules. This observation also led to better estimates of the relative weights of a variety of gases, including hydrogen and oxygen. Avogadro's hypothesis also helped to explain other oddities noticed by Gay-Lussac, such as the fact that equal volumes of gas under changing conditions (i.e., changes in temperature, pressure, or volume) behave identically.

There are several reasons that Avogadro's research was so long neglected. He was not well known as a meticulous experimentalist and he failed to support his paper with solid data, leaving his results open to interpretation. In addition, he published his results in relatively minor journals that were not well known and, therefore, were easily overlooked or discounted. On top of that, his language lacked clarity, leading to errors in interpretation of his results. And, finally, the results of his work, if believed, promised to overturn the work of other, better known researchers. Thus, it was simply easier to ignore his work and accept the status quo.

In addition to his scientific research, Avogadro was a devoted family man, having six sons with his wife Felicita, the Countess Avogadro. Unlike Avogadro, his children took an interest in law, government, and the military.

Two years after his death, a colleague again presented Avogadro's work to the scientific community, showing that it helped to explain and solve some outstanding problems at that time. Given a more careful reading, Avogadro's work was acknowledged as accurate and valuable, earning him proper recognition at last. Avogadro is now regarded as one of the founders of modern chemistry and his famous number is used and taught as one of the cornerstones of chemical theory.

P. ANDREW KARAM

Antoine Henri Becquerel
1852-1908
French Physicist

Henri Becquerel had already established himself as a respected French physicist when his discovery of radioactivity, in 1896, catapulted him into the ranks of the world's leading scientists. Although the discovery was unexpected, it was not random. Becquerel's background, experience, and particular circumstances positioned him for this historic event.

His background began with his ancestors. Becquerel's father, Alexandre Edmond Becquerel (1820-1891), and his grandfather, Antoine César Becquerel (1788-1878), had each been the physics professor at the Natural History Museum in Paris. Edmond Becquerel was especially interested in phosphorescence. He assembled a large collection of luminescent minerals for the Museum.

Edmond Becquerel's son, Antoine Henri (known as Henri), decided to follow the path that his father and grandfather had chosen. This path gave him the experience, and later the professional positions, that proved to be crucial for his future. Henri Becquerel entered the Paris Polytechnical School in 1872, where he earned an engineering degree. In 1878 he became his father's assistant at the Natural History Museum.

Henri Becquerel investigated the magnetic properties of different substances and the effects of magnetism on light. He also studied infrared spectra, luminescence, and absorption of light. For his researches he was awarded the doctorate in 1888. Becquerel was elected to the French Academy of Sciences in 1889. He was also named to the French Legion of Honor.

Becquerel became professor at the Museum in 1892, after his father's death. He was also appointed to his father's position at the National Conservatory of Arts and Trades. In 1895 Becquerel received his third professorship, this time at the Polytechnic School, where his father and grandfather had previously held the post. He continued to work at all three institutions simultaneously. He was named chief engineer of bridges and highways in 1894.

Early in 1896 Becquerel heard a talk that would forever change his life. A German physicist had just discovered invisible rays that could pass through opaque objects. Becquerel learned that the invisible rays seemed to come from the phosphorescent screen used in the experiments. It was natural to wonder whether other phosphorescent materials might also produce these rays (later called x rays).

Becquerel returned to the Museum and began testing the specimens in his father's collection of phosphorescent minerals. He soon found that uranium minerals emitted invisible rays, which he believed were a form of light. He published his findings during 1896-1897.

At first most scientists were not very interested in Becquerel's rays, since their journals were being flooded with reports of all sorts of invisible rays. Becquerel himself turned to the newly discovered Zeeman effect in 1897, since it related to his researches of magnetism's effects on light. He was drawn back to his rays after Marie Curie (1867-1934) and Pierre Curie (1859-1906) used them to trace and identify two new elements, polonium and radium (1898). Around that time Becquerel was recognized as having discovered a new property of matter, which Marie Curie named radioactivity.

Along with several other researchers, Becquerel found that part of the rays emitted by radium were beams of electrons (1900). In 1901, checking to see whether uranium's rays

might come from an impurity, Becquerel separated a radioactive substance from uranium, leaving the uranium inactive. Amazingly, the uranium eventually regained its original activity. This information helped Ernest Rutherford (1871-1937) and Frederick Soddy (1877-1956) to develop a revolutionary theory of radioactive transmutation in 1903.

Radioactivity became a popular subject for research. Later it would develop into the field of nuclear physics. Becquerel's earlier finding that rays from uranium could affect electrified bodies (the process was later called ionization) was essential for radioactivity studies. However, Becquerel no longer was in the forefront of the field, and he eventually returned to his earlier interests of phosphorescence and light spectra.

For his discovery of radioactivity Becquerel shared the 1903 Nobel Prize in physics with Marie and Pierre Curie. He was also honored by election as vice president, president, and permanent secretary of the Academy of Sciences.

Becquerel's first wife, Lucie-Zoé Jamin, was the daughter of a noted French physicist. She died soon after the birth of their son Jean in 1878, and Becquerel later remarried. After Henri Becquerel's death, Jean Becquerel was named to the physics chair at the Natural History Museum, the fourth generation of his family to hold that position.

MARJORIE C. MALLEY

Jöns Jacob Berzelius
1779-1848
Swedish Chemist

The most renowned chemist of the first third of the nineteenth century, Jöns Jacob Berzelius excelled as a theorist, experimenter, teacher, designer of laboratory equipment, and disseminator of chemical information. He invented the modern system of chemical notation; named such chemical concepts as isomerism, isomorphism, allotropy, and catalysis; and discovered the chemical elements of cerium, selenium, and thorium.

Descended from three generations of Lutheran clergy, Berzelius lost his father at age two and his mother at age nine and was raised by relatives. He began medical studies in Uppsala in 1796 and completed them in 1802, thereafter making a meager living as a district doctor to the poor while working as an unpaid assistant to the professor of pharmacy at the School of Surgery in Stockholm. His fortunes improved upon succeeding to the professorship in 1807, which in 1810 was renamed a chair in chemistry.

Meanwhile, Berzelius had become friends with Wilhelm Hisinger, a prosperous mine owner. Together they undertook original researches using galvanic piles (the earliest batteries) to decompose numerous salts into their respective acidic and basic components, thereby greatly expanding the knowledge and techniques of inorganic analysis. Utilizing the newly formulated atomic theory of the British chemist John Dalton (1766-1844), Berzelius used these analytical results to establish both atomic weights and equivalent weights (the effective multiples of atomic weights for elements in chemical reactions) for more than two dozen elements, to previously unattained degrees of accuracy.

Extending Antoine Laurent Lavoisier's (1743-1794) previous systematic reform of chemical nomenclature for naming elements and compounds, Berzelius also simplified the existing unsystematic set of chemical symbols. He represented each element with the first letter of its Latin name and used a second significant letter when necessary to distinguish two or more elements with names having the same initial letter—e.g., C for Carbon, Ca for calcium, and Cu for copper (Latin "cuprum"). With minor alterations, this system is still in use today.

A triumphant visit to England in 1812 and meetings with its most eminent scientists secured Berzelius's international reputation. The year 1814 saw the first edition of his new classification of minerals according to chemical properties, which quickly replaced previous systems based on physical descriptions. In 1818-1819 he made a second trip abroad, to France, where he collaborated in research with leading chemists and completed a volume on chemical proportions in inorganic reactions. This formed a second part of what ultimately became the six-volume *Textbook of Chemistry* completed in 1830, which went through five editions and became a standard reference work.

While in France Berzelius was elected secretary of the Academy of Science in Stockholm, which doubled his salary and provided him with new laboratory facilities. In 1821 he founded an annual report on current chemical research throughout Europe, which he published each spring until his death in 1848. Generally known by its German title, the *Jahresberichte,* this

became the single most important source of information for chemists of the era. In 1832 Berzelius resigned his other university obligations to concentrate on this work, and upon his belated marriage in 1835 was made a baronet.

Berzelius's most important theoretical contribution was his theory of electrochemical dualism. Classifying all elements as either electropositive or electronegative in character, he argued that all compounds resulted from combinations of these opposites and that a given element could substitute for another of the same electrical character in a compound. Initially this theory facilitated the explanation of a wide array of inorganic reactions, and aspects of it anticipated modern theories of polar bonding.

However, with the development of organic chemistry beginning in the 1830s, Berzelius's theory proved unable to explain how elements such as hydrogen and chlorine with supposedly opposite electrical characters could replace one another in organic compounds. Consequently the theory soon lost favor to the emerging rival substitution theories of chemical "types" and "radicals." Berzelius's reputation suffered as he refused to acknowledge the inadequacies of his system, used the *Jahresberichte* to violently attack his critics, and broke off long-standing friendships with Justus Liebig (1803-1873) and other chemists who opposed his views. However, his many lasting accomplishments and the work of his most prestigious students influenced chemistry for decades to come.

JAMES A. ALTENA

Friedrich Wilhelm Bessel
1784-1846
German Astronomer and Mathematician

Friedrich Bessel was one of the founders of modern astronomy. He made the first accurate measurements of the positions of stars and of interstellar distances, providing the data upon which an understanding of the dimensions of the universe could be based. He also made significant contributions to mathematics, especially the development of "Bessel functions."

Bessel was born in Minden, Brandenburg, in 1784. He showed interest in scientific and mathematics in school but ended his formal education in 1799, at age 14, to enter into an apprenticeship with an import-export company. His interest in the naval activities of the company, combined with his natural interest in science, led him to study the mathematics of navigation. This study developed further into an interest in astronomy. In 1804 Bessel published a paper in which he reported his calculation of the orbit of Halley's Comet, based on measurements that had been made in the seventeenth century. Several prominent astronomers recognized Bessel's ability, and he was given a position in the Lilienthal Observatory of astronomer J. H. Schroter soon thereafter.

Bessel's reputation spread rapidly, and in 1809 he was appointed director of Germany's first government observatory which was under construction in Königsberg. Since university professors were required to have doctorates, the University of Königsberg awarded a doctorate to Bessel in 1810 so that he could become the professor of astronomy at that university. He remained in this position until his death in 1846.

At Königsberg Bessel was able to devote his full attention to astronomy and mathematics. His extensive work in developing and refining methods for the accurate instrumental measurement of the positions of stars resulted in the founding of the division of astronomy known as astrometry. Bessel is credited with making the first accurate measurement of the position of a star. In 1818 he published the positions of 3222 stars, providing quantitative astronomy with an accurate data bank for the first time.

Bessel's careful measurements allowed him to observe very small variations in the positions of stars. One of the results was his discovery that stars appear to have minor elliptical motions. Bessel showed that this phenomenon, known as "parallax," results from the earth's motion around the sun, providing final proof of the heliocentric model of the solar system. Bessel predicted companion stars for Sirius and Procyon based on minute periodic deviations in the positions of these bright stars, and predicted that a planet, as yet unobserved, must exist because of slight deviations he detected in the orbit of Uranus. The development of more powerful telescopes led to the discovery of the companion stars of Sirius and Procyon, and the planet Neptune was found where Bessel had predicted it to be.

Bessel's work in related fields included the measurement of a value of 1/299 for the departure of the earth's curvature from a perfect sphere. The best known of his contributions to mathematics is his "Bessel functions." Bessel developed the mathematics of these functions to facilitate calculations involving three bodies—for example, planets or stars—that are simultaneously under the influence of each other's gravitational effects. Bessel also made significant contributions to the reform of university education, even though his own formal education had ended at age 14. His pioneering studies of mental chronometry and reaction times are further indications of his intellectual breadth and curiosity.

Bessel's work brought him considerable recognition within the scientific community. Among other awards, he received a prize from the Berlin Academy in 1815, was made a Fellow of the Royal Society in 1825, and received the Gold Medal of the Royal Astronomical Society.

J. WILLIAM MONCRIEF

Jean Baptiste Biot
1774-1862
French Physicist

Jean Baptiste Biot is perhaps best described as a polymath who made important contributions to acoustics, optics, and electromagnetic theory during a career that also included significant work in astronomy, geodesy, and many other fields. In 1803 he helped confirm that meteorites are of extraterrestrial origin, and in 1804 made a balloon ascent with French physicist and chemist Joseph Gay-Lussac (1778-1850). In 1856, in recognition of his writings on several subjects, Biot was elected to the Académie Française, an honor quite rare for a scientist. A dominant figure within the French university system, he was among the first to recognize and encourage the young Louis Pasteur (1822-1895) in his scientific career.

Biot was born the son of a treasury official, the first in his family to rise from the peasantry to a position of social standing. He served in the army during the French revolutionary period and his subsequent career reflected both the ideals of the revolution and the new institutions introduced in France at that time. He was among the first students to enter the Ecole Polytechnique and, following graduation and some minor academic posts, became Professor of Astronomy at the Collège de France when it was established by Napoleon Bonaparte in 1806. Remaining at this post until his retirement in 1849, Biot enjoyed the reputation of an outstanding teacher.

News of the discovery by Danish physicist Hans Oersted (1777-1851) of a connection between electricity and magnetism reached Paris

Jean Baptiste Biot. *(The Library of Congress. Reproduced by permission.)*

crystalline substances, but could be observed in certain liquids and even vapors. He concluded, correctly, that the rotation was a characteristic of the individual molecules. In 1816 he used the effect to argue for the chemical identity of sugar derived from sugar beets and sugar cane. In 1832 he began the study of another material of biological origin, tartaric acid. Four years later he published a paper devoted entirely to the rotation of polarized light by this substance and noted that it depended on color in a manner different from that of the substances he had previously investigated. Biot's investigation of optical rotation by tartaric acid and its salts was continued by Pasteur, in whose career Biot took a personal interest. In his first major research, Pasteur established the power of the optical rotation technique for future chemical research.

DONALD R. FRANCESCHETTI

in September 1820. By October 30 of the same year, Biot and Felix Savart (1791-1841) were able to deliver a paper at a meeting of the Academie des Sciences establishing the quantitative law for the magnetic force between currents flowing in different electrical circuits, now generally known as the Biot and Savart law.

The greatest contribution made by Biot, however, lies in the area of optics, particularly the behavior of polarized light. Until 1808, when Etiennne Malus (1775-1812) discovered that light beams reflected by a smooth surface would be polarized if the reflection occurred at the appropriate angle, the only clue to the polarizability of light was the double refraction, or splitting, of light beams that occurs in crystals of a low degree of symmetry. Biot became active in the field following the 1811 discovery by his colleague Dominique Arago (1786-1853) that light polarized by reflection could be split into two differently colored beams by certain crystals. Biot, an admirer of Isaac Newton (1642-1727), explained this result in terms of forces acting on the particles of light that comprised a light beam. The following year Biot observed that the rotation of the plane of polarization produced by a plate of quartz was dependent on the color, a process now known as rotatory dispersion.

In 1815 Biot showed that the rotation of the plane of light polarization was not restricted to

Ludwig Eduard Boltzmann
1844-1906
Austrian Physicist

Ludwig Boltzmann was an important figure in the development of physics as the science made the transition from the classical physics of Isaac Newton (1642-1727) to the physics of the twentieth century. His work helped to usher in the fields of statistical mechanics (used to describe the behavior of fluids) and helped to better explain the Second Law of Thermodynamics. He became embroiled in controversy as European scientists debated the validity of his work, eventually winning recognition and acceptance shortly after his death.

Boltzmann was born in Vienna, Austria, the son of a taxation official. He received a doctorate from the University of Vienna in 1866 at the age of 22, completing his graduate work on a kinetic theory of gases under the supervision of Josef Stefan (1835-1893). Following his graduation, Boltzmann continued working for a short time under Stefan as an assistant before moving to Heidelberg and Berlin, where he worked under Robert Bunsen (1811-1899), Gustav Kirchoff (1824-1887), and Hermann Ludwig von Helmholtz (1821-1894).

Appointed to a chair of theoretical physics at Graz University in 1869, Boltzmann taught for four years prior to accepting a chair in mathematics at the University of Vienna in 1873. Three years later, he returned to Graz, where he stayed until 1894, this time as an experimental physicist.

His next move was to return to Vienna, where he assumed the chair of theoretical physics vacated by the death of his former advisor, Stefan. Unfortunately for Boltzmann, however, Ernst Mach (1838-1916) joined the faculty at Vienna the following year. Mach was not only a scientific rival, but the two men were on unfriendly terms, which prompted Boltzmann to leave Vienna in 1900 to move to Leipzig.

During this time, Boltzmann continued his work on statistical mechanics, running into much controversy and dispute from his colleagues. At that time, with Newtonian mechanics reigning supreme, Boltzmann claimed that the motions of gases could be better described using probability and statistics rather than by calculating the properties of every individual atom or molecule. And, furthermore, not all physicists at that time even believed in atoms or molecules, leading to further disagreement. In addition, Newtonian mechanics did not depend on a particular direction in time; that is, one could just as easily calculate motions or activities in the past as in the future. Based on this view of physics, the behavior of any substance could be calculated perfectly, given an accurate knowledge of the starting conditions. Boltzmann refused to accept this premise, arguing his points tenaciously for many years.

Bolzmann's work—independent of Scottish physicist James Clerk Maxwell (1831-1879)—led to the discovery of the Maxwell-Boltzmann distribution law concerning the energy of atoms in a gas. Boltzmann was among the first to realize the importance of Maxwell's work in electromagnetic theory. Boltzmann was also responsible for discovering the constant that bears his name. The Boltzmann constant fugures in his seminal equation which expresses the relation between entropy and probability.

The continuing attacks on his work in statistical mechanics and thermodynamics took a physical and emotional toll on Boltzmann. Always prone to mood swings, he began to feel that he was unsuccessful in defending his theories and that his life's work was to amount to nothing. Boltzmann hanged himself while on vacation in Trieste. Ironically, the experiments that were to validate his work were to take place soon thereafter.

P. ANDREW KARAM

Marie Curie
1867-1934
Polish French Physicist and Chemist

Marie Curie was the first woman to be granted a Ph.D. in Europe and the first woman ever to win a Nobel Prize (she went on to garner a second). A woman of enormous personal character, Curie stands as one of the most significant and influential scientists of the twentieth century.

Curie was born Manya Salomea Sklodowska on November 7, 1867, in Warsaw, Poland. She was the youngest of five children. Both of her parents were ardent educators, and instilled in their children respect for physical work and a love of the countryside and nature.

Curie excelled at school and finished first in her class in the gymnasium in 1883. At the time, Warsaw University did not admit women, and a young graduate's only options were to go abroad to study or to get married. Curie hatched a plan with her sister Bronia whereby each would pay for the other's studies in France. While waiting her turn, Marie took a job as a governess. It was during this time that she decided on a career in science.

In 1891, at the age of 23, Marie Curie enrolled as a student at the Sorbonne in Paris. She had arrived at a critical time. It would be a while until French women were fully emancipated, but a number of technological innovations such as electricity, the telephone, and moving pictures were contributing to an explosion of interest in science, and scientific studies were generously subsidized. Curie was exposed to some of the finest scientific minds of the day, including the mathematicians Paul Appell and Henri Poincaré (1854-1912), and the physicist Gabriel Lippmann (1845-1921). Curie took her degree in physics in 1893, graduating first among her classmates, and ranked second for the degree in mathematics the following year.

Her initial plan had been to return to Poland to teach. But in the spring of 1894 she met 35-year-old Pierre Curie (1859-1906), who shared her love of and devotion to science. The two were married in 1895 and were soon an inseparable team of researchers. Their daughter Irène was born in 1897. A second daughter, Eve, followed in 1904.

The 1895 discovery of mysterious "x rays" by Wilhelm Conrad Röntgen (1845-1923) created a sensation among scientists and the general public alike. In 1896 Henri Becquerel (1852-

Marie Curie. *(The Library of Congress. Reproduced by permission.)*

1908) found that uranium also gave off rays, a phenomenon the Curies would later call radioactivity. Because uranium rays were new, and little attention had been paid to them in the wake of excitement over x rays, Curie decided to concentrate her doctoral thesis on uranium rays.

Working in a primitive space in the school where Pierre taught, she was surprised to find that pitchblende, a source of uranium, contained two unknown elements that were far more active than uranium. She named these elements polonium and radium. In 1903, the year Marie defended her Ph.D., she and Pierre, along with Becquerel, were awarded the Nobel Prize in physics for "joint researches in radiation phenomena." For the first time, light had been shed on the forces within the nucleus of the atom.

Pierre Curie died tragically in a street accident in 1906, and Marie took up the position he had obtained at the Sorbonne. Several years later her highly publicized love affair with a married friend and colleague, Paul Langevin (1872-1946), almost eclipsed the announcement that she had been awarded the 1911 Nobel Prize in chemistry for the discovery of radium. Fallout from the scandal put a substantial strain on her health.

Public obsession with her personal life was swept away with the advent of World War I. Curie distinguished herself during the war by procuring and equipping cars and personnel to make x-ray examinations available to soldiers at the front.

After the war, the dangers of radium and other radioactive substances, previously assumed to be harmless in small doses, became increasingly obvious. In 1934 Curie herself succumbed to leukemia, the result of a lifetime of exposure to radioactivity. She was buried beside her husband in a small cemetery in Sceaux. A year later, her daughter Irène Curie (1897-1956) and Irène's husband Frédéric Joliot (1900-1958) were awarded the Nobel Prize in chemistry for their discovery of artificial radioactivity.

GISELLE WEISS

John Dalton
1766-1844
English Chemist, Physicist and Meteorologist

John Dalton proposed the atomic theory of matter as a result of his investigations of the atmosphere. By viewing matter as made up of indivisible particles each with their own particular weight, Dalton offered a way to understand chemical reactions. This led to the rapid development of chemistry in the nineteenth century.

Dalton was born into a family of devout Quakers in a small village in the Lakes District of northwest England. His Quaker heritage, with its emphasis on education and interest in science, played a key role in Dalton's life. After completing all the schooling he could obtain in his village, Dalton left in 1781 to replace his elder brother as an assistant in a Quaker boarding school in Kendal, the principal town in the Lakes District. The school had a very good library as well as scientific equipment that allowed Dalton to continue his education by reading and experimentation.

While in Kendal Dalton was befriended by a fellow Quaker, John Gough. Although blind, Gough was highly educated and had a particular interest in the natural sciences. He tutored Dalton in mathematics, meteorology, botany, Latin, Greek, and French. Dalton also began taking on teaching duties in Kendal and giving public lectures on various scientific matters, a practice he would continue through out his life. In 1792 Dalton was appointed to the position of professor of mathematics and natural philosophy at New College in Manchester. He spent the next eight years teaching various subjects before resigning in 1800 to open his own school offer-

John Dalton. *(The Library of Congress. Reproduced by permission.)*

ing private instruction in mathematics, experimental science, and chemistry.

By 1805 Dalton had essentially produced an outline of his most important contributions, his two laws concerning the gaseous state and the atomic theory of matter. What led to these discoveries was Dalton's attempt to try to understand certain aspects of the atmosphere. In particular, why was the atmosphere a homogeneous mixture of gases instead of layers of gases arranged according to their weight, the heaviest at the bottom and the lightest on the top? Another question was how water vapor could be absorbed in the atmosphere. The key to these problems was Dalton's belief in the particulate nature of matter, a concept that had been accepted by many of his contemporaries.

Dalton's insight was that the mixing of gases need not be a chemical reaction and that gases existed independently of each other in the atmosphere. This mechanical explanation for the mixing of gases led Dalton to propose in 1801 his law of partial pressures. Dalton generalized that in any mixture of gases, each component acted independently. A further principle stated by Dalton was that at constant pressure all gases will expand equally given the same quantity of heat. A similar principle was discovered by the French physicist Jacques Charles (1746-1823), who is generally given credit for this instead of Dalton.

Dalton's unorthodox view that chemical attraction was not a force in the atmosphere was met with much criticism and disbelief. In attempting to find experimental proof for his mechanical concept of the mixing of gases, Dalton relied on the work of his friend William Henry (1774-1836), who studied the effect of pressure on the amount of gas that could be dissolved in water at constant temperature. Henry's experiments showed that the amount was related to the pressure, thus showing that mixing of gases in the atmosphere had to be a mechanical phenomenon. Dalton found in his own experiments that the nature of the substance played a role in terms of how much could be dissolved at a constant pressure. What distinguished one substance from another was its mass; thus, Dalton believed that it was the size of the particles (atoms) that was the crucial determinant of an element's chemical properties. In 1803 Dalton published these results along with his measurement of the relative weights of different gases, in the process producing the first atomic weight table.

Dalton quickly realized that the concept of elements as being ultimately made up of atoms, with each atom having a unique atomic weight, was a way to explain why compounds such as water have a constant composition. Dalton also explained how atoms could react with other atoms in more than one ratio by weight and produce a series of compounds, an observation that fueled acceptance of Dalton's atomic theory.

Dalton's theory appeared in print in *A New System of Chemical Philosophy,* published in two parts in 1808 and 1810. Dalton was to contribute little new after 1810 and spent the rest of his life developing the theory by making measurements of atomic weights and public lectures. Dalton lived the balance of his life in Manchester and received many honors, including election to the Royal Society (1822), the Royal Society medal (1826), and honorary degrees from Oxford (1832) and Edinburgh (1834). On his death in 1844 more than 40,000 people filed past his coffin and a public funeral was held.

MARTIN D. SALTZMAN

Humphry Davy
1778-1829
English Chemist

Humphry Davy is most famous for his discoveries of potassium, sodium, chlorine, and other elements using powerful voltaic batteries. Through his research he established the science of electrochemistry. On a more practical note, Davy also invented the miner's safety lamp.

Davy was born to a family of modest means in the remote coastal town of Penzance in Cornwall, England. A mediocre student, he preferred the rocky ocean cliffs to the classroom and spent his spare time writing poetry and reading philos-

HUMPHRY DAVY'S EXPERIMENTS WITH NITROUS OXIDE

~

Humphry Davy spent ten months experimenting with breathing nitrous oxide. He varied the concentration of the gas, increased the quantity, and tested its effects upon headache, indigestion, and dental pain. Changes in hearing, vision, thought, emotion, and desire plus physical changes in the pulse, blood, and breathing were observed and recorded. Ultimately, Davy was seeking to understand the chemical bases of passion and life. For all his experimental caution Davy almost asphyxiated (suffocated) himself at points.

Davy tested the gas publicly on others, often to great spectacle and hilarity. He always first administered regular air to the unknowing subject to check for what we now call the placebo effect. After having been administered the real gas, the subject was observed and—once effects had diminished—questioned as to his or her experiences. Some subjects reported despondent feelings; others reported feelings of ecstasy. A few jumped up and fled the room!

These experiments came under political and social attack. Political essayist Edmund Burke linked the gas and Davy's employer, Dr. Beddoes, to the destructive chaos of the French Revolution.

ophy. At age 16 he entered an apprenticeship as an apothecary to a local doctor, hoping some day to become a medical doctor himself. At age 18 he began studying chemistry and performing his own experiments.

The young Davy's chemical research so impressed a few local scientists that in 1798 at age 19 he was recommended for a position at Dr. Thomas Beddoes's Pneumatic Institution in Bristol, an institution dedicated to researching the medical uses of gases. It was there that Davy discovered through self-experimentation the intoxicating effects of breathing nitrous oxide.

Davy carefully researched the chemical and physiological properties of nitrous oxide and thereby secured his reputation as a chemist.

In 1801 Davy left Bristol for London to accept a position at the recently established Royal Institution. In the next 12 years he enjoyed tremendous success as both a lecturer and a chemical researcher. As a lecturer Davy's spectacular demonstrations and especially his inspired scientific discussions laced with religious, metaphysical, and patriotic references so thrilled audiences as to make chemistry all the rage among London's wealthiest and most fashionable classes. Even more successful were Davy's experiments using the voltaic battery. The battery, invented in 1799, was simply a concatenation of copper and zinc plates, separated by fluid, that generated a continuous current of electricity. Davy was convinced that the battery would revolutionize chemical understanding. In 1806 and 1807 Davy presented his most brilliant research on the battery and his theories of electrochemistry (a term he coined.) Through a series of rigorous experiments he demonstrated how the processes of chemical composition and decomposition correspond to electrical states of bodies. Based on his findings, he argued that electricity is identical to chemical attractive and repulsive forces and that indeed electricity is an essential property of matter.

Davy became most famous, however, for his discovery of potassium and sodium in 1807. Davy subjected the alkali potash (and then soda) to the strongest electrical powers he could elicit from three connected batteries totaling 600 double plates. The unusual lustrous globules that appeared at the negative pole of the battery proved to be a new chemical element. Across Europe chemists sought to duplicate Davy's experiments and to re-evaluate the order of chemistry. Meanwhile, Davy built larger and larger batteries and used them to discover numerous other chemical elements and to prove that chlorine, as he called it, is an element rather than an acid. Davy was knighted for his achievements in 1812.

Davy traveled widely in the following years while also continuing his scientific researches. Most importantly, he invented the miner's safety lamp in 1815 in response to disastrous explosions in British coal mines. He discovered that by encasing the lamp's flame in wire mesh it would not ignite methane gas trapped in a mine. As reward for his invention Davy was made a baron. Davy served as president from 1820 to 1827 of the Royal Society, Britain's most prestigious scientific institution.

In his final years the ailing Davy traveled, studied natural history, and wrote poetry and metaphysical treatises. Soon after turning 50 he died and was buried in Geneva, Switzerland.

JULIANNE TUTTLE

Michael Faraday
1791-1867
British Chemist and Physicist

Michael Faraday, British chemist and physicist, demonstrated electromagnetic rotation (the basis of electrical motors and dynamos), the laws of electrochemistry, and introduced the concept of electromagnetic fields.

Michael Faraday was born in London, the son of a poor blacksmith. As a child his education covered only the basics of reading, writing, and arithmetic. At the age of 14 he was apprenticed to a bookbinder for seven years. This occupation brought him into contact with books on science and introduced him to the idea that the subject was comprehensible. He spent the years of his apprenticeship reading any scientific works he could find and attending public science lectures. In 1812 Faraday began to seek employment in science. He wrote to Humphry Davy (1778-1829), a leading British scientist of the day, and enclosed with the letter the notes he had taken during several of Davy's lectures at the Royal Institution. He was hired as Davy's assistant, and in 1813 he was granted a full-time position as laboratory assistant at the Royal Institution, an institution designed to promote science among the public. He became director of the laboratory in 1825. During these years he became an extremely skilled experimentalist and began pursuing his own research in chemistry and electromagnetism. After being appointed director, Faraday initiated the Friday Evening Discourses, which became a prominent feature of the Royal Institution. Faraday was devoted to the idea of communicating science to the public and for close to forty years his very popular lectures brought experimental and theoretical physical science to a broad audience.

In 1821 Faraday conducted a series of experiments that demonstrated electromagnetic rotation. The previous summer Hans Christian Oersted (1777-1851) had discovered that an electrical current could produce a circular magnetic force. Faraday showed that this force would make a magnet rotate around a current-carrying wire. After a decade of pursuing other research

Humphry Davy. *(The Library of Congress. Reproduced by permission.)*

projects, Faraday returned to the study of electromagnetism in 1831. He was able to demonstrate the reverse of Oersted's 1821 discovery. His brilliant experiments uncovered the phenomenon of electromagnetic induction, the process by which changing magnetic conditions produce an electric current. His paper on this was the first in a 25-year series of articles entitled "Experimental Researches in Electricity" and established his reputation as a leading scientist. He showed in part that a moving magnet would produce an electrical current. This was a discovery of fundamental importance because it proved that the physical movement of the magnet could be converted into electricity. Faraday's experimental setup was the first dynamo, a simple machine using mechanical force to produce electricity.

In 1834 Faraday developed his two laws of electrochemistry, which claimed that the amount of chemical activity in matter is directly proportional to the amount of electricity passed through the matter, and that the masses of matter produced from an electrochemical reaction are proportional to their chemical equivalents. During his electrochemical studies, Faraday introduced such now-standard vocabulary terms as ion, cation, anion, electrode, cathode, and anode.

Faraday's final major contribution to science arose from many outside sources, including his religion. Faraday belonged to a small sect called

istry and physics, including groundbreaking work in the study of gases, the formulation of a law of combining volumes for chemical reactions, and the development and application of techniques for chemical analysis.

Gay-Lussac graduated from the Ecole Polytechnique in Paris in 1797. Accepting a position as an assistant to Claude Louis Berthollet (1748-1822) in 1801, he became a member of a group of young scientists who lived near Berthollet's private laboratory in the town of Arcueil. This group is now known as the Arcueil circle or the Laplacian school. The latter name is due to the influence of Pierre Simon Laplace (1749-1827), who believed that the application of Newtonian physics to chemical changes would lead to an understanding of chemistry comparable to the quantitative explanations for such physical phenomena as planetary motion. The members of this group, including Gay-Lussac, followed the leadership of Laplace, and their work contributed significantly to the foundation of physical chemistry, a branch of chemistry that applies the theories and methods of physics to chemistry.

In 1802 Gay-Lussac announced his first major scientific contribution—the law of thermal expansion of gases. According to this law, all gases, regardless of the properties of the individual gas, expand by the same fraction with the same increase in temperature. In 1804 Gay-Lussac made an ascent in a hydrogen balloon with Jean-Baptiste Biot (1774-1862) to an altitude of 13,000 feet. Later in the year Gay-Lussac made a second, solo ascent to 23,018 feet. The purpose of these ventures was to measure the variation of the earth's magnetic field and the chemical composition of the atmosphere with change in altitude. He determined that neither varied with altitude, at least at the height to which he rose.

While working with Alexander von Humboldt (1769-1859), Gay-Lussac determined the relative proportions of hydrogen and oxygen that react to form water, concluding that two volumes of hydrogen are required for one volume of oxygen. In 1808, on the basis of this discovery and on observations of other gaseous reactions, Gay-Lussac announced the law of combining volumes, now known as "Gay-Lussac's law." This law states that the volumes of gases that combine in chemical reactions always do so in simple integral proportions. Gay-Lussac's law was further explained in 1811 by Amedeo Avogadro (1776-1856), who proposed that equal volumes of gases (at the same temperature and pressure) contain the same number of

Michael Faraday. *(The Library of Congress. Reproduced by permission.)*

the Sandemanians who attempted to practice a simple, early form of Christianity. Because he believed that God had created a beautiful, orderly world, he was convinced that electricity and magnetism were produced by some single force. Whereas most scientists of the time believed electromagnetism to be caused by fluids of particles, Faraday introduced the concept of the "field," described by lines of force, in which the manifestation and energy of electricity and magnetism existed in the space around the magnet and not in the magnet itself. His concept of fields would later be fully clarified and expanded by James Clerk Maxwell (1831-1879) in his famous mathematical equations.

Despite his lack of mathematical training, Faraday made vital contributions to chemistry and physics in the nineteenth century. His expertise with experimentation led to numerous discoveries, and his somewhat unorthodox and idiosyncratic ideas led him to suggest radical theories which opened new doors of scientific investigation.

REBECCA B. KINRAIDE

Joseph Louis Gay-Lussac
1778-1850
French Chemist and Physicist

Joseph Gay-Lussac was a versatile scientist who made numerous contributions to chem-

and was elected to the Chamber of Deputies in 1831, 1834, and 1837. He also received a peerage from King Louis-Philippe.

J. WILLIAM MONCRIEF

Joseph Gay-Lussac. *(The Library of Congress. Reproduced by permission.)*

molecules. In other words, one volume of hydrogen contains the same number of molecules as one volume of oxygen. Since two volumes of hydrogen are required to react with one volume of oxygen to form water, twice as many molecules of hydrogen than oxygen are necessary in the formation of water.

Gay-Lussac, at times with the assistance of Louis-Jacques Thenard (1777-1857), explored the properties of potassium and iodine. He discovered boron, isolated and studied cyanogen, and developed a method for the analysis of organic molecules. He synthesized, isolated, and studied the properties of hydrogen chloride, hydrogen iodide, and hydrogen fluoride. His observations of the acidity of these compounds led to the refinement of the concept of acidity. Gay-Lussac also analyzed the composition of substances isolated from animals and vegetables, measured the solubility of salts as a function of temperature, and developed volumetric methods for chemical analysis. In 1815 he introduced a method of determining the vapor densities of liquids, and in 1821 demonstrated that wood soaked in borax is inflammable.

The results of Gay-Lussac's extensive scientific work were widely influential. Even after his death his observations and discoveries provided the basis for additional studies. He was respected by both his colleagues and fellow citizens,

Josiah Willard Gibbs
1839-1903
American Theoretical Physicist and Theoretical Chemist

J.Willard Gibbs is regarded as one of the greatest American scientists of the nineteenth century and one of the founders of modern physical chemistry. His theoretical and mathematical treatments of physical and chemical processes involving heat and work developed the science of thermodynamics into one of the most useful tools available to physicists and chemists.

Gibbs, whose father was a professor of sacred literature at Yale University, grew up in a highly intellectual environment. He developed into a shy, somewhat frail young man, deeply involved in intellectual pursuits. He graduated from Yale, then, in 1863, received the first doctorate of engineering awarded in the United States. In that same year he became a tutor at Yale.

After Gibbs's parents died and he and his two sisters inherited their estate, the three traveled together to Europe in 1866. They stayed for three years, giving Gibbs the opportunity to attend the lectures of some of Europe's most outstanding mathematicians and physicists.

Returning to the United States, Gibbs continued his work with engineering problems. As a result of an investigation of steam engines, he became interested in the science of thermodynamics, which involves the study of heat and work and their interchange. This interest became the basis of most of his lifelong scientific endeavors.

In 1871 Gibbs was named professor of mathematical physics at Yale and continued to devote his research efforts to the theoretical treatment of thermodynamic phenomena. In 1872 he published a theoretical treatment of the thermodynamics of the adsorption of substances on surfaces, and in 1876 produced his best known work, *On the Equilibrium of Heterogeneous Substances,* a general theoretical treatment of thermodynamics.

In 1878 Gibbs published his phase rule in *Transactions of the Connecticut Academy,* a journal that was so obscure that his work went unnoticed for 20 years. The phase rule relates the

Josiah Gibbs. *(The Library of Congress. Reproduced by permission.)*

number of intensive quantities (e.g., temperature and pressure) and the number of chemical components in a system with the number of phases of a substance (e.g., liquid or gas) that may exist simultaneously. Gibbs's phase rule has found considerable application in the development and improvement of industrial processes.

Perhaps his most outstanding contribution was the introduction of the concept of chemical potential or free energy, now universally called Gibbs free energy in his honor. The Gibbs free energy relates the tendency of a physical or chemical system to simultaneously lower its energy and increase its disorder, or entropy, in a spontaneous natural process. Gibbs's approach permits the calculation of the change in free energy that occurs in such a process—for example, a chemical reaction—and, consequently, enables prediction as to whether and how rapidly the process may be expected to occur. The use of this concept revolutionized the scientific understanding of those chemical and physical processes that involve changes in heat and work. Since virtually all chemical processes and many physical ones involve such changes, his work has significantly impacted both the theoretical and empirical aspects of these sciences.

Another of Gibbs's major contributions was the development of statistical mechanics and its application to thermodynamic systems. Statistical mechanics has subsequently found extensive application in other fields, including quantum mechanics.

Although the importance of Gibbs's accomplishments was immediately recognized in Europe, recognition came more slowly in America. This was partially due to the more empirical bias of American science, but it may also have been the result of his withdrawn nature. He remained a bachelor, living with his sister's family throughout his life.

In 1881 Gibbs received the Rumford medal from the American Academy in Boston and, in 1901, the Copley medal from the Royal Society in England.

J. WILLIAM MONCRIEF

Hermann Ludwig Ferdinand von Helmholtz
1821-1894
German Physicist, Philosopher and Physiologist

Hermann von Helmholtz's scientific career marked him as one of the foremost scientific minds of the nineteenth century. His landmark theories and ideas encompassed electrodynamics, fluid and thermo dynamics, mathematics, optics, and physiology.

Though interested in many aspects of knowledge in his youth, Helmholtz was not distinguished by any early scientific gifts. He leaned toward physics in his university years but began his scientific work in the field of medicine with medical school in Berlin in 1838. Lectures included attendance at the University of Berlin, where he found mentors in J. P. Müller in physiology and H. G. Magnus in physics. Helmholtz's medical thesis (1842) on the relation of animal nerve fibers and cells and heat led to his interest in the theory of the conservation of energy. Thus, early on this personal research netted the writing of a fundamental paper in classical physics titled "On the Conservation of Force" (1847), which would find application in several areas of experimental science. Yet his primary career was as a professor of physiology, which included several important academic positions: Königsberg (1849), Bonn (1855), and Heidelberg (1858).

Starting in 1850 Helmholtz would invent instruments for studying the human eye, leading to contributions to the understanding of the structure and mechanism of the eye. This was

(1858) and the physics of fluid flow (also 1858). He also worked out two basic, or primitive, equations based on the mathematics of fluid flow, showing the stability of vortex motion and the constancy of velocities of that motion. Later, he produced an important paper ("On Atmospheric Motion," 1888) dealing with the limited radiation in the polar regions, which meant sinking air over that area in relation to the general circulation of the atmosphere.

Helmholtz's renown as a theoretical physicist grew as his life progressed. He held the physics professorship of his late professor Magnus at Berlin (1871). By 1888 he had become the foremost scientific advisor in Germany, the proof of which was his appointment as the director of the new Physico-Technical Institute at Berlin, which was the first national scientific laboratory in Europe.

WILLIAM J. MCPEAK

Hermann von Helmholtz. *(The Library of Congress. Reproduced by permission.)*

expanded between 1857 and 1859 into investigations of physiological acoustics, the study of tone (combination and vowel), and important work on the mechanism of the bones of the middle ear, emphasizing the physical nature of the eardrum. He had also returned to nerve studies earlier (1850), successfully measuring the speed of the propagation of nerve impulses.

In regard to his physical study Helmholtz's 1847 paper proved that the work or energy done by a force was conserved, and that the total energy of particles interacting through some central force was dependent upon only the mass of the particles and their spatial separation. In this he had philosophically and physically defined the theory by just two concepts: matter and force (Helmholtz called it an "ultimate" force), and his theory was more comprehensive than that of other scientists studying the conservation of energy, such as Sadi Carnot (1796-1832), Julius Mayer (1814-1878), and James Joule (1818-1888). Helmholtz also did studies in energetics, or the transformation of energy of one type into another, which were important to the prevailing concept of a mechanical or dynamic theory of heat. Among the applications in which he proved the conservation of energy were electrostatics, electrodynamics, and thermodynamics.

Helmholtz provided seminal contributive papers to the fields of electricity and magnetism

William Henry
1774-1836
English Chemist and Physician

William Henry was a leading experimental chemist who helped establish the validity of Dalton's atomic theory. Best known for his investigation of gases, he formulated Henry's Law, which describes the relationship between mass and pressure for a gas dissolved in liquid. Henry also wrote the most influential chemistry textbook of his time, which stood as the standard for over thirty years.

Henry was born into a wealthy English family in Manchester. At age 10, he was injured by a falling beam and left with chronic, lifelong pain. This limited his play and led to his becoming an avid student. At age 16, he began his studies in medicine and in 1795 entered the University of Edinburgh, Scotland. A year later, however, he left to work in the family's manufacturing business. During this period, he did original research in chemistry. He returned to school in 1805 and earned his medical degree in 1807.

Henry was fascinated by the work of the French chemist Antoine Lavoisier (1743-1794). A generation earlier, Lavoisier had made chemistry a true scientific discipline, providing basic principles and challenging the existing vestiges of alchemy. Henry delivered lectures on the French chemist's work, and these evolved into *Elements of Experimental Chemistry*. First published in 1801, this textbook went through 11 editions in

30 years and introduced generations of chemists to Lavoisier's chemical nomenclature and the use of careful experimental measurement.

Henry was a friend of John Dalton (1766-1844), a friendship that was pivotal to both of their careers and to chemistry. Dalton is famous for his atomic theory, which holds that all the elements of matter are made up of indivisible, indestructible atoms. Dalton was brilliant and audacious in his thinking about chemistry, but clumsy and careless in the lab. As a teacher, Dalton also had little time or money for experimentation. Henry, on the other hand, had the time, the money, and the talent. He performed critical experiments that supported Dalton's theory and stimulated Dalton's thinking about atoms. Most of Henry's experiments were done on gases because they provided a simpler chemical model than other forms of matter. During this time, he formulated Henry's Law, his most famous achievement. Simply stated, this law holds that the mass of a gas that dissolves in a liquid is proportional to the pressure exerted by the gas on the liquid if the temperature is kept constant. Henry's Law is limited to less soluble gases and gases that do not react chemically with the liquid. Still, it is useful and continues to be employed to explain carbonation and to make calculations for safe diving.

Thanks in large part to Henry, the atomic theory came to be generally accepted. This led to a deeper and more subtle view of nature. Elements could be identified, ordered, and categorized according to their chemical properties. The monumental result of atomic theory was the creation of the periodic table of elements, which is one of the most important tools for understanding and manipulating materials.

Henry himself was slow to back atomic theory. Many historians believe that, had he become an early proponent, he would share credit with Dalton. While he had taken a leap in accepting Lavoisier as a young man, later in life Henry became cautious and did not take a stand when his experiments pointed to change (as with the composition of hydrochloric acid). He held onto old beliefs, such as insisting heat had mass. For society, Henry's reluctance had important results in medicine.

In 1824 Henry's bad health ended his career as an experimental chemist. Surgery on his hands made further work in the lab impossible. Henry turned his attention to medicine. He theorized that there was a chemical nature to disease, and that using heat to destroy the bad chemicals could alleviate illnesses and reduce the spread of diseases. During the 1831 cholera epidemic, he created an inexpensive and simple device that used heat to disinfect clothing. The device was effective and might have prevented illness and saved many lives, but, again, Henry failed to become an advocate. Heat sterilization was forgotten for another 30 years until Louis Pasteur (1822-1895) developed his germ theory of disease.

Along with chronic pain, Henry suffered from depression. In 1836, after a dozen years of being unable to work in the lab, he took his own life.

PETER ANDREWS

Caroline Lucretia Herschel
1750-1848
German-English Musician and Astronomer

Caroline Herschel was the first woman to gain wide recognition in astronomy. She assisted her brother William Herschel (1738-1822), regarded as the founder of modern quantitative astronomy, in his work, and discovered three nebulae, eight comets, and compiled several extensive collections of star and nebula positions.

One of six children, Caroline was born in Hanover; her father was an oboist, and later the bandmaster, of the Hanoverian Foot Guards Band. Her four brothers were trained as musicians, but, although her father managed to provide her with some education, her mother strongly opposed any formal education for her daughters. For much of her early life, she was principally engaged in housework and was basically a servant.

When the French occupied Hanover in 1757, her brother William emigrated to England. Her father died in 1767, and she served as her mother's housekeeper until, against strong objections from her mother, she joined William in England in 1772. William had done well with his music and had become an orchestra director and the organist at the chapel in Bath. Caroline managed their home and William taught her English and gave her voice lessons. She became an accomplished and successful singer.

William had become interested in astronomy and spent his evenings engaged in his hobby of building telescopes and observing the heavens. Caroline assisted him, learning the mathematics necessary to record observations and to make the necessary calculations. She even fed William when his hands were occupied for long periods in the construction of mirrors for his telescopes.

Caroline Herschel. *(The Library of Congress. Reproduced by permission.)*

death. During this period she helped educate William's son, John Herschel (1792-1871), who later made a name for himself in mathematics and astronomy.

Soon after William died in 1822, Caroline returned to Hanover where she completed a catalog of 2500 nebulae in 1828. She was widely respected by scientists throughout Europe, and many of them visited her in Hanover. She received the Gold Medal of the Royal Astronomical Society in 1828, and in 1835 she was one of the first two women elected to honorary membership in the Royal Society. She was elected a member of the Irish Academy in 1838 and was awarded the Large Gold Medal for Science by the King of Prussia in 1846. She died in 1848, shortly before her 98th birthday. In 1889 a minor planet was named Lucretia, Caroline's middle name, in her honor.

J. WILLIAM MONCRIEF

In 1781 William discovered the planet Uranus, an accomplishment that brought him recognition and a position as court astronomer for King George III. He and Caroline gave up their musical careers and devoted all of their time and energy to astronomy. She continued to perform most of the calculations on the data that resulted from his work, but also began to make her own systematic observations of the sky. Using a telescope that William had given her, Caroline discovered three new nebulae in 1783. In 1786 she submitted to the Royal Society a manuscript entitled *Index to Flamsteed's Observations of the Fixed Stars*. This volume was the result of her meticulous work checking and correcting the catalog of stars that had been published by the first Astronomer Royal, John Flamsteed (1646-1719). She also included 560 additional stars that Flamsteed had omitted. This, together with the discovery of her first comet that same year, brought her recognition as an astronomer in her own right, and she was given her own salary by King George III as William's assistant in 1787.

William eventually married in 1788. Though he and Caroline no longer lived in the same house thenceforth, Caroline continued to perform the mathematical calculations that he needed. She also discovered eight comets by 1797. After 1797, however, she produced no more work of her own until after William's

Urbain Jean Joseph Leverrier
1811-1877
French Mathematician and Astronomer

Urbain Jean Joseph Leverrier was born in St. Lô France in 1811. Little is known about his early life, but he studied briefly in the lab of famous chemist Joseph Gay-Lussac (1778-1850) prior to settling on celestial mechanics as a field of study. By the age of 25 he had been appointed as an astronomer and professor at the Polytechnique and, in 1846, was admitted to the French Academy.

Leverrier is best known for his work leading to the discovery of Neptune in 1846, a discovery he shares with John Couch Adams (1819-1892) and Johann Gottfried Galle (1812-1910). Like Adams and others, Leverrier was intrigued by the fact that Uranus's orbit did not match predictions based on accepted laws of physics. Solving the problem some time after Adams did, Leverrier published his results first. He approached a number of French astronomers, requesting that they look at the calculated position for the new planet, but met with no success. Part of the reason for this was that he was known as a rather unpleasant person.

Leverrier eventually convinced Galle to search for the new planet. Galle and his assistant found it in their first observing session at precisely the calculated position, beating a group of English astronomers who searched slightly less diligently. This led to a dispute between the English and

French over priority, eventually being resolved as a joint discovery shared by both Adams and Leverrier. Of this discovery, one of Leverrier's colleagues said, ". . . he discovered a star with the tip of his pen, without any instruments other than the strength of his calculations alone."

Immediately following the announcement of Neptune's discovery, the London *Times* announced "Leverrier's planet found" in large headlines. And, in recognition of his role in Neptune's discovery, Leverrier was awarded the Copley Medal of the Royal Society of London and other honors.

Because of the acclaim that resulted from this discovery, Leverrier was made a Senator by Louis Napoleon in 1852 and then Director-General of the Paris Observatory in 1854. However, these honors did not stop his research. Following his success with Neptune, Leverrier became interested in similar problems in predicting the orbit of Mercury. Taking a similar approach, in 1859 Leverrier predicted the existence of a planet, which he called Vulcan, closer to the sun than Mercury. Although several astronomers later reported unconfirmed sightings of Vulcan and Leverrier believed in its existence until his death, it has been determined that no such planet exists. In fact, Albert Einstein (1879-1955) showed the anomalous motions of Mercury to be due instead to an application of relativity theory, removing the physical and mathematical necessity for a new innermost planet altogether.

Leverrier was an unpopular director because of his continuing quest for greater and greater levels of efficiency. Because of this, Leverrier was removed from his position at the Paris Observatory in 1870. However, when his appointed successor died only three years later, Leverrier was re-appointed to the post, but with severe restrictions on his authority. He continued there, still searching for Vulcan, until his death in 1877 at the age of 66. He is remembered as not only the co-discoverer of Neptune, but as one of France's greatest astronomers.

P. ANDREW KARAM

Charles Lyell
1797-1875
English Geologist

Charles Lyell is considered by many to be the father of modern geology. His masterwork, *Principles of Geology,* published between 1830

and 1833, added scientific rigor to geologic interpretations of rocks and fossils. Lyell also became the first to propose the division of the Cenozoic era (the current geologic period) into epochs based solely on fossil evidence. This led to our current understanding of geologic time and the history of Earth.

Lyell was the eldest of ten children born to Scottish parents. His father was an active naturalist who had a large library, including many books on geology, which may have sparked Charles' interest in this subject. Lyell attended Oxford, studying mathematics, law, and geology.

Upon graduation he began a legal career, but he left it quickly to pursue his interest in geology. Although unskilled at first, Lyell quickly became expert through study and increasing practical experience. This, coupled with his knowledge of zoology, helped him to make sense of the rocks he studied in his travels.

Lyell was familiar with contemporary theories of the history of Earth and with various explanations for the geologic phenomena he observed. As he became more confident of his knowledge of geology, he became sure that the then-current reliance on catastrophic explanations for the formation of most of Earth's features was incorrect. Like most scientists of his time, Lyell believed Earth to be ancient; unlike his contemporaries, Lyell agreed with Scottish geologist James Hutton (1726-1797) that everyday phenomena, continuing over vast stretches of time, were sufficient explanation for mountains, oceans, and other large structures on Earth.

In 1830 Lyell began his best-known work, *Principles of Geology,* destined to become a three-volume exposition that formed the foundation of modern geology as a true science. In the first volume Lyell reintroduced Hutton's Uniformitarianism, the idea that geological change is the result of long-acting and relatively uniform processes. Lyell envisioned a "great year" in which all creatures rose, fell, and rose again. In other words, Lyell's Earth lacked directionality: it did not progress but cycled eternally. To illustrate, he wrote, "Then might those genera of animals return, of which the memorials are preserved in the ancient rocks of our continents. The huge iguanodon might reappear in the woods, and the ichthyosaur in the sea, while the pterodactyl might flit again through the umbrageous groves of tree-ferns." In this view Lyell was wrong, a fact he later admitted.

Lyell's second volume dealt with his observations on igneous and metamorphic rocks, noting that high temperatures can cause changes in rocks. Among the phenomena explained by this are the changes seen in sedimentary rock into which hot igneous rocks have intruded. Lyell's third volume dealt with paleontology and stratigraphy (the ordering and interpretation of layers of rock), in which he became the first to try to arrange more recent rocks in a coherent order, dividing them into epochs according to the fossils contained in various strata. This method of ordering and dating rock layers relative to one another is still used today.

Lyell's work inspired Charles Darwin, who carried *Principles of Geology* with him his voyage to South America that prompted him to formulate his theory of evolution. At one point Darwin commented on *Principles of Geology* that "The greatest merit of the *Principles* was that it altered the whole tone of one's mind, and therefore that, when seeing a thing never seen by Lyell, one yet saw it through his eyes." Lyell was knighted for scientific accomplishment at the age of 51, was named a baron at age 67, and died a year later, preceded in death by his beloved wife.

P. ANDREW KARAM

James Clerk Maxwell
1831-1879
Scottish Physicist

James Clerk Maxwell developed the mathematical theory of electricity and magnetism, and introduced statistical methods to the kinetic theory of gases and thermodynamics. Arguably the nineteenth-century scientist who exerted the greatest influence on twentieth-century science, his work had widespread significance in a variety of fields, including the development of relativity and quantum mechanics. The importance of Maxwell's work is ranked with that of Isaac Newton (1642-1727) and Albert Einstein (1879-1955).

Maxwell was born into a family whose original surname was Clerk; his father added the name Maxwell when he inherited the Maxwell estate. Maxwell's mother, who was 40 years old when he was born, died of abdominal cancer when he was eight. He was initially tutored at home, then attended Edinburgh Academy where he published his first scientific paper at age 14.

Maxwell entered the University of Edinburgh at age 16 and moved on to Cambridge at

age 19. He subsequently became one of the most influential members of the group of nineteenth-century scientists, now known as the Cambridge School, who provided leadership in the application of mathematics, especially calculus, to physical problems.

CHARLES LYELL
AND THE RETURN OF THE DINOSAURS

Charles Lyell, the father of modern geology, is widely accepted as the person who made geology into a predictive science. He did this by pointing out that "the present is the key to the past"; in other words that we can make inferences about the past because the same processes are taking place today and can be observed and measured. However, in one case, he appears to have taken a slightly too literal approach to Uniformitarianism because he felt that, at some time in the future, dinosaurs would return to replace man on Earth.

Lyell firmly believed in a strong Uniformitarianism—that geological change on Earth is the result of long-acting, relatively uniform processes. From this perspective the absence of dinosaurs from the world had to be explained, because this was something that had obviously changed. He resolved this dilemma by positing that Earth followed a cyclical form of history, not unlike the seasons, and that the classes of animals changed throughout the "great year." Mammals dominated the colder "seasons" of this great year, but when the calendar turned to summer again, "Then might those genera of animals return, of which the memorials are preserved in the ancient rocks of our continents. The huge iguanodon might reappear in the woods and the ichthyosaur in the sea, while the pterodactyle might flit again through umbrageous groves of tree-ferns."

This passage occasioned no end of ridicule among Lyell's contemporaries, even inspiring one of his colleagues, de la Beche, to draw a mocking cartoon showing "Professor Ichyosaur" lecturing a class about the skull of an extinct human.

In 1856 Maxwell became professor of natural philosophy at Marischal College in Aberdeen, Scotland, and in 1860 was appointed professor of natural philosophy at King's College in London. Between 1860-65 he wrote two papers that introduced his mathematical treatment of the field theory of electricity and magnetism. He also had a continuing interest in vision, color, color blindness, and geometric optics, and in 1861 showed the feasibility of color photography by demonstrating that photographs of the same subject, successively taken through filters of the three primary colors, could be combined to produce a colored image of the subject.

Among the most important contributions that resulted from Maxwell's application of mathematical methods to a variety of physical phenomena was the introduction of statistical methods to thermodynamics. In his treatment of the thermodynamics of gaseous systems he assumed that the amount of kinetic energy possessed by individual molecules is distributed statistically about an average energy that is related to the temperature of the system. This became the basis of a general statistical kinetic theory of gases and the statistical interpretation of thermodynamics.

Maxwell became a member of the Royal Society in 1860. In 1865, at age 34, he retired to the family estate to concentrate on his scientific work. The principal result was the full development of his field theory of electromagnetism, recognized as one of history's greatest shifts in scientific thinking. He published his results as *A Treatise on Electricity and Magnetism* in 1873. This theory treats electricity and magnetism as aspects of a single force—electromagnetism. Maxwell demonstrated that this force could be regarded as extending out through space as a field that did not require the presence of matter for its propagation. He showed that the rate of movement of this force through space is equal to the speed of light and, furthermore, that visible light itself is electromagnetic in nature and is a part of a broad range of electromagnetic radiation. The field and wave nature of his electromagnetic equations introduced an approach that would later be fundamental in the development of Einstein's special theory of relativity and the wave equations of quantum mechanics.

In 1871 Maxwell returned to Cambridge to become the first Cavendish Professor at that university. He died of colon cancer in 1879 at age 48, the same terminal age and disease as his mother.

J. WILLIAM MONCRIEF

Dmitri Ivanovich Mendeleyev
1834-1907
Russian Chemist

Dmitri Mendeleyev is best known for his development of the periodic table of the elements. His table is based on the patterns he observed when the elements are organized by increasing atomic weight. Using his table, he was able to accurately predict properties of previously unknown elements.

Mendeleyev was born in 1834 in Tobolsk, Russia. His father was a teacher who died by the time Dmitri was a teenager, and his mother opened a glass factory in order to support the large family. When Dmitri was old enough for college, his mother traveled with him across Russia to St. Petersburg, a journey of thousands of kilometers that they took largely on foot. There, in 1850, he enrolled in the Institute of Pedagogy to be educated as a teacher.

After graduating Mendeleyev continued his education at the University of St. Petersburg. He received an advanced degree in chemistry and was awarded a scholarship that allowed him to study in Europe. In 1860 he attended the Karlsruhe Conference in Germany. This conference was the first international meeting of chemists. One of the scientists Mendeleyev met there was Stanislao Cannizzaro (1826-1910), who had made accurate measurements of the atomic weights of the elements. These measurements were to be useful to Mendeleyev in his own research.

After returning to St. Petersburg Mendeleyev became a chemistry professor. He could find no textbook that was suited to his students' needs, so he decided to write his own. While working on his book, *Principles of Chemistry* (1868-1870), he began to look for a logical way of arranging the elements. He wrote the name of each element on a note card and listed its properties underneath. Then he began to arrange the cards in different ways, looking for patterns.

Eventually, Mendeleyev found that when he ordered the cards by increasing atomic weight, elements with similar properties appeared at regular, or periodic, intervals. Mendeleyev used his observations to make a table that reflected this pattern. This type of arrangement became known as a periodic table. (Mendeleyev's table forms the basis of the modern periodic table. However, the modern table is organized by increasing number of protons—atomic number—rather than increasing atomic weight.)

In March 1869 Mendeleyev presented his table to the Russian Chemical Society. Two years later, he published a revised and more detailed version. In the papers that accompanied these tables, Mendeleyev attempted to show that an element's physical and chemical properties were a function of its atomic weight. Mendeleyev called this relationship the periodic law.

As he was working on his table, Mendeleyev began to suspect that there were elements that had yet to be discovered. (In fact, at that time scientists had described only 69 of the 112 chemical elements known today.) He left blanks in his table to accommodate these elements and even made predictions of their properties based on his periodic law. Three of these—gallium, scandium, and germanium—were discovered within 20 years of the publication of Mendeleyev's first table. When the scientific community realized that his predictions were accurate, Mendeleyev soon became quite well known and was frequently invited to give lectures throughout Europe.

Mendeleyev was not as accepted in Russia, however. He was considered controversial because he allowed women to attend his lectures and because he openly expressed his criticism of the Russian government. Although he was denied admission to the Russian Academy of Sciences, he was made director of the Bureau of Weights and Measures in 1893. Throughout his life he continued to write about chemistry as well as other topics, including education and art. In 1906 Mendeleyev missed being awarded the Nobel Prize in chemistry by a single vote.

STACEY R. MURRAY

Albert Abraham Michelson
1852-1931
American Physicist

Albert Michelson was a renowned physicist whose work in optics inspired, among others, Albert Einstein (1879-1955). He made the first accurate determinations of the speed of light and helped to disprove the existence of "ether," previously thought to permeate all space.

Michelson was born in Strzelno, Poland, to Samuel and Rozalia Michelson. The Michelson family left Poland for the United States when Albert was three, his father eventually becoming a successful merchant in San Francisco. Michelson entered the U.S. Naval Academy at age 17, where he performed well in his academic studies but rather poorly in seamanship. He graduated in 1873 and was a science instructor at the Academy until 1879.

In 1878 Michelson began what was to be his life's passion—obtaining accurate measure-

ments of the speed of light. He first used crude homemade equipment to arrive at reasonable values. Knowing he would need to study optics in order to obtain the accuracy he desired, Michelson spent two years in Paris, Berlin, and Heidelberg. During this period, he resigned from the Navy to be able to concentrate on his research. Upon returning to the United States in 1882, he calculated that the speed of light was 299,853 km/sec, the most accurate measurement to be made until his own revision nearly 30 years later.

Accepting a professorship in physics at Case University in Cleveland, Ohio, Michelson began a collaboration with Edward Morley (1838-1923) to try to prove or disprove the existence of "ether," a substance that was thought to permeate space, allowing electromagnetic radiation to be transmitted. The Michelson-Morley experiment was elegant and decisive, using the wave-like properties of light in a device called an interferometer.

When light waves meet and are slightly out of phase (meaning the crests and troughs do not match each other exactly), observers see a series of bright and dark bands. The bright bands represent places where two crests meet and reinforce each other, while the dark bands show where a crest and a trough meet and cancel each other out. Someone holding two fingers very close together in front of a light can see these dark and light bands between their fingers if they look closely and carefully.

Michelson and Morley realized that as the Earth moved in its orbit it would be in motion across the ether. That meant that, if the ether existed, a light shining in different directions would be moving with different velocities with respect to the ether. They set up an "L"-shaped set of mirrors and bounced beams of light from a single source from the mirrors at either end of the "L." As the light returned to the origin, they looked for bands, because differences in the travel times of the beams of light would cause the beams to be out of synch with each other. When repeated experiments showed no bands, Michelson and Morley reluctantly came to the conclusion that "ether" did not exist. This conclusion not only rattled conventional science, but also set the stage for the later theory of relativity. For this experiment and his groundbreaking work on the speed of light, Michelson was awarded the Nobel Prize in Physics in 1907, the first American to receive this honor. It is important to note that from the perspective of Michelson and Morley this experiment was a failure, as they were attempting to show that ether existed. However, from their "failure" came a number of advances in our understanding of the physical universe, proving in the end more profitable than most experimental "successes."

Of Michelson's work, Einstein said: "My honored Dr. Michelson, it was you who led the physicists into new paths, and through your marvelous experimental work paved the way for the development of the theory of relativity."

Michelson commented in 1894: "The more important fundamental laws and facts of physical science have all been discovered, and these are now so firmly established that the possibility of their ever being supplanted in consequence of new discoveries is exceedingly remote. . . . Our future discoveries must be looked for in the sixth place decimals." The irony of this quote is that Michelson's own work set the stage for discoveries in relativity that would overturn our view of the universe. Also of interest, this statement was made shortly before the discoveries of x-rays and radioactivity, which led to studies of the atom, quantum physics, and other areas that are still being fruitfully explored today.

In addition to his scientific accolades, Michelson served as the first chair of the Physics Department at the University of Chicago and was president of the National Academy of Sciences from 1923-1927. He married twice during his life, having three children in each marriage, and died in 1931 at the age of 78.

P. ANDREW KARAM

Hans Christian Oersted
1777-1851
Danish Physicist

In 1820 Hans Oersted described the action of an electric current on a magnetized needle, demonstrating a connection between electric and magnetic forces unexpected by the majority of researchers in electricity and magnetism. His observation stimulated further work on electromagnetic phenomena by numerous investigators, including André Ampère (1775-1836) and Michael Faraday (1791-1867), leading to a grand synthesis in 1873 by English physicist James Clerk Maxwell (1831-1879). Maxwell's equations explained the behavior of light as an electromagnetic wave, led to the discovery of radio waves, and posed questions concerning

the velocity of light which would lead to Einstein's theory of relativity.

Oersted received a broad education in scientific subjects at the University of Copenhagen, receiving a degree in pharmacy in 1797. Two years later he received his doctorate for a thesis that argued for the importance of the philosophy of Immanuel Kant (1724-1804) for the study of nature. In 1801 Oersted built his own version of the Voltaic pile, invented the year before by Alessandro Volta (1745-1827), and used it to gain entrance to laboratories throughout the German-speaking world. At the same time, he was able to attend lectures on the "Naturphilosophie" being developed by a number of German scholars. Always interested in philosophical simplicity, Oersted was suspicious of the new chemistry introduced by Antoine Lavoisier (1743-1794) with its numerous chemical elements, preferring the "chemistry of opposites" proposed by Hungarian chemist J. J. Winterl. Drawing from Kant the notion that there are only two basic forces, attraction and repulsion, Oersted came to believe that the various forces apparent in nature were all versions of the same basic two.

By 1820 the majority of physicists believed that electricity and magnetism were unrelated, although they had some superficial similarities. Oersted, in contrast, believed that the same electric current that could produce light or heat or chemical effects depending on the size and surroundings of the wire that carried it must also be the source of magnetic effects, though not in the direction in the current. By sending a current down a wire passing by a magnetic compass, he was able to produce a slight deflection of the compass needle. Refining the experiment over the next few months, he was able to demonstrate that the lines of the magnetic field run in circles about a current carrying wire.

Building on Oersted's result, the French physicist Ampère , and later Biot and Savart, was able to determine the detailed magnetic interaction between electrical circuits. Ampère also provided an explanation for the existence of permanent magnets as materials with built-in electrical currents. Seeing that an electrical current could produce magnetic effects, the logical question arose as to whether a magnetic field could produce an electrical effect. English physicist and chemist Michael Faraday provided the correct connection about 10 years later by the discovery of electromagnetic induction, that is, the creation of a voltage by changing the magnetic flux through a surface. The results of Ampère and Faraday attracted the attention of the James Clerk Maxwell, who had the mathematical skills needed to express the experimental conclusions in the form of differential equations. Maxwell then noted that his equations could take the form describing transverse waves in empty space, waves that traveled with a speed related to the force constants appearing in the Coulomb electrostatic force law and the Ampère law for the magnetic force. Numerically, the speed of Maxwell's waves turned out to be three hundred million meters per second, the known speed of light in vacuum, thus suggesting that light was an electromagnetic wave. The creation of lower frequency radio waves, and Einstein's theory of relativity, needed to explain how the speed of Maxwell's waves could be the same as measured by all observers soon followed Maxwell's discovery.

Oersted remained fascinated with philosophy and at the time of his death was writing articles for a book entitled *The Soul in Nature*. While Oersted's bias against atomic theory and the growing list of chemical elements was certainly incorrect in hindsight, the search for a unified description of the forces of nature remains an important theme in modern physics.

DONALD R. FRANCESCHETTI

Friedrich Wilhelm Ostwald
1853-1932

German Chemist, Philosopher and Color Theorist

Co-founder of the discipline of physical chemistry, Wilhelm Ostwald (1853-1932) won the 1909 Nobel Prize in Chemistry for work on catalysis, electrochemistry, and solution theory. A prolific researcher with an engaging personality, he wrote 45 books and almost 500 articles, and edited six journals. His experimental results and techniques reshaped chemical theory by their integration with concepts and methods drawn from physics.

Born and raised in Riga, Latvia, Ostwald enrolled at the University of Dorpat (now Tartu in Estonia) in 1872, receiving his doctorate in 1878. After serving as a laboratory assistant and non-salaried lecturer, he became a professor at the Riga Polytechnic Institute in 1881, where he gained fame as a teacher and researcher. In 1880 he married Helene von Reyher, a surgeon's daughter; of their five children, Wolfgang became a noted colloid chemist, and Gretel wrote her father's biography. In 1887 the University of

Leipzig offered Ostwald the world's only professorship in physical chemistry. During his 19 years there, his laboratory became an international research center that pioneered the modern methods of professor-graduate assistant collaboration. At his retirement in 1906, Ostwald's students, eventually including three Nobel laureates, held 45 of almost 50 new chairs in physical chemistry.

During the later 1800s, organic chemistry dominated chemical research. Studies focused on static properties of chemical composition and structure—isolation and identification of chemical elements, analysis and synthesis of organic compounds, and theories of molecular shapes and functional groups. However, Ostwald's attention turned instead to unresolved problems involving dynamic chemical processes—reaction velocity, mass action (the effectively participating proportions of reactants), chemical equilibrium (the final balance between reactants and products), and affinity (the relative degree of attraction between various reactants).

Since chemical methods change the nature of a reaction, they cannot be used to measure the rates at which chemical processes occur. Ostwald realized that such processes can be studied instead by measuring changes in the physical properties that accompany chemical reactions in solutions, such as specific volume (mass per unit volume, the inverse of density), viscosity (resistance to fluid motion), refractive index (degree of deflection of a beam of light), rotation of polarized light, and electrical conductivity. Applying these techniques to hundreds of acid-base reactions, Ostwald obtained precise numerical values for chemical affinities, which previously could only be described qualitatively. This made it possible to predict and measure the speed and efficiency of chemical reactions.

In 1884 Ostwald's work garnered vital support from two fellow chemists. Svante Arrhenius (1859-1927), who became Ostwald's assistant, explained chemical reactivity as the result of dissociation of molecules in solutions into electrically charged particles or *ions*, while Jacobus van't Hoff (1852-1911) applied concepts from thermodynamics, or the physics of heat and temperature effects, to the analysis of chemical processes. Their ideas provided Ostwald theoretical explanations for his experimental results on affinities and reaction rates. In 1887, by then close friends, they founded the research journal *Zeitschrift für physikalische Chemie* to propagate their controversial new ideas, for which all three won Nobel prizes.

In 1888 Ostwald formulated his famous "dilution law," relating electrical conductivity and chemical activity in weak acid-base solutions to increasing dilution and ionic dissociation. During the 1890s, his researches shifted to the study of catalysis, or the acceleration of chemical reaction rates by use of a facilitating reagent that itself remains unchanged. Ostwald's findings related catalytic activity to ionic dissociation, electrical conductivity, and the energy states of chemical solutions, and made consideration of intermediate reaction stages and of time indispensable factors in chemical analysis. The use of catalysts is now part of virtually every industrial chemical process.

Beginning in 1891, unresolved problems involving contemporary atomic theories and thermodynamics led Ostwald to reject atomism for the radical theory of "energism," which posited that only energy exists, with matter being merely a manifestation of local energy complexes. Ultimately he advocated a "chemistry without matter," which sought to explain all chemical phenomena solely in terms of energy types and transformational processes. Initially given a respectful hearing, Ostwald's views were generally rejected after a celebrated debate with his opponents at the 1895 Lübeck conference.

Ostwald finally retracted his opposition to atomism in 1908, but by then he had recast energism as a general philosophy, arguing that the laws of thermodynamics proved the efficient application of all forms of energy to be the highest moral principle of mankind. From 1906 to 1914, working as a free-lance philosopher and scientific popularizer, he used his "energistic imperative"—"Do not waste energy; utilize it!"—to promote a wide variety of social reform schemes. When the outbreak of World War I dashed these hopes, Ostwald turned until his death in 1932 to a third career as a color theorist. In a laboratory on his estate "Haus Energie" in Grossbothen near Leipzig, he developed an original system of standardized color scales and patterns of harmonization, based on precisely measured gradations between shades of black, white and gray. While heavily criticized in German artistic circles, Ostwald's ideas have exerted considerable influence in German dye and fabric industries and in art schools in Britain and the United States.

JAMES A. ALTENA

Guiseppe Piazzi
1746-1826
Italian Astronomer

Guiseppe Piazzi was born in Ponte di Valtellina, Italy, in 1746. At the age of 18 he became a Theatine monk. While studying at colleges of that order throughout Italy, he developed a fascination with mathematics and astronomy. After serving as a professor of theology in Rome and, later, as professor of higher mathematics at the Academy of Palermo, Piazzi founded the government observatory of Palermo in 1791.

As the observatory's first director, he launched a lengthy project dedicated to updating existing star catalogues. Piazzi wanted to document precisely the exact number and astronomical position of several thousand stars. He soon demonstrated that most stars were in motion relative to the sun. His first compilation, published in 1803, included 6,784 stars. Eleven years later, he updated his list, publishing a second catalogue containing 7,646 stars. Both lists were awarded prizes by the Institute of France.

While working on his star maps, Piazzi made his greatest discovery. On January 1, 1801, he spotted a small object 470 miles in diameter in the constellation Taurus, between the orbits of Mars and Jupiter. The unusual object shone like a faint star, but behaved like a planet. Piazzi named it Ceres, after the Roman goddess of the harvest.

Piazzi's discovery caused great excitement in the astronomical community. For years scientists had been searching that same area of sky for a mystery planet. According to Bode's Law, a scientific formula for calculating distances between planets, there was a "missing" planet between Mars and Jupiter. But Piazzi's object was smaller in comparison to the planets, and its shifting motion ruled out the possibility of a new star. Unsure whether the strange heavenly object was a comet or a planet, Piazzi continued regular observations until mid-February 1801, when he fell ill. By the time he recovered a few days later, he had time to make just one more observation before the mystery object disappeared in the sun's glare.

What Piazzi had discovered was the first and largest asteroid (or "minor planet") with a mass 1/100,000 that of the earth. Alone, this single asteroid is equivalent to half the total weight of all known asteroids today.

Using data from Piazzi's three distinct observations, mathematician Carl Gauss (1777-1855) calculated Ceres' orbit with such accuracy that it was found one year later within 0.5 degrees of the predicted position. Ceres orbits the Sun every 4.6 years at an average distance of 257 million miles.

Soon afterwards, three other similar objects were discovered at the same distance. Those asteroids were named Pallas, Juno, and Vesta. Today many thousands of asteroids are known and catalogued. They are numbered in order of their discovery, for example 1 Ceres.

King Ferdinand wished to present Piazzi with a gold medal in commemoration of his discovery, but the astronomer requested that they money be used to purchase a much needed equatorial telescope. A decade before his death, Piazzi established a second observatory at Naples. Piazzi died in Naples on July 22, 1826. The one thousandth asteroid discovered was named Piazzia in his honor.

KELLI A. MILLER

William Ramsay
1852-1916
British Chemist

The most eminent British chemist of the late nineteenth century, William Ramsay won the 1904 Nobel Prize for his discovery of the so-called "noble" gases. A brilliant experimentalist rather than a theorist, and one who did his best work in collaboration with others, his research helped to establish the new discipline of physical chemistry in Britain. A champion of educational reform, an accomplished linguist, poet, and athlete who traveled widely, and deeply religious, his generous nature, humility, and lively sense of humor made him a beloved figure.

Of Scottish descent, Ramsay was the only child of devout Calvinists, and descended from three generations of workers in the dyeing industry on his father's side. In 1869 he first began chemical studies at university in Glasgow, going to Tübingen in Germany in 1870 to study organic chemistry and obtaining his doctorate there in 1872. He then returned to Glasgow and served six years as an assistant at a technical college. In 1880 Ramsay advanced to a professorship at University College, Bristol, a move that coincided with a shift in his research interests from organic to physical chemistry. In 1881 he became principal of the College and wed Margaret Buchanan, and their happy union produced a daughter and a son. Between 1882 and

William Ramsay. *(The Library of Congress. Reproduced by permission.)*

1887 he gained notice through several papers with Sydney Young, both on the thermal properties of solids and on the relation between rates of evaporation and molecular dissociation for substances at various pressures.

In 1887 Ramsay succeeded Alexander Williamson, the dean of British chemists, at University College, London, where he remained until retiring in 1912. His researches there initially focused on the behavior of nitrogen oxide compounds. Realizing that the current "hydrate theory" of solutions, which assumed that a solute and solvent form a weak chemical combination, could not explain the activity of electrolytes as freely moving charged particles, he became an advocate of the new "ionic" solution theory.

In September 1892 and April 1894, the renowned physicist Lord Rayleigh (John William Strutt; 1842-1919) published notices that two different methods of preparing elemental nitrogen, one from atmospheric air and the other from ammonia, consistently yielded products with very slightly differing densities. Ramsay, recalling Henry Cavendish's (1731-1810) famous 1785 experiment that isolated nitrogen from atmospheric air but left an unexplained minute gaseous residue, suspected that a previously unknown element awaited detection.

Between April and August 1894 Ramsay repeatedly passed quantities of atmospherically derived nitrogen over heated magnesium, which reacted to produce magnesium nitride. He thereby succeeded in isolating an unreacted gaseous fraction, which spectral analysis proved to be a new element, which he named "argon" (Greek for "inactive one"). Ramsay and Rayleigh's joint announcement of this discovery to the Royal Society of London created a sensation, since the new element showed no chemical activity and did not fit into the existing pattern of the recently established periodic table.

In March 1895 Ramsay read a report of the emission of an unknown gas from clèveite, a uranium-based mineral. He immediately bought the available supply of it in London, collected the gas by treating the mineral with sulfuric acid, and submitted it to spectral analysis. The result matched that of helium (from the Greek for "sun"), an element previously detected in the solar spectra but not known to exist on Earth. Between 1895 and 1898, with his assistant and later biographer Morris Travers, Ramsay painstakingly collected 15 liters of liquefied argon gas from atmospheric nitrogen. They repeatedly subjected this to fractional distillation, searching for minute traces of further elements. The resulting fractions were again subjected to spectral analysis, and between June and September 1898 yielded three new elemental gases: krypton ("hidden one"), neon ("new one"), and xenon ("strange one").

In later years until his death from cancer Ramsay, assisted by Frederick Soddy (1877-1956), investigated problems involving radioactive decay and elemental transmutation. These studies proved inconclusive, as they preceded necessary developments in atomic theory. However, a gaseous "emanation" from radium that Ramsay called "niton"—the molecular weight of which he determined from a minute sample in a remarkable experiment—was later proved to be yet another gas element, radon.

JAMES A. ALTENA

Samuel Heinrich Schwabe
1789-1875

German Pharmacist and Astronomer

Samuel Schwabe, a pharmacist by training and profession, is best known for his discovery of the 11-year cycles of solar activity and sunspot abundance. As an amateur astronomer, Schwabe's observations helped found the modern era of solar observations and sunspot research.

Schwabe was born in Dessau, Germany, in 1789, the son of a local councilor who was the personal physician to the duke. The oldest of 11 children, Schwabe studied pharmacology in Berlin and, when possible, attended additional lectures in astronomy and botany. Upon his grandfather's death in 1812, Schwabe returned to Dessau to assume responsibility for his pharmacy. There he plied his formal profession until selling the pharmacy in 1829 for enough money to allow him to pursue his scientific interests full-time.

Schwabe was one of the first astronomers to spend extensive time studying the sun. In 1610 Galileo (1564-1642), while observing the sun directly, risked blindness from the extreme brightness. His discovery of sunspots, contested by two others, came as a surprise because the face of the sun had been thought to be perfect in all respects. Another co-discoverer, the priest Christoph Scheiner (1573-1650), developed a safer way to observe the sun by projecting its image onto a screen. This technique is still in use today by amateur and professional astronomers alike. After early activity directed toward the confirmation of sunspots, however, the field languished until Schwabe's work.

Schwabe's interest in astronomy changed from theoretical to observational in 1825 when he won a telescope in a lottery. Later that year he ordered a larger refracting telescope and erected a small observatory from which he began making observations. A larger instrument was added shortly afterwards, with a focal length of six feet. Using these instruments, Schwabe made observations of Saturn (discovering eccentricity in the rings) and began searching for planets inside the orbit of Mercury. He never found new planets, but kept detailed notes of the locations and numbers of sunspots he observed, noticing over time that their numbers changed in a regular fashion. In 1843, based on his notes from 1825 through 1843, Schwabe announced his discovery of a 10-year sunspot cycle, but received little serious consideration. Only later, after independent observations by professional astronomers confirmed his claims, was Schwabe taken seriously. Further studies showed that the sunspot cycle, now known to be 11 years in length, was related to cyclic variations in the earth's magnetic field, adding to Schwabe's support.

Today, the 11-year sunspot cycle is accepted by all solar astronomers and is linked to periodic changes in solar activity. Because of its proximity and strength, these solar changes affect the Earth. Solar magnetic storms have caused widespread power outages, disrupted satellite communications, interfered with spacecraft (in fact, solar heating of the atmosphere led to Skylab's early demise), and have been correlated with weather patterns, rabbit populations, and more. While some of these associations are undoubtedly appropriate, others are still under study. Nonetheless, it is apparent and well accepted that the cycle first noticed by Schwabe is both real and important.

In addition to his astronomical discoveries, Schwabe served as founder and long-time president of a local Society for Natural History, sharing his interests in mineralogy and botany with his fellow members. He also published a two-volume book on local plants in which he described over 2400 specimens in some detail. Because of his contributions to astronomy and his wide-ranging interests, Schwabe was elected a Fellow of the Royal Society in 1868, having already received its Gold Medal in Astronomy in 1857.

Schwabe died in his hometown of Dessau in 1875, 20 years after his wife passed away. He left his scientific papers to the Society for Natural History and the local college.

P. ANDREW KARARN

William Smith
1769-1839
English Mineral Surveyor, Engineer and Geologist

William Smith grew up in the small village of Churchill in southern England, where he received his only formal education at the village school. At the age of 18 Smith took a job as an assistant to a surveyor named Edward Webb, and this experience led to his employment as a surveyor with the Somerset Coal Canal Company. Smith helped to engineer the canals, which were the transport highways for barges carrying the goods that were the lifeblood of the early part of the industrial revolution in England. He gathered information about the rock into which the canals were dug and paid particular attention to the fossils contained within the rock layers. Smith grew particularly interested in the vertical changes in the layers of rocks. The data he collected led him to recognize widespread regularity in rock successions and to realize that rock strata could be distinguished from each other by the fossils that they contained.

Smith kept detailed notes on the geology and through oral discussions of his discoveries generally made them known before he published them. His lack of formal schooling made publication difficult at first for Smith. However, in 1815 Smith published *A Map of the Strata of England and Wales, with Part of Scotland*, the first detailed geologic survey of England. This eight-foot-by-six-foot geologic map surpassed all others that had been made before, not only in

WILLIAM SMITH AND THE HOT SPRINGS OF BATH

~

In the era of William Smith (and Jane Austen), Bath was the busiest, most exciting destination in Britain. The streets were crowded with aristocrats, celebrities, the newly rich, and curious tourists, all eating, drinking, shopping, gambling, and attending parties, music, and theater. The main attractions, though, were the natural hot springs, famous since Roman times for their curative properties and widely prescribed as a medical therapy. Patients immersed themselves in the steamy baths and drank the sulphurous mineral water in copious quantities at the fashionable Pump Room.

When the hot springs showed signs of diminishing in 1808, city authorities turned anxiously to their resident expert, William Smith. Smith knew about groundwater from his experience building canals, and he was working at an attempted coal mine nearby that was constantly flooded. Smith noticed that one of the affected hot baths filled more quickly on Monday than on Saturday, which he attributed to the fact that the leaking mine was not pumped out on Sunday. He then devised a way to plug the hole, restoring the flow to the hot springs and halving the time required to fill the bath. The city praised Smith's knowledge and skill, but as a working man he nevertheless remained excluded from the high society that his efforts supported.

size but also in area covered and stratigraphic detail. Smith's map covered areas where there was little natural exposure of rock. This made the map particularly valuable, because it was extremely useful in the planning for future canals, quarries, mines, and as a guide to soil types. Smith realized that soil type is directly related to the underlying bedrock. By examining Smith's geologic map, it was now possible to predict throughout a large part of England what type of rock would be found near the surface and, most importantly, the successive layers of rock that lay beneath the surface of Earth. In addition to his famous map Smith also published several small volumes on the identification of strata by fossils between 1815 and 1820. In 1820 he moved to northern England, where he lectured on geology from 1824 to 1828 and subsequently (1828-33) worked as a land steward.

William "Strata" Smith, as he has become to be known, is most remembered for his revolutionary discovery that fossils are not randomly distributed in rocks, and that fossils in sedimentary rocks occur in a particular vertical order that is predictable. The idea that fossils were distinctive of individual stratum or groups of strata and that distinctive assemblages of fossils could be traced cross-country led geologists to realize that the sequences of fossils in England could be matched with similar sequences elsewhere in the world. Geologists took note of Smith's techniques and, using the principles of rock-type division and fossil correlation, quickly recognized major stratigraphic divisions throughout Europe and the eastern United States. The basic ideas of rock correlation and fossil succession that Smith recognized in the late eighteenth century are the fundamentals of rock correlation that are practiced to this day. In recognition of his contribution to the science of geology, the Geological Society of London awarded Smith the first Wollaston Medal in 1831, and he later received a pension from the British government.

STEPHEN A. LESLIE

Sir Joseph John Thomson
1856-1940
English Physicist

As the director of the Cavendish Laboratory at Cambridge University, J. J. Thomson was instrumental in many important experiments and advances that marked the transition from classical to modern physics at the turn of the twentieth century. He discovered the electron in 1897 and was awarded a Nobel Prize for Physics in 1906. With his student Ernest Rutherford (1871-1937), also a Nobel laureate, Thomson made many fundamental discoveries concerning the properties of ionizing radiation.

Thomson was born in 1856 in the north of England. He studied physics at Cambridge and, at only 28 years old, was greatly surprised to be simultaneously named director of the Cavendish Laboratory at Cambridge and professor of experimental physics. By his own description, the laboratory he took over used "string and sealing wax" as its equipment, presenting him with a considerable challenge. Under his supervision Cavendish became one of the world's preeminent nuclear physics laboratories.

Just three years later, in 1897, Thomson successfully demonstrated that cathode rays are streams of electrons. radiation. Further research in this area allowed Thomson to determine the mass of the electron to be less than one two-thousandth that of the hydrogen atom, until then the lightest bit of matter known to exist. This led to the realization that electrons are subatomic—units of matter smaller than the atom itself. These same experiments also showed that electrons were negatively charged.

Previous research using magnetic fields had shown that cathode rays (electrons) could be deflected by magnetic fields. This is the principle underlying the operation of televisions, computer monitors, and other CRTs (cathode ray tubes). However, Thomson was the first to place cathode rays in an electrical field by bringing oppositely-charged electric plates next to a beam of cathode rays. When he applied electric current to the plates, the beam deflected towards the positively-charged plate, indicating that the rays had a negative charge. Further experiments in both electric and magnetic fields showed the ratio of mass between hydrogen atoms (protons) and the cathode rays (electrons). For this work, Thomson was awarded the Nobel Prize in 1906.

Additional research by Thomson showed that the interaction between electrons and matter could produce x-rays and that, conversely, x-rays interacting with matter could produce electrons. Thomson also developed the first modern atomic model, albeit one that did not stand the test of time. In his "plum pudding" model, Thomson envisaged a sphere of positive changes with an equal number of negative charges (electrons) embedded within. This model was later supplanted by a number of others, leading eventually to the current model in which a cloud of electrons forms a shell surrounding a nucleus comprised of both protons and neutrons.

It may be argued that Thomson's mentoring of Rutherford was as important as his own discoveries in physics. Rutherford, who also went on to win the Nobel Prize for chemistry, was a pivotal figure in the formation of contemporary physics. Among his major discoveries, he proposed that radioactivity results from the disintegration of radioactive atoms, facilitated the development of today's model of atomic structure, and conducted other groundbreaking research into the nature of matter.

Thomson married and his son, George, also went into physics. George Thomson (1892-1975) was awarded the Nobel Prize in 1937 for research in electron diffraction by crystals. It is interesting to note that the work conducted by J.J. Thomson was based on the material properties of the electron while his son's work depended on the wave-like properties of electrons. The elder Thomson was knighted in 1908, and died in 1940 at the age of 84. He was buried near Isaac Newton (1642-1727) in Westminster Abbey in London.

P. ANDREW KARAM

William Thomson, better known as Lord Kelvin. *(The Library of Congress. Reproduced by permission.)*

William Thomson, Lord Kelvin
1824-1907
British Physicist

William Thomson, Lord Kelvin, conducted important work in the areas of thermodynamics and electrical theory. Strongly influenced by the more mathematical French style of physics, Kelvin's own work was heavily mathematical and was influential in encouraging other British physicists to follow this example. Kelvin's work in the study of heat led to the development of the Kelvin scale of temperature, the standard of measurement still used by scientists around the world.

Kelvin was born in Belfast, Ireland. When Kelvin was eight years old, his father, a professor of engineering, accepted a position teaching mathematics at the University of Glasgow, where Kelvin began his studies at age 10. At age 15 Kelvin won a gold medal from the University of Glasgow for his paper, "Essay on the Figure of the Earth."

Kelvin was deeply impressed by the French mathematical approach towards describing the physical world. He first read *The Analytical Theory of Heat* by Jean Baptiste Fourier (1768-1830), in which the author used rigorous mathematics to describe the flow of heat. Further reading of works by Pierre Simon Laplace (1749-1827), Augustin Fresnel (1788-1827), Joseph Louis Lagrange (1736-1813), Adrien-Marie Legendre (1752-1833), and other French scientists made

a continuing impact on Kelvin, leading to his relocation to Paris after graduation. There Kelvin worked with a number of influential physicists, eventually mastering this approach to physics. In Paris Kelvin was also introduced to the concept of applying similar mathematical techniques to a wide variety of problems, ranging from heat transfer to fluid dynamics, electricity, and magnetism. His embrace of mathematics as a means to describe the physical world helped to turn physics into the science it is today, using math to predict, describe, and explain the world and universe that surrounds us.

In 1846 Kelvin's father arranged for him to be a favored candidate for a vacant position at the University of Glasgow. Selected to fill the spot, Kelvin began to collaborate with George Stokes (1819-1903) on theories of turbulence and other problems in hydrodynamics. Like many of Kelvin's inquiries into thermodynamics and electricity, much of this research led to nothing. However, his researches in thermodynamics led to proposing an absolute scale for temperature measurement, now known as the Kelvin scale, with units of degrees Kelvin.

Although many of Kelvin's research efforts were later deemed unsuccessful, he accomplished a great deal by initiating lines of research that later proved fruitful to others. In addition, he had a knack for developing laboratory equipment that could be used commercially. This proved valuable when Kelvin agreed to participate in efforts to establish a transatlantic submarine telegraph cable. Kelvin's instruments helped the first cable to successfully transmit telegraphic messages. Further cables relied even more heavily upon Kelvin's inventions and, as a result, Kelvin was knighted in 1866 and became wealthy from the sale of his electrical devices. He was later made Lord Kelvin, Baron of Largs, for his service to Britain.

Kelvin also weighed in on the losing side of many scientific arguments. He opposed Darwin's evolutionary theories, Rutherford's ideas regarding radioactive decay, and the idea of atoms. He also devoted much time to arriving at wildly erroneous proposals for the age of the Earth. These errors, however, do not detract from Kelvin's legitimate place as one of the great physicists of the nineteenth century.

During the latter part of his life, Kelvin received numerous accolades and awards. He published over 600 scientific papers, was elected a Fellow of the Royal Society (from which he received two prestigious awards), was named President of the London Maths Society, and served three terms as President of the Royal Society of Edinburgh. He was also elected president of the British Association for the Advancement of Science in 1871. William Thomson, Lord Kelvin, died in 1907 at the age of 83.

P. ANDREW KARAM

Thomas Young
1773-1829
British Physicist, Physician and Egyptologist

Thomas Young's career straddles the turn of the nineteenth century. In some ways he was an old-style natural philosopher, dabbling in many fields—physics, physiology, medicine, linguistics, navigation, insurance—and more concerned with ideas than applications. Yet, ironically, this theoretical trend also put him ahead of his scientific contemporaries; his revival of the wave theory of light was ignored for a generation. He is best remembered for his double-slit experiment demonstrating the interference of light, an absolute measure of the elasticity of solids known as Young's modulus, his optical studies, and his contribution to the deciphering of the Rosetta stone.

Born at Milverton, Somerset, England, of Quaker parents, Young was a child prodigy. He

Thomas Young. *(The Library of Congress. Reproduced by permission.)*

studied medicine at London, Edinburgh, Göttingen, and Cambridge universities. Young's interest in science was criticized by medical colleagues as taking time away from his medicine, so he published some papers anonymously.

In 1801 he was appointed to the Royal Institution, a British scientific organization, as Professor of Natural History, but quickly found himself at odds with the Institution's practical goals, lecturing on "pure knowledge" rather than the expected topics of the industrial and social applications of science. Furthermore, Young's style of presentation was not entertaining; a colleague remarked that he "was worse calculated than any man I ever knew for the communication of knowledge." Young left the Royal Institution in 1803.

Yet, while Young's lectures had been poorly received, a printed collection (1807) showed the scope of his thought, including his theories and experiments on light. The double-slit experiment, which consisted of passing light through extremely narrow openings and observing the interference patterns produced, led Young to surmise that light must be a wave, and he used the analogy of sound waves to suggest that light consisted of longitudinal waves (a back-and-forth compression) in an imaginary substance dubbed the ether. Later work by Augustin Fresnel (1788-1827) suggested that light consisted of transverse waves (which undulate up and down, perpendicular to the direction of propagation), and Young accepted this correction.

Young's theories on light stepped on important toes. Humphrey Davy (1778-1829) wrote to a friend, "Have you yet seen the theory of my colleague Dr Young, on the undulations of an Ethereal Medium as the cause of Light? It is not likely to be a popular hypothesis after what has been said by [Isaac] Newton concerning it." Indeed, Young's work on light was left in obscurity until revived by Hermann von Helmholtz (1821-1894) and other European scientists a generation later.

Young's published works include papers on the mechanism of the focusing of the eye, capillary action, the regulation and flow of blood, treatment of diseases, and many other topics. He deduced that the eye needed only three color receptors to enable a full range of color perception. He was the first in mechanics to use "energy" in the modern sense (the product of a mass with the square of its velocity), and he introduced an absolute measurement in elasticity, Young's modulus (the ratio of the stress to the strain on a solid). Young also recognized, from experiments with friction, that heat was not an element in itself, but rather a minute vibratory motion of particles.

From 1811 Young was physician at St George's Hospital, London, until he retired from medicine in 1814 and turned his talents to insurance and Egyptology, helping to translate the Rosetta stone. In his personal life Young shed the plainness of his Quaker origins and mixed in popular social circles. He had by all accounts a happy, but childless, marriage. Helmholtz, who is chiefly responsible for reviving Young's work, said that he "was one of the most clear-sighted men who have ever lived, but he had the misfortune to be too greatly superior in sagacity to his contemporaries." Young's ideas had come too soon.

DAVID TULLOCH

Biographical Mentions

André-Marie Ampère
1775-1836

French mathematician and physicist who made important discoveries toward the understanding of electricity and magnetism. Ampère proposed, with Amedeo Avogadro (1776-1856), that equal volumes of gas at the same pressure and temperature contain the same number of particles. He is better-known for his work that shows that magnetic fields result from the movement of small electrical charges (now known as electrons) and the relationship between distance and magnetic field strength. He is memorialized by the Ampere, a unit measuring electrical current strength.

Thomas Andrews
1813-1885

Irish chemist who discovered the critical point—that temperature above which it is impossible to liquefy a gas no matter how much pressure is applied. Before Andrews discovered this it was believed that many gases could never be liquefied. This insight allowed scientists to liquefy oxygen, hydrogen, nitrogen, and other gases, which were then known as "permanent gases," for the first time by lowering their temperatures. Andrews also proved that ozone, which creates the sour smell near a spark source, is a form of oxygen.

Friedrich Wilhelm August Argelander
1799-1875

German astronomer who completed a survey of the heavens that included data on the positions and brightnesses of over 324,000 stars. His catalog, called the *Bonner Durchmusterung,* included more precise data on the brightnesses of stars than previous catalogs, aiding astronomers who studied variations in stellar brightness. He himself studied variable stars and devised a method for estimating their brightnesses. He also confirmed William Herschel's findings on the motion of the sun.

Svante August Arrhenius
1859-1927

Swedish physical chemist who developed the theory of electrolytic dissociation of molecules in solution. His theory explained the phenomena observed when certain substances dissolve; he proposed that they break up into electrically charged positive and negative ions. He applied this theory to acids and bases, arguing that acids dissociate to produce H^+ ions and bases OH^- ions. The Arrhenius equation resulted from his generalization of the effect of temperature on the rate of chemical reactions. He received the 1903 Nobel Prize in chemistry.

Baron Carl Auer von Welsbach
1858-1929

Austrian chemist who invented the gas lantern mantle, discovered an alloy of cerium and iron (mischmetal) used in cigarette lighter flints, and invented the osmium lamp. The gas lantern mantle, commonly used in camping lanterns, uses metals impregnated into cloth fibers to give a brilliant light when heated by burning gas. All of these inventions use the properties of metals for illumination of some sort.

Hertha Marks Ayrton
1854-1923

British electrical engineer who analyzed and improved the electric arc, which was used to produce bright light. Arc lamps were widely used for indoor illumination, as well as in searchlights and movie projectors. Ayrton patented some of her improvements and published a comprehensive book, *The Electric Arc* (1902). She also did research in hydrodynamics. Ayrton was the first woman elected to the Institution of Electrical Engineers (1899) and the first woman permitted to read her paper in person before the Royal Society of London (1904).

Johann Friedrich Wilhelm Adolf von Baeyer
1835-1917

German chemist who revolutionized the textile industry by providing ways to produce dyes synthetically. His discovery of barbituric acid led to the development of the barbiturate sedatives. He also was influential in the development of the theory of structure of organic molecules, proposing tautomerism and the strain theory of carbon rings. He received the Nobel Prize in chemistry in 1905.

George Phillips Bond
1825-1865

American astronomer who, in 1848, assisted his father William Bond in the discovery of Hyperion, the eighth satellite of Saturn, and an inner ring of Saturn. George Bond received his training in astronomy from his father and, in 1850, worked with him to produce the first photographs (daguerreotypes) of recognizable quality of astronomical objects, the moon, and a star. He succeeded his father as director of the Harvard Observatory in 1859.

William Cranch Bond
1789-1859

American astronomer who, with his son George Bond, discovered Hyperion, the eighth satellite of Saturn, and an inner ring of Saturn in 1848. They also took the first recognizable photographs of astronomical objects in 1850. A self-educated watchmaker who discovered a number of comets, William Bond became the first astronomical observer at Harvard Observatory in 1839 and its first director in 1847. He was elected an associate of the English Royal Astronomical Society in Great Britain in 1849.

Karl Ferdinand Braun
1850-1918

German physicist who was known for his improvements in the fields of radio, television, and electronics. Braun's first great work was to convert alternating current, which travels in two directions, to direct current, which travels in one direction, which helped to improve radio signals. In 1897, he completed his oscilloscope, a precursor to the modern television cathode ray tube. Braun also made improvements to the distance and strength of Marconi's radio signals, and patented his new system in 1899. What he developed would later be used in radio, television, and radar. Braun shared the Nobel Prize for Physics with Marconi in 1909.

David Brewster
1781-1868

Scottish physicist whose primary contributions to science were his researches into the polarization of light. His work aided in the development of the spectroscope, as well as his invention of the kaleidoscope, the stereoscope, and a new system for lighthouse illumination. His other scientific interests included such areas as photography, astronomy, and the religious implications of extra-terrestrial life. He was an educator, author, and an active member of Britain's most important scientific societies.

Robert Wilhelm Bunsen
1811-1899

German chemist who popularized and improved the bunsen burner, a small tabletop torch whose high-temperature, nonluminous flame was perfect for his pioneering spectroscopy experiments. It has since become standard laboratory equipment. Bunsen was the codiscoverer of two elements, cesium and rubidium. His other accomplishments included the development of an antidote to arsenic poisoning, an explanation for the action of geysers, and invention of the carbon-electrode battery, used for arc lights and electroplating.

Nicolas Sadi Carnot
1796-1832

French physicist whose discoveries created the study of thermodynamics. Carnot, in his short life, discovered the principles by which heat engines work. He was the first to show that the efficiency of a heat engine is related to the temperature difference across the engine and is inversely proportional to the temperature of the heat sink. Carnot's discoveries are used today to describe the efficiency of machines from engines to nuclear power plants.

Richard Christopher Carrington
1826-1875

English astronomer who discovered that the location of sunspots on the sun's surface follows a regular pattern throughout the solar cycle. He discovered that the sun rotates not as a rigid body, all at the same rate, but as a fluid, with parts near the equator moving more quickly than parts near the poles. He also made a catalog of stars in the northern sky near the pole star.

Thomas Chrowder Chamberlin
1843-1928

U.S. geologist recognized for advancing geological education and shaping the planetesimal hypothesis. This hypothesis states that a star passing near the sun broke into pieces that condensed and clumped together to become planets. Chamberlin became chief of the Wisconsin Geological Survey in 1881 and later became the president of the University of Wisconsin at Madison. Then, for 26 years, he was head of the University of Chicago geology department, developing it into a world leader in the field of geology. However, his most famous contribution came after retirement with the development of the planetesimal hypothesis.

Jean Antoine Claude Chaptal
1756-1832

French chemist who was one of the first scientists to apply the concepts of chemistry to industry. Chaptal was put in charge of a gunpowder factory after the French Revolution. Because France was involved in numerous wars at the time, the industrial manufacture of gunpowder using chemical methods was crucial in preventing the country's defeat. In 1807, Chaptal wrote the first book on industrial chemistry, *Chemistry Applied to the Arts*. He also played a major role in introducing the metric system to France.

Johann von Charpentier
1786-1855

German-Swiss geologist who first worked out the details of a European ice age. The ice age, earlier proposed by Swiss geologist Ignatz Venetez (1788-1859), was not accepted until Charpentier convincingly described the details to Louis Agassiz (1807-1873), who went on to bring the idea to popular and scientific acceptance.

Michel-Eugène Chevreul
1786-1889

French chemist, physicist, philosopher, and psychologist who, as a result of his work with the chemistry of fats, is regarded as one of the founders of modern organic chemistry. He placed the chemistry of fats on a firm scientific basis by explaining saponification; isolating oleic, palmitic, and stearic fatty acids; and improving methods for soap and candle manufacture. He also made important contributions to the chemistry of dyeing, the physics of color, and the psychology of color.

Rudolf Julius Emmanuel Clausius
1822-1888

German physicist considered, along with Lord Kelvin, to be one of the originators of the second law of thermodynamics. Clausius reanalyzed the earlier work of Carnot on the efficiency of steam engines, taking into account the conversion of heat energy into mechanical work. In 1865 he suggested that he ratio of heat energy exchanged between systems to absolute temperature be considered a change in a new function, the entropy, of the systems.

Ludvig A. Colding
1815-1888

Danish engineer and physicist who made important discoveries in hydrodynamics and designed the Copenhagen water supply system. Colding discovered the Law of Conservation of Energy independently of James Joule, then went on to make discoveries about the way water behaves. In particular, he determined the reasons for a flooding disaster in southern Denmark in 1872 (high winds at sea causing flooding ashore) and helped design drainage systems to divert groundwater to make land more useful.

Gustave Gaspard de Coriolis
1792-1843

French physicist who first derived the fictitious forces, known as the centrifugal and Coriolis forces, that appear to act on bodies in a rotating frame of reference like Earth. The Coriolis force—also known as the Coriolis effect or acceleration—acts perpendicular to the motion of the body in the rotating frame, causing a sidewise deflection. It plays a role in many dynamical problems on Earth's surface, including the circulation of cyclones, the trajectories of long-range ballistic missiles, and the precession of a pendulum.

Pierre Curie
1859-1906

French physicist who performed research in magnetism and radioactivity, awarded the 1903 Nobel Prize in Physics. Best known as the husband of Marie Curie, he was a distinguished scientist in his own right. His research into magnetic properties is fundamental for theories of magnetism, while his radiation research set the stage for future discoveries in atomic theory. Pierre Curie died prematurely, hit by a horse-drawn cart while crossing the street.

William Morris Davis
1850-1934

American geoscientist whose career and scientific contributions spanned the disciplines of meteorology, geology, and physical geography. In the latter field Davis's interest in geology, landform processes, and rain effects led to his defining the science of geomorphology , based on his theory of a continuing, progressive cycle of terrestrial erosion. His meteorological teaching and study at Harvard (1876-1912) produced *Elementary Meteorology* (1894), a standard college text for over 30 years. Davis's research included comprehensive field work and publications on the origins of limestone caverns and coral reefs and islands.

Warren De la Rue
1815-1889

English astronomer who pioneered the use of photography in astronomy. He also advanced the science of electricity, inventing the silver chloride cell battery and experimenting with platinum filament light bulbs and electrical discharges in gases. De la Rue's initial application of photography in astronomy was a series of stereoscopic photographs of the sun and moon, shown at the International Exposition of 1862. His photography of the solar eclipse of 1860 is regarded as photography's first major contribution to astronomy.

James Dewar
1842-1923

Scottish chemist and physicist known for having invented a double-walled, glass container that keeps hot liquids hot and cold liquids cold. The Dewar flask (or thermos bottle) slows the flow of heat in two ways: maintaining a vacuum between the walls prevents the conduction of heat; mirroring the glass on the vacuum side of the walls slows the radiation of heat. For Dewar, the flask provided a convenient tool for studying cold, liquefied gases. Dewar was the first to liquefy hydrogen and the first to maintain liquid oxygen long enough to discover its magnetic properties.

Giovanni Battista Donati
1826-1873

Italian astronomer who investigated stellar spectra and discovered Donati's comet in 1858. Donati was the first to observe the spectrum of a comet and used stellar spectroscopy to identify the chemical elements that make up the comet's body and tail. Since comets are thought to represent the primordial composition of the solar system, this technique is used today to gain better understanding of how the solar system formed.

Christian Doppler
1803-1853

Austrian physicist best known for research into the effects of motion on acoustic pitch (the Doppler effect). Doppler described the reason for the rising pitch of an approaching train whistle and the drop as the train passes. He suggested applying this principle to astronomical objects, where it has become a powerful tool for determining the motions of stars, verifying the existence of black holes, and measuring the size of the universe.

Henry Draper
1837-1882

American astronomer and professor of medicine who took the first photographs of the Orion nebula (1880) and the spectrum of a star (Vega, 1872). He built the instruments he used for these and other photographs, including images of the moon and the spectra of several planets, the sun, and a comet. His widow established a memorial fund that financed the Henry Draper star catalogues produced at Harvard College Observatory, which remain among the most cross-referenced tools for astronomers.

Johann Franz Encke
1791-1865

German astronomer known for his mathematical work in astronomy, especially on short periodic comets. Following the discovery of a comet with a very short orbiting period around the Sun (3.3 years), Encke developed (in 1819) the mathematical formulae needed to calculate the orbits of this and other short periodic comets. During his life Encke worked at a number of German universities, and he was a professor in Berlin until 1863, two years before his death.

Emil Fischer
1852-1919

German organic chemist who made significant contributions to our knowledge of the structure of sugars and proteins; awarded the Nobel Prize in chemistry in 1902. Fischer was the first to show that naturally occurring sugars existed as molecules with either right or left-handedness. In 1903 Fischer synthesized a class of molecules called barbituric acids. Derivatives of barbituric acids called barbiturates were subsequently used as sedatives.

Osmond Fisher
1817-1914

British geophysicist who was ahead of his time in suggesting that there were convection currents in Earth's interior. His model proposed that these currents rise under the oceans and fall under the continents. He also proposed that a large supercontinent was split apart when the Moon was torn from Earth, leaving the Pacific Ocean as a scar, and that the remaining pieces of continental crust floated apart.

Armand Hippolyte Louis Fizeau
1819-1896

French physicist who in 1849 was the first to measure the speed of light on Earth. Fizeau was also the first to use the Doppler principle to measure the speed of stars moving toward or away from the observer. This principle is still used quite extensively in astronomy to measure the velocity of galaxies, the motion of gas swirling into black holes, and the expansion of the universe.

Jean Bernard Léon Foucault
1819-1868

French physicist who demonstrated the earth's rotation using a pendulum. Foucault made important discoveries in many areas, including the understanding of crystals, inventing the gyroscope (used in inertial navigation systems today), and better processes for silvering telescope mirrors. Foucault's most famous discovery, however, was the pendulum named after him. Attached to a nearly frictionless pivot, the Foucault pendulum swings in a constant plane while the earth rotates beneath it, making the plane of the pendulum seem to rotate.

Joseph von Fraunhofer
1787-1826

German craftsman and scientist who studied and advanced the field of theoretical optics. While working as a lens grinder he learned mathematics and optical science. He applied this knowledge to the development of optical instruments, especially telescopes. Using a prism he discovered that the spectrum of the Sun's light contains dark lines. The discovery of these lines was the first step in analyzing sun and star light to reveal the elements with which those bodies are composed.

Augustin Jean Fresnel
1788-1827

French engineer, mathematician, and physicist. Though trained as a civil engineer, his interest was in physical optics. He studied the nature of light and became convinced that light was composed of undulations, or waves, rather than emissions, or particles. His mathematical papers arguing for this view convinced many other scientists and made him one of the founders of the wave theory of light. He was awarded a medal by London's Royal Society for his work.

Johann Gottfried Galle
1812-1910

German astronomer who first saw Neptune at the position predicted by Urbain Leverrier (1811-1877) and John Couch Adams (1819-1892). An astronomer at the Berlin Academy of Sciences, Galle was involved in creating a survey of stars in the zodiacal regions of the sky to assist in locating asteroids. He was asked by Leverrier to help locate Neptune, whose position Leverrier had calculated. Using the then unpublished survey, Galle and his assistant, Heinrich d'Arrest (1822-1875), discovered Neptune within the first few minutes of their search.

Thomas Graham
1805-1869

Scottish physical chemist who studied the process of the diffusion of gases and was responsible for the development of colloid chemistry. Graham in 1831 measured the rate of diffusion of gases through a small whole and found that it was inversely proportional to the square root of its molecular weight (Graham's law). In connection with his work on colloids, he was the first to describe the processes of osmosis and dialysis. His investigations of dialysis would lay the foundation for producing artificial kidney machines.

Charles Edouard Guillaume
1861-1938

Swiss physicist who developed a nickel-steel alloy (Invar), whose properties make it ideal for precision instruments and standard measures. Guillaume served as the director of the Bureau of International Weights and Measures, where he contributed toward the standardization of accurate scientific and commercial measurements. He was awarded the 1920 Nobel Prize in physics.

René-Just Haüy
1743-1822

French mineralogist and a founder of the science of crystallography. From 1784-1822 he published successive versions of his mathematical theory of crystal structure, which emphasized geometrical simplicity over experimental data. He also used his theory as the basis for a new classification of minerals, although he ultimately had to concede that some minerals have more than one crystal structure. Haüy held a variety of scientific teaching posts in French institutions, both before and after the Revolution.

John Frederick William Herschel
1792-1871

English scientist who worked in many different fields. Son of the astronomer William Herschel, John became one of the nineteenth century's most accomplished and respected scientists. His interests included astronomy, chemistry, mathematics, botany, geology, light, and photography. From 1833-1838 he made an important expedition to South Africa to observe the southern heavens. He was an esteemed member of Britain's most important scientific societies, and his work influenced many other important nineteenth-century scientists, including Charles Darwin.

Heinrich Rudolf Hertz
1857-1894

German physicist who discovered the electromagnetic radiation known as radio waves in 1888. Having studied with Hermann von Helmholtz, Hertz began experimenting with James Clerk Maxwell's theory of electromagnetic waves at Karlsruhe Polytechnic in Berlin (1885-1889). He verified the theory's prediction that a fourth type of radiant energy (other than visible light, infrared, and ultraviolet) could propagate in space between two objects. He discharged a spark, with its transport reflected by a second weaker spark at a distance, proving the progressive propagation in space of electromagnetic waves—radio waves called Hertzian waves. He went on to measure the length and velocity of electromagnetic waves and, by their characteristics of transverse form, vibration, reflection, refraction, and polarization, proved they were like light and heat energy-which proved that light was indeed an electromagnetic wave.

August Wilhelm von Hofmann
1818-1892

German chemist who discovered several important chemical compounds, founded the aniline dye industry, and developed a method of determining molecular weights. His systematic studies of aniline resulted in its extensive use in dyeing textiles. He discovered formaldehyde and allyl alcohol and developed the Hofman reaction, the standard reaction for converting amides into amines. A major contribution to chemistry was the development of a method of determining molecular weights of liquid organic compounds by measuring vapor densities.

Margaret Lindsay Huggins
1848-1915

Irish-English astronomer whose observations of the spectra of celestial objects, made in collaboration with her husband William Huggins, revolutionized astronomy and laid the groundwork for modern cosmology. The Huggins used spectroscopy to prove that the elements in stars are the same as those in the sun and the earth. They also proved that gases make up nebulae, discovered hydrocarbons in comets, and measured stellar velocities using Doppler shifts of spectra, providing data for theories concerning the origin and expansion of the universe.

William Huggins
1824-1910

English astronomer whose observations of the spectra of celestial objects revolutionized astronomy and laid the groundwork for modern cosmology. Huggins and his wife Margaret used spectroscopy to prove that the elements in stars are the same as those in the sun and the earth. They proved that nebulae consist of gases, discovered hydrocarbons in comets, and measured stellar velocities using the Doppler shift of their spectra, providing data for theories concerning the origin and expansion of the universe.

James Prescott Joule
1818-1889

English physicist who investigated the equivalence of mechanical, electrical, and heat energy, and who is often credited (together with Joseph Mayer) with formulating the law of conservation of energy. The unit of energy in the international system is named after Joule. Joule's family had acquired substantial wealth through a family-owned brewery, leaving him free to devote his time to scientific experiments, generally involving careful measurement of the heat energy released by different physical processes.

Friedrich Kekulé
1829-1896

German organic chemist who contributed to an understanding of the structure of molecules by the use of the concept of valence. Kekulé believed that each element had a fixed number of other atoms that it could combine with. This led to the development of structural formulas showing the way in which atoms are actually bonded to each other. Kekulé's most important contribution was his explanation of the unusual properties of benzene in 1865, in which he proposed that benzene existed as a ring of six carbon atoms with alternating single and double bonds.

Gustav Robert Kirchoff
1824-1887

German-Russian physicist who made fundamental contributions toward the understanding of electricity and light. Kirchoff developed the concept of a "black body," an idealized body that emits light whose color is based solely on temperature. This concept helped lead to the development of quantum theory in the early twentieth century. Kirchoff's work in electrical theory demonstrated that electrical and magnetic forces are related through a constant—the speed of light in vacuum—suggesting that electromagnetism and light are related.

Adolf Kolbe
1818-1884

German organic chemist who made critical contributions to the development of methods to synthesize organic compounds. Among his achievements was the synthesis of acetic acid—an organic compound from totally inorganic compounds—in 1845. This confirmed the concept that organic compounds did not need a vital force supplied by a living tissue in order to be made. In 1859 Kolbe discovered a method to make salicylic acid, a naturally occurring compound found in willow bark that had pain-relieving properties, cheaply and in large scale. A derivative of salicylic acid called aspirin was first produced in 1899 and has been in use ever since.

August Adolph Eduard Kundt
1839-1894

German physicist who performed work in acoustics, particularly the physics of standing waves and associated phenomena. Kundt may be best known for his work on air vibrations and standing waves. In 1866 he invented the "Kundt's tube," a glass tube in which dust is shown to collect at the nodes of standing waves, enabling the measurement of sound velocity in gases and solids. Today, his work is being used in the field of acoustic levitation, where sound waves are used to suspend droplets or small solid objects in ovens for more even heating.

Heinrich Friedrich Emil Lenz
1804-1865

Russian physicist who discovered the relationship between electrical resistance and temperature (also called Joule's Law) and first stated the law describing electrical inductance. Lenz began studying theology at Dorpat University, though shifted to chemistry and physics. He became a professor of physics at the St. Petersburg Academy of Science and was later named Dean of Mathematics and Physics. Lenz's work in electrical theory, especially his work on the effects of temperature on electrical resistance, were important in understanding electrical phenomena.

Sir Joseph Norman Lockyer
1836-1920

English astronomer who, in 1868, discovered helium, which he found in the sun's atmosphere; its existence on earth was not known until 27 years later. Lockyer began his spectroscopic observation of sunspots in 1866, and discovered that solar prominences are clouds of gas resulting from upheavals in the outer layer of the sun. In 1868 he developed a method of spectroscopic study of solar prominences that did not require the aid of an eclipse to block out the intense light of the sun.

William Edmond Logan
1798-1875

Canadian geologist and first director of the Geological Survey of Canada, 1842-69. Logan's geological career began in Wales, where he made detailed maps and established the origin of coal deposits. Once in Canada he personally made extensive field investigations of the colony's Precambrian and Paleozoic rocks, and he built the survey into an institutional success. After his exhibit of Canadian mineral resources in Paris in 1855, Logan was knighted and received the Geological Society of London's Wollaston Medal.

Ernst Mach
1838-1916

Austrian physicist associated with the speed of sound and its multiples (known as Mach 1, Mach 2, Mach 3, etc., in his honor). Mach showed that the characteristics of airflow over a moving object change dramatically as the object approaches the speed of sound. Mach's biggest impact was in the philosophy of science. He provided a clear distinction between physics and psychology and brought rigor to both disciplines. He upheld the principle, a tenet of scientific positivism, that no statement should be accepted by science until it has been verified by experimentation.

Etienne Louis Malus
1775-1812

French physicist who spent his brief career as an engineer in the French Army. While investigating the phenomenon of double refraction, the splitting of a light beam into two by certain crystals, he discovered that a beam with similar characteristics to each refracted beam is produced by reflection of sunlight from any smooth surface at a unique angle. He described the beams thus produced as "polarized," initiating a period of intense investigation of the properties of polarized light.

Julius Robert Mayer
1814-1878

German physiologist and physicist often credited with the first formulation of the principle of energy conservation, specifically the inter-conversion of electrical and magnetic energy, mechanical energy, and heat. Mayer's interest in the subject apparently began in 1840, when, serving as a ship's surgeon, he observed that the venous blood of sailors in the tropics was a brighter shade of red than in the northern latitudes, indicating less need to generate body heat through the oxidation of food.

William Hollowes Miller
1801-1880

British mineralogist who in 1839 developed a method using spherical trigonometry to describe how crystal faces were orientated about a crystal. A professor of mineralogy at Cambridge University, Miller used numbers derived from the intercepts of the crystal faces on a mineral's crystallographic axes to characterize the position of any crystal face. These numbers, known as Miller indices, define the position and orientation of a crystal face.

John Milne
1850-1913

English geologist who invented the first modern seismograph in 1880. Born and educated in England, Milne lived for 20 years in Japan where he invented the seismograph for measuring the arrival time and magnitude of waves traveling through the ground from earthquakes. Seismographs today are an indispensable tool used to better understand not only earthquakes, but the structure of the entire earth, as they show how seismic waves are changed by the various materials they pass through in the earth's interior.

Maria Mitchell
1818-1889

The first female astronomer in America. She grew up helping her father make astronomical observations for New England whaling ships. In 1847 she discovered a new comet. From 1849-1868 she worked at the U.S. Nautical Almanac Office to calculate the ephemerides of Venus before resigning to devote all of her time to her job as professor of astronomy and director of the college observatory at Vassar Female College. She was the first woman to be elected to the American Academy of Arts and Sciences.

Eilhard Mitscherlich
1794-1863

German chemist who made major advances in chemical crystallography, discovered a number of important chemical compounds, and first recognized catalysis. He discovered that chemicals with similar structure and activity may have similar crystalline forms (isomorphism) and that a chemical may have different crystalline forms (polymorphism). He discovered selenic acid, permanganic acid, and nitrobenzene and named benzene. Mitscherlich was the first to recognize what he called "contact action" (catalysis) and was first to recognize yeast as a microorganism.

Friedrich Mohs
1773-1839

German geologist best known for developing the Mohs Hardness Scale. Mohs presented his system for classifying minerals in *Outline of Mineralogy*, published in 1822. This same volume contained a description of his hardness scale, used to compare the hardness of various minerals as a way to help identify them. Still in use today, the Mohs scale ranges from hardness of 1 (talc) through 10 (diamond).

Edward W. Morley
1838-1923

American chemist who in 1887 performed, in collaboration with Albert Michelson, the crucial experiments showing the medium called ether does not exist. The ether was thought to be a massless substance that pervaded all of space, even a vacuum. The assumed presence of ether provided a way to rationalize how electrical and magnetic phenomena could act at a distance and to explain the propagation of light. Michelson proposed that, if there is an ether, it could be detected by using two beams of light; one would travel with the ether and the other across it. There should be a difference in time for these two beams. The lack of any difference made this one of the most famous negative results in history. The Michelson-Morley experiment played a role in Albert Einstein's development of his theory of special relativity in 1905.

Sir Roderick Impey Murchison
1792-1871

Scottish geologist who first described rocks of the Silurian system in England and conducted geologic mapping of Russia. One of the most distinguished geologists of the nineteenth century, Murchison was important in establishing the Permian and Devonian systems, and helped to put geologic time and events in the history of life on earth into a more understandable chronological and theoretical framework. He was named the director of the Royal School of Mines and director-general of the Geological Survey.

William Nicol
1768-1851

British geologist best known for inventing the Nicol prism for studying rocks and minerals under polarized light. The Nicol prism, still in widespread use today, made possible the use of petrographic microscopes for the study of sections of rock ground to near-transparency. This opened the door for the identification of small mineral grains and textures in rocks. Such knowledge gave new insights into the manner in which rocks, minerals, and large geologic features formed.

Alfred Bernhard Nobel
1833-1896

Swedish chemist who invented dynamite and established the Nobel Prize trust. After studying mechanical engineering in Russia and the United States, Nobel was employed at his family's factory in St. Petersburg, where he manufactured military devices. Nobel turned his attention to the development of safe packaging for dangerous chemicals (his brother was killed in an 1864 explosion), and in 1867 successfully cushioned nitroglycerin in organic material to produce dynamite. He also developed ballistite, a smokeless gunpowder. Nobel amassed considerable wealth as a manufacturer of explosives and, upon his death, bequeathed his fortune to the prestigious award foundation that bears his name.

Georg Simon Ohm
1787-1854

German physicist who quantified the current flow and the potential difference (or voltage) in an electric conductor and found these to be directly proportional to each other. Most conductors obey this empirical law that is named after him. Ohm identified the constant of proportionality in this law as a property of the conductor alone, known as the electric resistance, whose units are measured in Ohms.

Heinrich Wilhelm Matthias Olbers
1758-1840

German astronomer and physician who developed a method for calculating cometary orbits. He suggested a theory for why the tails of comets point away from the sun and discovered the second and third known asteroids—Pallas (1802) and Vesta (1807). Astronomers at the time assumed an infinite universe in which the night sky should be covered with stars, and his formulation of why, in that case, the night sky is dark became known as Olbers' Paradox.

William Perkin
1838-1907

English organic chemist who produced the first synthetic dyestuff in 1856 and founded the synthetic organic chemical industry. In attempting to produce quinine from aniline by oxidation, Perkin found that he produced a substance that when dissolved in alcohol produced a brilliant purple color and was able to dye many fabrics. By modifying the structure of aniline, Perkin and others were able to produce a whole range of colors. In 1874 Perkin retired from the chemical business to devote himself to basic scientific research. One of the products of this research was the synthesis of a substance named coumarin that had a vanilla-like smell. This became the basis for the beginning of the synthetic perfume industry.

Julius Plücker
1801-1868

German mathematician and physicist who discovered cathode rays and the idea of spectrum analysis. Plücker's work with cathode rays led to the discovery of x rays by Wilhelm Röentgen in 1895. Today, cathode rays are used in most television sets and computer monitors (also called cathode ray tubes, or CRTs). Spectrum analysis is an important laboratory research tool, widely used by astronomers to classify stars and to determine their chemical composition.

Jean-Louis Pons
1761-1831

French astronomer who discovered 37 comets during his life. Pons had only a basic education but, as a nonscientific assistant at the Observatory of Marseilles, received instruction from professional astronomers and developed skills that made him a respected astronomer in his own right. He discovered at least one comet yearly from 1801 until 1827, when he was forced to retire due to failing vision. Pons was honored with scientific awards from both Britain and France during his career.

Wilhelm Conrad Röntgen
1845-1923

German physicist who discovered x rays in 1895. Although other scientists had probably unknowingly observed x rays previously, Röntgen was the first to notice that the glass wall of his cathode-ray tube was emitting rays. He discovered the penetrating properties of x rays and took a number of photographs, including one of his wife's hand, showing for the first time the living skeleton. News of his discovery spread rapidly, and x rays were soon put to use in medicine, metallurgy, and physics. See long biography on p. 369.

Henry Augustus Rowland
1848-1901

American physicist who designed precise scientific instruments. A civil engineering graduate of Rensselaer Polytechnic Institute, Rowland researched the magnetic behavior of metal. As Johns Hopkins University's first physics professor, he demonstrated that electric charges in motion display magnetism. He refined values for units of electrical resistance and mechanical heat. The ruling engines and concave diffraction gratings he invented enabled scientists to assess an object's physical characteristics more accurately. Rowland won awards for the solar spectrum maps his apparatuses made possible.

Christian Friedrich Schonbein
1799-1868

German geochemist who discovered ozone in 1839. Ozone is formed in the presence of electrical current and can sometimes be smelled after an electrical short circuit or a nearby lightning strike. It is also formed by reactions using ultraviolet light in the upper atmosphere, where it helps to protect the earth's surface from harmful ultraviolet light. However, at the time Schonbein discovered ozone, it was regarded as only a laboratory curiosity.

George Julius Duncombe Poulett Scrope
1797-1876

English geologist who first realized that volcanoes build up in time as rock and ash are ejected from a volcanic vent, not by catastrophic uplifts. Based on his observations of volcanic rocks in Auvergne, France, he became an advocate of the enormity of geologic time. He also recognized that earthquakes precede volcanic activity, that water and heat in lava result in fluidity, and that volatiles are an important component in lava.

Pietro Angelo Secchi
1818-1878

Italian priest and astronomer who created one of the earliest systems for classifying stars. Secchi classified over 4,000 stars by their spectra. Secchi also helped develop photography as an astronomical tool. He studied solar prominences and observed markings on Mars that he interpreted as canals, sparking the interest of other astronomers. In 1890 he correctly identified dark spaces in photographs of star fields as clouds of dark interstellar matter.

Reverend Adam Sedgwick
1785-1873

British geologist who first established the existence of the Cambrian period of geologic time. One of Britain's elite geologists, the Reverend Sedgewick was the first to define the Cambrian period using fossil evidence unique to that period of time. He was also heavily involved in "The Great Devonian Controversy" that involved nearly all of England's geologists for over a decade as they struggled to understand the fossils and rocks from southern England.

Thomas Johann Seebeck
1770-1831

Russian-German physicist and chemist who first described the thermocouple. A thermocouple consists of two wires of different metals that are joined to form a closed loop. An electric current flows through the loop when the junctions of the wires are kept at different temperatures. This phenomenon, in which heat energy is converted to electrical energy in a closed circuit of different metals, is called the Seebeck effect. Today, thermocouples are used in some types of highly accurate thermometers.

Benjamin Silliman
1779-1864

American chemist and geologist who founded the *American Journal of Science and Arts* in 1818. Silliman is remembered through the mineral sillimanite, an aluminosilicate mineral found in rocks exposed to high temperatures.

Ernest Solvay
1839-1922

Belgian industrial chemist who invented an efficient low cost process to manufacture sodium carbonate (washing soda). Sodium carbonate is used in making ceramics, paper, as a cleaner, to soften water, and to make other sodium containing compounds. In the Solvay process ammonia and carbon dioxide are pumped through a saturated solution of sodium chloride (brine). A reaction occurs that produces sodium hydrogen carbonate, which precipitates out. Heating the sodium hydrogen carbonate converts it to sodium carbonate. The Solvay process displaced the LeBlanc process, which was energy intensive and produced massive amounts of pollution from the byproducts. All the products of the Solvay process are used or can be recycled.

Josef Stefan
1835-1893

Austrian physicist who, with his lab assistant Ludwig Boltzmann (1844-1906), showed the relationship between the temperature of a body and the amount of radiation it emitted. The Stefan-Boltzmann law, established in 1879, also describes the primary color emitted by a hot body. This law is used today by astronomers to determine the temperature of stars, the interstellar medium, and gases swirling into black holes based solely on the wavelength of light emitted by these objects.

George Gabriel Stokes
1819-1903

Irish physicist who developed Stokes' Law, which describes the motion of a sphere in a viscous liquid. Stokes worked with and described other problems concerning fluid flow, including the motions of and forces acting upon small bodies traveling through air and fluids such as water. Stokes's work helped set the stage for the field of fluid dynamics. Some of his equations are still in used in aircraft and ship design.

John William Strutt
1842-1919

English physicist who inherited the title of Baron Rayleigh at age 31 and is almost always referred to as Lord Rayleigh in scientific works. Despite appreciable wealth and social standing, Strutt worked steadily, doing research in every area of physics. He published 430 scientific papers and *The Theory of Sound*, a highly influential two-volume treatise on sound. He succeeded James Clerk Maxwell as Cavendish Professor of Physics at Cambridge and received the Nobel Prize in 1904 for the discovery of argon.

Eduard Suess
1831-1914

Austrian professor of geology at the University of Vienna who laid the foundation for the field of structural geology. Suess suggested that moving landmasses cause earthquakes and that sea level had changed through time. He recognized similarities between fossils from South America, Africa, Australia, and India and proposed that these areas were once connected as a large supercontinent he named Gondwana. Later geologists recognized that Gondwana included Antarctica.

John Tyndall
1820-1893

Irish physicist whose research described the scattering of light by particles. Tyndall's work showed why it is possible to view a beam of light passing through a cloudy liquid from the side. The Tyndall effect, as it is known, is used in water quality analyses and in looking for small (dust-sized) particles in space. Tyndall also helped to popularize the work of James Maxwell (1831-1879), who discovered that heat is related to the random motions of molecules.

Ignatz Venetz
1788-1859

Swiss civil engineer who demonstrated that a great ice sheet had once extended from the Central Alps across the Swiss plain to the Jura Mountains. He reasoned that alpine glaciers and large erratic boulders are remnants of this ice sheet and that moraines, piles of sediment deposited by glaciers, mark glacial advances and retreats. In 1836 he and Johann de Charpentier showed their evidence to Louis Agassiz, who further developed the idea of continental glaciation.

Hermann Karl Vogel
1841-1907

German astronomer who made pioneer spectroscopic studies of stars and refined an earlier star classification system. Working at the Potsdam Astrophysical Observatory, Vogel cataloged more than 4,000 stars and added sub-divisions to Pietro Secchi's classification scheme. He used spectroscopy to measure how fast the sun rotates and how fast some stars are moving away from Earth. He identified two stars as double stars on the basis of their spectra and calculated the sizes of both pairs of stars.

Alessandro Volta
1745-1827

Italian physicist who invented the electric storage battery. Volta's electrophorus replaced the Leiden jar for generating and storing electricity and led to modern electrical condensers. He showed that the galvanic effect (first noticed in the twitching of frogs' legs) could be generated with non-living objects, leading to the development of the storage battery. Volta's concepts of electrical tension and capacity are still in use today. He was memorialized by the term volt, denoting the unit of electromotive force.

John James Waterston
1811-1883

English physicist and astronomer best known for his research into the relationship between molecules and temperature. Overlooked for nearly 50 years, Waterston's research provided crucial insights that helped to define temperature in terms of average kinetic energy of molecules in a substance. This "kinetic" theory of heat helped displace the competing "caloric" theory and also showed that heat and light belonged in the same family of phenomena.

Wilhelm Eduard Weber
1804-1891

German physicist who invented the electrodynamometer and other devices for measuring electrical and magnetic effects. Weber directed the Göttingen astronomical observatory and was a professor of physics at Leipzig and Göttingen University. He conducted important research in electricity and magnetism with Johann Gauss, developed many useful devices for experimental physics, and wrote an important book on waves with his brother, Ernst Weber, a well-known physiologist and professor of anatomy.

Sir Charles Wheatstone
1802-1875

English physicist who invented the microphone, popularized (though did not invent) the Wheatstone bridge for measuring electrical resistance, and conducted important research in acoustics. Wheatstone also invented the concertina (a small accordion-like instrument played with bellows and keys), a form of telegraph, and was the first to explain the principle of the stereoscope. He was a professor of experimental philosophy in London and was knighted in 1868 for his contributions to physics.

Friedrich Wöhler
1800-1882

German chemist responsible for proving that organic compounds do not need a vital force supplied by a living tissue in order to be made. In 1828 Wöhler heated the inorganic compound ammonium cyanate and found that it was converted into urea. Urea is a nitrogen-containing compound that is produced by the metabolism of proteins in mammals and found in the urine. Wöhler proved that metabolic processes are chemical in nature by showing that benzoic acid was converted to hippuric acid. Another significant contribution was the discovery of calcium carbide, which reacts with water to produce acetylene. See long biography on p. 180.

William Hyde Wollaston
1766-1828

English scientist who made fundamental contributions to chemistry, physics, astronomy, biology, and medicine. Though trained as a physician, Wollaston's interests encompassed a wide variety of scientific fields. He discovered the elements palladium (1804) and rhodium (1805), and a technique to create malleable platinum. He also developed the industrial process for purifying transition metals, introduced the concept of equivalent weights in chemistry, pioneered the study of crystalline structure, observed the black lines in solar spectra, and invented the camera lucida.

Bibliography of Primary Sources

~

Books

Argelander, Friedrich and Eduard Schönfeld. *Bonner Durchmusterung.* 1859-62. This influential star catalog by the two German astronomers noted the position and brightness of nearly 325,000 individual stars, a monumental feat and one that kept the catalog useful well into the twentieth century.

Bessel, Friedrich. *Fundamenta astronomiae.* 1818. This catalog of more than 3,000 stars was one of the most important star catalogs produced in the nineteenth century and has been said to mark the beginning of modern astronomy.

Chaptal, Jean-Antoine-Claude. *Chemistry Applied to the Arts.* 1807. This was the first book on industrial chemistry.

Dalton, John. *New System of Chemical Philosophy.* Part I: 1808; Part II: 1810. This work grew out of lectures given by Dalton at the universities of Edinburgh and Glasgow in Scotland. In the work Dalton presented his pioneering atomic theory; interestingly, the atomic theory itself occupied only five out of the 916 pages of Part I.

Dreyer, John. *New General Catalog.* 1888. Dreyer's 1888 catalog noted more than 13,000 nebulae.

Faraday, Michael. *Experimental Researches in Electricity.* 3 vols. 1839-55. *Experimental Researches in Chemistry and Physics.* 1859. These two works summed up much of Faraday's pioneering efforts in the field of electricity and magnetism. They are often quoted by historians and provide a glimpse at what it was like to do science in the 1800s.

Herschel, John. *General Catalog.* 1864. This star catalog contained positions for 5,000 nebulae.

Lyell, Charles. *Principles of Geology.* 1830-33. In his landmark work Lyell proposed that the Earth was several hundred million years old, not a few thousand years old as most people had long believed. Lyell's breakthrough opened the door for many subsequent advancements, including the entire structure of modern geology, the theory of evolution, and concept of deep time in many branches of science.

Mendeleyev, Dmitri. *Principles of Chemistry* (English translation). 1868-1870. A textbook about inorganic chemistry written by the famed Russian chemist.

Smith, William. *A Map of the Strata of England and Wales, with Part of Scotland.* 1815. Here Smith provided the first detailed geologic survey of England. This eight-foot-by-six-foot geologic map surpassed all others that had been made before, not only in size but also in area covered and stratigraphic detail, and it was particularly useful in the planning for future canals, quarries, mines, and as a guide to soil types. Also, by examining Smith's geologic map, it was now possible to predict throughout a large part of England what type of rock would be found near the surface, and most importantly, the successive layers of rock that lay beneath the surface of Earth.

Periodical Articles

Arrhenius, Svante. "On the Dissociation of Substances Dissolved in Water." *Zeitschrift fur physikalische Chemie* I (1887): 631. Translated by H.C. Jones. A classic paper concerning electrolyte solutions, acidity, and the temperature dependence of rate constants.

Arrhenius, Svante. "On the Influence of Carbonic Acid in the Air upon the Temperature of the Ground." *Philosophical Magazine* 41 (1896): 237. A seminal paper on the greenhouse effect.

Curie, Pierre and Marie Sklodowska Curie. "On a New Radioactive Substance Contained in Pitchblende. Note by M. P. Curie and Mme. S. Curie, presented by M. Becquerel." *Comptes Rendus* 127 (1898): 175-8. Translated and reprinted in *The World of the Atom,* edited by Henry A. Boorse and Lloyd Motz. Vol. 1 (New York: Basic Books, 1966). This, the first of two landmark 1898 papers, announced the Curies' discovery of radium.

Curie, Pierre and Marie Sklodowska Curie. "On a New, Strongly Radioactive Substance, Contained in Pitchblende. Note by M. P. Curie and Mme. S. Curie, presented by M. Becquerel." *Comptes Rendus* 127 (1898): 1215-7. Translated and reprinted in *The World of the Atom,* edited by Henry A. Boorse and Lloyd Motz. Vol. 1 (New York: Basic Books, 1966). In their second classic 1898 paper the Curies announced the discovery of polonium.

Davy, Humphry. "On Some New Phenomena of Chemical Changes Produced by Electricity, Particularly the Decomposition of Fixed Alkalies, and the Exhibition of the New Substances which Constitute their Bases: and on the General Nature of Alkaline Bodies." *Philosophical Transactions of the Royal Society.* 1808. Davy here discussed the isolation of the alkali metals sodium and potassium.

Fresnel, Augustin J. "Mémoire sur la diffraction de la lumière." *Annales de Chimie et de Physique* 11 (1819): 246-96 and 337-78. The wave theory of light received important credibility with this article by Fresnel.

Gay-Lussac, Joseph Louis. "Memoir on the Combination of Gaseous Substances with Each Other" (English translation). 1809. In this article Gay-Lussac presented his law of combining gases.

Mendeleyev, Dmitri. "On the Relationship of the Properties of the Elements to Their Atomic Weights." *Zeitschrift für Chemi* (1869): 405. This article accompanied Mendeleyev's original periodic table of the chemical elements, the forerunner to the table still used today.

Mendeleyev, Dmitri. "The Periodic Law of the Chemical Elements." *Liebigs Annalen* 8 (1872): 133-39. This article contained Mendeleyev's second table of the elements.

Newlands, John A. R. "On Relations among the Equivalents." *Chemical News* 7 (7 February 1863): 70-72. first attempt to find relationships among elemental atomic weights and to determine equivalents or families of elements; he also set forth his law of octaves.

Ramsay, William. "An Undiscovered Gas." *Nature* 56 (1897): 378. Reprinted in *Classic Papers in Chemistry,* edited by David Knight. Second series. (New York: American Elsevier, 1970). In this article (from a speech) Ramsay expanded the periodic table to allow for noble gases and predicted the discovery and properties of the noble gas Neon (Ne).

Thomson, J.J. "Cathode Rays." *Philosophical Magazine* 44 (1897): 293. In this classic 1897 paper Thomson announced the discovery of the electron.

Volta, Alessandro. "On the Electricity Excited by the Mere Contact of Conducting Substances of Different Kinds." *Philosophical Magazine* (September 1800). In this paper Volta discussed his demonstration of the first electric battery.

Young, Thomas. "On the Theory of Light and Colours." *Philosophical Transactions* 92 (1802): 12-48. Young here described his wave theory of light.

Technology and Invention

Chronology

1803 First steam locomotive is built by British engineer Richard Trevithick.

1831 Cyrus Hall McCormick, an American, demonstrates his mechanical reaping machine.

1837 French artist Jacques Louis Mandé Daguerre makes the first photograph, or daguerreotype—a still life taken in his studio.

1844 Samuel Morse successfully transmits the first Morse code message over a telegraph circuit between Baltimore and Washington: "What hath God wrought?"

1844 Charles Goodyear of the United States patents vulcanization, a process for making rubber tough and flexible in all temperatures.

1846 Building on the work of French inventor Barthélemy Thimonnier and American Isaac Singer, Elias Howe of the United States develops the first practical sewing machine.

1856 English metallurgist Henry Bessemer develops what becomes known as the Bessemer process, a quick, inexpensive way of making steel.

1875 Alexander Graham Bell first transmits sound over electric cable; in the following year, he demonstrates his new telephone.

1879 Thomas Edison produces the first practical incandescent lightbulb.

1881 With his photographic gun for quickly recording a series of pictures, French physiologist Etienne-Jules Marey produces what may be considered the first motion-picture camera.

1883 Nikola Tesla, a Serbian-American electrical engineer, introduces the use of alternating current (AC), a highly efficient alternative to existing direct current (DC).

1885 German engineer Karl Friedrich Benz builds what may be considered the first true automobile, a vehicle in which engine and chassis form a single unit.

Overview:
Technology and Invention 1800-1899

What we think of as the modern world was born not in the twentieth century but in the technological innovations of the nineteenth century. Almost everything that is quintessentially modern—rapid transportation and communication, entrepreneurship and the market economy, crowded urban centers, the primacy of the individual, the mastery of nature—began in the nineteenth century. Even that icon of modernity, the computer, grew out of two nineteenth-century devices, the punch card system of Joseph Jacquard (1752-1834) and the analytical engine of Charles Babbage (1791-1871).

Copernicus (1473-1543) can be considered the ancestor of nineteenth-century technology because, with his contention that Earth revolves around the Sun, he founded the scientific revolution. His theory was revolutionary because it engendered a new way of thinking: the belief that humans could master nature through careful observation, systematic experimentation, and determined problem solving. The relationship between scientific inquiry and technological advances that the scientific revolution produced in the nineteenth century was circular. Sometimes technology led the way, with science unable to explain why an invention worked until many years later. For example, the invention of canning preceded an understanding of bacteriology. Other times science took the lead. For example, an understanding of electromagnetism preceded the invention the telephone. What science and technology shared was the desire to address the problems of everyday life and the confidence that those problems could be solved.

"It's a small world!" is a common exclamation these days, and rapid transportation and communication—two industries born in the nineteenth century—are what made the world shrink. The first locomotive powered by high-pressure steam was built in England in 1803 by Richard Trevithick (1771-1833), followed in 1825 by the first commercial railroad. By 1869 a railroad track joined the United States from east to west, shortening the transcontinental trek from months to days. Steamboats, with their capacity for carrying cannons and maneuvering through narrow rivers, allowed England to police its enormous empire. The internal combustion engine, invented by Etienne Lenoir

(1822-1900) in 1859 and improved by Nikolaus August Otto (1832-1891) in 1877, shortened distances even more by providing a source of power for cars and airplanes.

The communication industry saw comparable leaps during the nineteenth century. The telegraph, invented in 1835 by Samuel F. B. Morse (1791-1872), made it possible to send messages across wires sunk under oceans and strung across continents. Another form of signal transmission, the telephone, changed from a curiosity at the 1876 Centennial Exposition to a common sight in homes and businesses by the end of the century. Fast and cheap written communication was fostered by three advances in publishing: automated paper production, the steam-powered printing press, and the Linotype machine.

An economy characterized by entrepreneurship, the stock market, and international trade seems uniquely twentieth century, but in fact, these three features also characterized the nineteenth century. The century's first inventors, self-taught tinkerers such as Nicolas Appert (1749-1841), Oliver Evans (1755-1819), and Joseph Nicéphore Niépce (1765-1833), aggressively pursued patents and profits. Some, such as Isaac Singer (1811-1875) and Alexander Graham Bell (1847-1922), amassed fortunes from their inventions. Others, notably Charles Goodyear (1800-1860), died in debt. But they all shared the drive, determination, and single-mindedness that distinguish contemporary entrepreneurs. Because the cost of innovations like railroads, subways, and telegraph lines was enormous, no single person could afford to build them. A formalized stock market was created in response to this need. And as industrial output expanded during the nineteenth century, fresh markets had to be opened. Britain, the world's first industrialized nation, exported its manufactured goods to China, Australia, South America, and Canada and imported agricultural products to feed its industrial workers.

The modern world is an urban world. The migration from country to city began with the Industrial Revolution, which took farmers from their fields and put them to work in factories. This transition would have been impossible without the nineteenth-century technologies that made it possible to feed and house this new category of worker. Refrigeration and railroads kept workers supplied with fresh, nutritious food at the same time that subways, bridges, artificial lighting, and high-rise buildings made cities more livable and efficient. Railroads also opened new land for cultivation. No longer constrained to ship their goods to nearby market via rivers, farmers spread out into the vast expanses of the American West, Australia, and Argentina, where they took advantage of tools like Cyrus McCormick's (1809-1884) reaper to increase agricultural productivity.

An often-criticized feature of modern life is its narcissistic preoccupation with the self. The origin of this preoccupation can be found in Romanticism, a movement that flourished between the late 1700s and the mid-1800s. Romanticism glorified the individual, self-expression, and the imagination. The ability to observe oneself as a unique individual, to compare oneself to others, and to interpret one's own conception of the world was extended by several nineteenth-century inventions. The first was photography. By the 1860s photography was used both to depict idealized beauty and to document the horrors of war. It gave significance and permanence to the people and events it captured and offered a new outlet for the imagination. In the final decade of the century, motion pictures became another medium for creativity, while phonographs allowed people to listen to their favorite music in their own homes.

At the end of the twentieth century people are questioning the place of technology in their lives. Is it a useful tool or is it a threat to nature and the divine order of the universe? This same dilemma preoccupied people in the nineteenth century. Many objected to refrigeration on the grounds that God alone has the power to produce cold, and they warned that those who ate previously frozen food would pay with shortened lives. Telephones, electric lights, subways,

and suspension bridges were greeted with the same suspicion. Directions for turning on electric lights included the reassuring message, "The use of Electricity for lighting is in no way harmful to health, nor does it affect the soundness of sleep." Morse's first telegraph message, "What hath God wrought?" emphasized his reluctance to award himself first place as his invention's creator. Nevertheless, the technological advances of the nineteenth century were predicated on the belief that nature could be remade to suit human needs, and so it was. Nitroglycerin blasted a path for the railroad through the Sierra Nevada mountains, artificial lighting freed people from the limits of daylight, canning provided spring vegetables in the middle of winter, telephones brought distant voices into the same room, and machine guns ended hundreds of lives in a matter of minutes.

To describe the technological innovations of the nineteenth century, it is necessary to treat each one separately. In reality, however, there was no such separation, as can be seen in the case of steel. Without the large quantities of high-quality steel made possible by Henry Bessemer (1813-1898), there would have been neither railroad cars nor the rails they moved on. And the rapidly expanding cities that the railroads supplied with food and manufactured goods were equally dependent on steel for their bridges and subways. Technologies were also interdependent in a second way: development in one area inevitably spurred development in others. Charles Goodyear's vulcanized rubber did not find important markets until the invention of two unrelated technologies: electric lighting, with its need for insulated cables and wiring, and the automobile, with its need for tires. These connections weave across all the technologies discussed in this chapter, ultimately creating the fabric that is called modern life.

LINDSAY EVANS

French Inventor Jacquard Produces a Weaving Loom Controlled by Punch Cards (1801), Facilitating the Mechanized Mass Production of Textiles; the Punch Card System Also Influences Early Computers in the 1940s and 1950s

Overview

In 1801 Joseph-Marie Jacquard (1752-1834), a French weaver distressed by the poor working conditions plaguing laborers, revolutionized the weaving process with his invention of the Jacquard loom. A highly influential innovation in textile technology, his automated loom used punch cards to control intricate weaving of patterns and fabrics with more efficiency than human hands. Jacquard's idea of programming an automated machine using punch cards was used to design the first automated calculators and was eventually part of the earliest computers developed. His invention not only impacted the industrial revolution of his time, but also the twentieth century technological one.

Background

Late-eighteenth century Lyons, where Jacquard lived and worked, was well known as a silk weaving city. Working conditions, however, were less than appealing for the silk weavers. Laborers slaved from early in the morning to late at night, in crowded rooms, whole families often toiling over a single loom, their bodies pale and thin from lack of sunlight and proper nutrition, and their eyes expressionless from the monotony of their work. There were no holidays and no home life—little children grew pinched and old, and parents went too early to their graves. Joseph-Marie Jacquard became absorbed with thinking how he could ease the plight of the silk weavers by some kind of invention.

Others before Jacquard had attempted improvements to the design of the draw-looms used by silk weavers. To avoid the cost of employing a weaver's assistant, as well as the possible errors of said unskilled laborer, inventors endeavored to design loom modifications that would automatically perform the assistant's role. The weaver's assistant, known as a draw-boy, followed a squared and colored chart to control the figure-harness, an apparatus that allowed design elements to be woven in a fabric

Joseph-Marie Jacquard's 1801 loom used punched cards to control the lifting of certain warp threads. *(UPI/ Corbis-Bettmann. Reproduced by permission.)*

by pulling cords of silk in a particular sequence. The weaver then produced a pattern by throwing a shuttle through the shed, or passage, created by the movements of the figure-harness.

In 1725 Basile Bouchon devised a mechanism that could automatically select the cords to be drawn. His innovation incorporated a roll of paper perforated in rows according to the weaving pattern that passed round a perforated cylinder. Before each shuttle throw of the weaver, the cylinder was rotated to the next row of perforations. The presence or absence of perforations directed the movements of the loom's needles and silk cords. Three years later, a master silk-weaver named Falcon improved the mechanism by increasing the number of needles to multiple rows and by eliminating the paper roll in favor of a series of perforated rectangular cards strung together. Bouchon and Falcon's designs both superceded the slow and laborious process of setting up a pattern on the loom, a process that took a skilled worker up to two weeks and had to be repeated whenever a major change in pattern was desired. In 1775 Jacques de Vaucanson (1709-1782), an inventor famous for mechanical marvels, designed a new loom that improved on the ideas of Bouchon and Falcon. While impractical and never adopted due to a complicated sliding cylinder around which perforated program cards would be passed, Vaucanson's design featured a needles-and-hooks system that would eventually be perfected in Jacquard's solution.

The unfinished task of a completely mechanized loom was tackled by Jacquard with unparalleled enthusiasm. After each day's work as a laborer, he would spend hours designing improvements for his silk weaving machine. With a foundation in the designs of Bouchon, Falcon, and Vaucanson, Jacquard's design relied on a series of connected punch cards revolving on a prism, which allowed a complex pattern to be woven without any adjustments by an operator. The presence or absence of holes in each hardwood card controlled the weaving process and ensured that an identical pattern could be woven over and over, regardless of a weaver's skill or experience. Jacquard also devised a way to automate the shuttle process through an innovative hook and wire operation that relied on springs and steel rods to "read" the punched cards of his loom design.

In 1801 the Jacquard loom made its first public appearance at the Paris Industrial Exposition, and its inventor was awarded a bronze medal. Jacquard's invention attracted the interest of French leader Napoleon Bonaparte, who gave him a pension for his discovery on the condition that the patents revert to the city of Lyons. When Jacquard returned to Lyons to set up his looms, displaced laborers rebelled, fearing that their jobs would be eliminated with the gains made possible by Jacquard's loom. To quell the riots, city government set an example of Jacquard and his invention by ordering the destruction of one of his new looms on the public square. Lyons eventually forgave the inventor—especially in light of the prosperity the Jacquard loom brought to the city. Before Jacquard's death in 1834, several thousand looms from his design were in operation in Lyons, and the city awarded him a silver medal as well as the grand distinction of the Cross of the Legion of Honor. Six years after Jacquard's death, Lyons unveiled a monument to the inventor on the exact spot in the public square where his loom had been destroyed.

Impact

The mechanized mass production of weaving made possible by Jacquard's invention represented considerable financial savings for both producers and consumers. The Jacquard loom became an example of the benefits of using automation in a manufacturing environment. In addition, the innovation of Jacquard's loom—an automated machine controlled by holes punched in hardwood cards—became the inspiration for a series of inventions that eventually led to the development of the earliest computers of the twentieth century. His method of coding punch cards to program a machine (as Jacquard's machine demonstrated) was the starting point for modern computer technology.

Two inventions that never quite made it past the design stage were the work of British mathematician and inventor Charles Babbage (1791-1871) and his coresearcher Augusta Ada Byron, Lady Lovelace (1815-1852), daughter of the British poet Lord Byron. Babbage and his associate carried out a lifelong quest for a programmable machine after being inspired by the punched cards found in Jacquard's loom. As part of this quest, two significant inventions were conceived—the difference engine, capable of handling complicated mathematical problems, and the analytic engine (sometimes referred to as the analytical engine), a precursor of the modern computer. The difference engine was the researchers' first attempt at an automated computing machine. Inspired in the 1820s by Babbage's discovery of numerous errors while checking the navigation tables of the Royal Astronomical Society, the engine was an attempt to design a mechanical device that could automatically calculate and produce tables without the calculating or transcription errors of humans. After ten years of work, Babbage and Lovelace abandoned the project in favor of their analytic engine, the design of which resembles a modern computer.

While the difference engine could only be partially constructed (what remains of the engine is on display in London's Science Museum), it was the analytic engine that obsessed Babbage for the last 20 years of his life. Babbage's description of the components and function of the analytic engine marks his work as extremely forward thinking, displaying many features of the modern digital computer. The design called for a "store" for saving data (like a modern hard drive), an "input" stream of a deck of punched cards (like a modern disk), a "mill" for arithmetic operations (like a modern processor), and a printer to make a permanent record. Were it not for the limitations of technology in the mid-1800s (especially with regard to electricity and electronic devices), the analytic engine might have been built. Nevertheless, as a theoretical concept, the idea of the analytic engine and its logical design are of enormous significance in light of their impact on the development of modern computers.

In the late 1880s, nearly a century after the introduction of Jacquard's loom, his punched cards became a valuable element of the information age. The cards represented a basic binary code (i.e., hole vs. no hole corresponding to a binary notation of 0 or 1) that machines could easily comprehend. They could store complicated information that could be used over and over again without human interaction. Through the use of plugs and new punches, the cards could be edited and their data changed. They could store this data over a long time. Lastly, the cards could easily be copied so that many machines could reliably perform the exact same operations. American statistician and U.S. Census employee Herman Hollerith (1860-1929) developed the first statistical machine operating on these same principles. By the 1880s the collection, preparation, and tabulation of the enormous quantity of information generated by a countrywide census was a monumental undertaking that could only be improved by the introduction of a system specifically designed to tally huge quantities of data automatically. Hollerith perfected a system for encoding census returns onto punched cards and designed machinery to process the data received, demonstrating its effectiveness prior to the 1890 census by reorganizing record keeping systems in several large institutions. Hollerith's success, as well as that of his company, Tabulating Machine, and his application of punched cards was a significant advancement in the evolution of automated computing machinery. It is also important to note that Hollerith's Tabulating Machine Company eventually became the Computing-Tabulating-Recording Company, which in 1924 became International Business Machines, also known as IBM.

It wasn't until the 1940s that the first fully automated computer evolved from Jacquard's original concepts of 1801. In 1944 Howard Aiken (1900-1973), a Harvard University mathemati-cian, introduced what is considered to be the first digital computer. Constructed from adding machine parts, Aiken's machine was fed a roll of punched paper tape that represented an "instruction sequence," or program, used to solve a problem. The program was stored on the paper tape rather than in the computer (in 1945, a computer with program storage capabilities was constructed). Aiken's computer became rapidly obsolete, although the use of computer punch cards to store, process, and tabulate data continued well into the 1970s, when technological advancements surpassed the capabilities of the punch card systems and they, too, became obsolete.

Jacquard could never have imagined the significance of the improvements he made to a simple weaving machine. The punch cards he designed were only intended to ease the plight of French laborers, whose suffering he wished to see end. Jacquard would be astonished to experience the impact of his concepts as they have evolved in the twentieth century. The twentieth-century textile industry still relies on the Jacquard loom—it can be found in textile mills worldwide weaving cloth of intricate pattern and complex color combinations. Although its components have been modernized through the use of metals and plastics unavailable in Jacquard's lifetime, the basic design elements are identical to Jacquard's wooden loom of 1801. The twentieth century technological revolution owes a great debt to Jacquard's accomplishments, which have been built upon to create the extensive computer systems used daily by corporations, universities, and citizens around the world.

ANN T. MARSDEN

Further Reading

Books

Bolton, Sarah Knowles. *Lives of Poor Boys Who Became Famous.* New York: Thomas Y. Crowell & Co., 1885.

Cardwell, D.S.L. *Turning Points in Western Technology: A Study in Technology, Science and History.* New York: Science History Publications, 1972.

Cardwell, Donald. *The Norton History of Technology.* New York: W.W. Norton & Company, 1994.

Cortada, James W. *Historical Dictionary of Data Processing: Biographies.* Westport, CT: Greenwood Press, 1987.

Daumas, Maurice, ed., and Hennessy, Eileen B., trans. *The Expansion of Mechanization, 1725-1860.* Vol. 3 of *A History of Technology and Invention: Progress through the Ages.* New York: Crown Publishers, 1979.

Singer, Charles Joseph, ed. *The Industrial Revolution, c. 1750 to c. 1850.* Vol. 4 of *A History of Technology.* Oxford: Clarendon Press, 1954-84.

Usher, Abbott Payson. *A History of Mechanical Inventions.* 2nd ed. Cambridge, MA: Harvard University Press, 1954.

Van Ness, Robert G. *Principles of Punched Card Data Processing.* Elmhurst, IL: The Business Press, 1962.

Other

Dunne, Paul E. "Mechanical Aids to Computation and the Development of Algorithms: 19th Century Contributions and Their Impact on Elements of Modern Computers." WWW document. URL: http://www.csc.liv.ac.uk/~ped/teachadmin/histsci/htmlform/lect4.html

Tarses, Bonnie. "What's New!: Science & Nature." WWW document. URL: http://www.bess.net/whats_new/June2/science_and_nature/

Steam-Powered Railroad Systems Make Possible the Industrial Revolution and Fundamentally Alter the Transportation of Goods and People

Overview

The invention of the steam engine and the development of the railroad system were instrumental in creating the Industrial Revolution beginning in the late eighteenth century and continuing into the nineteenth century. Industrialization changed Western society radically, and the railroad was a primary tool of that change.

Background

The Industrial Revolution is defined as the change from the making of products by hand to their manufacture by machinery. Its development in Europe and then the Western Hemisphere between 1780 and 1880 triggered monumental social and political changes that altered Western society. It destroyed traditional ways of life and created new ones. This movement began in England, because England already had overseas markets, because a commercial class already existed, and because it supported a growing population that could supply workers for factories. One of the major inventions that drove this revolution in business and manufacturing was the steam-powered railroad system.

No single person invented the steam engine or the railroad. The first steam engine was built in 60 A.D. by Hero of Alexandria (fl. 60 A.D.), who used steam to make a ball spin around. It was a toy or a curiosity and did no practical work. Sixteen centuries later, in 1698, the first steam engine to do useful work was built in England by Thomas Savery (1650?-1715). Needing a device to drain water from coal and iron mines, he created an engine that condensed steam into a liquid. A partial vacuum and the escape of some of the steam moved a vessel. Valves were opened and closed to suck water up a pipe.

In 1712 Thomas Newcomen (1663-1729) improved this device by building an engine that used steam to move a piston up and down. In 1765 James Watt (1736-1819) used the heat and energy of these steam devices more efficiently by adding a separate cylinder and condenser that used one-fourth of the fuel that previous engines did. He patented his system in 1769 and later added a double action design that made his condensing engines practical to use in other applications, not just pumps.

Richard Trevithick (1771-1833) built the first high-pressure steam engine in 1801. He also invented a carriage that ran on rails. The idea of a car pulled along on rails had existed for centuries. The ancient Greeks had a rail system that helped to get their boats across obstacles. The men who rowed the boat pulled it along grooves in a limestone block. In the middle ages carts on rails were lowered into mines where men worked. There they were filled with ore and pulled to the surface by men, boys, or animals.

In 1803 Trevithick installed his engine on a flat car that could pull other cars. This became the first steam locomotive to run on rails. It was designed to pull loads of iron but was not a success as the loads proved to be too heavy for the rails. Still he had proved that such a device would work.

George Stephenson (1781-1848) created the beginnings of the first railway system in the world in Britain in 1814. His locomotive pulled eight coal cars filled with ore at four miles an hour. It was very slow, but it was a start and the speeds did improve. In the United States horse-powered railways had operated in mines and shipyards for years. In 1815 John Stevens (1749-1838) began building a steam-powered railroad system across New Jersey. He ran out of money and did not finish, but the idea of a railroad system had arrived. By 1820 the first locomotives using steam power and pulling public cars were operating successfully in England. By now some trains were made up of passenger carriages as well as freight cars. In 1825 a United States canal company built fourteen miles of wooden track in Pennsylvania and installed a locomotive on it that had been built in England. This first full-sized locomotive to be used in North America was not a success because it was too slow and the wooden track was too flimsy. The first modern railroad system began operation in England in 1825. It ran from the coal fields at Darlington in the North of England to the town of Stockton on the East Coast, close to ports where the coal could be loaded onto ships.

These inventions were a part of a general movement toward industrialization that occurred first in England, then spread to the rest of Europe and the United States and later to South America. Industrialization was possible because of growth in the population partially caused by improved sanitation and new medical discoveries. Another precipitator was the race between the various European nations to explore and claim far-flung lands. Many such places were considered strategically valuable, but more important was the fact that they had great quantities of the kind of raw materials needed to feed the factories back in Europe or America.

Impact

The Industrial Revolution changed the climate of society from a rural, agricultural economy whose goods were made by hand one at a time to an urban economy in which goods were manufactured in factories employing large concentrations of workers. From this came new cities, with their many advantages and numerous drawbacks, as well as far-flung markets. The advancement of knowledge and invention of new devices also led to many technical improvements in production, material, and transportation.

The steam-powered railway brought a revolution in transportation and accelerated the already developing industrialization of the Western world. Railroads answered the need to transport goods quickly to distant markets and to get the goods to ports where they could be taken by ship to even more distant markets overseas; railroads also brought raw materials to ports close to factories. Before the coming of the railroad, it was difficult to move some heavy industrial materials like iron, coal, or stone. The ability of the railroad and the steam boat to transport very heavy loads meant that more goods could be moved and more could be sold.

Operators of factories invested their profits in railroads to enhance their businesses. This was a good investment for the owners. Improvements were always necessary, and the expansion of railroads would serve mines and textile mills and bring more profit. When passenger cars were added, railroads were even more successful. It is clear that steam railroads accelerated industrialization, and industrialization in turn accelerated the building and improvement of railroads. The increased demand for coal and heavy manufactured goods was a guarantee of continued prosperity for the railroads. More factories were also needed to build more locomotives, rails, signals, switches, cars, and so on. More people had to be hired to build new track. Because of the spread of easy travel, it took less time for salesmen to sell goods, and quick sales meant quicker return on money that could then be invested in new lines or the manufacture of additional goods.

A single railroad could cost two million dollars—an enormous sum in the 1800s. Money was needed not only for the 2,500 men to build the structure but also the designers, engines, men who planned routes and decided where to build tunnels, embankments, and bridges. Therefore, construction of a new railroad required more capital than even a wealthy individual could handle alone. This necessitated the creation of corporations and stock companies to pool capital and resources to make the new railroad a reality. This sped up railroad construction but was not always advantageous to the citizens or the towns. The owners of the railroad and its right of way often become too powerful, controlling local government and monopolizing business and land holdings. The owners moved large amounts of manufactured goods and brought competition into an area from outside. For example, a company that produced wine had a general monopoly on the sale of local wine. When a new wine was brought in by the railroad, it changed sales in the area. If the imported wine was better or cheaper, the local wine producer would lose some of his profit to the new business. Railroads thus changed the way goods were advertised, priced, and sold.

The railroad has been called a fundamental innovation in American material life. It was a stimulus for the spread of U.S. population to the West and, in fact, created many small towns. Railroads were an efficient way to move men and supplies during the Civil War (1861-65). In the 1850s Congress began giving federal land grants to builders of new railroads. Northern businessmen had more money to spend on railroads than those in the South and, because the South was mainly agricultural, it lagged far behind the North in railroad mileage. Four times as many miles of railroads crisscrossed the northeastern part of the country as in the Southeast when the Civil War began in 1861. This advantage played an important part in the success of the North in the Civil War.

Long before the war, a rail connection to join the East Coast with the West Coast had been planned, but materials were unavailable during the war. The year after the war ended in 1865, the long-awaited connection was begun. Union Pacific built from the East and the Central Pacific began from the West. When the two railroads joined their tracks in Utah in 1869, the United States had a viable transportation route from the East Coast to the West.

The advent of railroad systems had numerous other effects. For instance, railroads made it possible for farmers to expand away from the banks of rivers and locate anywhere good farmland existed. Railroads also created the idea of small-town America. Furthermore, railroads stimulated the production of goods as well as propelling and spreading the idea of industrialization in the Western world. Railroads, the first major industry in the United States, made possible the growth of industries like coal, steel, flour mills, and commercial farming. They established cities like Chicago and had an impact on urban design. The finest minds and richest entrepreneurs were attracted to the engineering challenges of the railroad and to the legal and financial aspects of its operation.

LYNDALL B. LANDAUER

Further Reading

Dickinson, H. W. *A Short History of the Steam Engine.* New York: MacMillan, 1939.

James, Peter and Nick Thorpe. *Ancient Inventions.* New York: Ballantine Books, 1994.

Lane, Peter. *Industrial Revolution.* New York: Barnes & Noble, 1978.

Martin, Albro. *Railroads Triumphant.* London: Oxford University Press, 1992.

Musson, A. E. and Eric Robinson. *Science and Technology in the Industrial Revolution.* New York, Gordon & Breech, 1989.

Siegel, Beatrice. *The Steam Engine.* New York: Walker & Co., 1986.

Taylor, George Rogers. *The Transportation Revolution 1815-1860.* New York: Armonck, 1968.

Advances in Food Preservation Lead to New Products, New Markets, and New Sources of Food Production

Overview

At the beginning of the nineteenth century people relied on the same methods of preserving fresh food that their ancestors had used thousands of years before: drying, fermentation, salting, and chilling. By the end of the century a series of innovations had revolutionized food preservation, leading to new products, new markets, and new sources of food production.

Background

In order to survive primitive people had to preserve their precarious supply of food for as long as possible. One of the earliest preservation techniques was to dry meat, fish, or fruit in the sun or over a fire. Smoke added flavor to an often monotonous diet and also slowed spoilage. The ancient Egyptians and Romans stored wheat and barley in silos, where hot, dry air kept the grain usable for years. They used fermentation, a process activated by certain bacteria, to make wine and bread. Many cultures fermented dairy products into yogurt and cheese and fermented (or cured) meat into sausage. Salt, which suppresses the growth of harmful bacteria in addition to adding flavor, was another traditional preservative, particularly for fish and meat.

Freezing or chilling food in natural refrigerants such as ice and snow also inhibited the growth of bacteria. In warmer climates food was stored in caves or underground pits to slow the rate of spoilage.

Drying preserves food because yeasts, molds, and bacteria cannot grow without sufficient moisture. Several new methods of artificially drying food, or dehydration, were invented in Europe at the end of the eighteenth century. One called for immersing vegetables in hot water before drying them in hot air, a technique later used for drying meat. Small drying rooms originated in central Europe around this time. They were heated by a stove and filled with racks of drying trays. In the early 1800s a French inventor named Masson tried cutting vegetables into thin slices before drying them. His partner, Chollet, invented a mechanical warm air system for drying the vegetables, which were then compressed under high pressure. According to contemporary accounts the dried vegetables retained their flavor, and their condensed size simplified storage and transportation.

An English farmer, Downes Edwards, patented his process for preserving potatoes and other vegetables in 1841. He extruded cooked potatoes through a perforated disk, then dried the resulting "rice" on trays in gentle heat. A. E. Spawn, an American inventor, patented the first mechanized dehydration machine in 1886. Named the "Climax Fruit Evaporator," it consisted of trays of fruit that revolved slowly in an upward current of hot air.

Dehydration is the easiest way to preserve and store milk, one of the most nutritious foods and also one of the fastest to spoil. Milk was dried in the sun as early as the 1200s, when it was carried by Kublai Khan's soldiers as part of their rations. In the 1820s a more practical method was invented by Nicolas Appert (1749-1841). Appert also dehydrated meat and vegetable broth into solid cubes that could be reconstituted with water. Malted milk, a powdered combination of whole milk, ground barley malt, and wheat flour, was invented in Wisconsin in the 1880s. It quickly became popular as a sweet and nutritious drink.

Fermentation, the mechanism that creates yogurt, cheese, and butter, is so simple that it has changed little over the centuries. Scalded milk is exposed to special bacteria that convert lactose into lactic acid, resulting in a solid and less perishable product. Around the middle of the nineteenth century butter and cheese production began to shift away from farms as factory production made it possible to control environmental conditions more tightly and to take advantage of advances in refrigeration and transportation. The dry curing method of sausage-making perfected by the Romans became less common as other methods of preserving meat—especially refrigeration—became available in the second half of the century.

The most important advance in modern food preservation was canning. It was invented by means of trial and error and careful observation by Nicolas Appert, who opened his first factory in 1797. His three-step process is the same one used for home canning today. The food was placed in wide-necked bottles, sealed tightly, then heated in hot water. Appert theorized that heating destroyed or neutralized the "ferments" that make food spoil.

Appert's process, which he described in a pamphlet in 1810, was copied throughout Europe and the United States. Peter Durand obtained an English patent for preserving food in tin containers that same year. Soon after, Bryan Donkin and John Hall opened their own canning factory, and by 1814 they were shipping food to British military bases in tin cans, which had become the preferred food containers. Thomas Kessett received the first U.S. patent for canning and tin containers in 1825. Around that time the first sardine-canning factory was built on the coast of France, and other factories were constructed in Germany and Portugal.

Louis Pasteur's (1822-1895) research on spontaneous generation in the 1860s provided the scientific rationale underlying Appert's process. He found that heat kills microorganisms that cause food and beverages to spoil. He applied his research to the wine, vinegar, and beer industries, making it possible to transport these products without risk of deterioration. In the late 1800s John Tyndall (1820-1895), a British physicist, took Pasteur's heating process one step further. He heated food for 45 to 60 minutes to kill the active bacteria, then chilled it for 24 hours to kill any dormant bacteria spores. Milk pasteurization, introduced in the United States at the end of the nineteenth century, requires that milk be heated either to 145 degrees F (63 degrees C) for 30 minutes or to 161 degrees F (72 degrees C) for 15 seconds. Pasteurization kills disease-causing bacteria as well as those that lead to spoilage.

In 1856 Gail Borden (1801-1874), an American businessman who wanted to eradicate childhood illnesses caused by contaminated milk, patented a process for making condensed milk. Milk was heated in a vacuum pan that prevented outside air from entering, a device invented by the Shakers to condense fruit juice. There was not much demand for condensed milk until the Civil War (1861-65), when it became an important provision for soldiers.

Refrigeration preserves food by slowing the rate of growth of the bacteria. Several kinds of mechanical refrigeration were invented in the nineteen century. John Gorrie (1803-1855), an American physician, built the first ice-making and air-conditioning machine in 1844 but never found adequate financial backing for his invention. Around the same time Alexander Twining (1801-1884) patented an ice-making machine and opened the first factory to manufacture ice by vapor compression. The U.S. mechanical ice-making industry started to flourish in the 1870s, thanks to entrepreneurs such as Daniel Holden (1837-1924) and David Boyle (1837-1891). With both artificial and natural ice readily available, iceboxes became more common in American homes and factories, although they were not accepted in Great Britain and Europe. The first cold-storage warehouse was built in New York in 1865, followed by others near piers and railroad lines in major cities around the world. Refrigerated boxcars and steamships carried meat and produce from farms to cities and from one country to another. Keeping food cold during transport was especially important in the meat-packing industry. Meatpackers such as W. James Harrison (1816-1893), Gustavus Swift (1839-1903), and Philip Armour (1832-1901) found that it was cheaper to send frozen meat from centrally located slaughterhouses than to ship cattle to slaughterhouses near customers.

Impact

Advances in food preservation during the nineteenth century increased the healthfulness and variety of food available year-round. At the same time new food-processing technologies turned the pursuit of a constant supply of food from a small-scale, domestic activity to a large-scale, commercial industry, which led to further changes in people's lives at home and at work.

Among the first consumers of preserved food in the early 1800s were sailors, soldiers, and explorers, who were prone to malnutrition, especially scurvy, during long voyages without fresh food. Dried, compressed vegetables, in addition to their portability, minimized scurvy among soldiers in wartime and among miners in the Alaskan and Australian gold rushes. The invention of canning, refrigeration, and pasteurization prevented the loss of the valuable protein and vitamins in perishable foods and reduced the prevalence of some bacterial diseases. For those who could afford them, preserved foods made it easier to enjoy a nutritious, high-protein diet, regardless of season or location.

Changes in food preservation during the nineteenth century were linked to larger trends associated with the Industrial Revolution. Workers in urban manufacturing centers bought food instead of producing it themselves. Freed from the labor of food production and preservation, they were able to specialize in industrial work and were available for construction projects that required large work forces, such as building ships, trains, roads, and bridges. Thousands of additional jobs were created in the food preservation industry itself. This prosperous and growing urban market, in turn, encouraged farmers to specialize in items such as meat, eggs, dairy products, and fruit, to increase production, and to invest in new technology. Larger markets, product specialization, and the high cost of processing plants and refrigerated transport promoted consolidation in the food industry, as exemplified by the empires of Borden, Swift, and Armour. Faster transportation and improvements in refrigeration and canning meant that farms could be located farther from population centers, even as far away as Australia and Argentina. While cities became more densely populated, new farming areas were settled and cultivated. The result was the end of a predominantly farming economy and the beginning of a predominately manufacturing economy.

LINDSAY EVANS

Further Reading

Desrosier, Norman W., and James N. Desrosier. *The Technology of Food Preservation.* 4th ed. Westport, CT: AVI, 1977.

Prescott, Samuel C., and Bernard E. Proctor. *Food Technology.* New York: McGraw-Hill, 1937.

Robinson, R. K. *The Vanishing Harvest: A Study of Food and its Conservation.* New York: Oxford University Press, 1983.

The Steamboat:
First Instrument of Imperialism

Overview

American steamship inventor Robert Fulton (1765-1815) believed he had perfected a mechanism that would allow humankind to increase its productivity and create a better material life for all. He could not have imagined, however, that steamboats like his *Clermont* would evolve into the world's first intercontinental weapon and an instrument of Western imperialism.

Background

Western civilization's view of the world began to expand with the onset of the Commercial Revolution during the seventeenth and eighteenth centuries. This expansive increase in world trade was led by the English and the Dutch. It was based on the two economic ideas of mercantilism and monopoly. Mercantilism was based on the premise that there are only a finite number of resources available to the world community. A nation's power was linked to the number and amount of these resources it controlled. Worldwide economic competition was very expensive and the country that could concentrate its resources most effectively would be the nation most likely to succeed. Both England and the Netherlands created government-run monopolies to ensure their national welfare was protected. These monopolies increased the availability of goods to Europe and its colonies. The most successful of these organizations was the British East India Company.

The power of Europe would increase dramatically with the onset of the nineteenth-century Industrial Revolution, the greatest increase in productivity in the history of the world. Europe's ability to harness inanimate sources of power was the basis of this revolution. The steam engine perfected by James Watt (1736-1819) was the source of this power. It allowed machines with a hundred times the productive capacity of humans to operate without interruption. It was the Watt steam engine that served as the model for the power source of Fulton's *Clermont.*

European ultranationalism connected the drive to control resources and the social Darwinist philosophy of "survival of the fittest" to an unlimited, reliable source of energy. These tech-nological, economic, political, and philosophical factors set the stage for the "new imperialism" of the late nineteenth century.

Impact

Steam first became an active instrument of imperialism in south Asia under the control of the British East India Company. The British Empire was so large that timely communication was becoming an important problem. The British East India Company had been using steamboats to help maneuver large sailing ships into the narrow harbors of India. Steam technology was then used to link the British outposts in northern India. The Ganges River runs west to east across most of the northern part of south Asia. It provided a natural highway for steamboats to deliver goods, troops, and mail across the wide expanse of the northern frontier. In essence, steam power was used to shrink the size of northern India.

The East India Company next attempted to use steam to reduce the time of travel between the British Isles and India. However, the early steamboats were not up to the task. The water and weather of the world's oceans were too formidable for this new technology. The British developed an alternative plan: they created a route that extended from India, across the Persian Gulf, and through modern-day Iraq and Syria using the Euphrates River to the Mediterranean Sea. From there it was a short voyage across the Mediterranean, which at that time was a "British lake," through the Strait of Gibraltar, to England. This route had three important effects on the British imperial system: (1) it reduced travel and communication time between south Asia and Britain; (2) enlarged the British imperial presence by giving them a foothold in the Middle East; and (3) effectively blocked the southern expansion of one of Great Britain's rivals, Russia.

The use of steam as a weapon in imperial expansion came in 1824, when the East India Company found itself engaged in a war with the Burmese people. England had always considered Burma to be part of its Asian sphere of influence. Therefore, when Burma declared its national independence, the East India Company found itself bogged down in the first Western land war in Southeast Asia. Not unlike modern U.S. involvement in Vietnam, British troops suffered great losses as a result of the hit-and-run tactics of the Burmese guerrilla forces. The British searched for a way to train their superior firepower on the enemy. Building on the success of the Ganges River experience, the East India Company decided to try to use steam power to break the deadlock of the Burmese War. Steam-powered gunboats attacked and subdued Burmese war boats on the Irrawady River. This left their seat of government open to attack, prompting the Burmese king to end the fighting.

Steam power would play an even larger role in the Opium War of 1848-49. Once again England's policy of economic nationalism would drive Britain into a military confrontation, this time with China. By the early nineteenth century, the British system of trade was based upon the movement of three products—English wool, Indian cotton, and Chinese tea. Tea had become the most important consumer good in England. Virtually everyone in the empire consumed large quantities of tea. The result was a large trade deficit with China, which Britain tried to close by increasing its opium trade. Initially, the Chinese government accepted the increase and even took part in the profits. Eventually, everyone in China could see the great social disaster the trade had created, and the Chinese government informed Great Britain that it must stop dumping large quantities of opium into their nation. When the British refused, China destroyed the warehouses that held the dangerous drug. The British responded militarily and the Opium War began.

This was a classic confrontation between a formidable land power, China, and a naval superpower, England. Britain's military planners knew they could not carry on a prolonged land war in China. The lesson of the American Revolution was that any attempt to invade and occupy a vast expanse of enemy territory would lead to a military disaster. The introduction of the steamboat drastically altered this tactical situation. The British created a fleet of small, highly maneuverable steamboats in their naval yards in India. They mounted small cannons on the boats and sailed them to China. Since the weapons were placed on revolving turrets, they could be used to both attack Chinese fortifications and to engage enemy warships. The Chinese navy was no match for the advanced British technology, and in a short period of time the Royal Navy controlled all of the inland water-

ways. This left China's great inland fortifications defenseless, and the government agreed to a peace settlement. By the last quarter of the nineteenth century, Britain had used steamboats to conquer and control large areas of east Asia, and south Asia, as well as the Middle East.

The continent of Africa was a much different story. European interest in Africa predates the voyages of Columbus. Seven years before his famous voyage, the Portuguese had sent expeditions to explore the Congo. Again in the late sixteenth century, the Spanish sent explorers up the Zambezi River. This interest continued into the early nineteenth century, when the English attempted to explore the Niger River. Every expedition ended in failure. For over three centuries, disease, especially malaria, blocked exploration of the continent. It was the discovery of quinine as a preventive drug that allowed Europe to use the power of steam in Africa.

Once the medical problems were solved, steamboats proved again to be a valuable instrument of imperialism. The first application was Britain's successful attempt to gain control of the Niger River and the important product, palm oil. Steamboats were used to cover the great expanse of the Niger, collecting palm oil from government outposts. The outposts would gather the oil from African villages along the river. When on occasion relations became strained, gunboats were used to gain control of the situation. Both David Livingstone (1813-1873) and Henry Stanley (1841-1904) used steam power on their expeditions. By the late 1860s the British had perfected a steamboat small enough to be disassembled and carried past the dangerous rapids of many of Africa's rivers. Both France and Belgium also used steam in their colonizing attempts. Most notably Leopold II (1835-1909) of Belgium was able to penetrate into the heart of Africa by using steamboats to gain control of the Congo River.

For the first six decades of the nineteenth century steam power was used by the industrial nations as a tactical weapon. Iron paddlewheel boats lacked both the structural integrity and power to become a significant force in the world's major sea-lanes. This began to change in 1864, as a result of the creation of the Bessemer process. This allowed for the quick, inexpensive manufacture of steel. Steel was lighter, stronger, and more flexible than iron. These characteristics caused a major revolution in steam power. They allowed for the creation of oceangoing ships that would become the first intercontinental weapons. Steel boilers increased the productive capacity of the power plant. They allowed for an increase in pressure, which expanded power and speed, and at the same time allowed for the reduction in the size and weight of the boiler. Steel also made the use of the screw propeller a reality because these could now be manufactured strong enough to withstand the stress and pounding of the world's oceans. Propellers moved ships at a much faster rate than the paddle wheel. The combination of larger, faster ships, allowed the industrial nations to completely control the world's sea-lanes. England, France, Germany, the United States, and eventually Japan would be able to project their influence around the world. This accelerated the quest for colonies and led to the first modern arms race. It helped to create the environment that led to World War I in the second decade of the twentieth century.

RICHARD D. FITZGERALD

Further Reading

Diamond, Jared. *Guns, Germs, and Steel: The Fates of Human Societies.* New York: W. W. Norton & Company, 1998.

Headrick, Daniel. *The Tools of Empire: Technology and European Imperialism in the Nineteenth Century.* New York: Oxford University Press, 1981.

Hugill, Peter. *Geography, Technology, and Capitalism: World Trade Since 1431.* Baltimore: Johns Hopkins University Press, 1993.

Pakenham, Thomas. *The Scramble for Africa.* New York: Avon Books, 1991.

The Communication Revolution: Developments in Mass Publishing during the Nineteenth Century

Overview

James Watt's invention of the steam engine in 1765 brought rapid and expansive changes to many areas of culture including transportation, manufacturing, and science. Another major change resulting from the advent of steam power was in the area of mass publishing, where it was incorporated into the three major areas of production—paper making, typesetting and casting, and printing.

Background

One of the major advances in the printing industry was the invention of a device to produce printing paper quickly and inexpensively. This contribution came from Nicholas Louis Robert, who patented his automated paper-making machine in 1799. Robert's persistent efforts resulted in a machine that could produce a continuous sheet or roll of paper. His employer, St. Leger Didot, brought the plans for the improved machine to his brother-in-law, Gamble, who acquired a new patent for it in 1801. During the first decade of the nineteenth century, two brothers, Henry and Sealy Fourdrinier enlisted the services of an engineer, Bryan Donkin, to improve upon the idea. It was Donkin who actually received the financial benefits of the patent they received for an improved paper-making machine in 1807. By 1851, Donkin had designed 191 machines. In addition to the 83 machines that were put into production in England, over 100 were sold to mills in Europe. Another machine was sold for use in India, and by the end of the 1820s there were two in the United States. The first of these two was imported from England to Saugerties, New York, in 1827, while a second was built in Connecticut by George Spafford and James Phelps in 1829.

Machine paper making automated several processes previously done by hand—mixing pulp with water, coloring material, sizing, and distributing the fibers uniformly across the width of the paper form. The pressing and drying process was also automated: first, some of the water was removed from the paper mat by suction; then most of the water was removed as the roll, or "web," of paper moved through roller-presses and felt blankets; finally, all but a small percentage (about 5%) of the water was removed by passing the paper through steam-heated cylinders.

In addition to the "Fourdrinier," as the machine came to be known by modern paper makers, a cylinder-type machine was developed in 1809 by the Englishman John Dickinson. Around the same time, Thomas Gilpin of Brandywine Creek, Pennsylvania, built a cylinder machine, one which could produce a thirty-foot wide sheet at about sixty feet per minute.

Mass publishing also benefited from several inventions that improved the speed and accuracy of casting and composing type. A major problem for production printing had always been the slow and cumbersome way that type was set. Individual lead blocks of letters, numbers, or symbols had to be inserted by hand into long casting sticks and placed by a "composer" onto the printing form. In the 1880s two machines for typecasting and composing revolutionized prepress production.

Ottmar Mergenthaler (1854-1899) developed a system of casting type that streamlined the process to such an extent that his invention is often said to have been the most important contribution to the industry since Johannes Gutenberg (c. 1398-1468) developed moveable type. The German-born Mergenthaler, who had settled in Washington, D.C., in the 1870s, moved to Baltimore where he went to work to solve the problem of typesetting and casting. The result, the Linotype machine, was patented in 1884 and installed first at the *New York Tribune* in 1886.

The advantage of Mergenthaler's design was that one operator could perform the work of several people in a shorter period of time than handtype casting and composition. The Linotype machine consisted of a keyboard and a casting unit. An operator adjusted the machine for the size of type and length of the line of type needed. As the operator typed characters on the keyboard, metal in the casting unit, which had been heated to about 550 degrees Fahrenheit, stamped out the actual slugs. These matrices, or small pieces of stamped brass characters, moved from channels in the casting unit along a conveyor belt to the assembler box, or "composing stick," creating a whole line of type—thus the name Linotype.

Like the Linotype machine, Tolbert Lanston's Monotype machine consisted of a keyboard and casting unit. The system, however, differed in the way that the type was composed. The American-born Lanston devised a system of setting type based on precise mathematical calculations for the size and spacing of individual units, or characters. When the operator typed into the keyboard, perforations were made in a cylinder of paper that would then be used to cast the individual slugs and create accurately spaced characters of uniform sizes. Lanston's

machine entered the publishing industry in 1887, just a year after Mergenthaler's.

Innovations in the press room also improved the possibilities for mass communication. In the early 1800s Frederick Koenig adapted the steam engine to a cylinder-type printing press. After earlier unsuccessful attempts to automate the printing process by incorporating steam power, Koenig produced the first twin-cylinder steam-powered press in 1812. In 1814 Koenig's improved model was installed and put to work in the press room of the *London Times*. The power press increased the *Times's* output to 1100 sheets per hour, a 400% improvement. This increased rate of production allowed the *Times* to expand its distribution to over 30,000 papers per day by the end of the 1830s.

In 1830 Richard Hoe, at the age of eighteen, and his cousin Matthew Smith took over the Hoe Printing Company in New York City. Before retiring, Hoe's father, Robert, and Sereno Newton had developed and put into production an improved cylinder press. Hoe immediately set about trying to improve the rate of production beyond the 200 impressions per hour achieved by his father's machine. In 1837 he designed and produced the first large cylinder press used in the United States. Hundreds of these machines were in use in the printing industry within a decade.

Hoe continued to work out press designs to meet the ever-increasing needs of newspaper publishers. The result of his efforts was a "type-revolving machine," or rotary press. The first of these presses based on Hoe's patent was put into use at the *Public Ledger* in Philadelphia in 1847. The key feature in Hoe's design was an apparatus for securing the printing plates, or type forms, onto a large cylinder.

This type cylinder was surrounded by several impression cylinders, through which operators fed sheets of paper. This version of the rotary press was able to print 8,000 newspapers per hour. The increased output capability made the rotary press popular worldwide.

Other improvements followed. In 1865 William Bullock constructed the first press to print from a continuous roll (web) of paper. By 1871 Hoe and a partner, Stephen Tucker, had also built a web press. The first of these machines was put into production at the *New York Tribune*. The great advantage of Hoe and Tucker's web press was that it could print on both sides of the paper at once. This "perfecting press" allowed for even an higher rate of production. The *Tribune* could produce up to 18,000 newspapers per hour. In 1875 Tucker patented a rotating folding cylinder which allowed papers to be folded as they came off the press. The Hoe Company improved upon this feature in 1881 by creating a triangular folder. The folding apparatus and several other improvements incorporated by the Hoe company created the modern press as it was used in the printing industry until the advent of computerized printing in the last quarter of the twentieth century.

Impact

Enterprise was a key factor in the development of mass publishing. Creative, motivated inventors took advantage of the growing demand for educational and entertaining reading material by devising technological innovations that could serve that demand. To a certain extent, the expansion of mass education and the growing popularity of penny novels, newspapers, and periodicals contributed to this trend through their need for cheap, fast, and efficient methods of production.

The Fourdrinier paper-making machines not only made paper cheap and plentiful but changed the way that the industry operated. Previously, paper making had been a cottage industry, involving production by a lot of small companies. By the turn of the century, the industry was dominated by a few large companies that produced tremendous amounts of paper. These companies sold their inventory through a system of specialized wholesale distributors.

Improvements in typesetting and printing created advantages for book publishers, but the advantage of steam-powered printing to "jobbing printers"—those who produced all the nonbook print matter—was enormous. The costs of book printing were not necessarily lower as a result of power printing, but volume printing saw significant cost savings.

The impact of mass production was particularly important to the newspaper industry, where the inexpensive creation, composition, and distribution of information on a mass scale had to be repeated daily. As in the paper-making industry, the cost of purchasing power machinery resulted in a change in the industry's profile. By the end of the Victorian era, fewer, but larger-scale, printing companies were in business.

LISA NOCKS

Further Reading

Bruno, Michael H., ed. *Pocket Pal: A Graphic Arts Production Handbook*. Memphis, TN: International Paper Co., 1997.

Comparato, Frank E. *Chronicles of Genius and Folly: R. Hoe Company and the Printing Press as a Service to Democracy*. Culver City, CA: Labrynthos, 1979.

Feather, John. *A History of British Publishing*. London: Routledge, 1996.

Hart, Ivor B. *James Watt and the History of Steam Power*. New York: Schuman, 1949.

Moran, James. *Printing Presses: History and Development from the Fifteenth Century to Modern Times*. Berkeley: University of California Press, 1973.

Williams, Kevin. *Get Me a Murder a Day: A History of Mass Communication in Britain*. London: Arnold, 1998.

Advances in Photography during the Nineteenth Century

Overview

In August 1839, at a joint meeting of the French Academies of Sciences and Fine Arts, the astronomer François Arago (1786-1853) announced Louis Daguerre's (1787-1851) method of obtaining pictures by the interaction of light and chemicals. Daguerre's discovery instantly captured the imagination of the public everywhere. But the invention of photography is actually the work of three men. The combined efforts of Daguerre, Joseph Niépce (1765-1833), and William Talbot (1800-1877) altered for all time how people see themselves and the world around them.

Background

In ancient times the philosopher Aristotle (384-322 B.C.) described how during a partial eclipse of the sun, the gaps between the leaves of a tree cast images of the crescent-shaped sun on the ground. During the Renaissance, artists in Europe applied this optical principle to create the camera obscura, or darkened room, in which light passing through a small hole in a window covering projected an image on the opposite wall. By the seventeenth century the camera obscura was no longer a room but a portable box. It was an indispensable tool in working out accurate and proportionally correct renditions of buildings, landscapes, and people.

The desire of post-Renaissance society to portray things as seen by the eye rather than the mind contributed to a climate of scientific inquiry that emerged in the sixteenth century. Investigations into the appearance and structure of living things made it possible to portray organisms realistically. Discoveries in the physical sciences made painters more aware of the visual effects of the physical world, and stimulated their interest in mundane events in contemporary life. Among the middle class, these developments fed a craving for exact copying of nature, and for reality.

Images obtained in the camera obscura were fleeting, so inventors naturally turned to ways of making them permanent. Until the 1700s, the reactions of organic and mineral substances to heat and light remained largely the province of alchemists. Observations in the seventeenth century focused on substances such as silver nitrate, silver chloride, and ferrous salts. But in 1725, Johan Heinrich Schulze, a university professor, made the accidental discovery that exposing silver nitrate to sunlight darkened it.

In the eighteenth century, Dr. William Lewis and the chemist Joseph Priestley (1733-1804) studied and publicized the photochemical properties of silver halides. Thomas Wedgwood, an amateur scientist, and his associate Humphry Davy (1778-1829) demonstrated that it was possible to transfer pictures chemically by means of light. But they could not find a way to stop the action of light on the silver salts they used, and the result was darkened images.

In 1827, the French physicist Joseph Nicéphore Niépce exposed a pewter plate coated with bitumen, or asphalt, in the camera obscura for eight hours and managed to get an image in light and shade from the window of his workroom at Le Gras. Niépce, who described his camera as "a kind of artificial eye," tried without success to reduce the exposure time by increasing the sensitivity of his plates to light. A chance introduction to the scenic designer and Diorama owner Jacques Louis Mandé Daguerre resulted in their meeting in 1827 and forming a partnership to pursue the process of making images together.

After Niépce's death in 1833, Daguerre continued their joint work, discarding bitumen in favor of iodized silver plates. But the problem of lengthy exposure time continued to bedevil him until 1835, when he realized that he could use mercury vapor to bring out the latent, or hidden, image on the plate. A remaining problem was the action of light on the silver halides, which caused the image to darken into invisibility. But in 1837 Daguerre found that he could solve that problem by bathing the plate in a solution containing common table salt, a method he later replaced with hyposulfite of soda, a discovery made by the English scientist John Herschel (1792-1871). The result of this process was called the daguerreotype. The product was so delicate that it had to be encased in glass. And daguerreotypes were unique—they could not be duplicated. Still, their clarity of detail was breathtaking. Significantly, Daguerre gave his invention to the French government, which offered it free to the world.

Simultaneous with Daguerre's early discoveries, the English scientist and mathematician Willim Henry Fox Talbot found a way of making permanent images by exposing to light paper treated with alternate washes of sodium chloride and silver nitrate. Talbot's first experiments involved transferring leaf forms onto chemically sensitized paper. But in 1835, Talbot produced a one-inch-square negative paper image of his ancestral home. Talbot stabilized his early images with potassium iodide or salt, but like Daguerre, switched to hyposulfite of soda on the advice of John Herschel. Talbot called his images photogenic drawings, and tried without success to make a positive image from them by making a sandwich with silver-sensitized paper and exposing both sheets to light. Talbot's first images were reversed (a negative). In re-reversing the process, he obtained a positive copy.

The significance of Talbot's discovery was that in theory it would be possible to make multiple positives from a single negative. Like Daguerre, Talbot also hit on the phenomenon of latent development, in 1840. He sensitized his paper with a combination of chemical solutions, exposed it in the camera, removed what appeared to be a blank sheet of paper, and brushed the surface again with a bath of the same chemicals, during which the image appeared. This process reduced exposure time to 30 seconds on a bright day. Talbot called his invention a calotype, and he patented it.

It was John Herschel who persuaded Talbot to consider the term *photography* from the Greek *photos* for light and *graphein* for writing. Herschel also coined the terms *negative* and *positive*, and, 20 years later, *snap shot*. Herschel considered the three critical elements of photography to be very sensitive paper, a perfect camera, and a way to stop the further action of light. His solution to the problem of light action changed everything.

Talbot made his paper image process public in London in February 1839. The slowness of his method to gain popularity, especially in the United States, may have been due to people perceiving it as abstract, complex, and fuzzy, not clear like a daguerrotype. Moreover, Daguerre's invention had the sanction of the French government, whereas Talbot worked alone. Consequently, his fierce efforts to protect the rights to the use of his invention discouraged other people from using it—American photographers balked at paying license fees. In 1852, in response to requests from the Royal Academy and the Royal Society, Talbot relinquished control of his invention except in the case of portraiture.

The biggest problem for paper prints remained the lack of a stable printing medium. The surface of photographic printing papers was irregular, which made impossible the kind of detail that people had come to expect with daguerreotypes. Large-scale commercial printing finally became a reality with the development of the so-called colloidion or "wet plate" process in combination with the use of paper coated with albumen (egg white). The introduction of the colloidion process in 1851 made both daguerreotype and calotype processes almost instantly obsolete.

The first equipment used by Niépce, Daguerre, and Talbot was rectangular boxes modeled on the camera obscura that included a lens and a place to put the plate. Talbot's wife called his boxes mousetraps. Early photographic equipment was extremely cumbersome. For a single day's shoot, an amateur photographer alone easily lugged 100 to 120 pounds (45-54 kg) of equipment.

Impact

In England, France, and the United States, which were the three primary industrial powers with the wherewithal to develop the medium, photography found an immediately receptive public.

Portraiture satisfied people's desire for personal memoranda—every Victorian home had a family album. Visiting cards with small portraits on them called *cartes de visite* were introduced in 1851 and by 1861 had become an international craze. The photographer as explorer familiarized

people with their own lands and faraway places in a realistic manner, without their having to traverse seas or climb mountains. Roger Fenton (1819-1869) pioneered a new application of photography—war reportage—with his documentation of the Siege of Sebastopol during the Crimean War. Using a wagon fitted out as a darkroom, Fenton was the first to photograph still shots of battlefields, officers, and men during battle. The camera also provided an objective testament to the ruins of the Civil War, driving home the psychological difference between photography and other ways of making images.

People believed that photographs were authentic, that the people or things in them existed. William H. Jackson's (1843-1942) photographs detailing natural wonders were instrumental in the decision of the United States Congress to make the Yellowstone region a national park. The photographs of Jacob A. Riis (1849-1914) led a newly aware America to better living conditions for the poor in New York.

Once action studies were possible, they helped to understand movement. And inevitably, photography aspired not only to truth but also to beauty. As the science of photography adjusted to accommodate the art of picture making, artist-photographers came into being.

The substitution of gelatin plates for wet plates around 1879 freed the photographer of having to carry his darkroom with him. Gelatin dry plates could be stored for long periods and made instant photographs possible, shortening exposures to a fraction of a second. Modern photography still relies basically on gelatine emulsion. In 1888 George Eastman (1854-1932) introduced the first camera to incorporate a roll film, and made every tenth person in the United States a camera owner. "You press the button," ran the slogan for the Kodak camera, "we do the rest."

From the beginning, people sought ways of printing their photographic images. A variety of techniques made it possible to reproduce photographs in quantity, but they could not be printed on an ordinary press with type. All this changed with the invention of the halftone plate in the 1890s, which revolutionized the economy of news photography. Photographs could be reproduced in books, magazines, and newspapers, cheap and in limitless quantity.

At the close of the century, photographic technology had advanced to hand cameras and dry plates, enlargers and rapid printing paper, and more powerful lenses and high-speed shutters.

Although nascent, the development of successful color photography was still several years away.

GISELLE WEISS

Further Reading

Gernsheim, Helmut. *The History of Photography from the Earliest Use of the Camera Obscura in the Eleventh Century up to 1914.* London: Oxford University Press, 1955.

Newhall, Beaumont. *The History of Photography, from 1839 to the Present.* New York: Museum of Modern Art, 1964.

Rosenblum, Naomi. *A World History of Photography.* New York: Abbeville Press, 1984.

Cyrus McCormick
Invents the Reaping Machine

Overview

In 1831, Cyrus McCormick (1809-1884) developed the first device to reliably and effectively cut and gather grain in the field. Further refinements gave rise to the combine and other mechanical devices that have made it possible for a relatively small number of farmers to maintain and harvest thousands of acres of crops and to feed thousands of people each year.

Background

Agriculture, the domestication of plant and animal species for food production, is arguably the most important development in human history. Many have argued convincingly that the development of agriculture is what allowed nomadic hunter-gatherers to develop written language, cities, civilization, and more. However, for most of human history, growing grain and other crops has been wholly dependent on the availability of human labor for field preparation, planting, and harvesting. This placed severe constraints on not just farmers, but on society as a whole because workers in the fields were not available to operate factories, provide nonfarm services, or perform any of the other tasks upon which a technological society rests.

For thousands of years grain was harvested by people wielding scythes and other manual cutting devices. The grain was then bundled together by hand and brought to a central place where grains were separated from the stalks and chaff using technology that dated back to the earliest farmers. Although attempts were made to automate parts of this process, they were unsuccessful for a variety of reasons. This limited a farm's productivity to the amount of grain and the number of acres that could be maintained by a single person or family with occasional assistance.

McCormick built his first reaper in 1831, at the age of 22. In 1847 he opened his first factory, near the city of Chicago, then a small and unimportant town. He presented his reaper to the world at the 1851 Great Exposition in London, winning acclaim and increasing orders almost immediately. Already increasingly popular in the United States, McCormick's reapers spread overseas, helping to revolutionize agriculture in Europe and elsewhere.

In 1871, the reaper was mated with a mechanical thresher to form the first combine, a single machine that would cut wheat plants and thresh the grain at the same time. The combine was only the first of many mechanical farm implements to be developed, and the development of internal combustion engines made them even more efficient. The mechanization of agriculture has continued to the present day and, even today, agricultural engineers continue to work to automate and mechanize some of the most arduous and time-consuming tasks that remain.

Impact

The development of agriculture allowed a relatively few farmers to grow food for many. This gave nonfarmers the opportunity to invent forms of government, written languages, and to develop technologies that were not available to hunter-gatherers who spent their entire existence struggling to survive. Over the centuries and millennia, complex civilizations and technology were built around agriculture. In inventing the reaper, Cyrus McCormick set in motion events that would profoundly affect farm life and rural communities and that helped to make possible the technological and mechanized society in which most of the developed world finds itself. It has also led to the strange situation in which developed nations sometimes pay farmers to leave land fallow or to burn grain while in less developed countries, people starve. This, in turn, has led to interesting situations in global politics.

Farming, even with the assistance of machinery, is hard, dirty, back-breaking labor. It is highly dependent on weather conditions and,

THE GREAT EXHIBITION OF 1851

A landmark in the history of technology and the epitome of the Victorian spectacle, the Great Exhibition of the Works of Industry of All Nations took place in London in 1851. Held in the monumental Crystal Palace (designed for the event by Joseph Paxton), the exposition featured such sensational displays as a 2-ton steel ingot shown by Germany's Alfred Krupp alongside his 6-pounder cannon and cast steel cuirasses, Cyrus McCormick's Virginia grain reaper, Samuel Colt's revolvers, and Joseph Whitworth's exemplary lathes and assorted machine-tool equipment. The complete catalog of the 100,000 objects that formed the massive exhibition fills three volumes of 500 pages each.

Hundreds of exhibitors—showcasing everything from labor-saving devices to textiles, industrial chemistry discoveries, exotic musical instruments, agricultural displays, pottery, and furniture—vied for the attention of over six million visitors, some who paid more than one visit. Queen Victoria herself visited 34 times in the five-and-a-half months of the exhibition. By the time the doors closed on the Great Exhibition in October 1851, 6,063,986 had experienced the spectacle. Its wild success embodied the excess of the Victorian era as well as the hunger for invention and technology and the faith in commerce of the Industrial Revolution.

if using animals for heavy work, is also highly dependent on the ability to feed, water, and care for the animals. Regardless of any other factors, horses or oxen can fall ill, jeopardizing a harvest if they cannot be cured in time. By removing the dependence on less predictable animals, tractors and harvesting equipment have made planting and harvesting crops more predictable, giving farmers the ability to better plan their activities and the results in advance. In addition, by

reducing the manpower required to bring in a harvest of many crops, mechanization has helped reduce the cost and much of the uncertainty of the harvest time. Finally, by speeding the farming process and reducing the amount of work to a manageable level, machines such as McCormick's reaper have let a single person farm an incredibly large amount of land. However, mechanizing food-gathering has also led to attempts to breed foods that can better withstand machines, sometimes leading to less tasteful fruits and vegetables.

In general, societies without agriculture are primitive, societies with primarily manual agriculture (farming without machinery) are less developed, and societies with highly mechanized agriculture are more developed. This observation could be construed to indicate that more developed societies can afford machinery to perform heavy and dirty work, but this interpretation would appear to be false. Rather, history has shown that developing mechanized agricultural methods results in a vast reduction in the number of people required to feed a society. The workers no longer needed to till fields and harvest grain in turn can go on to work in factories, attend universities, and participate in all the other activities that allow a civilization to progress. Therefore, it seems likely that the invention of the mechanical reaper was the first step in the process leading to the society found throughout the developed world.

Using strictly manual methods, a single person can harvest an acre of wheat in 14 hours and can process a bushel's worth of grain in three hours. These same tasks can be accomplished by a single person in just over 30 minutes today, using mechanical farming equipment. This came at a price, however. By reducing the number of people needed to farm, many farming jobs disappeared, forcing a migration to the cities to find work. In the United States this has led to a steady decrease in rural populations and, according to many, an erosion in many of the values of joint sacrifice and hard work that seem to characterize farming communities.

Mechanizing agriculture also contributed significantly to the "green revolution" of the 1950s and 1960s. In fact, the green revolution rested on three developments: improved fertilizers, pesticides, and herbicides; better crops (albeit through conventional crop-breeding in pre-biotechnology days); and enhanced machinery. In fact, the green revolution is machinery-intensive. Machines till the fields, plant seeds, spray agricultural chemicals, and harvest the grains around the globe. One can argue, in fact, that farm machinery is a necessary prerequisite for the other elements of green revolution because, even if these activities could be performed without machinery, machines are necessary to harvest the resulting high crop yields before they rot in the field or on the vine. Although, even in the United States, much of food production remains dependent on manual labor, a great deal of it is amenable to mechanization, and machines are used extensively wherever possible.

In some ways, mechanizing agriculture has led to a near-embarrassment of riches in the developed world. In fact, many farmers are victims of their own efficiency in that their production techniques are so efficient that consistent bumper crops drive prices down, threatening to bankrupt the farmers. Governments around the world respond to this in similar ways. In the United States, governmental subsidies aimed at supporting the price of agricultural products resulted in paying farmers to leave land fallow, while other governmental agencies would sometimes purchase grain at artificially high prices, again to help support higher grain prices, and the grain would then be discarded or burned. These agricultural subsidies and food wastage led to vehement protests by many. Too, the obvi-

ous disparity between food surplus in the developed world and food shortage in much of the less developed world has often fueled tensions between these two parties.

This huge disparity in food supply has, in turn, affected global politics. Food shortages in the Soviet Union, China, North Korea, and other nations inimical to the United States led to American offers of food assistance, often with a political price tag attached. The United States also provided a great deal of food to the developing world, again for geopolitical and strategic advantage in many instances. To this must be added the frequent trade disputes in which the United States and many European nations are asked to reduce or remove governmental price supports, subsidies, protective tariffs, and other measures aimed at maintaining artificially high prices for domestic agricultural products.

P. ANDREW KARAM

Further Reading

Casson, Herbert. *Cyrus Hall McCormick: His Life and Work.* 1909. Reprint, 1971.

The Century of the Reaper, Cyrus McCormick. 1931. Reprint, 1972.

Diamond, Jared. *Guns, Germs, and Steel.* New York: W.W. Norton and Company, 1997.

Smith, Bruce. *The Emergence of Agriculture.* San Francisco: Scientific American Library, 1995.

Samuel Morse and the Telegraph

Overview

From the beginning of time, humans have sought ways to communicate quickly across distances. In ancient times, groups living far from each other would send messages via couriers traveling on foot, or by setting fires to create smoke signals. In the 1830s an artist-inventor named Samuel Morse (1791-1872) conquered the age-old problem of long-distance communication with his invention, known as the telegraph. Messages would no longer take days or even weeks to deliver, but could be transmitted across towns, across the country, even across the ocean, in a matter of seconds.

Background

The possibility of sending messages electronically was born as early as the 1700s. An experi-

menter in London was able to send an electrical impulse one-sixth of a mile along thread in 1727. A writer in a 1753 issue of *Scots Magazine* described a static electricity telegraph that could spell out messages over 26 wires, one for each letter of the alphabet.

The term "telegraph" was first coined by Frenchman Claude Chappe (1763-1805), from the Greek words *tele*, meaning far, and *graphein*, to write. What finally made the telegraph a possibility was the invention of the electric battery by Alessandro Volta (1745-1827) in 1800. Several battery-powered systems were soon created that could send messages short distances over a wire, but none were put into practical use.

In the early 1800s Morse traveled to London to study art. While there, he found he greatly missed his parents, and longed for a way to quickly get in touch with them. "I wish that in an instant I could communicate the information," Samuel wrote in one of his letters, "but three thousand miles are not passed over in an instant and we must wait four long weeks before we can hear from each other." Even at the tender age of 20, the first seeds were being planted for Morse's great invention.

In 1829, while once again traveling throughout the European continent, Morse became fascinated with the semaphore telegraph system designed by Claude Chappe (1763-1805). The device was comprised of platforms placed 15 miles apart. A man stood on each of the platforms, signaling to the next man down the line using wooden codes. In this way a message could be sent 150 miles in fifteen minutes.

While sailing home on the ship *Sully* in October 1832, Morse became involved in a discussion of electricity with his fellow passengers. They recounted the experiments of Benjamin Franklin (1706-1790), who had discovered that electricity could pass instantly over a wire. Franklin had used several miles of wire for a circuit and when he touched one end, it seemed to instantly create a spark at the other end. Suddenly, Morse had an idea. "If the presence of electricity can be made visible in any part of the circuit," he said, "I see no reason why intelligence may not be transmitted instantaneously by electricity." For the rest of the trip he thought through his idea. A current of electricity, passing along a wire, would be interrupted by a spark. The spark would be one sign, its absence would be another, and the length of time between sparks would be a third. Through a combination of three signs, dot, dash, and space, the signs could be made to represent letters and words. On the other end, the electric current would move a pencil that would then print the code. This way, messages could be transmitted clear across the country, even across the world, in a matter of seconds. His idea for the telegraph was thus born.

Impact

Morse's first telegraph receiver was a simple device, made of everyday household objects—a picture frame fastened to a table, the wheels of an old clock, and lead pieces which he himself melted. He hung a pencil at the end of a pendulum, and used a type rule to break the circuit. It worked on the very first try. But the world was not interested in his new invention, and Morse was out of money.

In the hopes of garnering support for his invention, Morse held a demonstration of his telegraph in the fall of 1837. At the time, he was teaching painting and sculpture at the University of the City of New York (now New York University). He used his classroom as his stage, inviting wealthy businessmen to view his contraption. When Morse sent his code over the circuit, a pencil hung above a paper at the other end of the wire began to write out a series of dots, dashes and spaces. But while many of the investors found this invention exciting, most feared it wouldn't be practical to use. All, except one young man.

In the room was a wealthy university student named Alfred Vail (1807-1859). His curiosity was piqued, and he asked Morse to explain his invention further. Vail's father and brother owned a large brass and ironworks factory, and he believed they could help develop some of the instruments needed for the telegraph and would be interested in a partnership. Morse offered them one-fourth interest in the telegraph. Morse also asked his friend, science professor Leonard Gale, to join them. Gale introduced Morse to Joseph Henry (1797-1878) and his work on electromagnets. Henry had constructed a working electromagnetic telegraph in 1831. Together, the team worked to improve the invention, which they named the American Electro-Magnetic Telegraph.

Morse improved on previous versions of the telegraph by designing a relay system, using a series of electromagnets, to open and close circuits along the wire. In this way, the current would be strong enough to travel long distances. He and his team also revised the code, in which a series of dots and dashes represented every letter of the alphabet and the numbers zero through nine. The most frequently used letters were assigned the shortest codes. For example, the code for "e" was a dot, while the code for "q" was dash-dash-dot-dash.

The partners applied for a patent in 1837, and Morse wrote a letter to the Secretary of the Treasury to list the advantages of his invention. "First," he wrote, "the fullest and most precise information can be almost instantaneously transmitted between any two or more points between which a wire conductor is laid."

"Second. The same full intelligence can be communicated at any moment, irrespective of the time of day or night, or state of the weather."

"Third. The whole apparatus will occupy but little space."

"Fourth. The record of intelligence is made in a permanent manner and in such form that it can at once be bound up in volumes, convenient for reference, if desired."

Morse took the telegraph to Washington, D.C., to ask the U.S. Congress for funding to test his invention. But the congressmen knew little of science, and were wary about giving money to a painter-turned-inventor. Morse then traveled abroad to secure patents, thinking the Europeans would be more amenable to his telegraph, but he was again turned down. Finally, in 1843, after nearly 12 years of hard work, Congress gave in and offered to fund testing of Morse's telegraph. He and his team were given just two months to lay a 40-mile telegraph line between Baltimore, Maryland, and Washington, D.C. They worked feverishly, stringing the wires on poles set about two hundred feet apart, and endlessly testing the mechanism to ensure that messages were being transmitted successfully.

On the morning of May 24, 1844, Morse sat in the United States Supreme Court building and sent the first official message to Baltimore. "What hath God wrought!" Alfred Vail, at the other end, quickly sent back the message. It had been received in the blink of an eye! Morse's telegraph was an instant hit.

By 1846 several private companies were using Morse's telegraph design to establish lines from Washington, D.C., to Boston, Buffalo, and beyond. People used the telegraph to quickly check on their loved ones, or send business messages, without having to travel several miles. By 1851 more than 50 competing telegraph companies were in operation, transmitting messages from town to town. The companies merged in 1856 to form the Western Union Telegraph Company.

In 1861 a line ran across the continent, and by 1866 messages could be sent across the Atlantic via a submarine cable, allowing for rapid communication between North America and Europe. Throughout the world, people began to rely on the telegraph as the fastest mode of communication.

Several refinements to Morse's invention were made over the years, including improved insulation methods, a duplex circuit which allowed messages to travel simultaneously in both directions, and in 1871 Thomas Edison invented the quadruplex, allowing for two messages to travel each way at once.

Morse's telegraph not only revolutionized the way people communicate, it led to later developments in signal transmission, such as the radio, telephone, and television. Today, instant communication is conducted via such modern inventions as cellular phones, faxes, and the internet. While information is now processed and transmitted much faster than Morse could have ever imagined, his invention helped lay the foundation for today's modern communications age.

STEPHANIE WATSON

Further Reading

Coe, Lewis. *A History of Morse's Invention and its Predecessors in the United States*. Jefferson, NC: McFarland, 1993.

Latham, Jean Lee. *Samuel F. B. Morse, Artist-Inventor*. Champaign, IL: Gouard Press, 1961.

Morse, Samuel F. B. *Samuel F. B. Morse, His Letters and Journals*. Boston: Houghton Mifflin, 1914.

Reid, James D. *The Telegraph in America: Its Founders, Promoters, and Noted Men*. Arno Press, 1974.

Charles Goodyear Discovers the Process for Creating Vulcanized Rubber

Overview

In 1839, a perpetually impoverished inventor who referred to a succession of debtors' prisons as his "hotels" rescued an ailing industry and made it a multimillion-dollar enterprise. Charles Goodyear (1800-1860) discovered a process for curing rubber, which transformed this remarkable but flawed natural substance from a curiosity fit for museums into the first of the modern plastics.

Background

During his second visit to the New World in 1493-96, Christopher Columbus (1451-1506) noted that native villagers in Hispaniola played a soccer-like game with a light and bouncy ball made from the milky, white sap of a tree. The Indians cured the sap, called latex, by smoking it to evaporate out the water before forming the latex into balls. Subsequent explorers from Europe learned that latex, which was both elastic and sticky, could be pressed not only into objects for games but also into usable articles such as waterproof cloth, inflatable bags, and molded bottles and boots.

In 1735, the French mathematical geographer Charles Marie de la Condamine (1701-1774) sent back samples of crude rubber from South America, and described its botanical nature and the products that could be made from it. The French called the material *cautchouc*, from the Mayan word for weeping wood, and excitedly advertised its springiness and resistance to water. It came to be called rubber after the British chemist Joseph Priestley (1733-1804) noted in 1770 that it was superior to breadcrumbs for rubbing out lead pencil marks.

The fact that latex hardened on drying spurred a search for solvents so that products could be made far away from where the rubber was collected. In 1763, two Frenchmen found that turpentine successfully dissolved rubber but left it sticky. Experiments with ether a few years later solved the problem of stickiness. Rubber was first used commercially by the English manufacturer Samuel Peal, who applied a solution of rubber in turpentine to waterproof cloth. In 1820 the Scottish chemist Charles Macintosh (1766-1843) found that immersing raw rubber in naptha produced a liquid rubber substance that could be brushed on sheets of cotton canvas. Two pieces of this rubberized cloth pressed together like a sandwich appeared to make an ideal waterproof raincoat, or mackintosh. The first rubber factory was founded in 1820 by English inventor Thomas Hancock (1786-1865), to manufacture footwear and clothing with rubber components. One of Hancock's innovations was a machine called a masticator that welded bits of waste rubber into solid masses that could be reused. Hancock and Macintosh eventually became partners.

The remarkable qualities of rubber were immediately obvious. Elastic, plastic, strong, durable, electrically inert, and water-resistant, it was a material that begged for exploitation. But at the time it was also intractable: Macintosh's raincoats, which worked so well in the London fog, melted in the heat of the American South. In the winter cold, rubber overshoes turned hard and brittle. And the smell of rubber was

unpleasant. Although demand for the light, pliant, waterproof material from Brazil was initially high, factories that sprang up in the 1830s in a wave of "rubber fever" quickly went under. Millions of dollars were lost in rubber ventures. Rubber products seemed destined to remain a marginal article of commerce.

In 1834, a bankrupt young inventor from Connecticut named Charles Goodyear tried to sell an improved airtight valve for life preservers to the Roxbury India Rubber Company. The company refused his invention on the grounds that it made no sense until someone could come up with a better rubber. Goodyear had always been enamored of rubber, and needed little encouragement. Although he was penniless and had no knowledge of chemistry, he set about to find the full range of rubber's potential. During this time, Goodyear, who had a wife and small children, was in and out of debtors' prison, where he cajoled the prison guards into letting him conduct experiments in the prison kitchen.

Goodyear plunged into a trial-and-error orgy of mixing natural latex rubber with anything he could find, including witch hazel, castor oil, and ink. Because rubber was naturally adhesive, he wondered what he might do to absorb the stickiness. He tried adding two drying agents to his rubber, first magnesia, then both magnesia and quicklime, with limited improvement. His first success occurred in 1836 when he treated rubber with nitric acid and bismuth and copper nitrates—the so-called acid-gas process—which made the rubber as smooth and dry as cloth and appeared to improve its resistance to heat. He found a financial backer to begin production of his material, but bad economic times ended the scheme, and Goodyear was reduced to fishing to keep his family alive. In 1837 he accepted a friend's contract to manufacture nitric-acid-cured rubber mailbags. But in the summer heat the mailbags crumbled, and Goodyear realized that the nitric acid had only cured the surface of the rubber. He redoubled his attempts to solve what he called the "riddle of rubber."

Five years into his investigations, and the year he turned 40, Goodyear accidentally spilled some raw latex mixed with sulfur on a hot stove. Instead of melting like ordinary rubber, the substance charred evenly, like leather. He applied extremes of heat to the mixture, but it did not melt. Nor did hanging the charred material outside overnight in the cold Massachusetts winter cause it to shatter. It remained pliable. In a later test, Goodyear saw that along the edge of the charred rubber was a border that was perfectly cured. His happy accident would quickly revolutionize the rubber industry.

In chemical terms, what Goodyear had managed to do was to link together the long, linear chains of molecules in rubber. Rubber is derived from the sap of the rubber tree *Hevea braziliensis*. When natural rubber is heated with sulfur, the sulfur forms cross-links between the chains that are like a bridge, or the rungs of a ladder. The more sulfur is added, the greater the degree of cross-linking. In natural rubber, the chainlike molecules can slip past one another, or around each other, which is what causes the rubber to melt when it gets warm, and to break apart when it gets cold. If car tires were made of natural rubber, the friction of the road surface acting on them would reduce them to useless glop. But the cross-links act like anchors to prevent slippage. They are also what allows rubber to return to normal after it has been stretched (a property called elasticity). A rubber band, which is only moderately cross-linked, is elastic. A car tire, which is extensively cross-linked, is hard and bouncy rather than elastic.

Impact

Although Goodyear now knew that heat and sulfur changed the properties of rubber in a dramatic way, he had no idea how much heat he needed to apply or for how long. Many months of personal hardship passed, but in the end he came up with a formula that guaranteed uniform results. Overnight, what Goodyear called "vegetable leather" and "elastic metal" because of its durable properties became a useful commodity. It was stretchable, tough, waterproof, and could be used to stick things together. Above all, it was moldable, which made it ideally suited to mass manufacturing. Used first to make the ruffled fronts then fashionable in men's shirts, vulcanized rubber was soon shaped into other items of clothing, harnesses, bottle stoppers, frames for photographic plates, cigarette holders, and rubber dental plates. Consumption of rubber increased from 38 tons in 1825 to 8,000 tons in 1870.

Solid rubber tires, kinder to roads and to grasslands, appeared in 1867 for use in steam road vehicles, and in 1869 for bicycles. Rubber belting was used as conveyors to handle seeds and grain, and then minerals and ores, and to drive machinery and vehicles. With the discovery of the incandescent electric lamp, electricity would need to be distributed, stimulating demand for insulated cables and wiring.

But vulcanized rubber was not without disadvantages. It had to be gathered by hand, and prices were kept high worldwide by a small cartel of rubber "barons." Moreover, the material was not corrosion-resistant, which made it vulnerable as an insulator for submarine cables. Only once vulcanized rubber began to be used in machinery and particularly in automobile tires did the modern rubber industry become a reality. So effective was Goodyear's formula that, despite some technical improvements, the vulcanization process has changed very little since his day.

An appealing feature of Goodyear's recipe for cured rubber was its simplicity. Unfortunately, it also made the formula very easy to copy. Goodyear patented his invention in the United States, but had no money left to file for British patent protection. He proposed a British-American joint venture to the firm of Charles Macintosh & Co., but his agent's handling of the matter was maladroit. Consequently, four years after Goodyear's own discovery, Thomas Hancock, the managing director, reverse-engineered some samples the agent had left behind and promptly filed a British patent himself, only weeks before Goodyear got around to filing his own patent. It was Hancock who, on the suggestion of a friend, named the process vulcanization, after Vulcan, the ancient Roman god of fire.

Goodyear spent the rest of his life embroiled in patent suits—he prosecuted 32 infringement cases all the way to the Supreme Court—at one time hiring Daniel Webster at $15,000 for two days' work to defend him. At the Great Exhibition of 1851, he borrowed heavily to underwrite a Vulcanite Court, every part of which, including furniture, musical instruments, and six-foot-diameter balloons filled with hydrogen, was fashioned from vulcanized rubber. Six million people visited his exhibit.

Goodyear's life was laced with irony. Jailed in Paris for 16 days in 1855 for nonpayment of debts, he received the Cross of the Legion of Honor from French emperor Napoleon III while in prison (his son brought the medal to his cell). By 1860 an industry founded on his rubber patents was employing 60,000 people and making over $8 million a year; yet Goodyear died $200,000 in debt. At one time or another he pawned his children's schoolbooks and his wife's jewelry to pay bills. His autobiography netted him more money than he ever made from his 60 patents. His vision was unabashedly monomaniacal, though curiously enough, some of the ideas Goodyear had for uses of rubber that must have appeared madcap a century ago enjoy application today, including packaging for food, rubber paint, car springs, wheelbarrow tires, water beds, inflatable life rafts, and frogman suits. Had he been able to, he would have used rubber to remake the world. And in a way, he did. Vulcanized rubber paved the way to synthetic materials.

GISELLE WEISS

Further Reading

Books

Fenichell, Stephen. *Plastic: The Making of a Synthetic Century.* New York: HarperBusiness, 1996.

Goodyear, Charles. *Gum-Elastic and Its Varieties, with a Detailed Account of Its Applications and Uses, and of the Discovery of Vulcanization.* New Haven: Pub. for the author, 1853.

Periodicals

Kauffman, George G. "Charles Goodyear—Inventor of Vulcanisation." *Education in Chemistry* (November 1989): 167-170.

Other

"Charles Goodyear and the Strange Story of Rubber." http://www.goodyear.com/us/corporate/strange.html

Invention of the Sewing Machine

Overview

When the first cave dwellers attempted to keep warm by covering themselves with animal skins, they likely recognized the advantage of joining several pelts together to form larger "garments." Eventually, one of them may have noticed a sliver of bone on the ground and thought to use it as an tool to string together multiple pelts.

Despite such speculation, the important fact is that from the dawn of humanity until the present day all fabric attachment processes have had the same thing in common: the needle. It has undergone many changes, not only in appearance, but through all the stages of iron, steel, and other metals that lend themselves to high levels of refinement.

Background

The needle is an indispensable tool that enables the production of highly sophisticated clothing for much of the civilized (no longer naked) world. However, its facility was long limited to the skill of the individual using it. We have only to look at the surviving articles of embroidery and needlework from centuries ago to recognize the abilities of the people who used their needles to such advantage.

With population increases and more and more clothing needed for daily use, it was only a matter of time until someone would try to find a way to move the needle mechanically—thus much faster and consistently.

Elias Howe. *(Archive Photos. Reproduced by permission.)*

The earliest mention of a sewing machine is in a 1790 patent issued to British inventor Thomas Saint. Actually, Saint didn't use a needle per se, but an awl that pierced a hole over which a needlelike rod hooked the thread through to the underside of the fabric and moved it forward to repeat the operation. When the cycle was repeated, a second loop was created that formed a chain and effectively locked the stitch in place. Its use was mainly for leather and canvas and could not be adapted for clothing fabric weights. Even though the basic idea was good, Saint's machine never went beyond what we presently refer to as a prototype.

The next important step toward a machine that would sew fabrics mechanically was taken in 1830 by French tailor Barthelemy Thimonnier. He was encouraged to develop his machine by the French army, which needed new uniforms and ordered 80 machines to help produce them. Factory workers were so frightened that the sewing machines would put them out of work that an angry mob of tailors destroyed all 80 of the units, which effectively delayed further research on mechanical sewing for some years.

The next major innovator of the sewing machine was an American named Walter Hunt (1796-1859), who in 1834 devised a machine that utilized both an eye-pointed needle and a moving shuttle. He failed to patent it at that time and when he tried later to get it registered, he was rejected on the basis of "product abandonment."

This activity was followed by another American named Elias Howe (1819-1867), who put together a unit that contained many of the mechanisms that Hunt had used earlier. Although Howe patented his device, the machine was not accepted in the United States. He took it across the Atlantic, where it had a better reception and where he was persuaded to sell some of his patent rights to British investors for the paltry sum of 250 pounds sterling ($1,250).

Howe moved to England and worked toward adapting his machine for use on leather and other heavier materials. Personal misfortunes beset him and, when he returned home, he found his wife dying. He endured years of financial hardship and was shocked to find that in his absence, other people in the United States were manufacturing and selling sewing machines in violation of his patent. After many years in court, he prevailed and in 1854 began receiving royalties on his invention. Until his patent expired in 1867, he received royalties on all sewing machines produced in the United States.

Throughout the years of litigation, there were numerous manufacturers (some of them inventors) who added refinements to Howe's basic model. The largest and most successful name in the patent pool was Isaac Merrit Singer (1811-1875). After more than 10 years of working as a machinist, Singer had secured employ-ment at a machine shop in Boston in 1851. One day, he was asked to repair a sewing machine that had been manufactured by Lerow and Blodgett. Eleven days later, he came up with a much improved machine that he patented and sold through his own company then called I. M. Singer & Company. The business was incorporated in 1863 and was then called the Singer Manufacturing Company.

Impact

During this productive era, many other sewing machine manufacturers emerged, but none as prominent and successful as Singer. His treadle-operated unit was the first practical sewing machine for domestic use, and he soon bundled his various patents to build the machines on a mass-production basis.

To further his cause, he went on to demonstrate the first electric-powered machine at the Philadelphia electric exhibition in 1885. This created an unprecedented market for his product and Singer was the first major manufacturer to make it available to every home in America by what would become a widespread consumer practice: Buy on credit, pay by the month.

The sewing machine saga involves the names of many now relegated to obscurity, but who nonetheless produced most of the "add-ons" that have made the machines so successful. Allen Benjamin Wilson patented the rotary bobbin in 1850. Reece Machinery Company of the United States pioneered buttonhole machines before the end of the nineteenth century. John Barran of England developed a multi-layered cloth cutter that was an important adjunct to the factory machines, as was the Hoffman press, which provided machine pressing of sewn fabrics instead of the time-consuming hand operations.

For those who have not looked closely at an electric sewing machine, the operation may appear complicated. However, in actual practice, it is a relatively simple progression of steps. The basic nineteenth-century format has been embellished and refined, but the results are readily apparent. Generally, sewing machines use two sources of thread: one for the needle which pierces the fabric; the other, from the bobbin beneath, which secures the thread in a straight line of locked stitches.

Here's how it works. The needle carrying the upper thread moves down through the material and below a metal plate that looks like a small platform with one or more holes milled for specific penetration. Under this plate is a cylindrical bobbin around which is wound the same ply of thread as the needle carries. When the needle moves below the plate, it picks up the thread from the bobbin and forms a loop. The size of this loop is determined by a tension device on the visible (upper) part of the machine. The loops are repeated and form what is aptly called a lockstitch.

In the early days of mass-produced articles of clothing, most machine sewing was augmented by hand-finishing work. In the more expensive "design" clothing, hand sewing was still much in demand and there was always a market for skilled seamstresses and men's tailors. However, as years passed, the United States became a dominant supplier in the ready-to-wear industry and consequently, many Europeans came to this country to work in the factories (and sometimes sweatshops) centered mainly in New York City. They brought their willingness to work in exchange for the religious and economic freedoms they could not find in their native countries.

Along with their conscientious work habits, they contributed a healthy spirit of enterprise, which engendered numerous improvements in the sewing machine industry as a whole. The machines were made more effective with refinements such as automatic buttonhole makers, machine-sewn hems and collars, machine embroideries, mechanical pleat-folders, and patterns which could be mass-produced by blades that cut through multiple layers of cloth. This latter development led to a spreading machine, which spread fabric from long bolts into layers that could contain hundreds of plies of fabric, depending on the thickness and density of the weave.

As the sewing machine was further improved, another major industry developed to keep pace with the machines—weaving. With the increased demands for longer and longer lengths of fabrics, mills were built—particularly in the British Isles—to spin, dye, treat, and weave more and more cloth. However, the supply of wool and cottons in Britain was soon diluted by the construction of American mills on the east coast, where imports of cotton, woolens, and silks increased as the shipping industry took giant steps forward. The New England region of America soon challenged the Old England market in the production of fabrics for commercial and consumer use.

By the end of the nineteenth century electric sewing machines were used wherever electric power was available. However, a large portion of the world's population was still without even electric light for their homes, farms, or remote villages. For this substantial market, the original hand- or treadle-operated sewing machines were a viable product. In places where electric power is still lacking or unavailable, the foot-treadle machine remain in use today.

GERALD F. HALL

Further Reading
Brandon, Ruth. *Singer and the Sewing Machine: A Capitalist Romance.* Kodansha International, 1996.

Elisha Graves Otis Produces the First Passenger Elevator with Safety Locks, Facilitating the Growth of High-Rise Buildings

Overview

A descendent of James Otis, a British immigrant who arrived in North America in the seventeenth century, Elisha Graves Otis was born in 1811. He grew up in Vermont before moving to the state of New York to work as a master mechanic. Working for the Bergen Company in Yonkers, he installed a "safety hoist" in a building in 1852 as a means to prevent the sudden fall of the elevator. Five years later, he designed and constructed the first safety elevator in a New York City building. By 1861, following additional improvements to his system, Otis's sons formed the Otis Elevator Company. Together with other innovations in building practices, the elevator became the cornerstone for the effective development of the skyscraper, the first of which rose in New York City. The elevator thus changed the urban landscape and, consequently, the way businesses worked in cities.

Background

Elevators date as far back as Roman times. Engineering texts from the first century B.C. describe the use of platforms that employed pulleys and were operated by humans, animals, or even water power. Rome's Coliseum, for example, had 12 elevators that were used to hoist gladiators and wild animals to the stage level. Such tools, however, were used primarily to lift construction material and their existence is documented all the way to the nineteenth century. In some palaces, dumbwaiters were installed to bring the monarch's food more quickly then through the stairway. By then, in England, the steam machine was used to power certain lifting platforms, and a variation, the hydraulic pump (in which the fluid in the cylinder was thrust by steam), was also tested. Regardless of their potential use, these elevators all suffered from a major disadvantage—the platform might break loose from its attachment, thereby harming or killing passengers or the operators below. It comes as no surprise, then, that most elevators remained confined to use as freight lifts in factories. Otis's innovation, the elevator brake, however, would change this situation.

Elisha Graves Otis suffered from ill health throughout his life, also affecting his early attempts at establishing a business. In 1845 he moved to Albany, New York, where he worked as a master mechanic in the Tingley Bedstead Company. While there, he invented a railway safety brake and other devices to improve the running of turbine wheels.

In 1852 he moved to Yonkers, New York, to organize and install machinery for another bedstead company, which required the use of a hoist to transport equipment to the factory floor. Otis became concerned with the equipment's safety problem and devised a pair of spring-loaded pieces of metal that would engage into the cog-shaped rail if the rope gave way. So successful was the device that Otis soon received three unsolicited orders for similar systems. He set up his own shop and formally went into business for himself in 1854, selling his safety system for $300 apiece.

No new orders followed, however, prompting Otis in 1854 to promote his invention at the Crystal Palace Fair in New York City during its second season. There, facing a large crowd, Otis ascended in a drum-operated platform. Suddenly, he took out a knife and slashed the rope, and the safety system engaged automatically. Otis then announced (as he would in several such

Elisha Graves Otis demonstrating his elevator. *(UPI/ Corbis-Bettmann. Reproduced by permission.)*

demonstrations), "all safe, gentlemen, all safe." New orders for Otis's device soon followed, totaling $13,488 by 1856. But it was not until 1857 that the first passenger elevator began operation, in the New York City Haugwout department store. At that time, the steam-powered elevator traveled at a leisurely 40 feet per minute (4.8 miles per hour). Otis, however, did not live to see the further success of his machine; he died of diphtheria at the age of 49 in 1861.

Otis's sons Charles and Norton took over the business, which at the time faced a strong deficit, and built a new factory and devised new elevator models that put the company back in the black. The hydraulic elevator replaced the vapor-powered one, as it could rise to greater heights at higher speeds (steam used winding drums while early hydraulics relied on a ram type, based on a plunger and cylinder). Later, the roped or geared hydraulic elevator would become capable of speeds up to 600 feet per minute and rise to 30 or more stories. Between 1880 and 1900, all major 10- or 12-story buildings in New York City used hydraulic elevators. In parallel to new methods of propulsion, other improvements—including enclosures, doors, stronger wires, and warning bells—were also added. The first electric elevator was installed in 1889, and five years later the first push-button machine went into service. Otis's invention now became an industrial tool essential to all kinds of businesses, thus changing the urban landscape.

Impact

The advent of the elevator in buildings reversed many trends and created new ones. For instance, hotels that had had trouble filling upper levels now found that they could charge more for rooms with a view. At the same time, the need for business space that followed the economic boom in North American cities in the 1870s meant that downtown locations were at a premium. The appearance of the iron-structured building, combined with the existence of the elevator, cleared the way for the appearance of the first skyscrapers, since metal structures allowed a better distribution of the building's weight and could include the elevator shafts without unduly stressing the building's internal structure. Insurance companies were among the first to advocate the construction of large, tall buildings that would reflect their financial power and success. The Equitable Company, for example, by commissioning a building with two elevator shafts—built by Otis Tufts, a competitor of the Otis Elevator Company—was able both to double the size of the average business building and to rent out the upper floors successfully.

Otis Elevator Company, however, remained a leader in the field. It rose to many challenges, including the design of a special diagonal elevator system (which remains in use to this day) for Paris's Eiffel Tower, inaugurated in 1889. Other challenges included devising how many elevators would be necessary for different types of buildings and how fast these machines should go.

By 1900 "elevatoring" was no longer an inventor's craft, but a complicated science that required both heavy engineering knowledge and an understanding of the business requirements of each building, especially as the number of floors increased. (In 1893 the average business "skyscraper" was 20 stories tall; ten years later, it was reaching 25.) Initially, architects suggested a simple rule of doubling the number of elevators every time the number of floors doubled. Soon, however, they realized that this rule did not solve the problem, for people still had to wait for the elevator. The advent of the new electric-traction elevator made further increases in speed and numbers of floors possible. Instead of huge hydraulic shafts required for elevator operation, the electric drum could be mounted at the top of the shaft

without taking up rental space. The use of electricity also allowed better regulation of speed.

Safety problems, which continued to affect the elevator industry in the early years of its existence, were eventually ironed out. Otis's initial safety system turned out to work well only for slow-moving vehicles. Using the elevator shaft as a safety system in which compressed air would slow a falling car did not always work; figures released for the years 1909 to 1911 suggested that over 2,600 people had been killed in elevator accidents. Through the installation of new brake systems that clamped around the guiding rails, the elevator slowed down without injuring its occupants through sudden jerk movements.

In parallel to elevator safety improvements, a new jump in skyscraper design followed. In 1913 the Woolworth building was completed. It had 60 floors and 26 elevators and the safety system tested on inauguration day functioned perfectly. (The rope was cut and the bucket of water left in the car reportedly did not spill a drop.) In fact, the cars went so fast in regular traffic that New York City building codes had to be changed to allow for greater speeds. These speeds increased again in 1931 with the inauguration of the Empire State Building, whose 73 elevators went as fast as 1,200 feet per minute. There is a limit, however, to how fast one can travel comfortably in an elevator car. Several Mitsubishi models used in Japan since the mid-1990s average 2,200 feet per minute and take 12 floors to slow down to a stop—any faster and the passengers would likely fall sick or even injure themselves as a result of gravitational forces.

Further improvements followed but, aside from the fully automatic elevator that did not require an operator, most improvements have been a matter of architectural taste as well as a function of the passengers' comfort. The proliferation of elevators has, in turn, spawned a whole new behavioral process. People waiting for an elevator and express their impatience by pushing the button repeatedly. Then, for a few seconds or even minutes at a time, unacquainted users find themselves together in very tight confines. Unless they are riding in a glass paneled elevator (common in certain hotels and shopping malls), passengers typically occupy spaces near the controls or other corners and avoid eye contact or small talk. As for elevator phobia, it existed from the very beginning of safety elevators, being exacerbated by numerous accidents in the first years of operation. While elevators have in fact become one of the safest means of transportation, claustrophobia as well as the fear of crime in public elevators remain strong. These machines have also come to be associated in popular culture with various terrors and pleasures, ranging from fires and elevator-shaft falls to humorous skits and romantic encounters.

The sturdiness of elevator systems remains quite astounding, and some machines are in better condition than the buildings in which they operate. The hydraulic elevator operating in New York City's 34 Grammercy Park building, for example, was installed in 1883 and still runs today—at 51 feet per minute. Virtually as omnipresent as the elevator itself, the Otis Corporation continues to dominate the elevator market from its headquarters in Connecticut, where it tests new models in an 11-story tower.

GUILLAUME DE SYON

Further Reading

Books

Douglas, George H. *Skyscrapers: A Social History in America.* Jefferson, NC: McFarland, 1996.

Landau, Sarah Bradford, and Carl W. Condit. *Rise of the New York Skyscraper 1865-1913.* New Haven, CT: Yale University Press, 1996.

Strakosh, George R. *Vertical Transportation: Elevators and Escalators.* New York: John Wiley, 1967.

Periodicals

Dale Jackson, Donald. "Elevating Thoughts from Elisha Otis and Fellow Uplifters." *Smithsonian Magazine* (November 1989): 211-34.

Klaw, Spencer. "All Safe, Gentlemen, All Safe!" *American Heritage* (August/September 1978): 40-45.

Other

www.elevator-world.com/magazine/magazine.html-ssi

English Inventor Henry Bessemer Develops Process to Produce Inexpensive Steel

Overview

In 1856, Henry Bessemer (1813-1898) developed a new method for manufacturing steel. The Bessemer process made possible the manufacture of large amounts of high-quality steel for the first time. This, in turn, provided steel at relatively low cost to various industries. By revolutionizing the steel industry, the Bessemer process helped to spur on the Industrial Revolution. Within a few decades, foundries were making railroad track, bridge girders, locomotives, armor plating, and other steel-based products.

Background

Iron has been known to man for several thousands of years; the earliest iron implements found thus far are from Egypt and were made about 3000 B.C.. However, iron is a relatively soft and brittle metal on its own. By about 1000 B.C. the Greeks had discovered that heating iron would help to harden it, but it was still brittle and not very useful for any tasks requiring strength. Other cultures heated iron ore and charcoal together, making what we now call wrought iron.

During the fourteenth century, pig iron was developed. Pig iron is made by heating iron

Henry Bessemer's invention of the pneumatic conversion process in 1856 made steel a commercially viable material. *(Archive Photos. Reproduced by permission.)*

bars, coal coke, and limestone together in a fire or furnace. First the iron ore is covered with the coke and limestone and heated for up to a week, allowing the carbon to diffuse into the surface of the iron. Then, the metal is hammered and folded in order to mix the carbon throughout the iron in the same way that kneading bread distributes the yeast throughout the dough. Unfortunately, this process was time- and labor-intensive and produced steel of varying quality.

Heating these ingredients together in a crucible was the next innovation. By allowing some degree of mixing during the heating, this process produced a more uniform quality of steel, although still in limited quantities. Even so, this method allowed one city in England to increase its steel production from about 200 tons annually to over 20,000 tons per year within a century.

The next breakthrough came in 1856 when Henry Bessemer developed the process for steel-making named for him. In fact, unknown to Bessemer, this method of making steel was developed nearly simultaneously by the American William Kelly (1811-1888). However, Bessemer filed his United States patent application first and has received the majority of the credit. In the Bessemer process, air is continuously forced through the steel while it is contained in the crucible. This burned the carbon present in the steel, raising the temperature, and removing many of the impurities that would otherwise impair the quality and strength of the final metal. In addition, Bessemer's converters (now also called blast furnaces because of the large volumes of air that are blasted through the molten steel) ran continuously rather than in batches as was the case with crucible steel. This produced larger quantities of steel, another improvement. In this process, some carbon remains in the steel, helping to make the steel both stronger and more flexible than the original iron.

The Bessemer process has proved nearly as durable as the steel it produces. After nearly 150 years it is still the primary method of steel manufacture in the world. Other manufacturing processes are used, but mostly for specific types of steel requiring different properties. In addition, although steel today is similar to that of the 1860s, a number of specialty steels have been developed. Some examples of these are stainless steel, tool steel, spring steel, and special alloys used in extreme environments (such as jet engines, nuclear reactors, inside the human body, and so forth).

Impact

The impact of the availability of inexpensive, high-quality steel can hardly be overstated. The nineteenth century in Europe was the time of the Industrial Revolution. The development of the steam engine into a useful device, the development of the railroad, and the internal combustion engine all occurred during this century. These devices were made more efficient by Bessemer's steel. And this new steel also made other innovations possible. The importance of Bessemer's steel soon became apparent in the areas of industry and commerce, civil engineering, and the military.

Industry was the first and most obvious beneficiary of the new steel. Steel production in Britain increased from about 50,000 tons to over 1.3 million tons annually in just 25 years. At the same time, the price of this steel dropped to half of its previous level. Much of this extra steel went to industry. Some of the steel was used for infrastructure, for making the machines that made the goods that were sent to market. Using steel instead of iron, wood, or other materials helped to increase the lifespan of manufacturing devices, in the end making the manufacturing process more efficient. In addition, the greater strength of steel made some devices possible that simply could not be constructed before. Steel blades or cutting surfaces helped make saws last longer, reducing the cost of manufacturing wood products, while steel tools and parts helped make possible high-speed lathes. Also, industry could use steel in the products it made for consumers. Steel was harder than iron, less expensive than wrought iron, and more durable than wood, stone, or glass. While not the perfect material for all uses, its versatility was impressive.

All of these attributes make steel an important part of any national economy. The availability of durable and relatively inexpensive steel products to consumers encourages purchasing, thereby fueling the supply-and-demand market economy. High-quality, low-cost manufactured goods can be exported, bringing foreign currency into a country. Steel is one of the cornerstones of a healthy industrial economy, and steel plants are among the first major purchases made by developing nations as they become industrialized. It is also worth noting that, in the 1990s alone, a number of trade disputes revolved around allegations of "dumping" steel into foreign markets at artificially low prices to gain economic advantage or, in some cases, to try to hurt another nation's steel manufacturing industry. That such practices occur is yet another indication of the economic importance of steel to nations.

Steel was also important in the making of many industrial empires. In the United States, Andrew Carnegie made his fortune selling steel to the developing nation. Steel from Carnegie's furnaces went into the trains traveling to the American frontier, it built the rails they rode on, and the guns the passengers carried. In Germany, the industrial empire of the Krupp family was built, in part, on steel and steel products. Best known for supplying weapons to Nazi Germany, Krupp also built farming equipment and vehicles for many years until they were split up by the Allies after World War II.

Steel has also had a major impact on civil engineering and architecture. Without steel there would be no skyscrapers, no suspension bridges, no railroads, no reinforced concrete, and no modern highways. The strength and relative lightness of steel have made all of these things possible. It is safe to say that no large city would look the same without steel.

One of Bessemer's reasons for developing an improved steel manufacturing process was the need for more guns by British troops during the Crimean War. Among the first products he made with his new steel were guns, along with railroad track. Guns are, of course, directly useful by the military and, by allowing more rapid troop mobilization, the railroad helped to revolutionize warfare. During the American Civil War the superiority of the Northern rail system combined with the North's industrial strength helped the northern states win the war.

Other uses to which steel was quickly put were armor for vehicles, stronger gun barrels for artillery pieces, motor vehicles, naval ships, engines and turbines, and, later, tanks and parts of airplanes. As in so many other areas, it is hard to conceive of modern warfare without steel.

Although steel is being supplanted in some areas by polymers, ceramics, composites, and other materials, it still plays a vital role in modern society. While some automobiles are now made with polymer body panels, for example, steel is used for horizontal surfaces and for reinforcement within the doors because of its superior strength and ability to protect passengers during an accident. The computer and keyboard on which these words are typed are composed largely of plastic, silicon, and copper, but the assembly line on which they were manufactured is made with steel. Steel may, at some point in time, be replaced with other materials, but that time is not likely to be in the near future. In the meantime, there can be no doubt that the world we live in would be profoundly different without the availability of inexpensive and high-quality steel.

P. ANDREW KARAM

Further Reading

Diamond, Jared. *Guns, Germs, and Steel: The Fate of Human Societies.* New York: W.W. Norton and Company, 1999.

Institute of Materials. *Sir Henry Bessemer, F.R.S.: An Autobiography.* Ashgate Publishers, 1989.

Kent, Zachary. *Andrew Carnegie: Steel King and Friend to Libraries.* Enslow Publishers, Inc., 1999.

The Advent of Mechanical Refrigeration Alters Daily Life and National Economies throughout the World

Overview

People have taken advantage of natural refrigeration for thousands of years. Caves, holes dug in the ground, springs, ice and snow, and evaporative cooling have all been used to cool food and drinks. Natural refrigeration, however, has limitations. Its availability depends on location and weather conditions, and it has never been adequate to chill large quantities for long periods of time. The scientific study of thermodynamics

and chemistry that began in the seventeenth century, accompanied by advances in manufacturing technology, led to the birth of the mechanical refrigeration industry in the nineteenth century.

Background

An Egyptian fresco from 2500 B.C. shows slaves fanning water jars, an early record of human efforts at cooling. The people of ancient Egypt and India knew how to make ice by exposing jars of water to the clear night air. While both societies credited supernatural forces for this phenomenon, it was a combination of evaporative cooling through the porous jars and radiational cooling into the night sky that chilled the water and froze its surface.

A Chinese poem to the Great One of Cold written around 1175 B.C. describes harvesting ice and storing it in a cave. One thousand years later ancient Greeks and Romans collected snow from the mountains and kept it in pits covered with straw and branches. Wealthy Romans used snow to chill water and wine and for cold baths called *frigidaria*. Alexander the Great had thirty pits filled with snow so that his troops could drink cold wine during the siege of the Indian capital of Petra.

Chilling preserves food by slowing both the growth of harmful microorganisms and the rate of metabolism and cellular respiration of the food. Long before this relationship was understood, there is evidence that Iron Age (beginning in Europe and the Middle East c. 1200 B.C.) communities stored food underground. The low temperature and humidity in caves maintained the freshness of seeds and grains while preventing losses from mold, fungus, and insects. Before mechanical refrigeration crops such as potatoes, apples, and cassava were often stored underground or in aboveground structures covered with straw and soil where they stayed fresh for as long as six months.

It has been known since antiquity that certain chemicals lower the temperature of water or snow. Chemical refrigeration was not common until the 1500s, however, when it became fashionable for the Italian nobility to chill wine in a solution of water and saltpeter (ammonium nitrate). Later that century British scientist Francis Bacon (1561-1626) furnished the royal family with ice by mixing saltpeter and snow. Robert Boyle (1627-1691), one of the founders of thermodynamics, also studied various salts as freezing agents and the effect of cold on animals, vegetables, and minerals. His *Experimental History of Cold* was the first scientific study of refrigeration.

As experimentation with chemical refrigeration continued during the eighteenth century, the invention of the mechanical pump and thermometer provided the technology for new areas of research. William Cullen (1710-1790), a professor of medicine in Scotland, found that evaporative cooling increases in a vacuum and that volatile liquids like ether produce even lower temperatures.

Edward Nairne (1726-1806) and John Leslie (1766-1832) discovered that sulfuric acid absorbs water vapor, an effect that produces cold. John Dalton (1766-1844) observed that air cools or heats its surroundings depending on whether it is expanding or contracting. These and other experiments were the foundation of mechanical refrigeration in the nineteenth century.

Before mechanical refrigeration became practical, the first half of the nineteenth century saw the rise of the natural ice industry. Norway was a major exporter of ice, sending large quantities to Europe and England. In the United States Frederic Tudor (1783-1864) built a business that eventually shipped ice to every major port in South America, Asia, and Australia. The demand for ice grew with the century, boosted by the brewing industry in the 1860s. By 1872 the United States exported 220,000 tons of ice a year. A specialized technology for harvesting and storing ice arose, including circular ice saws, ice houses, and iceboxes for homes and businesses. Artificial ice began to overtake natural ice in the 1890s as water near cities became polluted. An unusually mild winter in 1890 emphasized the unreliability of natural ice. At the same time, mechanical refrigeration was increasingly adopted for ice making, cold storage, and breweries.

Two types of refrigeration machines evolved during in the nineteenth century. The first worked by compressing either air or a vapor such as ethyl ether and ammonia. When the air or vapor expanded again, it produced cooling. The first vapor-compression system was invented by Oliver Evans (1755-1819), who published a theoretical description of a closed vapor-compression machine in 1805 but renounced his rights to the invention. In 1834 his friend Jacob Perkins (1766-1849) took out an English patent on Evans's design, which was used to build the first ice-making machine.

John Gorrie (1803-1855), a Florida physician, invented the air-compression refrigerator in 1844 because he believed that heat and humidity were responsible for "...the mental and physical deterioration of the native inhabitants...." He used his machine to make ice and cool the bedrooms of his malaria and yellow-fever patients. Gorrie's ultimate goal was to cool entire cities, but he was unable to obtain financial backing for his ice-making machine and died disappointed and in debt.

Apparently without knowledge of these earlier designs, the American engineer Alexander Twining (1801-1884) began experimenting with ether evaporation and condensation in 1848. He built a prototype freezing machine in 1850 and opened the first commercial ice-making plant using vapor refrigeration five years later. His attempt to open a second plant in New Orleans was stymied by the Civil War (1861-65), which gave his rivals, Ferdinand Carré (1824-1894) and W. James Harrison (1816-1893), the opportunity to gain a foothold in the southern states.

Charles Piazzi Smyth (1819-1900) shared Gorrie's belief in the need for mechanical cooling in hot climates, particularly in the hospitals of India that he visited for the British government. Around the same time as Gorrie, Smyth designed and built a compressor that could be powered by humans or oxen on a treadmill. Air-cycle compression systems were further advanced in the 1860s by Alexander Kirk (1830-1892), who installed ice and refrigeration machines throughout the British empire. Air-cycle refrigeration systems, the predominant design for ships and hospitals until the 1890s, had relatively low thermal efficiency. However, air, unlike ether and ammonia, was free, non-toxic, non-flammable, and always available.

W. James Harrison began experimenting with vapor-compression refrigeration in 1854 after studying the patents of Perkins, Gorrie, and Twining. He moved from Australia to England, where he worked with the engineering firm Siebe & Company on the production of a refrigeration machine. The first models were sold to a brewery and a petroleum company. His next goal was to export frozen Australian meat. After years of failure and bankruptcy, Harrison finally achieved success with Thomas Mort (1816-1878) and Eugène Nicolle (1824-1895?), owners of the Fresh Food and Ice Company in Sydney.

One of the earliest successful pioneers in commercial liquid-vapor refrigeration in the United States was David Boyle (1837-1891). Boyle saw the potential in refrigeration after making $8,000 selling cold lemonade. Unable to buy a satisfactory machine, he designed his own ammonia-compression system in 1873, a design that was manufactured until 1905.

Breweries, which require cool temperatures for fermentation and storage, greatly increased the market for refrigeration in the second half of the nineteenth century. Carl Linde (1842-1934) brought a rigorous scientific approach to the design and construction of brewery refrigeration systems. After studying the thermodynamics of refrigeration, he built a machine with double the efficiency of existing plants. He then patented his design, which was installed in breweries in Germany, Great Britain, and the United States.

The second type of refrigeration, aqua-ammonia absorption, grew out of experiments by Edward Nairne (1726-1806), John Leslie (1766-1832), and John Vallance (1801-1850) in which they used sulfuric acid to absorb water. Edmund Carré (1822-1890?), the first to commercialize the absorption method, sold ice-making machines to several French cafes in the 1850s. His brother Ferdinand, who went on to obtain more than 50 patents in the field of refrigeration, improved the method by switching from acid to ammonia. Two ammonia absorption machines that were smuggled into southern ports during the Civil War became the prototypes for absorption refrigeration machines throughout the United States. In 1877 Edmund Carré installed one of his brother's machines in a steamer that carried sheep carcasses from Buenos Aires, Argentina, to Marseilles, France.

Daniel Holden (1837-1924) improved one of Carré's original machines by making ice with distilled water instead of river water. The result was crystal-clear ice, an important attraction for customers. Holden built both absorption and compression plants and patented a combination of the two in 1877.

By the end of the century 400 aqua-ammonia absorption plants had been built, mostly in the southern United States. They were thermodynamically suited for ice making, but less so for cold storage and shipboard refrigeration. Absorption plants, as well as air-cycle machines, lost favor once high-efficiency engines for ammonia compression plants were invented in the 1890s.

Impact

The development of refrigeration during the nineteenth century propelled scientific and technological progress in related fields. Advances were made with refrigerants, insulating materials, refrigerated boxcars, steamships, warehouses, and compressors. Because refrigeration was a competitive and potentially profitable business, the drive for improvements was constant, culminating in the development of small home refrigerators at the turn of the century.

Refrigeration also had an immediate impact on people's lives. Instead of spoiling during seasonal gluts, food could be refrigerated and consumed out of season. Properly refrigerated or frozen food was less likely to carry disease-causing bacteria. Even the kind of food people ate changed. Former delicacies such as cold drinks, ice cream, and imported meat and fish became more widely available. A dependable supply of ice also made life in hot climates more comfortable.

Refrigeration was a factor in the continued growth of urban centers in Europe and the United States whose populations depended on the dairy products, meat, and produce brought to them by refrigerated boxcars and steamships. Refrigeration was also fundamental to the meat-exporting economies of Australia, New Zealand, and Argentina.

LINDSAY EVANS

Further Reading

Books

Donaldson, Barry. *Heat and Cold: Mastering the Great Indoors; A Selective History of Heating, Ventilation, Air-Conditioning and Refrigeration from the Ancients to the 1930s.* Atlanta, GA: American Society of Heating, Refrigeration and Air-Conditioning Engineers, 1994.

Woolrich, Willis Raymond. *The Men Who Created Cold; A History of Refrigeration.* New York: Exposition Press, 1967.

Periodicals

Woolrich, Willis Raymond. "The History of Refrigeration: 220 Years of Mechanical and Chemical Cold, 1748-1968." *ASHRAE Journal* 11 (1969): 31-39.

American Edwin L. Drake
Drills the First Oil Well (1859)

Overview

In 1859, Edwin L. ("Colonel") Drake (1819-1880) helped dig the world's first petroleum well. This launched the era of relatively cheap and abundant energy. In a very short time, petroleum was powering the industrial world in the form of internal combustion engines, jet turbines, and many power plants. In the twentieth century, petroleum has furnished raw materials for plastics, roadways, fertilizers, and more. Petroleum has become so important to a country's economic and military power that petroleum extraction and distilling facilities have become prime military assets and targets and, in fact, secure access to petroleum was a major factor in events leading to the Persian Gulf War of 1990 and 1991. Petroleum, however, seems to be a mixed blessing. Of primary concern to most environmentalists is the environmental risk that accompanies all aspects of petroleum extraction, processing, and use, while economists and industrialists are concerned about the leverage that petroleum-producing nations have over national economies. All of this notwithstanding, petroleum continues to be the most important resource on Earth for billions of people, a position it is likely to retain for some time to come.

Background

Petroleum was known to man from prehistoric times in the form of a sticky, black substance that appeared in the occasional seep. It was known to burn, to lubricate, and thought to possibly have some medicinal properties. Kerosene was developed in the mid-nineteenth century to help light lamps, but there was no reliable and abundant supply of petroleum from which to make kerosene, so the price remained relatively high. Europeans began actively extracting petroleum in the vicinity of natural seeps, and, in the 1840s and 1850s, Americans were beginning to do the same. However, manual digging of shafts near oil seeps was of limited utility in many cases.

In 1858, Edwin Drake was offered the job of helping to arrange for property rights to drill for oil near Titusville, Pennsylvania. Accomplishing this, he next began to make plans to drill for oil using technology already well established for water and salt extraction. Although called "Colonel" in correspondence, apparently to impress the local public, Drake was not a military man. Neither was he an engineer, a businessman, or geologist. He was simply in the right place at the right time. Drake struck oil in 1859, launching an oil boom in western Pennsylvania that quickly spread to Ohio, Texas, California, and around the world.

Petroleum was initially used primarily for kerosene, bitumen (for roads), and lubricating oil. When distilled to produce these heavier "fractions," much of the lighter products were discarded or burned as waste. These products included gasoline and natural gas. In part, the lighter fractions were discarded because they were too volatile and too flammable—simply put, they were too dangerous for the day's technologies to make use of. However, with the advent of the gasoline internal combustion engine in the 1880s, this began to change.

Another major step in the utilization of petroleum came in the pre-World War I years when Winston Churchill, then the First Lord of the Admiralty, started the process of converting the British Navy from using coal to oil. Other nations followed suit, making access to petroleum a matter of national security for the developed nations of the world. As petroleum became ever more useful and ubiquitous, exploration increased, leading to the discovery and exploitation of major new oil fields and leading to fantastic wealth for nations, corporations, and businessmen involved in its recovery. By the end of World War II, petroleum was firmly ensconced as a resource of prime importance militarily and economically.

In the latter part of the twentieth century, the rate of discovery of new petroleum reserves began to slow at the same time that awareness of the environmental impacts of widespread petroleum use began to make themselves felt. Petroleum exploration began to go to greater lengths as the continental shelves, the American far north, and other increasingly remote regions were explored and tapped. At the same time, the petroleum industry found ways to recover a higher percentage of petroleum at existing sites, helping to extend their productive lifespans. And, on the environmental front, some spectacular disasters and accidents led to increasing regulations and engineering standards designed to minimize the potential for future large-scale accidental releases to the environment.

Impact

With the exception of electrical power generation, the increasing utilization of petroleum is arguably the most important and wide-ranging development in recent human history. These impacts are in the following areas: (1) Petroleum has made possible the engines that power virtually all modes of transportation on the land, by sea, and in the skies. (2) Petroleum byproducts include plastics, asphalt, and fertilizers that have sparked revolutions in food production, ground transportation, and the industry of materials. (3) The process of discovering, extracting, process-

ing, and using petroleum has resulted in undeniable environmental concerns.

Petroleum replaced coal and wood as the fuel of choice for vehicles in the early twentieth century. In warships, this was because its use added speed, provided more rapid acceleration, and required less manpower than coal, making ships more effective. Aircraft were unable to truly fly until engines harnessed the high power output made possible by using gasoline as a fuel. Automobiles and most other modes of ground transportation also utilize either gasoline or diesel fuel. This litany of transportation uses of petroleum fractions is intended to show the extent to which modern transportation depends on petroleum. In fact, only a relative handful of vehicles (some nuclear-powered military ships and submarines, rockets, and some experimental automobiles) do not use some sort of petroleum directly for fuel. In turn, these vehicles have opened the world to quick, efficient, and relatively inexpensive transportation. Transportation, in turn, has contributed a great deal to the "shrinking" of the world, allowing, for example, a family in Europe to eat fresh Chilean strawberries in February, to vacation in the United States in the summer, and to send holiday gifts to relatives in Australia.

In addition, petroleum has contributed to the mechanization of agriculture by fueling the tractors and combines that till fields, plant and harvest grain, and transport food around the world. In conjunction with modern fertilizers, mechanized farm equipment has made it possible to feed many more of the world's population than would otherwise be the case. This, in turn, may be contributing to possible crises as the world's population continues to grow and to use ever-increasing amounts of energy.

Fertilizers are one product made from petroleum. Plastics are another. The first plastic, bakelite, was made in 1909 by the inventor Leo Baekelund (1863-1944), and was somewhat of a novelty at first. It was not until the 1960s that plastics became more widely used, and developments in the 1980s and 1990s led to an amazingly wide range of plastics and other polymers with properties that began to match those of metals and other more traditional materials.

The preceding impacts have been largely positive in nature. In the minds of many, the negative impacts resulting from the widespread use of petroleum outweigh these benefits. However, it must be stressed that some of these negative environmental impacts may or may not come to pass; although many vocal critics make claims with a high degree of certainty, the scientific evidence is still ambiguous and is likely to remain ambiguous for years or decades to come. While it is undeniable that man has had a measurable impact on the global environment, the scientific jury is still out as to the magnitude, seriousness, and duration of this impact. That being said, it is still necessary to explore the environmental impacts of petroleum use.

The first impacts were noted soon after Drake's well first began production. Since oil underground is under high pressure from the rocks covering it, it was not uncommon for it to spray out, and many oil regions had oil saturating the ground, streams, and lakes. Even today, with modern technology and techniques, oil extraction requires transportation of large amounts of equipment to remote areas, housing for personnel, and all the facilities necessary to support and entertain tens to hundreds of people for the life of the oil field. Since no technology is perfect, there are occasional leaks and fires that can affect the local ecosystem.

Once extracted, the petroleum must be transported, and this has thus far proved to be the greatest risk. The *Exxon Valdez* is only the most visible oil tanker accident, and there have

been occasional failures of oil pipelines, too. These have resulted in the contamination of areas ranging from a few hundred square meters to thousands of square kilometers. While long-term studies are still underway, it appears as though ecosystems are more resilient than previously thought. In addition, it must be pointed out that petroleum is a natural product that was discovered because it seeps to the Earth's surface (or into the oceans) naturally, so small-scale oil releases are not uniquely a product of industrial society. However, there is no denying that large-scale petroleum spills place a stress on the environment. Finally, burning petroleum releases exhaust gases that are known to cause respiratory problems, and it may also have long-term implications for the Earth's climate. While the case for global warming is strong, it is not yet definitive because of normal changes in solar activity, and global temperature is not yet fully understood. The answer to these questions may be better known in the near future.

P. ANDREW KARAM

Further Reading

Yergin, Daniel. *The Prize: The Epic Quest for Oil, Money, and Power.* New York: Simon & Schuster, 1991.

The Internal Combustion Engine

Overview

Physicists call the internal combustion engine a "prime mover," meaning it uses some form of energy (e.g., gasoline) to move objects. The first reliable internal combustion engines were developed in the middle of the nineteenth century and were almost immediately put to use for transportation. The development of the internal combustion engine helped to free men from the hardest manual labor, made possible the airplane and other forms of transportation, and helped to revolutionize power generation.

Background

In 1698, Thomas Savery (c. 1650-1715), a British military engineer, built the "Miner's Friend," a device that used steam pressure to pump water out of flooded mines. A few years later, Thomas Newcomen (1663-1729) would

expand upon Savery's design and create the first true engine. Newcomen's engine, unlike both Christiaan Huygens (1629-1695) and Savery's, used a piston that was attached to the engine itself. It could therefore produce continual (though hardly smooth) power.

Three conditions present during the nineteenth century encouraged the development of the internal combustion engine. The main condition was the demand for power presented by the Industrial Revolution. Second, physicists were beginning to understand the key concepts upon which the internal combustion engine was built. Third, the fuel needed to power the engine was becoming more available.

Between 1700 and 1900 scientists developed the field of thermodynamics, which gave inventors the tools to calculate the efficiency and power output of different types of engines. These calculations suggested that the internal combustion engine was potentially far more efficient than the steam engine (which, in contrast, was an external combustion engine, meaning it ignites the fuel outside of the engine itself).

The most important event in the early history of the internal combustion engine occurred in 1859 at the hands of Belgian inventor Jean-Joseph Etienne Lenoir (1822-1900). The Lenoir engine was both durable (some of them worked perfectly after 20 years of use) and, more importantly, reliable. Earlier versions of the engine were of poor quality and would stop operating for no reason. The Lenoir engine delivered continuous power and operated smoothly. In 1862, Lenoir invented the world's first automobile.

During the 1860s, Nikolaus Otto (1832-1891) began playing around with the Lenoir two-stroke and Alphonse Beau de Rochas's (1815-1893) theoretical four-stroke engines. Otto was a grocery salesman; he had no technical education or experience. In 1866, Otto—with the help of Eugen Langen (1833-1895), a German industrialist—developed the successful, but heavy and noisy Otto and Langen Engine. He continued to experiment with engines. In 1876 he released the "Silent Otto," the world's first four-stroke engine. In addition to being quieter than previous engines, the Silent Otto was also far more fuel efficient.

Otto's engine set the standard for the times. In fact, the fundamental design of modern engines remains identical to Otto's. As thermodynamics had predicted, the internal combustion engine was far more fuel efficient than the steam engine. Internal combustion engines that were quieter, cheaper to operate, and less bulky than steam engines began to appear in industrial plants throughout northern Europe.

In order for the internal combustion engine to make use of liquid fuels, it must first convert the liquid into a vaporous state. The next challenge for engine makers was to come up with a way to make this change happen. Between 1880 and 1900, different processes were invented to accomplish this task. Three methods were developed between 1885 and 1892: carburetion, hot bulb vaporization, and the diesel engine.

In carburetion, a device called a carburetor mixes air with vapors from the liquid fuel. The carburetor then delivers the mixture into the engine. A spark or flame inside the engine ignites the mixture. This is the function of the carburetor in today's automobiles. By comparison, the hot bulb engine sprayed gasoline onto a hot surface next to the cylinder, and then drew the evaporating fuel into the engine in vapor form. With the hot bulb engine it was possible to use less volatile fuels such as kerosene. A third method is the diesel compression engine. Rather than using an external heat source to ignite the gas, as in the first two methods, German engineer Rudolf Diesel (1858-1913) invented a process in which the gas ignites itself. Diesel had a strong background in math and science, and he knew that when a gas is compressed, its temperature increases to the point where the fuel ignites.

Impact

By the turn of the century, internal combustion engines had become integral to Western life. Industrial plants throughout Europe and America used them extensively, and the gateway for the large-scale automobile production of the 1900s opened.

In the area of transportation, the gasoline internal combustion engine and its variants (primarily the diesel engine) have been adapted for use in travel by sea, land, and air. At sea, a great number of smaller ships were, and continue to be, powered by diesel engines, speeding the movement of people and goods between any places connected by water. This has served to make trade more rapid and less expensive. Combining sea transportation with more efficient land transportation of goods makes these advantages even more significant. In turn, enhancing trade tends to lead to greater prosperity and a higher standard of living for both parties, not to mention the formation of new jobs.

Airplanes also owe their existence to the development of the gasoline engine. Many inventors had attempted powered flight at the end of the nineteenth century, but it wasn't until low-weight, high-output gasoline engines were available that the field of aviation was established. In fact, gasoline engines dominated aviation for the first half of the twentieth century and even today play an important role in private, commercial, and military aviation.

Also to be considered is the impact on farming and food production. Tractors and other modern farming equipment, usually running on diesel or gasoline engines, play a significant role in the abundance of food in the developed world and in parts of the developing world. The use of tractors to till, plant, and harvest as well as to pull heavy loads has helped to increase the amount of land a single farmer can work, as well as increasing the yield per hectare. This dual increase in the efficiency of individual farmers results in more food at lower prices. In the developed world this means not only more and cheaper food available for its citizens, but more food available for export to all nations.

The diesel engine is an outgrowth of the internal combustion engine, as mentioned previously. Diesel engines are powerful, require less maintenance, and use less highly refined fuel than gasoline engines. These factors make them less expensive, and they have become the engine of choice for rail travel, large boats and small ships, and trucks. Diesel engines are also widely used for electrical power generation, especially as emergency backup power supplies for installations such as hospitals and nuclear power plants. In both capacities, diesel engines have proven themselves dependable and inexpensive to maintain and operate.

The final impact that must be discussed is the environmental impact of the internal combustion engine. All internal combustion engines operate by burning some form of hydrocarbon and discharging exhaust gases. These hydrocarbons are typically derived from petroleum, and they burn to form carbon dioxide, carbon monoxide, and water. Although hydrogen engines have been developed that burn hydrogen and produce water vapor as an exhaust gas, they are uncommon as of this writing.

From the perspective of fuel, petroleum reserves are finite and are becoming ever-more difficult to discover and extract. The process of extraction invariably results in some environmental impact, not only at the drilling site, but along the transportation route. Since most petroleum is recovered in regions distant from refineries and industrial nations, much of it is transported by ocean-going tanker ships which sometimes cause spills with potentially serious results.

Once burned in engines, hydrocarbon fuels release many gases, most of which have contributed to air pollution. Until banned in the United States, many fuels also contained lead compounds, which were implicated in cases of lead poisoning. Even without lead, however, carbon dioxide, the primary combustion exhaust gas, seems to be produced in sufficiently high quantities that atmospheric levels have been noted to be increasing globally. Since carbon dioxide is known to help trap solar heat, there is a great deal of speculation that widespread use of internal combustion engines is causing temperatures to rise worldwide with potentially catastrophic results. However, it must be stressed that data that have been interpreted to show global warming are subject to many different readings, and not all scientists believe that global warming is actually occurring. In addition, it must be remembered that, for most of the history of the Earth, temperatures have been much higher than at present. So, even if global warming is occurring, it may or may not be due to burning fossil fuels in internal-combustion engines.

TODD JENSEN AND P. ANDREW KARAM

Further Reading

Combs, Harry. *Kill Devil Hill*. Boston: Houghton Mifflin Company, 1979.

Hardenberg, Horst O. *The Middle Ages of the Internal-Combustion Engine 1794-1886*. Detroit: Society of Automotive Engineers, 1999.

Roberts, Peter. *Veteran and Vintage Cars*. London: Drury House, 1967.

The Mass Production of Death: Richard Jordan Gatling Invents the Gatling Gun and Sir Hiram Maxim Invents the Maxim Machine Gun

Overview

At the end of the nineteenth century, a new military technology appeared on the scene that would fundamentally change the way warfare was conducted, and which would lead to some of the most tremendous slaughters of human beings ever witnessed. That technology was the machine gun, and it changed warfare by making it possible for a handful of men to kill thousands in only minutes.

Background

For centuries, battles had been conducted between two massed armies, with the goal of the attacking army being to break the defensive line of the other. While bows and arrows, cavalry, and even artillery could be used to weaken the line of the massed units, warfare was still a matter of those two lines moving ever closer to one another until the moment of the charge, when the attackers would rush forward to try and overwhelm the weakened defensive positions.

Even with the development of infantry troops carrying rifles and muskets, the deciding factor in any battle was that moment when the combatants would close for hand-to-hand combat. With the development of the machine gun, all basic strategies and tactics of warfare had to be changed fundamentally, because now that moment of massed attack could only amount to foolishly heroic suicide.

While projectile weapons that could fire more than one round at a time had existed in one form or another for centuries, it was only with the development of the Gatling gun in 1862 and the Maxim gun some 20 years later that the first two true machine guns were brought into combat in any widespread way. The development of these weapons hinged upon two developments in cartridge technology: the cased round that incorporated its own percussive cap, and the development of slow-burning smokeless powder.

At the outbreak of the American Civil War in 1861, most projectile weapons were still muzzle-loaders; that is, powder was poured down the barrel of the rifle, a round was rammed down with a ramrod, a percussion cap was placed on a "nipple" at the far end of the barrel, and when the hammer fell upon the cap, a spark was thrown into the breech of the barrel, igniting the powder, which caused an explosion that propelled the bullet forward. This was a time-consuming and awkward process, and the effectiveness of any multiround weapon was severely hampered by the process of having to load it. During the American Civil War, however, cartridge rounds were developed. These were bullets as we know them today, with a copper or brass casing that held the powder charge and the bullet, and with a percussion cap built into the base of the casing. It was this development that made the Gatling gun possible.

Impact

Richard Jordan Gatling (1818-1903), the inventor of the Gatling gun, was, for most of his life,

Richard Jordan Gatling. *(The Library of Congress. Reproduced by permission.)*

involved with the development of farm equipment. The son of a North Carolina plantation owner, Gatling was living in Cincinnati during the outbreak of the war, and came up with the idea for his weapon, according to legend, while watching the wheel of a paddleboat turn. The design of the weapon, however, also owes at least some of its inspiration to the sowing and seeding machines that would have been familiar to Gatling.

The Gatling gun had six to ten barrels arranged in a circular pattern, which rotated around a central pivot when a crank was turned. On top of the gun was a hopper, which fed bullets into the barrels as the barrels rotated. As the barrels rotated, bullets would be fed into the barrels, locked, fired, and extracted as the barrels moved around the pivot. In this way, the Gatling gun could fire up to 350 rounds per minute, with some experimental models achieving a rate of fire of over 1,000 rounds per minute. However, the weapon was extremely prone to jams and was large, requiring at least a three- to four-man crew. Initial versions were incapable of traversing fire—it was essentially an artillery weapon and, despite the potential it held, saw almost no use in combat save for some limited use during the Spanish–American War.

The two main problems with the Gatling gun were the complicated process through which rounds were loaded, fired, and extracted, and the fact that it was a hand-cranked weapon. In 1880 Sir Hiram Steven Maxim (1840-1916), another American inventor, came up with the idea for a weapon that would use the force generated by the bullet's recoil to operate the loading, firing, and extraction process. Maxim had previously designed a number of inventions dealing with gas illumination, as well as a process for treating the filaments in electric lights. In 1880 he came to England and, again according to legend, was attending a trade show when someone said to him, "If you want to make real money, invent something for these fool Europeans to kill one another with." Maxim's answer to that was the Maxim gun, the first recoil-operated machine gun.

The major problem that Maxim had to solve concerned the powder that was used in cartridge rounds. Black powder, which was the common form of gunpowder at the time, burned too quickly to allow gas pressures to build up and generate the kind of force needed to operate a recoil weapon. Maxim's solution to this problem was the invention of a new type of gunpowder that incorporated nitrocellulose, called Maximite. With this new powder, a fired round would generate enough power to force back a sliding breech bolt, and the spent casing would be carried back with it, to be ejected when the bolt reached the back of the breech. This backward motion of the bolt would also compress a large spring, which would then force the bolt forward. As it moved forward, the bolt would force another round from a belt of ammunition into the barrel; the round would be fired, and the whole process repeated again. When Maxim demonstrated his new weapon for His Royal Highness the Duke of Cambridge in Hatton Garden in 1884, it was capable of firing 600 rounds per minute without jamming, and could be fed rounds from continuous belts that could be linked together to provide virtually uninterrupted fire.

This devastating weapon was slow to be adopted by European military forces, however, and it wasn't until 1888, after it was made the official machine gun of the German army, that the Maxim was adopted in any large-scale way by a military force. During the last decade of the nineteenth century and the first decade of the twentieth century, the true power of the Maxim gun was demonstrated in a number of incidents during European colonial actions, and would come to its apex as it dominated the battlefields of World War I.

An incident from the British colonial campaign in Egypt provides an illustration of how even a few machine guns could change the entire course of a battle. During the Battle of Omdurman on September 2, 1888, a force of 26,000 British and Egyptian troops met 40,000 Arab troops. Under normal circumstances the British troops, even though they were armed with rifles, could have expected a tough battle, and might very well have been overwhelmed by the superior force. However, the British were armed with six Maxim guns, and at the end of the day these made a decisive difference. Over 10,000 dervishes were killed as they attempted to charge and overwhelm the British positions, compared to only 20 British and 28 Egyptian casualties. Over three-quarters of the Arab causalities were attributed to the Maxim guns, and as the British writer Sir Edward Arnold put it, "The battle was won by a quiet, scientific gentleman living in Kent." In 1884 Maxim had become a British subject; in 1901 he was knighted, presumably for the service his weapon had given to the British colonial armies.

It was during World War I, however, that the devastating power of the machine gun was truly demonstrated. In fact, World War I might very well be called the machine gun war. In the intervening years since 1888, Maxim had marketed his machine gun to all the armies of Europe, and as war broke out in 1914, the forces of both sides were armed with the new weapon. Set up in trench emplacements, and later mounted on aircraft and tanks, European troops for the first time faced the power of the weapon they had used so effectively against native populations. They obviously learned their lessons well; with emplacements set up every few hundred yards along a trench, it was possible to create overlapping fields of traversing fire, and hundreds of men could be killed before they even stepped a few feet out of their trench in an attempt to charge and overwhelm the enemy line. With machine guns in the trenches of both sides, the traditional strategy of charging to overwhelm an enemy position could only mean almost certain death or wounding to those making the charge. The machine gun was largely responsible for the decimation of an entire generation of young British, French, and German men who fought in the war.

After World War I the tactics of warfare changed almost completely, due in large part to the introduction of the machine gun in that war. Instead of massed armies that used overwhelming force to break an enemy line, tactics now relied upon small, often motorized, mobile units armed with rapid fire weapons. Such units could quickly move in and control a piece of ground while large armored units pushed forward, clearing the way for more infantry to advance—the *blitzkrieg* approach, employed with stunning success by the German army during World War II. This strategy and its variations are still used today.

In addition to affecting combat tactics, the machine gun totally changed the scale and violence of warfare and exerted a profound psychological impact on its participants. Post-traumatic combat stress, called "shell shock," emerged as a new category of battle injury during World War I, afflicting many soldiers who witnessed the horrific spectacle of mechanized mass killing on the battlefield.

PHIL GOCHENOUR

Further Reading

Ellis, John. *The Cultural History of the Machine Gun.* Baltimore: Johns Hopkins University Press, 1975.

Hallahan, William H. *Misfire: The History of How America's Small Arms Have Failed Our Military.* New York: Charles Scribner's Sons, 1994.

The Development of the Automatic Writing Machine: The Typewriter

Overview

Prior to the nineteenth century, almost all letters, business records, and other documents were written by hand. The only practical alternative was to have them printed on a printing press—an expensive process if only a few copies were needed. Thus, almost all documents, whether business, legal, or personal, were handwritten. During the 1860s, three American inventors, Christopher Latham Sholes (1819-1890), Samuel W. Soulé, and Carlos Glidden, developed an automatic writing machine called the typewriter. Within two decades, a modified version of this machine would soar in popularity and revolutionize business practices around the world.

Background

Various kinds of automatic writing machines were invented in the early nineteenth century. Many were large and difficult to use, and most printed words much more slowly than a person could write by hand. It was not until the second half of the century, however, that the first practical and commercially successful typewriter was invented.

Christopher Latham Sholes worked as a port official in Milwaukee, Wisconsin, in the 1860s. This job gave him time to pursue a second career as an inventor. Sholes had previously worked as a newspaper editor, and so he was familiar with printing presses and steel type. (A *type* is a rectangular piece of steel with a raised letter or symbol on one end.) Sholes worked on his inventions at a workshop owned by C.F. Kleinsteuber, where a group of inventors developed new machines.

In 1864, Sholes and Samuel W. Soulé, another inventor who frequented Kleinsteuber's, were granted a patent for a page-numbering machine. This machine could print consecutive numbers for the pages of a book or for a set of railway tickets. In 1867, Carlos Glidden, another regular at the workshop, showed Sholes an article in the journal *Scientific American* that described a writing machine. This machine was called the pterotype and had been patented by John Pratt in England the previous year. Glidden suggested that Sholes could modify his page-numbering machine to print letters as well as numbers.

Sholes, Glidden, and Soulé soon set about building such a machine. Their initial attempt consisted of a telegraph key connected to a single type with piano wire. This machine could only print the letter "W." Their first working model of the typewriter, however, could print all 26 letters, but it used piano keys to move individual types in a somewhat clumsy design. Their second model, patented in 1868, was much improved and could produce printed text faster than writing by hand. Despite this success, the men had difficulty raising the money necessary to manufacture their typewriter. Later that year, a businessman named James Densmore bought a share of the typewriter patent in exchange for taking over the previous costs involved in developing the machine.

Densmore first tried to interest the Automatic Telegraph Company in the typewriter, but Thomas Edison (1847-1931), who worked there as a mechanic, claimed that he could build a better model for less money. (Edison went on to invent the phonograph, the microphone, and the motion-picture camera, among many other inventions.) Finally, in 1873, Densmore arranged for Sholes to sign an agreement with the gun manufacturing company E. Remington and Sons for $12,000.

Remington had manufactured weaponry during the American Civil War, and it was now

OTHER PRACTICAL (AND LESS THAN PRACTICAL) WRITING MACHINES

~

As typewriter sales began to rise, dozens of companies eager to make a quick profit jumped into the industry. Some of the machines were designed to suit customers with specific needs. For example, the Virotyp typewriter could be worn on the user's wrist like a watch, and the Trebla typewriter was small enough to be carried in a pocket. The Electric Blick typewriter allowed users to type on paper up to 3 feet (91 cm) wide. Other typewriters seemed more suited to separate gullible purchasers from their money. For instance, Cary Writing Gloves were offered as an alternative for people intimidated by machinery. This "typewriter" consisted of a pair of rubber gloves with raised letters on the fingertips, knuckles, and other portions of the hand. The user was instructed to wear the gloves and coat them with ink. Then he or she could "type" by pressing the appropriate letters against a piece of paper.

looking for items to produce during peacetime. The company had already added sewing machines to its inventory, and it put its first commercial typewriters up for sale in 1874. To type on a Sholes & Glidden, as the machines were called, the user pressed keys labeled with letters or symbols, similar to those on a computer keyboard. When a key was pressed, a type bar swung upward and hit an inked ribbon, printing the letter or symbol on the type on a sheet of paper. This upward striking mechanism kept the paper hidden from the typist, so that he or she could not see the words as they were typed. Such machines came to be known as blind writ-

ers. Sholes & Glidden typewriters sold for about $125 (about the cost of a home computer system in today's dollars). The first five years the typewriter was produced, only 5,000 were sold.

Impact

During the Industrial Revolution, machines came to be used to manufacture products faster than could be done by hand. Large factories replaced smaller workshops. Improved travel and communication between cities increased trade. The railroad, the telephone, and the telegraph each contributed to the growth of business. As business and industry grew, so did the amount of written papers. Simple record keeping soon became more than one person could handle just with pen and paper.

When the typewriter first appeared, however, the machine was regarded as little more than a novelty. For example, the Sholes & Glidden made an appearance at the Centennial Exposition in 1876 in Philadelphia, but unlike Alexander Graham Bell's (1847-1922) telephone, it made little impact. In fact, not a single typewriter sold. Many people initially thought the typewriter was a form of printing press, such as those used to produce newspapers and books. They had difficulty imagining how they might use the typewriter in their daily lives. It seemed to be too much trouble to use a machine just to write a business letter or fill out an invoice.

The public eventually overcame its reluctance after an improved version of the typewriter, the Remington Model 2, appeared in 1878. After several years, the Remington 2 became a tremendous success. In 1881, Remington sold a total of 1,200 typewriters for the year. By 1888, however, the company was selling 1,500 typewriters each month. As the Remington Model 2 began to catch on, many other manufacturers jumped at the chance to make a profit. In order to avoid infringing on Remington's patents, they had to come up with different—often drastically different—designs. By the end of the nineteenth century, there were typewriters with wheels of type, strips of type, and type shuttles. There were keyboards that had keys arranged in straight rows, curved rows, and even circles. By 1909, there were 89 manufacturers of typewriters in the United States alone. By 1905, the number of U.S. patents relating to typewriters exceeded 2,500.

As the typewriter came to be an accepted business tool found in most offices, several changes occurred to its basic design. The original Sholes & Glidden typewriter could only be used to write capital letters. The Remington 2, however, could type lowercase letters as well as capitals. Each bar of type on this machine contained two letters, a lowercase and an uppercase. The letter that printed was controlled by a shift key. When the typist pressed the shift key, an uppercase letter would be typed. When he or she released the shift key, a lowercase letter would be typed.

At about the same time as the Remington 2 came on the market, other manufacturers were producing double-keyboard typewriters. Double-keyboard typewriters had a separate key for each uppercase and lowercase letter. For instance, these typewriters had both an "S" key and an "s" key. As touch-typing came to be widely accepted, double-keyboard typewriters grew less popular and went out of fashion; they were slower because typists had difficulty using so many keys.

A second change to typewriter design was brought about when Remington decided to make its own typewriter ribbons rather than purchasing them. In response, John Thomas Underwood, a ribbon manufacturer, founded his own typewriter company. He bought a new model of typewriter that allowed the typist to see the letters on the paper as they were being typed. The type bars on this machine swung forward and hit the ribbon from the front rather than swinging upward and hitting the ribbon from the bottom. This type of machine became known as a visible writer (as opposed to the blind writers being produced by Remington). The Underwood Number 5 was an immediate success upon its release, and by 1908 all major typewriter manufacturers had switched to visible writing machines.

Another design feature was the use of the QWERTY keyboard, established by Christopher Sholes. The QWERTY keyboard takes its name from the first six letters of its top row of letters. Remington's competitors frequently claimed that Sholes created a confusing keyboard so that typists would be forced to type slowly. Slow typing, they argued, would prevent Sholes's faulty keys from jamming. However, Sholes actually designed his keyboard to prevent jamming and to improve typing speed. Sholes had noticed that keys often locked when two adjacent type bars were pressed one after the other. He took the most common letter pairs (such as *th* and *ed*) and placed them so that their type bars were not next to each other. Because jamming with such an arrangement would be less likely to occur, typists could type as fast as they were able. Other keyboard designs have been proposed, but none has ever become popular—even on keyboards where jamming is no longer an issue.

Thomas Edison invented the first electric typewriter in 1872. Electric typewriters use power from an electric motor to hit the type bars against the paper. Consequently, the typist does not have to press the keys as hard, which results in faster typing. Edison's machine evolved into the ticker-tape printer once used by the New York Stock Exchange. The ticker-tape printer allowed stock prices to be printed on a moving tape by means of a telegraph. Electric typewriters did not come into popular use until 1960. These machines used a type ball that replaced the individual type bars. (Type balls had initially appeared in the 1880s but did not catch on at the time.)

Besides greatly increasing office productivity, the invention of the typewriter played an important role in opening up new fields of employment for women. During the 1800s, the only jobs available to most women were in shops or factories, where hours were long and conditions were often unsafe. The Young Woman's Christian Association in New York was concerned about the work conditions of women who labored in sweatshops. They thought that typewriters might offer a solution. In 1881, they offered a typing class to eight women, who were immediately offered jobs upon graduation. Soon, business schools across the United States and England were offering typing programs. Office work offered poor women a way to avoid factory work and allowed middle-class women a way to become independent. In 1880, only 5 percent of clerical office workers were female. By 1900, this figure had climbed to 75 percent. However, the typewriter eventually came to be seen as a symbol of gender inequality in the workplace. Women were traditionally employed in low-skilled positions such as secretaries and assistants while men worked as managers and executives.

Today, the typewriter has been almost entirely replaced by the personal computer. However, the keyboards of computers reflect many of the design elements of typewriters. For instance, they still make use of the QWERTY keyboard and the shift key.

STACEY R. MURRAY

Further Reading

Books

Adler, Michael H. *The Writing Machine*. London: George Allen & Unwin Ltd., 1973.

Flatow, Ira. *They All Laughed...From Light Bulbs to Lasers: The Fascinating Stories Behind the Great Inventions That Have Changed Our Lives*. New York: HarperCollins, 1992.

Hooper, Meredith. *Everyday Inventions*. New York: Taplinger Publishing Company, 1972.

Richards, G. Tilghman. *The History and Development of Typewriters*. 2nd ed. London: Her Majesty's Stationery Office, 1964.

Other

"Mavis Beacon Teaches Typing: A Short History of Typing." The Learning Company, Inc., 1999. http://www.mavisbeacon.com/history.html

Rehr, Darryl. "The Typewriter." http://www.popularmechanics.com/popmech/spec/9608SFACM.html

Alexander Graham Bell
Patents the First Telephone (1876)

Overview

The invention and patent of the telephone by Alexander Graham Bell (1847-1922) in 1876 stands as one of the world's most important innovations. The telephone grew into an indispensable part of everyday life and became instrumental in the development of the modern world. It changed the way people communicated forever and fostered the rise of big business, city life, and changed people's perceptions of community. Today, a world without the telephone is as unimaginable as one without electricity, automobiles, or television.

Background

Most children have strung a thread between two empty soda cans and "invented" their own telephones. This crude contraption proves the telephone is a relatively simple device. In hindsight, it is hard to believe it took so long for the telephone to come into existence. In fact, the two major scientific advances required for its invention—electromagnetism and induction—had been understood for years before Alexander Graham Bell devised what he called the "harmonic telegraph."

Hans Oersted (1777-1851), a Danish scientist, introduced electromagnetism to the world in 1820, while English scientist Michael Faraday (1791-1867) published his discovery of induction in 1831. With these technological advances, either person could have invented the telephone. Over the next 45 years, numerous scientific advances took place in intellectual circles around the world, but still no one put all the pieces together. The reason for this is as much societal as scientific. The leap of faith required to assume human voices could be transmitted over an electromagnetic device was unfathomable to most scientists and researchers of the period.

Perhaps the scientific community needed a person like Alexander Graham Bell, who was not a trained scientist or an electrician, to make the intellectual jump necessary to invent the telephone. Although not formally trained, the young Scottish immigrant was the third generation of Bells to center his life on the concept of sound, and he displayed a passionate interest in the education of the deaf. In fact, Bell had never invented anything before devising the telephone.

Impact

In 1876, America's 100th anniversary, Bell, then professor of vocal physiology at Boston University, traveled to Philadelphia to display his harmonic telegraph at the national Centennial Exposition. Bell didn't want to go, but his fiancée, Mabel Hubbard, realizing the importance of publicizing his invention, persuaded him to go.

At the exposition, Bell's table was hidden in a dark corner of a small upstairs gallery. A distinguished group of judges, including Emperor of Brazil Dom Pedro, evaluated each exhibit entry. The emperor held the hearing device to his ear while Bell, 500 feet (152 m) away, recited one of Hamlet's soliloquies into the "magneto" telephone. When the Emperor tried it, he was startled. He exclaimed, "I hear, I hear! My God! It talks!" The judges fought to be next in line to hear Bell. A witness claimed that other scientists made so much commotion that exposition police thought the building caught fire.

Eventually Bell and his assistant, Thomas Watson (1854-1934), publicized their invention before many other audiences. First Watson would talk to Bell and then after considerable protest he would sing *Yankee Doodle* and *Auld Lang Syne*. The telephone had begun its journey from novelty to necessity. Within a year, Bell installed 230 telephones and he and his financial partners established the Bell Telephone Company (which would later become AT&T). Bell realized that he would only have 17 years to use his original patent, so they sought quick profits and charged high rates. Initially, the telephone developed primarily as a tool of business and commerce.

Even before President Rutherford B. Hayes had the first telephone installed at the White House in 1879, orders had begun to pour in from residential customers. Since the Bell Telephone Company was a moneymaking venture from the start, its leaders decided to sell the company. Bell tried to sell the company to Western Union, then the largest corporation in America, for $100,000.

Western Union's president William Orton flatly refused Bell's offer, stating, "What use could this company make of an electrical toy?" Orton's judgment has since been deemed one of the worst business decisions of all times. Instead, Western Union hired inventors Thomas Edison (1847-1831) and Elisha Gray (1835-1901) to develop a new phone system. The competition between Bell and Edison fueled further innovation. Western Union opened the first crude switchboard in New Haven, Connecticut, in 1878 and published the first telephone directory a few weeks later, a single page containing 50 names. The Bell Company sued Western Union for patent violation, while Western Union countered by labeling Bell a fraud. The company stated that its scientist, Elisha Gray, was actually the true inventor of the telephone.

Bell's performance during the subsequent patent trial was outstanding. He was passionate and forceful in his beliefs. Western Union had little choice but to abandon the suit. The next day, Bell Telephone's stock rose from $300 to $1,000 a share. This episode was the first of hundreds of suits in which Bell would be forced to defend his invention. Throughout the 1880s, Bell had over 600 court battles with people who challenged his patent. The constant legal pressures took their toll on Bell, and he eventually left the business and moved into his family retreat in Canada. Bell, however, continued to think about new innovations. He began designing flying machines and dreamed of wireless communications.

In the early years, after waging the many patent battles and squeezing out its competition, AT&T symbolized corporate greed, poor quality, and terrible customer service. The company's rebirth in the early twentieth century, however, under the leadership of financier J.P. Morgan and business executive Theodore Vail, made AT&T a model for the modern corporation.

Employment at the phone company in the late 1800s gave females an outlet outside the traditional women's jobs in the schoolhouse and home. The first woman operator, Emma M. Nutt, was hired in late 1879 to work in Boston. Throughout the grueling 12-hour shifts, women handled hundreds of calls each hour and worked the switchboard with both hands simultaneously. They were forced to follow a strict dress code, and any behavioral transgressions were noted in records kept by company officials. Even with the rigid policies, female operators effectively and efficiently took over the day-to-day operations of the company.

AT&T (commonly called "Ma Bell" by customers) and the telephone became virtually synonymous. Both the company and invention which made it famous hold lofty positions in American history. AT&T was the largest corporation in the world for much of the twentieth century, employing over one million people. At the time of its court-mandated breakup in 1984, the company's assets totaled $155 billion, more than General Motors, Mobil, and Exxon combined.

Through its Western Electric subsidiary, AT&T formed alliances with companies around the globe to manufacture telephone equipment. By 1914 locations included Antwerp, London, Berlin, Milan, Paris, Tokyo, Buenos Aires, and many others. Thus, the telephone deserves credit in the development of a global community. AT&T was a leader in opening foreign markets, including China, east Asia, South America, and Europe.

AT&T's scientific innovations, through Bell Labs (now the independent Lucent Technologies), resulted in an exhaustive list of inventions. In addition to the company's important work spreading phone service across the nation and then around the world, Bell Labs invented the transistor, which replaced vacuum tubes, in 1948. Widely regarded as one of the most important inventions of the twentieth century, the transistor won Bell Labs the Nobel Prize in

1956. AT&T's R&D lab was also instrumental in developing cellular wireless technology (1947), the computer modem (1957), communications satellites (1962), and commercial ISDN long-distance network services (1988). The electronic switching systems AT&T installed in 1965 after years of research permitted a vast increase in phone traffic and paved the way for the Information Age. These advances in switching technology allow the Internet to exist today.

AT&T played a role in the growth of the U.S. military-industrial complex dating back to the First World War when it expanded domestic military communications and installed telephone lines in France. Western Electric and Bell Labs completed projects for the military throughout World War II. AT&T made important advances in radar technology, which later became the chief means of transmitting long-distance phone calls and television signals after the war. In the 1950s and 1960s, AT&T worked on satellite communications and launched its first satellite in 1962.

Culturally, the telephone's impact has been immense. By the early 1900s, the telephone was already considered an indispensable part of life for most individuals and businesses. Telephones connected rural and farm areas with the growing cities and urban centers. AT&T also created the distinction between local and long-distance phone calls, which has become a staple of modern telecommunications. The separation between the two markets facilitated the rise of the regional Bell Companies, the "Baby Bells," and ultimately to the breakup of the parent company.

Once people became more familiar with the telephone and the convenience it offered, they began demanding phone service over greater geographical areas. Long-distance service, however, barely existed and, in fact, frustrated would-be inventors for many years. The technology to send a human voice over great distances was not available. No significant breakthroughs were made until a signal amplifier was invented nearly 40 years after the advent of the telephone. Alexander Graham Bell himself placed the first New York-to-Chicago call, 600 miles (965 km) away. Bell was so happy with the event that he allowed himself to be photographed with his invention for the first and only time.

AT&T continued to develop its long-distance service. In 1915, the 68-year-old Bell was once again summoned to make the first transcontinental telephone call with his old friend Thomas Watson. AT&T executives worried about the call between the two old men turning into a public relations nightmare, so they attempted to choreograph the call by providing each man with a script. The headstrong Bell, however, threw the script aside after the initial "Hoy, Hoy, Watson," and Watson replied "Hoy." Later in the conversation, Bell evoked their initial conversation by saying, "Mr. Watson, come here, I want to see you." Watson then replied, "Mr. Bell, I will, but it would take me a week now." When Bell died in 1922 at age 75, AT&T paid tribute to its founders by stopping phones across the nation from ringing for one full minute.

The telephone has led to the development of several cultural icons. The Yellow Pages, more widely read than the Bible, were published to help customers use their phones more often and more effectively. AT&T began the use of the telephone as a service tool. Initially, the phone served as a means to get weather and time reports. Today, one can receive almost any information over the phone, from sports scores and soap opera updates to movie listings and bank information. The ubiquitous image of teenagers on the phone in movies and television mirrored the real life development in the 1950s when disposable income and a population explosion made phones readily available for teens to use.

Technology has advanced to the point where cellular telephone service is an everyday part of life. In Europe, many people only have a cellular phone. In the United States, cellular telephones allow people to take calls anywhere and be connected 24 hours a day.

The telephone continues to influence popular culture. Telecommunications companies spend billions of dollars in marketing and advertising each year. And, as a result, these companies have some of the strongest brand names in the world. News released by the telephone giants makes the front page of financial papers around the world.

With over 2 million shareholders, AT&T has always been one of the most widely held stocks in the world. Thus, the company's fortunes continue to have an impact on people everywhere. It is a corporate giant that produces major headlines with every significant action. However, unlike many other large corporations, AT&T, due to Bell's original invention, has become an ingrained part of American popular culture.

BOB BATCHELOR

Further Reading

Brooks, John. *Telephone: The First Hundred Years*. New York: Harper & Row, 1976.

Bruce, Robert V. *Bell: Alexander Graham Bell and the Conquest of Solitude*. Boston: Little Brown, 1973.

Fischer, Claude S. *America Calling: A Social History of the Telephone to 1940*. Berkeley: University of California Press, 1992.

Garnet, Robert W. *The Telephone Enterprise: The Evolution of the Bell System's Horizontal Structure, 1876-1909*. Baltimore: Johns Hopkins University Press, 1985.

Kleinfield, Sonny. *The Biggest Company on Earth: A Profile of AT&T*. New York: Holt, Rinehart, and Winston, 1981.

Smith, George David. *The Anatomy of a Business Strategy: Bell, Western Electric, and the Origins of the American Telephone Industry*. Baltimore: Johns Hopkins University Press, 1985.

Stern, Ellen and Emily Gwathmey. *Once Upon A Telephone: An Illustrated Social History*. New York: Harcourt Brace, 1994.

Wasserman, Neil H. *From Invention to Innovation: Long-Distance Telephone Transmission at the Turn of the Century*. Baltimore: Johns Hopkins University Press, 1985.

Artificial Gas and Electrical Lighting Systems Are Developed That Change Living and Work Patterns

Overview

With the use of gas and electricity, lighting systems in the nineteenth century provided more versatile illumination in both interior and exterior applications. These new lights extended the day at home, at work, and at play as people could perform more activities beyond the hours of light provided by the Sun. Artificial light supplanted natural light so that people increasingly relied on the technology of lighting in organizing their lives.

Background

At the beginning of the nineteenth century men like Humphry Davy (1778-1829), head of London's Royal Institution, demonstrated that an electrical current sent across a gap in a circuit created a bright white spark of light. Electrical technology was in its infancy at that time, and little practical application resulted from this new electrical phenomenon. Yet, at a time when most illumination consisted of the dim light from candles or oil lamps—which required high maintenance by replacing candles or trimming wicks and refilling lamps—the prospects of brighter, easier-to-use lighting systems using electricity or gas appealed to a broad constituency.

Two developments competed with electricity as sources of light in the nineteenth century: gas illumination and kerosene lamps. The discovery in 1781 by Archibald Cochrane that burning coal produced a coal gas, and the successful use of coal gas in lighting by William Murdock (1754-1839) in the period from 1780 to 1810 led to the technology of gas lighting. Murdock's system and a new gas lamp design by Francois Ami Argand, who was familiar with Antoine Lavoisier's new chemical theories of combustion and the role of oxygen in burning, combined to produce a brighter gas light—an attractive new and widely used lighting technology. This system became more deeply entrenched with the utilization of the incandescent gaslight invented by Carl Auer von Welsbach (1858-1929) in 1886. Welsbach gas mantles provided a soft light especially suitable for people's homes, offices, and shops. Although gas illumination presented problems of supply, fire danger, and switch-on, this system competed successfully with electrical lighting into the early years of the twentieth century.

The discovery of vast reserves of petroleum in the eastern United States in the 1860s revived the oil lamp. Especially in those areas not served by gas or electric distribution systems, the kerosene lamp provided an adequate and relatively economical source of illumination, especially in the home. The kerosene lamp had to be refilled and the wick had to be trimmed—disadvantages that doomed its use once gas or electric lighting systems were available. However, for much of the last third of the nineteenth century, such oil lamps served a wide population.

In the 1870s the production of a steady electrical current with improved dynamos, or

Gas lighting had a tremendous impact on nineteenth-century city and town life. *(Library of Congress.)*

generators (especially the design of Zenobe Gramme), allowed lighting by electricity to gain a place in lighting system technology of that time. The first widely used technique was arc lighting, in which current passed through a circuit with two carbon rods separated by a small gap; this produced a bright white glowing spark. With the work of Americans such as Charles Brush and Elihu Thomson (1853-1937), this very bright light saw application in public places such as city streets, parks, factories and large stores, lighthouses, exhibition halls/arenas, theaters, and railway stations. These first commercially viable electric lighting systems established the necessary elements of a new lighting technology. But even though arc lighting was successful as a new and very bright artificial light source, it had its drawbacks as well. The carbon rods had to be replaced daily; the lights left a dirty carbon ash residue; and the intensity of the arc light made it unsuitable for smaller areas such as homes. So inventors in the 1870s sought to subdivide the electric arc by creating an incandescent electric light whose filament produced a soft glow in an enclosed glass tube, thereby eliminating the chief problems with the arc light.

Two pioneers developed a commercial incandescent lamp: Joseph Wilson Swan (1828-1914) of Britain and Thomas Alva Edison (1847-1931) of the United States. They realized that a successful lamp needed a high-resistance filament surrounded by a vacuum to reduce the rate of combustion in the bulb. Finding a suitable filament required a substantial effort in directed trial and error methods; the availability of the effective Sprengel vacuum pump solved the problem of securing a satisfactory vacuum within the lamp. By 1879 both men had produced a carbon-filament lamp suitable for commercial application and had begun marketing this new form of artificial light. Edison especially foresaw the importance of research and development for the continual improvement of this new technology and relied on his Menlo Park Laboratory staff to create an electrical lighting system, adding generators, distribution systems, central power stations, and even metering devices to his basic light bulb. Further, his marketing acumen made his new system more widely accepted than those of his competitors. Edison's insights and prescience regarding electricity and its utilization chiefly in an urban setting allowed for the expansion of electric lighting to a wide variety of applications and its eventual dominance in lighting technology for the twentieth century.

Impact

Improved gas and newly developed electric lights transformed life in the latter third of the nineteenth century. Artificial light appeared in homes, factories and offices, stores and streets, various public buildings and arenas, and churches. These lamps freed people from the limited hours of natural light and extended their days and their activities.

In the home, the new lights replaced candles and oil lamps. Reading and writing were much easier. Various home-centered events from small gatherings to large parties could take place more safely at night. Home cleanliness improved because the brighter lights drew attention to dirt and dust that was less obvious with earlier illumination. Electricity provided a cleaner, safer source of light with low maintenance; it no longer was necessary to clean glass globes or to trim wicks. In addition, this new technology allowed the separation of public and private lighting spaces. Illumination of rooms in a house would be softer and gentler than lights used outside and around the home.

Interior house design changed with electric lights as well. Earlier, unadorned windows and light colored walls helped to brighten otherwise dim rooms lighted by candle or oil. But the brighter effect of gas or electric lights meant that households used drapes and curtains more extensively to soften that new light. Likewise, interior decor, from wall colors to furniture, was more muted so that the light was diffused rather than reflected. With less reliance on natural light, people used Tiffany lampshades to filter electric lamps or stained glass windows to vary the effect of light on the interior from both natural and artificial sources. Even windowless rooms had more appeal with the accessibility of this innovative lighting technology from gas or electricity.

The worker in a store, office, or factory experienced a change in operations due to the new lights. Stores used exterior lighting to highlight a sign or storefront; their expanded interiors were brighter from display windows to display cases. Nighttime sales hours were possible, and staff found night schedules part of their work week. In addition, factory laborers found their hours extended. Especially during the short winter days, factory owners used interior lighting to lengthen the work day. At the same time night shift work became more widespread. The adage of working from dawn to dusk had less relevance by century's end.

In public spaces, gas and electric lighting made city streets safer and public buildings more accessible. First electric arc lights and then incandescent lamps illuminated public squares and parks, urban streets and arenas/exhibition halls, houses of worship, and art galleries and museums. Theaters profited from electric lighting as well: the expensive limelight, used earlier in the century, gave way to the intense blue-white brightness of arc lights as the new spotlights in auditoriums. Instead of using dangerous open flame lamps for stage lights, theater operators welcomed enclosed incandescent lamps as a more convenient and safer replacement. These many new applications occurred mainly in urban centers, and people who lived in such locales experienced a transformation in lifestyle with the advent of electric lighting.

What began as a laboratory curiosity in 1800 with Humphry Davy's demonstration of a simple arc light ended in 1900 with fully developed electric lighting systems, using both arc lights and incandescent lamps. Because of their softer illumination and ease of use and maintenance, incandescent lamps dominated the marketplace. The nineteenth century saw the transformation of lighting technology and with it a significant change in the way people lived. Light no longer was restricted to the Sun's rays, and the impact of artificial light varied from the drudgery of longer working hours for industrial laborers to safer streets and a wider variety of nighttime activities for many. For much of the latter third of the century, gas illumination competed successfully with electric lights as the accepted lighting technology of the era. But the cleaner, more varied aspects of electric lighting along with heavy promotion by innovators such as Thomas Edison established electric lighting as the dominant system early in the twentieth century.

H. J. EISENMAN

Further Reading

Bowers, Brian. *Lengthening the Day*. New York: Oxford University Press, 1998.

Friedel, Robert D. and Paul Israel. *Edison's Electric Light: Biography of an Invention*. New Brunswick, NJ: Rutgers University Press, 1986.

O'Dea, William T. *The Social History of Lighting*. London: Routledge and Paul, 1958.

Schivelbusch, Wolfgang. *Disenchanted Night*. Berkeley: University of California Press, 1988.

Stoer, G.W. *History of Light and Lighting*. Eindhoven, Netherlands: Philips Lighting, 1988.

Use of Electric Power Becomes Widespread

Overview

It is difficult to envision a world without electricity, especially in the developed world. Electric lights have allowed us to extend our waking hours beyond the constraints of daylight; electricity powers our factories, computers, stereos, and air conditioning units, pumps our water, and helps us iron our clothes. Yet, until the last part of the nineteenth century, electricity was more a scientific curiosity than a useful phenomenon, and it was not until the second quarter of the twentieth century that electrical power became available to most of the population of the United States, Europe, and parts of Asia. By the end of the twentieth century, most of the world's population had at least some access to electricity, although many areas had limited, unreliable, or no access. This widespread use of electricity was made possible, in large part, by the development of relatively efficient means of generating and using alternating current (AC) which, unlike direct current (DC), can be transmitted over long distances without excessive losses due to resistance in the power transmission lines.

Nikola Tesla in his laboratory. *(Corbis-Bettmann. Reproduced by permission.)*

Background

Electricity was first noticed and recorded by the ancient Greeks, who noted that rubbing a piece of amber with wool could cause the amber to shock the person holding it. However, its properties remained a mystery until scientists—including Benjamin Franklin (1706-1790)—began investigations in the eighteenth century. By 1831, Michael Faraday (1791-1867) had discovered electric induction and Hippolyte Pixii (1808-1835) created the first dynamo for generating electrical current the following year. In the 1860s and 1870s a number of scientists and inventors came up with improvements to Pixii's dynamo, but electricity remained confined to the laboratory because there was no real use for it. Lacking a use, there was also no reason to develop efficient means of production or transmission of the electrical energy.

This began to change in the 1880s with the development of both DC and AC electric motors and generators. The primary forces behind these developments were Serbian-American inventor Nikola Tesla (1856-1943) and the American Thomas Edison (1847-1931). Tesla first developed an AC motor and generator in 1883, bringing it with him to the United States the following year. He worked briefly for Edison, later leaving to work on his own. In 1885 he sold his inventions to George Westinghouse (1846-1914), who launched the first large-scale attempt to generate AC electrical power.

In so doing, Westinghouse and Tesla ran afoul of Edison, who was promoting his own DC electrical power systems. The struggle was acrimonious, with Edison proclaiming the dangers of alternating current (pointing out that it was used in electric chairs) while Tesla and Westinghouse pointed out that DC could not be transmitted for long distances without substantial inefficiencies. In fact, the greater efficiency of AC generation and transmission eventually won the day, resulting in one of Edison's few defeats. In addition, it turns out that AC motors can be made lighter, more efficient, and less expensively than DC motors, adding to their advantage.

By the year 1900 the electrification of the United States, Europe, and Japan was beginning, and, by the 1930s, much of the rest of the world was beginning to construct power-generation stations. This trend has continued through the present and, even today, increasing use of electrical power is the norm in virtually every nation on Earth. It is nearly impossible, with the exception of small populations in parts of South America, New Guinea, and Africa, to find people whose lives have not been touched in some way by electricity, whether in the form of products, electric lights, or radios. Regardless of the many different means of generation, electrical power seems likely to remain the most widely used form of energy on Earth for decades to come.

Impact

Electric motors have helped to make mechanical power portable and accessible. Prior to the development of electric motors, mechanical power consisted of muscles (human or animal), gravity (in water wheels and similar devices), wind, or steam (in the form of large steam engines). In many cases mechanical power was impractical because of physical or geographic constraints. For example, gravity-powered devices such as water wheels require running water nearby, making them impractical in the desert. Muscle power could work well for many projects, but had its limitations, too. Horses, for example, could be teamed together to pull heavy loads, but required relatively level surfaces and enough room to operate. Other sources of mechanical power suffered similar limitations.

Electric motors began to change this. Turning a saw in a sawmill no longer required a nearby stream or river when an electric generator could provide sufficient current to turn an electric motor continuously, regardless of weather or water supply. Virtually unlimited power was available at any time simply by running wires or bringing in a generator to wherever work needed to be done. As well as making mechanical power portable, electric motors could out-perform both man and animal. Lifting and moving heavy loads could be done more easily and more reliably than before without the resulting risk to life and health.

In addition to these advantages, electric motors were more easily controlled than teams of either humans or animals. Since the torque and power output of an electric motor can be precisely controlled, the mechanical power provided by these motors could be harnessed and controlled with a much higher degree of precision than before. While this level of control is relatively unimportant for cutting lumber or pulling heavy loads, it is vital for the development of precision objects. Too, the fact that electric motors can be made almost arbitrarily small

allows this same precision at almost any scale, making possible near-microscopic motors and mechanical devices that are beginning to play an important role in modern technology.

Electric motors are also important in operating pumps. While seemingly mundane, pumps are vital to modern society. For example, pumps supply water to boilers that make steam for electric generators to power our civilization. Massive pumps supply drinking and fire-fighting water to cities while smaller pumps supply gasoline to our automobile engines. Pumps of another sort, called compressors, run our air conditioning systems, smaller versions of which circulate coolant through high-performance supercomputers to keep their electronics from melting. Other electrical pumps circulate blood through heart-lung machines during open heart surgery, helping surgeons to repair damage that would otherwise be fatal. Virtually every system designed to move a fluid, whether liquid or gaseous, relies on pumps or fans, most of which are powered by electrical motors similar to those first designed by Nikola Tesla.

Powering these motors, of course, is electricity from a generator of some sort. While there have been many alternate methods of energy production in the last few decades, the overwhelming majority of the world's electrical power comes from dynamos similar to those invented in the latter part of the nineteenth century. It is also interesting to note that, from an engineering and theoretical standpoint, electric motors and generators are nearly identical and, in fact, "motor-generators" exist that operate equally well as motors or generators, depending on the direction of electrical current flow.

In any event, all of the benefits accrued from electric motors would fail to exist if there was no reliable way to generate electrical power in the first place. However, electrical power generation reaches much further than simply providing a power supply for motors. Everything that makes modern life comfortable for inhabitants of the developed world is dependent in part or in whole on the presence of relatively cheap and reliable electrical power.

Much of the impact of electricity on society came from the rural electric cooperatives, first in Japan and Germany and later in the United States and Europe. These cooperatives banded together to build power stations, run lines, and power the farms. Grain crushers, threshers, milking machines, food storage and preservation, and other machines became possible, removing much of the drudgery from farm life while making farms more efficient.

At the same time, widespread electrical power was making life in the cities healthier and more comfortable. Refrigeration made it possible to keep food fresh for several days while freezers kept food fresh for weeks. This made daily trips to the market a thing of the past, which, along with household devices, began to free women from their traditional roles as homemakers. Electric elevators and escalators have made high-rise buildings possible, completely changing the layout and skyline of the world's major cities. Meanwhile, computers, wholly dependent on an uninterrupted supply of precisely conditioned electrical power, have had (and will continue to have) a profound effect on virtually all aspects of technological society. Lest one argue that these impacts are limited to the developed world, with its plenitude of electrical power, it is worth noting that many developing and less developed nations are actively pursuing expansion of their domestic electrical power systems.

The negative side of this increased use of electrical power is the environmental toll it has exacted. Burning fossil fuels has led to air and water pollution, along with possible long-term effects on global climate. Damming rivers for hydroelectric power production has flooded parts of America's Grand Canyon, forced the rescue of Egyptian temples and statues, and necessitated the relocation of millions of Chinese peasants. And, of course, the extraction of fossil fuels for energy production has led to strip mining, land subsidence, acid mine drainage, and other environmental insults. In addition, by enhancing so many aspects of our lives, electrical power helps to enhance human health and prolong life. Ironically, this has resulted in steady and rapid population growth in many parts of the world, further straining the terrestrial environment and threatening to reduce the quality of life for so many.

P. ANDREW KARAM

Further Reading

Conot, Robert. *A Streak of Luck: The Life and Legend of Thomas Alva Edison.* Seaview Books, 1979.

Hunt, Inez and Wanetta Draper. *Lightning in His Hand: The Life Story of Nikola Tesla.* Sage Books, 1964.

Elegant Spans: Suspension Bridges

Overview

The suspension bridge using iron and/or steel as its structural material was a new form of spanning space developed in the nineteenth century. Used for pedestrian, vehicular, canal, and railroad traffic, this new bridge form celebrated the versatility and strength of iron and steel as structural elements. The resultant monumental technology created elegant and efficient bridges which captured the imagination of the public as well as the artistic community.

Background

Spanning space by suspending or hanging a bridge surface from towers dates to ancient times when people in Tibet and Peru used the design as foot bridges in the Himalayas and Andes. However, the modern form of the metal suspension bridge is a product of nineteenth-century designers in the United States, France, and Britain as a response to the increased transportation needs of industrialism.

The American James Finley, from Uniontown, Pennsylvania, patented his design of the first modern suspension bridge in 1808. Incorporating metal cables or chains into his bridges, Finley used an empirical method of design to construct spans as long as 250 feet (76 m) and influenced other bridge builders in the first few decades of the nineteenth century.

In the early 1800s Finley's design diffused to Europe where Thomas Telford (1757-1834) and Samuel Brown in Britain and Claude-Louis-Marie-Henri Navier in France built on Finley's original work to create longer and stronger suspension bridges. Relying on the superior metallurgical practices of Britain, Brown and Telford used iron bars to produce their suspension chains and were able to improve on the length, design integrity, and elegance of Finley's pioneering work. For example, Telford's 1826 Menai Straits Bridge, appearing structurally substantial and strong, spanned 580 feet (177 m), a major feat for the nineteenth century.

During the 1820s, Louis Navier, a well-known French engineer and theoretician, visited Brown and Telford in Britain and studied their suspension bridges. Navier combined his theoretical approach to engineering problems with his knowledge of these British bridges and the successful French wire cable designs to shape a new era of suspension bridge design and construction in France for almost three decades. Both the Americans and French substituted wire cables for iron bars because they could produce quality iron wire rather inexpensively; iron wire's strength, dependability, and low cost made it a better choice than the iron bars used by the British. French interest in suspension bridges diminished after a bridge failure in the town of Angiers in 1850. This accident allowed Americans to provide the next major achievements in suspension bridge design and construction.

The use of iron cables became a trustworthy practice and a mature style through the developmental efforts of Charles Ellet, Jr. (1810-1862), and John Roebling (1806-1869). Both men were versed in French suspension bridge activities, including knowledge of the analysis, construction, and design of these bridges. Ellet, who built the first significant wire cable bridge in America at Philadelphia in 1842, and Roebling were rival bridge builders in mid-nineteenth-century America; Ellet's early death in 1862 left Roebling as the chief champion of wire suspension bridges in the period following the Civil War in the United States, a time of rapid economic and industrial growth.

Roebling, a German emigré to the United States, translated his extensive knowledge of suspension bridges into a patent (1847) for spinning wire cable. Used first in several canal aqueducts, the technique, along with a special anchorage method for securing the wire cables, became a hallmark of Roebling's bridge work. A major disagreement between Ellet and the directors of the Niagara Gorge Bridge project gave Roebling the opportunity to complete the road and railway bridge in 1855 with a span of 821 feet (250 m); the success of this bridge enhanced Roebling's reputation as a leading suspension bridge builder. His Cincinnati Bridge, completed in 1867 with a span of 1,057 feet (322 m), matched European designs in elegance and sophistication and served as the model for Roebling's most famous design, the Brooklyn Bridge.

With a central span of almost 1,600 feet (488 m) and its massive twin Gothic arch towers, Roebling's Brooklyn Bridge represented the

culmination of the nineteenth-century suspension bridge design. In 1869 John Roebling died of complications from a site inspection accident and his son, Washington Roebling, who was well versed in his father's methods and techniques, supervised the remaining work of building the bridge. Completed in 1883, the bridge was the longest suspension bridge in the world, a record it held for many years.

Impact

The Brooklyn Bridge symbolized the impact of the industrial revolution and its attendant age of progress. At a time when the Western world, and the United States in particular, was being transformed by technology, this bridge mirrored the attributes of this new era.

In scale alone, the Brooklyn Bridge was a major achievement. The bridge spanned almost 3,500 feet (1,067 m) with towers reaching 276 feet (84 m) above water level. The New York City area had never seen a structure this high nor long, and it dominated the landscape of lower Manhattan and Brooklyn in the last two decades of the century. Before skyscrapers of the twentieth century dwarfed its size, the bridge symbolized the monumental nature of a new construction technology using iron and steel. The size and scale of bridges and buildings accelerated as rapidly as the technology which produced them.

The growing industrial economy of the Western world increased the development of canals and railroads in the nineteenth century, and the suspension bridge was an efficient and economical means of spanning space in these transportation networks. This bridge design was attractive for those settings requiring an uninterrupted passage, especially for ships, or for locales where an especially deep gorge made it impossible to construct a supporting pier. In addition, the weights of canal aqueducts and railroad lines required stronger bridges so metal became increasingly important as a construction material. By blending strength and economy, the suspension bridge was a natural solution to these transportation needs and contributed to the transportation revolution of the time.

Roebling's techniques for spinning wire cable and erecting towers with the caisson method were innovations in bridge technology. These techniques, which allowed for the widespread use of the suspension bridge, stand as significant American engineering contributions to civil engineering in the nineteenth century—a time when American engineers gained a reputation for blending ingenuity and empiricism into successful and striking projects. Results of this kind reflected the vernacular style in which functionalism, a design based on form following function, determined the aesthetics of a structure. The blending of innovation with functionalism created a unique American motif characteristic of the industrial age.

As a prototype for suspension bridges, the Brooklyn Bridge represents the impact of that bridge style on the social and cultural fabric of America. By providing a regular and dependable connection between Brooklyn and Manhattan, this bridge allowed for a population expansion in the greater New York City area. This growth helped to transform New York City into a thriving, teeming metropolis that became a modern city by the turn of the century. Roebling's design of a pedestrian walkway across the bridge provided a place for walkers and for relaxation on benches, as well as a venue from which to enjoy city views. The multi-purpose function of the bridge, and others like it, has made it attractive to vehicular and pedestrian traffic since its inception. Cities such as New York and San Francisco, surrounded by water, continue to depend on a series of suspension bridges as foundations to their transportation and commercial networks.

Along with the impact on transportation technology and the quality of urban life, the aesthetic qualities of the suspension bridge captured the attention of the artistic community from writers and poets to painters and photographers. Georgia O'Keeffe, John Marin, and Joseph Stella memorialized the bridge on canvas with paintings that celebrate the interplay of forms and shapes typical of suspension bridge geometry with its many cables against a backdrop of towers and urban skylines. The American photographer Walker Evans created a series of dramatic photographs of the bridge from above and below its deck and Hart Crane's poem, "The Bridge," reflects power of the bridge on the creative imagination. This monumental technology became a worthy subject for art as both symbol and artifact of a world which was becoming increasingly a creature of science and technology.

The modern suspension bridge grew out of the newly developed structural technology of iron and steel. As a handmaiden to the growth of railroads, harbors, ports, highways, and urban centers, this elegant style remains closely identified with the power and influence of industrialism in shaping the lives and landscape of the last half of the nineteenth century and beyond. It allowed the spanning of previously unbridgeable space with a physical object whose aesthetic and monumental profile continues to capture the imagination and admiration of both the artistic and technical communities. This bridge design stands as a celebration of the ingenuity of engineers and the elegance and artistry of an efficient and economical means of closing wide and deep spaces in cities, along waterways, and in steep terrain. The pioneering genius of men like John Roebling laid the foundation for a timeless design which reflected a new industrial age of technological progress.

H. J. EISENMAN

Further Reading

Kranakis, Eda. *Constructing A Bridge: An Exploration of Engineering Culture, Design, and Research in Nineteenth Century France and America.* Cambridge, MA: MIT Press, 1997.

Latimer, Margaret, Brooke Hindle, and Melvin Kranzberg, eds. *Bridge to the Future.* New York: The New York Academy of Sciences, 1984.

Trachtenberg, Alan. *Brooklyn Bridge: Fact and Symbol.* Chicago: University of Chicago Press, 1979.

Vogel, Robert M. *Building Brooklyn Bridge.* Washington, DC: Smithsonian Institution, 1983.

The Invention of Automobiles

Overview

In the latter part of the nineteenth century, Gottlieb Daimler (1834-1900) and Karl Benz (1844-1929) developed a gasoline-powered automobile, a significant improvement over the existing steam-powered devices. Henry Ford (1863-1947), in 1913, introduced the assembly line, lowering automobile production costs sufficiently that most families could afford their own car. This also heralded the start of mass production, which had significant impacts in virtually all areas of manufacturing throughout the world. The internal combustion engine eventually revolutionized transportation, industry, and farming, and has had significant impact on many environmental issues.

Background

Transportation had changed very little between the time of the Romans and the early 1800s. People walked, rode horses, or rode in slow vehicles pulled by horses. At sea, people relied upon wind and muscle power. The first invention that began to make travel more efficient was the steam engine.

The first useful device using steam power was a pump for dewatering mines, introduced in 1698 by Thomas Savery (c. 1670-1715) in England. The first steam-powered transportation appeared in 1769, a carriage that would carry up to four people at a speed of slightly over 2 miles (3.2 km) per hour. This was the first automobile. It was not until Daimler and Benz, in 1885 and 1886, married the recently invented gasoline engine to a chassis, wheels, and a steering device that automobiles began to be useful, if expensive.

Although early automobiles look very little like modern machines, their basic design remains surprisingly similar. Both early and current cars have a gasoline-powered engine, four wheels (although Benz's 1885 automobile was three-wheeled), steering, and braking mechanisms all attached to a chassis with seats and a fuel tank. Obviously, the automobile has evolved over the decades, but the essentials remain the same.

Automobiles remained expensive until Henry Ford introduced the assembly line in 1913, changing cars from hand-crafted machines to mass-produced transportation appliances. With the advent of mass production, the cost for assembling a single car dropped so much that they became financially accessible to many families. Since that time, cars have become ubiquitous in all developed nations and in many less-developed countries as well, with concomitant changes in lifestyle, transportation, and environment.

Impact

It may be argued that the internal combustion engine and the automobile are among the most significant inventions in human history. The engine itself revolutionized all forms of transportation, each of which in turn had far-reaching impacts on society. As a form of transporta-

tion, the automobile has had an impact on nearly every person on Earth. Finally, the environmental impact of the automobile and its variants has been, and will likely continue to be, significant. Each of these areas will be discussed in further detail in the rest of this article.

The invention and popularization of the automobile has changed society in many countries in almost unimaginable ways. The automobile has made possible the mobile lifestyle common to most Americans. Until the advent of the automobile it was almost unheard of for working-class families to take extended family vacations, for people to work 20 or more miles from their home, to routinely visit friends or family more than a few miles away, and so forth. It is argued by some that the entire phenomenon of the suburbs is due to the automobile.

The automotive infrastructure dominates the landscape in many parts of the developed world and, increasingly, in the developing world. Parking lots, shopping malls, roads, and highways—not to mention car part shops, automobile dealerships, car washes, and repair shops—exist almost solely due to the ubiquity of automobiles in everyday life in the developed world. The movement of population away from city centers towards the suburbs also depends in large part on the presence of the automobile. In addition, many developing nations have or are building national road networks and are beginning to resemble the developed nations in this respect.

As a direct result of the rise of automobiles in the United States (in contrast to Europe and parts of Asia), the use of mass transportation has declined sharply and, in many cities, is almost nonexistent. Public train systems and subways exist in only a few cities and, in the majority of American municipalities, mass transportation must be heavily subsidized in order to exist at all. By comparison, Europe and parts of Asia, although developed, have not embraced the automobile with quite the fervor as Americans and have been more reluctant to establish suburban lifestyles or to abandon mass transportation. Because of this, many nations continue to have effective nonautomotive transportation networks and less land devoted to parking lots and roads. Another mitigating factor in this trend may be the relatively compact size of many nations compared to the United States, making mass transportation more convenient.

The ability to mass-produce automobiles and, later, other products has changed society as well. First developed as a way of making automobiles affordable to the general public, mass production quickly spread to virtually all forms of manufacture. It is hard to realize that, until this innovation, everything purchased was made by hand, much of it laboriously. Although the first assembly lines still utilized manual assembly, they set the stage for the highly automated assembly lines of today. This, in turn, has led to a relative abundance of inexpensive goods of high, consistent quality. In some industries, manufacture is now performed primarily by machines with little or no human intervention. A far cry from Henry Ford's first Model T assembly line, these factories are nonetheless a direct descendent of Ford's first plant.

In the environmental arena, the automobile has also had a dramatic impact on society and, potentially, the world. The construction of hundreds of millions of cars takes a tremendous amount of iron, aluminum, plastic, and rubber, not to mention the lead in car batteries, copper in wiring systems, and so forth. These materials must all be mined, transported to the manufacturing plant, turned into automobiles, and transported to the dealer for purchase.

To use the cars still requires burning petroleum products (except in rare instances), which must be extracted from the Earth, transported to a refinery, and made into fuel. All of these steps carry with them the potential for adverse environmental impact such as spills, fires, and more. In addition, every transportation step also involves the use of fossil fuels.

The combustion of petroleum (and many other substances) releases carbon dioxide, carbon monoxide, smoke, and other byproducts. Although the introduction of catalytic converters has reduced automotive emissions significantly in some parts of the world, these are not widely used in developing nations. As a result, air quality in many large cities is abysmal, leading to respiratory problems for the very young, the elderly, and the ill. In addition, although the scientific community remains deeply divided over the issue, there is the possibility that the release of large amounts of carbon dioxide into the atmosphere may lead to widespread atmospheric and oceanic warming that could result in equally widespread climate changes. These changes could result in the polar ice caps melting, which would probably swamp many large cities, including Amsterdam, Dakka (Bangladesh), Washington D.C., and parts of Buenos Aires (Argentina), to name a few. However, it must be stressed that data regarding global warming is, as of this writing, still subject to great debate and interpretation. This is further complicated by the fact that the Earth's average temperature is continually changing, and the Earth at this time is cooler than is typical throughout the history of the planet.

All of these environmental changes must, of course, be weighed against the benefits that have come from the invention of the automobile. There is no doubt that today's world is an improvement in many ways over that of a century ago and that much of this improvement is due to the relative ease of transporting goods and people, enhanced farming techniques, and many of the other benefits noted above.

P. ANDREW KARAM

Further Reading

Gott, Philip G. *Changing Gears: The Development of the Automotive Transmission*. Society of Automotive Engineers, 1991.

Womack, James P., Daniel T. Jones, and Daniel Roos. *The Machine That Changed the World*. Rawson Associates Press, 1990.

Quest for Sound:
Thomas Edison's Phonograph

Overview

Thomas Edison (1847-1931) has gone down in history as one of the great inventors of the late nineteenth and early twentieth centuries. He was responsible for developing the first electric light bulb, as well as the early motion picture industry. But arguably his greatest, and certainly his fondest, invention was the phonograph, which was not only groundbreaking, but laid the foundation for the future music recording industry.

Background

When he was just 11 years old, Edison would disappear into his family's cellar, neglecting his

Thomas Edison and his phonograph. *(The Granger Collection. Reproduced by permission.)*

schoolwork, for hours on end. His family would discover him experimenting with all sorts of chemicals, creating strange and mysterious concoctions. It was the earliest spark of genius in a man who would come to personify invention.

In July 1877, while trying to develop a new and better transmitter for the telephone—invented the previous year by Alexander Graham Bell (1847-1922), Edison first discovered that he was able to create an impression on a piece of wax paper that could actually record sound. Soon after he wrote in his diary, ". . . there is no doubt that I should be able to store up and reproduce at any future time the human voice perfectly."

Edison was not the first to record sound; a French inventor had done so 20 years before. What made Edison's discovery unique was that he was also able to play it back. Soon after, the word "phonograph" began to appear in Edison's notes. The first phonograph was a simple apparatus. To make a recording, a person would speak into a funnel-like mouthpiece that directed sound into a thin metal diaphragm. When the sound waves struck the diaphragm, they caused it to vibrate. As the diaphragm vibrated, it caused an attached needle to also vibrate, which then made indentations on a thin piece of tinfoil. To play back the recording, the procedure was reversed, making the needle and diaphragm vibrate, and sending back the sound.

Edison's first recorded words were memorable but not profound: "Mary had a little lamb, its fleece was white as snow, and everywhere that Mary went, the lamb was sure to go." Nevertheless, when Edison heard this simple nursery rhyme played back in his own voice, he said he "was never so taken aback in all my life."

Impact

When the public first heard of Edison's invention, they thought he was a magician. In fact, he was dubbed "the Wizard of Menlo Park," named for the New Jersey town where he lived and worked. Hordes of reporters descended upon Edison's home and laboratory. Huge audiences came to see traveling demonstrations of his new and wondrous device. He was invited to Washington, D.C., to showcase his new phonograph to an astounded Congress. While there, he paid a late night visit to President Rutherford B. Hayes, who excitedly awoke his wife so that she, too, could witness the miraculous phonograph.

But the early excitement of the phonograph faded quickly as Edison became distracted by another of his great inventions—electricity. In 1881 Alexander Graham Bell, who had long wondered why he had not come up with the idea for the phonograph, decided to take up the project where Edison had left off. Edison's early recordings were of poor quality, could only play for a minute or two, and deteriorated after only a few uses. Bell thought he could do better. With prize money awarded by the French government, he set up a lab in Washington, D.C., to begin work. Bell's assistants—his cousin Chichester Bell, a chemist, and Charles Tainter (1854-1940), a scientist—produced their own version of the phonograph, which they named the graphophone. Instead of using tin foil, which could only be played two or three times, the graphophone instead used the more durable wax-covered cardboard. Edison's clunky hand crank was also replaced with a motor driven by a foot treadle.

Bell suggested that he and Edison work together and pool their patents to create one superior product, to which Edison bitterly replied, "Under no circumstances will I have anything to do with Graham Bell" and his band of "pirates." Instead, Edison challenged his competitor by creating a better and cheaper machine than Bell's. In June 1888 Edison's improved version of the phonograph was completed.

When Edison had first begun work on his phonograph, he came up with a list of things his invention could be used for. High on that list was taking office dictation, which he believed would be the machine's primary function. He was to be proven wrong. As early as 1890, music became the main use for the phonograph. In concert halls and sing-alongs, people gathered to hear their favorite tunes. In penny arcades, phonograph parlors, and train stations, people lined up to drop a nickel in the slot and hear Gilbert and Sullivan songs, operatic overtures, and vaudeville comedy monologues. Bell's graphophone also gained similar popularity during this time.

But even as the phonograph grew in appeal, it was still limited by the fact that cylinders could not be mass produced. This meant that Edison's workers had to painstakingly mold each individual cylinder, while performers three floors above sang the same song, over and over, to make each individual recording.

In the 1890s Edison was once again distracted from his work on the phonograph; this time, by the newly emerging motion picture. While he was away, his rival Bell began marketing the graphophone for its first home use, with a new spring motor to make it lightweight and inexpensive. At around the same time another Edison rival was gaining prominence in the recording industry. The Berliner Gramophone, invented by Emil Berliner (1851-1929), would become the first record player to use discs instead of cylinders.

In 1901 Berliner founded the Victor Talking Machine Company, and the next year opened a record factory. Now, for the first time, records could be easily mass-produced. Berliner also attempted to appeal to public taste when deciding what type of music to record. Previously, Edison made his decisions based on his own taste, which ran toward sentimental ballads like "I'll Take You Home Again Kathleen." Victor's catalog offered a wide variety of artists, including the famed opera singer Enrico Caruso. The company was also responsible for creating the world's first home entertainment system—the Victrola, which featured a cabinet in which to store records, and hid the big horn, making it look more like an elegant piece of furniture. The public went wild for this new invention.

Edison worked to meet Victor's challenge by introducing a new, longer, four-minute cylinder and by developing a machine that would compete with the Victrola—the Amberola. But it was now clear that the public wanted discs, and Edison would have to follow demand if he was to continue making money.

He did so by working around the clock to create an even better disc machine than his competitor. Edison's goal was to "produce a disc machine free from a mechanical tone, distortion of the original sounds and free from those irritating scratchy sounds now in all disc machines." In 1915 Edison's disc phonographs were finally ready to challenge the Victrola. While they sold reasonably well, they were never able to reach Victor's sales. Although the fidelity was better in Edison's product, Victor still had a larger catalog of artists.

Then, in 1920, a new invention emerged called radio. Soon, everyone was tuning in to the static-filled broadcasts and tuning out their phonographs. Edison's son encouraged his father to enter the radio business, but he resisted the radio invasion, calling it merely a "fad." By 1929, faced with plummeting sales, Edison was forced to admit defeat and close his phonograph company forever.

Edison's phonograph had not only created thousands of jobs, but it led to the development of the recording industry as we know it today. Future improvements would include the LP, or long playing record, in the 1940s, stereophonic sound in the late 50s, and, in 1982, the compact disc. Edison would go on to become a pioneer in the early motion picture industry. However, it was the phonograph that would be his most prized invention, and the one for which he retains the sobriquet "the Wizard of Menlo Park."

STEPHANIE WATSON

Further Reading

Books

Baldwin, Neil. *Edison: Inventing the Century.* New York: Hyperion, 1995.

Bruce, Robert V. *Alexander Graham Bell and the Conquest of Solitude.* Boston: Little, Brown and Company, 1973.

Israel, Paul. *Edison: A Life of Invention.* New York: John Wiley & Sons, 1998.

Other

The Edison Effect: The Phonograph. Produced by Jon Wilkman. 50 min. A&E Home Video, 1995. Videocassette.

Herman Hollerith's Punched Card Tabulating Machine Automates the 1890 U.S. Census

Overview

By successfully automating the calculation of the 1890 United States census with an electromechanical punch-card device, inventor Herman Hollerith (1860-1929) laid the foundation for the next century's explosion of information-processing machines, technologies, systems, and businesses, including IBM, the world's largest computer corporation. Hollerith's use of standardized punched cards to represent information, and his addition of electricity to mechanical tabulation, greatly increased the effectiveness and range of applications of tabulating and, later, computing machines.

Background

The explosive growth of industry and population throughout the nineteenth century was matched and in some ways exceeded by growth in the amount of information those industries and people generated. There was an important and expanding need to manage and manipulate the statistics derived from that information, both for social/political and economic purposes. For example, while the 1850 census allowed for 60 different statistical entries regarding race and sex, there were more than 1,600 such entries in the 1890 census. The world was in danger of drowning in data. There was an increasing demand not only for more accurate records-keeping but also for faster calculations and tabulations of those records.

For the most part the statistics were recorded and tabulated by hand, although mechanical calculating devices that used gears and levers to

record numbers had been employed with varying degrees of success throughout the century. (Indeed, simple aids to calculation had been used for thousands of years; the abacus, with its movable beads, is a good example of such a device, and has been in existence for more than 5,000 years.) Mechanical calculating devices were primarily useful for simple addition, and still required a great deal of human labor, including the painstaking recording and transcription of data before and after the calculating process itself. Still, it was clear that mechanical assistants could make simpler all manner of arithmetical operations.

Early in the nineteenth century, English inventor Charles Babbage (1792-1871) built the first of several mechanical calculating machines, which he called difference engines. Completed in 1822, Babbage's first difference engine was intended to handle a variety of calculating and computing tasks, and to print the results of its computations either on paper or on printer's plates. Babbage continued to refine and enhance his engines until his death in 1871. While Babbage's machines were marvels of design and operation, they remained difficult to operate, and less precise than desired.

Still, the importance of mechanical aids for computation could not be denied, and by the 1880s, research into computing machines was proceeding at a rapid pace. Various approaches were taken to increase the effectiveness of mechanical calculators, none more successful than the innovations introduced by American inventor Herman Hollerith. It occurred to Hollerith—as it had to others—that since statistics are patterns of information, they could perhaps be encoded in ways similar to those used in mechanical looms to weave patterns of thread into fabric. That process, using cards with holes punched in them to guide the loom's workings, had been developed in France more than a hundred years before by Joseph Marie Jacquard (1752-1834). The Jacquard loom had revolutionized the textile industry; now Hollerith sought to adapt the technology to weaving patterns of statistics.

By May 1883, now working for the United Sates Patent Office in Washington, Hollerith had roughed out his idea. Central to the idea—and one of the keys to Hollerith's great contribution—was the use of electricity to automate the recording of information.

Holes punched in strips of paper would represent different types of information. The paper would be fed into a mechanical device comprised of a hand press which pushed metal pins into contact with the paper strip. Those pins which passed through punched holes would make contact with a container of mercury, completing an electric circuit; where no hole was encountered, no contact would be made. Completed circuits in turn triggered mechanical counters which recorded the appropriate information. There had been mechanical calculators before Hollerith, but his introduction of electricity into the equipment dramatically increased their efficiency—and potential.

Another of his innovations was the replacement of strips of paper with punched cards. These stiff cards proved far easier to sort and manipulate than paper, further enhancing the speed of calculating operations. Over the next three decades Hollerith and his company would work to standardize the size and format of punched cards. Standardization was vital to the spread of tabulating—and, later, true computing—as an industry.

Hollerith machines were used to calculate the 1890 United States census. His Tabulating Machine Company accomplished the calculation of population in less than six months, and the rest of the census information in under two years, far more rapidly than any census had ever been completed. Also attracting attention was the fact that Hollerith's machines performed their tasks far more economically than could human calculators. The census was completed more than $5 million below its budget.

Impact

The impact of Hollerith's innovation was both immediate and long-lasting. The immediate impact was simply an increased demand for his machines by governments and industries throughout the world. In 1897, for example, Hollerith systems were used to calculate the Russian census. For that task, more than 100 million punched cards were employed.

While census calculation formed the heart of Hollerith's business throughout the 1890s, other applications and uses for electromechanical computation were clear. Virtually any undertaking that relied on numbers or statistics could automate its calculations, saving money and increasing efficiency. Hollerith's tabulating systems found ready use in the calculation of railroad freight shipments, agricultural output, insurance and banking figures, and so on.

Many of these applications required features unavailable in the first Hollerith machines, including multiplication and division. Hollerith continued to innovate, adding these and other mathematical features, as well as introducing printed output of results. With the widespread arrival of electrical power Hollerith added that as well, producing true electrical calculators rather than electrochemical ones. He and his engineers continued to automate various aspects of their equipment, notably the punching of holes in cards, and the sorting of those cards.

The uses of Hollerith tabulators—and machines from competing companies—continued to spread. United States President Theodore Roosevelt foresaw ways in which social reform and government policies could be shaped and directed as a result of statistics generated by the census, as well as other information sources. In addition to industry and government, the sciences also found many uses for mechanical calculators.

The business of calculation grew as well. Hollerith perceived not only the need for large-scale calculating and computing services, but also the all-but-unlimited potential demand for such services as data entry, machine maintenance and servicing, equipment upgrades, and so on. One of his business innovations, which influenced the computer industry for most of the century, was his policy of leasing, rather than selling his tabulating machines. Leases tended to be more affordable than outright purchase, and had the added advantage of insuring Hollerith a continuing market for services and cards.

In 1911 Hollerith's Tabulating Machine Company merged with the International Time Recording Company and two other companies to form the Computing-Tabulating-Recording Company (CTR) with Hollerith as president. He continued to work with the firm for another decade, retiring in 1921. He died eight years later.

CTR changed its name in 1924, becoming International Business Machines (IBM.) As the world's largest supplier of calculating and tabulating machines, as well as the punched cards on which information was recorded, this offspring of Hollerith's original firm shaped the role of information management for governments, businesses, and institutions throughout the twentieth century.

As other technologies emerged, IBM—and its competitors—incorporated them into their products. The arrival of electronics, first in the form of vacuum tubes, later in the form of transistors and printed circuits, hastened the transformation of calculating machines into true computers, able to accomplish billions of calculations of all types in seconds. While still primarily statistical machines, computers played increasingly important roles throughout society. The twentieth century, in fact, came to be referred to as the Information Age, a fact largely made possible by our ability, thanks to calculating, tabulating, and computing machines, to process and manage information.

With the expansion of the computer's abilities during the second half of the twentieth century, as well as their decreasing size and cost, computers became perhaps the world's single-most-important

technology. Complex modern society and business would be all but unmanageable without these descendants of early tabulating machines. Everything from systems of traffic lights to telephone networks to inventories at stores and factories came to be automated by computers. The arrival, late in the 1970s, of small, relatively inexpensive personal computers extended the usefulness of computers to individuals.

The marriage of computers with telecommunications created, by the last decade of the century, a global network of information, commerce, communication, entertainment, and education unlike anything the world had seen—all barely a century after Hollerith used punched paper to calculate the United States census.

KEITH FERRELL

Further Reading

Books

Austrian, Geoffrey D. *Herman Hollerith: Forgotten Giant of Information Processing.* New York: Columbia University Press, 1982.

Babbage, Charles. *Passages From the Life of a Philosopher.* New Brunswick, NJ: Rutgers University Press, 1994.

Periodicals

Heide, Lars. "Shaping A Technology: American Punched Card Systems 1880-1914." *IEEE Annals of the History of Computing* 19 (October-December 1997): 28-41.

Kistermann, Friedrich W. "The Way to the First Automatic Sequence-Controlled Calculator: The 1935 DEHOMAG D 11 Tabulator." *IEEE Annals of the History of Computing* 17 (Summer 1995): 33-49.

Reid-Green, K.S. "The History of Census Tabulation." *Scientific American* (October 1989): 78-83.

Capturing Life Onscreen: The Invention of Motion Pictures

Overview

Motion pictures combined three earlier technologies. Early nineteenth-century experimenters knew how to make drawings appear to move, by passing them rapidly before the eye. Magic lanterns and shadow puppets were ways to project silhouettes onto a wall or screen. Photography allowed the capture of realistic images of people, animals, and their surroundings. Motion picture pioneers competed to find a way to make photographs seem

to come alive, and project them for display to an audience.

Background

The perceptual phenomenon called *persistence of vision* was known to the ancient Egyptians, but was first described by Peter Mark Roget (1779-1869) in 1824. You can demonstrate it yourself by looking briefly at a light, and then closing your eyes. The image of the light remains on your retinas for 1/20 to 1/5 of a second before it fades away. Early in the twentieth century, psychologists showed that the brain also has a perceptual threshold, in which images that flash by quickly will seem to be continuous. Together these phenomena make it possible to produce the illusion of motion using a series of closely spaced images that change by degrees.

The first motion picture devices were mechanical, and the images were simple drawings. The thaumatrope, from the Greek for "wonder turning," was a cardboard disk with images that seemed to merge and move when it was spun on a piece of thread. Several similar "optical toys" were developed in the early 1830s, including the Wheel of Life, the phenakistoscope, and the stroboscope. A device with pictures on a strip surrounding a rotating drum was marketed as the zoetrope in the 1860s.

Light-projected images do not require technology; they are familiar to anyone who has ever made finger shadows into a bunny on the wall. Shadow puppet shows using ornate paper silhouettes were performed in China, India, and Java for over 1,000 years. Magic lanterns, using candles or oil lamps to illuminate and project images drawn or painted on glass slides, were popular in sixteenth- and seventeenth-century Europe. In 1666 the diarist Samuel Pepys wrote of a visitor who showed him "a lantern with pictures in glass to make strange things to appear on a wall, very pretty." Gears, rotary disks, and sliding panes of glass were used to provide special effects.

The Frenchman Emile Reynaud combined the magic lantern and the mechanical motion picture device, attaching a projecting lens to the Zoetrope. The result was bright sharp projections of drawings that appeared to move. In 1892, he presented the popular *pantomimes lumineuses* to packed houses at his Theatre Optique. Unfortunately for Reynaud, the projection of moving photographic images, which would sweep away his invention, was only a few years away.

Cameras enabled the photographic reproduction of real-life scenes in the middle of the nineteenth century. The world's first permanent photographic image was made by Joseph Nicéphore Niepce (1765-1833) in 1826. The British scientist William H. F. Talbot (1800-1877), in the course of inventing a way to print photographs on paper, devised the first negative-positive photographic system. In 1849, the Langenheim brothers of Philadelphia introduced positive images fixed onto glass plates. These transparencies paved the way for the projection of photographs.

A story which some regard as apocryphal holds that in 1872 Edward Muybridge (1830-1904) was hired to use photography to capture the motion of a racehorse, in order to win a $25,000 bet for California millionaire Leland Stanford. Stanford wagered that when a horse galloped, at some point all four legs left the ground simultaneously. Whether or not the bet actually took place, it is known that six years later, after many failed experiments, Muybridge set up a row of 12 cameras beside the Palo Alto racetrack. As the horse galloped past, it tripped wires attached to the cameras' shutters. In fact, all four legs did leave the ground at once, and Muybridge later produced an 11-volume study of animal locomotion, using as many as 24 cameras to take his photographic sequences. But he didn't produce a realistic motion picture. When his photographs were affixed to the rim of a wheel, and the wheel was spun, the animal seemed to run in place.

What was needed was a way for one camera to take multiple pictures in succession. In that case, objects in the image would be seen to change position in relation to the camera. The cumbersome glass plates early cameras used for each exposure made rapid shooting impossible. Movies were made possible by the invention of flexible film, consisting of photographic emulsion deposited on a celluloid base. This process was discovered by Hannibal Goodwin, a minister in Newark, New Jersey, in 1887. It was patented by George Eastman (1854-1932), the local photographic manufacturer soon to be a multimillionaire.

Impact

The first movie camera using the new films, called the kinetograph, was developed in 1889. It was a product of the prolific Thomas Edison Laboratories in West Orange, New Jersey, from which the electric light bulb, phonograph, and tickertape machine had also emerged. The moving picture research team was led by the Englishman William Kennedy Laurie Dickson. In 1891 the Edison Lab patented the Kinetoscope. This was a wooden box with a peephole through which an individual could view short Kinetograph film loops.

Edison's films were made in the world's first film studio, a box-like contraption in West Orange called the Black Maria. The room was mounted on a turntable so it could be rotated to let the sun in through a skylight. The short films produced there included the first motion picture kiss, between Broadway stars May Irwin and John Rice in 1896. Peepshow parlors soon sprung up around the country, where people could peer into the kinetoscope for a nickel, collected by an automated coin slot of Edison's devising. The peepshow was one of the most popular entertainment fads of the time. Believing that projecting his films for mass audiences would undercut this lucrative business, Edison lost interest in advancing the technology.

As a result, the ingenious Dickson quit and went to work with Woodville Latham and his sons Gray and Otway. The Lathams were committed to the future of motion picture projection. One of their later developments, the Latham loop, was a way to leave slack in the film as it went through the projector. This reduced the stress on the celluloid, preventing it from tearing, and allowed longer films to be made. But the Lathams and Dickson were not the only ones working on motion picture projection during the 1890s. Competing teams presented several successful demonstrations during 1895 and 1896.

On November 1, 1895, Max and Emil Skladanowsky used their movie projector, which they called the bioscope, to present a variety program at the Berlin Wintergarten. They showed several very short circular films of children dancing, kangaroos boxing, and other subjects. Their technology was never developed past these brief repetitive loops, although the term "Bioscope" was used in South Africa for decades to mean "cinema."

The first successful showing of a movie to a paying audience took place in Paris on December 28, 1895. In 20 minutes, the brothers Auguste and Louis Lumière (1862-1954 and 1864-1948, respectively) showed 33 customers 10 short films. One sequence, called *"La Sortie des Usines,"* showed workers leaving the Lumières' factory in Rue Saint-Victor, Lyons. The street has since been renamed rue du Premier-Film.

Most historians credit the Lumières' demonstration as the birth of cinema. They had developed a portable hand-cranked movie camera that could shoot, print, and project motion pictures. With it they shot 15- to 20-second sequences all over the world, freeing cinema from the primitive studio and the look of the staged tableau. Their naturalistic film sequences included military parades, comic pieces, scenic landscapes, and living portraits. Their work began to hint at the scope of this new medium.

In 1896, inventors Norman Raff and Thomas Armat modified the Kinetoscope into a projector, paid Thomas Edison (1847-1931) a fee to use his name for credibility, and started advertising "Edison's latest marvel, the Vitascope." On April 23 of that year, at Koster and Bial's Music Hall in New York City, they showed a selection of the Edison kinetograph movies. The program included scene of waves crashing toward the camera on the beach, causing people sitting in the first few rows to duck. Later showings of "The Kiss" outraged audiences, who thought the larger-than-life image was lascivious. The enormous amount of publicity that these shows received left many with the impression that Edison invented the movie projector, including the officials of Macy's, who put up a plaque to that effect when they built their department store over the old 34th street site of Koster and Bial's.

The motion picture pioneers were inventors, not artists. Many were not fully aware of the potential of their devices. The first audiences, including many of the crowned heads of Europe, were amazed by the realistic motion alone. It didn't occur to them to demand high drama or glamorous film stars. Edison took out only American patent rights for his camera and projector, not believing the extra $150 to cover England and France was worth it. He was already intent on new projects. The Lumières' agent was driven out of America in 1897 by the heavy duties of the protectionist Dingley Tariff Act.

By 1900, there were dozens of small film companies throughout the United States and Europe. Inventors and technicians were giving way to theatrical artists, trick photographers, and even magicians. They began to explore cinema as an art form in its own right. In particular, the French illusionist George Méliès (1861-1938) is regarded by many as the father of the narrative film. He is credited with the first use of techniques such as fades, double exposures, time-lapse photography, and artificial lighting effects. Between 1896 and 1906, his Star Film Company produced more than 500 short films, including fantasies and documentaries. Motion pictures were no longer a novelty; by the turn of the century, they had become an industry.

SHERRI CHASIN CALVO

Further Reading

Giannetti, Louis and Scott Eyman. *Flashback: A Brief History of Film*. Englewood Cliffs, NJ: Prentice-Hall, 1986.

Nowell-Smith, Geoffrey. *The Oxford History of World Cinema*. Oxford: Oxford University Press, 1996.

Parkinson, David. *History of Film*. New York: Thames and Hudson, 1996.

Robinson, David. *From Peep Show to Palace: The Birth of American Film*. New York: Columbia University Press, 1996.

Sklar, Robert. *Film: An International History of the Medium*. New York: Harry N. Abrams, Inc., 1993.

Wenden, D.J. *The Birth of the Movies*. New York: E.P. Dutton, 1975.

The First Subways

Overview

The growth of railroads in the middle of the nineteenth century coincided with improvements in tunneling techniques. Underground railway service began in London in the 1860s, as a way to help ease street traffic problems. However, the coal-fired locomotives of the day created air pollution problems in the early subway tunnels. The success of the first electric subway, which opened in London in 1890, encouraged other cities to develop systems of their own.

Background

In the early nineteenth century, city traffic was becoming unmanageable. Narrow, twisting lanes and dead-end streets built for foot traffic and mounted riders were being confronted with increasing volumes of coaches, carriages, and omnibuses. The congestion became a major civic issue in population centers such as London, Paris, New York, and Boston.

In London, the growth of the Thames dock system had led to extensive waterfront development on both sides of the river. To get from one side to the other required a ferry, or a roundabout trip via the London Bridge. Produce withered and dairy spoiled before it could be delivered to its customers. Another crossing was a necessity, and the powerful Duke of Wellington, with military defense considerations in mind, supported a tunnel rather than an additional bridge.

Yet no one had ever been able to build a tunnel in the soft silt of a riverbed. The ancients had dug passages and channels into rock, for tombs, aqueducts, and sewers. Rock was difficult to cut, but in the seventeenth century engineers learned to use gunpowder to help blast it out. And once rock tunnels were cut, they tended to stay put, because their walls were self-supporting. Trying to dig under soft earth or watery silt was a different matter.

In mining tunnels, for example, the walls were braced with wooden frames, but the miners had to dig ahead of the frames in order to progress further. It was not uncommon for soil and loose rocks to suddenly pour in, destroying both the tunnel and the miners in it. In the early 1800s, a true soft-ground tunnel had never yet been built. In 1809, after a few attempts had failed, a learned committee was formed to judge proposals for building one. A prize of 500 British pounds was offered. The judges rejected dozens of proposals and finally declared the entire idea "impracticable."

Civil engineer Marc Isambard Brunel (1769-1849) thought the committee's conclusion was ridiculous. In 1818, he patented the *tunneling shield*. Miners stood inside a number of what were essentially huge hollow cast-iron drill bits, which were pushed forward through the earth six inches (15 cm) at a time by hydraulic presses. As the shield crept forward, it provided the supportive structure to hold the tunnel up. Meanwhile, behind the protection of the shield, each successive six-inch section of tunnel wall was bricked over.

With Wellington's support, Brunel gained the approval of the British government for his scheme. He broke ground for the Thames Tunnel in 1825. Building the tunnel was an arduous process. Financial directors insisted on piecework wages, paying by the amount of ground covered or number of bricks laid, rather than by the hour. This led workers to take dangerous shortcuts. A collapse in 1827 led to the project's being abandoned for a decade. Finally the tunnel was opened in 1843, and 50,000 people passed through it in the first 24 hours. Although

A picture of the London subway, popularly called "The Tube," which marked the second generation of that city's pioneering mass transit system. *(Photograph by Neil Strassberg; Archive Photos, Inc. Reproduced by permission.)*

it was only 400 yards (366 m) long, it was regarded as one of the wonders of the world.

Impact

For 23 years the Thames Tunnel served as a pedestrian passage. But its arrival in the 1840s coincided with the height of enthusiasm for the new railroads, which were expected to solve all transportation problems. Rails had been used for horse-drawn mining carts and other rough-terrain hauling for hundreds of years. The invention of the steam engine by Thomas Newcomen (1663-1729) in 1712 and later improvements by James Watt (1736-1819) led to the development of the steam locomotive, capable of pulling heavy trains over great distances. Society was transformed in the early decades of the nineteenth century, as people and goods traveled more easily between cities and the countryside.

The Thames Tunnel had pointed the way to using underground transportation to ease city congestion. Railway lines crisscrossed England, but they all stopped at the edge of London. Proposals to get them into the city failed for lack of space. Underground railways were the inevitable solution, and were championed by Sir Charles Pearson, sometimes called the father of the subway. In 1854, the newly formed Metropolitan

Railway Company received the Royal Assent to build an underground line in London.

A major concern about running locomotives underground was preventing the passengers from suffocating. The tunnels of the time were very poorly ventilated, and coal-fired locomotives were a pollution problem even in the open air. The Metropolitan Railway's chief engineer, John Fowler (1817-1898), attempted to assuage these concerns by promising to build a "fireless locomotive" that would build up enough steam outside the tunnel to get through it without burning any more fuel. So there would be no smoke, and even the steam would be condensed into a cold water reservoir rather than being released in the tunnel. This promising idea turned out to be impossible to implement, and the Metropolitan retreated to a standard coal-fired locomotive with a few minor modifications.

This meant it was all the more important to have well-ventilated tunnels, and so they were made shallow, with "open cuts" to the surface at intervals to allow air exchange. Brunel's shield was not appropriate for such shallow work, and so another tunneling technique called *cut and cover* was used. First an open trench was dug, and the sides reinforced. Then the trench was roofed over, and the surface restored, except at the open cut sections.

Cutting a 3-mile (4.8 km) trench across the ancient city of London was like stepping into a time machine. Excavators unearthed fossilized invertebrates, Roman water pipes, and ruins left by the Celts, Saxons, Danes, and Normans. The excavators were, of course, not archaeologists, but laborers and businessmen in a hurry, so one can only speculate at the knowledge that was lost.

On January 9, 1863, the world's first subway train pulled out of the Bishop's Road Station. It was packed with dignitaries, although the Prime Minister, the elderly Lord Palmerston, declined, writing that he preferred to remain above ground as long as possible. The next day the "Underground" opened to the public, and 30,000 passengers tried it out. Within a few years, companies were vying for the right to open subway lines.

Early London subways had first, second, and third class carriages, and the cars were attractively done up in Victorian style, with plush seats, brass hardware, and polished paneling. Air quality, though, was still a problem, even with the open cuts. Additional holes drilled to the surface were not sufficient to ventilate the tunnels, and served mainly to startle pedestrians with sudden blasts of smoke. Railway officials issued claims that coal and sulfuric acid fumes were therapeutic for asthmatics and bronchitis victims. Sometimes the smoke was so thick that the trainmen could not see their signals. Eventually, it became necessary to install large, expensive exhaust fans.

It had been hoped that the shallow tunnels would help with ventilation, but this was not the case. While it was understandable to have assumed that digging a shallow tunnel would be less expensive than digging a deep one, that hadn't worked out either. Shallow tunnels interfered with streams, water and sewer lines, and gas mains, and incurred the cost of buying all the property through which the trench would run. A deeper tunnel would avoid all these concerns. But it would be impossible to run steam locomotives in them.

Fortunately, refinements of Brunel's deep tunnel-digging methods coincided with the development of electric railroads. On November 4, 1890, the Prince of Wales, later to be King Edward VII, presided over the opening of the City and South London, the city's first "tube." The tubes had only one passenger class, and were pollution-free. Their success sparked further subway development in London and elsewhere. On the European continent, the first subway line opened in Budapest in 1896. The first subway in the United States, a one-and-a-half mile (2.4 km) line in Boston, opened in 1897. Work on the Paris Metro began in 1898, and it opened in 1900.

Subway development in New York got a slow start due to the machinations of the famously corrupt politician William Marcy "Boss" Tweed, who was safeguarding the interests of the omnibus and streetcar lines that paid for his patronage. There was also a great deal of sensational publicity in New York about the ventilation problems in the early London underground. A prototype pneumatic subway had a brief run in the 1870s, and a number of elevated railroad lines were built. The first underground line of the Interborough Rapid Transit Company, or IRT, opened in 1904.

SHERRI CHASIN CALVO

Further Reading

Bobrick, Benson. *Labyrinths of Iron: A History of the World's Subways.* New York: Newsweek Books, 1981.

Cudahy, Brian J. *Change at Park Street Under: The Story of Boston's Subways.* Brattleboro, VT: S. Greene Press, 1972.

Day, John Robert. *The Story of London's Underground.* London: London Transport Executive, 1979.

Trench, Richard and Ellis Hillman. *London Under London.* London: John Murray Publishers Ltd., 1994.

Safe Enough to Kill:
Advances in the Chemistry of Explosives

Overview

Warfare took a deadly step forward with the invention and development of powerful explosives in the nineteenth century. Guncotton and a safer explosive, cordite, supplied a "smokeless" propellant that made the battlefield visible. Big explosions became more practical when Alfred Nobel stabilized nitroglycerin by converting it to dynamite. But while mass destruction was made easier, so was mining, the drilling of oil wells, and reshaping the land for roads, railroads, and construction. The understanding of nitrogen chemistry that came from research into explosives built a basis for creating new fertilizers and medicines. While Nobel's hope that dynamite would end war by making it too horrible to engage in was proven wrong, his prizes have provided the twentieth century's most visible recognition for contributions to peace, as well as to science and literature.

Background

Explosives research emerged from advances in the general understanding of chemistry. Scientists such as Jöns Berzelius (1779-1848) were eager to figure out how new materials might be created by combining familiar materials. In fact, Berzelius, who discovered three elements, began the practice of using compact symbols for elements, so that familiar formulations such as H_2O for water and $C_6H_{12}O_6$ for glucose. This contribution helped chemists to reveal and clearly articulate how chemicals combined and rearranged to form new compounds. Explosives became a serious area of chemical research when Ascanio Sobrero (1812-1888), who had studied under Berzelius, discovered nitroglycerin in 1847. Sobrero slowly added glycerin to a mixture of nitric and sulfuric acids and observed that adding a small amount of heat to the new compound created a big bang. He was horrified by his discovery, and, while he reported the powerful effects of nitroglycerin, he never tried to put it to use.

Others were less cautious. Even though nitroglycerin is highly unstable and extremely dangerous to manufacture and use, its power proved irresistible. One person who was attracted to it was Alfred Nobel (1833-1896). Nobel had grown up with weapons. His father moved the family to Russia to create underwater mines and machine tools for the Tsar. With the end of the Crimean War, the family's business was having difficulties, and Nobel saw the advantage of developing a better form of nitroglycerin. One of his first contributions was the invention of a detonator for the explosive. He began to manufacture nitroglycerin in his native Sweden, but when a factory blew up, killing several people including his brother, Nobel was banned from rebuilding his factory. Undeterred, Nobel set up a laboratory on a barge in the middle of a lake. There he had a lucky accident. In 1866 some nitroglycerin leaked into a container filled with diatomaceous

Alfred Nobel. *(The Library of Congress. Reproduced by permission.)*

earth (kieselguhr). The combination, called dynamite, was stable. It could not be jarred into action, but once ignited with a blasting cap (a smaller explosive, also invented by Nobel), it had the full power of the original nitroglycerin.

Luck and Alfred Nobel are connected with another development in the chemistry of warfare, the invention of smokeless powder. Battlefields at the beginning of the nineteenth century were overhung with a cloud of black smoke. Gunpowder creates a dirty explosion, and the combined firings of many guns and cannons often meant that troops were fighting blind. Unless they had favorable winds, generals could not effectively direct their troops. An accident set off a change in this situation in 1845. Christian Schönbein (1799-1868) did basic research in chemistry and is known for the discovery of ozone. Schönbein had been forbidden from conducting experiments in his wife's kitchen, but he was doing so anyway. While he was combining nitric acid and sulfuric acid, he spilled the mixture. He needed to clean it up before his wife came back, so he grabbed her cotton apron, mopped up the mixture, and set the apron next to the stove to dry. Unexpectedly, the apron disintegrated. Schönbein had created guncotton.

Unlike Sobrero, Schönbein had no qualms about putting his invention to use. In fact, he rushed guncotton into production as a "smokeless" (in reality merely lower smoke) alternative to gunpowder. Like nitroglycerin, however, guncotton proved to be deadly to manufacture and use. Ultimately, after the explosion of a few factories, it fell out of favor. It was two British scientists who took on the problem of creating a safer alternative.

Frederick Abel and James Dewar (1842-1923) were practical men, and their work in explosives was deliberate, not accidental. While both were interested in fundamental aspects of chemistry, they relied on defense work to provide a steady income. Together, they set out to create a smokeless propellant that had the same stability as dynamite. The combination was not obvious and the work was dangerous and disappointing. Dewar actually met with Nobel to discuss the project. (Nobel had produced his own smokeless powder, ballistite.) It was soon afterward that Abel and Dewar mixed nitroglycerin, guncotton, and vaseline to create cordite. Cordite first exists as a liquid that can be extruded and dried to create cords (hence its name). These cords, like dynamite, are highly stable and do not present the same dangers as guncotton. Cordite's burn rate can be fine tuned by modifying the shape of the grains. The cords also can be measured out and conveniently cut to the desired length to create exactly the right size of explosion. The Spanish-American War (1898) was the last major conflict fought without smokeless powder.

Impact

Dynamite and cordite did not introduce the world to massacres. The American Civil War's Battle of Antietam (1862), for instance, left 23,000 soldiers dead, wounded, or missing in a single day. But dynamite and cordite did allow for massive stockpiling of the first weapons of mass destruction. When the "powder keg" was ignited during World War I, eight and a half million people were killed. The world was horrified by what was incorrectly called "the war to end all wars." Today, explosives are as popular as ever. In 1995 almost five billion tons of explosives and blasting agents were produced in the United States alone.

Explosives also found peaceful uses. Nobel, who studied in the United States, was inspired in his work by the possibilities he saw in taming the American West. In fact, the efficient road, railroad, and tunnel systems that connect people today would not have been possible without dynamite. Dynamite has been essential to unlocking resources. Nobel himself benefited from the use of his explosive to improve oil production in Russia, and E.A.L. Roberts received a patent in the U.S. for using dynamite to unlock reserves of pools of oil. All types of mining have been facilitated as well, making the coal, stone, and metals that are essential to modern life available at affordable prices.

The explosive reshaping of our planet through the creation of roads, tunnels, and dams, not to mention strip-mining and the burning of coal, has also led to cases of environmental devastation. Today in many developed nations, limits—such as environmental impact statements—have been placed on development activities and the use of natural resources. But through most of the history of modern explosives, the users have been heedless of the environmental consequences of their use. As a result, there have been many cases where species and ecosystems have been made extinct by explosive-facilitated development.

A less apparent effect of research into explosives has been progress in nitrogen chemistry. Gunpowder, guncotton, nitroglycerin, cordite, and dynamite all depend on nitrogen compounds. As chemists have worked to harness the explosive power of these materials, they have come to a better understanding of the chemical properties of nitrogen. Nitrogen is the most abundant chemical in the air (in the form of N_2, nitrogen gas). It is also the basis for fertilizers, which helped turn aside nineteenth-century predictions of worldwide starvation by dramatically improving agricultural yields. In the early 1900s Fritz Haber (1868-1934) developed a Nobel Prize-winning process for "fixing" atmospheric nitrogen, creating a synthetic source of material for the production of both fertilizers and explosives. The work of this chemist (who was also a weapons developer involved in gas warfare) would not have been possible without the earlier work of chemists exploring nitrogen compounds for their explosive properties. Nitrogen chemistry has also led to the creation of new medicines. Nitroglycerin itself has found widespread pharmaceutical use as a vasodilator—it dilates, or enlarges, blood vessels—and is taken by people with angina pectoris to relieve their chest pains.

An indirect effect of nineteenth-century advances in the chemistry of explosives came from the wealth accumulated by Alfred Nobel. The production and sale of dynamite was highly profitable but controversial. Nobel was called "the merchant of death." Perhaps concerned about his legacy, Nobel at his death left more than nine million dollars (an enormous sum at the time) in trust to finance yearly awards for achievement in chemistry, physics, medicine, literature, and peace. (Economics was not provided for in Nobel's will but was added later.) These have become the most prestigious awards in their disciplines and now have a cash value of more than one million dollars each. They have motivated brilliant and powerful people and made the public aware of significant achievements.

PETER J. ANDREWS AND SCOTT BOHANON

Further Reading

Asimov, Isaac. *Isaac Asimov's Biographical Encyclopedia of Science and Technology.* New York: Doubleday and Co., 1976.

Brown, G. I. *The Big Bang: A History of Explosives.* Stroud, UK: Sutton Publishing, 1998.

Williams, Trevor. *Alfred Nobel: Pioneer of High Explosives.* London: Priory Press, 1974.

Biographical Sketches

Nicolas Appert
1749-1841
French Inventor

Nicolas Appert was born in 1749 in Châlons, which lies in the Champagne district of France. He developed an industrial method for canning food and beverages that is still used today.

Appert, who came from a long line of farmers and innkeepers, was born at his family's inn. As a boy, he learned to cook and to cork champagne bottles. He learned more about food and wine by working at other inns and as a master chef for a duke and duchess.

In 1784 Appert used an inheritance to open a candy and grocery shop in Paris. The following year, he married Elisabeth Benoist, with whom he had five children. When the French Revolution broke out in 1789, Appert contributed money to the revolutionary army and represented his political district at Louis XVI's execution. His prominence led to his arrest on false charges during the chaotic period known as the Terror.

As his grocery business expanded into wholesale produce in the 1790s, Appert began to experiment with preserving food beyond the harvest season. He was dissatisfied with traditional preservation techniques such as drying or adding sugar, salt, or vinegar because he believed they spoiled the taste and healthfulness of food. Appert used champagne bottles, which were thick enough to withstand the pressure of bubbles, for his first experiments. He filled them with prepared food, leaving room at the top for expansion, corked and bound them, and then heated them in boiling water. He hypothesized that "heat destroys or at least neutralizes the fermentation that changes the quality of animal and vegetable substances." Louis Pasteur (1822-1895) later credited Appert with being the first to understand the basic principle of pasteurization.

By 1802 Appert had moved from his Paris workshop to a large property in Massy that combined factory and farm so that food could be preserved where it was grown. One of the factory kitchens was equipped with large brass pots with compartments for holding meat or chicken and faucets on the bottom for pouring off broth. Another was reserved for dairy products and a third for fruits and vegetables. There was a room for washing and rinsing jars, another for labeling and packaging, and a laboratory for testing new products. Appert supervised the factory, which employed up to fifty people during harvest season, and continued to maintain his shop in Paris.

Appert enlarged his markets throughout the early 1800s. He traveled to port cities in France and asked naval authorities to test his preserved meats and vegetables as a remedy for malnutrition during long ocean voyages. After a year on board ship, the food was praised for its perfect preservation and acclaimed as "advantageous for sailors." Appert also continued to enlarge his range of products. He experimented with red wine, which spoiled during exportation, by resealing and heating several bottles and then storing them on a Caribbean-bound ship. Two years later, when the bottles were returned, a wine connoisseur compared them to unpreserved wine Appert had kept in his cellar. He found the preserved wine superior in every dimension.

In an attempt to win support for his research, Appert took samples of his food to the Society for the Encouragement of National Industry. He was awarded 12,000 francs with the stipulation that he either patent his process or publish a detailed description. Appert, who believed he was serving humanity by sharing his discoveries, wrote *The Art of Preserving All Animal and Vegetable Substances for Several Years* in 1810. Reviews were positive, and two revised editions as well as German, English, Italian, and Swedish translations followed. Despite his success, Appert struggled financially because of the high cost of his equipment. His prospects were beginning to improve when the Prussian and Austrian armies invaded France in 1814 and destroyed his factory at Massy. One year later, after Napoleon's defeat at Waterloo, the British turned his property into a hospital.

Forced to start over at age 64, Appert received a subsidy from the government to build a new factory in Paris that used tin cans instead of glass jars. He designed his own tinsmith where up to 1,500 cans per month were made to his specifications. He won a gold medal for his canned products in 1820 and a 2,000-franc prize from the Society for the Encouragement of National Industry in 1824. Appert's last invention was a process for making gelatin, dried bouillon, and concentrated milk in an autoclave, a kind of pressure cooker.

When the government stopped subsidizing the rent on his Paris factory, Appert moved yet again and built a new factory at the age of 78. He received a small award from the Ministry of Commerce and Transportation in 1832, but was refused the prestigious Legion of Honor award. He retired in 1836 to Massy. He died alone in 1841 at the age of 91 and was buried in a pauper's grave.

LINDSAY EVANS

Alexander Graham Bell
1847-1922
Scottish-American Inventor

Alexander Graham Bell is best known for his invention of the telephone in 1876. However, he was also the second president of the National Geographic Society, invented a number of other useful devices, founded the first telephone company, and was a noted humanitarian. In short, in spite of the tremendous impact the telephone has had on society, Bell would have been a remarkable and important figure even without the invention.

Bell was born in Scotland in 1847 to a mother who was nearly deaf and a father who was an expert in speech. These facts, coupled

Alexander Graham Bell. *(AT&T Laboratories. Reproduced by permission.)*

with his grandfather's prominence as a speech expert, ensured Bell's interest in sound and speech from an early age, and he spent much of his life helping to teach the deaf to communicate.

One of Bell's first jobs, however, was teaching elocution in England. A job he excelled at, it often struck many as ironic that a Scot was teaching English to the English. At the same time, he provided instruction to the deaf, helping to teach the fundamentals of his father's "Visible Speech" system that helped show the positions of tongue, lips, and teeth for various sounds and words.

Bell graduated from the University of Edinburgh and then moved to the United States with his father, becoming a professor of vocal physiology at Boston University and specializing in the mechanics of speech. This led to his forays into inventing as he worked to develop devices that would help the deaf to enter the hearing world. Among the devices Bell invented or tried to invent was the phonoautograph, which was to make sounds visible but which was difficult to interpret.

Bell admitted that, had he known more about electricity, he might never have set out to invent the telephone because he would have realized it to be impossible. However, he seems to have known enough to succeed and not enough to fail and, in 1876, after a great deal of work and experimentation, the pieces fell into place. Bell's invention was demonstrated at Philadelphia's Centennial Exposition in 1876, meeting with instant acclaim.

Not content to rest on his laurels, Bell went on to investigate problems associated with human flight and, later, invented the hydrofoil. Another important invention was the audiometer, a device that permits testing hearing acuity in a variety of frequencies. This device allowed for early testing of students' hearing, revealing that many children who had been labeled as "slow" or "inattentive" were simply hard of hearing.

In addition to his technical achievements, Bell was a devoted family man and a dedicated humanitarian. Bell's wife Mabel, a deaf woman, started as one of his speech students after he moved to the United States. Eventually they fell in love and married, in spite of early objections by Mabel's parents. Mabel and their two daughters were an important part of Bell's life until his death.

Finally, Bell did his utmost to contribute to society as much as possible. He remained a teacher of the deaf his whole life, helping to promote Visible Speech and to encourage programs to teach the deaf to speak. He was also a factor in Helen Keller's life, leading her to her tutor, Annie Sullivan, and becoming a good friend over the years. And, as a prominent early member of the National Geographic Society, Bell became the Society's second president. From that position he was able to transform a dry and relatively unpopular scientific magazine into one of the most successful and widely read magazines in history. Bell's relationship with the National Geographic Society continued beyond his tenure as president when his daughter married Gilbert Grosvenor, another man destined to become an important force in the society.

Bell died at his home in Nova Scotia in 1922. When he was buried, all telephone service in the United States was stopped for one minute to honor the man, his work, and his life.

P. ANDREW KARAM

Karl Friedrich Benz
1844-1929
German Engineer

It is difficult to say who actually "invented" the automobile. In part, this is because it is hard to define what exactly constitutes an automobile. The first self-propelled road vehicle was probably

Karl Friedrich Benz. *(The Library of Congress. Reproduced by permission.)*

the steam-powered tractor driven in the 1760s by Nicolas Cugnot (1725-1804), a French army engineer who designed it to pull artillery. However, the first person to manufacture and sell liquid-fueled machines similar to what we call cars today was undoubtedly Karl Benz.

Born in Mannheim, Germany, in 1844, Benz shared with his father, one of the early locomotive engineers, a passion for the mechanical sciences. At school he delved into chemistry, the new field of photography, and was especially intrigued by clockworks. In his autobiography he remembered "the marvelous language that gear wheels talk as they mesh with one another."

At age 16 Benz entered a polytechnic school, where he was inspired by efforts to find a more user-friendly substitute for the steam engine. Upon graduation, he was employed by an engineering firm, but found the work unchallenging and left after a short time. Gottlieb Daimler (1834-1900), with whom Benz would later forge the automotive alliance Daimler-Benz in 1926, became the new senior engineer soon after Benz's departure.

In 1871 Benz opened a factory to make what he called "tin working machinery." The business failed financially, due in part to a stock market crash. Undaunted, he borrowed the funds to start a new business, designing and producing an engine for stationary use. This business prospered. His investors didn't mind if he devoted some time to his dream project: a self-propelled vehicle powered more efficiently than the heavy, clumsy steam carriages that had developed after Cugnot's tractor.

Benz put a small engine under the seat of a two-wheeled horse cart, removed the shafts and placed a smaller third wheel at the front, steered by a lever. A local newspaper covered the story: "A velocipede driven by Ligroin gas, built by the Rheinische Gasmotorenfabrik of Benz & Cie. was tested early this morning on the Ringstrasse, during which it operated satisfactorily."

During another, unauthorized, morning test, Benz's wife and two sons "borrowed" the machine for a joyride while papa Benz slept. The trio drove it the astounding distance of 70 miles from Mannheim to Pforzheim and back, having to get out and push on steeper parts.

The machine was proving practical. Improved models in the three-wheel format came along, but Benz was determined to design a four-wheel model that would allow more capacity and greater stability at speed. A big problem involved safely steering two wheels without dragging one mercilessly at every turn. By 1893 he had such a car. He called it the Victoria to celebrate his victory over the tricky steering challenge, not to honor a British queen as is sometimes supposed.

More models followed with different seating arrangements, sizes, and shapes. By 1902 a Benz advertisement showed a front-engine car with a steering wheel rather than a tiller. The ad also highlighted words that sound rather like modern car ads—reliable, fast, durable, elegant.

Then there was trouble in the factory. Both Benz and employee Marius Barbaroux claimed design credit for the front-engine Parsifal model. The friction was relieved only when Benz resigned and Barbaroux became chief engineer.

BROOK HALL

Sir Henry Bessemer
1813-1898
English Inventor and Engineer

The early manufacture of steel in England is invariably discussed with one of the nineteenth-century's greatest inventors: Sir Henry Bessemer. Born on Jan. 19, 1813 in Charlton, Hertfordshire, England, Bessemer showed great promise early in his youth. According to his account of his life in *An Autobiography* (1905), he was interested in mechanical and chemical processes as early as his seventh year. He recalled making his first machine, a contraption that produced small, white pipeclay bricks for model making. Since his father was a typefounder, Bessemer also had early access to molten metals, which he used for a variety of experimental castings. In his autobiography he mentions walking around the English countryside with a favorite dog, picking up lumps of clay, and molding them into different shapes. Later, at home, he would make molds from the shapes and cast them into type metal.

By the time he was 17 years of age, he was familiar with all sorts of metal works, including foundries, large melting furnaces, and the combination of various metals to produce alloys. At this time his father decided to move the family business to London, a move that delighted the young Bessemer, who claims he never tired of walking the interesting streets full of shops, galleries, and squares.

During the ensuing years he patented numerous inventions that brought him both wealth and fame. Among them were the movable stamps for dating deeds and other important documents, the manufacture of "gold" powder from brass, and an advanced design of sugarcane-crushing machines. Much of his income was generated by the artistic decor of the times, in which gilding and heavy, florid trims were popular. The powdered brass was used in paints to produce the "gold" look that was so prevalent in this era.

Although he is credited with numerous developments during and after the Crimean War (1853-56), his work with metals in that era was one of the achievements that spread the Industrial Revolution from Europe to the rest of the world. The war effort inspired him to work on producing a stronger type of iron that would withstand greater pressures and temperatures. When he found that the excess oxygen he was using in his furnace seemed to remove the carbon from the pre-heated iron pigs he was using, he carried purification further by blowing air through the melted cast iron, rendering it hotter and easier to pour.

Although Bessemer is credited with being the first to mass-produce steel, there were many others who contributed ideas, adaptations, and advanced methods of removing impurities from pig iron. These included William Kelly (1811-

1888), an American who conceived of and developed—independently but concurrently with Bessemer—the same purification process used by Bessemer; Robert Forester Mushet, and Goran Goransson. Their improvements resulted in a revolution of the construction industry and provided low-cost steel to replace the perishable iron formerly used on railroads and other basic industries.

Bessemer's work led to the open-hearth process in the late 1860s, which eventually replaced much of the original furnace operations. However, even in his later years (he was still working at age 70), Bessemer went on discovering and developing new processes and machinery. He is credited with building the first solar furnace as well as an astronomical telescope. The latter was not for public use but for his own entertainment. Also, the prestige that London enjoys today as a world center for diamonds is due in part to the set of machines designed by Bessemer for polishing the valuable stones.

The British government honored Bessemer with a knighthood in 1879 and a fellowship in the Royal Society. He died in London on March 15, 1898.

GERALD F. HALL

Ada Byron, Countess of Lovelace. *(Doris Langley Moore Collection. Reproduced by permission.)*

Augusta Ada Byron, Countess of Lovelace

1815-1852

English Mathematician

Augusta Ada Byron, Countess of Lovelace, was a pioneer in the field of computer science. The work she did on a paper describing Charles Babbage's analytical engine revealed her understanding of the machine we now call the computer. Her remarkable insight far exceeded an understanding of the technology available in her lifetime.

In 1815 Augusta Ada Byron was born in London to Annabella Millbanke, an amateur mathematician, and the renowned poet Lord Byron (George Gordon). Byron inherited her father's creative traits and her mother's love of mathematics.

Byron's parents separated soon after her birth, and her mother raised her with the help of private tutors. Byron's tutors fostered her early interest in systems as well as her desire to understand how things operated. Along with traditional subjects, tutors taught the young girl

mathematics and astronomy. Math and science became her favorite subjects.

As a teenager Byron outgrew her governess tutors. She continued to teach herself mathematics. Correspondence with informal tutors including the mathematician Mary Somerville (1780-1872) improved her mathematical skills. However, social prohibitions limited her potential to develop into a great mathematician. In the early nineteenth century the universities did not enroll women as students. The higher level of mathematical instruction available to men at Cambridge University was denied to Byron.

In 1833 Byron met Charles Babbage (1792-1871), a mathematician and pioneer in the field of computer science. The friendship that formed between these two individuals led to Byron's work in the field of computer science.

Babbage designed a machine with the capability of calculating complex numerical problems. He also imagined a machine that when programmed could do any type of calculation. His advanced ideas would have produced an early version of the computer that would use modern computing ideas. However, Babbage never produced his computer because the technology needed to construct the machine was not available in the early 1800s. What was available was Byron's willingness and ability to describe and write programs for Babbage's conceptual machine called the analytical engine.

In 1842 L. F. Menabrea, an Italian mathematician, wrote a descriptive paper on the analytical engine. It became Byron's task to translate this paper from French into English. Babbage gave her permission to add her own notes to the paper. The product of this task became Byron's noteworthy contribution to the field of mathematics.

Byron translated Menabrea's words and added her ideas to the paper. The finished document she published was three times the length of the original paper. Included was a sample set of instructions, or a program, for the machine. She also noted that these types of machines were unable think for themselves, rather their functioning depended on their programmers. Byron's work on this paper showed her insight into the future of computers, as she revealed an understanding of the concept of a programmed machine that was beyond her time. Along with Babbage, she understood the basics of a machine she would never have the opportunity to actually work on.

This publication was Byron's only major mathematical contribution. It was not until after her death that her role as author of this visionary piece became public. Since societal views held it improper for women to write technical papers, Byron had signed only her initials to her work. Thirty years after the paper was published, Byron's full name appeared as the paper's author.

The mathematical community forgot Byron's work on this paper until the second half of the twentieth century. During the late 1970s, mathematicians and computer scientists reexamined her work. They noted her exceptional ability to see the technology of the future so clearly and acknowledged her early efforts to write computer programs. The United States Department of Defense honored her contribution to computer science in 1979 when they named their high-level computer language ADA.

HEATHER M. MONCRIEF-MULLANE

Andrew Carnegie
1835-1919
American Businessman and Philanthropist

Andrew Carnegie embodied the "rags-to-riches" American dream. A poor Scottish immigrant, he amassed great fortune as a financier, then built an empire in the steel industry. Later in life, Carnegie devoted his time to giving away his money and became history's most prominent philanthropist.

Born in Dunfermline, Scotland, on November 25, 1835, Carnegie was the son of a weaver. Like the millions of immigrants pouring into the United States, he and his family searched for a better way of life. They settled in Allegheny, Pennsylvania (now Pittsburgh). With little formal education, Carnegie went to work at age 13 as a bobbin boy in a cotton mill. Later, Carnegie became one of the city's best telegraph operators and advanced through a series of jobs with Western Union.

Carnegie's work with the Pennsylvania Railroad, however, served as his apprenticeship into the business world. Thomas Scott, the superintendent of the railroad, hired Carnegie in 1852 to be his personal telegrapher and secretary. Scott became Carnegie's mentor and friend and taught Carnegie about the world of high finance. Carnegie invested in many companies, including an oil business, a bridge manufacturer, and a railroad sleeping-car company. By 1863, Carnegie earned enough on his investments to declare himself rich.

In 1865, at age 30, Carnegie resigned from the railroad. For the next seven years, Carnegie accumulated great wealth as a stock speculator, but he yearned to start his own company. Recognizing the expanding market for iron and steel, he established the Carnegie Steel Company in 1872.

Carnegie used his knowledge from the railroads in his new venture. His manufacturing strategy revolved around control. He introduced sophisticated cost-accounting systems and kept detailed records of operations, always trying to lower production costs while increasing output. Carnegie used the records to ferret out inefficiencies and find processes that were successful. Carnegie was the first manufacturer to analyze his workers in this manner. Other business leaders adopted Carnegie's management techniques, which served as the foundation of a uniquely American system of manufacturing.

Carnegie was also an early advocate of vertical integration, which led him to acquire the sources of raw materials needed in his mills and sell the finished products. Thus, he controlled the entire production process in the steel industry. Carnegie bought out his raw material suppliers, then acquired a fleet of ships to transport goods. In the 1890s, Carnegie established a sales office and began producing finished goods.

As a result, Carnegie's company became the world's largest steel manufacturer. By 1900, the United States surpassed Great Britain as the leading steel producer. Carnegie's steel served as the backbone of the nation's expanding railroad system and allowed the vast expansion in American cities.

Carnegie solidified his control over the steel industry, then began turning over daily operations to close associates, like Charles M. Schwab and Henry Clay Frick. Carnegie spent more time traveling around the world and interacting with leading scholars and intellectuals.

There was a downside to Carnegie's control, which led to one of the nation's most infamous labor battles. In 1892, union steelworkers in Homestead clashed with Pinkerton strikebreakers hired by Frick. The ensuing gunfight killed one Pinkerton guard and injured many more. The governor of Pennsylvania, at Frick's request, called in the state militia to restore order. Under this protection, strikebreakers were able to smash the union. Many historians believe Carnegie's philanthropy was his way to atone for destroying the union in Homestead.

In 1901, at age 65, Carnegie sold the company to J. P. Morgan for $400 million, making Carnegie the richest man in the world. He devoted the rest of his life to philanthropy and writing. Carnegie believed that the rich have a moral obligation to give back to humanity. He gave numerous personal gifts and established trusts centered primarily on peace studies and education. Carnegie created many philanthropic and educational organizations in the United States, including the Carnegie Corporation of New York, and several more in Europe.

Carnegie had a lifelong interest in establishing free public libraries, which were previously a rarity. He subsequently spent over $56 million to build 2,509 libraries throughout the English-speaking world. During his lifetime, Carnegie gave away over $350 million. He died in Lenox, Massachusetts, on August 11, 1919.

BOB BATCHELOR

Sir George Cayley
1773-1857
English Inventor and Aviator

Sir George Cayley was born in Yorkshire, England, in 1773. He led a privileged childhood at his parent's estate, Brompton Hall, and was primed to take over the baronetcy. As a child,

Cayley's greatest fascination was with the scientific world. He kept notebooks full of sketches he made of plants and animals, was an avid study in mathematics and navigation, and loved to tinker with all sorts of gadgets. His great passion was flight, however, and he dreamed of one day taking man into the air to soar like the birds.

In the 1780s, the world had a new fascination with flight, thanks to the work of balloonist brothers Jacques-Etienne and Joseph-Michel Montgolfier (1745-1799 and 1740-1810, respectively). But Cayley wasn't interested in this type of ascent. He wanted to fly with the help of mechanics, and was convinced that it was only a matter of time before he would create a real flying machine.

His first device was modeled after a toy helicopter invented by Frenchmen Launoy and Bienvenu in 1784. His version had feathers for propellers and used an airscrew to achieve mechanical flight. Cayley carefully studied the flight of birds, for clues on how to apply the same dynamics to his machine. He discovered that birds twisted their wings to help them fly long distances, and envisioned that the same effect could be created in a fixed-wing, flying machine with the use of cambered wings.

In 1799 he sketched out his idea on a silver disc. On one side, he illustrated the aerodynamic force on a wing, on the other, the design for his fixed-wing airplane. Five years later, he had built the world's first workable model glider, which consisted of a paper kite used as a wing, with an adjustable tail mounted on the end that could be used for horizontal and vertical control. In 1809, he built a full-sized version which could fly unmanned.

Cayley's next hurdle was to find a way to power his flying machine. In this, he was constrained by the technology of his day. Steam engines were too heavy for flight, so he invented his own hot-air engine, using a gunpowder motor. In 1809-10, he published his theories on aviation in the article "On Aerial Navigation." His article was not well received in the scientific or general communities, but Cayley was not ready to give up.

He was willing to put his dream on hold, however, and spent the next three decades pursuing other occupations. In 1832, Cayley became involved in public affairs and was the parliamentary representative of Scarborough. He cofounded the British Association for the

Advancement of Science in 1831 and of the Regent Street Polytechnic Institution in 1838.

In 1843, when he was 70 years old, Cayley finally returned to his true love, and his drawings of a helicopter-like machine were published in *Mechanics' Magazine*. His design had four circular lifting rotors, which could be adjusted to form wings, as well as two pusher propellers.

In 1849, Cayley built his first full-sized glider. The only problem was, it could only fit a small boy. That boy (whose name has not been recorded) became Cayley's test pilot, and the first person in history to fly—if only briefly. In 1853 he built a new and improved glider, and convinced his reluctant coachman to make the first flight. The machine and coachman soared 900 feet (274 m) across a Brompton dale before crashing.

At the time of his death in 1857, Cayley had never received public recognition for his work, but today, he is credited as a pioneer in the field of aeronautics. His work paved the way for Orville and Wilbur Wright's (1871-1948 and 1867-1912, respectively) famous flight at Kitty Hawk, and for later advances in aviation.

STEPHANIE WATSON

Louis Jacques Mandé Daguerre
1787-1851
French Artist and Inventor

Louis Jacques Mandé Daguerre was born in 1787 to a middle-class family in Cormeilles, near Paris. He was an accomplished scenic designer who created the Diorama and invented the daguerreotype, the first practical method of making photographs.

Daguerre's artistic talent was evident at an early age. He served apprenticeships with a local architect and a stage designer in Paris. At 28 he was appointed scenic designer of the Paris Opéra. Two years later, he cofounded the Diorama, a theater in which enormous, lifelike murals and special lighting effects created the illusion of changing scenes. Audiences flocked to see famous sights such as the tomb of Napoleon, an alpine village, and Canterbury Cathedral.

To obtain the exact perspectives that were crucial for making these scenes appear real, Daguerre relied on a *camera obscura*. The camera, used by painters for centuries, was a box with a lens on one end and a mirror at a 45-degree angle on the other. The mirror reflected

Louis Jacques Mandé Daguerre. *(The Library of Congress. Reproduced by permission.)*

an image onto a glass on the top of the box, where it could be copied onto translucent paper. In time, Daguerre began to experiment with making the reflected images permanent.

Daguerre bought the lenses for his camera from Vincent Chevalier, a Parisian optician. Another of Chevalier's customers was Nicéphore Niépce (1765-1833), who had invented a method of recording the camera's reflected image on chemically treated paper and stone plates. After hearing about this invention from Chevalier, Daguerre wrote to Niépce. Over the next few years, the two men met in Paris and exchanged many letters before finally signing a partnership agreement. They collaborated via letters written in a number code devised by Daguerre to guard the secrecy of their experiments. In one letter, Daguerre suggested substituting silver iodide for the asphalt substance Niépce was using, a critical innovation that would shorten the time required to create an image from eight hours to several minutes. Daguerre also designed a new lens that produced sharper images.

After Niépce's death in 1833, Daguerre maintained a partnership with Niépce's son Isidore but conducted his research independently. Daguerre continued to improve his silver-iodide method by treating the exposed silver-iodide plate with mercury vapor. He gave credit to Nicéphore Niépce for the original invention but took credit himself for perfecting the process, which he named the daguerreotype in 1838. Daguerre's work impressed the Académie des Sciences so strongly that the French government offered to buy his invention. Eminent scientists of the day traveled to Daguerre's studio to see demonstrations. One of them, Samuel F. B. Morse (1791-1872), the American inventor of the telegraph, marveled at the daguerreotypes' "exquisite minuteness of...delineation."

In 1839 Daguerre's Diorama, his only source of income, burned to the ground. His supporters convinced the French government to grant a generous annual pension to both Daguerre and Isidore Niépce in return for their publishing the technical details of both the original research and the daguerreotype. Daguerre, although described as timid and embarrassed as a speaker, gave demonstrations and classes and wrote a brochure that became an international bestseller. A company was created to manufacture the equipment for making daguerreotypes, with one-half of the profits going to the manufacturer and the rest shared by Daguerre and Isidore Niépce. As the daguerreotype grew popular around the world, others made improvements that shortened the exposure time to forty seconds by 1841.

Daguerre retired to Bry-sur-Marne, a small village outside Paris. Behind the altar of the local church, he painted a mural that gave the impression of leading into an immense cathedral. He died of a heart attack in 1851.

LINDSAY EVANS

Gottlieb Wilhelm Daimler

1834-1900

German Mechanical Engineer

German mechanical engineer Gottlieb Wilhelm Daimler was born on March 17, 1834, in Schorndorf, Wurttemberg, Germany. He was an obviously gifted youngster, and his father arranged for him to enter into a prestigious, classical education. His report cards soon showed that he was little attracted to such subjects as Latin but showed great aptitude for subjects like geometry. He persuaded his parents to let him change his scholastic endeavors toward the technical side.

Young Daimler's first job was in the foundry of a gunsmith's factory. He was very proficient at

Gottlieb Daimler. *(The Library of Congress. Reproduced by permission.)*

this hot, dirty work but soon found it lacking. He enrolled at a technical institute. While most engineering students were then busy trying to exploit and improve the power of steam engines, Daimler was looking for a different type of engine all together.

After visiting France and England, including the 1862 London World Fair, Daimler was exposed to the latest European engineering advances. He took a job at the mills of an old family friend where he devised systems, tools, and power units to mechanize the old factory. Later, he joined (as senior engineer) the Karsruhe Engineering Works, where Karl Benz (1844-1929) had labored earlier. Here, he hired Wilhelm Maybach (1946-1929), who helped Daimler modify the Otto Atmospheric Engine, considered the premier engine of the time.

The two continued to work together when Daimler became technical director of the Deutz Engine Works, owned by Nikolaus A. Otto (1832-1891) and Eugen Langen (1833-1895). They made many advances, but were not satisfied with the still-heavy, slow-revving stationary engines. Daimler's vision was of a light, high-speed engine that could run on benzene.

In 1882, Daimler established his own business in Cannstatt. Maybach followed his boss to the new factory. Here, they designed a new high-speed internal combustion engine that used gasoline as fuel. Three years later, in 1885, Daimler installed one of his engines on a bicycle he had designed. This resulted in a motorcycle remarkably similar to the modern one. Shortly thereafter, the Daimler engine was installed in boats, tram cars, and a dirigible. It wasn't long before this new engine was installed in a four-wheeled vehicle (1889). In 1900, one of the most significantly advanced cars ever produced was developed by Maybach and Daimler's son, Paul. Unfortunately, the construction of the first Mercedes, as it was called (after the daughter of Daimler's sales agent on the French Riviera), came too late for Gottlieb Daimler to see. He died on March 6, 1900, in Cannstatt.

GERALD F. HALL

Thomas Alva Edison
1847-1931
American Inventor

Thomas Alva Edison is perhaps the most famous inventor in American history. He held a world record 1,093 patents for inventions such as the incandescent electric lamp, the phonograph, and the motion-picture projector. Edison also played a pivotal role in bringing the modern age of electricity to the world.

As a child, Edison developed hearing problems that left him partially deaf. He attended school on and off for five years, but he had difficulty hearing and his teachers considered him slow. To compensate, Edison became an avid and inquisitive reader.

Edison quit school at age 12, and took a job selling newspapers and snacks on the railroad. By that time, the rail line was using a telegraph to control the movements of its trains. Edison learned how to use the telegraph and in 1863 became an apprentice telegrapher, replacing one of many operators who went to fight in the Civil War.

Initially, messages received on the Morse telegraph were inscribed as a series of dots and dashes on a piece of paper that had to be decoded and read. The transformation of telegraphy to an auditory system left the partially deaf Edison at a disadvantage. He spent six years as a travelling telegrapher, devoting much of his time to improving upon the telegraph itself. By January 1869 Edison had made enough progress on a telegraph capable of transmitting two messages simultaneously on one wire, and a printer which converted electrical symbols to letters, that he

Thomas Alva Edison. *(The Library of Congress. Reproduced by permission.)*

was able to pursue a career as a full-time inventor in Newark, New Jersey. There, he continued working on the automatic telegraph system.

In 1876 Edison moved his operation to Menlo Park, New Jersey, where he made some of his most significant discoveries, including a carbon-based conductive system in which an electrical current could be changed according to the amount of pressure it was under. In 1877 Edison began experiments that used that same pressure system to amplify and improve the sound quality of the telephone, which Alexander Graham Bell (1847-1922) had patented the previous year. By the end of 1877, Edison had developed the carbon-button transmitter that is still used in telephone speakers and microphones.

The phonograph is considered Edison's most original discovery. In the summer of 1877 he was trying to come up with a machine that would transcribe the sound of a human voice as it came over a telephone line. Building on earlier theories that each sound, if it could be graphically recorded, would produce a distinct shape resembling shorthand, Edison used a stylus-tipped carbon transmitter to make impressions on a piece of waxed paper. When he pulled the paper back beneath the needle, the tiny indentations generated sound. Edison unveiled his first phonograph in 1877, but it took 10 years for it to become a commercial success.

Edison spent five years trying to come up with a safe, inexpensive electric light that would replace the gaslight. In September 1882 he turned on the lights to the world's first permanent, commercial central power system, located in lower Manhattan. It was years before incandescent lighting powered by central stations replaced gas lighting, but isolated plants for hotels, theatres, and stores were an instant hit. Edison quickly garnered a reputation as the world's greatest inventor.

In 1887 Edison moved his workshop to West Orange, New Jersey, where he built the world's first industrial research laboratory. The first major project at the new lab was the commercialization of the phonograph. It was during this time that Edison began work on the first movie projector. He succeeded in building a working camera and a viewing instrument, but synchronizing the sound and motion proved next to impossible, so he gave up and the silent movie was born. The original Kinetoscopes, as the viewing machines were called, had peepholes that allowed one person at a time to view the moving pictures. Rival inventors soon edged out the Kinetoscope with screen-projection systems that allowed for group screenings.

Edison spent his career inventing devices that could satisfy real needs and that could be used by everyone. He, more than any other inventor, laid the foundation for the modern electric world.

CATHERINE M. CRISERA

Robert Fulton
1765-1815
American Inventor and Engineer

Robert Fulton is best known as the inventor of the first operational steamboat. However, he also developed an early submarine, canal designs, and patented machines for sawing marble and twisting hemp into rope. As one of America's first civil and military technologists, Fulton helped shape the worldview and ideals of modern mechanized society.

Fulton was born on November 14, 1765, in Little Britain Township, Pennsylvania. He spent most of his life in Lancaster, then the largest inland city in the American colonies. Lancaster was a hub of intellectual and technological activity during the American Revolution. Many historians believe the stimulation of this environment had a great impact on the young Fulton.

Robert Fulton. *(The Library of Congress. Reproduced by permission.)*

After the Revolution Fulton worked in Philadelphia as a portrait painter, and in 1786 traveled to England to study art; there he was a student of American painter Benjamin West. However, enamored by the marvels of industrial technology, Fulton turned to engineering instead during the early 1790s. While in England Fulton also took up residence with socialist Robert Owen (1771-1858), poet Samuel Taylor Coleridge (1772-1834), and chemist John Dalton (1766-1844), who together greatly influenced Fulton's worldview. They all believed that it was within the capacity of society to perfect the human condition. This could be achieved by developing strategies that were based on the interconnection among economics, science, and the humanities.

Fulton's vision of the future rested upon the free and inexpensive movement of material goods. Domestically, he envisioned the population centers in the United States and Europe linked by an extensive network of canals. Small, uniformly constructed steamboats would transport people and goods throughout the nations of the Western world. Science and technology would improve the efficiency of transportation and increase the availability of products, which in turn would result in a better life for all. Fulton published these ideas in his first major work, *The Treatise on the Improvement of Canal Navigation*.

Fulton held the same view for the global market. He believed that the idea of international trade based on the concept of freedom of the seas would create international agreements similar to the modern North American Free Trade Act (NAFTA), which would facilitate the free flow of goods between nations. Oceangoing steamboats were the backbone of his plan and served a twofold purpose. First, they would lead to the creation of a global marketplace. Secondly, if every nation had steamboat technology, the ability of the English navy to control the sea-lanes would be curtailed. Fulton believed the acceptance of the concept of total freedom of the seas would allow the nations of the world to disarm and to concentrate their wealth on social expenditures for the betterment of humankind. He fervently believed that the British navy was the main obstacle to this utopian view. His first major marine engineering project was to construct a submarine that would neutralize the British warship.

His first chance at significant government funding for this project came from France during the Napoleonic Wars (1799-1815), when they were facing a British naval blockade. Fulton constructed an operational model run on human power, and on two occasions he unsuccessfully tried to use it against the British navy. Both times English intelligence became aware of his plan and the Royal Navy was able to move their ships out of harm's way.

The British then sent a secret agent to Fulton with a proposal to fund further research. The navy wanted to use the research information to create a series of countermeasures against future submarine attacks. His ethic of freedom of the seas was strained, then torn when he decided to take the money from the very organization he had originally wished to neutralize. The British government cut off funding when they became convinced that submarine warfare was not a possibility. They paid Fulton about $70,000 in today's money and allowed him to bring a fully operational Watt steam engine back to the United States.

When Fulton returned to America, he aggressively looked for money to develop an operational steamboat. He finally received backing from the wealthy New Yorker Robert Livingston. Combining his knowledge of earlier research and hard work, he successfully launched his first steamboat, the *Clermont,* in 1803. This revolutionized transportation in the United States and Europe. It helped propel the westward movement across North America and set the stage for Western expansion into the rest of the world.

RICHARD D. FITZGERALD

Charles Goodyear
1800-1860
American Inventor

American inventor Charles Goodyear made important contributions to the practical application of rubber and its related industries. His discovery of the process of vulcanization, by which raw rubber could be made into a strong, malleable material, became useful for a large number of common products, most famously the rubber tire.

Goodyear's father was a New Haven, Connecticut, hardware inventor, manufacturer, and merchant specializing in farm tools, but also purveying items as diverse as pearl buttons. While attending public school, young Charles spent much time at his father's store, factory, and farm. He showed an interest in studying for the clergy, but his father saw a budding businessman and arranged for Charles to learn the hardware trade at a firm in Philadelphia. He did and, upon returning to New Haven and entering into partnership with his father, contributed to the success of their business, especially on the sales and merchandising side. He married a New Haven woman, Clarissa Beecher, in 1824.

In 1826 Charles and his wife moved back to Philadelphia to open his own hardware store, stocking mainly his father's products. By 1830 both Charles and his father were bankrupt, primarily because they were too generous in extending credit to their customers. Charles, although he had health problems on top of his financial ones, did not use the bankruptcy laws to assuage the pain. He was able to pay off some of his creditors by giving them interests in new Goodyear inventions. This was inadequate, however, and Goodyear was to suffer imprisonment for debt more than once before he died.

In 1834 he called on a company that dealt in India rubber goods, thinking a better valve might improve their inflatable life-preserver (and save the Goodyears from financial ruin). He devised such a valve, but the rubber company manager, more impressed by the ingenuity of its designer, told Goodyear of a better way to make big bucks. The rewards, he said, would flow if he could solve the rubber industry's big prob-

Charles Goodyear. *(The Library of Congress. Reproduced by permission.)*

best friend. In an animated discussion with a group of interested gentlemen in his laboratory, Goodyear accidentally dropped a blob of the rubber-sulfur mixture on top of a red-hot stove. The pancake did not melt, but was transformed into a strong, pliable, resilient, unsticky (albeit slightly charred) material. He had discovered the process that would later be called vulcanization (named for Vulcan, the Roman god of fire).

Of course, the process needed development and refinement, which Goodyear undertook on borrowed money, most of it never repaid. Many people made fortunes from rubber, or in the case of lawyers, the litigation about its patents and processes. Goodyear seems to have piled up only debts until the day of his death. He did, however, receive accolades. In France he was awarded the Grand Medal of Honor and the Cross of the Legion of Honor.

BROOK HALL

lem: During the summer, rubber became sticky, melted, and decomposed.

Goodyear was inspired by the challenge and began to experiment with rubber. His first tests were made in a Philadelphia jail. Experiments with magnesia looked good in the winter of 1834-35, but deflated his hopes in the summer.

By 1837, then back in New Haven, Goodyear was relying on the charity of others, even to feed his family. Two New Yorkers helped him continue his experiments in that city. One gave him a room; the other, a druggist, supplied rubber and chemicals. That year he obtained Patent No. 240 and began to manufacture sample articles including rubber clothes. In his *Gum Elastic and Its Varieties,* Goodyear provided the following description of himself in the words of another: "If you meet a man who has on an India rubber cap, stock, coat, vest and shoes, with an India rubber money purse without a cent of money in it, that is he."

A year later, Goodyear met Nathaniel Hayward, who had discovered that sulfur was good for taking stickiness out of rubber. His process involved the combining of rubber with a sulfur and turpentine mixture, then applying Goodyear's patented acid-metal process.

This set the stage for Goodyear's greatest discovery. Heat, rubber's old enemy, became its

Herman Hollerith
1860-1929
American Inventor, Businessman and Statistical Engineer

Herman Hollerith's invention of a machine able to tabulate information encoded in the form of holes punched in paper cards dramatically speeded up the 1890 United States census, and laid the foundation for the explosion of information processing in the twentieth century. The business Hollerith founded in 1896, the Tabulating Machine Company, later became a major component of the International Business Machines Corporation (IBM).

The child of German immigrants, Hollerith was born in Buffalo, New York, on February 29, 1860. He entered college at age 15, attending both City College of New York and Columbia School of Mines. He graduated from the School of Mines in 1879 with a degree in engineering.

That same year Hollerith moved to Washington, D.C., to accept a position as a special agent in the United States Census Office. The Census Office was responsible for counting the country's population every 10 years, and for deriving employment, income, and other statistics from the census reports. Hollerith proved a gifted statistician, and within a few months his $600 annual salary was raised to $800. His experiences in the tabulation of the 1880 census would alter the course of his life. A brief stint at

Herman Hollerith. *(The Library of Congress. Reproduced by permission.)*

the United States Patent Office would likewise prove valuable to him later on.

In 1882 Hollerith left the Census Office and joined the faculty of the Massachusetts Institute of Technology as an engineering instructor. While teaching mechanical engineering, among other subjects, Hollerith turned his mind to the question of increasing the efficiency with which census information could be tabulated.

Beginning in 1884, Hollerith devoted himself to his invention and to building a business around it. He applied for and was awarded patents for his devices, even as he continued to make improvements on them. By 1887 Hollerith machines were being used to calculate death statistics in both Baltimore and New York. He entered his machines in an 1889 competition to select the equipment that would tabulate the 1890 census. Hollerith won the competition and the census contract.

His tabulating machines accomplished the 1890 census in record time, completing the basic population count in under six months and the entire census in two years, coming in more than $5 million under budget. Hollerith and his business were firmly established. In 1897, with his business continuing to grow, he formed the Tabulating Machine Company.

Over the next decade Hollerith would continue to introduce innovations to his system, as well as expanding his commercial ventures. In addition to tabulating census information for countries around the world, Hollerith machines found use in virtually all types of industries, helping keep track of financial information, railroad shipments, insurance policies and mortality estimates, and wage and payroll figures.

In 1911, faced with an increasingly challenging competitive environment, Hollerith merged the Tabulating Machine Company with three other businesses to form the Computing-Tabulating-Recording Company (C-T-R). Hollerith served as the company's first president, a position he held only briefly. From 1911 until his retirement in 1921, Hollerith remained involved with C-T-R as a consulting engineer, and one of the company's directors. Three years after Hollerith's retirement, the company changed names once more, becoming in 1924 International Business Machines, known throughout the world as IBM.

Following his retirement, Hollerith continued to pursue various inventions, as well as agricultural experiments, a long-standing hobby of his. He had made more than $1 million from the sale of his stock in C-T-R, but despite various plans for other businesses, he did not start another company. Married since 1890, he devoted the last years of his life to his family and to continuing improvements to the punched cards that had created his fame and fortune. He died on November 15, 1929.

KEITH FERRELL

Joseph Marie Jacquard
1752-1834
French Inventor

Joseph Jacquard is the inventor of the weaving loom that still bears his name. The Jacquard loom revolutionized the textile industry and is the basis for the modern automatic loom.

Little is known about the formative years and education of Joseph Marie Jacquard. He spent the first years of his professional life as an apprentice in bookbinding, type-founding, and cutlery shops. It is believed that his parents had some connection to the weaving industry. Upon their deaths Jacquard inherited a small piece of property, which afforded him the opportunity to leave his apprenticeship and begin a series of experiments with weaves that contained patterns and designs. Unsuccessful, he lost his inheritance and was forced to return to type-founding and cutlery work.

Jacquard did not completely abandon his dreams and in 1790 he conceived of the idea for his famous loom, but his work was cut short by the onset of the French Revolution. The war lasted until 1793, during which time Jacquard fought on the side of the Revolutionaries in the defense of his hometown, Lyon.

In 1801 Jacquard introduced a loom for weaving net that was an improved version of work done by three previous loom inventors. He was sent to Paris to demonstrate it, where he received a bronze medal from the French government as well as a patent for this first invention. Along with the honor came a small pension that allowed Jacquard to study at the Conservatoire des Arts et Métiers, where in 1804-1805 he perfected a mechanism for pattern weaving.

The mechanism, known as the Jacquard loom or the Jacquard attachment, was incorporated into special looms to control individual yarns. The device utilized interchangeable punched cards that controlled the weaving of the cloth so that any desired pattern could be created automatically. It enabled looms to produce fabrics with intricate woven patterns such as tapestry, brocade, and damask, and it was later adapted to the production of patterned knitted fabrics.

Using the Jacquard attachment, a given pattern is made of a predetermined series of threads that are either raised or not raised according to the holes on the punched cards. As a punched card moves into place on the loom, the weaving needles pass through the holes in the card and specific threads are raised to make a section of the desired pattern. Where there are no holes, the needles are simply pushed back off the card and no threads are raised. By adding several Jacquard attachments to one loom, a weaver can produce patterns that are both very intricate and of considerable size.

In 1806 the Jacquard loom was declared public property, and Jacquard was given a pension and royalty on each machine. But his invention was not well received by weavers, who feared that its labor-saving capabilities would take away their jobs. Weavers in Lyon burned machines and physically attacked Jacquard in protest. Eventually, the advantages of the Jacquard loom brought about its general acceptance, and by 1812 there were 11,000 of them

in use in France. Jacquard received a gold medal and the Cross of the Legion of Honor in 1819. By 1820 his invention had reached England, and then it was quickly spread to the rest of the world.

Jacquard's punched card system introduced the concept of storing information for controlling data processing in a machine. In 1834, the year of Jacquard's death, these punched cards were adopted by noted English inventor Charles Babbage (1792-1871) as an input-output medium for his proposed analytical engine, the first automatic digital computer. Similarly, in 1880 American statistician Herman Hollerith (1860-1929) developed a machine capable of reading and then sorting data represented by a pattern of holes punched in cards. Using Hollerith's machine, it took just six weeks to process the 1890 United States census results—one-third the time required in 1880. Jacquard's punched cards were also used as a means of inputting data into early digital computers, but they were eventually replaced by electronic devices.

CATHERINE M. CRISERA

Margaret E. Knight
1838-1914
American Inventor

As a woman and an inventor, Margaret Knight is significant due to the number of inventions she produced and for the number of patents she received during her lifetime. Her most notable work was her development of the first machine capable of making square-bottomed paper bags.

Margaret Knight was born in York, Maine, on February 14, 1838. Mattie, as her parents and friends called her, had woodworking tools as favorite toys, which she enjoyed putting to use to make things. She later said of her childhood, "the only things I wanted were a jack knife, a gimlet (a tool for boring holes) and pieces of wood." She received some education through secondary school but no formal education beyond this. As a child and an adult, Knight used her creative mind and her interest in mechanical things to produce many inventions.

Knight's family moved to New Hampshire during her childhood. Like many other young people during the early years of the industrial era in the United States, her brothers became employees of a local cotton textile mill. It is believed that Knight conceived the idea of her first invention in this mill.

At age 12, Knight visited her brothers at work. She watched the heavy steel-tipped shuttles move on the large looms. While watching, she saw a loom malfunction and a shuttle fly out and hit a worker. She designed a safety device to prevent this type of accident. Her shuttle restraining device turned the entire machine off when something malfunctioned. When used, it kept shuttles from falling out of the looms.

As an adult, Knight went to work in a paper bag shop in Springfield, Massachusetts. The bags being made were weak, narrow bags with an envelope shape. Like others, Knight worked to develop a better design. Knight's efforts to improve the bag design led to her improving the machine used to make paper bags. The new part she developed enabled the machines to fold square-bottomed bags. Unlike their flat predecessors, these new bags were superior in strength, and their ability to stand made them more practical. Knight took out her first patent for this invention in 1870.

Knight is best known for her inventions in the paper bag industry. She acquired two more patents in 1871 and 1879 for further improvements she made to the paper bag machine. Grocery stores around the world still use her bag design.

Knight invented three domestic items in the 1880s. In 1883 she designed a dress and skirt shield. The next year she developed a clasp for holding robes. Then in 1885 she designed a spit; a spit is a long, pointed tool used to skewer meat for cooking.

In 1890 Knight turned to a new field for her inventions. This time she focused on the making of shoes. Before shoes were sewn together, a person or machine used a pattern to cut them out of large pieces of material. Knight designed several machines that improved this process of shoe cutting. In just four years she acquired six patents for her inventions in this field.

A few years passed before Knight began working with motors and rotary engines to produce her next set of inventions. In the early 1900s she created a number of components for rotary engines and motors. Her first was patented in 1902, her last in 1915. The sleeve-valve automobile engine was her most notable invention in this category. Knight's lack of education prevented her from fully understanding the

mechanics behind engines and motors and limited her understanding of her own work.

When she died in 1914, Knight had acquired at least 27 patents and made approximately 90 inventions. Her largely self-taught abilities and great interest in machinery made her an extraordinary American inventor of the industrial era.

HEATHER M. MONCRIEF-MULLANE

Auguste Marie Louis Nicholas Lumière
1862-1954
French Inventor

Louis Jean Lumière
1864-1948
French Inventor

Auguste and Louis Lumière, inventors and experts in the realm of photography, were the inventors of a camera and projector apparatus called the Cinématographe, which became the basis for contemporary cinematic projection. The brothers gained additional distinction for creating the first efficient color-photography process, known as the Autochrome plate, and are commonly considered the founders of modern cinema.

Auguste Lumière was born in Besançon, France, on October 19, 1862, with Louis following on October 5, 1864. Their father, Antoine Lumière, was an accomplished portrait painter who switched media to deal in photographic manufacturing and supplies. Auguste and Louis developed an early fascination with the photographic equipment their father produced for his business in Lyon, France. After attending technical school in Lyon with great success, Auguste and Louis worked for Antoine's business. In 1894 Louis developed a new process for photographic-plate preparation; he then opened his own photographic-plate manufacturing plant. By 1895 the Lumière factory was one of the most successful of its kind in Europe, producing over 10 million plates per year.

It was an event of the previous year, however, that would lead to the Lumière brothers' greatest triumph. In 1894 their father visited Paris for a demonstration of Thomas Edison's Kinetoscope, a peep-show apparatus that allowed a film loop to run continuously between a shutter and an incandescent lamp. Antoine urged Auguste and Louis to invent an improved version of Edison's device. The impressive result, a single-piece apparatus, contained a projector, printer, and camera in the same machine. Patented in 1895, the Cinématographe was markedly unlike its ponderous predecessor. Movable and hand-operated with a claw foot to advance the film—and with a reduced number of frames needed per second—the new camera allowed the brothers to leave the studio and take to the Parisian streets to film all varieties of daily life.

Although initially reserved for private showings to specialists, the Cinématographe debuted in the Grand Café on the Boulevard des Capucines in Paris on December 28, 1895, thus heralding the birth of cinema. The public reaction was spectacular, and Auguste and Louis were soon showing 20 of their short comedies and documentaries per day. Within four months of the Cinématographe's debut, the Lumières opened theaters in New York, London, Berlin, and Brussels. They also showed the first news reel, which was of the French Photographic Society Conference. By 1897 the brothers enjoyed worldwide renown; their small group of titles had expanded to over 700, and they were able to send cameramen all over the globe in search of interesting subjects to film.

After the Paris Exposition of 1900, the Lumières distanced themselves from the creative process of filming to focus on the production of the Cinématographe. Recipients of extensive acclaim and awards for their invention, the brothers continued to pursue cinematic manufacturing until Louis's death in 1948 and Auguste's in 1954.

MEGAN MCDANIEL

Guglielmo Marconi
1874-1937
Italian Physicist and Inventor

Guglielmo Marconi, a physicist and inventor, was responsible for pioneering a new method of communication known as radio telegraphy. A recipient of the 1909 Nobel Prize for physics, Marconi's revolutionary conception of long-distance transmissions continues to underpin contemporary applications of wireless technologies.

Marconi was born on April 25, 1874, in Bologna, Italy, to an Italian father and an Irish mother. Privately tutored at his father's estate in Pontecchio, near Bologna, Guglielmo was an inquisitive child who quickly developed an interest in atmospheric and static electricity. He read extensively from his father's scientific library and was particularly fascinated by the way Benjamin Franklin (1706-1790) had proved lightning to be electricity. After studying in Bologna and Florence, Marconi became a physics student at the technical school in Leghorn, where he studied the work of Heinrich Hertz (1857-1894) and Oliver Lodge (1851-1940).

In 1894 Marconi began experimenting by gathering a few simple devices: a Morse key to send signals, an induction coil, a spark discharger, and a radio wave-detecting instrument called a coherer. By repeated trial and error, Marconi found that he could achieve a signal span of 1.5 miles. Marconi found a more receptive audience for his experiments in London and in 1896 obtained a patent for his invention there. Together with his assistant, engineer William Preece, Marconi continued to conduct trials with great success. Using kites and balloons to achieve higher elevations for his antennae, he increased the range of his signals up to nine miles.

Many scientific groups expressed skepticism about the development of wireless telegraphy because of their belief that the earth's curvature would prevent long-distance transmittals over 150 to 200 miles in length. In 1901 Marconi disproved such naysayers when he received signals sent to Newfoundland from England. Scientists across the globe marveled at this intellectual—as well as oceanic—leap, which was the first crucial step in the development of twentieth-century radio communication. Shortly thereafter, Marconi made a pivotal discovery while on board the U.S. oceanliner *Philadelphia*. During this trip, he received transmittals from 700 miles away during the day but from 2,000 miles away at night. He concluded that, since a large degree of radio waves are conveyed by being reflected in the upper atmosphere, nighttime conditions allow for more effective communication. Put another way, sunlight—by absorbing radio waves in the lower regions of the atmosphere—prevents waves from reaching the climate in which transmission is most far reaching. By 1910 Marconi had refined his antennae to the extent that he could receive 6,000-mile transmittals in Buenos Aires from Ireland, and by 1922 radio broadcasting had exploded from a mere handful of stations to over 590.

Until his death in 1937, Marconi was a tireless experimenter who was perpetually improving upon the wireless-telegraphy discoveries that are still vital to current long-distance radio technology. In addition to receiving the Nobel Prize, which he shared with the German physicist Ferdinand Braun (1850-1918), Marconi was given numerous awards for his scientific advancements. He was a signatory at the 1919 Paris peace conference. He also held a distinguished position in the Italian senate and was elected president of the Royal Italian Academy.

MEGAN MCDANIEL

Cyrus Hall McCormick
1809-1884
American Inventor

Cyrus McCormick developed an agricultural reaper that mechanically cut crops. Farmers needed to harvest wheat soon after it ripened, and McCormick's reaper enabled them to cut the same amount of wheat in a day that required one person two weeks using hand tools. The mechanical reaper inspired the improvement of other farm machinery, encouraged farmers to migrate to unsettled western land, and allowed workers to shift from agricultural to business or industrial employment. Most significantly, the reaper helped farmers produce enough food to meet population demands.

A native of Rockbridge County, Virginia, McCormick was the son of Robert and Mary Ann (Hall) McCormick. His father was an inventor who patented blacksmith and farm tools. Attending school infrequently, McCormick worked with his father. The elder McCormick's unfulfilled goal was to develop a reaping machine. Other inventors were striving to accomplish this too. Reapers needed devices to isolate grains on stalks, position them for cutting with a blade, and collect the pieces.

Cyrus McCormick introduced a horse-drawn reaper for wheat harvesting in July 1831 and received a patent three years later. A year prior to McCormick's legal claim, Obed Hussey, a Maine inventor, patented a reaper. Competition with Hussey and other inventors resulted in McCormick seeking patent rights, although he did not market his reaper immediately because he wanted to refine its design and was distracted by duties at the McCormick's iron works. Technical aspects of McCormick's reaper outlined in the patent not only modernized American agriculture but also were incorporated in all reapers afterwards.

Cyrus McCormick. *(The Library of Congress. Reproduced by permission.)*

Beginning in 1837 when an economic panic threatened his financial stability, McCormick assembled reapers at the family's blacksmith shop at Walnut Grove, Virginia, for local consumers. By 1843 he sold rights to manufacturers in New York and Ohio to make reapers. These producers, however, did not create machines of the quality upon which McCormick insisted. To increase production while maintaining standards, he established a factory in Chicago in 1847. When his first patent expired the next year, McCormick sued competitors that threatened his business. Some of his revisions to the reaper were protected by patents, but his original design entered the public domain and could be copied.

McCormick's reaper was the first of its type that was commercially viable. He promoted his reaper at exhibitions. His reapers also were sold by salesmen who convinced farmers to purchase McCormick's machinery by traveling through rural neighborhoods and sharing testimonials. They also demonstrated the benefits of other farm tools such as cultivators. McCormick encouraged payment plans to assist farmers to buy machinery. All customers received performance guarantees.

The company became the country's largest manufacturer of agricultural technology. McCormick began selling his reaper in Europe in 1851, displaying it at the Crystal Palace in London. During the American Civil War, the reaper had a significant impact in that it freed men from farm labor to fight in the war. Although the foundation structure of the reaper was retained, McCormick constantly sought design enhancements such as adding the twine binder. McCormick overcame such obstacles as the 1871 Chicago fire and hostile farm organizations protesting equipment prices. He was elected to the French Academy of Sciences in 1879. In the 1880s McCormick sold 50,000 reapers annually and hired 1,400 employees.

At the time of McCormick's death, six million harvesters had been manufactured by his company. Wealthy from profits, McCormick invested in railroad and mining stock. He donated funds to the Presbyterian church and schools. A Democrat, McCormick unsuccessfully ran for Congress in 1864. Merging with five farm implement producers in 1902, McCormick's business became the International Harvester Company. McCormick had married Nettie Fowler in 1858, and they had six children, some of whom directed the company and participated in reaper centennial celebrations.

ELIZABETH D. SCHAFER

Ottmar Mergenthaler. *(The Library of Congress. Reproduced by permission.)*

Ottmar Mergenthaler
1854-1899
German-American Inventor

Ottmar Mergenthaler revolutionized the distribution of data. His Linotype machine expanded the length of newspapers, which made more information available to the reading public.

Ottmar Mergenthaler was born in 1854 in what would eventually become part of the nation of Germany. His family background prepared him well for the life of an inventor. It created an early love of education and instilled the qualities of self-discipline and curiosity. Ottmar began his education at the school where his father taught. He was exposed to a strong German curriculum, which included an emphasis on science and technology. At home the Mergenthaler children were expected to help with the daily chores. This helped him to develop a disciplined mind and spirit.

When he reached the age of 14 he was expected to begin the training that would prepare him for a teaching career. Ottmar informed his parents that he was not interested in teaching, but that a life in technology was more to his liking. He and his family finally agreed on an apprenticeship with his uncle in watch- and clock-making. Ottmar was a very successful apprentice, but he was still not satisfied. On his own initiative he took classes in mechanical drawing and drafting in the evenings and on Sundays. This training would pay off later in his career by allowing him to understand the workings of highly technical machines.

Upon completion of his apprenticeship Ottmar decided to immigrate to the United States. This decision would be the turning point in his life. His cousin ran a company that made electrical instruments in Washington, D.C. He traded part of his future labor for passage and arrived in America ready to begin his new life. Washington was the location of many important government agencies. The best inventors in the country lived and worked there. This was Ottmar's true training grounds, where he was exposed to some of the best technological minds in the country.

In 1873 the nation began to experience the first great economic downturn of the second Industrial Revolution—which began in the latter half of the nineteenth century. By 1876 most of the business community was feeling the effects of the economic problems. Ottmar took advantage of the situation and made himself available as a consultant for the development of new technologies. One day an inventor came in looking for assistance with his new writing machine. Ottmar solved the problem, but he also began thinking about the practical application of this new concept. His work would eventually lead to the invention of the Linotype machine.

Before Mergenthaler's invention, all the type for a newspaper had to be done by hand. This process was long, tedious, expensive, and resulted in newspapers of under 10 pages in length. Ottmar's machine was controlled by a keyboard similar in concept to the one found with most personal computers today. A linotype operator could produce 4-7 lines of print a minute with his 90-character keyboard.

This innovation caused the greatest information explosion since Johannes Gutenberg (c. 1398-1468) invented movable type. Newspapers doubled and tripled in size. Most importantly, it allowed the average American to become a more informed citizen. Mergenthaler's success with the linotype coincided with the onset of the Progressive Movement in America. Great social critics, such as Upton Sinclair, were now able to get their vivid, in-depth accounts of political corruption and social poverty printed in great detail in newspapers and magazines across the country. The American people responded to this movement because they became better informed. It also made the corporations that ran the newspapers more powerful. More people purchased newspapers because of their increased quality, which gave the papers great power over the shaping of public opinion.

Ottmar Mergenthaler died in 1899 in Baltimore, Maryland. His greatest invention, however—the Linotype machine—remains a very good example of technology's impact on society. It helped set the stage for the great social movements of the twentieth century.

RICHARD D. FITZGERALD

Samuel Finley Breese Morse
1791-1872
American Inventor and Painter

Samuel Morse was a successful portrait painter who turned to invention in midlife. He developed the electric telegraph and the communication system known as Morse Code.

Morse's education, while not offering a strong scientific background for an inventor, did influence the directions his future work would take him. He first attended Phillips Academy in Andover, where he was not considered to be a

Samuel Morse. *(The Library of Congress. Reproduced by permission.)*

strong student. His parents then sent him to Yale. There he developed an interest in painting miniature portraits and attended lectures on electricity, which at the time was a topic that was still not understood well. After graduating, Morse traveled to England to continue studying painting.

When Morse returned from England, he earned respect as a painter. He first attempted to sell the historical canvases that he had developed a preference for while studying in England, but, finding that such works were of little interest to American buyers, he decided to paint portraits. Working first as an itinerant painter, he settled in New York and developed a sizable reputation for his portraits. In 1826 he helped found the National Academy of Design.

While making a trip related to his artistic career, Morse would find himself beginning to change the direction of his life's work. In 1829, after experiencing the successive deaths of his parents and wife, he returned to Europe to further his artistic studies. On his return voyage in 1832, Morse's interest in electricity was renewed during a conversation with fellow passengers, including chemist Charles T. Jackson (1805-1880), about the newly discovered electromagnet. During the voyage, he conceived the idea of sending messages over a wire using electricity. Although the idea for an electric telegraph had already been suggested before 1800, Morse thought that his proposal was the first. His first working model of the telegraph was probably made by 1835. Morse gave up painting in 1837, a decision influenced by both disappointments in his artistic career and his interest in developing the telegraph.

With his work on the telegraph impeded by his lack of a scientific background, Morse turned to others for assistance. While teaching art at the University of the City of New York (later to be called New York University), Morse received aid from a colleague who showed him a detailed description of an earlier, alternative model. Chemistry professor Leonard Gale helped him improve both his electromagnet and his battery. Gale introduced Morse to Joseph Henry (1797-1878), who shared his knowledge of electromagnetism with him, enabling Morse to invent an electromagnetic relay system, which made long-distance transmission possible by renewing the electric current along a line containing a series of relays. Morse filed an intent to patent in 1837 and by 1838 had developed the system of dots and dashes known as Morse Code. Both Morse's device and his code received improvements made by a new partner, Alfred Vail (1807-1859). After a long struggle to receive financial support from Congress, Morse, with the aid of Vail and Ezra Cornell (1807-1874), built a telegraph line between Baltimore, Maryland, and Washington, D.C., sending the historic message, "What hath God wrought!"

The telegraph brought Morse fame and wealth. When the United States government refused to buy the rights to the telegraph, he and his partners formed their own business, the Magnetic Telegraph Company. Morse was confronted by considerable litigation over patent rights to the telegraph, but his patent rights were upheld by the Supreme Court. He became increasingly involved in politics and was an opponent of abolition. He helped found Vassar College and was a noted philanthropist. Morse preferred not to be remembered for his portraits, but, ironically, as his reputation as an important artist has grown, his fame as the inventor of the telegraph has lessened, particularly since the advent of the telephone, radio, and television.

GARRET LEMOI

Joseph Nicéphore Niépce
1765-1833
French Inventor

Joseph Nicéphore Niépce, born in 1765 in Chalon-sur-Saône, France, was the first to

Joseph Nicéphore Niépce. *(Corbis Corporation. Reproduced by permission.)*

make negative photographic images on paper and positive photographic images on metal plates. He also invented a method of making multiple copies of existing pictures.

As a young man, Niépce served as an officer in the French army under Napoleon Bonaparte. After poor health forced him to resign from the military, he settled in Nice, where he married and became a government administrator. He returned to his family's estate in Chalon-sur-Saône in 1801 and, with his brother Claude, devoted his life to experimentation and invention.

The brothers' first invention, patented in 1807, was an internal combustion engine they called the *pyréolophore*. The engine used the same piston and cylinder system as a modern gasoline engine and worked well enough to propel a boat upstream. Because the lycopodium powder that fueled the engine was expensive, however, the engine never achieved commercial success. The Niépce brothers entered government competitions to invent a hydraulic ram and to find a substitute for indigo, a blue dye. Although their ideas received positive responses from the government and the scientific community, they lost the competitions.

Nicéphore Niépce next became interested in lithography, a new method of printing copies of drawings inked on stone. Since he could neither draw nor find the kind of stone required for lithographs, Niépce experimented with copying images on chemically treated paper using a *camera obscura*. (Painters had been using this simple camera—a wooden box with a lens at one end and a mirror at the other—for more than 200 years to help them draw more accurately.) In 1816 he wrote about his first successful experiment to his brother, who was in Paris promoting the pyréolophore. With a camera he had produced a faint negative image on white paper of a birdhouse outside his window.

By 1824 Niépce had discovered that bitumen of Judea, a kind of asphalt, is sensitive to light. Niépce took advantage of this property to invent a process of copying existing images onto a bitumen-coated plate. This process was later perfected by his nephew and used by printers for decades. Niépce's goal, however, was to capture new images on the bitumen-coated plate. On September 16, 1824, he wrote to his brother that he had finally obtained "a picture from nature as good as I could desire."

Louis Daguerre (1787-1851), a Parisian painter who was also experimenting with photography, heard about Niépce's success and wrote to him, asking for more information. Niépce, however, preferred to keep his work secret and replied evasively. After receiving a second request the following year, Niépce agreed to meet Daguerre in Paris on his way to London, where his brother was now living. In London Niépce showed what he called his heliographs to every British scientist he could meet and even sent examples to Windsor Castle. After five months of discouraging responses from the English, Niépce received another enthusiastic letter from Daguerre and decided to return to France.

Niépce sent a heliograph to Daguerre, who responded with detailed suggestions. Niépce, in turn, offered a partnership to Daguerre, who accepted immediately and proposed publishing the process as Niépce's discovery. They carried out their research separately, communicating via letters in a number code devised by Daguerre to maintain secrecy. They experimented with electric currents, iodine fumes, solar microscopes, and various exposure times. This fruitful collaboration was cut short by Niépce's death on July 5, 1833. Daguerre formed a new partnership with Niépce's son, Isidore. They were granted a government pension in 1839 in return for disclosing the technical details of both the original invention and Daguerre's new invention, the daguerreotype.

LINDSAY EVANS

Alfred Bernhard Nobel
1833-1896
Swedish Chemist, Engineer and Industrialist

As the inventor of dynamite, Alfred Nobel amassed great wealth producing instruments of destruction. He left the majority of his fortune to establishing the Nobel Foundation, among whose prizes is the prestigious award for the promotion of peace.

Nobel's education received an important push when his father found success as a manufacturer of explosive mines and machines tools in St. Petersburg. The family left Stockholm to join his father in Russia, enabling the young Nobel to study under private tutors. Fluent in several languages, Nobel showed great interest in engineering and chemistry. For a year he studied in Paris under the noted chemist Theophile Jules Pelouze (1807-1867); while most regard Christian Schonbein (1799-1868) as the inventor of guncotton, some attribute the honor to Pelouze. Nobel then traveled to the United States and worked for four years under John Ericsson (1803-1889), who built the ironclad warship the *Monitor.*

Nobel began experimenting with explosives on the family estate in Sweden after his father's Russian business, unable to make a peacetime profit once the Crimean War ended, went bankrupt. At the time, black powder, a form of gunpowder, was the dominant explosive used in mines. Even though Ascanio Sobrero (1812-1888) had recently invented the liquid compound known as nitroglycerin, a much more powerful explosive, it was much too volatile for widespread use. Despite the danger, Nobel's father devised a method for the large-scale production of nitroglycerin and Nobel built a factory for its manufacture. To deal with the problem of nitroglycerin's volatility, Nobel sought to control the detonation of the explosion. His first important invention was a practical detonator for nitroglycerin involving a container of nitroglycerin and a plug containing black powder, the igniting of the less-volatile powder setting off the explosion of the nitroglycerin. Nobel then developed an improved detonator that used a mercury fulminate blasting cap that could be exploded by either shock or moderate heat. These inventions not only marked the beginning of the wealth Nobel was to acquire making explosives, but introduced the modern era of explosives as well.

After the accidental death of a brother and several others at Nobel's nitroglycerin factory, he built several more factories and continued his research into nitroglycerin. He found that a porous clay known as kieselguhr would absorb the nitroglycerin, making its manipulation and transportation much safer. Nobel took the name of this new product from the Greek *dynamis* meaning "power," calling it dynamite. Nobel's 1867 invention of dynamite, more powerful than gunpowder and relatively safe to use in blasting, brought him a fortune.

Following the invention of dynamite, Nobel continued to research explosives and enlarged his manufacturing interests. In 1875 he invented a more powerful form of dynamite, variously called blasting gelatin, saxonite, gelignite, and Nobel's Extra Dynamite. In 1887 Nobel introduced ballistite, which would become the precursor of another smokeless explosive powder, cordite. (Nobel also invented such nonexplosive products as artificial silk and leather.) In 1894 Nobel bought an ironworks in Bofors, Sweden, which became the basis for the widely known Bofors arms factory.

In great contrast to what the bulk of his career would suggest, Nobel was a devoted humanitarian and philanthropist during his lifetime. In his fiercely contested will, he bequeathed most of his wealth to create the Nobel Foundation, whose awards honor those who have rendered an intellectual service to humanity. The prizes reflect his interest in the areas of chemistry, physics, physiology, and literature. Nobel's inspiration to establish the famous peace prize was most likely influenced by his friendship with the Austrian pacifist Bertha von Suttner. The Nobel Foundation awarded its first monetary prizes in 1901.

GARRET LEMOI

Norbert Rillieux
1806-1894
American Inventor and Engineer

Norbert Rillieux is best known for helping to mechanize the process of sugar refining. He invented an evaporator that converted sugarcane juice into crystalline sugar. His evaporator worked more efficiently than previous methods and contributed to the availability of inexpensive sugar.

Rillieux was born in New Orleans, Louisiana, in 1806. His mother was of mixed race, and his father was a white plantation owner. Rillieux showed a gift for learning, which became even more apparent when he attended college at l'Ecole Centrale in Paris, France. Rillieux earned a degree in engineering and, after graduation, became a teacher at his former school.

While in France, Rillieux became interested in making the process of sugar refining more efficient through the use of machinery. As a youth, he had observed the manual process of sugar refining on plantations in Louisiana. In this laborious and dangerous task, African-American slaves tended fires used to heat sugarcane juice in open vats. As the liquid boiled down, it was transferred to smaller vats with the aid of ladles. Eventually, the liquid would boil away completely, leaving behind crystals of sugar that were often brownish and of poor quality. In part because sugar refining was such a lengthy and labor-intensive process, white crystalline sugar was quite expensive, and only the rich could regularly afford it.

Rillieux designed a much more efficient method of refining sugar in which the quality of the crystals could be better controlled. His method was based on an invention that he called a multiple-effect vacuum pan evaporator. In this machine, sugarcane juice was heated in a series of sealed vats. Some of the air in the vats was pumped out, reducing the pressure and creating a partial vacuum. As a result of the reduced pressure, the juice boiled at a lower temperature, which meant that less fuel was needed. Pipes led the steam and the heated juice from one vat to the next. Steam from the first vat helped to boil the liquid in the second vat, and so on. This "recycling" of the steam's heat resulted in even lower fuel needs.

In 1834 Rillieux left Paris and returned to New Orleans, where he built and tested his evaporator on a sugar plantation. This test was unsuccessful, but Rillieux was able to make changes to his designs based on the results. In 1846 he was granted a patent for an improved version. His invention was adopted by the sugar industry within a few years, and thousands of evaporators were eventually installed in Louisiana and the Caribbean. Through the use of Rillieux's pan evaporator, the production of sugar increased and its price decreased. Sugar was no longer a luxury, and it soon became the common household item that it remains today.

After the success of his evaporator, Rillieux turned his attention to the design of a sewage and drainage system for New Orleans. The city, which lies at the marshy mouth of the Mississippi River, had long had a sanitation problem. City politicians rejected Rillieux's plans, although a system quite similar to the one he had designed was eventually adopted. Rillieux left New Orleans in the 1850s as restrictions on free African-Americans continued to increase in the years before the Civil War.

Rillieux returned to Paris, where he continued to work as an engineer and write articles for scientific journals. In addition, he became interested in the history of Egypt and worked on deciphering ancient Egyptian hieroglyphics for nearly 20 years. In 1881 Rillieux returned to the sugar industry and designed a modified system for heating multiple vats with vapor. Although he lost a patent battle for this design, the basic principle behind it was eventually adopted in the manufacture of soap, glue, and paper, in addition to the refining of sugar.

STACEY R. MURRAY

Isaac Merrit Singer
1811-1875
American Inventor

Isaac Singer was an American inventor, who is best known for making the mechanical sewing machine much more practical and useful. In addition to the sewing machine, he developed a rock-drilling machine, a metal- and wood-carving machine, and more. As a businessman, Singer also introduced several innovations that were to have a significant effect on consumer sales and marketing.

Singer, a native of Pittstown, New York, worked with machines and inventions from an early age. Apprenticed to a machinist in his teens, Singer invented a rock-drilling machine at the age of 28 and a machine for carving metal and wood at 38. Two years later, working for a customer in Boston, Singer saw his first sewing machine. Within two weeks he had not only repaired the broken machine but had designed a vastly improved version that he patented and sold through his own company.

Singer did not actually invent the sewing machine. That distinction instead belongs to France's Barthélemy Thimonnier. In fact, two other Americans had produced sewing machines before Singer. The first, Walter Hunt, failed to patent his machine, which contained many important features. The second, Elias Howe, however, did patent his sewing machine. Howe's machine incorporated an eye-pointed needle (a needle in which the eye is in the point of the nee-

dle rather than in the opposite end) that created an interlocking stitch with a second thread on a shuttle on the other side. Singer's, however, was the first sewing machine also to incorporate an overhanging arm containing the needle, which allowed virtually any part of the fabric to be sewn in nearly any orientation.

Singer became embroiled in a patent infringement suit with Howe almost immediately but, not wanting to wait for resolution through the courts, began manufacturing sewing machines anyway. Despite losing the suit, he filed patents for a dozen other innovations and was able to continue building his machines. By 1860 Singer's company was the world's largest manufacturer of sewing machines.

Singer continued to improve the sewing machine for the next few years, in particular working on ways to power the machines better. The first machines were powered by a hand crank, but this forced tailors to work with only one hand. Inventing the foot treadle freed both hands for sewing and was considered a great improvement.

Singer turned out to be as innovative in business as with machinery. He was among the first manufacturers to use the concept of installment payment plans for purchasing items, giving his company increased sales while placing relatively expensive pieces of equipment within the financial reach of many more families and small businesses. This move had a profound impact on the consumer-sales sector that continues to this day.

That modern industrial sewing machines—and most home machines too—continue to display the same fundamental features as Singer's original models is a tribute to his design. Specifically, the eye-pointed needle, interlocking stitches, the overhanging arm, and foot or (knee) controls are all standard features on most of the world's sewing machines to this day.

Singer's first company, I.M. Singer & Company, was superseded by the Singer Manufacturing Company, which was owned by Singer and his partner, Edward Clark. Shortly afterwards, Singer retired to England, where he lived until his death in 1875 at the age of 63. His company survived for another 113 years, pioneering such innovations as the first electric-powered sewing machine (1885) and the first mass-produced sewing machine (1910). Sewing machines with the Singer brand name continue to be produced under a different parent company.

P. ANDREW KARAM

Richard Trevithick
1771-1833
English Engineer

Richard Trevithick was a mechanical engineer and inventor who created the first high-pressure steam engine and built the first steam railway locomotive. Despite the importance of his inventions, Trevithick failed to profit by them.

Trevithick was born in Cornwall, England. Not considered a good student, he attended the village school, and throughout his life remained barely literate. His father managed a coal mine, and thanks to an ability to solve engineering problems, Trevithick received his first job as an engineer at several Cornish ore mines at the age of 19.

Because of his association with coal mines, Trevithick was early exposed to the steam engines made by James Watt (1736-1819). Watt had invented a steam engine that employed steam at a low pressure. A cautious inventor, Watt believed that high-pressure steam was too dangerous to harness. Trevithick, however, took high-pressure steam and allowed it to expand within the cylinder, which resulted in an engine that was smaller and lighter than Watt's version but without any loss of power. He demonstrated, contrary to contemporary belief, that machinery could be powered by using pressures higher than the atmosphere's. After building working models of both stationary and locomotive engines, Trevithick built a full-scale engine for hoisting ore. These engines were used at the Cornish mines and because they released steam were called "puffer whims." (Trevithick brought increased power to his engine by releasing waste steam into the smokestack, increasing the fire's draft and temperature.) In 1802 he and his cousin Andrew Vivian took out a patent for high-pressure engines.

Trevithick used what he had learned building high-pressure steam engines to build a steam-driven locomotive. He drove his first steam carriage, which carried passengers, on Christmas Eve in 1801. In 1803 Trevithick drove a second carriage through the streets of London and built the first steam railway locomotive in South Wales. In 1804, that same locomotive, the *New Castle,* carried 10 tons of iron and 70 men along 10 miles of tramway. He built a similar locomotive at Gateshead in 1805, and in 1808, Trevithick's third locomotive, the *Catch-me-who-can*, ran on a track in London. Tre-vithick abandoned these locomotive projects because cast-iron rails proved unable to support the weight of his engines. He demonstrated, however, that smooth metal wheels moving over a smooth metal track could actually create enough friction to move weight. Unfortunately, George Stephenson (1781-1848) and his son Robert Stephenson (1803-1859), who used many of Trevithick's discoveries, were to become known as the "parents" of steam-powered railway transportation.

Trevithick then turned his attention to other designs. He adapted his engine to drive an iron-rolling mill as well as to propel a barge that used paddle wheels. He designed dredging and threshing machines, also powered by his steam engine. (He attempted, but failed, to dig a tunnel under the Thames River using a dredger.) The success of these engines was related to improvements that Trevithick made in the design and construction of boilers.

Trevithick's career was marred by a series of poor business decisions. An untrustworthy partner and the Thames dredging failure contributed to the 1811 bankruptcy of his London business. His engines were later ordered for use in the Peruvian silver mines, and in 1816 he left for Peru and Costa Rica with the prospect of mineral wealth and the intention of linking the Atlantic and Pacific Oceans by rail. This venture also failed, and in order to return to England Trevithick borrowed money from one of the Stephensons, who by this time had become rich from railway profits. Trevithick died in poverty and, only after the men in his workshop contributed money for his funeral, he was buried in an unmarked grave.

GARRET LEMOI

George Westinghouse
1846-1914
American Inventor and Industrialist

George Westinghouse is best known for founding his electric company and promoting the adoption of alternating current for electric power transmission in the United States. He also made significant improvements in railroad safety.

Westinghouse's career inclinations were apparent early on. Born in New York, he worked in his father's factory, where he learned about mechanics. After serving on the Union side during the Civil War, he received his first patent for a rotary steam engine in 1865.

Westinghouse's first significant accomplishment involved the railroad. By the end of the Civil War, railroad cars were stopped by brakemen who, stationed along the length of the train, turned hand brakes on each car at the signal of the engineer. Westinghouse saw the need for a safe means of braking trains, one involving a single braking system for the entire train. He first experimented with a system involving pressure and steam. After becoming aware of a Swiss tunnel being excavated using compressed air, he developed a train brake that employed an air pump powered by the train's engine and that was directed by a single valve controlled by the engineer or brakeman. Pipes ran the length of the train to each car where mechanical devices activated the brakes. The invention was a great success, and in 1869 Westinghouse formed the Westinghouse Air Brake Company.

Further testing showed some problems with his braking system, and by 1872 Westinghouse had changed his approach. Instead of using compressed air to activate the brakes, he now activated them using a drop in air pressure. The principles developed by Westinghouse to brake trains remain in use today. Seeing the advantages of standardizing air-brake equipment among cars belonging to different train lines, Westinghouse became one of the first to advocate the practice of standardization.

Westinghouse also was concerned with the era's inadequate railroad signaling. Combining his own inventions with patents he had acquired from other inventors, Westinghouse developed an electrical and compressed air system of signals. He founded the Union Switch and Signal Company in 1880.

Another interest of Westinghouse's was natural gas. He drilled a gas well on his Pittsburgh estate that exploded and, after it was lighted, became known as Westinghouse's Torch. Intending to promote the use of natural gas as a power source, he worked to engineer a safe transmission network. In so doing, he developed a leakproof piping system, an automatic cutoff regulator, and a gas meter. He founded the Philadelphia Company, which provided gas service to customers.

Perhaps Westinghouse's most important achievement was to change the United States over from the use of direct current (DC) to alternating current (AC) for the transmission of electricity. The great advantage of AC power is that, by using a transformer, voltage can be easily increased (at the power plant for long-distance transmission) or decreased (at the customer's end for practical use). While the electrical system then in development in the United States used DC power, in 1881 Lucien Gaulard of France and John Gibbs of England demonstrated a successful AC system. Importing a set of Gaulard-Gibbs transformers and a Siemens AC generator, Westinghouse established an electrical system in Pittsburgh. With the help of electrical engineers, he perfected the transformer and introduced a constant-voltage AC generator.

In 1886 he created the Westinghouse Electric Company, later renamed the Westinghouse Electric & Manufacturing Company, which began producing transformers, electric motors, and dynamos. Although the advocates of DC power, including Thomas Edison (1847-1931), fought strongly, they were unable to suppress the widespread use of AC power. In 1893 Westinghouse won contracts to build the electrical generating station at Niagara Falls and to light the World's Columbian Exposition in Chicago. During the economic panic of 1907, Westinghouse lost control of his electric company and most of his wealth.

GARRET LEMOI

Biographical Mentions

~

Clément Ader
1841-1925

French aviation pioneer who experimented with aircraft beginning in 1882. He designed and later built several machines with bat-shaped wings. He attempted to fly one, the steam-engine-powered *Eole,* in 1890. Although he took off, he failed to remain airborne. Later craft also failed, even though Ader later claimed he had succeeded. He is credited, however, with inventing the French term for airplane, *avion.*

Sir William George Armstrong
1810-1900

English inventor and engineer who made significant developments in the fields of hydraulic engineering, shipbuilding, and artillery production. Early in life Armstrong developed an interest in hydraulics and electricity, subjects upon which he became a popular lecturer. While studying the instruments and equipment being produced at Newcastle's High Bridge Works, he built his first hydraulic machine, which he later used to power cranes, hoists, dock gates, and bridges. During the Crimean War, the British War Office commissioned him to design and make submarine mines. This led him into the field of artillery, and in 1855 he designed a lightweight gun for field use, which fired lead projectiles rather than cast-iron balls. Later, he developed an interest in shipbuilding, and worked to furnish ships with more complex hydraulic equipment.

Joseph Aspdin
1779-1855

English bricklayer and inventor who patented Portland Cement. Concrete, a mixture of cement, gravel, and sand, had existed since the time of the Romans. However, the cement portion was often of uneven quality. Aspdin discovered that the key to making hydraulic cement (cement that would harden when mixed with water) was roasting a mixture of clay and limestone powder in a furnace. Mixing this cement with sand, gravel, and water allowed the lime and clay to form a kind of gel that coated the sand and gravel, cementing them together when it dried. The hardened mixture reminded Aspdin of the prized building rock quarried on the Isle of Portland, thus the term "Portland Cement." Modern concrete contains only about 11 percent cement by weight, sand and gravel constituting over 65 percent of the weight and water most of the rest.

Charles Babbage
1791-1871

English mathematician considered the inventor of the first mechanical computer. Babbage devised plans for two machines, a difference engine (1823) to perform mathematical calculations and an analytical engine (1834), the forerunner of a modern computer. Because these mechanical devices required precise machining of geared wheels exceeding the capabilities of mid-nineteenth century technology, they were never completed as working devices. But Babbage's basic ideas persisted in the mathematical world and influenced later computer development.

Alexander Bain
1810-1877

Scottish inventor who designed the first method for transmitting images, or facsimile. Bain worked as a clockmaker in London, where in 1845 he invented and was granted a British patent for his electric clock. That same year he patented a method for using electricity to control the operation of steam railway engines with the use of a steam valve. His most famous invention was the forerunner of the fax machine, a chemical telegraph that could be used to transmit text and images over distances.

Erastus Brigham Bigelow
1814-1879

American inventor and industrialist best known for developing the power carpet loom and for his role as a founder of the Massachusetts Institute of Technology (MIT). Bigelow invented a number of power looms for weaving everything from lace to figured fabrics and carpeting. In 1843 he, with his brother, built a gingham mill around which the town of Clinton, Massachusetts, developed. In 1851 he completed a power loom for the manufacture of Brussels and Wilton carpets. This loom greatly improved carpet manufacture throughout Europe and the United States.

Louis Braille
1809-1852

French educator who developed a system of printing and writing by which the blind could write in relief and read by touch. That system, braille, which consists of a six-dot code in various combinations, was officially named after him in 1834. Blinded at the age of three, he studied at the Paris Blind School, where he later became an instructor. He was an accomplished musician, and developed a separate code for reading and writing music and mathematics.

John Watkins Brett
1805-1863

English engineer who helped develop Britain's inland telegraph system. Brett, along with Sir Charles Tilston Bright and Cyrus Field, established the Atlantic Telegraph Company.

Isambard Kingdom Brunel
1806-1859

English engineer who designed England's Great Western Railway and built several of the era's great steamships. Brunel began his career as an engineer on the Thames River Tunnel project, headed by his father, Marc Brunel. Isambard Brunel went on to design the River Avon suspension bridge, still in use today. One of Brunel's greatest achievements was the design and construction of the London-to-Bristol rail line, better known as the Great Western Railway, distinguished by its low-arch bridges and two-mile tunnels. His innovations in rail gauges helped reform England's locomotive industry. Brunel was also a pioneer in steam navigation, and he designed three of the world's great steamships—the *Great Western* (1838), the *Great Britain* (1845) and the *Great Eastern* (1858), the largest steam vessel of its time.

Marc Isambard Brunel
1769-1849

British civil engineer who built the world's first underwater tunnel, the Thames Tunnel in London. Born in France, Brunel fled to America as a royalist refugee in the aftermath of the French Revolution. He served as chief engineer of New York City, and built a canal between Lake Champlain and the Hudson. In 1799 he settled in England, and began construction of the Thames Tunnel in 1825. He was knighted in 1841.

William Seward Burroughs
1855-1898

American inventor of the calculator. The earliest seeds of Burroughs's passion for invention were planted while tinkering in his father's machine shop as a young child. At age 15 he came up with the idea for a device that would calculate the sum of numbers. In 1881, he began working on his invention, and three years later had it patented, but his machine still needed work. It wasn't until 10 years later that Burroughs had designed a fully functional version of his adding machine, which printed its results on paper. His invention proved to be a great success, but not in Burroughs's lifetime.

Ferdinand Carré
1824-1894

French engineer who invented the first refrigeration machine. In 1859 he received a French patent for an aqueous ammonia absorption system for making ice. His invention remained popular until the early 1900s, when it was replaced by refrigeration systems that use the liquid vapor compression cycle. Carré also conducted research in the field of electricity. He invented an electric light regulator and the Carré machine, a mechanism for producing high voltages.

(Louis-Marie) Hilaire Berniguad, comte de Chardonnet
1839-1924

French chemist who invented rayon, the first synthetic fiber to come into common use. During a silkworm epidemic that threatened the French silk industry, Chardonnet realized there was a market for artificial silk. Expanding upon the previous work of Swiss chemist George Audemars and Sir Joseph Swan of England,

Chardonnet experimented with cellulose-based fibers. He treated cotton with nitric and sulfuric acids, then dissolved the mixture in alcohol and ether, forming fibers he called rayon. His work was patented in 1884 and he began to manufacture rayon in 1891. Chardonnet went on to study several other subjects, including ultraviolet light and telephony.

Samuel Colt
1814-1862

American firearms manufacturer who popularized the revolver. His single-barreled pistols and rifles featured a cartridge cylinder rotated by cocking a hammer. In 1855, Colt built and maintained the world's largest private armory. Although his guns were not initially well-received, his pistols were the most widely used in the American Civil War. His Colt Industries plant took the manufacture of interchangeable parts and the assembly line further than any industrialist before him.

George Henry Corliss
1817-1888

American inventor and entrepreneur who revolutionized the design and construction of the steam engine. At age 25 Corliss opened a boot store and patented his own boot-stitching machine. Two years later, while employed as a draftsman for a Rhode Island engineering firm, he worked on improvements for the steam engines the firm manufactured. In 1856 he founded his own company, the Corliss Engine Company, and began selling his new engines for use in cotton mills. His patented Corliss valve, a mechanism for regulating steam flow in engines, was among his most significant design contributions.

John Frederic Daniell
1790-1845

English chemist who invented the Daniell cell battery, as well as the dew-point hygrometer. Daniell's career began in the sugar-refining industry, to which he made several improvements. He also did significant work in the lighting industry, before making a major contribution to the field of meteorology with his invention of the hygrometer, an instrument that measured relative humidity, or the dew point. Daniell was probably best known for his invention of the Daniell cell, a battery that had more lasting power than the previous zinc-copper voltaic battery. In his lifetime, Daniell was also active as an educator, philosopher, and writer.

Rudolph Diesel
1858-1913

German engineer who devised an internal combustion engine named after him. Based on his theoretical engineering education in Munich and his association with German refrigeration engineer Carl von Linde, Diesel recognized the potential for an internal combustion engine without any external ignition system for its operation. After many years of developmental work, he marketed a high-pressure, high-efficiency durable engine successfully employed in railroad locomotives, trucks, ships, and automobiles.

Edwin Laurentine Drake
1819-1880

American petroleum engineer who pioneered a new method of oil drilling. In his early years, Drake was a jack-of-all-trades, working as a hotel clerk, railway express agent, and conductor before finding a position with a Pennsylvania oil company. While there, he studied the use of cable tools in salt-drilling, and came to the conclusion that the same method could be used to drill for oil. Drake pioneered the use of an iron pipe in drilling, which kept the bore hole from filling up. He was the first to strike petroleum at its source, at a depth of 69 feet (21 m). His work led to oil-drilling expeditions around the globe, and to the development of petroleum engineering.

John Boyd Dunlop
1840-1921

Inventor of the pneumatic tires (patented in 1888). Dunlop first began manufacturing tires for bicycles in 1900, but soon began producing tires for automobiles (1906). To protect his investment, Dunlop began buying rubber plantations on the Malay Peninsula but eventually sold the land back to Malaysian investors. He continued manufacturing tires, and the company he founded, Dunlop Rubber Co., Ltd., is still in business.

George Eastman
1854-1932

American industrialist who devised a carefree system of photography. Eastman began by manufacturing dry plates for the emerging field of photography in America (1880) and recognized the potential for amateur photography with his marketing of the simple Kodak box camera (1888). With the Kodak camera, the use of celluloid film rolls (1889) and the child-friendly Brownie (1900), Eastman dominated the marketplace in photography and shared his resulting wealth by making several philanthropic gifts.

Gustave Alexandre Eiffel
1832-1923

French engineer who pioneered steel bridge designs and constructed the tower named after him. After chemistry studies, he worked as a railway engineer and later headed his own engineering firm. In 1885 he designed the internal metal structure of the Statue of Liberty, sculpted by Bartoldi and given to the United States. Two years later, he supervised the construction of the famous 900-foot-tall tower, which was inaugurated at the 1889 Universal Exposition in Paris. He spent the remainder of his life studying aerodynamics.

Charles Ellet
1810-1862

American civil engineer who designed and built wire suspension bridges. Influenced by European engineers, Ellet wrote pamphlets addressing transportation technology. The federal government contracted him to survey the Mississippi River, which led to his suggestion of creating levees for flood control. Ellet established a railroad across Virginia's Blue Ridge Mountains and the James River and Kanawha Canal in that state. Commanding a fleet of Civil War ram-boats he produced, he secured Memphis, Tennessee, for the Union in 1862, dying from a battle wound.

John Ericsson
1803-1889

Swedish-born American engineer who influenced the design of warships. Building the first successful screw propeller in 1836, he immigrated to the United States three years later. He helped produce the *Princeton*, a unique steam-powered warship with engines and boilers below the water surface. Ericsson is best known for his production of the USS *Monitor*, an ironclad that withstood the Confederate ship *Virginia*, reinforcing Union support for the navy. Ericsson also refined structural plans for additional ironclads.

William Fairbairn
1789-1874

Scottish engineer who revolutionized the construction of steam engines through his invention of a riveting machine. Fairbairn moved to Newcastle upon Tyne in 1804, where he was apprenticed to an enginewright at North Shields. From there he moved to London, where his engineering studies helped him produce such inventions as a steam excavator and a sausage-making machine. By 1817 he had established a facility in

Manchester, where he made machinery for water wheels and cotton mills. His reputation grew and, before long, he was actively working on ships, bridges, locomotives, and other important contributions to the British economy. He was made a baronet in 1869.

Cyrus West Field
1819-1892

American financier best known for sponsoring the laying of the first telegraphic cable across the Atlantic Ocean. After a number of failed attempts, Field arranged for Queen Victoria to send the first transatlantic message to President James Buchanan in August 1858. The cable broke three weeks later. It took eight more years for Field to finally complete his project, when in 1866 a single ship laid out 2,700 miles of telegraphic cable that stretched from Newfoundland to Ireland.

Benoit Fourneyron
1802-1867

French hydraulic engineer who first patented a design for hydraulic turbine installations. Though the hydraulic (water-driven) turbine had been invented by Pierre-Simon Girard in 1775, Fourneyron's contribution involved the development of an installation to make use of the turbine's mechanical power. Fourneyron went on the build over 100 of these installations in various parts of the world, where they were used for power for a variety of industrial applications.

Richard Jordan Gatling
1818-1903

American inventor best known for his invention of the Gatling gun, a crank-operated rapid-fire machine gun that he patented in 1862. Working at the outbreak of the Civil War, Gatling believed a gun that could quickly fire large numbers of cartridges without stopping would reduce the number of soldiers required on the battlefield. The son of a wealthy planter, he also invented seed-sowing machines that did much to revolutionize the agricultural system in America.

Heinrich Geissler
1814-1879

German inventor who developed the Geissler tube and the Geissler mercury pump. Geissler learned glassblowing and settled in Bonn, where he made scientific instruments. The Geissler tube is a sealed glass tube containing a near-vacuum; the rarefied gas within shows the passage of electricity through it. These tubes became the forerunner of the cathode ray tube, later used in televisions and computer monitors.

Henri Giffard
1825-1882

French engineer who flew the first practical airship. When Giffard first became interested in aeronautics, balloons were the only means to fly, depending on wind for movement and steering. Existing engines were too heavy, so Giffard designed a 3-hp motor that weighed some 90 lbs and installed it on a 144-ft long, torpedo-shaped balloon. He flew the machine successfully at up to 6 mi an hour at the Paris Hippodrome on September 24, 1852, but could not make full turns. He continued working on the problem, but was stricken with blindness and eventually committed suicide.

Zénobe Théophile Gramme
1826-1901

Belgian electrical engineer who invented the direct current (DC) electrical generator and motor. Gramme produced the first successful DC generator in 1869 and, after two years of making improvements, began manufacturing them in 1871. They were initially sold for use in electroplating as well as electric lighting. Later, working with Hippolyte Fontaine, he showed that an electrical generator (also called a dynamo) could be used as a motor when the flow of electrical current was reversed.

Sir Robert Abbott Hadfield
1858-1940

British metallurgist who developed manganese steel. Manganese steel is extremely hard and resistant to wear, properties that make it an excellent material from which to make railroad tracks and machinery for breaking apart rock. Earlier forms of manganese steel had been too brittle to be of practical use. Hadfield eliminated this problem by increasing the percentage of manganese. He also experimented with adding silicon to steel. In 1925 he wrote a reference book titled *Metallurgy and Its Influence on Modern Progress.*

Joseph Henry
1797-1878

American physicist whose experiments with electricity led to the development of electromagnets, which in turn made possible the electric motor. In honor of this work, the unit of inductance is called the henry. As first secretary of the Smithsonian Institution, Henry resisted congressional pressure to turn that organization into a passive repository of knowledge, insisting that the Smithsonian also support original scientific research.

Richard March Hoe
1812-1886

American inventor and industrialist who invented the first successful rotary printing press. His first press was used by the *Philadelphia Public Ledger* beginning in 1847, and a later, improved model—the Hoe web perfecting press—was first used by the *New York Tribune*. The web printing press formed the basis for newspaper printing for many years.

Elias Howe
1819-1867

American inventor who patented the first sewing machine in the United States. As an apprentice, Howe mastered machinery processes. He observed his wife sewing to create a machine that duplicated human motions. Howe patented the machine in 1846, but American manufacturers refused to buy it. A London factory owner purchased his machine but stole Howe's profits. Returning home, Howe successfully sued rival inventors who had appropriated his design. He later established a sewing machine company that thrived.

John Wesley Hyatt
1837-1920

American inventor best known for his discovery of the process for making celluloid, the first artificial plastic. In 1870, while searching for an ivory substitute for the manufacture of billiard balls, Hyatt combined nitrocellulose, camphor, and alcohol, and heated the mixture under pressure. When it cooled and hardened, the resulting substance was celluloid, a discovery that facilitated the development of plastics. Hyatt is also credited with inventions in sugarcane refining and water purification.

William Kelly
1811-1888

America inventor and metallurgist who developed improved methods of making steel. Kelly was the first to force air through molten cast iron, a process that removed much of the carbon and produced a less brittle and more widely useable metal. Unfortunately, Henry Bessemer received a U.S. patent for a similar process before Kelly could build more than seven of his converters, prohibiting Kelly from capitalizing on his invention.

Friedrich Koenig
1774-1833

German inventor and printing press designer who produced the first cylinder flatbed press. Koenig spent his early adult years developing models of a new printing press that would produce many more copies than those in use at the time—and at far greater speeds. The platen presses of the era were hand-operated and time-consuming. Koenig's press was controlled by a series of rollers, operated in turn by a system of gear wheels. The rollers both raised and lowered the platen, providing the to-and-fro movement of the press bed as well as the inking of the form. After several unsuccessful trials in 1811, a working version of Koenig's press was placed on active duty at the *London Times* in 1814 and Koenig took his place in printing history.

Alfred Krupp
1812-1887

German industrialist and armaments magnate, known as the Cannon King, who was an important contributor to the growth of the German military-industrial complex, as well as the development of modern warfare. Introduced in 1847, Krupp's cast-steel cannons and other armaments were sold to over 45 war ministries worldwide. Under his leadership, the family business also expanded and found prosperity in manufacturing railway tires, springs, and axles of high-grade steel.

Samuel Pierpont Langley
1834-1906

American aviation pioneer who first worked as an astrophysicist before becoming secretary of the Smithsonian Institution in 1887. While there, he studied solar radiation and popularized scientific knowledge through magazine articles. Beginning in 1890, he used his knowledge of aerodynamics to design and construct powered aircraft models and test them from a houseboat on the Potomac River. A full-scale piloted machine failed to fly in 1903, but the hired pilot survived. Langley claimed later that his machine could have beaten the Wright brothers' aircraft; the design, however, was structurally too weak to withstand aerodynamic forces.

Tolbert Lanston
1844-1913

American inventor who developed the Monotype machine in 1887. The Monotype, used for type-forming and type-composing, became commercially available in 1897 and revolutionized the process of printing. The Monotype (and its variants) was used for decades to set print for newspapers, magazines, books, and other printed materials.

Lewis Howard Latimer
1848-1928

American inventor and engineer who invented a durable and inexpensive carbon filament for lightbulbs in 1882. Latimer was originally trained as a draftsman and drew many of the designs for Alexander Graham Bell's first telephone. Later, he became the first African-American electrical engineer to work for the Edison Electric Light Company. He oversaw the installation of carbon-filament lighting in several major cities during the 1880s, and he published a manual of electrical engineering titled *Incandescent Electric Lighting* in 1890.

Georges Leclanché
1839-1882

French chemist and engineer who invented the dry cell battery. Leclanché gave up his career as an engineer to perfect his battery, which he completed in 1866. The Belgian telegraph service adopted Leclanché's battery in 1868, and it soon came into general use. The dry cell battery was much superior to the wet cells of the day, which were similar to present-day automobile batteries in that they could spill corrosive liquid if not handled carefully.

Jean-Joseph Etienne Lenoir
1822-1900

Belgian inventor and engineer who developed the first usable internal combustion engine. Lenoir was able to convert a steam engine to run on a coal gas and air mixture, making a practical engine. He later modified his device to run on liquid fuel, similar to today's internal combustion engines. He used his engine to propel a land vehicle in 1860 and a boat in 1886.

Ferdinand-Marie, vicomte de Lesseps
1805-1894

French diplomat and engineer who supervised construction of the Suez Canal and started work on a sea level canal across Panama. Lesseps began his career as a diplomat, holding several posts between 1825 and 1854. He directed the construction of the Suez Canal between 1860 and 1869 and, in 1881, began work on the French Panama Canal, later abandoned. He was charged with fraud and sentenced to five years in prison (later suspended) for his part in the Panama Canal debacle.

Otto Lilienthal
1848-1896

German aviation pioneer who built and flew experimental gliders. Beginning in 1891, Lilienthal used data gathered from the observation of bird flight to design his first contraption. Two years later, he carried out controlled glides to a distance of approximately 750 feet from a hill outside Berlin. He died when his glider stalled in midair and crashed. Although not all of his data was accurate, it inspired the Wright brothers to proceed with their own glider experiments.

Carl Paul Gottfried von Linde
1842-1934

German engineer who developed a process for converting large volumes of gas into liquid. He founded a factory that used this process to produce liquefied air. By 1895 he was able to separate oxygen from liquid air. Liquid gases soon became important to the manufacture of steel and to the refrigeration industry. Earlier in his career, Linde invented refrigerators that used either methyl ether or ammonia as coolants. He used these machines to perform calculations involving the efficiency of heat transfer.

Mahlon Loomis
1826-1886

First person to use radio waves to transmit telegraph messages. Mahlon Loomis overcame ground and weather obstacles by using aerials, which were held aloft by kites. His first radio wave telegraphy traveled 14 miles (22.5 km) between two mountains near his home in West Virginia. Loomis was granted the world's first wireless patent on July 20, 1872.

Hiram Stevens Maxim
1840-1916

American-born British inventor who revolutionized warfare in 1884 with the invention of the world's first automatic, portable machine gun. The Maxim recoil-operated machine gun was used extensively and with devastating effect in World War I. Maxim's many other inventions include a simple steam-powered flying machine, an electric pressure regulator, smokeless gunpowder, a pneumatic gun, carbon filaments for lightbulbs, vacuum pumps, gas motors, and an automatic steam-powered water pump. He was knighted in 1901.

John Loudon McAdam
1756-1836

Scottish engineer/inventor whose name is known worldwide as the father of modern road building. In 1819 McAdam appeared before Britain's House of Commons to report on his system of road building and mending. His macadam design helped establish a new standard of road engineering and became popular in much of England and the United States. Smooth, solid roads replaced gravel-on-soil roads, facilitating travel and communications and opening many areas to new influences.

Elijah McCoy
1844-1929

Canadian-born mechanical engineer and inventor. The son of former slaves from Kentucky, he revolutionized the industrial machine industry with his lubricator cup, which allowed small amounts of oil to drip into a machine as it worked. The creation of the lubricator cup dramatically sped up production in factories and on farms, as workers no longer had to stop at regular intervals to apply lubricants by hand.

William Murdock
1754-1839

Scottish-born British engineer and inventor who pioneered the use of coal gas for lighting. In 1792 Murdock generated gas from coal and lit his Cornwall home. His work with gas lighting culminated in its commercial application. In 1802 Murdock installed an exterior lighting system at Boulton and Watt's Soho factory (interior lights were added in 1803) and in 1805 he installed lights in the Phillips and Lee spinning mill in Manchester.

James Nasmyth
1808-1890

Scottish inventor who designed the first powerful steam hammer (patented in 1842). James Nasmyth manufactured more than 100 steam locomotives, a series of small high-pressure steam engines, hydraulic presses, an assortment of steam-driven pumps, and other useful machines. He did all these things before the age of 48, when he retired from active business and devoted himself to astronomy. In 1874, he wrote *The Moon: Considered as a Planet, a World, and a Satellite.* Nasmyth died in 1890 in London, England.

Paul Nipkow
1860-1940

German engineer and inventor who developed a mechanical scanning device (the Nipkow disk) used in early televisions. The Nipkow disk consisted of a spinning metal disk with a spiral pattern of holes drilled in it. Invented in 1884, it was used in televisions until 1932, then replaced by electronic scanning devices.

Elisha Graves Otis
1811-1861

Designer of the safety hoist elevators. Born in Halifax, Vermont, in 1811, Elisha Graves Otis was a hard-working New Englander with exceptional talents in improving old machines and inventing new ones. Working as a master mechanic in 1852 for a company in Bergen, New Jersey, he was sent to Yonkers, New York, to operate a new factory and install its machinery. It was there that he designed and installed a safety hoist, the first elevator with backup devices to keep it from falling in case the lifting chains or ropes failed to support the weight of the hoist. On March 23, 1857, the first passenger elevator in history was installed in a New York store. Otis died in 1861 but the company he founded is still in business.

Nikolaus August Otto
1832-1891

German engineer who built the first four-stroke internal combustion engine. The principle of a four-stroke engine (intake, compression, combustion/power, exhaust) had been patented by the Alphonse Beau de Rochas several years earlier, but Otto's engine was the first to be constructed. The four-stroke engine ran more efficiently than previous types and helped make internal combustion engines lighter in weight and more powerful. This type of engine is still known as the Otto cycle.

Alexander Parkes
1813-1890

British chemist and inventor who perfected the extraction of metal from lead. In 1850 he patented the Parkes process for using zinc to desilver lead. In the 1850s Parkes's experiments with cellulose fibers and nitric acid led to the discovery of cellulose nitrate. This led to his invention of parkesine, the world's first plastic, which was patented in 1855. Parkesine (later known as xylonite) was used in the production of ornaments, knife handles, and fishing reels.

Charles Algernon Parsons
1854-1931

British engineer and inventor who in 1884 invented a reaction turbine that revolved through the use of steam. In Parson's design steam was permitted to expand in a number of stages, performing useful work at each stage. In 1897 he adapted his compound steam turbine for maritime use. Parsons also invented a geared turbine and nonskid automobile chains.

Joseph Paxton
1801-1865

British architect and horticulturist best known for his design of the Crystal Palace, an iron and glass structure erected in London's Hyde Park in 1851. A botanist, self-taught gardener, director of works, and park designer, Paxton was inspired by the tropical greenhouses and conservatory he constructed in gardens at Chatsworth. His Crystal Palace, which substituted transparent surfaces for solid walls and roofs, was a revolutionary new concept in architecture.

Hippolyte Pixii
1808-1835

French engineer and instrument maker who created the first electrical generator. In 1832, following Hans Christian Oersted's discoveries, Pixii constructed a hand-cranked machine that rotated a magnet past a bar of iron wrapped with wire. This produced alternating current, which he converted to direct current. Although superseded by more efficient devices using electromagnets, Pixii's device was the first bona fide electrical generator constructed, and it paved the way for subsequent devices.

George Pullman
1831-1897

American industrialist and inventor of the Pullman railroad car. Pullman recognized the marketability of a comfortable, elegant sleeping car in the age of railroad expansion. He created his own company in 1867 to manufacture these cars and built the company town of Pullman in present-day Chicago, complete with church, community center, recreation facilities, shops, and other amenities, to create an uplifting living environment for his workers and their families.

William John Macquorn Rankine
1820-1872

Scottish engineer and scientist who made technical and scientific advances in thermodynamics, steam engines, and applied mechanics. Rankine was chair of engineering at Glasgow University for many years and wrote standard textbooks on steam power, shipbuilding, and applied mechanics. His scientific research added a great deal to the fields of thermodynamics and theories of elasticity and waves. He was elected a member of the Royal Society in 1853 in recognition of his contributions to science and engineering.

John Rennie
1761-1821

Scottish civil engineer and architect who won fame as a bridge builder. Rennie constructed three bridges over the Thames River in London: the Waterloo Bridge (1811-17), the first in London to be built with a flat roadway; the Southwark Bridge (1814-19), the first metal bridge over the Thames; and London Bridge (1824-34), dissembled in 1968 and subsequently relocated to Lake Havana, Arizona. Rennie also designed the London docks and built dams, canals, and harbors.

John Augustus Roebling
1806-1869

German-American engineer and industrialist who pioneered the construction of suspension bridges in the United States. Born and educated in Germany, Roebling emigrated to America, where he designed and built several well known bridges, including one that spanned Niagara Falls and the Brooklyn Bridge over the East River. An important Roebling innovation was that of replacing the bulky chains used in German bridges with lighter, stronger woven steel cables, which his company also manufactured.

Henry Shrapnel
1761-1842

British inventor and artillery officer whose name is synonymous with the exploding fragmentation shell he invented in 1784. The shrapnel shell was adopted by the British army in 1803 and used for the first time in warfare by the British Army against the Dutch in Suriname in 1804. It contained small lead or iron balls that also came to be known as shrapnel, and a timed fuse designed to explode the shell in the air above enemy troops.

Ernst Werner von Siemens
1816-1892

German electrical engineer and inventor who made important contributions to the electrical supply industry in Germany, Britain, and Russia. Siemens also developed the telegraphic system in his native Prussia and invented the dial telegraph (1846). In 1857 he invented an apparatus that included an electromagnetic generator. In January 1867 Siemens's experiments led to his independent discovery of the self-excited dynamo (also discovered in England by Charles Wheatstone).

Germain Sommeiler
1815-1871

French engineer who designed and built the first major railway tunnel. The Mount Cenis Tunnel (completed in 1870), which passed through a section of the Alps, was the first significant tunnel to be built for the purpose of train traffic. In building it, Sommeiler set the stage for future tunnels that would carry car and truck traffic not only through mountains, but beneath rivers and, eventually, beneath the English Channel.

William Stanley, Jr.
1858-1916

American inventor who designed the induction coil, or transformer, that revolutionized the transmission of electricity. This device allowed the transmission of electricity over long distances via alternating current (AC), rather than the impractical direct current (DC) in use at that time. Inventor and industrialist George Westinghouse hired Stanley as his chief engineer at his Pittsburgh factory in order to design the induction coil. During his lifetime, Stanley was granted 129 patents covering a wide range of electronic devices.

Charles Proteus Steinmetz
1865-1923

Polish-American engineer who made significant contributions to the field of electrical engineering. Steinmetz studied in Germany but was forced to leave due to his socialist activities. Relocating to the United States, he subsequently developed lightning arrestors for electrical transmission lines, arrived at a law for determining magnetic hysteresis, and invented a simple notation for mathematically describing alternating current circuits. Steinmetz consulted for General Electric for many years before becoming a professor at Union College in 1902.

George Stephenson
1781-1848

English railroad engineer who built some of the world's first railroad engines. Stephenson received little formal education, but became an enginewright and, in 1814, introduced his first steam locomotive. This machine was able to pull 30 tons of material more rapidly than a horse-drawn cart. Stephenson built many more engines, his most famous being the Rocket, which ran at the impressive speed of 36 miles per hour (58 kilometers per hour).

Louis Sullivan
1856-1924

American architect who strongly influenced modern U.S. architecture and architects, especially Frank Lloyd Wright, with his building designs. Sullivan is best known for his unique skyscraper style emphasizing the vertical thrust of buildings using steel or iron frame skeleton construction. Among his most distinguished buildings are the Auditorium Building (1866-89) and Carson Pirie Scott department store building (1899-1904) in Chicago and the Wainwright Building in St. Louis (1890-91).

Joseph Wilson Swan
1828-1914

British chemist and inventor who invented a primitive electric lamp (1860) and the carbon-filament incandescent lightbulb (1880), which he developed independently of Thomas Alva Edison. Swan also experimented with photographic printing, patenting the carbon process of printing (1864) and inventing the dry photographic plate (1871) and bromide photographic paper (1879). In 1883 Swan patented the process of squeezing nitrocellulose through holes to generate fibers, thus creating the first feasible artificial silk.

William Henry Fox Talbot
1800-1877

English chemist and pioneer of photography who invented an early photographic technique called the talbotype, or calotype. This process used a sheet of paper coated with silver chloride that, when exposed to light for one minute, created a negative from which multiple images could be made. Had his method been announced a few weeks earlier, Talbot (not French inventor Louis Daguerre) would have been known as the father of photography. Talbot's *Pencil of Nature* was the first book to contain photographic illustrations.

Frederick W. Taylor
1856-1915

American engineer who pioneered scientific management as a feature of mass production. Taylor's experiences at Midvale Steel Company, from shop laborer to chief engineer, exposed him to inefficiencies in the way work was done. Using time studies and job analysis, he strove to eliminate wasted efforts and time by shop workers. Many workers and managers resisted his methods, but the methods were instrumental in rationalizing industrialism in the age of mass production.

Thomas Telford
1757-1834

Scottish civil engineer who has been described as the father of structural engineering. Telford began his career as a journeyman stonemason in England, and in 1793 was assigned to build the Ellesmere Canal in Wales. Between 1802 and 1812 he constructed more than 1000 miles of roads and 1200 bridges in Scotland, before returning to England to build roads. By 1814 Telford had become the most distinguished civil engineer in Britain. Telford's notable works include the Caledonian Canal (1803-23), the London to Holyhead road, his Waterloo Bridge at Betws-y-Coed (1818), his chain-link suspension bridge over the Menai Straits (1819-26), and the St. Katherine's Docks in London (1824-28).

Lewis Temple
1800-1854

American inventor who in 1848 developed an improved toggle harpoon that revolutionized the technology of whaling. An African-American born into slavery in Richmond, Virginia, Temple became a shipsmith and whalecraft manufacturer. In 1829 he arrived in Massachusetts and, by 1836 had his own waterfront blacksmith shop, where he specialized in whalecraft. Temple's toggle substantially improved the efficiency of the whale hunt and the innovative device became the standard harpoon of the whaling industry by the middle of the nineteenth century.

Nikola Tesla
1856-1943

Croatian physicist and electrical engineer who invented a number of important electrical devices, including the AC motor and generator. Tesla was born in Croatia and moved to the United States in 1884, where he worked in Thomas Edison's laboratory in Menlo Park, New Jersey. Leaving the Edison Works to pursue his own work, Tesla was instrumental in developing alternating current as a more efficient means of electrical power. He was memorialized with the unit of magnetic force, the tesla.

Sydney Gilchrist Thomas
1850-1885

English inventor and metallurgist who, with his cousin Percy Gilchrist, developed an improved iron-smelting process that removed phosphorus. The Gilchrist-Thomas process made superior iron by removing phosphorus impurities that resulted in brittle, poor-quality iron. This development made it possible to use many of the large European iron deposits that contained phosphorus, effectively doubling the potential world's ability to produce steel.

Elihu Thomson
1853-1937

American engineering scientist and inventor in electrical technology. Thomson devised and marketed a successful electric arc lighting system, developed an alternating-current motor, and held almost 700 patents relating to electrical devices or processes. He helped establish an industrial research laboratory at General Electric, a company formed by consolidation of his Thomson-Houston Electric and the Edison Electric companies. With his many technical talents, Thomson helped make electrical technology a consumer product.

Robert Whitehead
1823-1905

British engineer and inventor who developed the first self-propelled torpedo in 1866. Whitehead's torpedo was driven by a compressed-air engine and carried a 15kg explosive charge. It also featured a self-regulating device for maintaining a constant preset depth. In 1872 Whitehead established a firm in the Adriatic port of Fiume (also called Rijeka) to manufacture his torpedoes. By 1881 he had customers worldwide, from Britain to Russia to Argentina.

Joseph Whitworth
1803-1887

British mechanical engineer and inventor whose attention to quality and precision gained him a worldwide reputation for superior machine tools. Whitworth developed and perfected tools for many industries; in 1835 alone, his patents included a knitting machine, several planing machines, a vertical drill, a gear-cutting machine, and a screw-cutting lathe. Perhaps his most important innovation was the creation of a machine tool capable of measuring to an accuracy of one hundredth-thousandth of an inch.

Granville T. Woods
1856-1910

American engineer and inventor who received approximately sixty patents. Considered the most prolific nineteenth-century African American inventor, Woods innovated and improved transportation and communication technology. Most of Woods's inventions were railroad related. His telegraph system relayed messages between moving trains as well as stations to prevent collisions. He also patented an egg incubator, telephone transmitter, and electric car. He later established the Woods Electric Company, selling such inventions as automatic air brakes to George Westinghouse.

Bibliography of Primary Sources

Books

Appert, Nicolas. *The Art of Preserving All Animal and Vegetable Substances for Several Years.* 1810. Appert shared his discoveries concerning food preservation in this 1810 book.

Bell, Alexander Graham. *The Deposition of Alexander Graham Bell.* 1908. Published by his own company (American Bell Telephone), this work is most detailed explanation by Bell of his work in telephony.

Bessemer, Henry. *An Autobiography.* 1905. Bessemer's memoir is his only account of his pioneering work with steel and is the source of most information written about him since his death.

Fulton, Robert. *A Treatise on the Improvement of Canal Navigation.* 1796. Fulton offered detailed ideas about canal navigation in this work.

Goodyear, Charles. *Gum-Elastic and Its Varieties, with a Detailed Account of Its Applications and Uses, and of the Discovery of Vulcanization.* 1853. In this work Goodyear described his accidental discovery of vulcanized rubber.

Talbot, William. *Pencil of Nature.* 1844-46. This landmark book was the first to be illustrated with photographs.

Periodical Articles

Carnegie, Andrew. "Wealth." 1889. This article by Carnegie outlined what came to be called the "Gospel of Wealth," which was Carnegie's belief that the rich had a moral obligation to spend their money for the betterment of society. In the article Carnegie wrote, "A man who dies rich dies disgraced."

General Bibliography

Books

Agassi, Joseph. *The Continuing Revolution: A History of Physics from the Greeks to Einstein.* New York: McGraw-Hill, 1968.

Anderson, E. W. *Man the Aviator.* London: Priory Press, 1973.

Anderson, E. W. *Man the Navigator.* London: Priory Press, 1973.

Asimov, Isaac. *Adding a Dimension: Seventeen Essays on the History of Science.* Garden City, NY: Doubleday, 1964.

Bahn, Paul G., ed. *The Cambridge Illustrated History of Archaeology.* New York: Cambridge University Press, 1996.

Barrow, Sir John. *Sketches of the Royal Society and Royal Society Club.* London: F. Cass, 1971.

Basalla, George. *The Evolution of Technology.* New York: Cambridge University Press, 1988.

Benson, Don S. *Man and the Wheel.* London: Priory Press, 1973.

Boorstin, Daniel J. *The Discoverers.* New York: Random House, 1983.

Bowler, Peter J. *Biology and Social Thought, 1850-1914.* Berkeley: University of California Press, 1993.

Bowler, Peter J. *The Norton History of the Environmental Sciences.* New York: W. W. Norton, 1993.

Brock, W. H. *The Norton History of Chemistry.* New York: W. W. Norton, 1993.

Bruno, Leonard C. *Science and Technology Firsts.* Edited by Donna Olendorf, guest foreword by Daniel J. Boorstin. Detroit: Gale, 1997.

Bud, Robert, and Deborah Jean Warner, eds. *Instruments of Science: An Historical Encyclopedia.* New York: Garland, 1998.

Bynum, W. F., et al., eds. *Dictionary of the History of Science.* Princeton, NJ: Princeton University Press, 1981.

Bynum, W. F. *Science and the Practice of Medicine in the Nineteenth Century.* New York: Cambridge University Press, 1994.

Cardwell, D. S. L. *From Watt to Clausius: The Rise of Thermodynamics in the Early Industrial Age.* Ames: Iowa State University Press, 1989.

Carnegie Library of Pittsburgh. *Science and Technology Desk Reference: 1,500 Frequently Asked or Difficult-to-Answer Questions.* Washington, D.C.: Gale, 1993.

Coleman, William. *Biology in the Nineteenth Century: Problems of Form, Function, and Transformation.* New York: Wiley, 1971.

Collier, Bruce. *The Little Engines That Could've: The Calculating Machines of Charles Babbage.* New York: Garland, 1990.

Crone, G. R. *Man the Explorer.* London: Priory Press, 1973.

Dellenbaugh, Frederick Samuel. *Books by American Travellers and Explorers from 1846 to 1900.* New York: G. P. Putnam's Sons, 1920.

Dibner, Bern. *Darwin of the Beagle.* New York: Blaisdell Publishing Company, 1964.

Dibner, Bern. *Ten Founding Fathers of the Electrical Science.* Norwalk, CT: Burndy Library, 1954.

Dunn, L. C. *A Short History of Genetics: The Development of Some of the Main Lines of Thought, 1864-1939.* Ames: Iowa State University Press, 1991.

Elliott, Clark A. *History of Science in the United States: A Chronology and Research Guide.* New York: Garland, 1996.

Ellis, Keith. *Man and Measurement.* London: Priory Press, 1973.

Erlen, Jonathan. *The History of the Health Care Sciences and Health Care, 1700-1980: A Selective Annotated Bibliography.* New York: Garland, 1984.

Fearing, Franklin. *Reflex Action: A Study in the History of Physiological Psychology.* Introduction by Richard Held. Cambridge: MIT Press, 1970.

Fink, Karl J. *Goethe's History of Science.* New York: Cambridge University Press, 1991.

Fox, Robert, and Anna Guagnini. *Laboratories, Workshops, and Sites: Concepts and Practices of Research in Industrial Europe, 1800- 1914.* Berkeley: University of California Press, 1999.

French, Roger, and Andrew Wear. *British Medicine in an Age of Reform.* New York: Routledge, 1991.

Galton, Sir Francis. *English Men of Science: Their Nature and Nurture.* London: Cass, 1970.

Gascoigne, Robert Mortimer. *A Chronology of the History of Science, 1450-1900.* New York: Garland, 1987.

Gasking, Elizabeth B. *The Rise of Experimental Biology.* New York: Random House, 1970.

Good, Gregory A., ed. *Sciences of the Earth: An Encyclopedia of Events, People, and Phenomena.* New York: Garland, 1998.

Graham, Loren R. *Science in Russia and the Soviet Union: A Short History.* New York: Cambridge University Press, 1993.

Grattan-Guiness, Ivor. *The Norton History of the Mathematical Sciences: The Rainbow of Mathematics.* New York: W. W. Norton, 1998.

Greene, Mott T. *Geology in the Nineteenth Century: Changing Views of a Changing World.* Ithaca, NY: Cornell University Press, 1982.

Gregor, Arthur S. *A Short History of Science: Man's Conquest of Nature from Ancient Times to the Atomic Age.* New York: Macmillan, 1963.

Gullberg, Jan. *Mathematics: From the Birth of Numbers.* Technical illustrations by Pär Gullberg. New York: W. W. Norton, 1997.

Harman, P. M. *Energy, Force, and Matter: The Conceptual Development of Nineteenth-Century Physics.* New York: Cambridge University Press, 1982.

Hellemans, Alexander, and Bryan Bunch. *The Timetables of Science: A Chronology of the Most Important People and Events in the History of Science.* New York: Simon and Schuster, 1988.

Hellyer, Brian. *Man the Timekeeper.* London: Priory Press, 1974.

Hodge, M. J. S. *Origins and Species: A Study of the Historical Sources of Darwinism and the Contexts of Some Other Accounts of Organic Diversity from Plato and Aristotle On.* New York: Garland, 1991.

Holmes, Edward, and Christopher Maynard. *Great Men of Science.* Edited by Jennifer L. Justice. New York: Warwick Press, 1979.

Hoskin, Michael. *The Cambridge Illustrated History of Astronomy.* New York: Cambridge University Press, 1997.

Howsom, Colin, ed. *Method and Appraisal in the Physical Sciences: The Critical Background to Modern Science, 1800-1905.* New York: Cambridge University Press, 1976.

Huxley, Julian. *Essays in Popular Science.* New York: A. A. Knopf, 1927.

Kelman, Peter, and A. Harris Stone. *Mendeleyev: Prophet of Chemical Elements.* Englewood Cliffs, NJ: Prentice-Hall, 1970.

Knight, David M. *Sources for the History of Science, 1660- 1914.* Ithaca, NY: Cornell University Press, 1975.

Lankford, John, ed. *History of Astronomy: An Encyclopedia.* New York: Garland, 1997.

Lincoln, Roger J., and G. A. Boxshall. *The Cambridge Illustrated Dictionary of Natural History.* Illustrations by Roberta Smith. New York: Cambridge University Press, 1987.

Martin, Ernest G. *The Story of Our Bodies: The Science of Physiology, Organs, and Their Functions in Human Beings.* New York: P. F. Collier & Son Company, 1930.

Maulitz, Russell Charles. *Morbid Appearance: The Anatomy of Pathology in the Early Nineteenth Century.* New York: Cambridge University Press, 1987.

McGrath, Kimberley A., ed. *World of Invention.* Second ed. Detroit: Gale, 1999.

McGrath, Kimberley A., ed. *World of Scientific Discovery.* Second ed. Detroit: Gale, 1999.

Parker, Geoffrey, ed. *The Cambridge Illustrated History of Warfare: The Triumph of the West.* New York: Cambridge University Press, 1995.

Porter, Roy. *The Cambridge Illustrated History of Medicine.* New York: Cambridge University Press, 1996.

Rehbock, Philip F. *The Philosophical Naturalists: Themes in Early Nineteenth-Century British Biology.* Madison: University of Wisconsin Press, 1983.

Reingold, Nathan, ed. *Science in America Since 1820.* New York: Science History Publications, 1976.

Rothenberg, Marc. *The History of Science in the United States: An Encyclopedia.* New York: Garland, 2000.

Routledge, Robert. *A Popular History of Science.* New York: G. Routledge and Sons, 1881.

Rudwick, M. J. S. *The Meaning of Fossils: Episodes in the History of Paleontology.* New York: American Elsevier, 1972.

Sachs, Ernest. *The History and Development of Neurological Surgery.* New York: Hoeber, 1952.

Sarton, George. *The History of Science and the New Humanism.* New Brunswick, NJ: Transaction Books, 1987.

Sarton, George. *Introduction to the History of Science.* Huntington, NY: R. E. Krieger Publishing Company, 1975.

Schneiderman, Ron. *Computers: From Babbage to the Fifth Generation.* New York: F. Watts, 1986.

Scott, Wilson L. *The Conflict Between Atomism and Conservation Theory, 1644- 1860.* New York: American Elsevier, 1970.

Singer, Charles. *A History of Biology to About the Year 1900: A General Introduction to the Study of Living Things.* Ames: Iowa State University Press, 1989.

Smith, Roger. *The Norton History of the Human Sciences.* New York: W. W. Norton, 1997.

Spangenburg, Ray, and Diane K. Moser. *The History of Science in the Nineteenth Century.* New York: Facts on File, 1994.

Stiffler, Lee Ann. *Science Rediscovered: A Daily Chronicle of Highlights in the History of Science.* Durham, NC: Carolina Academic Press, 1995.

Stwertka, Albert, and Eve Stwertka. *Physics: From Newton to the Big Bang.* New York: F. Watts, 1986.

Swenson, Lloyd S. *Genesis of Relativity: Einstein in Context.* New York: B. Franklin, 1979.

Travers, Bridget, ed. *The Gale Encyclopedia of Science.* Detroit: Gale, 1996.

Watkins, George, and R. A. Buchanan. *Man and the Steam Engine.* London: Priory Press, 1975.

Webster, Charles. *Biology, Medicine, and Society, 1840-1940.* New York: Cambridge University Press, 1981.

Whitehead, Alfred North. *Science and the Modern World: Lowell Lectures, 1925.* New York: The Free Press, 1953.

Winsor, Mary P. *Starfish, Jellyfish, and the Order or Life: Issues in Nineteenth-Century Science.* New Haven, CT: Yale University Press, 1976.

Young, Robyn V., ed. *Notable Mathematicians: From Ancient Times to the Present.* Detroit: Gale, 1998.

Index

~

Numbers in bold refer to
main biographical entries